Rockets and People

Volume II: *Creating a Rocket Industry*

Boris Chertok

Asif Siddiqi, Series Editor

The NASA History Series

National Aeronautics and Space Administration
NASA History Division
Office of External Relations
Washington, DC
June 2006
NASA SP-2006-4110

Library of Congress Cataloging-in-Publication Data

Chertok, B. E. (Boris Evseevich), 1912–
　　[Rakety i lyudi. English]
　　Rockets and People: Creating a Rocket Industry (Volume II) / by Boris E.
　　　Chertok ;
　　[edited by] Asif A. Siddiqi.
　　p. cm. — (NASA History Series) (NASA SP-2006-4110)
　Includes bibliographical references and index.
　　1. Chertok, B. E. (Boris Evseevich), 1912–　2. Astronautics—
Soviet Union—Biography.　3. Aerospace engineers—Soviet union—
Biography.　4. Astronautics—Soviet Union—History.
I. Siddiqi, Asif A., 1966-　II. Title.　III. Series.　IV. SP-2006-4110.
TL789.85.C48C4813 2006
629.1'092—dc22

　　　　　　　　　　　　　　　　　　　　　　　　　　　　2006020825

For sale by the Superintendent of Documents, U.S. Government Printing Office
Internet: bookstore.gpo.gov　Phone: toll free (866) 512-1800;　DC area (202) 512-1800
Fax: (202) 512-2250 Mail: Stop IDCC, Washington, DC 20402-0001

ISBN 0-16-076672-9

I dedicate this book
to the cherished memory
of my wife and friend,
Yekaterina Semyonova Golubkina.

Contents

Series Introduction

In an extraordinary century, Academician Boris Yevseyevich Chertok lived an extraordinary life. He witnessed and participated in many important technological milestones of the twentieth century, and in these volumes, he recollects them with clarity, humanity, and humility. Chertok began his career as an electrician in 1930 at an aviation factory near Moscow. Thirty years later, he was one of the senior designers in charge of the Soviet Union's crowning achievement as a space power: the launch of Yuriy Gagarin, the world's first space voyager. Chertok's sixty-year-long career, punctuated by the extraordinary accomplishments of both Sputnik and Gagarin, and continuing to the many successes and failures of the Soviet space program, constitutes the core of his memoirs, *Rockets and People*. In these four volumes, Academician Chertok not only describes and remembers, but also elicits and extracts profound insights from an epic story about a society's quest to explore the cosmos.

Academician Chertok's memoirs, forged from experience in the Cold War, provide a compelling perspective into a past that is indispensable to understanding the present relationship between the American and Russian space programs. From the end of the World War II to the present day, the missile and space efforts of the United States and the Soviet Union (and now, Russia) have been inextricably linked. As such, although Chertok's work focuses exclusively on Soviet programs to explore space, it also prompts us to reconsider the entire history of spaceflight, both Russian and American.

Chertok's narrative underlines how, from the beginning of the Cold War, the rocketry projects of the two nations evolved in independent but parallel paths. Chertok's first-hand recollections of the extraordinary Soviet efforts to collect, catalog, and reproduce German rocket technology after the World War II provide a parallel view to what historian John Gimbel has called the Western "exploitation and plunder" of German technology after the war.[1] Chertok describes how the Soviet design

---

1. John Gimbel, *Science, Technology, and Reparations: Exploitation and Plunder in Postwar Germany* (Stanford: Stanford University Press, 1990).

team under the famous Chief Designer Sergey Pavlovich Korolev quickly outgrew German missile technology. By the late 1950s, his team produced the majestic R-7, the world's first intercontinental ballistic missile. Using this rocket, the Soviet Union launched the first Sputnik satellite on 4 October 1957 from a launch site in remote central Asia.

The early Soviet accomplishments in space exploration, particularly the launch of Sputnik in 1957 and the remarkable flight of Yuriy Gagarin in 1961, were benchmarks of the Cold War. Spurred by the Soviet successes, the United States formed a governmental agency, the National Aeronautics and Space Administration (NASA), to conduct civilian space exploration. As a result of Gagarin's triumphant flight, in 1961, the Kennedy Administration charged NASA to achieve the goal of "landing a man on the Moon and returning him safely to the Earth before the end of the decade."[2] Such an achievement would demonstrate American supremacy in the arena of spaceflight at a time when both American and Soviet politicians believed that victory in space would be tantamount to preeminence on the global stage. The space programs of both countries grew in leaps and bounds in the 1960s, but the Americans crossed the finish line first when Apollo astronauts Neil A. Armstrong and Edwin E. "Buzz" Aldrin, Jr. disembarked on the Moon's surface in July 1969.

Shadowing Apollo's success was an absent question: What happened to the Soviets who had succeeded so brilliantly with Sputnik and Gagarin? Unknown to most, the Soviets tried and failed to reach the Moon in a secret program that came to naught. As a result of that disastrous failure, the Soviet Union pursued a gradual and consistent space station program in the 1970s and 1980s that eventually led to the Mir space station. The Americans developed a reusable space transportation system known as the Space Shuttle. Despite their seemingly separate paths, the space programs of the two powers remained dependent on each other for rationale and direction. When the Soviet Union disintegrated in 1991, cooperation replaced competition as the two countries embarked on a joint program to establish the first permanent human habitation in space through the International Space Station (ISS).

Academician Chertok's reminiscences are particularly important because he played key roles in almost every major milestone of the Soviet missile and space programs, from the beginning of World War II to the dissolution of the Soviet Union in 1991. During the war, he served on the team that developed the Soviet Union's first rocket-powered airplane, the BI. In the immediate aftermath of the war, Chertok, then in his early thirties, played a key role in studying and collecting captured German rocket technology. In the latter days of the Stalinist era, he worked to develop long-range missiles as deputy chief engineer of the main research institute,

---

2. U.S. Congress, *Senate Committee on Aeronautical and Space Sciences, Documents on International Aspects of the Exploration and Uses of Outer Space, 1954-1962, 88th Cong., 1st sess., S. Doc. 18* (Washington, DC: GPO, 1963), pp. 202-204.

the NII-88 (pronounced "nee-88") near Moscow. In 1956, Korolev's famous OKB-1 design bureau spun off from the institute and assumed a leading position in the emerging Soviet space program. As a deputy chief designer at OKB-1, Chertok continued with his contributions to the most important Soviet space projects of the day: Vostok, Voskhod, Soyuz, the world's first space station Salyut, the Energiya superbooster, and the Buran space shuttle.

Chertok's emergence from the secret world of the Soviet military-industrial complex, into his current status as the most recognized living legacy of the Soviet space program, coincided with the dismantling of the Soviet Union as a political entity. Throughout most of his career, Chertok's name remained a state secret. When he occasionally wrote for the public, he used the pseudonym "Boris Yevseyev."[3] Like others writing on the Soviet space program during the Cold War, Chertok was not allowed to reveal any institutional or technical details in his writings. What the state censors permitted for publication said little; one could read a book several hundred pages long comprised of nothing beyond tedious and long personal anecdotes between anonymous participants extolling the virtues of the Communist Party. The formerly immutable limits on free expression in the Soviet Union irrevocably expanded only after Mikhail Gorbachev's rise to power in 1985 and the introduction of *glasnost'* (openness).

Chertok's name first appeared in print in the newspaper *Izvestiya* in an article commemorating the thirtieth anniversary of the launch of Sputnik in 1987. In a wide-ranging interview on the creation of Sputnik, Chertok spoke with the utmost respect for his former boss, the late Korolev. He also eloquently balanced love for his country with criticisms of the widespread inertia and inefficiency that characterized late-period Soviet society.[4] His first written works in the *glasnost'* period, published in early 1988 in the Air Force journal *Aviatsiya i kosmonavtika* (Aviation and Cosmonautics), underlined Korolev's central role in the foundation and growth of the Soviet space program.[5] By this time, it was as if all the patched up straps that held together a stagnant empire were falling apart one by one; even as Russia was in the midst of one of its most historic transformations, the floodgates of free expression were transforming the country's own history. People like Chertok were now free to speak about their experiences with candor. Readers could now learn about episodes such as Korolev's brutal incarceration in the late 1930s, the dramatic story behind the fatal space mission of Soyuz-1 in 1967, and details of the failed and abandoned

---

3. See for example, his article "Chelovek or avtomat?" (Human or Automation?) in the book M. Vasilyev, ed., *Shagi k zvezdam* (Footsteps to the Stars) (Moscow: Molodaya gvardiya, 1972), pp. 281-287.

4. B. Konovalov, "Ryvok k zvezdam" (Dash to the Stars), *Izvestiya*, October 1, 1987, p. 3.

5. B. Chertok, "Lider" (Leader), *Aviatsiya i kosmonavtika* no. 1 (1988): pp. 30–31 and no. 2 (1988): pp. 40–41.

Moon project in the 1960s.[6] Chertok himself shed light on a missing piece of history in a series of five articles published in *Izvestiya* in early 1992 on the German contribution to the foundation of the Soviet missile program after World War II.[7]

Using these works as a starting point, Academician Chertok began working on his memoirs. Originally, he had only intended to write about his experiences from the postwar years in one volume, maybe two. Readers responded so positively to the first volume, *Rakety i liudi* (Rockets and People) published in 1994, that Chertok continued to write, eventually producing four substantial volumes, published in 1996, 1997, and 1999, covering the entire history of the Soviet missile and space programs.[8]

My initial interest in the memoirs was purely historical: I was fascinated by the wealth of technical arcana in the books, specifically projects and concepts that had remained hidden throughout much of the Cold War. Those interested in dates, statistics, and the "nuts and bolts" of history will find much that is useful in these pages. As I continued to read, however, I became engrossed by the overall rhythm of Academician Chertok's narrative, which gave voice and humanity to a story ostensibly about mathematics and technology. In his writings, I found a richness that had been nearly absent in most of the disembodied, clinical, and often speculative writing by Westerners studying the Soviet space program. Because of Chertok's storytelling skills, his memoir is a much needed corrective to the outdated Western view of Soviet space achievements as a mishmash of propaganda, self-delusion, and Cold War rhetoric. In Chertok's story, we meet real people with real dreams who achieved extraordinary successes under very difficult conditions.

Chertok's reminiscences are remarkably sharp and descriptive. In being self-reflective, Chertok avoids the kind of solipsistic ruminations that often characterize

---

6. For early references to Korolev's imprisonment, see Ye. Manucharova, "Kharakter glavnogo konstruktora" (The Character of the Chief Designer), *Izvestiya*, January 11, 1987, p. 3. For early revelations on Soyuz-1 and the Moon program, see L. N. Kamanin, "Zvezdy Komarova" (Komarov's Star), *Poisk* no. 5 (June 1989): pp. 4–5 and L. N. Kamanin, "S zemli na lunu i obratno" (From the Earth to the Moon and Back), *Poisk* no. 12 (July 1989): pp. 7–8.

7. *Izvestiya* correspondent Boris Konovalov prepared these publications, which had the general title "U Sovetskikh raketnykh triumfov bylo nemetskoye nachalo" (Soviets Rocket Triumphs Had German Origins). See *Izvestiya*, March 4, 1992, p. 5; March 5, 1992, p. 5; March 6, 1992, p. 5; March 7, 1992, p. 5; and March 9, 1992, p. 3. Konovalov also published a sixth article on the German contribution to American rocketry. See "U amerikanskikh raketnykh triumfov takzhe bylo nemetskoye nachalo" (American Rocket Triumphs Also Had German Origins), *Izvestiya*, March 10, 1992, p. 7. Konovalov later synthesized the five original articles into a longer work that included the reminiscences of other participants in the German mission such as Vladimir Barmin and Vasiliy Mishin. See Boris Konovalov, *Tayna Sovetskogo raketnogo oruzhiya* (Secrets of Soviet Rocket Armaments) (Moscow: ZEVS, 1992).

8. *Rakety i lyudi* (Rockets and People) (Moscow: Mashinostroyeniye, 1994); *Rakety i lyudi: Fili Podlipki Tyuratam* (Rockets and People: Fili Podlipki Tyuratam) (Moscow: Mashinostroyeniye, 1996); *Rakety i lyudi: goryachiye dni kholodnoy voyny* (Rockets and People: Hot Days of the Cold War) (Moscow: Mashinostroyeniye, 1997); *Rakety i lyudi: lunnaya gonka* (Rockets and People: The Moon Race) (Moscow: Mashinostroyeniye, 1999). All four volumes were subsequently translated and published in Germany.

memoirs. He is both proud of his country's accomplishments and willing to admit failings with honesty. For example, Chertok juxtaposes accounts of the famous aviation exploits of Soviet pilots in the 1930s, especially those to the Arctic, with the much darker costs of the Great Terror in the late 1930s when Stalin's vicious purges decimated the Soviet aviation industry.

Chertok's descriptive powers are particularly evident in describing the chaotic nature of the Soviet mission to recover and collect rocketry equipment in Germany after World War II. Interspersed with his contemporary diary entries, his language conveys the combination of joy, confusion, and often anti-climax that the end of the war presaged for Soviet representatives in Germany. In one breath, Chertok and his team are looking for hidden caches of German matériel in an underground mine, while in another they are face to face with the deadly consequences of a soldier who had raped a young German woman (Volume I, Chapter 21).[9] There are many such seemingly incongruous anecdotes during Chertok's time in Germany, from the experience of visiting the Nazi slave labor camp at Dora soon after liberation in 1945, to the deportation of hundreds of German scientists to the USSR in 1946. Chertok's massive work is of great consequence for another reason—he cogently provides context. Since the breakup of the Soviet Union in 1991, many participants have openly written about their experiences, but few have successfully placed Soviet space achievements in the broader context of the history of Soviet science, the history of the Soviet military-industrial complex, or indeed Soviet history in general.[10] The volumes of memoirs compiled by the Russian State Archive of Scientific-Technical Documentation in the early 1990s under the series, *Dorogi v kosmos* (Roads to Space), provided an undeniably rich and in-depth view of the origins of the Soviet space program, but they were, for the most part, personal nar-

---

9. For the problem of rape in occupied Germany after the war, see Norman M. Naimark, *The Russians in Germany: A History of the Soviet Zone of Occupation, 1945-1949* (Cambridge, MA: The Belknap Press of Harvard University Press, 1995), pp. 69–140.

10. For the two most important histories of the Soviet military-industrial complex, see N. S. Simonov, *Voyenno-promyshlennyy kompleks SSSR v 1920-1950-ye gody: tempy ekonomicheskogo rosta, struktura, organizatsiya proizvodstva i upravleniye* (The Military-Industrial Complex of the USSR in the 1920s to 1950s: Rate of Economic Growth, Structure, Organization of Production and Control) (Moscow: ROSSPEN, 1996); and I. V. Bystrova, *Voyenno-promyshlennyy kompleks sssr v gody kholodnoy voyny (vtoraya polovina 40-kh – nachalo 60-kh godov)* [The Military-Industrial Complex of the USSR in the Years of the Cold War (The Late 1940s to the Early 1960s)] (Moscow: IRI RAN, 2000). For a history in English that builds on these seminal works and complements them with original research, see John Barber and Mark Harrison, eds., *The Soviet Defence-Industry Complex from Stalin to Khrushchev* (Houndmills, UK: Macmillan Press, 2000).

ratives, i.e., fish-eye views of the world around them.[11] Chertok's memoirs are a rare exception in that they strive to locate the Soviet missile and space program in the fabric of broader social, political, industrial, and scientific developments in the former Soviet Union.

This combination—Chertok's participation in the most important Soviet space achievements, his capacity to lucidly communicate them to the reader, and his skill in providing a broader social context—make this work, in my opinion, one of the most important memoirs written by a veteran of the Soviet space program. The series will also be an important contribution to the history of Soviet science and technology.[12]

In reading Academician Chertok's recollections, we should not lose sight of the fact that these chapters, although full of history, have their particular perspective. In conveying to us the complex vista of the Soviet space program, he has given us one man's memories of a huge undertaking. Other participants of these very same events will remember things differently. Soviet space history, like any discipline of history, exists as a continuous process of revision and restatement. Few historians in the twenty-first century would claim to be completely objective.[13] Memoirists would make even less of a claim to the "truth." In his introduction, Chertok acknowledges this, saying, "I . . . must warn the reader that in no way do I have pretensions to the laurels of a scholarly historian. Correspondingly, my books are not examples of strict historical research. In any memoirs, narrative and thought are inevitably subjective." Chertok ably illustrates, however, that avoiding the pursuit of scholarly history does not necessarily lessen the relevance of his story, especially because it represents the opinion of an influential member of the postwar scientific and technical intelligentsia in the Soviet Union.

Some, for example, might not share Chertok's strong belief in the power of scientists and engineers to solve social problems, a view that influenced many who sought to transform the Soviet Union with modern science after the Russian Revo-

---

11. Yu. A. Mozzhorin et al., eds., *Dorogi v kosmos: Vospominaniya veteranov raketno-kosmicheskoy tekhniki i kosmonavtiki, tom I i II* (Roads to Space: Recollections of Veterans of Rocket-Space Technology and Cosmonautics: Volumes I and II) (Moscow: MAI, 1992) and Yu. A. Mozzhorin et al., eds., *Nachalo kosmicheskoy ery: vospominaniya veteranov raketno-kosmicheskoy tekhniki i kosmonavtiki: vypusk vtoroy* (The Beginning of the Space Era: Recollections of Veterans of Rocket-Space Technology and Cosmonautics: Second Issue) (Moscow: RNITsKD, 1994). For a poorly translated and edited English version of the series, see John Rhea, ed., *Roads to Space: An Oral History of the Soviet Space Program* (New York: Aviation Week Group, 1995).

12. For key works on the history of Soviet science and technology, see Kendall E. Bailes, *Technology and Society under Lenin and Stalin: Origins of the Soviet Technical Intelligentsia, 1917-1941* (Princeton, NJ: Princeton University Press, 1978); Loren R. Graham, *Science in Russia and the Soviet Union: A Short History* (Cambridge: Cambridge University Press, 1993); and Nikolai Krementsov, *Stalinist Science* (Princeton, NJ: Princeton University Press, 1997).

13. For the American historical discipline's relationship to the changing standards of objectivity, see Peter Novick, *That Noble Dream: The 'Objectivity' Question and the American Historical Profession* (Cambridge, UK: Cambridge University Press, 1988).

lution in 1917. Historians of Soviet science such as Loren Graham have argued that narrowly technocratic views of social development cost the Soviet Union dearly.[14] Technological hubris was, of course, not unique to the Soviet scientific community, but absent democratic processes of accountability, many huge Soviet government projects—such as the construction of the Great Dnepr Dam and the great Siberian railway in the 1970s and 1980s—ended up as costly failures with many adverse social and environmental repercussions. Whether one agrees or disagrees with Chertok's views, they are important to understand because they represent the ideas of a generation who passionately believed in the power of science to eliminate the ills of society. As such, his memoirs add an important dimension to understanding the *mentalité* of the Soviets' drive to become a modern, industrialized state in the twentieth century.

Chertok's memoirs are part of the second generation of publications on Soviet space history, one that eclipsed the (heavily censored) first generation published during the Communist era. Memoirs constituted a large part of the second generation. In the 1990s, when it was finally possible to write candidly about Soviet space history, a wave of personal recollections flooded the market. Not only Boris Chertok, but also such luminaries as Vasiliy Mishin, Kerim Kerimov, Boris Gubanov, Yuriy Mozzhorin, Konstantin Feoktistov, Vyacheslav Filin, and others finally published their reminiscences.[15] Official organizational histories and journalistic accounts complemented these memoirs, written by individuals with access to secret archival documents. Yaroslav Golovanov's magisterial *Korolev: Fakty i Mify* (Korolev: Facts and Myths), as well as key institutional works from the Energiya corporation and the Russian Military Space Forces, added richly to the canon.[16] The diaries of Air Force General Nikolay Kamanin from the 1960s to the early 1970s, published in

14. For technological hubris, see for example, Loren Graham, *The Ghost of the Executed Engineer: Technology and the Fall of the Soviet Union* (Cambridge, MA: Harvard University Press, 1993).

15. V. M. Filin, *Vospominaniya o lunnom korablye* (Recollections on the Lunar Ship) (Moscow: Kultura, 1992); Kerim Kerimov, *Dorogi v kosmos (zapiski predsedatelya Gosudarstvennoy komissii)* [Roads to Space (Notes of the Chairman of the State Commission)] (Baku: Azerbaijan, 1995); V. M. Filin, *Put k 'Energii'* (Path to Energiya) (Moscow: 'GRAAL', 1996); V. P. Mishin, *Ot sozdaniya ballisticheskikh raket k raketno-kosmicheskomu mashinostroyeniyu* (From the Creation of the Ballistic Rocket to Rocket-Space Machine Building) (Moscow: 'Inform-Znaniye', 1998); B. I. Gubanov, *Triumf i tragediya 'energii': razmyshleniya glavnogo konstruktora* (The Triumph and Tragedy of Energiya: The Reflections of a Chief Designer) (Nizhniy novgorod: NIER, four volumes in 1998-2000); Konstantin Feoktistov, *Trayektoriya zhizni: mezhdu vchera i zavtra* (Life's Trajectory: Between Yesterday and Tomorrow) (Moscow: Vagrius, 2000); N. A. Anifimov, ed., *Tak eto bylo—Memuary Yu. A. Mozzhorin: Mozzhorin v vospominaniyakh sovremennikov* (How it Was—Memoirs of Yu. A. Mozzhorin: Mozzhorin in the Recollections of his Contemporaries) (Moscow: ZAO 'Mezhdunarodnaya programma obrazovaniya, 2000).

16. Yaroslav Golovanov, *Korolev: fakty i mify* (Korolev: Facts and Myths) (Moscow: Nauka, 1994); Yu. P. Semenov, ed., *Raketno-Kosmicheskaya Korporatsiya "Energiya" imeni S. P. Koroleva* (Energiya Rocket-Space Corporation Named After S. P. Korolev) (Korolev: RKK Energiya, 1996); V. V. Favorskiy and I. V. Meshcheryakov, eds., *Voyenno-kosmicheskiye sily (voyenno-istoricheskiy trud): kniga I* [Military-Space Forces (A Military-Historical Work): Book I] (Moscow: VKS, 1997). Subsequent volumes were published in 1998 and 2001.

four volumes in the late 1990s, also gave scholars a candid look at the vicissitudes of the Soviet human spaceflight program.[17]

The flood of works in Russian allowed Westerners to publish the first works in English. Memoirs—for example, from Sergey Khrushchev and Roald Sagdeev—appeared in their English translations. James Harford published his 1997 biography of Sergey Korolev based upon extensive interviews with veterans of the Soviet space program.[18] My own book, *Challenge to Apollo: The Soviet Union and the Space Race, 1945-1974*, was an early attempt to synthesize the wealth of information and narrate a complete history of the early Soviet human spaceflight program.[19] Steven Zaloga provided an indispensable counterpoint to these space histories in *The Kremlin's Nuclear Sword: The Rise and Fall of Russia's Strategic Nuclear Forces, 1945-2000*, which reconstructed the story of the Soviet efforts to develop strategic weapons.[20]

With any new field of history that is bursting with information based primarily on recollection and interviews, there are naturally many contradictions and inconsistencies. For example, even on such a seemingly trivial issue as the name of the earliest institute in Soviet-occupied Germany, "Institute Rabe," there is no firm agreement on the reason it was given this title. Chertok's recollections contradict the recollection of another Soviet veteran, Georgiy Dyadin.[21] In another case, many veterans have claimed that artillery general Lev Gaydukov's meeting with Stalin in 1945 was a key turning point in the early Soviet missile program; Stalin apparently entrusted Gaydukov with the responsibility to choose an industrial sector to assign the development of long-range rockets (Volume I, Chapter 22). Lists of visitors to Stalin's office during that period—declassified only very recently—do not, however, show that Gaydukov ever met with Stalin in 1945.[22] Similarly, many Russian sources note that the "Second Main Directorate" of the USSR Council of Ministers managed Soviet missile development in the early 1950s, when in fact, this body

17. The first published volume was N. P. Kamanin, *Skrytiy kosmos: kniga pervaya, 1960-1963gg.* (Hidden Space: Book One, 1960-1963) (Moscow: Infortekst IF, 1995). Subsequent volumes covering 1964-1966, 1967-1968, and 1969-1978 were published in 1997, 1999, and 2001 respectively.

18. Sergei N. Khrushchev, *Nikita Khrushchev and the Creation of a Superpower* (University Park, PA: The Pennsylvania State University Press, 2000); Roald Z. Sagdeev, *The Making of a Soviet Scientist: My Adventures in Nuclear Fusion and Space From Stalin to Star Wars* (New York: John Wiley & Sons, 1993); James Harford, *Korolev: How One Man Masterminded the Soviet Drive to Beat America to the Moon* (New York: John Wiley & Sons, 1997).

19. Asif A. Siddiqi, *Challenge to Apollo: The Soviet Union and the Space Race, 1945-1974* (Washington, D.C.: NASA SP-2000-4408, 2000). The book was republished as a two-volume work as *Sputnik and the Soviet Space Challenge* (Gainesville, FL: University Press of Florida, 2003) and *The Soviet Space Race with Apollo* (Gainesville, FL: University Press of Florida, 2003).

20. Steven J. Zaloga, *The Kremlin's Nuclear Sword: The Rise and Fall of Russia's Strategic Nuclear Forces, 1945-2000* (Washington, DC: Smithsonian Institution Press, 2002).

21. G. V. Dyadin, D. N. Filippovykh, and V. I. Ivkin, *Pamyatnyye starty* (Memorable Launches) (Moscow: TsIPK, 2001), p. 69.

22. A. V. Korotkov, A. D. Chernev, and A. A. Chernobayev, "Alfavitnyi ukazatel posetitelei kremlevskogo kabineta I. V. Stalina" ("Alphabetical List of Visitors to the Kremlin Office of I. V. Stalin"), *Istoricheskii arkhiv* no. 4 (1998): p. 50.

actually supervised uranium procurement for the A-bomb project.[23] In many cases, memoirs provide different and contradictory information on the very same event (different dates, designations, locations, people involved, etc.).

Academician Chertok's wonderful memoirs point to a solution to these discrepancies: a "third generation" of Soviet space history, one that builds on the rich trove of the first and second generations, but is primarily based on *documentary* evidence. During the Soviet era, historians could not write history based on documents since they could not obtain access to state and design bureau archives. As the Soviet Union began to fall apart, historians such as Georgiy Vetrov began to take the first steps in document-based history. Vetrov, a former engineer at Korolev's design bureau, eventually compiled and published two extraordinary collections of primary documents relating to Korolev's legacy.[24] Now that all the state archives in Moscow—such as the State Archive of the Russian Federation (GARF), the Russian State Archive of the Economy (RGAE), and the Archive of the Russian Academy of Sciences (ARAN)—are open to researchers, more results of this "third generation" are beginning to appear. German historians such as Matthias Uhl and Cristoph Mick and those in the United States such as myself have been fortunate to work in Russian archives.[25] I would also note the enormous contributions of the Russian monthly journal *Novosti kosmonavtiki* (News of Cosmonautics) as well as the Belgian historian Bart Hendrickx in advancing the state of Soviet space history. The new work has opened opportunities for future research. For example, we no longer have to guess about the government's decision to approve development of the Soyuz spacecraft, we can see the original decree issued on 4 December 1963.[26] Similarly,

23. Vladislav Zubok and Constantine Pleshakov, *Inside the Kremlin's Cold War: From Stalin to Khrushchev* (Cambridge, MA: Harvard University Press), p. 172; Golovanov, *Korolev*, p. 454. For the correct citation on the Second Main Directorate, established on December 27, 1949, see Simonov, *Voyenno-promyshlennyy kompleks sssr*, pp. 225-226.

24. M. V. Keldysh, ed., *Tvorcheskoye naslediye Akademika Sergeya Pavlovicha Koroleva: izbrannyye trudy i dokumenty* (The Creative Legacy of Sergey Pavlovich Korolev: Selected Works and Documents) (Moscow: Nauka, 1980); G. S. Vetrov and B. V. Raushenbakh, eds., *S. P. Korolev i ego delo: svet i teni v istorii kosmonavtiki: izbrannyye trudy i dokumenty* (S. P. Korolev and His Cause: Shadow and Light in the History of Cosmonautics) (Moscow: Nauka, 1998). For two other published collections of primary documents, see V. S. Avduyevskiy and T. M. Eneyev, eds. *M. V. Keldysh: izbrannyye trudy: raketnaya tekhnika i kosmonavtika* (M. V. Keldysh: Selected Works: Rocket Technology and Cosmonautics) (Moscow: Nauka, 1988); B. V. Raushenbakh, ed., *Materialy po istorii kosmicheskogo korablya 'vostok': k 30-letiyu pervogo poleta cheloveka v kosmicheskoye prostranstvo* (Materials on the History of the 'Vostok' Space Ship: On the 30th Anniversary of the First Flight of a Human in Space) (Moscow: Nauka, 1991).

25. Matthias Uhl, *Stalins V-2: Der Technolgietransfer der deutschen Fernlen-kwaffentechnik in die UdSSR und der Aufbau der sowjetischen Raketenindustrie 1945 bis 1959* (Bonn, Germany: Bernard & Graefe-Verlag, 2001); Christoph Mick, *Forschen für Stalin: Deutsche Fachleute in der sowjetischen Rüstungsindustrie 1945-1958* (Munich: R. Oldenbourg, 2000); Asif A. Siddiqi, "The Rockets' Red Glare: Spaceflight and the Russian Imagination, 1857-1957, Ph.D. dissertation, Carnegie Mellon University, 2004.

26. "O sozdaniia kompleksa 'Soyuz' " (On the Creation of the Soyuz Complex), December 4, 1963, RGAE, f. 298, op. 1, d. 3495, ll. 167-292.

instead of speculating about the famous decree of 3 August 1964 that committed the Soviet Union to compete with the American Apollo program, we can study the actual government document issued on that date.[27] Academician Chertok deserves much credit for opening the doors for future historians, since his memoirs have guided many to look even deeper.

The distribution of material spanning the four volumes of Chertok's memoirs is roughly chronological. In the first English volume, Chertok describes his childhood, his formative years as an engineer at the aviation Plant No. 22 in Fili, his experiences during World War II, and the mission to Germany in 1945–46 to study captured German missile technology.

In the second volume, he continues the story with his return to the Soviet Union, the reproduction of a Soviet version of the German V-2 and the development of a domestic Soviet rocket industry at the famed NII-88 institute in the Moscow suburb of Podlipki (now called Korolev). He describes the development of the world's first intercontinental ballistic missile, the R-7; the launch of Sputnik; and the first generation probes sent to the Moon, Mars, and Venus.

In the third volume, he begins with the historic flight of Yuriy Gagarin, the first human in space. He discusses several different aspects of the burgeoning Soviet missile and space programs of the early 1960s, including the development of early ICBMs, reconnaissance satellites, the Cuban missile crisis, the first Soviet communications satellite Molniya-1, the early spectacular missions of the Vostok and Voskhod programs, the dramatic Luna program to land a probe on the Moon, and Sergey Korolev's last days. He then continues into chapters about the early development of the Soyuz spacecraft, with an in-depth discussion of the tragic mission of Vladimir Komarov.

The fourth and final volume is largely devoted to the Soviet project to send cosmonauts to the Moon in the 1960s, covering all aspects of the development of the giant N-1 rocket. The last portion of this volume covers the origins of the Salyut and Mir space station programs, ending with a fascinating description of the massive Energiya-Buran project, developed as a countermeasure to the American Space Shuttle.

It was my great fortune to meet with Academician Chertok in the summer of 2003. During the meeting, Chertok, a sprightly ninety-one years old, spoke passionately and emphatically about his life's work and remained justifiably proud of the achievements of the Russian space program. As I left the meeting, I was reminded of something that Chertok had said in one of his first public interviews in 1987. In describing the contradictions of Sergey Korolev's personality, Chertok had

---

27. "Tsentralnyy komitet KPSS i Sovet ministrov SSSR, postanovleniye" (Central Committee KPSS and SSSR Council of Ministers Decree), August 3, 1964, RGAE, f. 29, op. 1, d. 3441, ll. 299-300. For an English-language summary, see Asif A. Siddiqi, "A Secret Uncovered: The Soviet Decision to Land Cosmonauts on the Moon," *Spaceflight* 46 (2004): pp. 205-213.

noted: "This realist, this calculating, [and] farsighted individual was, in his soul, an incorrigible romantic."[28] Such a description would also be an apt encapsulation of the contradictions of the entire Soviet drive to explore space, one which was characterized by equal amounts of hard-headed realism and romantic idealism. Academician Boris Yevseyevich Chertok has communicated that idea very capably in his memoirs, and it is my hope that we have managed to do justice to his own vision by bringing that story to an English-speaking audience.

Asif A. Siddiqi
Series Editor
October 2004

---

28. Konovalov, "Ryvok k zvezdam."

Introduction to Volume II

As with Volume I, Boris Chertok has extensively revised and expanded the material in Volume II from the original Russian text. In this volume, Chertok takes up his life story after his return from Germany to the Soviet Union in 1946. At the time, Stalin had ordered the foundation of the postwar missile program at an old artillery factory northeast of Moscow. Chertok gives an unprecedented view into the early days of the Soviet missile program. During this time, the new rocket institute known as NII-88 mastered V-2 technology and then quickly outgrew German technological influence by developing powerful new missiles such as the R-2, the R-5M, and eventually the majestic R-7, the world's first intercontinental ballistic missile. With a keen talent for combining technical and human interests, Chertok writes of the origins and creation of the Baykonur Cosmodrome in a remote desert region of Kazakhstan.

He devotes a substantial portion of Volume II to describing the launch of the first Sputnik satellite and the early lunar and interplanetary probes designed under legendary Chief Designer Sergey Korolev in the late 1950s and early 1960s. He ends with a detailed description of the famous R-16 catastrophe known as the "Nedelin disaster," which killed scores of engineers during preparations for a missile launch in 1960.

Working on this project continues to be an extraordinary honor and pleasure. I owe a debt of gratitude to many for their hard work in bringing these stories to the English-speaking world. As before, I must thank historian Steve Garber, who supervised the entire project at the NASA History Division. He also provided insightful comments at every stage of the editorial process. Similarly, thanks are due to Jesco von Puttkamer for his continuing support in facilitating communications between the two parties in Russia and the United States. Without his enthusiasm, sponsorship, and support, this project would not have been possible.

Many others at NASA Headquarters contributed to publication of these memoirs, including NASA Chief Historian Steven J. Dick, Nadine J. Andreassen, William P. Barry, and others.

Heidi Pongratz at Maryland Composition oversaw the detailed and yet speedy copyediting of this book. Tom Powers and Stanley Artis at Headquarters acted as invaluable liaisons with the talented graphic design group at Stennis Space Center. At Stennis, Angela Lane handled the layout with skill and professional grace, Danny Nowlin did an expert job proofreading this book, and Sheilah Ware oversaw the production process. Headquarters printing specialists Jeffrey McLean and Henry Spencer professionally handled this last and crucial stage of production.

As series editor, my work was not to translate, a job that was very capably done by a team at award-winning TechTrans International, Inc., based in Houston, Texas. Their team included: Cynthia Reiser (translator), Lydia Bryans and Laurel Nolen (both editors), Alexandra Tussing and Alina Spradley (both involved in postediting), Trent Trittipo, Yulia Schmalholz, and Lev Genson (documents control), Daryl Gandy (translation lead), Natasha Robarge (translation manager), and Elena Sukholutsky.

I would also like to thank Don P. Mitchell, Olaf Przybilski, Peter Gorin, Dr. Matthias Uhl, and T. V. Prygichev for kindly providing photographs for use in Volume II. Finally, a heartfelt thank you to Anoo Siddiqi for her support and encouragement throughout this process.

As the series editor, my job was first and foremost to ensure that the English language version was as faithful to Chertok's original Russian version as possible. At the same time, I also had to account for the stylistic considerations of English-language readers who may be put off by literal translations. The process involved communicating directly with Chertok in many cases and, with his permission, taking liberties to restructure paragraphs and chapters to convey his original spirit. I also made sure that technical terms and descriptions of rocket and spacecraft design satisfied the demands of both Chertok and the English-speaking audience. Finally, I provided many explanatory footnotes to elucidate points that may not be evident to readers unversed in the intricacies of Russian history. Readers should be aware that all of the footnotes are mine unless cited as "author's note," in which case they were provided by Chertok.

Asif A. Siddiqi
Series Editor
June 2006

A Few Notes about
Transliteration and Translation

THE RUSSIAN LANGUAGE IS WRITTEN using the Cyrillic alphabet, which concists of 33 letters. While some of the sounds that these letters symbolize have equivalents in the English language, many have no equivalent, and two of the letters have no sound of their own, but instead "soften" or "harden" the preceding letter. Because of the lack of direct correlation, a number of systems for transliterating Russian (i.e., rendering words using the Latin alphabet), have been devised, all of them different.

Russian Alphabet	Pronunciation	US Board on Geographic Names	Library of Congress
А, а	ă	a	a
Б, б	b	b	b
В, в	v	v	v
Г, г	g	g	g
Д, д	d	d	d
Е, е	ye	ye* / e	e
Ё, ё	yŏ	yĕ* / ĕ	ë
Ж, ж	zh	zh	zh
З, з	z	z	z
И, и	ē	i	i
Й, й	shortened ē	y	ĭ
К, к	k	k	k
Л, л	l	l	l
М, м	m	m	m
Н, н	n	n	n
О, о	o	o	o
П, п	p	p	p
Р, р	r	r	r
С, с	s	s	s
Т, т	t	t	t
У, у	û	u	u
Ф, ф	f	f	f
Х, х	kh	kh	kh
Ц, ц	ts	ts	ts
Ч, ч	ch	ch	ch
Ш, ш	sh	sh	sh
Щ, щ	shch	shch	shch
Ъ	(hard sign)	"	"
Ы	gutteral ē	y̦	y̦
Ь	(soft sign)	̦	̦
Э, э	ĕ	e	ĭ
Ю, ю	yû	yu	iu
Я, я	yă	ya	ia

* Unitially and after vowels

For this series, Editor Asif Siddiqi selected a modification of the U.S. Board on Geographic Names system, also known as the University of Chicago system, as he felt it better suited for a memoir such as Chertok's, where the intricacies of the Russion language are less important than accessibility to the reader. The modifications are as follows:

- the Russian letters " ь " and " ъ " are not transliterated, in order to make reading easier;
- Russian letter " ё " is denoted by the English "e" (or "ye" initally and after vowels)—hence, the transliteration "Korolev", though it is pronounced "Korolyov".

The reader may find some familiar names to be rendered in an unfamiliar way. This occurs when a name has become known under its phonetic spelling, such as "Yuri" versus the transliterated "Yuriy," or under a different transliteration system, such as "Baikonur" (LoC) versus "Baykonur" (USBGN).

In translating *Rakety i lyudi,* we on the TTI team strove to find the balance between faithfulness to the original text and clear, idiomatic English. For issues of technical nomenclature, we consulted with Asif Siddiqi to determine the standards for this series. The cultural references, linguistic nuances, and "old sayings" Chertok uses in his memoirs required a different approach from the technical passages. They cannot be translated literally: the favorite saying of Flight Mechanic Nikolay Godovikov (Vol. 1, Chapter 7) would mean nothing to an English speaker if given as, "Now you see it, now you don't." The jargon used by aircraft engineers and rocket engine developers in the 1930s and 1940s posed yet another challenge. At times, we had to do linguistic detective work to come up with a translation that conveyed both the idea and the "flavor" of the original. Puns and plays on words are explained in footnotes. *Rakety i lyudi* has been a very interesting project, and we have enjoyed the challenge of bringing Chertok's voice to the English-speaking world.

TTI TRANSLATION TEAM
Houston, TX
October 2004

List of Abbreviations

AFU	Antenna Feeder System
AKT	Emergency Turbine Contact
AMS	Automatic Interplanetary Station
APR	Automatic Missile Destruction
AS	Automatic Station
ASSR	Autonomous Soviet Socialist Republic
AVD	Emergency Engine Shutdown
AVD-APR	Emergency Engine Shutdown and Emergency Missile Destruction
AVDU	Emergency Engine Unit Shutdown
BDU	Strapon Propulsion Unit
BESM	Large Electronic-Computation Machine
BKIP	On Board Power Switchboard
BMP	Armed Fighting Vehicle
BN	Ballistic Normal
BON	Special Purpose Brigade
BS	Ballistic Staged
EKR	Experimental Cruise Missile
EPAS	Apollo-Soyuz Experimental Flight
FED	Feliks Edmundovich Dzerzhinskiy
FIAN	Physical Institute of the Academy of Sciences
FTI	Physical-Technical Institute
FTU	Photo-Television Unit
GAU	Main Artillery Directorate
GAI	State Automobile Inspection
GAZ	Gorky Automobile Factory
GDL	Gas Dynamics Laboratory
GIPKh	State Institute of Applied Chemistry
GIRD	Group for the Study of Reactive Motion
GKAT	State Committee for Aviation Technology
GKOT	State Committee for Defense Technology
GKRE	State Committee for Radio Electronics

GKS	State Committee for Ship Building
GOKO	State Committee for Defense
Gosplan	State Planning Commission
Gossnab	Main Directorate for State Procurement
GSKB	State Special Design Bureau
GSKB Spetsmash	State Special Design Bureau for Special Machine Building
GSO	Approximate Solar Orientation
GTsKB	State Central Design Bureau
GTsP	State Central Firing Range
GULAG	Main Directorate of Labor Camps
IKI	Institute of Space Research
IP	Tracking Station
KB	Design Bureau
KBV	Traveling Wave Coefficient
KD	Contact Sensor
KDI	Design Development Test
KDU	Correction Engine Unit
KIK	Command-Measurement Complex
KIS	Control And Testing Station
KN	Winged Normal
KR	Winged Staged
KRL	command radio-link
KRZ	Kiev Radio Factory
KS	Staged Winged
KUNG	All-Purpose Standard Clearance Body
LII	Flight-Research Institute
LIPAN	Academy of Sciences Instrumentation Laboratory
LKI	Flight-Development Test
LMZ	Leningrad Metal Works
LVMI	Leningrad Military-Mechanical Institute
MEI	Moscow Power Engineering Institute
MGU	Moscow State University
MIFI	Moscow Engineering and Physics Institute
MIIGAiK	Moscow Engineering Institute of Geodesy, Aerial Surveying and Cartography
MIK	Assembly and Testing Building
MNII	Naval Scientific-Research Institute
MNIIEM	Moscow Scientific-Research Institute of Electromechanics
MOM	Ministry of General Machine Building
MPSS	Ministry of the Communications Systems Industry
MVTU	Moscow Higher Technical School
NDMG	Unsymmetrical Dimethylhydrazine
NII	Scientific-Research Institute

NIIAP	Scientific-Research Institute of Automation and Instrumentation Building
NII Avtomatiki	Scientific-Research Institute of Automatics
NIIIT	Scientific-Research Institute of Current Sources
NIIP	Scientific-Research and Test Firing Range
NIIPM	Scientific-Research Institute of Applied Mathematics
NIP	Ground Measurement Point
NIR	Scientific-Research Work
NIRA	Scientific Institute of Reactive Aviation
NISO	Scientific Institute for Aircraft Equipment
NKVD	People's Commissariat of Internal Affairs
NTS	Scientific-Technical Council
OKB	Experimental Design Bureau
OPM	Department of Applied Mathematics
Ostekhbyuro	Special Technical Bureau
PGU	First Main Directorate
PIK	Floating Measurement Complex
POS	Tin And Lead Alloy
PS	Simple Satellite
PSO	Constant Solar Orientation
PTR	Programmed Current Distributor
PVRD	Ramjet
PVU	Programmed Timing Device
RKK Energiya	Energiya Rocket-Space Corporation
RKKA	Workers' and Peasants' Red Army
RKO	Radio Control Orbit
RKS	Apparent Velocity Regulation/Control
ROKS	Aircraft Coordinate Radio Locator (22)
RNII	Reactive Scientific-Research Institute
RUP	Radio-Control Ground Station
RVGK	Supreme Command Reserve
RVSN	Strategic Rocket Forces
SAS	Emergency Rescue System
SB	Special Bureau
SKB	Special Design Bureau
SOB	Tank Emptying System
SOBIS	Tank Depletion System and Synchronization
SOZ	Startup Support System
SPVRD	Supersonic Ramjet
SUBK	Onboard Complex Control System
SUK	Solar Heading Indicator
SVA	Soviet Military Administration
TASS	Telegraph Agency of the Soviet Union

TGU	Third Main Directorate
TNA	Turbopump Assembly
TOGE	Pacific Ocean Hydrographic Expedition
TP	Engineering Facility
TsAGI	Central Aerohydrodynamics Institute
TsAKB	Central Artillery Design Bureau
TsIAM	Central Scientific Institute for Aviation Motor Construction
TsKB	Central Design Bureau
TsNII	Central Scientific-Research Institute
TsNIIAV	Central Scientific-Research Institute for Artillery Armaments
TsNIIChernmet	Central Scientific-Research Institute for Black Metallurgy
TsNIIMash	Central Scientific-Research Institute of Machine Building
TsSKB	Central Specialized Design Bureau
TU	Technical Condition
UD	Administration
VDNKh	Exhibitions of Achievements of the National Economy
VEI	All-Union Electrical Institute
VISKhOM	All-Union Institute of Agricultural Machine Building
VKP(b)	All-Union Communist Party (Bolsheviks)
VNIIEM	All-Union Scientific-Research Institute of Electromechanics
VNIIT	All-Union Scientific-Research Institute of Current Sources
VPK	Military-Industrial Commission
VSNKh	All-Russian Council of the National Economy
VV	Explosive Matter
ZIM	V. M. Molotov Factory
ZIS	Stalin Factory

Chapter 1

Three New Technologies, Three State Committees

During World War II, fundamentally new forms of weapons technology appeared—the atomic bomb, radar, and guided missiles. Before I resume my narrative, in this chapter, I will write about how the Soviet Union organized work in these three new fields through a system of three "special committees" organized at the highest levels.

WORLD WAR II FORCED US TO LEARN QUICKLY. Despite evacuations, relocations, reconstruction, building from scratch, and losing factories in the Ukraine and Byelorussia, after two years of war, our aircraft, artillery, tank, and munitions industries were producing such quantities of guns, tanks, and airplanes that the course of the war was radically altered. We overcame the mortal danger of total defeat during the first two years of the war. Beginning in mid-1943, we became hopeful that we would not only save our country, but would also defeat Nazi Germany. However, to achieve this superiority in manpower, the heroism of soldiers and officers was not enough.

According to the most optimistic calculations, a year-and-a-half to two years of war lay ahead of us. Despite the human losses—from prewar repressions, the deaths of scientist-volunteers in the militias in 1941, and all those who starved to death during the siege of Leningrad—the Soviet Union retained its intellectual potential, enabling it not only to improve the weapons it had, but also develop fundamentally new weapons.

Setting up operations to deal with the new challenges required the recruiting of scientists released from their wartime work routine and necessitated the introduction of a new system of research and development. Soon, the People's Commissars recognized (and then prompted the members of Stalin's Politburo to grasp) the need to coordinate all the basic operations in these fields at the state level, conferring on them the highest priority. But priority over what? Over all branches of the defense industry?

The experience of war had taught us that conventional weapons attain new levels of capability and become much more effective when combined with modern systems, for example, when aircraft are equipped with radar, when anti-aircraft batteries fire according to the precise target indications of radar fire control systems rather than the readings of antediluvian sound rangers, when missiles use radio guidance, when airplanes could carry atomic bombs, and on and on—the prospects were limitless. During the war it was still too early to limit the production of conventional weapons, but they had to be upgraded according to new trends. That being the case, where were the resources to come from?

There remained the tried and true "mobilization economy" method, that is, take everything you could from all the branches of industry responsible for producing conventional civilian goods.[1] In addition, after the defeat of Germany, we could restructure conventional weapons production to benefit new fields and also use the potential of captured German technology.

During the war, the aircraft, artillery, and tank industries' mass production process had become highly developed and had accumulated tremendous organizational experience. But what should be the path for new technologies? Should the new industries be entrusted to individual People's Commissariats?[2] Even before we began our work on rockets in Germany, scientists—nuclear and radio engineers—had sensed and had convinced high-ranking officials that such problems required an integrated systematic approach not only in the field of science but also in terms of management. The challenge required a special supervisory agency headed by a Politburo member, who would report directly to Stalin and who would be authorized, unhindered by bureaucratic red tape, to make rapid decisions on the development of the new technology that would be binding for everyone, regardless of departmental subordination.

THE FIRST SUCH GOVERNMENTAL AGENCY TO BE ESTABLISHED WAS FOR DOMESTIC RADAR TECHNOLOGY. With radar, the senior leadership had the most clarity as to its "why and wherefore." On 4 June 1943, on the eve of the great battle of Kursk, the State Defense Committee (GOKO) issued a decree signed by Stalin "On the Creation of the GOKO Radar Council."[3] Stalin appointed G. M. Malenkov as Council Chairman.[4] This decree, which appeared during the most trying wartime period, was the most critical governmental resolution for our radar development. By forming this council, supervision over the development of this new branch of technology and the implementation of an extensive set of measures in what had previously been isolated organizations was concentrated in the hands of a single governmental

---

1. Broadly speaking, "mobilization economics" in the Soviet context meant massive state diversion of industrial resources to wartime needs, as happened during World War II.

2. People's commissariats were governmental bodies equivalent to industrial ministries. After 1946, all Commissariats were renamed ministries.

3. GOKO—*Gosudarstvennyy komitet oborony.*

4. Georgiy Maksimilianovich Malenkov (1902–88) was one of the top government administrators during the Stalin era. In 1953, he succeeded Stalin as Chairman of the USSR Council of Ministers, serving in that position until 1955, when he was effectively ousted by Nikita Khrushchev.

agency. However, no matter how perfect the organizational structure, it is the leaders who determine the success. Amazingly, all three new fields were blessed with true leaders, all engineer-scientists.

The most brilliant figure in the history of domestic radio engineering was Radar Council Deputy Chairman Aksel Ivanovich Berg. He was a top-level scientist, military chief, and bold government official combined in one person. I first met Aksel Berg in late 1943. At Factory No. 293 in Khimki we were trying to develop the Aircraft Coordinate Radio Locator (ROKS) system for the flight control of the BI fighter.[5] My deputy for radio engineering, Roman Popov, said that without Aksel Ivanovich's help, nothing we were doing would work. He mustered the courage to invite him to Khimki.

At that time, Berg occupied the post of Deputy People's Commissar of the Electrical Industry. He was also Malenkov's deputy on the Radar Council, and a month earlier he had been selected as a corresponding member of the Academy of Sciences. In person, Aksel Ivanovich in no way matched the mental image that I had formed in my high school days of this respected scientist with the title of professor. I had spent my last two years in high school sitting long into the night in the Lenin Library striving to grasp the theoretical fundamentals from Professor Berg's book *Radio Engineering*.[6] Fifteen years had passed since that time. Rather than an elderly professor, it was a seaman with the rank of Vice Admiral who came to see us in Khimki. Berg quickly went over the naïve proposals of these young air defense enthusiasts, gave us practical advice—not at all professorial—and promised us real assistance. He made good on his promises, although we never finished ROKS because of other circumstances.

Twenty-five years later, I saw 75-year-old academician Berg at a meeting of our Academy of Sciences department. He was still as vibrant and unique as he had always been.

Festive celebrations were held for Berg's 70th birthday in 1963 and later his 75th birthday in 1968. His unusual biography became available to the scientific community at the time. Aksel Berg's father was a Swede and his mother an Italian. No matter how hard the pseudo-patriotic biographers tried, they could not find a drop of Russian blood in him. During World War I, the 22-year-old Berg was a submarine navigator, becoming a submarine commander after the Revolution. Following the civil war, Berg graduated from the Naval Academy, stayed on there as a radio engineering instructor, and attained the academic title of professor and the military rank of captain first class.

How could the vigilant security services resign themselves to the fact that a

---

5. ROKS—*radioopredelitel koordinat samoleta.*

6. More recent editions were published as A. I. Berg, and I. S. Dzhigit, *Radiotekhnika i elektronika i ikh tekhnicheskoye primeneniye* [*Radio Engineering and Electronics and Their Technical Applications*] (Moscow: AN SSSR, 1956).

person of obscure nationality and a former tsarist officer was training Red Navy commanders? To be on the safe side, they arrested this already well-known professor and author of the most current work on the fundamentals of radio engineering. However, sober heads prevailed and they released Berg and conferred on him the rank of rear admiral. Berg never lost his sense of humor. He had a simple explanation for his elevation in rank: "They accused me of being a counterrevolutionary conspirator. Over the course of the investigation the charge was dropped, but I held onto the first part of the accusation and tacked on 'admiral'."[7]

In March 1943, Berg was recalled from the Naval Academy and appointed deputy people's commissar of the electrical industry. Remaining in that office until October 1944, Aksel Ivanovich managed the daily operations of the Radar Council and of the entire radio industry, which was part of the People's Commissariat of the Electrical Industry.

In June 1947, the Radar Council was converted into Special Committee No. 3, or the Radar Council under the USSR Council of Ministers. M. Z. Saburov, Chairman of the USSR *Gosplan*, was appointed council chairman.[8] A. I. Shokin, who would later become deputy minister of the radio electronic industry and then minister of electronics industry, managed the committee's day-to-day activity.

Berg organized and became the director of the head Central Scientific-Research Institute No. 108 (TsNII-108) under the Radar Committee.[9] From 1953 through 1957, he occupied the high-ranking post of USSR deputy minister of defense. Berg infused the working environment with new and creative plans. He immediately proposed radical designs and unwaveringly rejected slipshod work. Among scientists, Aksel Ivanovich possessed a vibrant individuality. In spite of years of repression, he did not hesitate to express his sometimes very blunt opinions on matters of technical progress and economic policy. During the postwar years, he very boldly spoke out in defense of cybernetics as a science, despite the fact that officially, just like genetics, it had also been persecuted.[10] Berg, who had developed methods for calculating the reliability of systems that contained a large number of elements, even got involved in debates with our chief designers.

The Radar Committee was abolished in August 1949, and its responsibilities were divided among the Ministry of Armed Forces and the ministries of the various branches of the defense industry. In 1951, drawing on the personnel from the

---

7. The word for counterrevolutionary in Russian is *kontrrevolutsionnyy,* and the word for rear admiral is *kontr-admiral,* hence the play on words.

8. *Gosplan—Gosudarstvennaya planovaya komissiya* (State Planning Committee)—founded in 1921 by the Council of People's Commissars, was in charge of managing allocations for the Soviet economy.

9. TsNII—*Tsentralnyy nauchno-issledovatelskiy institut.*

10. For works on the ideological battles over genetics and cybernetics in the Soviet Union, see Nikolai Krementsov, *Stalinist Science* (Princeton, NJ: Princeton University Press 1997); Slava Gerovitch, *From Newspeak to Cyberspeak: A History of Soviet Cybernetics* (Cambridge, MA: The MIT Press, 2002).

abolished committee, under the aegis of Lavrentiy Beriya, the Third Main Director-ate (TGU) was created under the USSR Council of Ministers.[11] The Third Main Directorate was entrusted with the task of missile defense. Ryabikov was appointed the direct chief, and Kalmykov, Vetoshkin, and Shchukin were appointed his deputies.[12]

By this time, Korolev and his deputies—Vasiliy Mishin, Konstantin Bushuyev, and I—had already had the opportunity to develop a closer relationship with Valeriy Kalmykov. In 1948, he was director of Scientific-Research Institute No. 10 (NII-10) of the Ministry of the Shipbuilding Industry, where Viktor Kuznetsov worked.[13] Kuznetsov had been appointed the chief designer of gyroscopic command instruments for all of our rockets.

At the beginning, Kalmykov received us very cordially and personally led us on a tour of the laboratories, demonstrating the mockups and newly developed operational detection and ranging systems. He was most interested in thermal detection and ranging in the infrared range. He demonstrated one project, a thermal detector, aiming it from the laboratory window at distant factory smokestacks that were barely perceptible by the naked eye. The effect was impressive. Kalmykov was very well-liked, not only as the director of a giant institute, but simply as a friendly, intelligent person with a good sense of humor, a quality he demonstrated over tea, pulling Vitya Kuznetsov's leg about his stay in Berlin in 1941 as a "prisoner" of the Germans at the beginning of the war.[14]

In 1954, Kalmykov was appointed minister of the radio engineering industry. I often had to meet with him, in the different setting of his office or at the test range. His unfailing tact, competence, and friendly nature (which not every minister is able to maintain, even if he possessed those qualities before his appointment) facilitated decision-making on the most convoluted interdepartmental, organizational, and technical matters. Among the very many ritual farewells that have taken place over the last several decades at Novodevichye Cemetery, I recall with great sorrow my final goodbye to Valeriy Dmitriyevich Kalmykov.[15] The successes of the radio electronic industry were of decisive importance for the subsequent evolution of rocket-space technology. That is why I felt it necessary to make this digression into history.

---

11. TGU—*Tretye glavnoye upravleniye.* The Soviet government initiated the air defense project in August 1950 and organized the TGU the following February to manage the program.

12. Valeriy Dmitriyevich Kalmykov (1908–74), Sergey Ivanovich Vetoshkin (1905–91), and Aleksandr Nikolayevich Shchukin (1900–) later became high-level managers in the Soviet military-industrial complex.

13. NII—*Nauchno-issledovatelskiy institut.*

14. *Author's note:* In the summer of 1941, V. I. Kuznetsov was sent to Berlin on a temporary assignment. When the war started, like all Soviet citizens in Germany, he was interned and later made a long trip through neutral countries to return to the USSR.

15. Kalmykov died in 1974 at the age of 65.

From the author's archives.

In 1947, Sergey Korolev created one of the most innovative management mechanisms in the early Soviet missile program—the Council of Chief Designers. This photo, a still from a rare film from the postwar years, shows the original members of the Council and Boris Chertok at a meeting. From the left, Chertok, Vladimir Barmin, Mikhail Ryazanskiy, Korolev, Viktor Kuznetsov, Valentin Glushko, and Nikolay Pilyugin (standing).

THE LEADERSHIP OF THE ATOMIC PROBLEM or, as it was sometimes called, the "uranium project," followed a slightly different script. While military and defense industry leaders took the initiative in gathering specialists and organizing the Radar Committee, in the case of atomic weaponry, it was the scientists and physicists who advocated for centralization from the very beginning, as was the case in the United States and Germany. However, because of their modesty, having been brought up working on laboratory-sized projects, they did not always dare to take away the country's essential vital resources. As early as 1942, I. V. Kurchatov was entrusted with managing the scientific aspects of the problem at the recommendation of Academician A. F. Ioffe. Stalin personally supervised the operations. But as the scale of operations expanded, a small governmental staff was required.

At first, Deputy Chairman of the Council of People's Commissars M. G. Pervukhin was in charge of organizing atomic projects.[16] He was simultaneously the People's Commissar of the Chemical Industry. Soon, it became apparent that the

---

16. The Council of People's Commissars was the equivalent of the governmental cabinet in the Soviet system. In 1946, it was renamed the USSR Council of Ministers.

expenses and scale of the projects required new efforts from a half-starved people and a country that had not yet recovered from wartime ravages. In addition, following the Americans' example, the highest degree of secrecy needed to be ensured. Only the department of the all-powerful Lavrentiy Beriya could provide such a regime.[17]

On 20 August 1945, the State Defense Committee passed the decree for the organization of a special committee under GOKO, which would be also called Special Committee No. 1. According to the decree, the Special Committee comprised the following members:

1. L. P. Beriya (Chairman)
2. G. M. Malenkov
3. N. A. Voznesenskiy
4. B. L. Vannikov (Deputy Chairman)
5. A. P. Zavenyagin
6. I. V. Kurchatov
7. P. L. Kapitsa
8. V. A. Makhnov
9. M. G. Pervukhin (Deputy Chairman)

The decree stated:

"The Special Committee under GOKO shall be entrusted with the management of all projects researching the nuclear energy of uranium, as well as the construction of nuclear power plants and the development and production of an atomic bomb."[18]

The document was long and very detailed. It relieved Beriya of his duties as the people's commissar for internal affairs, but to make up for it he received absolutely unlimited authority to create the nuclear industry. In connection with this, he was soon named first Deputy Chairman of the Council of People's Commissars. This same decree entrusted B. L. Vannikov, the People's Commissar of Ammunition to be Beriya's first deputy in the Special Committee. Vannikov organized and headed the First Main Directorate (PGU), which in fact meant he was the first nuclear minister of the USSR.[19]

Besides all the other advantages that Beriya had over conventional ministers, he had at his disposal an unknown number of workers, laboring without pay—the

---

17. Lavrentiy Pavlovich Beriya (1899–1953) was the feared manager of the Soviet security services. Between 1938 and 1945, he headed the NKVD, the predecessor to the KGB.

18. The GOKO decree No. 9887ss/op, issued on August 20, 1945 was first published in V. I. Ivkin, "Posle Khirosimy i Nagasaki: s chego nachinalsya yadernyy pariter" [After Hiroshima and Nagasaki: The Origin of Nuclear Parity], *Voyenno-istoricheskiy Zhurnal* [Military-Historical Journal], 4 (1995):65–67.

19. PGU—*Pervoye glavnoye upravleniye*. The PGU was the management and administrative branch of the Special Committee for the atomic bomb.

inmates of the "GULAG Archipelago" and an army of the internal troops of the People's Commissariat of Internal Affairs (the NKVD) numbering many thousands.[20]

Beriya's deputy, Boris Lvovich Vannikov, was a very colorful figure. Not very tall, quite energetic, typically Jewish in appearance, sometimes rudely cynical, sometimes very blunt, and friendly and amicable when necessary, he possessed quite exceptional organizational skills. In 1941, he held the post of People's Commissar of Armaments, and right before the war he was arrested. He was kept in solitary confinement at Lubyanka Prison, in the same building where the office of the all-powerful People's Commissar Beriya was located. Who would have surmised that four years later he would be Beriya's deputy for the creation of nuclear weaponry? While Vannikov was in prison, his position was filled by the 33-year-old director of the Bolshevik Factory in Leningrad, Dmitriy Fedorovich Ustinov.

The war required just as much effort and heroism from industry as it did from the army. A story, which sounded like it might even be true, was in circulation to the effect that two months into the war, when enormous lapses were discovered in supplies of shells, mines, and even cartridges, Stalin asked Beriya about Vannikov's fate. He was quickly given some medical treatment to make him at least look healthy after his stay in Lubyanka Prison and delivered to Stalin, who, as if nothing had happened, offered Vannikov, an "enemy of the people," the post of People's Commissar of Ammunition and asked him "not to hold any grudges over what had happened."

Thus, Vannikov and Ustinov, who had replaced him, worked in tandem almost throughout the entire war.[21] During the war, Vannikov's tremendous contribution was to eliminate problems in ammunitions production and delivery. Therefore, it was not the least bit surprising that Stalin and Beriya, despite Vannikov's past and his Jewish ethnicity, put him in charge of all operations for the development of the atomic bomb as head of the First Main Directorate.

By late 1947, when we began our campaign in Moscow to bring in specialists from various enterprises and institutes for our work on rocketry, we often ran up against the all-powerful, super-secret, but very broad-based personnel recruiting system, which snatched the tastiest morsels right out of our mouths. This was Vannikov's atomic system already at work. He was using Beriya's staff on his own behalf. In 1947, Kurchatov was the all-powerful scientific chief of the field. He was director of the Academy of Sciences' Instrumentation Laboratory (LIPAN).[22] Today, the enormous Kurchatov Atomic Energy Institute stands on the former site of LIPAN.

During those first years of the rocket industry's formation, Korolev—who is

---

20. The Main Directorate of Correctional Labor Camps (*Glavnoye upravleniye ispravitelno-trudovykh lagerey*, GULAG) was a vast system of prison labor camps spread throughout the remote areas of the Soviet Union. The NKVD—*Narodnyy komissariat vnutrennykh del* (People's Commissariat of Internal Affairs)—was the precursor of the KGB.

21. Vannikov was the commissar of ammunition (1942–46), and Ustinov was the commissar of armaments (1941–46).

22. LIPAN—*Laboratoriya izmeritelnykh priborov akademii nauk.*

often compared with Kurchatov in history-themed journalism in terms of his accomplishments—could in no way be compared with him in terms of power and resources. And in terms of material support for the laboratories, and the scientists' and specialists' standard of living, we in the missile industry looked like "poor relatives" compared with the nuclear scientists. Until the last few years, in terms of their services and utilities, the comfortable standard of living, cultural and social amenities, child-care and medical services, and supplies of fresh produce and household goods, there was absolutely no comparison between the closed atomic cities and the "rocket towns" built at Kapustin Yar, Tyuratam, and Plesetsk and the numerous ground measurement stations (NIPs) located throughout the country.[23] When our professional collaboration with the nuclear scientists began in 1952, we discovered with some envy what limitless resources they had for production, experimental facilities, residential construction, and other goods in short supply. Korolev took the fact that we were "lagging behind" very hard, and often complained to Ustinov, who, he felt, underestimated our work. Now, many years later, one can see that it was not Ustinov's doing at all. The country wasn't capable of creating such comfortable conditions for everyone working in the three fields of nuclear, missiles, and radar.

We in the rocket industry worked together with the Ministry of Defense and with army personnel, but our facilities were built not by GULAG prisoners, but by military builders; the corresponding main directorates of the Ministry of Defense supervised the operation and acceptance of our work. In other words, we dealt with soldiers and officers who themselves led a semi-destitute existence.

STATE COMMITTEE NO. 2, OR SPECIAL COMMITTEE NO. 2, as it was sometimes called, was second according to numeric designation, but it was the third one to be organized after the atomic and radar committees. It was created by special decree of the Central Committee and Council of Ministers dated 13 May 1946, No. 1017-419. This decree is the document that marked the beginning of the organization of large rocket technology operations in the Soviet Union. Naturally, this decree came out too early to mention cosmonautics or the use of outer space for peaceful or scientific purposes. It discussed the organization and distribution of responsibilities among ministries and enterprises for the development of rockets for purely military purposes and for the use of the contingent of German specialists.

The reader will find it useful to spend a little time perusing the full text of the decree of the USSR Council of Ministers dated 13 May 1946, cited below. Studying this text will facilitate the understanding of many subsequent events in the history of the establishment of rocket technology and of the role of specific individuals in this history.[24]

---

23. NIP—*Nazemnyy izmeritelnyy punkt.*

24. This text of this decree, which Chertok presents, was first published openly in 1994 in a book published by the Russian Strategic Rocket Forces. See "Voprosy reaktivnogo vooruzheniya." In I. D. Sergeyev, ed., *Khronika osnovnykh sobytiy istorii raketnykh voysk strategicheskogo naznacheniya* (Moscow: TsIPK 1994), pp. 227–234.

To be returned within 24 hours to the USSR Council of Ministers
Administration (U.D.) special group[25]

SECRET
(SPECIAL FILE)

USSR COUNCIL OF MINISTERS
DECREE No. 1017-419 *top secret*
13 May 1946, Moscow, Kremlin
<u>*On Questions of Reactive Armaments*</u>

Considering the creation of reactive armaments and the organization of scientific-research and experimental work in this field a vital task, the USSR Council of Ministers
DECREES

I.

1. To create a Special Committee for Reactive Technology under the USSR Council of Ministers with the following members:

G. M. Malenkov	– chairman
D. F. Ustinov	– deputy chairman
I. G. Zubovich	– deputy chairman, having been relieved of his duties at the Ministry of the Electrical Industry
N. D. Yakovlev	– Committee member
N. I. Kirpichnikov	– Committee member
A. I. Berg	– Committee member
P. N. Goremykin	– Committee member
N. E. Nosovskiy	– Committee member

2. To entrust the Special Committee for Reactive Technology with the following responsibilities:

a) Supervise the development of scientific-research, design, and practical operations for reactive armaments; review and submit plans and programs directly for the approval of the Chairman of the USSR Council of Ministers; develop scientific research and practical operations in the aforementioned field; and also specify and approve quarterly needs for monetary appropriations and material and technical resources for reactive armaments projects;

b) Track the completion status of the scientific research, design, and practical operations assigned by the Council of Ministers to the ministries and departments involved with reactive equipment;

c) Cooperate effectively with the appropriate ministries and departmental directors to ensure the timely fulfillment of the aforementioned assignments;

3. The Special Committee shall have its own staff.

4. To establish that the work fulfilled by the ministries and departments on reactive

---

25. UD—*Upravleniye delami.*

armaments shall be monitored by the Special Committee for Reactive Technology. No institutions, organizations, or individuals shall have the right to interfere with or ask for information concerning the work being conducted on reactive armaments without the special permission of the Council of Ministers.

5. The Special Committee for Reactive Technology must submit its plan of scientific research and experimental operations for 1946-1948 to the Chairman of the USSR Council of Ministers for approval. Its top-priority task will be the reproduction of V-2 (long-range guided missiles) and Wasserfall (surface-to-air guided missiles) rockets using domestic materials.

II.

6. The following shall be designated as the head ministries for the development and production of reactive armaments:

a) Ministry of Armaments—for missiles with liquid-propellant rocket engines;

b) Ministry of Agricultural Machine Building—for missiles with solid-propellant rocket engines;

c) Ministry of Aviation Industry—for cruise missiles.

7. To establish that the primary ministries involved with subcontractor production and tasked to carry out scientific research, design, and experimental operations, and also to fulfill orders for the head ministries approved by the Committee shall be:

a) Ministry of Electrical Industry—for ground-based and onboard radio control equipment, tuning equipment and television mechanisms, and radar stations for target detection and ranging;

b) Ministry of Shipbuilding Industry—for gyroscopic stabilization equipment, resolvers, naval radar stations for target detection and ranging, shipborne launcher stabilization systems, homing missile warheads for use against undersea targets, and for [other] instruments;

c) Ministry of Chemical Industry—for liquid propellants, oxidizers, and catalysts;

d) Ministry of Aviation Industry—for liquid-propellant rocket engines for long-range rockets and aerodynamic research and rocket tests;

e) Ministry of Machine Building and Instrumentation—for mountings, launch equipment, various compressors, pumps and equipment for them, as well as other accessory equipment;

f) Ministry of Agricultural Machine Building—for proximity fuses, munitions, and gunpowder.

III.

8. In the interests of fulfilling the tasks entrusted to the ministries, the following directorates shall be created:

in the Ministries of Armaments, Agricultural Machine Building, and the Electrical Industry—Main Directorates for reactive technology;

in the USSR Ministry of Armed Forces—a Directorate of reactive armaments within the structure of the GAU and a directorate of reactive armaments within the structure of the Navy;[26]

---

26. GAU—*Glavnoye artilleriyskoye upravleniye* (Main Artillery Directorate).

11

in the Ministries of Chemical Industry, Shipbuilding Industry, and Machine Building and Instrumentation—directorates of reactive technology;

in the Gosplan of the USSR Council of Ministers—a department of reactive technology headed by a deputy chairman of Gosplan.

9. The following scientific-research institutes, design bureaus, and test ranges for reactive technology shall be created in:

a) Ministry of Armaments—Scientific-Research Institute of Reactive Armaments and Design Bureau using the facilities of Factory No. 88, taking all its other programs and distributing them among the other Ministry of Armaments factories;[27]

b) Ministry of Agricultural Machine Building—Scientific-Research Institute of Solid-propellant Reactive Projectiles using the facilities of State Central Design Bureau No. 1 (GTsKB-1), a design bureau using the facilities of the Ministry of Aviation Industry NII-1 Branch No. 2, and the Scientific-Research Test Range for Reactive Projectiles using the facilities of the Sofrinsk Test Range;[28]

c) Ministry of Chemical Industry—Scientific-Research Institute of Chemicals and Propellants for Rocket Engines;

d) Ministry of Electrical Industry—Scientific-Research Institute with a design bureau for radio and electronic control instruments for long-range and surface-to-air missiles using the facilities of the NII-20 telemetry laboratory and Factory No. 1. Task Comrade Bulganin with reviewing and making a decision on the issue of transferring Factory No. 1 of the Ministry of Armed Forces to the Ministry of Electrical Industry so that the responsibility for this factory's program will rest with the Ministry of the Electrical Industry;

e) USSR Armed Forces Ministry—GAU Scientific-Research Reactive Institute and State Central Test Range for Reactive Technology for all of the ministries involved with reactive armaments.

10. It shall be the responsibility of the Ministries of Armaments (Ustinov), Agricultural Machine Building (Vannikov), Electrical Industry (Kabanov), Shipbuilding Industry (Goreglyad), Machine Building and Instrumentation (Parshin), Aviation Industry (Khrunichev), Chemical Industry (Pervukhin), and the Armed Forces (Bulganin) to approve the structures and staff of the directorates, NIIs, and design bureaus of the corresponding ministries.

IV.

11. The following work on reactive technology in Germany shall be considered top-

---

27. This organization eventually became Scientific-Research Institute No. 88 (NII-88), which was the seed of the Soviet missile and space industry.

28. GTsKB—*Gosudarstvennoye tsentralnoye konstruktorskoye byuro*. GTsKB-1 later became NII-1, and finally the Moscow Institute of Thermal Technology, the developer of modern-day Russian mobile ICBMs such as the Topol. The NII-1 Branch No. 2 was later successively known as KB-2 and GSNII-642. Currently, it is known as GNIP OKB Vympel and develops ground and launch equipment for the Russian space program.

priority tasks:

a) The complete restoration of the technical documentation and models of the V-2 long-range guided missile and Wasserfall, Rheintochter, and Schmetterling surface-to-air guided missiles;

b) The restoration of the laboratories and test rigs with all the equipment and instrumentation required to perform research and experimentation on V-2, Wasserfall, Rheintochter, Schmetterling, and other rockets;

c) The training of Soviet specialists who would master the design of V-2, surface-to-air guided missiles, and other rockets, testing methods, and production processes for rocket parts, components, and their final assembly.

12. Comrade Nosovskiy shall be named director of operations for reactive technology in Germany and shall reside in Germany. He shall be released from other work not related to reactive armaments. Comrades Kuznetsov (GAU) and Gaydukov shall be appointed as Comrade Nosovskiy's assistants.

13. The Reactive Technology Committee shall be responsible for selecting the necessary number of specialists with various backgrounds from the corresponding ministries and sending them to Germany to study and work on reactive armaments, keeping in mind that each German specialist shall be assigned a group of Soviet specialists so that the latter may gain experience.

14. The ministries and departments shall be forbidden to recall, unbeknownst to the Special Committee, their employees working on committees studying German reactive armaments in Germany.

15. The Ministries of Armaments, Agricultural Machine Building, Aviation Industry, Electrical Industry, Chemical Industry, Machine Building and Instrumentation, and the USSR Armed Forces shall have one month to prepare and submit for the approval to the Special Committee for Reactive Technology specific plans for design, scientific-research, and experimental operations in Germany on reactive armaments, specifying assignments and deadlines for each design bureau.

Comrades Ustinov, Yakovlev, and Kabanov shall be sent on assignment to Germany with a group of specialists for 15 days in order to familiarize themselves with the work being conducted on reactive armaments in Germany, with a view toward preparing a plan for impending operations.

16. The USSR Ministry of Armed Forces shall be tasked with forming a special artillery unit in Germany to master, prepare, and launch V-2 rockets.

17. The transfer of the design bureaus and German specialists from Germany to the USSR by the end of 1946 shall be predetermined.

It shall be the responsibility of the Ministries of Armaments, Agricultural Machine Building, Electrical Industry, Aviation Industry, Chemical Industry, and Machine Building and Instrumentation to prepare facilities for the placement of the German design bureaus and specialists. The Special Committee for Reactive Technology shall submit proposals on this matter to the USSR Council of Ministers within a month.

18. The Special Committee for Reactive Technology shall be permitted to pay a higher salary to German specialists recruited for work involving reactive technology.

19. It shall be the responsibility of the USSR Ministry of Armed Forces (Khrulev) to allocate the following items in support of all the Soviet and German specialists involved in work on reactive armaments in Germany:

free rations per norm No. 11—1000 units;

supplementary rations per norm No. 2—3000 units;

vehicles: passenger cars—100 units;

trucks—100 units;

provide fuel and drivers.

20. It shall be the responsibility of the USSR Ministry of Finance and the Soviet Military Administration in Germany to allocate 70 million marks to finance all of the operations conducted by the Special Committee for Reactive Technology in Germany.

21. The Special Committee for Reactive Technology shall be granted permission to order various special equipment and hardware in Germany for the laboratories of the scientific-research institutes and for the State Central Test Range for Reactive Armaments as reparations. The Special Committee jointly with Gosplan and Ministry of Foreign Trade shall be charged with specifying a list of orders and their delivery dates.

22. The Special Committee shall be assigned to submit proposals to the USSR Council of Ministers concerning a business trip by a commission to the U.S. to place orders and procure equipment and instruments for the laboratories of the scientific-research institutes for reactive technology, having stipulated in these proposals that the commission be granted the right of procurement by public license for a sum of 2,000,000 dollars.

23. Deputy Minister of Internal Affairs Serov shall be responsible for creating the requisite conditions for the normal operation of the design bureaus, institutes, laboratories, and factories involved with reactive technology in Germany (food supply, housing, transportation, etc.).

The USSR Ministry of Armed Forces (Khrulev) and SVA Supreme Commander Sokolovskiy shall be responsible for assisting Comrade Serov as needed.[29]

V.

24. The Special Committee for Reactive Technology shall be responsible for taking inventory of all the equipment, tools, hardware, as well as materials and models of reactive technology brought back to the USSR by the various ministries and departments and also for redistributing them among the appropriate ministries and departments in accordance with the tasks assigned them.

25. The USSR Ministry of Armed Forces (Bulganin) shall be tasked with making proposals to the Council of Ministers concerning the site for and construction of the State Central Test Range for reactive armaments.

26. The Special Committee for Reactive Technology shall be responsible for submitting for approval to the Chairman of the USSR Council of Ministers its policy on awarding bonuses for the development and creation of reactive armaments, as well as

---

29. SVA—*Sovetskaya voyennaya administratsiya* (Soviet Military Administration).

proposals for paying a higher salary to particularly highly qualified employees in the field of reactive technology

27. The Special Committee for Reactive Technology shall be permitted to consider the scientific-research institutes and design bureaus recently established by the Ministries of Armaments, Agricultural Machine Building, Aviation Industry, Electrical Industry, Machine Building and Instrumentation, Chemical Industry, and the USSR Armed Forces as equal to the scientific institutions of the USSR Academy of Sciences in terms of salaries and the provision of industrial and food supplies in accordance with USSR Council of People's Commissars decree No. 514, dated 6 March 1946.

28. The Ministry of Aviation Industry (Khrunichev) shall be responsible for transferring 20 specialists in the fields of engines, aerodynamics, aircraft construction, etc. to the Ministry of Armaments.

29. Minister of Higher Education Kaftanov shall be responsible for arranging for engineers and scientific technician to be trained in the field of reactive technology at institutions of higher learning and universities and also for retraining students close to graduating who majored in other specialties for a reactive armaments specialty, ensuring that the first graduating class from technical institutions of higher learning yields at least 200 specialists in the field of reactive armaments and at least 100 from universities by the end of 1946

30. The Special Committee for Reactive Technology shall be entrusted, jointly with the Ministry of Higher Education, with selecting 500 specialists from the scientific-research organizations of the Ministry of Higher Education and other ministries, retraining them, and sending them to work in ministries involved with reactive armaments.

31. In an effort to provide housing for the German reactive technology specialists transferred to the USSR, Comrade Voznesenskiy shall be tasked with providing 150 prefabricated sectional Finnish-style houses and 40 eight-apartment log houses per the order of the Special Committee for Reactive Technology.

32. Work for the development of reactive technology shall be considered the most important governmental task and it shall be the responsibility of all ministries and organizations to prioritize reactive technology assignments.

USSR Council of Ministers Chairman I. Stalin
USSR Council of Ministers Adminstrator Ye. Chadayev

Lev Gaydukov, Georgiy Pashkov, and Vasiliy Ryabikov prepared the main text of the decree with the direct involvement of Marshal Nikolay Yakovlev and Minister Dmitriy Ustinov.[30] The draft decree affected dozens of leading ministries and

---

30. Lev Mikhaylovich Gaydukov (1911–98) supervised recovery operations in Germany in 1946–47. Georgiy Nikolayevich Pashkov (1911–93) was a senior official in *Gosplan* responsible for the new missile industry. Vasiliy Mikhaylovich Ryabikov (1907–74) was Ustinov's first deputy in the Ministry of Armaments.

departments, determined the fates of many thousands of people, and demanded truly heroic efforts for the creation of a new field of technology and industry from a people bled dry by four years of war. Nevertheless, the text of the decree was concurred at all echelons with an urgency appropriate to wartime. As Gaydukov related many years later, only about 20 days elapsed from the first handwritten outline to the final text viewed by all the ministers and Malenkov himself. Stalin, to whom Malenkov reported, read and signed the draft without comments. The long and comprehensive document was essentially a strategic decision. In terms of its historic significance, it was comparable to the decree on the nuclear problem that preceded it.

Georgiy Malenkov, who headed the Special Committee for Reactive Technology, remained a member of the Special Committee on the Atomic Problem. His closeness to Stalin and the knowledge and experience he had gained preparing and issuing all the "atomic" decrees aided the development and rapid passage through the state and Communist Party bureaucracy of all the decisions implementing the "rocket" decree of 13 May 1946. The 13 May decision served as the basis for subsequent ones defining dozens of particular issues for decrees and prompted an avalanche of orders within each ministry and department. Ustinov, the most enterprising and decisive of the ministers, without waiting for the appearance of the main decree, issued his own order in May 1946 for Artillery Factory No. 88 to begin studying the drawings of rockets arriving from Germany.

Ustinov's order of 16 May 1946 announced the organization of the State Head Scientific-Research Institute No. 88 (NII-88), which was specified as the primary scientific-research, design, and experimental design facility for missile armaments with liquid-propellant rocket engines. NII-88 was created using the facilities of Artillery Factory No. 88, located in the suburban Moscow town of Kaliningrad near the Podlipki station.

After meeting with us in Germany, Ustinov and the other ministers quickly issued their orders in furtherance of the decree of 13 May on personnel assignments, having obtained concurrence from the All-Union Communist Party of the Bolsheviks (VKP[b]).[31] On 9 August 1946, as ordered by Ustinov, Korolev became chief designer of "Article No. 1"—the long-range ballistic missile.

On 16 August a decree of the Council of Ministers and Ustinov's subsequent order made L. R. Gonor director of NII-88. Gonor would develop and Minister Ustinov would approve the structure of the head institute, which would contain a special design bureau (SKB).[32] Department No. 3 was part of the SKB. Gonor

---

31. VKP(b)—*Vsesoyuznaya kommunisticheskaya partiya (bolshevikov)*, was the official designation of the Soviet Communist Party between 1925 and 1952, after which it became the *Kommunisticheskaya partiya sovetskogo soyuza* (KPSS) (Communist Party of the Soviet Union [CPSU]).

32. SKB—*Spetsialnoye konstrukturskoye byuro.*

issued his own order to appoint Korolev chief of the NII-88 SKB's Department No. 3.

The Ministry of Armaments headed by D. F. Ustinov received the leading role in the strategic decree. This was not coercion from above, but the result of Ustinov and his first deputy Ryabikov's initiative when they visited the Institute RABE in 1945. Both of them had already foreseen that rocket technology was the future for the entire industry. The decree was prepared after the special commission headed by Marshal Yakovlev visited Berlin, Nordhausen, and Bleicherode in February 1946. We in Germany, of course, had no way of knowing about this decree that determined our future fate.

Sergey Ivanovich Vetoshkin, our direct chief within the Ministry, and later in the Committee, scrutinized our affairs very carefully in Bleicherode. An artilleryman through and through, he understood that the time had come to reeducate himself. An intelligent man, kind and modest, with a great sense of responsibility, he tried first and foremost to gain an understanding of this completely new field of technology. Every free minute he could find away from commission meetings he would very politely address any one of the old hands in Bleicherode and request, "Please explain this to me—a mechanic who doesn't understand much about electricity…" asking for an explanation of how the gyroscopes worked or the *mischgerät*.[33] In short, each answer required a lecture. On returning from Germany, Sergey Ivanovich was one of the leaders in the ministry office, and then in the new committee, who helped us daily.

Somewhat unexpectedly, Malenkov was named chairman of Committee No. 2. He was already chairman of the Radar Committee and a member of Committee No. 1. Evidently, from Stalin's viewpoint, things were going so well there that he could throw Malenkov into another new field—missile production. However, Minister of Armed Forces N. A. Bulganin soon replaced Malenkov as Committee chairman.[34] Neither Malenkov nor Bulganin played a special role in establishing our field. Their prominent role boiled down to looking through or signing draft decrees that the committee office prepared with the active support of or on the initiative of Ustinov, Yakovlev, and the chief designers.

Right from the beginning, Ustinov and Vetoshkin, who was appointed chief of the Seventh Main Directorate within our ministry, paid special attention to rocketry and even displayed infectious enthusiasm, which was unusual for leaders.[35] Unfortunately, Ryabikov, one of our first patrons in the Ministry of Armaments, was soon transferred from our field of rocket technology to "air defense and radar" to head

---

33. The *mischgerät* was an amplifier that received signals from the gyroscopes on the V-2 rocket.

34. Bulganin replaced Malenkov in May 1947.

35. The Seventh Main Directorate was one of several "main directorates" within Ustinov's Ministry of Armaments. Soviet ministries typically had between six and a dozen such directorates, that is, functional units, assigned to fulfill specific tasks. Other directorates in the Ministry of Armaments focused on non–rocket-related weapons.

the Council of Ministers' Third Main Directorate. However, in 1955, Ryabikov once again returned to deal with problems of long-range missiles. They appointed him chairman of a new special committee for rockets and also chairman of the state commission for testing the first R-7 intercontinental missiles.

ALONG WITH THE HEAD INSTITUTE OF NII-88, a number of other organizations in other ministries played important roles in the early development of Soviet missiles. OKB-456, headed by Chief Designer Valentin Glushko, was charged with developing liquid-propellant rocket engines and their serial production.[36] The OKB was created using the facilities of aviation Factory No. 84. Before the war, Factory No. 84, located in Khimki on the outskirts of Moscow, had specialized in the production of Li-2 transport aircraft, a copy of the famous American DC-3 airplane produced by Douglas. In 1938, the OKB headed by Viktor Bolkhovitinov was relocated from Kazan to this factory. When completing my final thesis in 1939, I returned to Bolkhovitinov's OKB at Factory No. 84. Soon thereafter, next to this large series-production factory, Bolkhovitinov built his new experimental Factory No. 293, and his OKB relocated there as well.

After his return from Germany, Glushko was faced with setting up a factory where the entire "Bolkhovitinov team"—Isayev, Chertok, Mishin, Bushuyev, Raykov, Melnikov, and many others—had worked before him. They joked that Glushko had exiled the native Khimki-ites to Podlipki.

MINISTRY OF ARMED FORCES FACTORY NO. 1 was designated as the lead factory for control systems and renamed NII-885. N. D. Maksimov was appointed its director and Mikhail Ryazanskiy its first deputy director and chief designer. In the beginning, Nikolay Pilyugin was the deputy chief designer for autonomous control systems. During the war, the factory that was later to be the site of NII-885 had specialized in the production of remote-controlled electric motors and magneto generator field telephones. To make a call the user had to crank the handle. The factory's production and technology culture, equipment, and staff were so far removed from those of rocket instrumentation that Ryazanskiy and Pilyugin complained spitefully that, "Korolev will transform artillerymen into missile specialists, Glushko will train aviation to use his beloved liquid-propellant rocket engines, and we are going to provide them all with control technology, using telephone cranks as our main component."

Vladimir Barmin was appointed head developer of the ground-based launching complex and fueling and transport equipment, with Viktor Rudnitskiy as his first deputy. Their organization was called GSKBSpetsMash and was located at the Kompressor Factory site, which had been the head enterprise for the production

---

36. OKB stood for both *Osoboye konstruktorskoye byuro* (Special Design Bureau) and *Opytno-konstruktorskoye byuro* (Experimental-Design Bureau). In the case of OKB-456, it was the latter.

of *Katyusha* guards' mortars, the vehicle-mounted multibarreled solid-fuel rocket launchers.[37]

Of the six main chief designers, Viktor Kuznetsov and his associates were probably more fortunate. He returned to the shipbuilding NII organization, which held him in high esteem, and to a well-equipped laboratory. At that time the organization was developing gyroscopic navigation systems for ocean-going ships and had created a unique gyroscopic stabilization system for a tank gun for mobile use. But Kuznetsov did not like administrative work and had no aspirations for the director's chair. The position of chief designer suited him completely, and he was a true chief in his field. He had no fear of theoretical mechanics equations and an excellent command of the theory of gyroscopic systems, but at the same time sensed a design's adaptability to the manufacturing process and loved to delve into the fine points of production.

Once, I dropped in on Kuznetsov at home (at that time he lived on Aviamotornaya Street) and was amazed by the abundance of all sorts of electronic radio parts, bundles of wires, and fitting tools scattered about the room and on the desk. Viktor explained that he loved to unwind with a soldering iron in his hands. It turns out that he had assembled a homemade television and a unique television tube with a particularly high degree of clarity. This was at that time when televisions with tiny screens had just barely begun to appear in Muscovites' apartments.

A missile system, even the first—and by modern conceptions such an elementary system as the A4 (R-1)—contained current converters in its control system—motor generators, or, as we sometimes called them *Umformers*.[38] These assemblies transformed 24 volts of direct current into 40 volts of alternating current with a frequency of 500 hertz to supply power for gyroscopic instruments. They tasked Ministry of the Electrical Industry's NII-627 to manufacture these assemblies. Andronik Gevondovich Iosifyan headed this NII. He was responsible for manufacturing electric motors, trimming capacitors, and polarized relays for control-surface actuators. Several years later Andronik, as Korolev intimately liked to refer to him, took on a much larger challenge. He was appointed chief designer of onboard electrical equipment for a wide range of rockets. NII-627 was already a ready-made scientific production facility that specialized in servo drive technology and all sorts of low-power electrical machines. The small Moscow Mashinoapparat Factory was designated as the series-production facility for the onboard electrical equipment.

The Moscow Prozhektor Factory was charged with the development and manufacture of all of the ground-based electrical equipment. Aleksandr Mikhaylovich Goltsman was appointed chief designer of these systems. Chief designer Mark Izmaylovich Likhnitskiy, who had worked in the Leningrad fuse NII, was assigned

---

37. GSKB Spetsmash—*Gosudarstvennoye spetsialnoye konstruktorskoye byuro spetsialnogo mashinostroyeniya* (State Special Design Bureau for Special Machine Building).

38. *Umformer* is the German word for transformer.

to develop fuses for the warheads. The Ministry of Higher Educational Institutions was tasked with setting up special departments and training rocket technology specialists.

A WORD ABOUT THE CUSTOMER FOR ROCKETS—The Special Committee reserved a special role for the Ministry of Defense's Main Artillery Directorate (GAU). Artillery Marshall Nikolay Yakovlev continued to be in charge of it. The Main Artillery Directorate was designated the primary customer for long-range ballistic missile systems. To this end, they created a special Fourth Main Directorate in the Main Artillery Directorate headed by General Andrey Sokolov. Using the facilities of an institute of the Academy of Artillery Sciences, a special military institute, NII-4, was created under the Main Artillery Directorate to work on problems of the military application of missiles. General Aleskey Nesterenko became the institute's first chief. General Lev Gaydukov was named Nesterenko's deputy. Gaydukov had supported all of our undertakings in Germany; had managed to get Stalin to bring in Korolev, Glushko, and other formerly imprisoned missile specialists for our work; and had headed the Institute Nordhausen. He was already well acquainted with those of us who would be creating his new rocket technology. Why not then entrust him with one of the defining leadership posts in the new Main Artillery Directorate missile organizations? But the war had ended, and many combat generals were left without jobs appropriate for the well-earned high ranks that they had gained in combat. Soon thereafter, Nesterenko was relieved of his directorship at the NII-4 institute of the Academy of Artillery Sciences, and some time later General Sokolov was put in charge. He had been the first of the Soviet military specialists to "domesticate" Peenemünde in 1945.

Lieutenant Colonel Georgiy Tyulin, also a member of our "German" company, became the chief of the theory of flight department in the Main Artillery Directorate.

In late 1946 Lieutenant-General Vasiliy Ivanovich Voznyuk, who had commanded major guards' mortar subunits during the war, was appointed chief of the State Central Test Range (GTsP), which technically still did not exist.[39] Colonel Andrey Grigoriyevich Karas became the chief of staff of the State Central Test Range. He would later become the chief of the Defense Ministry's Central Directorate of Space Assets, the precursor to the Russian military space forces.

Voznyuk and Karas were very colorful figures in the history of the test range at Kapustin Yar and during the first years of our rocket technology in general. During the early days of our new assignments, these combat generals had to grapple with such a multitude of problems that they recalled the most difficult battles of World War II as heroic but simple work. Their work was complicated by the necessity to

39. GTsP—*Gosudarstvennyy tsentralnyy polygon*. The official name of the Kapustin Yar, the first Soviet long-range missile testing facility, was GTsP-4.

deal appropriately with "those civilians," that is, the chief designers, to cede to a few chiefs from Moscow, and to report not to the commander of an army or an army group but to the Central Committee and additionally to General Ivan Serov of state security.

They had to make time to resolve a plethora of domestic issues, to look after the housing and amenities for the officers, their families, and thousands of construction workers assigned to the projects. But they also had to gain an understanding of the new technology. All of the newly created organizations were expected within a very short time to determine their structure, fill out their staff, and begin the necessary construction. A mass of organizational, scientific and technical, and social problems crashed down on everyone. In spite of the very difficult postwar economic situation in the country, this newly created field, like the atomic industry, was appropriately prioritized in the *Gosplan* and Ministry of Finance to receive supplies, funds for capital construction and reconstruction, and production and laboratory equipment.

Here I feel it is fitting to make an observation in defense of the centralized state "bureaucratic" planning and coordinating apparatus. The competence of the officials of Committee No. 2 and their effective efforts not to shirk from making decisions rendered us quick and energetic assistance in setting up our operations. The decisions to recruit new firms for the work and drafting Council of Ministers' decrees and similar matters were resolved with the urgency that had not been lost since wartime.

OF THE THREE NEW TECHNOLOGIES—RADAR, ATOMIC, AND MISSILES—atomic technology was the most science-intensive. Perhaps because of this, Special Committee No. 1 included two academicians: Igor Kurchatov and Petr Kapitsa.

Malenkov headed two of the three Special Committees (radar and missiles), created in 1945–46; Beriya headed the third (atomic). Both Malenkov and Beriya reported directly to Stalin, who attentively, strictly, and in a very demanding manner monitored the execution of the scientific, technical, and production tasks assigned to the committees. Stalin's supervision was anything but detached. Stalin inserted his corrections and additions into drafts of decrees that had already been accepted. One such Stalin initiative was the top secret decree dated 21 March 1946, "On Awards for Scientific Discovery and Technical Achievement in the Use of Atomic Energy and for Cosmic Radiation Research Projects Contributing to the Solution of This Problem."

This decree called for large monetary awards to be granted to individuals who solved specific scientific and technical problems. It stipulated prizes of one million rubles for the directors of the work and would confer on them the titles of Hero of Socialist Labor and Stalin Prize laureate. At government expense they would be granted, in any region of the Soviet Union, ownership of a villa, a furnished dacha, a car, double pay or salary for the entire period of time they worked in that field, and the right to free transportation (for life for the individual and wife or husband and for the children until they came of age) within the USSR by rail, water, or air

21

transport. Large monetary awards were stipulated not only for the directors but also for the primary scientific, engineering, and technical employees who were involved in the work. The individuals who had distinguished themselves the most were presented with orders and medals of the USSR. No one but Stalin could dare offer such bountiful generosity. For the atomic scientists and everyone associated with them, this decree was unexpected. Scientists of all ranks, engineers, and technicians were so accustomed to working for nearly nothing, to living poorly and sharing the adversity of the entire populace, that the blessings promised by the decree shocked them at first.

Stalin wasn't just looking after the senior science staff. At his instruction, beginning in the second half of 1946, wages were increased one-and-a-half to two times for all employees in the atomic industry. Budgetary expenditures on science, in particular on the Academy of Sciences, were tripled in 1946 compared with 1945 and then doubled again in 1947!

If 13 May 1946 (THE DAY THE DECREE WAS ISSUED) is considered the beginning of broad-scale missile technology operations in the USSR, then it was eight months behind the corresponding date for nuclear technology. This proved to be sufficient time to train government officials on the basis of the nuclear experience to prepare and issue decrees that had been worked out in minute detail to solve the most vital strategic, military, and technical problems.

The State Defense Committee (GOKO), which was created at the very beginning of World War II, held all the strings to control the economy. It created an original centralized military-industrial and transport management system, which supported the development of weapon prototypes and the production of all types of military hardware. Under peacetime conditions at the very beginning of the Cold War, centralization of the political and economic authority made it possible to effectively use the wartime experience for organizing operations.

After the war, State Defense Committee functions were transferred to the Council of Ministers.[40] The industrial ministries, formed from the people's commissariats, received a great deal of independence. However, solving the new and very complex scientific and technical problems called for the formation of the special committees described previously. These committees allowed the higher political leadership and Stalin personally to manage the solution of complex problems that required enormous material expenditures, scientific leadership, and participation of various branches of industry.

The complex government mechanism controlling the entire defense industry, as well as all of the branches of industry composing the country's economy, was under the supervision of the Communist Party Central Committee. All of the decrees affecting the life of the country, its science, and its defense were made on behalf of

---

40. The GOKO was a temporary body established to operate only under wartime conditions. The Council of Ministers was the cabinet-level body managing Soviet industry and society.

the Council of Ministers and the Central Committee of the All-Union Communist Party of the Bolsheviks. To be sure, one should mention that Beriya, unlike Malenkov, tried to keep the Party apparatus from participating in generating decisions on matters within the scope of the Special Committee that he headed and the First Main Directorate subordinate to it.

After Beriya was overthrown, tried, and shot in 1953, stories leaked of his leadership methods. On one occasion, the Ministry of Aviation Industry had received instructions from Beriya to prepare a governmental decree to reassign one of its factories producing aircraft instruments to the First Main Directorate. The minister dared to inform Beriya that the decree must be issued in concordance with the defense department of the Party Central Committee. "What is the Central Committee to you?" shot back Beriya in indignation. "Stalin *is* the Central Committee and I will report this to him."

In the mid-1950s, the interests of the three Special Committees became intertwined. They began to move atomic explosives from airborne bombs into missile warheads. A massive campaign was underway to "missilize" the infantry forces, navy, and air force. Radio electronic systems from auxiliary facilities were converted into the primary means of determining the effectiveness of anti-aircraft defense and, later, anti-missile defense. It was time to rethink the traditional division of the military into the three branches of the armed forces: the infantry, navy, and air force.

A scientific theory for a systemic approach to the management of complex hierarchical systems did not yet exist, but the organizers of industry, having cast aside their departmental differences, decided to consolidate the management of the country's entire military-industrial complex. And so the special committees were dissolved and the managerial coordination of all the defense ministries was transferred to a new agency—the Commission on Military-Industrial Issues under the USSR Council of Ministers, or the VPK.[41] I will write about this governmental agency later.

Here I would like to say a kind word about the mangers and bureaucrats during that period—the staff members of all the special committees, the defense departments of the Party Central Committee, the people's commissariats' main directorates, and later the ministries, *Gosplan*, and military chiefs—with whom, in one way or another, I had the occasion to come into contact during the period from 1945 through 1955, the period during which the three technologies came into being. The overwhelming majority of the governmental and party officials who made up the large managerial machine of the military-industrial complex were at their core dedicated to their cause and competent organizers. They were a necessary component of the driving force behind the creative process for the birth of a new technology.

---

41. The full name of this body was the Commission on Military-Industrial Issues, but it was more commonly known as VPK—*Komissiya po voyenno-promyshlennym voprosam* (Military-Industrial Commission). Officially formed in December 1957, the VPK was the top management body for the entire Soviet defense industry. Commission members typically included the ministers of various branches of the defense industry (including the rocket industry).

Chapter 2
The Return

I spent 21 months in Germany. The majority of the Soviet specialists who worked at Institutes RABE and Nordhausen spent considerably less time there, 6 to 12 months. Korolev himself was in Germany for about 15 months. The future chief designers of future new Soviet technology, including Valentin Petrovich Glushko, Nikolay Alekseyevich Pilyugin, Viktor Ivanovich Kuznetsov, Vladimir Pavlovich Barmin, Mikhail Sergeyevich Ryazanskiy, and almost all of their first deputies and future leading specialists and researchers, designers, process engineers, and military testers—several thousand people in all—for over a year had simultaneously undergone retraining, recertification, the difficult "breaking in" process, and getting to know one another. Many of us acquired good friends that we would have for years to come.

A plethora of new scientific and technological difficulties arose during the creation of these large and complex technical systems. One of them was totally unforeseen. It required the development of new "system-oriented" interrelationships among the people creating all the elements of a large system. This factor, a purely human one, had exceptionally great significance after our return and indeed from the very beginning of our activity in 1947.

We returned almost two years after victory, but during a difficult and complex time. Caught up in a new field of creative activity opening up boundless prospects, we made the most optimistic plans for future rocket technology. Having lost touch with the postwar reality of Moscow, before our return to the Soviet Union, we had virtually no experience with the everyday cares that were normal for Soviet people at that time. Finding ourselves plunged into this new atmosphere in the first months of 1947, we were forced to expend time and energy readapting to our native land.

After returning from comfortable Thuringia, not everyone was able to find quarters in conditions that were reasonable even by the postwar standards of that time. My family—there were four of us now—returned to the NII-1 superstructure, building No. 3 on Korolenko Street in Sokolniki. Here we occupied two adjacent rooms. Yevgeniy Shchennikov's family, which also had four members, occupied the other two rooms. He was an official of the Russian Federation Council of Ministers. The apartment had no bathtub and no shower. It had one toilet and one sink for everyone. The latter was also the kitchen sink for the small common kitchen. The apartment had a wood-burning stove, for which the wood had to be carried up from a shed in the courtyard to the fifth floor, and, of course, there was no elevator. After our fashionable Villa Frank in Bleicherode, these circumstances required psychological adaptation. Yet, many envied us. First, we had an average of six square meters per person, and, second, we had good neighbors. Our wives immediately became friends, and our children were still friends a half-century later.

A year passed before Korolev received a separate apartment in the factory building, not far from the main entrance. Almost all of 1947 he spent nights on a couch in the old apartment on Konyushkovskaya Street. After his arrest in 1938, his wife Kseniya Vintsentini and daughter had been left with one tiny room.

Many lived wherever they could, "catch as catch can." In other words, they were registered at the factory dormitories so that their passports were in order, but they lived without a residence permit with relatives or friends or rented rooms in dachas on the outskirts of town. In Podlipki, where our new NII-88 rocket center was located, only the old staff workers of the former artillery factory had separate apartments. The newly hired young specialists and workers were housed in barracks that had been built in abundance. However, we were not the least bit depressed! Even when we were living and working for many months under arduous conditions—verging on the impossible—at the Kapustin Yar test range, we saw things with humor and optimism.

It was more difficult to adjust to the country's general atmosphere of a stifling ideologically repressive system. While enthusiastically working for some time as victors in another country, which previously had been under even harsher repressive control, we were sure that the postwar life in our country would be much more democratic. These same hopes were shared by the military intelligentsia, including the many combat officers who had experienced the crucible of war.

During the war, people faced death and performed feats under the motto "For the Fatherland!," "For Stalin!," or "For the tears of our mothers!" At the rear they labored heroically under the motto "Everything for the front, everything for Victory!" We had triumphed at the cost of countless lives, with real heroism and genuine unity of the people in the face of a common mortal danger. But now, once again, they were demanding heroism, this time in the workplace.

Hope for a better life, faith in the wisdom of the "greatest leader of the peoples," and constant ideological Communist Party pressures proved so strong that in spite of all the sacrifices made during the war, people were prepared to endure postwar difficulties and to accomplish new feats for the even greater consolidation of military might and for new accomplishments and triumphs in Soviet science and technology.

There was a wave of triumphant euphoria, of genuine nationwide exultation, but instead of being caught up by this enthusiasm and releasing the powerful spirit of free creative initiative, against all logic and common sense, Stalin and his entourage intensified their regime of repression. A new series of reprisals followed. A campaign of ideological repression against the intelligentsia intensified. The government carried out resettlements, with the massive exile of entire ethnic groups, a process that had begun during the war. And former prisoners of war, officers, and millions of young Soviets, who had already undergone all manner of torments, were subjected to totally inexplicable repressions for being forced by the Germans to work in Germany.

During one of my first encounters with Isayev after returning from Germany, he asked, "Do you remember the walking skeletons at the Dora camp that the Americans didn't take with them, but left behind for us just because they flat out refused, and demanded to be handed over to the Soviet authorities?"

"Of course I remember. You don't just forget things like that."

"Well, all of them, who by some miracle survived the German camps, have now been sent to our camps. Sure, our camps are different from the German camps. Ours don't have crematoria and they don't trust the prisoners to be involved with the production of missiles or things like that!"

Applications for employment and admission to institutes of higher learning and technical schools contained such questions as: "Were you or any of your relatives held captive or on territories occupied by Nazi forces? Have you or any of your relatives been repressed? Have you or any of your immediate relatives been abroad? If yes, when and for what reason?"

Fifty years later I am trying and cannot find a satisfactory answer for myself to the question of why all the strata of postwar Soviet society—the army, scientists, intelligentsia from the applied sciences and humanities, the working class united by labor unions, and the poverty-stricken peasantry—made no historically significant attempts to change the state system or to stop the repression of millions of innocent people and the political suppression of any dissent. Stalin, Roosevelt, and Churchill were idols of the masses who had struggled with Hitler's Germany. After victory, only Stalin remained. Up until 1953, there was no internal opposition whatsoever to his dictatorship. If, in the late 1940s or early 1950s, a poll similar to the ones nowadays had been taken to determine Stalin's popularity, I am sure he would have rated much higher than the subsequent leaders of the Soviet Union and contemporary Russia.

While working in Germany, we had understood that after the war, international scientific cooperation would be of utmost importance for the development of domestic science and technical progress. We dreamed that instead of the confrontation that had begun to emerge, the interaction of the scientists from the victorious countries would be a natural continuation of the military alliance. In late 1946, Korolev, who had returned from some meeting in Berlin, smiled enigmatically at Vasiliy Kharchev and me, "Get ready to fly across the ocean." Alas! Until the very day he died, neither Korolev, nor any one of his closest associates was ever "across the ocean."

In autumn of 1947, many of the specialists returning from Germany, among them Korolev, Pobedonostsev, Kosmodemyanskiy, Ryazanskiy, and I, began to give

27

lectures for the higher engineering programs organized at the N. E. Bauman Higher Technical Institution. There, the entire "elite" of the still quite young rocket industry had been assembled to retrain military and civilian engineers. We were supposed to pass on the experience and knowledge we had acquired in Germany. I was assigned to teach the course "Long-range Missile Control Systems." Korolev prepared the first systematized work for these courses, "Fundamentals for the Design of Long-Range Ballistic Missiles."[1] This was the first real engineering manual for designers in our country.

In these courses it was impossible to avoid mention of history and German achievements. Aside from the *Katyusha,* we still did not have our own combat rockets. Our first "almost domestic" R-1 rocket was to fly only a year later in autumn 1948. In spite of that, the administrator who supervised the higher engineering courses, averting his eyes, asked that we "remove mention of the Germans' work from the lectures to the extent possible." Preparing a cycle of lectures, I conscientiously described the A4 missile's control system and the basic history of its development. At Pobedonostsev's recommendation, one of the publishing houses accepted this book for open publication, and by the middle of 1948 it had already been submitted for printing. Pobedonostsev unexpectedly called me in and said that the "powers that be" had really lit into him for agreeing to be the editor of my book. The publishing house had already received the order to scratch the printing job and to destroy all the printed copies of the manuscript.

"You in particular need to be circumspect and cautious now. If you have a type-written copy, hide it, and I will report that everything was destroyed!"

Alas, I had nothing to hide. I had handed over all the copies to the publishing house. I very much regretted that soon thereafter I had to part ways with Pobedonostsev. They transferred him to the managerial staff and to teach at a recently established industrial academy to train leadership cadres for the Ministry of Armaments.[2]

THE SUBURBAN MOSCOW RAILROAD STATION with the poetic name Podlipki was located 20 kilometers from the Yaroslavskiy station. That is where our special train from Germany arrived. The A4 missiles that we had assembled in Thuringia were housed in the airfield hangars on approximately the same site where the spaceflight Mission Control Center is now located. During the war it was the site of one of the

---

1. The second and main part of these lectures has been published. See "Osnovy proyektirovaniya ballisticheskikh raket dalnego deystviya" ["Fundamentals for the Design of Long-Range Ballistic Missiles"]. In M. V. Keldysh, ed., *Tvorcheskoye naslediye akademika Sergeya Pavlovicha Koroleva: izbrannyye trudy i dokumenty* [*The creative legacy of academician Sergey Pavlovich Korolev: Selected works and documents*] (Moscow: Nauka, 1980), pp. 208–290.

2. Yuriy Aleksandrovich Pobedonostsev (1907–73) served as the Chief Engineer of NII-88 from 1946–49. In May 1950, he was transferred to the Scientific Department of the Academy of the Defense Industry.

air defense airfields where the fighter aviation defending Moscow was based. For the first years we used this airfield for its real purpose.

Truthfully, when we first saw the future missile factory in Podlipki, we were horrified. There was dirt and primitive equipment, and even that equipment had been ransacked. Compared with the aviation industry from whence we had transferred, this seemed like the Stone Age to us. There was no need even to compare it with the conditions in Germany. There was no comparison. Korolev and his entourage began a stubborn struggle to establish a production culture. I must say that Minister of Armaments Dmitriy Ustinov gave us vigorous support in this. He did a great deal to establish the rocket industry and understood very well that rocket technology required new conditions and a more elevated culture and technology than artillery, which was the basis for the formation of our industry. But proper credit must also be given to artillery technology and to the industrial and process engineers who took part in the solution of our problems with wartime enthusiasm.

We had to create our own laboratory facilities and debug and test the missiles that had been brought in. Based on the Germans' experience, we knew that even if a missile had been tested somewhere but was then transported to a different site, during subsequent tests it might not fly. The German missiles failed in large numbers right on the launch pad if thorough tests and checks had not been conducted to the end. For that reason we paid particular attention to debugging the missile tests. In particular, in my department we developed a testing/simulating bench, where we debugged all the test automatics, and in place of a "live" missile there was a set of onboard equipment with the appropriate indicator lights simulating operations during the launch phase of the trajectory.

In Germany, using Institute Nordhausen resources and then at NII-88 in Podlipki, two missile series of 10 units each were prepared. We assembled series "N" in Germany at the Kleinbodungen factory and also performed the horizontal tests there, using the process previously employed at Mittelwerk. We assembled the "T" series in Podlipki at the NII-88 experimental factory from assemblies and parts that we had prepared in Germany.

The engines for the T series had undergone firing tests in 1946 in Lehesten, but we retested them. The pairing of the engines with the turbopump assemblies and steam gas generators required tests and the recording of data to precisely determine parameters. OKB-456 in Khimki headed by Valentin Glushko performed all of these procedures.

The control system hardware for both rocket series underwent retesting at NII-885 before it was sent to the test range. Mikhail Ryazanskiy and Nikolay Pilyugin supervised this work. A complex problem was solved at Naval Scientific-Research Institute No. 1 (MNII-1) of the Ministry of the Shipbuilding Industry.[3] Here,

---

3. MNII—*Morskoy nauchno-issledovatelskiy institut.*

under the leadership of Viktor Kuznetsov and Zinoviy Tsetsior, the *Gorizont*, *Vertikant*, and *Integrator* gyroscopic instruments were almost completely reassembled. The conventional bearings that they had been fitted with at the Zeiss factory in Jena were replaced with precision bearings, the rotors were balanced to reduce vibrations, and the command potentiometers were adjusted. The latter were perhaps the most delicate elements of the command gyroscopic instruments.

All of the ground equipment gave us a lot of trouble. The *Viktoriya* system was designed to perform lateral flight correction. In Germany we had not managed to come up with all the parts necessary to outfit it in its nominal form. Therefore, at NII-885, under the supervision of Mikhail Borisenko, workers not only performed restorative work but also partially developed and fabricated missing assemblies and antennas for the ground control station and thoroughly tested out its joint operation with the onboard receiver. For this they even conducted special aircraft tests at the Kapustin Yar State Central Test Range (GTsP) before we arrived there for the rockets' first launches.

Under the supervision of Vladimir Barmin and his deputy Viktor Rudnitskiy at the Kompressor factory, workers repaired and checked out all of the ground-based launching and fueling equipment. The ground-based electric equipment was completed, retested, and shipped to the test range by the Prozhektor Factory. Aleksandr Goltsman was in charge there. He was one of the chief designers who had not been with us in Germany.

The individuals responsible for the reproduction of the onboard electrical equipment were Andronik Iosifyan, chief designer of the Moscow Electromechanical Scientific-Research Institute (MNIIEM), and Nikolay Lidorenko, chief designer of the Scientific-Research Institute of Current Sources (NIIIT).[4] The explosives for the warheads made use of domestic development under the supervision of NII-46 Chief Designer Mark Likhnitskiy. NII-20 of the Ministry of the Communications Systems Industry (MPSS) directed development of the telemetry systems.[5] Grigoriy Degtyarenko and Special Purpose Brigade (BON) officer Captain Kerim Kerimov, who had both undergone training in Germany, supervised the preparation and operation of this system.[6] Thus, aside from the six "really chief" designers (Korolev, Glushko, Pilyugin, Ryazanskiy, Barmin, and Kuznetsov), there were at least four more who were not "not so chief" but were also chief designers (Goltsman, Iosifyan, Lidorenko, and Likhnitskiy).

In September 1947, on our special train, we set out for Kapustin Yar, where the Ministry of Defense had created the State Central Test Range for the testing of

---

4. MNIIEM—*Moskovskiy nauchno-issledovatelskiy institut elektromekhaniki;* NIIIT—*Nauchno-issledovatelskiy institut istochnikov toka.*

5. MPSS—*Ministerstvo promyshlennosti sredstv svyazi.*

6. BON—*Brigada osobogo naznacheniya.* The BON was the artillery brigade assigned to operate captured German missiles in the postwar era.

rocket technology. We traveled in comfort in our two-berth compartments. I was in the upper berth, and Viktor Kuznetsov was in the lower one. Only Korolev, as the technical director of the State Commission had a deluxe compartment with a small boardroom. NII-88 director Lev Robertovich Gonor traveled in a separate compartment.

We were not involved in the test range site selection—military officials did this on their own. Kapustin Yar was an old village in the lower reaches of the Volga River, on a flood plain that was usually not covered with water. This was the area between the Volga and Akhtuba Rivers. Further along the firing line were the uninhabited Volga steppes. Lieutenant General Vasiliy Ivanovich Voznyuk was appointed chief of the test range.

I met General Voznyuk for the first time during the hot summer of 1947 in the NII-88 director's office. Gonor invited Korolev, Voskresenskiy, and me to a meeting with the chief of the country's first state rocket test range. When we entered, a broad-shouldered lieutenant-general of above-average height stood up to meet us. His chest was decorated with row after row of service ribbons and the Gold Star of a Hero of the Soviet Union. He gave each of us a firm handshake and wore a teasing, kind smile as he studied us, looking us straight in the eyes.

"Well, well. I thought General Gonor had officers, but I see that you all are running around in undershirts quenching your thirst with Borzhomi mineral water. Out there I've still got only barren steppe, the temperature is over 40°C (104°F), there is no good water, no roads, and nowhere to live. I still don't know what you're planning on building, where you're going to build it, where it's coming from, where it's going to, or what you're going to fire it with." Smiling broadly, Voznyuk said, "Help me to gain some understanding of this," happily downing yet another glass of mineral water that Gonor poured for him. We explained our understanding of the test range's missions to Voznyuk as best we could.

"This will not be Peenemünde, and we have no pretensions of building a Schwabes Hotel," joked Gonor. "To begin with, we will be arriving on our special train and will be living in it. And then we will help design firing test rigs, a rocket processing hangar, and launch pads."

Military construction workers who had gained considerable experience on rush jobs during the war carried out the construction at the test range. It started literally from scratch. The officers were housed haphazardly in a small town of adobe huts. The soldiers lived in tents and dugout huts. The task of providing electricity to all of the test range facilities could be compared to a military operation.

But in September 1947, despite all of General Voznyuk's energy, the test range was still not ready for tests. The first thing that we had to do was to place one of the rockets on a test rig and conduct integrated firing tests. The second thing was to equip the launch pad and assembly and testing building. We were supposed to have a concrete platform on which the launch pad would be installed and an assembly and testing building where the rockets would be tested in the horizontal position before they were brought out for launch. This building was called the "engineering facility." We needed several cinetheodolite tracking stations, which were supposed to film the rockets' launch and flight.[7] The test range was supposed to have a rather large meteorological service because the launches needed to be conducted under good weather conditions in order to observe and film them. A synchronized time service was needed so that all the test range services would use a synchronized time system.

To begin with, efforts were focused on completing the test rig. This was a large three-tiered rig, the design of which drew from the experience in Peenemünde and Lehesten. The rocket was secured to the rig in a gimbal ring brought from Peenemünde. Our job was to equip it with everything it needed and to set up all of the launch and fueling equipment. The firing rig was quite far from our special train. It was next to the airfield, where airplanes landed on an unpaved airstrip. And the launch pad was further away, approximately three kilometers. Here they also began to build the command bunker. But missile launch control would be initiated not from the bunker but from the German armored fighting vehicle, the *Panzerwagen*, which were reminiscent of modern infantry armored fighting vehicles (BMP); the *Panzerwagen* was widely used by the German military for V-2 launches.[8]

A large wooden structure, cold and drafty, was built to serve as the assembly and testing building. There, we began the horizontal tests on the rocket before it was hauled out to the firing rig, which was being finished with the help of a round-the-clock all-hands rush job by the military construction workers under the supervision of Marshal Vorobyev.

A state commission appointed by governmental decree managed and monitored the conducting of the first long-range ballistic missile launches in the USSR.

The members of the commission were:

1. N. D. Yakovlev—Chairman, also Artillery Marshal and Head of the Main Artillery Directorate
2. D. F. Ustinov—Deputy Chairman, also Minister of Armaments
3. I. A. Serov—First Deputy Minister of Internal Affairs
4. S. N. Shishkin—Deputy Minister of Aviation Industry
5. N. I. Vorontsov—Deputy Minister of the Communications Industry
6. V. P. Terentyev—Deputy Minister of Shipbuilding
7. M. P. Vorobyev—Marshal, Commander of the Infantry Engineering Troops
8. M. K. Sukov—Head of the Main Directorate of the Oxygen Industry Under the Council of Ministers
9. S. I. Vetoshkin—Head of the Main Directorate of Reactive Armaments of the Ministry of Armaments
10. P. F. Zhigarev—Deputy Commander-in-Chief of the Armed Forces

---

7. Cinetheodolites are optical cameras that record the position and movement of objects in flight.
8. BMP—*Bronemashina pekhoty.*

Members of the State Commission were housed, and conducted their almost round-the-clock activity, in two trains: in Special Train No. 2, where we lived, and in Special Train No. 1, which was reserved for the military. The state commission approved by decree industrial representatives who were allowed to participate in operations; it also appointed technical management for the testers. Korolev was appointed technical director of testing. His deputies, all chief designers, were members of the Council of Chiefs. They were V. P. Glushko, V. P. Barmin, M. S. Ryazanskiy, and V. I. Kuznetsov. Pilyugin was not included in the technical management, because the decree had named Ryazanskiy chief designer of the guidance system and Pilyugin as his deputy. During the flight tests in 1947, Pilyugin had two duties at the firing range. First, he was chief of Electrical Department No. 1 both at the engineering facility and at the launch site during electrical testing of the missile. Second, during launch he served as a firing department operator. I was also on the roster as a firing department operator.

It bears mentioning that the organizational structures for the launches were developed at General Tveretskiy's Special Purpose Brigade (BON) back in Germany and were applicable to troop operations, with provisions that took into consideration the need for personnel training. Technical management required that each military unit concerned with technology have monitors or industrial representatives who worked with the military personnel.

The State Commission had to approve two organizational structures, one for military personnel and one for civilian personnel. During work, no one thought about who was where in the organizational hierarchy. Everyone worked harmoniously. I cannot remember a single "who's in charge here?" conflict. Special groups were created in the vast mixed military-industrial staff to support missile preparation and launch. These included analytical groups, groups for science experiments, instrumentation, meteorology, communications, medical assistance, and all the services supporting the critical functions of the special trains and the hundreds of individuals involved in testing.

German specialists occupied an entire railroad car in our special train. Helmut Gröttrup was in charge of the German "firing squad." He brought almost all of the leading specialists from Gorodomlya. In addition to them, Glushko wanted to have his own German engine specialists from Khimki.

ON 14 OCTOBER, the missile was finally brought out to the almost completed firing rig. The only difference between the rig version of the missile and the combat version was that the "Heck," or tail section, had been removed from it. This was done in keeping with the German way of testing at Peenemünde. It took days to connect the ground-based electrical control and measurement networks, to test them, sort them out, and eliminate problems that inevitably appeared in a large and complex electrical system assembled for the first time and in a hurry. Barmin and Rudnitskiy received personal instructions from Marshal Yakovlev to monitor and be responsible for the fueling process.

Shown here are the leading participants who oversaw the historic first A4 (V-2) launches from Kapustin Yar in the fall of 1947. Sitting huddled on the ground are (from left to right): M. I. Likhnitskiy, N. A. Pilyugin, G. A. Tyulin, N. N. Khlybov, and S. S. Lavrov. In the middle row (left to right) are: M. S. Ryazanskiy, V. P. Barmin, S. P. Korolev, S. I. Vetoshkin, L. M. Gaydukov, and V. I. Kuznetsov. Standing at the back (from left to right) are: unknown (face obscured), V. P. Glushko, D. D. Sevruk, B. Ye. Chertok, M. I. Borisenko, L. A. Voskresenskiy, unknown, and V. A. Rudnitskiy.

The engine was started up directly from the *Panzerwagen* by the firing squad, which included Captain Smirnitskiy and industry "operators" Voskresenskiy, Pilyugin, Ginzburg, and me. No matter what we did, however, we couldn't get the engine to start up. The "lighters"—the special electrical devices that ignited the fuel—kept getting knocked out during the very first firing, and the engine did not start up. For the most part the defects were in the electrical starting system. First one relay would fail, then another... All of these incidents were heatedly discussed in the *bankobus* during State Commission sessions.[9] We testers had to report on each operation to the State Commission. There, in Kapustin Yar in 1947, was the birth of the term *bobik*, which later became part of the missile field vernacular. Since then, testers have called a failure that requires several hours to identify and eliminate a *bobik*. The source of this folklore was an anecdote that Ginzburg told, very appropriately, in the *bankobus* after the engine's latest failure.

It was on perhaps the third day of our sufferings, after we had spent several sleep-

---

9. *Author's note:* The term *bankobus* was formed by combining two words, "bank" (in the sense of a collective discussion) and "bus." We met in a dilapidated bus that had been pulled up close to the rig so that we could have some sort of shelter from the wind and rain.

less nights attempting to start up the engine, that an aggravated Serov addressed us in the presence of the entire commission:

"Listen, why are you doing this to yourselves? We'll find a soldier. We'll wind some twine onto a long stick, dunk it in gasoline, the soldier will insert it into the nozzle, and you'll have your ignition!"

The idea was "splendid," but in spite of the fact that it was Colonel General Serov's, no one fell for it. We continued to discuss the causes of the latest *bobik*. It was cramped in the *bankobus* and everyone was chain-smoking. Thank goodness there was a strong draft through the broken windows.

"Why was there was no ignition this time? Have you analyzed it?," Serov meddled once again.

Korolev said that Pilyugin could give a report, adding, "His circuit failed." Pilyugin explained, "Yes, we found the cause. A relay in the ignition circuit didn't trip."

"And who is responsible for that relay?" asked Serov.

"Comrade Ginzburg," responded Pilyugin after a brief pause.

"Show me this Ginzburg," said Serov menacingly. Pilyugin, who was leaning on Ginzburg's shoulder, surreptitiously pressed him into the crowd that was huddled around, and answered that he could not point Ginzburg out because he was at the rig replacing the relay. I should say that over that entire time no harm came to any of us, although the "Sword of Damocles" was constantly hanging over each of us.

Finally, on the night of 16–17 October, from one of the armored vehicles that served as the command post where Pilyugin, Smirnitskiy, Voskresenskiy, Ginzburg, and I were located, we started up the engine! The feeling of triumph was extraordinary! For the first time, a liquid-propellant rocket engine had been started up at the State Central Test Range in Kapustin Yar. Tired and worn out, we barely managed to crawl out of the armored vehicle. I pulled an ordinary soldier's flask filled with pure alcohol out of my pocket and treated the entire crew of our armored vehicle. And that was the first toast that we raised to the successful launch of our rocket, albeit still only on the rig.

After the test-firing, we did not conduct any more tests on that rig. Instead of spending more time on that, we switched to preparing and launching rockets from the launch pad.

In those days, we didn't drive to the launch pad over a luxurious concrete road as they do today. We drove along dusty roads in American Jeeps, and our favorite hymn was the song, "Eh, roads, dust and fog..." The autumn weather tormented us a great deal, and the most popular people then were the meteorologists. There were two reasons for this: first, we waited for them to give us permission for launch; and second, there were a lot of young women in this service, which relieved our difficult workaday routine somewhat.

The launch team in our military unit was staffed primarily by servicemen from the Special Purpose Brigade formed in Germany. Its personnel had worked with us at the Institutes RABE and Nordhausen practically all of 1946, and each officer knew his job. The most highly trained specialists from industry were included on the launch team. Engineer Major Ya. I. Tregub was in charge of the launch team on

From the author's archives.

Shown here are Korolev and his principal associates during a break of the A4 (V–2) rocket tests in Kapustin Yar in 1947. From left to right are N. N. Smirnitskiy, L. A. Voskresenskiy, S. P. Korolev, Ya. I. Tregub, and an unknown associate.

behalf of the military and L. A. Voskresenskiy on behalf of industry. The assistant commander of the launch team was Engineer Major Rafail Vannikov, the son of the first minister of the atomic industry Boris Vannikov.

During the first launch, technical director Korolev was in the armored vehicle. He had the last word on the operation. At Ustinov's insistence, a German specialist, Corporal Fritz Viebach, was there as controller and consultant.

THE FIRST LAUNCH WAS EXECUTED ON 18 OCTOBER 1947 AT 10:47 A.M. It was a series T rocket. During the launch, I was in the armored vehicle and was thus denied the opportunity to delight for the first time in the spectacle of a launching rocket, an event that never leaves anyone indifferent. The weather was quite decent, and we were able to monitor the launch phase using test range systems. The rocket flew 206.7 kilometers and deviated to the left by almost 30 kilometers. They didn't find a large crater at the impact site. Subsequent analysis showed that the rocket disintegrated upon entry into the dense layers of the atmosphere.

They also used a series T rocket for the second launch. It was conducted on 20 October. During the launch phase, the rocket deviated significantly to the left of its plotted course. No reports were received from the calculated site of impact, and the test range observers announced rather tongue-in-cheek, "It went toward Saratov."[10]

---

10. Saratov is a large industrial center about 800 kilometers southeast of Moscow on the banks of the river Volga.

After a couple of hours, the State Commission promptly convened. At this meeting Serov reprimanded us:

"Imagine what would have happened if the rocket had reached Saratov. I won't even begin to tell you; you can guess what would have happened with all of you."

We quickly grasped that it was much farther to Saratov than the 270 kilometers that the rocket was supposed to fly, and so we were not very alarmed. Then it turned out that the rocket had successfully covered 231.4 kilometers, but had deviated to the left by 180 kilometers. We needed to find out why. And then, as annoying as it was for us, Ustinov decided to seek advice from the Germans. For the analysis, they enlisted the services of German specialists at the firing range who were in a separate "German" railroad car in our special train. Before this, Dr. Kurt Magnus, a specialist in the field of gyroscopy, and Dr. Hans Hoch, an expert in the field of electronic transformations and control, had been sitting around at the test range without anything in particular to do. Ustinov said to them, "This is your rocket and your instruments; go figure it out. Our specialists don't understand why it went so far off course."

The Germans sat down in the laboratory car, which was part of the special train, and began to experiment with a complete set of all the nominal control instruments. Dr. Magnus suggested testing the gyroscopic instruments on the vibration table. We put the gyroscope on the vibration table, connected it to the *mischgerät*—the amplifier-converter that received commands from the gyroscopic instruments—switched on the control-surface actuators, and thus simulated the control process, exposing it to vibrations under laboratory conditions. They succeeded in showing that in a certain mode, vibration could cause detrimental interference to the legitimate electrical signal. Dr. Magnus showed that the *mukholapka,* that is, the device that picked up the current from the gyroscope potentiometer, reacted to frequencies close to 100 hertz and began to "dance" and apply interference to the legitimate signal.[11] Dr. Hoch explained that the process of differentiation in the amplifier-converter amplifies the interference such that it jams the legitimate signal. As a result, the rocket veers away from the assigned course in any direction and could even dive into the ground. Former Corporal Viebach, a participant in many combat launches, confirmed that in Germany there had been similar instances during test and combat launches when they had not been able to explain the true causes of the large deviations. Gröttrup joked about this, "If Dr. Magnus and Dr. Hoch had worked with us in Peenemünde during the war, British losses during our bombardment of London would have been considerably greater."

The solution proved to be simple: we needed to put a filter between the gyroscopic instrument and the amplifier-converter that would allow only legitimate signals to pass and would cut off detrimental noise generated by vibration. Dr. Hoch himself designed the filter right then. He found everything he needed among our

---

11. *Mukholapka* literally means "fly foot."

From the author's archives.

Conditions at Kapustin Yar were difficult for even the most seasoned war veterans, with weather oscillating from extreme heat to unbearable cold. Shown here in their rugged attire are the armored vehicle crew for the first A4 (V-2) launches in the fall of 1947. From left to right are: A. M. Ginzburg, B. Ye. Chertok, N. A. Pilyugin, L. A. Voskresenskiy, N. N. Smirnitskiy, and Ya. I. Tregub. All of these men would later reach senior engineering or military positions in the Soviet space program.

spare parts. We placed the filter on the next rocket, and the effect was immediately evident. Lateral deviation was slight.

To celebrate, Ustinov ordered that all the German specialists and their assistants be given what were for that time enormous bonuses—15,000 rubles each and a jerrican of alcohol for all of them. They, of course, couldn't cope with it all and generously shared it with us. We celebrated the successful launch together. The authority of the German specialists, whom up until then only the "technicians" had respected, immediately rose in the eyes of the State Commission.

During the merrymaking in the German railroad car, having enjoyed a good mutton pilaf, I boasted to Dr. Magnus that in April 1945, in Adlershof, I had found a report, authored by him, on the development of a new type of gyroscope. The report had been approved by Dr. Schüler, and the title page had been stamped *Geheim*, that is, "Secret."

Magnus, who had seemed tipsy, gave a start and immediately sobered up.

"Where is that report now?"

"I saved it, in violation of my instructions. But I can't give it to you because that would now be my second gross record-keeping violation."

"On your instructions, Dr. Hoch and I are developing proposals for a new control system that would be much more reliable than the one on the A4. That report

From the author's archives.

On the thirtieth anniversary of the first launch of the first Soviet V-2, several veterans reunited to celebrate in the event in 1977. Standing in front of the memorial are, from left to right: General A. G. Karas, Ye. V. Shabarov, Gen. V. A. Menshikov, B. Ye. Chertok, and [initials unknown] Kolomiytsev. At the time, Karas was commander of the Soviet military space forces.

would be very useful to us."

I never submitted the report for declassification, and it got lost in the chaos among my books. Two years later, Magnus and Hoch reproduced the report's contents and it became part of the design of the G-1 rocket now under our "secret" stamp.

In 1953, Magnus returned to Germany, where he pursued a brilliant scientific career. He established a department and then an institute of mechanics at the Munich Technical University. In 1971, in West Berlin, Kurt Magnus' monograph *Gyroscop:. Theory and Applications* was published.[12] In 1974, the monograph was translated into Russian by the Mir publishing house and became a reference book for three generations of specialists.[13] Magnus also established an institute of mechanics at the Stuttgart Technical University. The Russian Academy of Navigation and Motion Control elected Professor Magnus as an honorary member. In September 2002, I was invited along with other Russian scientists to Stuttgart Technical University to celebrate the 90th birthday of this distinguished Doctor of Technical Sciences. Dr. Sorg, who officiated at the festive gathering, reported that, to his great regret, Dr. Magnus was ill and would not be able to attend the celebration in his honor.

Having been granted the opportunity to deliver the first congratulatory speech, I told the attendees about Magnus' work in the Soviet Union and about the episode at the Kapustin Yar test range in 1947. I asked that they pass on my gifts to the birthday boy: his 60-year-old report approved by Dr. Schüler and stamped *Geheim;* a commemorative medal issued for the 90th birthday of academician S. P. Korolev; and a commemorative souvenir from the Energia Rocket-Space Corpora-

---

12. Kurt Magnus, *Kreisel: Theorie und Anwendungen* [*Gyroscope: Theory and Applications*] (Berlin: Springer-Verlag, 1971).

13. K. Magnus, *Giroskop* [*Gyroscope*] (Moscow: Mir, 1974).

From the author's archives.

Shown here are engineers responsible for the guidance and control systems during the historic first A4 (V–2) launches from Kapustin Yar in the fall of 1947. Sitting in the front row are A. M. Ginzburg, V. I. Kuznetsov, M. S. Ryazanskiy, N. A. Pilyugin, B. Ye. Chertok, and M. I. Borisenko.

tion.[14] Without letting me leave the podium, Magnus' protégé, President of the Institute of Mechanics, Professor G. Sorg, reminded the assembled crowd that I was also 90 years old and therefore was being awarded a model of the gyroscope. The attendees were delighted.

NOW LET'S RETURN TO THE EVENTS OF 1947 IN KAPUSTIN YAR. For everyone, military and civilian, the work was hard. The most unpleasant procedure was waiting for a clear sky during cold, rainy weather in the damp tents at the launch site. The food was quite satisfactory and our mood was optimistic, although the living conditions were like military field conditions.

On 7 November, on the occasion of the 30th anniversary of the Great October Revolution, Minister Ustinov invited the senior technical staff and certain members of the State Commission for an airplane ride over Stalingrad. We took off in an 18-seat Douglas from an unpaved area right by the special train. The cloud cover was very low, and we flew to Stalingrad at an altitude of no more than 100 meters. We crossed the Volga and suddenly found ourselves over the ruins of Stalingrad.

---

14. Magnus was also one of several from the German rocket experts brought to the Soviet Union in 1946 who published memoirs of their times there. See Kurt Magnus, *Raketensklaven: deutsche Forscher hinter rotem Stacheldraht* [*Rocket Slave: German Scientists Behind the Red Barbed Wire*] (Stuttgart: Deutsche Verlags-Anstalt, 1993).

Ustinov emerged from the cockpit and shouted, "Look! They're already restoring the city. Let's fly to the Barrikady Factory that Gonor defended."

We pressed up against the windows, and the airplane banked sharply, suddenly climbed steeply, and turned once again. About 20 meters from the airplane the tall factory smokestack flashed by.

"Dmitriy Fedorovich has taken over the controls," commented Vetoshkin, the color gone from his face.

The airplane rocked violently. It was flying at very low altitude and a collision with Stalingrad seemed unavoidable. Gonor managed to shout, "What is he doing? We're about to crash into the Barrikady Factory." And with the next lurch of the plane he flew out into the aisle. Korolev looked angry and somber. Glushko looked straight ahead, calm and unruffled. Marshal Yakovlev could not contain himself, and, barely able to stay on his feet, he headed for the cockpit. We could not hear what he said when he confronted Ustinov, but the rocking stopped. Once again we crossed the Volga, and, after 20 minutes of calm flight, we taxied safely up to our special train.

In all we launched 11 German rockets and 5 of them reached their target. The reliability of the rockets was roughly the same as what the Germans had experienced during the war. Of the 11 rockets launched, 5 had been assembled in Nordhausen and 6 at Factory No. 88. But the assemblies and parts were all German. And they all proved to be equally unreliable.

The launch of an A4 rocket in the fall of 1947 was in some ways the fruit of our 18-month activity in Germany. The intense work in Germany during the period from 1945 through 1946 with the help of German specialists enabled us to save enormous resources and time for the formation of our domestic rocket technology. The flight tests in 1947 showed that Soviet specialists, both military and civilian, had mastered the fundamentals of practical rocket technology and had gained the experience needed to make an accelerated transition to a now independent development of this new, promising field of human endeavor.

Many years later, at the site of the first launch in 1947, an R-1 rocket was erected as a monument. In its outward appearance it was an exact copy of an A4. Enriched by the experience of the A4 tests, on our return from Kapustin Yar we immediately switched over to the task of developing domestic rockets, as the saying goes, without pausing to catch our breath. In the process of preparing for and conducting launches, we had discovered too many defects. Each of these defects, each negative observation and accident during launch needed to be thoroughly analyzed and a decision made as to what modifications were necessary for the creation of our own domestic R-1 rocket.

The tests also yielded other results that were certainly positive. First, combining all the services at the test range into a single collective during the process of the flight tests allowed both individuals and organizations to adjust to each other. The organizational experience of conducting such complex activities sometimes proves to be as valuable as the scientific and technical achievements.

Second, the participation on the State Commission of high-ranking military officials and the directors of a number of ministries definitely influenced their "rocket world view." Now it was not just the chief designers and all of their compatriots but also those individuals on whom we were directly dependent, who understood that a rocket was not simply a guided projectile. A rocket complex was a large, complicated system that required a new systematic approach during all the stages of its life cycle, such as design, development, fabrication, and testing. Given such an approach there should not be primary and minor tasks. In the system, everything should be subordinate to the interests of achieving a single final goal.

In this regard, I recall this episode, which later became an edifying anecdote, from the State Commission sessions. While analyzing the latest in a series of unsuccessful launches, it was determined that the most probable cause was the failure of one of the multicontact relays in the primary onboard distributor. Exercising his rights as the highest ranking minister and Deputy Chairman of the State Commission, Ustinov addressed Deputy Minister Vorontsov, who was in charge of rocket technology at MPSS. "How was it that your people didn't look through and check each contact?" Vorontsov was offended and retorted, "There are 90 relays on board and 23 on the ground. You can't look after every single one. Is it really that great a calamity, after all, one relay failed!" What a commotion erupted! The indignation reflected the gradual internalization of a new systematic thinking into our world view.

Third, at the test range, directors and specialists from various levels worked and lived together. In the future they would be implementing a national program on an enormous scale. Here they were not only developing an understanding of each other's difficulties but they were strengthening amicable relations; real friendships developed regardless of departmental affiliation. In the work that was to last for years to come, this was enormously important.

Finally, during the process of the first range tests, an unofficial agency became firmly established—the Council of the Chief Designers headed by Sergey Pavlovich Korolev. The authority of this council as an interdepartmental, nonadministrative, but scientific and technical governing body had critical importance for all of our subsequent activities.

Chapter 3

From Usedom Island to Gorodomlya Island

A total of over 200 German specialists came to NII-88 from Germany. With families, it was nearly 500 people. Among the new arrivals were highly qualified specialists—scientists and engineers who had worked with us at the Institutes RABE and Nordhausen and at the Montania factory. The German collective included 13 professors, 33 Ph.D. engineers, and 85 graduate engineers. As soon as they arrived in the Soviet Union, 23 German specialists were sent to Khimki to work at OKB-456 to help set up production of engines for the A4 rockets. OKB-456 Chief Designer V. P. Glushko was personally involved with their job placement.

The majority of the Germans were at the disposal of NII-88 director L. R. Gonor. They spent some time at health and vacation resorts in the vicinity of Podlipki. Beginning in the spring of 1947, they began to house the Germans in quickly repaired and newly constructed homes on Gorodomlya Island in Lake Seliger. Before the war, this lake had been known as the best lake for fishing and the most beautiful lake in central Russia, thus the most favored by tourists. At the time, Gorodomlya Island was closed to tourists; it was the location of a center for biological research in the fight against foot-and-mouth disease and anthrax. In 1947 the entire island was given to NII-88.

The organization of German specialists housed on Gorodomlya Island was given the status of NII-88 Branch No. 1; thus, formally, the entire staff was subordinated to NII-88 Director Gonor. At first, F. G. Sukhomlinov, who had previously worked in the offices of the Ministry of Armaments, was appointed director of the branch. Soon, however, P. I. Maloletov, the former wartime director of Factory No. 88, replaced him.

The former director of the Krupp Company's ballistics department, Professor Woldemar Wolf, was appointed director of the German contingent. Engineer/designer Blass was appointed his deputy. The German collective included prominent scientists whose works were well known in Germany: Peyse, thermodynamics expert; Franz Lange, radar specialist; Werner Albring, aerodynamics expert and pupil of Ludwig Prandtl; Kurt Magnus, physicist and prominent theoretician and gyroscope specialist; Hans Hoch, theoretician and specialist in automatic control; and Kurt Blasig, Askania Company specialist in control surface actuators.

The vast majority of German specialists in NII-88 at that time were not former associates of von Braun in Peenemünde. They were introduced to rocket technology at the Institutes RABE and Nordhausen, while working with us. Wernher von Braun had this to say about the German specialists that we had brought in to work with us: "... the USSR nevertheless succeeded in acquiring the chief electronics specialist Helmut Gröttrup... But he was the only important catch from among the Peenemünde specialists."

By mid-1947 more than 400 persons, including 177 Germans, were working on Gorodomlya Island in NII-88 Branch No. 1. Among the German specialists were 5 professors, 24 Ph.D.'s, 17 graduate engineers, and 71 "engineer practitioners."

Initially, the German specialists were combined into "collective 88." In August 1947, the Germans carried out a reorganization, and "collective 88" was named "Department G." The Germans themselves selected graduate engineer Gröttrup to be director of Department G; they also appointed him chief designer of new long-range ballistic missile designs.

The German specialists brought in from Germany worked at other locations in addition to NII-88 at Lake Seliger. For this reason, it is worth addressing their legal and material status in our country. It was practically the same in various organizations, because it was determined by orders coming from the top in the corresponding ministries. All of the specialists that had been brought to the USSR along with their family members, were provided with foodstuffs on a par with those of Soviet citizens, in accordance with the ration card system that existed in our country until October 1947. Upon arrival in the Soviet Union, they were housed in buildings that were quite comfortable. If the distance was sufficiently great, the specialists were transported from their place of residence to work and back on buses. Residences on Gorodomlya Island had undergone high-quality restoration, and the living conditions were quite decent for those times. In any case, specialists with families received separate two- and three-room apartments. When I arrived on the island, I could only envy the way they lived, because in Moscow my family and I lived in a communal four-room apartment, in which we occupied two rooms with a total area of 24 square meters. Many of our specialists and workers still lived in barracks, where they did not have the most elementary conveniences.

The German specialists received fairly high salaries, depending on their qualifications, academic titles, and degrees. Thus, for example, Drs. Magnus, Umpfenbach, and Schmidt each received 6,000 rubles per month, Gröttrup and Willi Schwarz received 4,500 rubles each, and graduate engineers received, on average, 4,000 rubles each. For the sake of comparison I can cite the monthly wages of the primary leading specialists of NII-88 (in 1947): Korolev (chief designer and department chief)—6,000 rubles; Pobedonostsev (the institute's chief engineer)—5,000 rubles; and Mishin (Korolev's deputy)—2,500 rubles. My monthly salary was 3,000 rubles. The average salary of the German specialists in the Ministry of Aviation Industry, to which OKB-456 was subordinate, also exceeded that of Soviet specialists. OKB-456 chief designer V. P. Glushko received a salary of 6,000 rubles per month in

1947–1948. In that same OKB-456, German specialist Dr. Oswald Putze, deputy chief of engine production, received 5,000 rubles per month. Glushko's deputy V. A. Vitka had a salary of 3,500 rubles. The Germans were permitted to transfer money to their relatives in Germany. On a par with all the Soviet specialists who worked at NII-88 and OKB-456, in addition to the aforementioned salaries, the Germans were given incentives in the form of large monetary awards for completing phases of work within the scheduled deadlines.

On weekends and holidays they were permitted to make excursions to the regional center of Ostashkov and to Moscow to go to shops, markets, theaters, and museums. Therefore, life on the island surrounded by barbed wire could not in any way be considered comparable to the status of prisoners of war.

The case of Ursula Shaefer, who left Bleicherode and ended up on Gorodom-lya Island on Seliger Lake, was unusual. The wives of the German specialists were not elated by the presence of a beautiful woman living alone in the rather closed German community. Frau Schaefer appealed to the administration with a request to find her husband, who was being held as a prisoner of war somewhere in the Soviet Union. The appropriate agencies actually looked for her husband in one of the POW camps. It turned out that he was an anti-fascist and quite possibly even the organizer of a new German party among the prisoners. They released him from the camp and sent him to his wife.

By that time, however, while he was being processed out of the camp and making his way to the island, his charming wife had abruptly changed her political orientation; among the German community, she turned out to be the most ardent supporter of the crushed fascist regime. The State security authorities on the island were in a complete state of confusion over it—such a beautiful woman and suddenly a true, unadulterated Nazi. What was to be done with her? Then her husband showed up, virtually a communist. They asked him to exert some influence over his unruly wife. It seems that he was unsuccessful in that venture. To get themselves out of harm's way, our security agencies sent them both to East Germany ahead of schedule.

Officially all the German specialists were referred to as "foreign specialists" in correspondence and were combined into "collective 88." The Germans themselves were divided into specialized structural subdivisions.

The NII-88 management had drawn up a thematic plan of work for the German collective for 1946 and early 1947 that included consultations for issuing a set of A4 rocket documentation in Russian, compiling diagrams of the A4 and surface-to-air guided missile research laboratories, studying issues related to boosting the A4 rocket engine, developing the design for an engine with a thrust of 100 tons, and preparing to assemble rockets that were made of German parts and had been outfitted with equipment at the Institute Nordhausen.

Probably the most vital stage of this period was the development of proposals for the A4 rocket launch program. Launches were scheduled for autumn 1947 at the State Central Test Range in Kapustin Yar. The German specialists, among whose ranks were those who had participated in combat firing, as well as specialists in measurements and ballistics, were tasked with obtaining as much information as possible about the rockets with a minimum number of launches. Basically, the idea was for a program of no more than 10 to 12 launches. The Germans handled the work successfully, while Hoch and Magnus, as I have already mentioned, helped to determine the cause of the A4's pronounced deviation during the second launch.

In June 1947, the NII-88 director held a meeting on the prospects and organization of the German specialists' subsequent work. Six months of experience had shown that the German specialists, who were not fully staffed, were virtually isolated from our newly formed production technology. They had no contact with our recently initiated network of cooperation on engines, control systems, and materials and were not capable of developing new rocket complexes. Nevertheless, at Gröttrup's recommendation, they were given the opportunity to test their creative powers and to develop the design of a new long-range ballistic missile. The missile design was assigned the designation G-1 (later the designation R-10 also appeared). Gröttrup was named project director and chief designer of the new missile.

The newly formed department in "collective 88" received the same rights that all of the institute's other scientific-research departments enjoyed. It consisted of branches for ballistics, aerodynamics, engines, control systems, missile testing, and a design bureau. The institute's chief engineer, Yuriy Aleksandrovich Pobedonostsev, became the immediate director of the department, as well as of other NII-88 departments. As Pobedonostsev's deputy for control systems, I was to supervise the work of the German specialists on the new control system. The chief of the NII engine department, Naum Lvovich Umanskiy, was assigned to help them with engines, Viktor Nikolayevich Iordanskiy with materials, and Leonid Aleksandrovich Voskresenskiy with testing, and so on.

THROUGHOUT 1947 AND 1948, I visited the "German" island many times. Usually after these business trips I had difficult and confidential conversations with Pobedonostsev and Gonor. It seemed obvious to me that the group of specialists, being completely out of the information loop, could not, in our system-oriented times, develop a design for a new rocket system that would fit in with the design, production, and most importantly, armament infrastructure being established in the Soviet Union.

Occasionally when speaking his mind, Pobedonostsev ruefully tried to explain, "Boris Yevseyevich! I can't believe you still don't realize that our security agencies are never, under any circumstances, going to allow the Germans to be involved in true joint work! They are under double scrutiny—ours (as specialists) and that of the state security agencies, who see in each of them a fascist who has gone over to the U.S. intelligence services. And anyway, no matter what they come up with, it won't be in step with our current trend in ideology, which dictates that everything created recently or previously in science and technology be done without any foreign influence."[1]

I had similar frank conversations with others. NII-88 director Lev Robertovich Gonor was a general and one of the first Heroes of Socialist Labor, but as a result of his Jewish parentage, he too could not withstand the rising turbid wave of the "struggle against foreign and cosmopolitan influence." Soon he too was removed from his job and then arrested on charges of complicity in a "Zionist" conspiracy. I will describe his fate later.

For the sake of fairness I must mention that the Germans, judging by the specialists with whom I was in close contact, adjusted quickly. In almost two years of working in vanquished Germany and interacting with Germans from different social groups, not once did I sense either anti-Semitism or a spirit of German chauvinism. At that time, I thought that it was the result of discipline, cowardice, and submission to the victors. But after visiting the Federal Republic of Germany in 1990, 1992, and 2002, I once again detected no traces of anti-Semitism, or what we referred to as revanchism.

Beginning in 1948, on orders of higher Communist Party authorities, all mass media outlets and especially liberal arts institutions, institutes, cultural organizations, and educational institutions, mounted a struggle against what they called "cosmopolitanism." As part of this campaign, they organized active searches for the Russian authors of all inventions, discoveries, and the latest scientific theories, without exception. A widely known joke circulated: "Russia should also be declared the birthplace of the elephant."[2]

But we should give credit to the directors of branches in the defense industry, such as Ustinov, Malyshev, Ryabikov, Kalmykov, Vetoshkin, and to their many like-minded associates—a fear of "cosmopolitanism" and "foreign influence" was not in their nature. Korolev did not maintain close contact with Germans for completely other, purely personal reasons. He was one of the founders of rocket technology in our country and had to drink a full cup of humiliation beginning with his arrest in 1938, only to find after his release in 1944 that many of the ideas that he had hatched had already been implemented by others and that, in many regards, the German rocket specialists had gone significantly farther than his most forward-

---

1. Here, Chertok is referring to the broader cultural trends of *Zhdanovshchina* ["time of Zhdanov"] and "anti-cosmopolitanism" campaign promoted by the Soviet Communist Party in the late 1940s and early 1950s, when many fields of intellectual inquiry were hostage to ideological interference and distortion. One of the central dimensions of these campaigns was to negate any and all Western influence on the development of Russian and Soviet science and technology. Another was to demonize Jews in the Soviet Union.

2. For more on the anti-cosmopolitanism campaign, see Gavriel D. Ra'anan, *International Policy Formation in the USSR: Factional 'Debates' During the Zhdanovshchina* (Hamden, CT: Archon Books, 1983).

thinking plans. Once he had finally obtained the position of Chief Designer, it offended him to be testing a German A4 rather than his own rocket and to design a domestic R-1, which by government decree was an exact copy of the A4. Being by nature a commanding and ambitious person who was easily hurt, he could not conceal his feelings when they hinted to him that "you're not making your own rocket, you're reproducing a German one." On this topic, Minister Ustinov, who initiated the program for the exact reproduction of the German A4 rocket as practice for the production process, had serious conflicts with Korolev on more than one occasion.

AFTER THE AFOREMENTIONED ENCOUNTER AT THE MEETING IN THE NII-88 DIRECTOR'S OFFICE IN JUNE 1947, the German collective was tasked with the independent design for a ballistic missile with a range of at least 600 kilometers. Korolev did not sympathize with this work assigned to the Germans, because he justly considered that priority in the development of this rocket should belong to his staff, that is, the NII-88 Special Design Bureau (SKB) Department No. 3. Suddenly it turned out that almost all of the NII-88 scientific-research departments under the supervision of Pobedonostsev, his co-worker at RNII until 1938, would be working not only for him, but also for the newly appointed chief designer of the G-1, Helmut Gröttrup, Wernher von Braun's closest associate.[3]

We had already begun developing the design for a rocket with a range of 600 kilometers back at the Institute Nordhausen. Tyulin, Mishin, Lavrov, Budnik, and many other Soviet specialists had participated in the project there. The majority of them were now working under Korolev's supervision. In 1947, Korolev's department, already at work on the R-1 rocket, was working at full speed to design a rocket with a range of 600 kilometers, with the designation R-2. Out of consideration for the continuity of the technology, Korolev's design called for the maximum use of the available parts stock for A4 and R-1 missiles. This also included requirements not to exceed the A4 diameter and to use the same engine, after having Glushko's OKB-456 boost its performance characteristics. At Korolev's initiative, the government approved the inclusion of the R-2 rocket in the NII-88 work schedule, although earlier they had envisioned developing the R-3 with a range of up to 3,000 kilometers immediately after the R-1. Korolev had quite correctly assessed the difficulty of such a qualitative leap and decided that they should first try their hand at an intermediate version. However, it was the engine specialists such as Glushko who had the decisive word as to the possible deadlines for developing a rocket with twice the range of the A4.

Here it is fitting to note the differences between the two leading luminaries of our domestic rocket technology, Korolev and Glushko, in their attitudes to the

---

3. RNII—*Reaktivnyy nauchno-issledovatelskiy institut* (Reactive Scientific-Research Institute). RNII was one of the founding organizations of Soviet rocketry. Chertok describes the history of RNII in detail in Chapter 26.

German specialists. Korolev simply, and sometimes even demonstratively, ignored everything that had to do with the work the German collective was doing on Lake Seliger. Not once did he visit Gorodomlya Island, nor did he associate with Gröttrup or with the other leading German specialists. In contrast, Glushko placed the German specialists at OKB-456 in positions of responsibility in engine production. He dealt with them personally and had the same expectations of them as he did of his own subordinates. However, the Germans still were not cleared to work on the new design for the new engines. After the A4 rocket engine technology was restored and production of its domestic analog, the RD-100 engine, had been mastered, the German engine specialists were simply no longer needed.

ENGINE MODIFICATION WORK AT OKB-456 BEGAN IMMEDIATELY UPON THEIR ARRIVAL FROM GERMANY. For reasons of secrecy, the Germans who worked at NII-88 on Gorodomlya Island were not informed about the work that their German colleagues were conducting per Glushko's instructions at OKB-456. However, at both places, people understood that the A4 rocket engine could be upgraded. According to calculations, its thrust on the ground could be increased to 35 to 37 metric tons by increasing the turbopump assembly's revolutions per minute and raising the pressure in the chamber.

They had already discovered significant reserves in the engine's layout and design during the A4 engine firing tests in Germany. The firing tests in Lehesten, initiated by Isayev and Pallo in 1945, continued under Glushko's supervision. They confirmed the feasibility of boosting the engine from a thrust of 25 metric tons to 35 metric tons. With the A4's structural mass of around 4 metric tons, this was sufficient to hurl an 800- to 1,000-kilogram warhead 600 kilometers instead of the 270 to 300 kilometers that had been attained!

However, increasing the range required a considerably greater amount of propellant and oxidizer. That meant larger tanks and a larger structural mass, which could nullify the gains achieved by boosting the engine. They studied several alternative versions, but in each of them they searched for reserves in structural volume and mass that would make maximum use of rigging that was fabricated and already available at the in-house factory. In early 1947, it was already evident that they needed to introduce a fundamental change into the design of the future long-range missile. Rather than the entire missile, only the nose section containing the warhead would fly to the target. This immediately eliminated the problem of the missile's body strength during entry into the atmosphere—one of the A4 rocket's weakest points.

The issue as to whose idea it was to have a separating nose section is debatable to this day. Beginning with the R-2, all modern long-range ballistic missiles have had a separating nose section. For a modern designer, it is incomprehensible why the Germans had the entire A4 enter the atmosphere and then were surprised that it disintegrated without reaching its target. But in 1947, the idea of nose section separation, like other daring proposals introduced during work on the design of the

The Collection of Olaf Przybilski.

A rare photograph of the Gorodomlya group of Germans while at Kapustin Yar during the fall of 1947 when the Soviets tested the A4 (V–2) missile. From left to right are Karl (Viktor) Stahl, Dr. Johannes (Hans) Hoch, Helmut Gröttrup, Fritz Viebach, and Hans Vilter.

R-2 rocket, was not immediately and unequivocally approved. All the new issues dealing with the separating nose section for the R-2 rocket were tested, first on a modification of the R-1 rocket known as the R-1A and then on an experimental version of the R-2 known as the R-2E .

Overtaking the project of Korolev—who was busy preparing for the A4 tests, organizing R-1 production, and practically fighting to establish his doctrine at NII-88—the Germans brought their G-1 (or R-10) design before the NII-88 Scientific-Technical Council (NTS) for discussion in September 1947.[4]

Director of operations Helmut Gröttrup presented the main report. NII-88 director Lev Gonor conducted the meeting. Participating in the discussion were Chief of the Main Directorate for Rocket Technology within the Ministry of Armaments Sergey Vetoshkin; Chief Engineer of NII-88 Yuriy Pobedonostsev; rocket technology pioneer Mikhail Tikhonravov; Chief Designers Ryazanskiy, Pilyugin, and Kuznetsov; head of the N. E. Bauman Moscow Higher Technical Institute Nikolayev; Chief Designer Isayev; Director of the USSR Academy of Sciences Institute of Automation Trapeznikov Professor Kosmodemyanskiy; Korolev's deputies Mishin and Bushuyev; and me, NII-88's deputy chief engineer. Korolev himself did not attend the meeting.

---

4. NTS—*Nauchno-tekhnicheskiy sovet.*

Gröttrup, Professor Umpfenbach, and Drs. Hoch, Albring, Anders, Wolf, and Shaefer traveled from Gorodomlya Island to Podlipki to defend the G-1 design. In his opening remarks, Gonor reported that the design had been developed with the participation of NII-88 radio engineering specialist Dmitriy Sergeyev and Naum Umanskiy, who specialized in the improvement of liquid-propellant rocket engines.

In his report Gröttrup said, "A rocket with a range of 600 kilometers should be a stage for the subsequent development of long-range rockets, and it is precisely our design that makes it possible to develop rockets with an even greater range of effectiveness." Reminding his audience that Soviet specialists were developing a rocket with the same range, making maximum use of A4 parts, he proposed, "From here on out it would also make sense to develop both designs simultaneously, but completely independently of one another until the test articles are fabricated and test launches are conducted."

The main features of the G-1 design were the following:

- Retaining the A4 dimensions while reducing the dry weight and significantly increasing the volume for propellant
- Greatly simplifying the onboard control system by transferring as many control functions as possible to ground-based radio systems
- Simplifying the rocket itself and the ground systems as much as possible
- Increasing accuracy
- Separating the nose section during the descent portion of the trajectory
- Cutting the launch preparation time cycle in half
- Using two load-bearing tanks—alcohol and oxygen—in the design

In 1941, when von Braun invited his teacher Hermann Oberth to Peenemünde, Oberth noted the faulty design of the A4 rocket tanks. As early as the 1920s, Oberth had written in his books that propellant tanks should be a load-bearing part of the rocket design. Structural stability, he argued, should be maintained by increased pressure, the pressurization of the tanks. Why, then, was von Braun not using such a productive idea? Although faulty from the point of view of Oberth and of any modern rocket designer, the load-bearing layout of the A4 did not require prolonged testing and verification. The A4 structural optimization was dictated not by mass, but by a time factor. The war was going on and the time required to develop a combat rocket played the decisive role. Pressurized tanks were not adopted at the time. Gröttrup's design for the G-1 and Korolev's design for the R-2 both used the concept of load-bearing tanks.[5]

The layout of the A4 engine was also changed significantly. The turbine that

---

5. *Author's note:* Disputes as to whose idea it was are pointless; Academician Rauschenbach demonstrated this in his book. See B. V. Raushenbakh, *German Oberт, 1894-1989* [*Hermann Oberth, 1894-1899*] (Moscow: Nauka, 1993). The volume was translated and published in English as Boris V. Rauschenbach, *Hermann Oberth: The Father of Space Flight* (Clarence, NY; West-Art, 1994).

turned the pumps feeding alcohol and oxygen was driven by gas taken directly from the engine's combustion chamber. A new radio control system provided a high degree of firing accuracy. The engine was shut down in one step when the rocket reached a specific trajectory point and speed, which was measured from the ground via radio. The speed was not only measured, but also corrected via radio during the straight segment of the trajectory. Regulating the speed by controlling the engine's thrust was a very progressive idea. The weak point of this proposition was the necessity for control via radio from the ground.

We first developed the apparent velocity regulation (RKS) system for a rocket in 1955, but did not put it to practical use until 1957, on the first R-7 intercontinental missile.[6] However, this system was purely autonomous and did not require the presence of a radio measurement system for rocket speed during flight. Currently, all liquid-propellant rockets, both for combat purposes and launch vehicles, have autonomous RKS systems.

Helmut Gröttrup expressed his confidence in the great merit of the design, which contained fundamentally new ideas and proposals. "The confidence with which we have put forth our design for discussion is based on the knowledge and experience of our colleagues. Accumulated experience provides the basis for the development of a rocket, which at first glance seems unrealistic; the range has been doubled without increasing the rocket's size, and in spite of a significant reduction in the number of control instruments, the striking accuracy has been increased tenfold."

The main difference between the G-1 design and that of the A4 and R-1 rockets (and our competing R-2 design) was the probabilistic error value, which was on a different order of magnitude than what we had in mind. Instead of the *Gorizont* and *Vertikant* free gyroscopes, the design called for a simple and inexpensive single-degree-of-freedom gyroscope, the theory for which Dr. Kurt Magnus had already developed in detail in 1941. The control loop as a whole was theoretically designed by Dr. Hans Hoch.

Pneumatic control surface actuators replaced hydraulic ones under the rationale that "pneumatic energy on board doesn't cost anything." Classic Askania control surface actuators, on the other hand, required heavy storage batteries and electric motors. The number of electrical instruments, connectors, and cables on board was sharply reduced. As a result of all these measures, the A4's structural mass was reduced from 3.17 to 1.87 metric tons, and, in so doing, the mass of the payload explosives was increased from 0.74 to 0.95 metric ton. Taking advantage of the newly freed space, they increased the propellant mass. The new design for the rocket layout featured a nose section that separated from the body at the end of the launch phase, smaller tail fins, and a body fabricated primarily of light alloys.

In conclusion, Gröttrup cited an estimate for the increase in the rocket's combat effectiveness: to completely destroy a 1.5- by 1.5-kilometer area from a range of 300 kilometers, 67,500 A4 rockets would need to be launched, while from a range

---

6. RKS—*Regulirovaniye kazhushcheysya skorosti.*

of 600 kilometers, only 385 G-1 rockets would be required. These estimates seem absurd from today's nuclear standpoint, but they show how unreal Hitler's hopes had been for the destruction of London using the V-2 "vengeance weapon."

The general assessment of the reviewers who had first studied the design in groups by discipline was positive. In particular, Mishin's speech was interesting. He referred to the Soviet work that had begun with his participation at the Institute Nordhausen. "Development of the proposed conceptual design [for an advanced A4] began in Germany. Around August 1946 they tasked us with assessing the possibility of modernizing the A4 rocket in order to attain a 600 kilometer range. We worked on this problem jointly with department No. 6 (Sömmerda) and department No. 3 (Institute RABE)."

Mishin could not resist describing the rival design to the G-1 of which he was the primary author. "We could see two ways to create such a rocket. The first way was to create a rocket based on existing designs and the experience gained operating them, taking into consideration the actual feasibility of realizing this rocket in metal. The second way was to create a rocket based on fundamentally new principles that, in and of themselves, require experimental testing. Meanwhile, existing designs would be used to an extremely limited degree, requiring a radical restructuring of production."

In conclusion, responding to statements and criticism, including some polemics from Mishin, Gröttrup defended the idea of the forward-looking proposals. "We are approaching our task to create a rocket with a 600-kilometer range from the following standpoints. This rocket is not the end of the evolution of rocket science. That means that we need to design new rockets so that they will also find application in the future evolution of rockets. Therefore, we have adopted a large number of new engineering solutions that could promote the further evolution of rocket technology."

In my evaluation, I supported the idea of simplifying the onboard control system (housing the instruments in a single location, the aft compartment) and recalled that: "the rocket of today has several tens of thousands of wires, thousands of two-way make-before-break contacts, and dozens of relays, potentiometers, etc. The operation of all of this equipment, even with well-trained personnel, is extremely intricate, both because of the complexity of the electrical system itself, and because all of the instruments are concentrated not only in the instrument compartment, but in other parts of the rocket and ground equipment... This new design offers a real and critical simplification of all the rocket's electrical equipment. This provides not only an advantage in weight (although ultimately, this advantage is not so important), but also an enormous operational gain... It seems to me that this is one of the great merits of the design."

Responding to the numerous critical remarks on the lack of calculations and theoretical foundations, Gröttrup made a statement referring to the experience of Peenemünde. In this mission statement, he said:

"Using our method to evaluate the design it is quite sufficient to present theoretical principles. During the design process, we can update and confirm the theoretical prin-

ciples via experimentation. Ours is an industry that demands an article be fabricated within specific deadlines and, of course, we are not in a position to conduct theoretical work on a large scale.

Therefore, as development progresses, we derive theory from experimentation. Essentially, theory should help to find the right direction for the experiment. Scientific-research institutes should provide the requisite textbooks for fundamental physics research. Many cases show that an experiment leads more rapidly to the objective and gives better results than theory. As one can easily understand, the second possible method requires some time. We don't have much time to develop our rockets, considering the work that is going on in the U.S. Nor is this method more reliable. When design theory and experimentation cooperate closely, the end result is reliable and complete. This method—based only on theory—has only one advantage: it makes it easier for the customer to evaluate the design. But I think that this advantage is less important than the considerable failure to meet deadlines."

Gröttrup's view was essentially the design doctrine for complex rocket systems of that period, but its main features still apply today. Today instead of simply criticizing the speaker for not presenting enough theoretical research, people would ask, "But where are the simulation results?" Alas, at that time they did not yet have modern simulation methods, nor did they do mathematical modeling using real equipment.

In this regard, Korolev's point of view concerning the procedure for evaluating rockets to make production decisions is also interesting. Immediately upon his return from Germany, Korolev began to pester the upper management to speed up rocket flight tests. In February 1947, Korolev prepared a memorandum for the upcoming discussion of the future plan of operations for rocket technology at the government level. Korolev wrote:

"It would be erroneous to think that the realization of the domestic R-1 rocket is a matter of simply copying German technology, of just replacing the materials with domestically produced materials. Besides replacing materials and restoring the entire manufacturing process for the rocket components and parts, we should keep in mind that the Germans did not bring the A4 rocket to that degree of perfection that is required of a product that has been accepted as an [operational] armament.

Our experience studying German rocket technology shows that to solve this problem, i.e., to achieve the final optimization of the A4 rocket, the Germans expended enormous manpower and resources. In addition to experimental design work, at numerous institutions on a broad scale, the Germans conducted scientific-research work of both an applied and problem-solving nature

It is also well known that a significant number of the Germans' rockets broke up in the air, and the causes for this were not determined with any degree of certainty. In many cases they did not manage to achieve the required flight trajectory and accuracy. There were many well-known cases of failures during launch due to defects in the control instruments, propulsion system assemblies and mechanisms, etc

So far, we have not succeeded in conducting tests in flight on the previously assembled

German production models and, consequently, we do not even have a complete under-standing of the design.

All of this and many other issues must be extensively studied and tested in our scien-tific-research facilities, institutes, factories, on test benches, and at test ranges during the development and fabrication of the first batch of domestic R-1 rockets.

To do this, first of all, we need to conduct flight tests on existing A4 rockets that have been lying in storage for a long time at the NII. This will give us the necessary practical experience and will generate a whole series of new tasks for everyone working in the field of long-range rockets

Right now we need to start equipping the launch pads and flight paths at the test range to conduct flight tests and we need to build a test rig near the test range…"

Decisions were made based on Korolev's memorandum. We set up experimental rig tests and conducted A4 rocket flight tests at the State Central Test Range in Kapustin Yar. I have already talked about this in the preceding chapter. It never even occurred to anyone to argue with Korolev or try to prove that the experiments should not be conducted and that we should focus on theoretical designs and then determine the fate of the R-1 rocket.

But in the case of the G-1 design, in spite of the Germans' sufficiently convinc-ing arguments, the NTS decided not to hurry with decision-making. Moreover, there were not only technical issues but others as well that the majority of us did not utter out loud. Here is an excerpt from the NTS decision:

"The report on the G-1 rocket design contains a number of interesting, fundamen-tally new designs for the rocket's individual structural assemblies. On the whole the design merits approval. Of particular interest is the rocket control system used in the design, which solves the problem of improving the grouping capability compared with the A4 rocket. However, the reports and the subsequent discussions show that many critical control system assemblies have not yet been optimized and do not meet the requirements of the draft plan… The idea of separating the warhead from the body of the rocket is a new one and deserves approval, as does Mr. Gröttrup's proposal to conduct experimental optimization of the payload on A4 rockets… The load-bearing propellant tanks con-structed of light alloys might substantially lighten the structure of the G-1 rocket's middle section compared with the A4… The design of the G-1 (R-10) propulsion system makes it possible to simplify the general layout of the propulsion system, to reduce its weight and its dimensions… Driving the turbine with gases from the combustion chamber certainly requires experimental testing… Before the development of the rocket's detailed design, individual experimental models of the aforementioned G-1 assemblies need to be fabricated and tested under test rig conditions… We need to speed up in every possible way the more detailed development of the control system as a whole and its fundamental assemblies all the way to the mockup phase, and subject the design of the radio equipment to an authoritative expert review… We also need to expedite follow-up on the theoretical and experimental principles of the design and speed up its further development in draw-ings so that at the next regularly scheduled NTS plenary session we can once again hear *a presentation of the rocket's draft plan.*"[7]

In theory, Gröttrup and his staff had no reason to protest the NTS decision. But in reality not only the NTS, but the management of the institute and Ministry of Armaments, at whose insistence this project had been implemented, found themselves in a very difficult situation.

Sergey Vetoshkin's position was revealing in this regard. In the Ministry of Armaments he was the chief of the main directorate which had authority over NII-88, and he was basically Minister Ustinov's right-hand man, managing the development of rocket technology.[8] I became acquainted with Vetoshkin back in Germany. He had flown in as a member of Marshal Yakovlev's commission. We had the highest regard for his genuine sophistication and intelligence, his ability to listen attentively to advocates of the most diametrically opposed technical points of view, his kindness, and his striving to delve, not just nominally, but into the essence of the most complex scientific and technical problems, and finally, his amazing capacity for work and unselfish devotion to our cause.

I also felt that he was well disposed toward me from our initial acquaintance. Time and again he candidly expressed his views and prognoses on the development of our technology and was also intent on getting my candid, rather than formal, observations.

One of these conversations took place soon after Gröttrup's defense of his design described above. Vetoshkin and I were squeezed into the aft single-seat cabin of a Po-2 airplane that served us at the test range in Kapustin Yar. When neither time nor automobiles were available to get from the special train where we lived to the launch site and back, sometimes we availed ourselves of this "air taxi."

On this particular occasion after takeoff, being to a certain extent an "aviator" because of my previous work, I noticed an unusually vigorous rocking of the aircraft's wings. Usually pilots would do that at low altitude to greet someone. I happened to glance at the wings that the pilot was "waving" so intensely, and I saw that the ailerons for roll control were clamped in control surface locks. These control surface locks were supposed to be latched on the ailerons and rudders after landing to prevent buffeting by the wind. In his haste, our pilot, evidently, forgot to remove them before takeoff and took off with them still clamped to the ailerons. I decided to keep quiet until we landed and not upset Vetoshkin. Thankfully the entire flight only took 10 to 12 minutes. The pilot made a long approach into the wind to the landing area near our special train and we touched down successfully. When we

---

7. The "draft plan" *(eskiznyy proyekt)* of a project typically denoted a document (usually several volumes long) that substantiated in detail the overall design of the system in question. Once designers signed off on the draft plan, they would then produce subsequent technical documentation for production to experimental workshops.

8. This was the Seventh Main Directorate of the Ministry of Armaments, one of many in the ministry overseeing weapons development.

had gotten out of our cramped cabin, I showed Vetoshkin the control surface locks, which did not look at all like they belonged on an airplane, and congratulated him on our successful landing, telling him that we could have ended up in the hospital because of that. Sergey Ivanovich decided to point this out to the pilot, but when we showed him the ailerons he smiled, unfazed, and said, "That's nothing, we've flown with worse."

After that, Vetoshkin asked me to drop by his compartment for some frank conversation over a glass of strong tea. After the "blowout" at the launch pad, after one more failed rocket launch attempt, this was very tempting. Over tea in the warm compartment he asked me straight out, "Boris Yevseyevich, you started all of this activity in Germany. You organized the Germans' work. You know better than I do what they are capable of. And now they are here with us designing a new rocket, with your help, incidentally. How do you envision the future course of this work? You and I heard them out at the NTS. There was quite a bit of criticism, and it was all useful and interesting. But the main issue that continues to haunt me and that Dmitriy Fedorovich castigated me out about is—what to do with the design of the [G-1] rocket? After all, the Germans can't create it by themselves on that island."

The issue was not a simple one. Lately, especially after the meeting with the Germans at the NTS in September, I had been mentally scrolling through all sorts of alternatives for the subsequent process of combining our operations in order to utilize the creative potential of the specialists we had brought in from Germany. It was not just the official, but also the moral, weight of responsibility for their fate that haunted me. Nevertheless, I did not see any real prospects for the German collective to work effectively on the design they had proposed. Out of political and security considerations, no one would allow us to create a mixed Soviet-German collective at NII-88 like the one we had in Germany. But even if they did give us permission, whose design would be developed there and who would be the chief designer? That Korolev would work under Gröttrup was absolutely out of the question. And if Gröttrup worked under Korolev? This too was unrealistic, because Korolev would immediately announce, "Why? We can handle it ourselves." In other words, we needed to set up two parallel design bureaus conducting parallel work. But this was beyond the powers of our institute and our subcontractors, especially because Ryazanskiy and Pilyugin would not implement the new ideas contained in the G-1 design, not because the Germans had proposed it, but because they also wanted to be the authors of their own developments and systems. Both Ryazanskiy and Pilyugin, with whom I had very good relationships, viewed the A4 and its domestic reproduction, the R-1, as practice, above all for technology, production, and setting up a domestic control systems industry. Then they dreamed of creating their own systems. In this regard, they shared Korolev's general attitude. In other words, we needed to use the Germans' experience and those ideas that they expressed in our subsequent work, and then, unless relevant decisions were conveyed from the very top, gradually send them home. Those were approximately the thoughts that I expressed to Vetoshkin.

He agreed with what I had said, but alluding to Ustinov's opinion, he said that the availability of a creative staff of German specialists should serve as a stimulus for our work. "After all, it is still unclear precisely what rockets we are going to need. We have no one to fight against using A4 rockets. And even if we double its range, it doesn't matter, because nobody needs it in a war. But we will certainly make it. Otherwise there will be no industry. And without factories, all the science in the world won't help us."

I left Vetoshkin, having thanked him for the tea, sugar, cookies, and frank conversation. Having crossed over to my two-berth compartment, I woke up Viktor Ivanovich Kuznetsov, who would later become a twice-recognized Hero of Socialist Labor and academician. A bust of Viktor Ivanovich now stands near his institute on Aviamotornaya Street. Over quite "allowable" portions of "Blue Danube"—that's what we called the 70% alcohol tinted with manganese crystals that we filled the rockets with—I told Viktor about my conversation with Vetoshkin and asked for his opinion. Soon thereafter a terribly worn out Voskresenskiy, who had just arrived from the launch pad, knocked at our door and entered.

The conversation continued among the three of us. Voskresenskiy expressed some really prophetic thoughts: "Sergey (as he called Korolev) wants to be the autocratic master of the problem. I have studied him better than you have. And he will be able to do it. For him the Germans have already done their job, and he doesn't need them any more. But the authorities are afraid of Korolev. They need a counterweight, and so for the time being we will pretend that we are interested in the German design. No matter what clever thing the Germans might propose, Sergey, Mikhail, and Nikolay will still do things their way. So there is no need to mess around. We have to be up early tomorrow, the weather is supposed to be good, let's say goodnight."

When we returned to Podlipki in late 1947 after the A4 launches, I once again had conversations with Pobedonostsev on that same subject.

In the winter of 1948—I don't remember if it was January or February—a group of colleagues, including my deputy for radio engineering Dmitriy Sergeyev, and I set out for the island to—in institute Director Gonor's parting words—"check how the implementation of the NTS decision was going." During these business trips sometimes you got on friendlier terms with people than during the everyday hustle and bustle on the job. I really liked Sergeyev, a "kindred spirit" and talented radio engineer always filled with a lot of new ideas. He was really fascinated with the proposals for G-1 radio control, which embodied new principles that were substantially different from what had been done in Peenemünde, but he had redone a lot of things and it was difficult to determine what had actually been done without his prompting or direct involvement.

During the aforementioned trip to the island, a meeting and difficult conversation with Gröttrup were unavoidable. In Bleicherode, I was "tsar, god, and military commander" to him. The moment he was boarded onto the railroad car bound for the Soviet Union, he understood that my authority had ended, and our interaction during our meetings in Podlipki and at Lake Seliger was usually rather dry and formal.

But this time Gröttrup was very happy about my arrival and announced that, whether I liked it or not, he had a lot of unpleasant things to tell me. The gist of the rather long speech that he unleashed upon me was that, in spite of the NTS's favorable decision regarding his design, he could not meet a single request listed on that document.

The testing that they had been faulted for omitting had not begun and was not scheduled on the island, in Podlipki at NII-88 itself, or in Khimki at Glushko's design bureau. In their small, closed collective, alienated and artificially isolated from Soviet science and the Soviet OKBs, they continued to work on their design, which would be criticized again, because not a single one of its fundamentally new proposals would undergo experimental testing.

"They do not give us the opportunity to use your wind tunnels. We want to set up experiments on rigs to test our new propulsion system layout, but we can't. And how can we prove that a turbine can actually be driven by taking gases straight out of the combustion chambers? That isn't the kind of system you can corroborate with analysis. You need an experiment. The radio system needs test range and aircraft tests. But we aren't capable of making the latest equipment here."

I do not recall all of the criticisms now, but the list was sufficiently convincing. Next, Gröttrup switched to a calm, confidential tone. Although more and more he was convinced that he was being deceived, he asked that I, a Soviet citizen whom he trusted, tell him candidly what the future held for their work.

It was 1948; could I candidly tell him everything that I thought? I did not dare tell Gröttrup what I had told Pobedonostsev, Vetoshkin, and Gonor about the Germans' work prospects. My reason was both professional and based in concern for Gröttrup's well-being. I did not think that I had the right to kill his hope for at least a partial realization of the idea he had conceived. Gröttrup was an engineer genuinely committed to his work. He had lost his homeland, at least for a long time, so he assumed. Now, except for his family, his only pleasure and goal in life was the interesting, risky, next to impossible, but exceedingly fascinating task of creating the rocket that they had not been able to, had not had time to come up with in Peenemünde. Even if it was for the Russians. To hell with them. But this would be Gröttrup's and his collective's creation. Half of Germany was under Stalinist Russia anyway. That meant that this rocket could benefit not only the Russians, but the Germans as well. In my mind, that was Gröttrup's reasoning. I must honestly admit that I liked him, both as a person and as a talented engineer. He just had that "divine spark."

During that winter visit and one more subsequent visit to the "German island," I acquired detailed knowledge of the work being conducted on the control system. Besides Sergeyev, who himself was actively involved in developing a radio control system, Kalashnikov also worked with me. He was my department deputy at the institute and the lead for electro-hydraulic control surface actuator development.

We confirmed that, in spite of the very primitive production equipment, the system's main new instruments were manufactured and undergoing testing. These instruments included a summing gyroscope proposed by Drs. Magnus and Hoch, which had a spherical gyro wheel and an electric spring, an amplifier-converter with magnetic amplifiers instead of the vacuum tubes that were used in the A4 rocket's *mischgerät*, a program mechanism, and pneumatic control surface actuators. Of the ground equipment, they were finishing fabricating the launch console and launch control system test panel.

Engineer Blasig, experienced from working at Askania, was developing the pneumatic control surface actuator. We criticized this project more than any other. Kalashnikov especially loved to argue with Blasig. A staunch proponent of hydraulic drives, Kalashnikov would not tolerate even the thought of using pneumatic control surface actuators on rockets. It is worth mentioning that the subsequent development of both Russian and foreign actuator drive mechanism technology proved us right. For a variety of reasons, all large rockets, ours and the Americans', used only hydraulic drives in various layouts and designs.

By late 1948, according to all indices, the G-1 design met the requirements of the draft plan. By this time we had returned from Kapustin Yar enriched by our experience from the range tests on the first series of R-1 rockets.

Right before the New Year, on 28 December 1948, the large NII-88 NTS gathered once again to discuss the G-1 design. This time it was not Gonor who conducted the session, but acting NII-88 Director Aleksey Sergeyevich Spiridonov. Gröttrup's team of specialists, who had arrived to defend their design, included Drs. Wolf, Umpfenbach, Albring, Hoch, Blass, Müller, and Rudolph. Bushuyev, Lapshin, Isayev, Glushko, and I were to review the design from the Soviet side.

Right off the bat Gröttrup decided to take the bull by the horns and announced that "the majority of the design elements could be considered suitable only after thorough check-out and testing…" The new rocket in its draft plan featured additional advantages compared with the attributes reported more than a year before. The primary parameter, the range, was not 600 kilometers anymore, but 810! The maximum targeting error was ±2 kilometers for azimuth and ±3 kilometers for range.

They had thought through some of the more innovative design elements in much greater detail and more thoroughly. In particular, the warhead separated from the rocket as a result of the difference in aerodynamic forces. Two solid-fuel braking rockets were incorporated on the body for reliability. A single load-bearing tank divided into two chambers by an intermediate plate was used for both components. It is worth mentioning that this design proposal was not subsequently used in Korolev's rocket designs. Many years later, V. N. Chelomey made use of it.[9] The idea of using the turbine exhaust gases to pressurize the alcohol tank was new.

---

9. Vladimir Nikolayevich Chelomey (1914–84) was a prominent Soviet designer of naval cruise missiles, ICBMs, space launch vehicles, and spacecraft. These included the UR-100 ICBM (and its various modifications), the Proton launch vehicle, the Soviet ASAT and ocean reconnaissance satellites, and the Almaz piloted military space station.

As he began addressing modifications to the propulsion system design, Gröttrup did not miss an opportunity to upbraid the critics: "We performed theoretical calculations in considerably greater detail than in Peenemünde, but of course, it would have been much better if, instead of excessively detailed theoretical research, we could have performed experiments on a test rig."

In spite of the criticism at the first NTS aimed at the radio control system, Gröttrup, who had enjoyed Sergeyev's genuine help and consultation over the past year, announced, "A purely autonomous control system is not feasible. We envisioned using instruments on the ground that had already undergone numerous tests, specifically radar." The Germans did not have documentation on our radar, and the control department that I directed made all of the primary ground radio equipment for the design development. Also among the proposals were further simplifications in the ground launching and fuelling equipment.

In the conclusion of his report Gröttrup said, "It seems to me that we can acknowledge that we have found a solution to the problem posed, and that the R-10 [G-1] rocket, in addition to having an increased range, also has other significant advantages over the A4: a streamlined and inexpensive manufacturing process; simplicity of maintenance; and reliability in operation... Even if the rocket was not attractive as a weapon, it would be needed as an object for the testing of the aforementioned innovations (separating nose section, load-bearing tanks, improved liquid-propellant rocket engine turbine, and new control), which are vital for the future development of a long-range ballistic missile..."

By way of discussion, all the disciplinary groups reported their findings after a preliminary study of the design of the G-1. On the whole, all of the findings were positive and amiable. The control group ended up having the most negative remarks, which I was forced to read out. I considered the most serious of these to be such system vulnerabilities as: the unreliability of the pneumatic control surface actuators at low temperatures, the transfer of the last electrical operations before launch from an automatic system to a human being, the lack of an operator error protection circuit in the preparation automatics, and an increase compared with the A4 in the number of "air-to-ground" pneumatic connections. Nevertheless, the control group approved the draft plan just as the other groups had. Everyone noted that in terms of scope it surpassed the requirements for the draft plan and that it was time to make the transition from designs to the realization of all of the stipulated experimental work.

ONE OF THE FUNDAMENTALLY NEW FEATURES IN THE CONTROL SYSTEMS DESIGN PROCEDURE WAS THE USE OF *BAHNMODEL*, the German term for trajectory simulators. In modern terms, this was the first time we had used an electromechanical analog simulator. This simulator was, of course, nothing like modern electronic machines, but for the first time it made it possible to simulate equations of the rocket's motion relative to its center of mass with variable coefficients and to obtain solutions for these equations, taking into account the characteristics of the individual instruments connected to the simulator. The simulator's inventor, Dr. Hoch, announced that it was now possible to conduct a preliminary checkout of the A4 rocket hardware before launches. There was no such simulator in Peenemünde. At that time, the Germans, and then we, used an elementary simulation involving a "Häusermann pendulum," a simple instrument named after its inventor.

These days, when an engineer is designing a rocket motion control system, simulation is the main way of selecting the system's parameters at the beginning and of performing a control check of the actual instruments at the end of the process. Electronic analog and digital simulators have attained such a degree of sophistication that when they are used to solve differential equations of the highest order, the results are more credible than the analytical calculations of the most distinguished mathematicians. Today simulation is viewed not as a desirable process of design and of the subsequent optimization of the control systems of any class of rocket, but rather but as necessary and mandatory. In this sense rocket technology instigated the evolution of a new and progressive method for the development of complex systems and has had considerable influence on many other fields of science and technology.

At least two ideas that were brought to the point of engineering realization and experimental testing belong to Dr. Hoch. These included one of the first electro-mechanical simulators in the Soviet Union and a simulating gyroscope. The latter development was a joint effort with Dr. Magnus.

Unfortunately, Dr. Hoch's very productive work was cut short. His reputation extended beyond the confines of our NII and reached the organization where air defense rocket control systems were being developed. Sergey Beriya, son of the all-powerful Lavrentiy Beriya, had been appointed chief designer there.[10] Without asking for approval, the leaders of this organization could transfer anyone from anywhere to work there. They transferred Dr. Hoch to work for young Beriya. According to rumors that reached us, he had settled down there quite well, was having great success on the job, and had asked to become a full-fledged Soviet citizen. But suddenly he ended up in the hospital, where he died after an operation as a result of purulent appendicitis.

THE DISCUSSION PROCESS IN THE NTS WAS NOT WITHOUT A CURIOUS DISPUTE. The person who caused a ruckus was an NII-88 consultant on issues of motion stability who was head of the department of celestial mechanics at Moscow State University, professor of mathematics from the N. Ye. Zhukovskiy Academy, Engineer Colonel N. D. Moiseyev. He was an exceptional polemicist, a brilliant lecturer, and vocal about his militant intolerance toward those displaying "dissident tendencies" in science.[11]

---

10. This was the Special Bureau No. 1 (SB-1) organization, established in 1947 in Moscow to develop air defense weapons. In 1950, SB-1 became Design Bureau No. 1 (KB-1), one of the most secret Soviet weapons design organizations. It was tasked with developing a foolproof air defense system around Moscow to protect against American strategic bombers.

11. In other words, Moiseyev supported the Communist Party's position on strong ideological control over Soviet science.

This time he plunged into a debate, first with the design reviewer regarding automatic stabilization systems and then with Dr. Hoch in connection with the praise directed at the new simulator. Hoch believed that the simulator could do in several hours what mathematicians took months to calculate. And even after years of incredibly complex work the mathematicians would be less reliable. In response the reviewer wrote, "…elements of the control system are presented in metal. The control system proposed in this design is new and original."

Regarding this same section of the design Moiseyev, on the other hand, declared, "The section devoted to the analysis of stability during the active and passive flight segments was not satisfactorily developed… The procedure used in the German studies of freezing the variable coefficients and analyzing the signs of the real components of the roots of characteristic equations was inadmissible, as the research of Soviet scientists has shown." Here he was speaking not about Soviet scientists in general, but specifically about Moiseyev's work on the "theory of technical stability."

Later he said, "I offered Dr. Hoch an example of linear differential equations with variable coefficients. About a week has passed since I gave him this example. However, so far we do not have a solution from Dr. Hoch for this simple example using the *Bahnmodel*…. As a stability theory specialist, I believe that… coefficient freezing and all of that is the kind of thing that, in 1948, is simply not worth writing about in scientific reports to be submitted to serious scientific institutions."

Regarding this position, Viktor Kuznetsov, who was well aware of the danger of overestimating the value of multistage theoretical calculations when working with gyroscopic systems, could not refrain from making this ironic statement: "Professor Moiseyev said that the theoretical grounds were insufficient. For us designers, on the contrary, what's important is the experimental method, which no calculation can replace, and the availability of such a method is a great achievement. Sergeyev, who had spent many days working on the island during the development of the design, was more blunt in his speech. "I think it is better to use the *Bahnmodel* than to write very complex equations as is the way in 1948 and end up with no rocket." Shapiro, another notable Moscow professor from another military academy, the artillery academy, supported Moiseyev's opponents. "Considering that we are taking hundreds of aerodynamic coefficients with insufficient accuracy, I think that we need to have a sense of proportion and understand that mathematical methods must be in keeping with the degree of accuracy of those parameters, particularly aerodynamic ones, that we know."

Moiseyev's attacks did not disturb Dr. Hoch. He responded that in his report to the group he had already cited an example of a solution for the system of equations that convincingly showed the insubstantial effect of the coefficients' variability. But the main advantage to his method was the use of the actual equipment, which could not be described with precise equations in theoretical studies. "If you look at any electrical instrument, you will see tolerance values for all of its resistors. I cannot order production to fabricate resistors with absolute precision."

And as far as the examples that Professor Moiseyev proposed for solution on the

Bahnmodel, he commented, "Unfortunately they transported this precise measuring instrument as if it were a sack of nails, so it doesn't have the accuracy it once did… I would like to remind you of a case in which Helmholtz once theoretically proved to Otto Lilienthal that human flight was altogether impossible."[12]

After long-winded squabbling with one another, in his final wide-ranging speech Professor Moiseyev decided to divide the blame between his Soviet and German opponents. "Dr. Hoch is undoubtedly an intelligent worker and he undoubtedly is conscientious in his work. One can clearly sense this from the diligence with which he has approached the solution of the problem… This department's advocacy of a simplistic treatment of theory and allowance for theoretical weakness is a gross error, and this error has its political significance… Comrade Shapiro is profoundly and fundamentally misguided, demonstrating his complete ignorance of elements of stability theory… The fruits of the simplification efforts of comrades Kuznetsov, Sergeyev, and Shapiro were immediately apparent. The authors of the design picked up their simplistic approvals and also began to defend what they had been forced to admit at NTS group reviews."

I am giving such a detailed account of the polemics at the NTS because at that time, even scientific problems that were far removed from politics and ideology, such as matters of rocket stability, could acquire political overtones.[13]

A sudden and unwelcome was not unimaginable…. a German scientist in a new secret rocket institute not only argues with a Russian professor and colonel, but even receives support from Soviet scientists, including Professor Shapiro, who was also a colonel and a Jew. The accusation that Moiseyev advanced ("this error has political significance"), at that time could unceremoniously turn into an affair that could end not only in loss of work, but also in an investigation by the security agencies to see if there were not something along the lines of a conspiracy there.

But by and large, the institute's staff of engineers and scientists did not support Moiseyev's line or similar attempts to insert political ideology into purely engineering problems. The general course of the discussion was friendly, but Glushko, Pobedonostsev, Bushuyev, and Mishin had consulted beforehand with Korolev and had a sense of the mood in the ministry. They were sure that the [German] rocket design on the whole could not be implemented.

In his closing remarks, Gröttrup expressed himself unequivocally. "Without

---

12. The account here is somewhat garbled and probably refers to an episode involving Herman Ludwig von Helmholtz (1821–94), the great German scientist, who pronounced in the 1870s that human flight powered by muscles was probably an impossibility. Otto Lilienthal (1848–96) was the famous German pioneer in the human conquest of the air, whose book *Der Vogelflug als Grundlage der Fliegekunst [Birdflight as the Basis of Aviation]*, published in 1889, greatly influenced the Wright Brothers' early designs.

13. Here, Chertok is once again referring to the broader cultural trends of *Zhdanovshchina* ["time of Zhdanov"] and anti-cosmopolitanism promoted by the Soviet Communist Party in the late 1940s and early 1950s, when many fields of intellectual inquiry were hostage to ideological interference and distortion.

experimentation it is impossible to develop this design... The experiments are not simple, because in some instances we are dealing with tests of designs based on completely new principles. For this reason, if these experiments are to be conducted at an accelerated pace now, which I and all of the specialists working on this design would like very much to see, I request that the delivery of materials and equipment be increased accordingly so that experiments can be conducted... We need to completely change the method that we have used to develop this rocket up until now and switch from theoretical and design work to broad experimentation."

Nominally the council's ensuing decision was quite favorable. It contained all the necessary requests for experimental optimization and for the acceleration of all activities

The council's favorable decision was, however, little consolation. Scrapping the two-year project—a constituent part of NII-88's plan—was impossible both practically and for formal reasons. A great deal of resources had been expended for the development of the G-1 (R-10) design, which was the basis for work at Branch No. 1. At the same time, there was neither enough engineering nor enough production manpower to realize a design being developed simultaneously with plans being executed under Korolev's direction.

The further development of rocket technology required the concentration of efforts in a single decisive area. The conditions that had been created at that time had already made the R-10 design unfeasible. However, work on the design continued throughout 1949.

In October 1949, our institute had already conducted range tests on the R-2E rocket—an experimental version of the R-2 developed by Korolev's OKB—at a range of 600 kilometers. At Branch No. 1, work on the [G-1] design that had seen so much effort poured into it were gradually curtailed. The German specialists were still hearing many promises to begin the experiments, but they lost faith and began to understand the futility of their activity.

Air defense guided missiles occupied a special place in the work of the German specialists. The goal of this work was an attempt at modernizing the *Wasserfall* and *Schmetterling* rockets. Chief designers Sinilshchikov and Rashkov conducted this work at the main base in Podlipki.[14] However, with the transfer of air defense projects to the Ministry of Aviation Industry, where they entrusted rocket development to well-known chief designer S. A. Lavochkin, and the development of the entire control complex to the new KB-1 organization, it no longer made sense to continue

---

14. Chief Designer Yevgeniy Vasilyevich Sinilshchikov (1910–90) headed the NII-88 SKB's Department No. 4 responsible for reproducing the *Wasserfall* missile. Chief Designer Semyon Yevelyevich Rashkov headed Department No. 5 responsible for reproducing the *Schmetterling*.

these operations at the Ministry of Armaments.[15]

During this same period and also under Gröttrup's leadership, on the island they were studying ideas for developing the R-12 (G-2) rocket with a firing range of 2,500 kilometers and a warhead with a mass of at least 1 metric ton. They intended that this rocket be developed immediately after the R-10 was put into production. The propulsion system for this rocket was to be constructed as an assembly of three R-10 engines, for a total thrust of over 100 metric tons. For the first time, this design called for doing away with gas-jet control fins. Loss of thrust in the propulsion system—because of the gas-dynamic resistance of the control fins standing in the stream of the hot gases—was thus eliminated, consequently increasing control reliability. One should note that no such proposals had been made while we were working in Germany. Eight years later we completely did away with graphite gas-jet control fins on the famous R-7 intercontinental missile. In the R-12 design, the Germans proposed that control be carried out by changing the thrust of the engines arranged along the periphery of the tail section at an angle of 120 degrees. More than 20 years later, we implemented a similar idea on the N-1 "lunar" rocket. If I am not mistaken, these were the only examples of the control of a heavy rocket using that method. But the R-12 did not go any further than a paper report, and work on the N-1 was curtailed in 1974 after four failed launches. All modern liquid-propellant rockets are controlled by special control engines, jet nozzles, or hydraulic drives rotating the primary engines relative to the rocket's body.

In addition to the detailed conceptual design of the R-10 rocket with a range of 800 kilometers and the proposal for the R-12 rocket with a range of 2,500 kilometers, the Germans had performed preliminary calculations for more forward-looking designs, such as the R-13 (G-1M) rocket, which was an R-10 body augmented by a propulsion system from an A4, the G-4 (R-14) ballistic missile, and the G-5 (R-15) cruise missile, with a range of 3,000 kilometers and a payload of three metric tons. All of these developments were in the stage of layout drawings and calculation of basic parameters. In terms of depth of developmental work they were inferior to the Peenemünde A9 and A10 designs and to the Sänger intercontinental rocket-bomber. The Germans conducted this work without having the opportunity to consult with Soviet specialists. Our similar work on long-range plans was strictly classified, and we did not have the authority even to discuss these subjects with the Germans.

IN THE SAME PERIOD, WE WERE MOVING AHEAD WITH RESEARCH ON THE R-3 MISSILE, one of the most important stages in the development of Soviet long-range

15. In August 1951, the Soviet government transferred all tactical air defense and winged missile projects from the Ministry of Armaments to the Ministry of Aviation Industry. As part of the move, projects from NII-88 were moved to aviation-based organizations such as OKB-301 led by Semyon Aleksandrovich Lavochkin (1900–60) and KB-1 led by Amo Sergeyevich Yelyan (1903–65).

missiles in the postwar era. Work on the R-3 plan began, under Korolev's supervision, as early as late 1947. The intention was to conduct broad-scale research for the development of a rocket with a range of at least 3,000 kilometers. To this end, four basic rocket design layouts were accepted for review: normal ballistic (BN), staged ballistic (BS), normal winged (KN), and staged winged (KS).[16] Primary attention was devoted to work on the BN layout. As for the winged layouts, simultaneously with work being conducted at NII-88, work on them had begun to a significant extent under the influence of Sänger's report as early as 1945 at NII-1 under Bolkhovitinov and was developed on a broader scale when Keldysh became the NII-1 director.

Work had been conducted on the R-3 design as part of the cooperation that had already taken shape during 1947. Korolev's KB was in charge of developing the conceptual design. Engines were being developed simultaneously in two organizations, at OKB-456 by Chief Designer Valentin Glushko and at NII-1 of the aviation industry by Aleksandr Polyarnyy.[17] Control system design as a whole was assigned to NII-885 headed by Mikhail Ryazanskiy and Nikolay Pilyugin. A competitive version of the radio-control system using a gyro-stabilized platform was being developed at the same time under the supervision of Boris Konoplev at NII-20 (in the radio unit) and at NII-49 (in the gyroscope unit).

Korolev personally supervised all of the work on the R-3 plan. He took on the responsibility for the content of the first volume of the draft plan, "Principles and Methods for Designing Long-Range Rockets." The entire design, which consisted of 20 volumes, not counting the tens of volumes and reports by subcontracting organizations, was completed in June 1949. In this work, I devoted a great deal of attention to the development of a stellar navigation system, on astro-correction for autonomous control systems, and above all, for winged versions that required control along the entire flight path. I will describe this in greater detail later.

On 7 December 1949, NII-88 held a meeting of its scientific-technical council, during which it examined the draft plan of the R-3 rocket, engines and control system. This meeting was held a year after the discussion of Gröttrup's R-10 design; it finally shut off the prospects for the development of the German version.

The R-3 draft plan was approved on the whole, but at the same time the council noted the tremendous complexity of the problem that had been posed and its "scale, which was unusual for our field." These words from Korolev's memorandum show his understanding of the need for a systemic approach and a concentration of great effort on a common targeted objective. In his memorandum, upon completion of the R-3 draft plan, referring to the organization of operations, Korolev concisely formulated the organizational principles for operations on such a scale:

---

16. BN—*ballisticheskaya normalnaya;* BS—*ballisticheskaya sostavnaya;* KN—*krylataya normalnaya;* and KR—*krylataya sostavnaya.*

17. Aleksandr Ivanovich Polyarnyy (1902–91) was a liquid propellant rocket engine designer who had worked together with Korolev in the early 1930s at the amateur GIRD group.

"[We must] conduct a set of large-scale measures in various fields of industry, resulting in a significant improvement in quality in the field of technology associated with the development of the R-3;

[We must] set up operations so that individual organizations and groups would not be actively working on the R-3 rocket, but R-3 development would be handled by the country's best, with as great a workforce as is required...

In order to attract the very best technical staff, a number of material conditions need to be provided. One of the most important of these is the provision of housing and material support... To a significant degree, [we must] expand and strengthen the experimental base for the new technology, find the capital investment needed to reequip it... Entrust the country's appropriate scientific and technical organizations with a whole complex of work and full responsibility for the solution of problems concerning the development of the R-3 rocket...

[We must] combine all the specialized organizations currently working on rocket technology into a single agency."

These positions, which embodied the mission as seen not only by Korolev, but also by his colleagues on the famous Council of Chief Designers, essentially defined the requirements on a national scale for the subsequent rocket technology development program. Common sense, however, suggested that even 3,000 kilometers was not the range that our rockets needed. The R-3 design was a jumping point for a program of long-term operations. It took five years before such a program began to be realized on a national scale, when work was launched on a broad scale to develop the R-7, the first intercontinental missile with a thermonuclear warhead.

IN 1950 THE NATURE OF THE WORK CONDUCTED BY NII-88 BRANCH No. 1 CHANGED. The Ministry of Armaments officially decided to halt further work on the design of long-range rockets in the German workforce. This decision was prompted by the perfectly understandable pessimistic moods of the Germans, the lack of faith in the purpose of their further work, and their loss of creative enthusiasm. The gap between the problems posed in 1947 and the actual capabilities for solving them was so obvious by 1950 that promises to correct the situation inspired little of the confidence required for work. As I mentioned earlier, for further productive work on the development of rockets, the main thing was that we needed to allow the German specialists to participate in joint work in all areas of our cooperation. But this would have involved "revealing state secrets." The island's isolation led to an ever increasing gap between the German scientists' level of knowledge and experience and that of the specialists from the "mainland."

To keep the collective busy, they came up with a list of minor odd jobs of various disciplines, which for one reason or another were not suitable to be performed on the main center of NII-88 at Podlipki. Among these projects were control system instruments, measurement instruments, and the optimization of the *Bahnmodel*. The latter very timely work, unfortunately, did not receive the proper development because of the departure of the primary author, Dr. Hoch, to another organization.

In October 1950, all work at Branch No. 1 of a secret character was terminated and the future stay of the German specialists in such a location and with its associated status lost any meaning. Earlier, on 13 August 1950, the USSR Council of Ministers made a decision concerning the further use of the German specialists. This decision regulated the conditions for the return of the German specialists to Germany. Those desiring to leave in the next two years could exchange their savings in rubles for German Democratic Republic marks; a worker would receive 75% of his wages and each member of his family, 25%. The trip home, transport of baggage to the border, and special passenger and freight cars were provided.

This decision obliged the Soviet Supervisory Commission in Germany to provide free passage to the German specialists and their families on German territory and ensure living quarters and work for them. Soviet organizations were allowed with mutual consent to continue using German specialists to complete projects.

The decision to send German specialists to the German Democratic Republic was made at the governmental level. They were sent in several groups. The first batch was sent from NII-88 Branch No. 1 in December 1951, the second in June 1952, and the last special train left for the German Democratic Republic in November 1953. As is appropriate for the captain of a sinking ship, Gröttrup and his family left the island last. We received only fragmentary and random information about the subsequent fate of the German specialists.

IN 1990, AT A CONFERENCE OF THE INTERNATIONAL ASTRONAUTICAL FEDERATION IN DRESDEN, Mishin met up with German aerodynamics specialist Dr. Albring. Their encounter was very warm. Albring reported that Gröttrup had passed away and that in the Federal Republic of Germany his wife Irmgardt had published a memoir about working in the Soviet Union.[18]

In spring 1991, while in the Federal Republic of Germany, I got to know Dr. Werner Auer, the leading specialist in space gyroscopic instruments. He turned out to be a protégé of Professor Magnus. In the foreword to Magnus' famous *Gyroscope: Theory and Application,* published in 1971, the editor of the translation wrote, "This is a fundamental monograph in which the author exhaustively elucidates the main aspects and applications of modern gyroscopic theory, its methods and most significant results, in particular those that belong to the author himself."

Every time necessity compels me to take this well-published book off the shelf, I remember the two jolly, young Drs. Magnus and Hoch, in 1947 intently working in the laboratory car of the special train at the Kapustin Yar test range, trying to find what caused the A4 rocket's great deviations during the second launch. At that time the mood of the German specialists was splendid. At any rate, it was better than during all of the subsequent periods of our joint work.

It was not until 1992 that I was able to learn the fate of the Gröttrup family after

---

18. The original German language monograph was published in English as Irmgard Gröttrup, *Rocket Wife* (London: Andre Deutch, 1959).

their departure from Moscow for the German Democratic Republic in 1953. In March 1992, the newspaper *Isvestiya* published an abridged version of my memoirs about our postwar activity in Germany and subsequent work of the German specialists in the USSR.[19] This series of articles was under the general headline: "Soviet Rocket Triumphs Had a German Origin." Boris Konovalov prepared the articles for publication, but he did not coordinate the headline with me. I called Editor-in-Chief Igor Golembiovskiy and expressed my displeasure with the headline's tendentiousness. He was surprised at my reaction, but promised to correct the situation. And so the last of the six articles in the series appeared under the headline "American Rocket Triumphs Also Had a German Origin." An *Izvestiya* reader in Hamburg with whom I was not acquainted came across the surname of her friend, Gröttrup, in the series of articles and asked her if the article was about her father. It turned out that Ursula Gröttrup really was Helmut Gröttrup's daughter. Ursula wanted to find out more details about her father and decided to travel to Moscow. She flew in to Moscow on 7 August 1992 and stayed with a Russian friend of her German acquaintance. This Muscovite woman, who had a beautiful command of German, also arranged a meeting for Ursula Gröttrup and me.

This was Ursula Gröttrup's story. She was eight years old when the Gröttrup family left Moscow for the German Democratic Republic in 1953. Her parents had intended to start to work in the new Germany because many friends had gone there and wrote that her father would be guaranteed good work.

But on the platform at the Berlin train station, instead of East German state security agents or the Soviet guards that had protected them for almost eight years, they were surrounded by young people who turned out to be agents of the U.S. and British intelligence services. They were holding passports for the Gröttrups (which later turned out to be fake) and used them to prove to the Berlin authorities that the Gröttrups had expressed the desire to live in West Berlin. They were taken directly from the train station and driven to West Berlin, where they were placed in one of the American residences.

After the initial processing of Ursula's parents, the Americans announced that they would create the necessary conditions for their work, but in Cologne rather than West Berlin. There was only one autobahn for the journey from West Berlin to Cologne through the German Democratic Republic and it was strictly controlled by the East German border guards. Apparently, they feared that Gröttrup, who did not have the necessary documents, might be detained, resulting in an operational failure of the mission, not to mention diplomatic unpleasantness. For that reason they did not put Gröttrup in a German vehicle, but in a station wagon with U.S. military license plates. These vehicles were not subject to inspection or control. Before their

---

19. *Izvestiya* correspondent Boris Konovalov prepared these publications, which had the general title "U Sovetskikh raketnykh triumfov bylo nemetskoye nachalo" [Soviets Rocket Triumphs Had a German Origin]. See *Izvestiya*, March 4, 1992, p. 5; March 5, 1992, p. 5; March 6, 1992, p. 5; March 7, 1992, p. 5; and March 9, 1992, p. 3.

departure, the raven that Gröttrup's wife Irmgardt had tamed before their departure from Seliger Island—and was carrying in a large cage—became a source of contention. The Americans demanded that she get rid of the bird, but she firmly refused, declaring that she was not going anywhere without the raven.

In Cologne they housed the Gröttrups in a separate villa guarded by U.S. soldiers. Released into freedom in the interior chambers of the villa, the raven wasted no time before it decorated the rich décor and smashed a precious vase.

Instead of work in West Germany, the Americans offered Gröttrup a contract to work on rockets in the United States. He said that he had to consult with his wife. Irmgardt Gröttrup declared that she had had enough rocket technology in Russia, that she was not going to leave Germany for anywhere, and that she did not need America. No amount of persuasion changed their minds; the Gröttrups absolutely refused to go to the United State. Six hours later they were simply put out on the street in front of their luxurious residence, along with the raven.

Literally out on the street without any means, they lived in poverty for almost a year. Finally, after a series of odd jobs, Gröttrup managed to get a good job in a department of the Siemens firm in Munich. This was during the beginning of the big boom in computer development. Gröttrup proved to be a capable engineer in this field, and soon he was in charge of more than 400 scientists and engineers. He worked a lot and earned a good living.

Soon Gröttrup appointed a young and very talented engineer as his deputy. Suddenly this deputy was arrested and charged with being a Soviet spy. At the trial Gröttrup vouched for his deputy, but they did not believe him, especially because he himself had worked for the communists for nine years. Insulted by the distrust, Gröttrup turned in his resignation to the Siemens firm and found himself once again unemployed. Friends and acquaintances helped him find work at a firm that manufactured machine tools for printing currency and all sorts of automatic machines for the banking industry. Already enriched from his experience in computer technology, here he developed the first automatic machines capable of counting paper currency, scanning credit cards, exchanging currency, and so on.

He once again prospered and his family lived well. "Father spent the whole day at work, and in the evening at his desk he wrote, calculated, and invented. Mother spent money, was very eccentric and lively. She used to tell unbelievable stories about her life in the Soviet Union." Her father warned Ursula that a lot of it was not quite that way.

Gröttrup could not avoid a merciless killer—cancer. He died in 1981. His wife, having obtained the freedom to follow her whim, published her diaries in 1958 under the title *The Possessed and the Powerful in the Shadow of the Red Rocket.*[20]

Not long before my encounter with Gröttrup's daughter I had the opportunity

---

20. Irmgardt Gröttrup, *Die Besessenen und die Mächtigen im Schatten der roten Rakete* [*The Possessed and the Powerful in the Shadow of the Red Rocket*] (Stuttgart: Steingrüben Verlag, 1958).

to acquaint myself with these diaries. It turned out that Frau Gröttrup's imagination was focused not on exaggerating her husband's role or that of the German specialists in Soviet rocket history, but on describing utterly incredible events from the time she had spent in Moscow, her association with certain high-ranking officials, and Soviet figures who had fallen in love with her. In addition, she described her participation in rocket launches at the test range in Kapustin Yar. But she had never been there, and everything that she described in her "diaries" was pure, unadulterated fiction. I told all of this to Ursula.

It turned out that her mother had died just three years before our meeting. Ursula agreed without protest that her mother had invented a great deal—that was her nature. Rather than write what actually happened, she could write how she wanted things to be. Alas, German readers have no way of figuring out which events in these rather lively narratives are true and which ones are pure fabrication.

Finding myself in the Federal Republic of Germany in September and October 1992, at the initiative of German television, I once again met with Gröttrup's daughter, who had already told the German television audience about her father against the backdrop of the country estate in the village of Trebra, where I had settled the Gröttrups in 1945 after their transfer from the American zone.

What then, on the whole, was the role of German operations in establishing our rocket and space technology? Rather than the work that the German specialists performed while they were in the Soviet Union, their greatest achievement should be considered what they managed to accomplish in Peenemünde and other places where they were developing rocket technology before their surrender in 1945.

The creation of a powerful scientific-research base such as Peenemünde, the development of the A4 rocket system and its mass production, the beginning of work on forward-looking long-range ballistic, cruise, and multistage missiles, and the development of various types of air defense missiles, in particular, the *Wasserfall*—all these achievements would serve as the foundation, the virtual launching pad for our subsequent work and that of the Americans.

The organization of rocket development in Germany during the war was an example of how a government, even in a difficult situation, was capable of concentrating its resources to solve a large-scale scientific and technical problem.

Doctrine that depended on the effectiveness of an unmanned rocket-powered bomber against important strategic objectives was a miscalculation for the Germans during the war years. For us, with the emergence of nuclear weapons, it became a real hope for preserving peace by creating the equal threat of a reciprocal nuclear strike. When missiles and nuclear weapons united into a single force, their use was virtually dominated by the two parties that had become locked in the Cold War struggle: the USSR and the United States. This arrangement has maintained peace on our planet for a long time and continues to do so. Thanks to its integration with nuclear weaponry and to its intense technical development, the Germans' "vengeance weapon" was converted into a real threat of a terrible retribution for all humankind, if the latter should lose its senses.

The Germans' technical experience, of course, saved us many years of creative work. After all, Korolev, during his captivity in Kazan, was the only one thinking about ballistic missiles. And even then he proposed making solid-propellant ballistic missiles, because he did not believe that liquid-propellant engines could provide the requisite colossal power.[21] But we saw that the Germans had real liquid-propellant engines with 30 metric tons of thrust and designs for engines with thrust up to 100 metric tons. This taught us not to fear scale. Our military leaders stopped looking at a rocket as a projectile, for which all you had to do was come up with a little bit better "powder" and then everything would be fine. When you think about it, that was precisely what had served as the basis of our prewar doctrine for the development of the celebrated *Katyusha* solid-propellant rocket-propelled projectiles of Petropavlovskiy, Langemak, Tikhomirov, Kleymenov, Slonimer, and Pobedonostsev.

In Germany we learned that a single organization or even a single ministry was incapable of dealing with rocket technology. The development of missiles required strong, nationwide cooperation. And the main thing was that we needed high-quality instrument building, radio engineering, and engine building infrastructure.

The fact that after a devastating war we mastered and surpassed German achievements over a very short time was enormously significant for the general rise of the culture of technology in our country. The development of rocket technology was an exceptionally strong stimulus for the evolution of new scientific fields such as computer technology, cybernetics, gas dynamics, mathematical simulation, and the search for new materials.

From the standpoint of the "human factor," as they say these days, in Germany we learned how important it was to have a solid intellectual nucleus of specialists from various fields. The unity generated in Germany was preserved even after our return to the USSR, although we were spread out over various ministries. And this was not just words or slogans, but in actual fact, despite the sometimes complicated personal relationships between the chief designers, their deputies, ministers, military, and governmental officials. Before the historical day of 4 October 1957, foreign publications wrote to the effect that the Russians were using German experience and German specialists to develop their rockets. All of these conversations and stories ended after the world saw the first artificial Earth satellite. The famous R-7 rocket, the first intercontinental missile, free of the "birthmarks" of German rocket technology, inserted this satellite into orbit. Its development was a leap in new quality and enabled the Soviet Union to take the lead in cosmonautics.

---

21. In 1944–45, Korolev proposed a series of long-range ballistic and winged missiles known as the D-1 and D-2 to his superiors. According to his plans, they were to use solid propellants. These proposals were never approved.

Chapter 4
Institute No. 88 and Director Gonor

The government decree of 13 May 1946 made the Ministry of Armaments State Union Scientific-Research Institute No. 88 (NII-88) the rocket industry's primary scientific-technical, design, and production facility. After many transformations, this organization exists to this day, but since 1967, it has been called the Central Scientific-Research Institute of Machine Building (TsNIIMash).[1]

I wrote earlier that during the summer of 1946, a high-ranking government commission headed by Artillery Marshall Yakovlev visited Bleicherode. Commission deputy chairman (and also Minister of Armaments) Ustinov and acting *Gosplan* representative Pashkov, before leaving Moscow, had evidently arranged with the personnel office of the VKP(b) Central Committee that, upon finishing our work in Germany, Pobedonostsev, Mishin, Voskresenskiy, Budnik, Chizhikov, and I would be transferred from the aviation industry's NII-1 to the new NII-88 under Ustinov. The decision for our transfer was formulated in paragraph 28 of the 13 May decree, but we did not yet know that such a decree existed.

This same commission also foreordained Sergey Pavlovich Korolev's transfer to NII-88 as department chief. When they offered him this post and the duties of head of the development of long-range ballistic missiles in Germany, he did not yet know that he would end up at NII-88, not under the authority of the director, but under Special Design Bureau (SKB) Chief Karl Ivanovich Tritko.

The day after our arrival from Germany, after riding the commuter train to Podlipki, I reported to NII-88 for the first time. I couldn't enter the grounds without a pass, so I stopped by the office of State Security Colonel Ivashnikov, deputy director for personnel and security. "I have an order on your assignment, but rules are rules. Go get the forms and fill them out like you're supposed to, and bring two photos. After you hand in the forms they'll give you a temporary pass, and then it's up to the director."

---

1. TsNIIMash—*Tsentralniy nauchno-issledovatelskiy institut mashinostroeniya.* TsNIIMash, originally known as NII-88, remains the leading R&D institute of the Russian space program. It also has supervision over the Russian Flight-Control Center (TsUP) at the Moscow suburb of Korolev.

And so my first day on the job was spent filling out forms in duplicate and writing an autobiography. It wasn't until the following day that I appeared before my immediate chief, NII-88 Chief Engineer Pobedonostsev. His office was located in the old building of the former artillery Factory No. 88 plant management. This pre-Revolutionary building was reminiscent of a monastery with its solid construction and thick walls. Pobedonostsev was very happy that I had finally showed up. He complained that people were always coming to him with problems in my field, and he already had a full plate. With that, he led me off to introduce me to Director Gonor.

Until then, I had seen Gonor only once in Germany, when he arrived as a member of Marshal Yakovlev's commission. At that time, he was in the uniform of a major general of the engineer artillery service. A Hero of Socialist Labor star, a Stalin Prize laureate medal, and three Orders of Lenin distinguished him from the many other combat generals.

Now, as we entered his large office appointed with heavy antique style furniture, he was also in his general's uniform, but of his many decorations he wore only the Hero star. We former aviation workers, still coming into our own as missile specialists, had an admittedly skeptical attitude toward artillerymen, and believed that, among all the men of arms, only Ustinov understood us. Nevertheless, we were going to have to work directly with Gonor and apparently for a long time. For this reason, I made up my mind to be obedient and to prepare myself to listen to instructions. Instead, what followed were simple questions: "How was your trip getting here? How are your living quarters? Have you been to your department yet?" He was clearly happy that I wasn't requesting an apartment in Podlipki. He lit up a Kazbek cigarette and offered the pack to me, to the obvious dissatisfaction of the nonsmoking Pobedonostsev. Shifting to current business matters, Gonor made it clear that I should quickly organize the work of the department, which was constantly acquiring new specialists, and that I should help Pobedonostsev sort out the job placement of the Germans. He said that intensive construction was underway for their resettlement on the island of Gorodomlya, and for the intervening three months they would live in health and vacation resorts in the vicinity nearby. There were very many transportation and domestic issues to be worked out.

"However," complained Gonor, "we're having more trouble with your friends than with the Germans." But he did not expand on that subject. When we returned to Pobedonostsev's office, he explained that Korolev and Mishin, especially the latter, had mounted an attack on Gonor from the very start, trying to circumvent their subordination to SKB Chief Tritko. But Ustinov had approved the NII-88 structure; everything had been done with the approval of the Central Committee office, and Gonor did not have the right to change anything. "We can work with him. He's a reasonable and sensible individual, and Sergey is picking a fight for no good reason where he should be biding his time." This was the first time I heard disapproval about Korolev's aggressive behavior.

Henceforth, I developed a completely professional, business-like relationship with Gonor, although once he even ordered that I be severely reprimanded. In that instance, the matter concerned a fire that occurred one Sunday in my department. The fire was trivial, but the times were not. The experienced Gonor signed the order even before the cause was determined, and called the duty officer at the ministry over the "Kremlin" telephone network and reported "The fire was extinguished and the guilty parties have been punished." After forcing me to sign my name to the order, he explained that, "It is better to receive a severe reprimand from me than to wait for a ministerial order removing you from the job. Should one of your people make a mistake, punish them yourself and quickly, in order to report that the 'cause has been determined and the guilty parties have been punished.' You gain time that way." There was no reason to be offended. This was a lesson in administrative leadership.

Gonor's fate was tragic after his appointment as NII-88 director. In this regard, I will allow myself to digress from the chronology to talk about him in greater detail. Like Ustinov, Gonor graduated from Leningrad Military Mechanical Institute, or *Voyenmekh*.[2] In general, this institute was the forge that produced the production and technology intelligentsia for the People's Commissariat of Armaments. Gonor received an appointment to the Bolshevik Factory in Leningrad, where he rapidly advanced from foreman to chief engineer. Thus, he actually became deputy to Ustinov, who was director of the factory. The personal qualities of the chief engineer contributed to the fulfillment of Stalin's immediate task for the mastery of new artillery systems for the navy. For this, the Bolshevik Factory, Ustinov, and Gonor received the first Orders of Lenin.

In 1938, they transferred the thirty-two-year-old Gonor from Leningrad to Stalingrad, as director of another large artillery factory, Barrikady.[3] Barrikady specialized in the production of 406-mm guns for battleship gun turrets, superpowerful infantry guns, and 122- to 305-mm howitzers. The factory had failed the reconstruction, and Gonor was supposed to salvage it from collapse. He managed to do that.

There, in Stalingrad during World War II, Gonor showed true heroism, and in the summer of 1942, he was among the first six military-industrial leaders to be awarded the title Hero of Socialist Labor.[4] To this day, Director Gonor is recalled with greater affection at the Stalingrad Barrikady Factory than at TsNIIMash—the former NII-88. During the Battle of Stalingrad, the Barrikady Factory was completely destroyed, and they transferred Gonor to Sverdlovsk to artillery Factory No. 9, which was being created under the auspices of *Uralmash*.[5] For the defense of

---

2. In Russian this institute is called the *Leningradskiy voyenno-mekhanicheskiy institut* (LVMI), hence *voyenmekh* for short.

3. The Barrikady Factory was also known as Factory No. 221.

4. The Hero of Socialist Labor was the highest award for civilians during the Soviet era.

5. The Ural Heavy Machine Building Factory (*Uralmashzavod*) was one of the largest mining and metallurgical enterprises existing during the Soviet period. Since its founding in 1933, it produced a huge array of industrial equipment for the Soviet economy and military.

Stalingrad and his subsequent activity during the war, Gonor was awarded the Stalin Prize first degree, yet another Order of Lenin, an Order of Kutuzov first degree, and the rank of major general of the engineer troops. On 24 June 1945, he attended the famous Victory Parade and the festive reception at the Great Palace of the Kremlin. But his most joyful assignment was his return to Leningrad in 1945 as director of the Bolshevik Factory, where he had begun his career path.

It seems that, in search of a future NII-88 director, Ustinov was playing a game of solitaire—only with people; he proceeded from the premise that, first of all, he should be an individual who was unconditionally devoted to him personally. Second, he should be a capable organizer, who had gone through a good school of production, "through thick and thin." And third, his candidacy should be supported by the apparatus of the Central Committee and perhaps even by Stalin himself. Postwar 1946 marked the recurrent rise of anti-Semitic sentiments as per directives from the top. But for the time being these were strategic appeals to the masses, who during the war had been driven by anti-German sentiments but were rarely anti-Semitic. In the defense industry, and in particular in the atomic industry, Stalin and Beriya not only tolerated, but protected talented Jews such as Khariton, Zeldovich, and many others.[6] They were guarded almost like members of the government.

Ustinov took a risk. He bet on Gonor and won. For the forty-year-old engineer general, who had outstanding accomplishments and capabilities, Stalin's confidence, and Ustinov's patronage, a brilliant future was opening up as the director of the first Soviet rocket center. At Gonor's disposal were missile specialists, whom Ustinov had persuaded to transfer to work for him. Gonor received an assignment as early as 1947 to begin flight tests on the German A4 missiles and, in 1948, to create the domestic R-1 missile. The government authorized the recruitment of many for this goal, including young specialists, those newly demobilized from the army, and scientists from institutes of the Academy of Sciences and institutions of higher learning; they would be able to work while simultaneously holding their former jobs. After becoming director, Gonor immediately created a scientific-technical council made up of scientists who had already made a name for themselves in our country.

As an artilleryman, Gonor used to associate with a very tight circle of scientists and military chiefs. Now dozens of individuals whom he hadn't known before, but who were extremely influential people, were asking for permission to visit his institute and look at the rockets. The shops and interior of the old artillery factory were completely unsuitable for meetings and for displaying the new technology. We urgently needed to construct clean assembly shops, a tower for vertical testing of the rockets, and demonstration laboratories where we wouldn't be ashamed to bring high-ranking guests and to show that less than a year after the decree was issued, we already had an institute. After all, Dornberger was able to create the now world-

---

6. Yuliy Borisovich Khariton (1904–96) and Yakov Borisovich Zeldovich (1914–87) were two pioneering physicists who played key roles in the development of the Soviet atomic bomb.

famous Peenemünde center from scratch.[7] Moreover, Gonor had to remember that NII-88 was the head institute of the new field. He had to combine the ideas and production output of engine specialists, guidance specialists, chemists, metallurgists, and mechanical engineers.

During the war, directors such as Gonor truly did accomplish great feats on the production front. A factory's director and chief engineer worked under the threat of a military tribunal for failing to meet an armament production goal. Directors of his rank were accustomed to giving completely of their physical and spiritual strength and their professional knowledge on the job. Incompetence was simply not allowed. They were monitored rigidly from above and watched constantly by the factory workers from below. The workers could forgive even a strict boss for being demanding if he was as demanding of himself, was interested in everything that affected his workers' living conditions, and showed sensitivity and humanity. Not every director possessed these qualities.

Now Gonor had to show his competence in a completely new field. Here he could not count on his reserve of knowledge and rich production experience. On more than one occasion during business meetings with him, he asked me to explain many problems of missile guidance that were incomprehensible to him. Yet, he was very helpful during the creation of our first integrated testing laboratory with operational test launching and onboard equipment, which included a large demonstrational light board that simulated the missile's launch process.

By late 1947, this laboratory had already become our pride and joy, and, for Ustinov, it served as an occasion to invite high-ranking leaders of the army, who had participated in the most recent session of the USSR Supreme Soviet, to NII-88.[8] For the first time, I found myself in the role of speaker at a gathering of such illustrious military leaders such as Marshals Zhukov, Rokossovskiy, Konev, Bagramyan, Vasilevskiy, Govorov, Sokolovskiy, and Voronov and army generals, whom I shall no longer risk listing here.[9]

Beginning early in the morning, laboratory chief Emil Brodskiy and I checked out the entire testing laboratory, and still during every routine cycle one glitch

7. Walter Dornberger (1895–1980) played a key role in the development of the German A4 (V-2) missile. In 1937, he was appointed military commander of the research station at Peenemünde and in subsequent years served as the effective manager of the project.

8. The Supreme Soviet (*Verkhovnyy sovet*), formed in 1936, was the highest legislative body in the Soviet Union, and the only one empowered to pass constitutional amendments. In practice, the Supreme Soviet was more of a rubber stamp "parliament," approving all decisions that came down from the Politburo.

9. Georgiy Konstantinovich Zhukov (1896–1974), Konstantin Konstantinovich Rokossovskiy (1896–1968), Ivan Stepanovich Konev (1897–1973), Ivan Khristoforovich Bagramyan (1897–1982), Aleksandr Mikhailovich Vasilyevskiy (1895–1977), Leonid Aleksandrovich Govorov (1897–1955), Vasiliy Danilovich Sokolovskiy (1897–1968), and Nikolay Nikolayevich Voronov (1899–1968) were a few of the most important military commanders who served the Soviet armed forces during World War II.

or another occurred. It was Murphy's Law. The laboratory—which had not been designed for such a large number of guests—was cramped when the brass, decked out with all their orders and medals, filled the room.

Ustinov began the explanations. With difficulty, Gonor and Korolev squeezed through the crowd to join me standing by the console. Both of them wanted to intercept Ustinov's initiative to contribute to the presentation. But suddenly he said: "And now our specialist Comrade Chertok will demonstrate the missile launching process."

During Ustinov's speech the marshals and generals had clearly begun to get bored, and I immediately switched to the demonstration, while providing commentary:

"The launch system is automated. Your attention, please! I am setting the switch on start! Look at what is happening on the light board. I am monitoring the process according to the message boards, and if I make a mistake, the system will not go into an erroneous launch. The automatics will reset everything into the initial position."

Actually, being nervous, I did something wrong, and Brodskiy didn't have time to correct me; the lights on the light board suddenly went out.

"I have just demonstrated that the system is foolproof. And now we will repeat the attempt to launch the missile."

Now I was ready to start again; Brodskiy understood my error and was watching my every move like a hawk. The steam gas generator lit up on the light board, I added the turbo pump assembly, and then the ignition began to glow. There went the preliminary, then the primary! With gusto I explained that the liftoff contact had been tripped and now "look, the engine is producing a full thrust flame—we have flight! In sixty seconds, without our intervention, the engine will shut down." Everything went splendidly.

Nevertheless, instead of expressing his gratitude as we expected, Marshal Rokossovskiy loudly exclaimed with a cunning smile:

"But regarding 'foolproof protection,' you were just pulling our leg." I was taken aback, but Ustinov kept a cool head.

"No, Comrade Marshal, the entire demonstration was free of deception. I personally have checked out the entire system a number of times both here and at the test range."

The marshals broke into smiles and began to exit the laboratory. They still needed to get a look at the rocket in the assembly shop. I told Brodskiy, "When I was sitting in the armored car during our first launch, my back was dry, and now I'm soaked."

He burst out laughing and said, "Me too."

Those were the kind of guests that director Gonor had to receive. But in this case Ustinov personally took on the role of host. To be sure, he later gave Gonor a dressing down because there was mud on the road to the assembly shop. What was he to do? We were well into autumn, and instead of snow, a light rain fell incessantly. But, in contrast to the factories that some of the marshals had occasionally visited during the war, workers in the assembly shop were already working in white lab coats. White lab coats at an artillery factory! What nonsense. Gradually there came a turning point in the psychology of the factory workers.

Gonor was far more demanding with regard to the factory than he was toward the scientific and design elite. He was right in his element with production, the mastery of new technological processes and the installation and reconfiguration of equipment. During the prewar and war years, directors of his level went through a sort of "production academy" and found themselves in situations that no formal education at institutes of higher learning could ever have anticipated.

In 1947, Gonor identified two tasks for the institute. First, we should master the technology for the clean assembly and testing of missiles from parts manufactured by us and shipped from Germany. This was the assignment of that very new assembly shop where the white lab coats had appeared for the first time. Second, we should begin to implement fabrication of missiles from domestic materials according to drawings that the Special Design Bureau had belatedly begun to issue. The most important among them were the drawings for the R-1 missile that Department No. 3, headed by Korolev, was issuing.

That year, Gonor traveled with us out to the State Central Test Range in Kapustin Yar to participate in the tests on the German missiles and the following year, in 1948, to participate in the tests on the first series of R-1 missiles. Here, he was the first to appear before the State Commission when production defects were discovered in the missiles. But the most difficult thing for him was supporting the lifestyle of all the big shots who did not want to depend on test range chief General Voznyuk and counted on the all-powerful, rich director of NII-88.

Relations between Gonor and Korolev were complicated. Formally, Gonor was not Korolev's immediate boss. Special Design Bureau chief Tritko, Gonor's former compatriot at the Barrikady Factory in Stalingrad, still stood between them. But due to Korolev's nature and his ambition, he could not endure two artilleryman bosses. Conflicts arose, often over irrelevant and immaterial matters. Korolev sometimes went over the heads of Tritko and Gonor to Vetoshkin, Ustinov, and other chief engineers on problems of design, new proposals, and relationships with contracting chief designers. Such behavior irritated some. On a number of occasions, knowing about the special relationship that Pobedonostsev and I had with Korolev, Gonor appealed to us with the request: "You know his character better than I. Have a little talk with him. Why must we have these quarrels?" We would have little success trying to smooth out conflicts over Korolev's demands—demands that he be granted greater independence, allowed to create his own experimental shop, granted privileges in the selection of specialists, and so on. After all, there were also many other chief designers of various air defense missiles that were zealously monitoring the actions of Gonor, Pobedonostsev, and Chertok. They might view any assistance rendered to Department No. 3 as an infringement of their interests. Complaints were making their way to the Party Committee and even to the local Party Municipal Committee in Mytishchi.[10]

---

10. Mytishchi is a suburb of Moscow, about twenty-two kilometers northeast of Moscow, close to Podlipki (Kaliningrad) where NII-88 was based.

Because the government considered rocket development especially important, using wartime experience, it sent a VKP(b) Central Committee organizer—instead of an elected Party committee secretary—to manage the NII-88 Party organization. Gonor was supposed to find a common language with this man, but this was considerably more difficult than at the factories during the war, when everyone was united by a single production program and a single motto: "All for the front, all for victory."

At that time, directors mingled with workers at Party conferences, at various meetings of Party and managerial leaders, and later at Party meetings in the departments. On these occasions, directors could also come under reverse scrutiny of the workers. At such gatherings, a director's duties included not just giving speeches in which he framed pressing problems, but he was also obliged to criticize the actions and behavior of senior management. Typically, Gonor was accused of not being demanding enough with regard to Korolev, who was not a Communist Party member. Gonor was sufficiently wise not to press his luck when it came to criticism from the top, especially because the general Party atmosphere was becoming increasingly oppressive. A campaign of anti-Semitism, no longer local, but widespread, was spreading under the slogan of "struggle against rootless cosmopolitans." The more accomplished and honored the campaign's latest victim, the more effective it seemed was the victory of the ideological champions of the general Party line.

During the war, Gonor, had been a member of the presidium of the Soviet Anti-Fascist Jewish Committee.[11] When word emerged about the "accident" with committee head Mikhoels, Gonor blurted out during one of its business meetings, "This is a very great misfortune. Bear in mind that now a purge will begin in our ministry.[12] Our institute is too visible. Our subject matter is very enviable and promising. Ustinov won't be able to protect us." And indeed, in August 1950, Gonor was removed from his post as NII-88 director and sent off to be the director of an artillery factory in Krasnoyarsk.[13]

Later, in January 1953, during the infamous "Doctors' Plot," Gonor was arrested.[14] Almost simultaneously, the security services also arrested our protector Artillery

---

11. The Soviet Jewish Anti-Fascist Committee was formed in 1942 with the support of the Soviet government to help foster support for the Soviet war effort and also to establish contacts with supporters outside the USSR.

12. Theater actor Solomon Mikhaylovich Mikhoels (1890–1948) became chairman of the Soviet Anti-fascist Jewish Committee in 1941. He was killed in an automobile crash in 1948 while visiting Minsk. Evidence suggests that Stalin was directly involved in staging this "accident."

13. This production facility, Factory No. 4 Named After K. Ye. Voroshilov, was an important manufacturer of artillery, mortar, sea mines, and bombs.

14. In January 1953, certain Kremlin physicians, mostly Jewish, were arrested on charges of medically mistreating and murdering various Soviet leaders. The Doctor's Plot served as a pretext for a broader society-wide anti-Jewish campaign that was interrupted only by Stalin's death. See Joshua Rubenstein and Vladimir P. Naumov, eds., *Stalin's Secret Pogrom: The Postwar Inquisition of the Jewish Anti-Fascist Committee* (New Haven: Yale University Press, 2001).

From the author's archives.

Shown here are (left) Maj.-General Lev Gonor (1906–69), the first director of NII–88, the institute tasked with developing long-range ballistic missiles in the postwar years. Gonor and Korolev (right) had a complex relationship during Gonor's tenure as NII–88 director. In the early 1950s, Gonor spent time in prison as part of a wave of anti-Semitic persecution.

Marshall Yakovlev and a number of his GAU colleagues. They were charged with deliberate sabotage during the production of the new automatic antiaircraft guns designed by Grabin.[15] They were all saved by the death of Stalin. Gonor was completely rehabilitated. The government returned all of his awards and appointed him director of the Central Institute of Aviation Motor Construction (TsIAM) branch in Turayevo, located in the Moscow suburb of Lyubertsy.[16]

It is difficult to explain what sort of logic governed our high-ranking officials in many similar cases. Consider the course of events: a specialist in the field of artillery production technology became the director of what was at that time the largest missile scientific-research center in Europe and, perhaps, in the world. Four years of managerial work in the field of rocketry provided the wise and experienced Gonor with a great deal of valuable knowledge, connections, and contacts and would have enabled him to be used to great advantage specifically in that field.

---

15. Vasiliy Gavrilovich Grabin (1900–80) was a famous wartime designer of guns and cannons. Chertok describes his work at length in Chapter 27.

16. TsIAM—*Tsentralnyy institut aviatsionnogo motorostroyeniya*. TsIAM was (and still is) one of the leading Soviet/Russian research institutes doing fundamental research on aviation propulsion.

But instead of bringing Gonor back to the missile industry, the Central Committee's defense industry department decided that the aircraft engine building base needed to be reinforced with experienced personnel. And so Gonor was once again to start from square one and learn the technology for state-of-the-art aviation engine building. But his health had already been undermined. He developed gangrene of the extremities, and his fingers had to be amputated. On 13 November 1969, Gonor died at the age of sixty-three. His name is practically forgotten in Kaliningrad, the town outside Moscow for which he did so much in the most difficult early postwar years.[17]

MORE THAN LIKELY, NOT WITHOUT PROMPTING FROM THE PARTY CENTRAL COMMITTEE MANAGEMENT, Ustinov approved the NII-88 structure so that the position occupied by Korolev in the official hierarchy was not all that high. He was just a department chief. And in 1947, the new NII already had more than twenty-five departments. From his very first days on the job at the new NII, Korolev's quest for personal authority and to broaden his sphere of activity caused conflicts with administrative and Party leadership.

In Germany, Korolev had been the chief engineer of the Institute Nordhausen and Glushko, Ryazanskiy, Pilyugin, Kuznetsov, and many other civilian and military specialists were under his authority. After Korolev returned to the Soviet Union, it was decided that he not be given such freedom and authority. Now Glushko, Ryazanskiy, Barmin, Kuznetsov, and Pilyugin stood considerably higher than Korolev on the official "table of ranks" because they were directors or "first deputy" directors of Soviet enterprises or institutes with experimental factories.

Structurally, NII-88 consisted of three major units:
- a special design bureau (SKB);
- a unit comprising scientific-research and design departments for various disciplines; and
- a large experimental factory.

K. I. Tritko was appointed SKB chief. He was the former chief engineer of the Barrikady artillery factory. Tritko was a typical administrative director of a wartime artillery factory. He had never come into contact with rocket technology or science before being assigned to NII-88. The SKB consisted of design departments headed by chief designers of rocket systems with the following tasks:

Department No. 3 (Chief Designer S. P. Korolev) was responsible for designing the R-1 and R-2 long-range ballistic missiles and for reproducing the German A4 missile.

Department No. 4 (Chief Designer Ye. V. Sinilshchikov) was responsible for

---

17. Kaliningrad is the Moscow suburb now known as Korolev where NII-88 (TsNIIMash), along with RKK Energiya, and a number of other Russian defense enterprises are still located.

designing long-range surface-to-air guided missiles with a homing head (R-101) and for modifying the captured *Wasserfall* missile (the Germans had not yet completed the acceptance process for it to become operational.).

Department No. 5 (Chief Designer S. Ye. Rashkov) was responsible for designing the R-102 medium-range surface-to-air guided missile and for reconstructing the German *Schmetterling* and *Rheintochter* missiles.

Department No. 6 (Chief Designer P. I. Kostin) was responsible for designing R-103 and R-110 solid- and liquid-propellant unguided surface-to-air rockets, with a range at altitude of up to fifteen kilometers, using as a basis the German solid-propellant *Typhoon* rocket, which had also not been optimized to the point of acceptance as an operational armament.

Department No. 8 (Chief Designer N. L. Umanskiy) was the special department involved in liquid-propellant rocket engines using high boiling point oxidizers for surface-to-air missiles. It had a test station and an experiment shop.

Department No. 9 (Chief Designer A. M. Isayev) was the department involved with liquid-propellant rocket engines for surface-to-air missiles. This department was created in 1948, incorporating the personnel transferred from NII-1. Two years later Department No. 9 absorbed Department No. 8. I had something to with that.

I shall digress in order to describe my involvement in Isayev's fate. Isayev left the Institute RABE in late 1945, and returned to his "home" Factory No. 293 in Khimki. By this time, the factory had become a branch of Ministry of Aviation Industry's NII-1.

Let me remind the reader that NII-1 was created from NII-3, the former RNII in Likhobory. To this day, the main building of this historical institute, where so many "enemies of the people" worked, displays an inscription that in days gone by concealed that institution's activity: "All-Union Institute of Agricultural Machine Building."

In fact, the building really *was* erected for the Institute of Agricultural Machine Building. But in 1933, when, at Tukhachevskiy's insistence, the Leningrad Gas Dynamics Laboratory (GDL) and the Moscow Group for Reactive Motion (GIRD) merged, they were given the main building and referred to as the Reactive Scientific-Research Institute (RNII).[18]

In autumn 1947, when I returned from Kapustin Yar and was deeply involved in setting up NII-88, Isayev tracked me down. He was in a terribly gloomy mood. He told me that our beloved patron, Viktor Bolkhovitinov, who was the NII-1 institute's scientific chief was not getting along with the ministry brass, had given up on the whole future of rockets, and was returning to the field of aviation as head of the design department at the N. Ye. Zhukovskiy Air Force Academy. A new

---

18. GDL—*Gazodinamicheskaya laboratoriya*; GIRD—*Gruppa issledovaniya reaktivnogo dvizheniya*.

director, Mstislav Keldysh, was coming to NII-1 from the Central Aero-Hydro-dynamic Institute (TsAGI).[19] "He doesn't know anything about liquid-propellant rocket engines, and there is nothing for me to do there."

I told Isayev about the prospects for NII-88 and went on and on about what Minister Ustinov—whom Isayev did not yet know—Vetoshkin, and Director Gonor thought of our work. "And of course you know Pobedonostsev very well. He will certainly support your transfer!"

Isayev knew how to make radical decisions. In both his personal and professional life, he was not afraid of changing course if he had come to the conclusion that he was on the wrong one. "Blow my brains out! Why didn't I think of that before? Why was I dragging my feet, what was I waiting for?"

The following day, I persuaded Pobedonostsev to take on Isayev and together we visited Gonor in his office. He approved our proposal and immediately telephoned Vetoshkin. Having received Vetoshkin's approval, Gonor requested that I convince Isayev and, so that there weren't any misfires, tell him that the matter had already been approved by the NII-88 director and the Ministry of Armaments.

Gonor told us that "Isayev himself must appeal to the Ministry of Aviation Industry so that they don't accuse us of luring specialists away from the aviation industry over and above the quota allowed by the [1946] decree."

Isayev energetically sprung into action, and as a result, in 1948, the two ministers issued an order transferring Isayev's entire staff from the Khimki branch of NII-1 (Factory No. 293) to NII-88.

This decision was important for Isayev's subsequent fate—as well as that of many of his colleagues. At NII-88 an experimental facility was created for Isayev. He rapidly took over work on low-thrust liquid-propellant rocket engines using high boiling point components for surface-to-air missiles, medium-range missiles, and, subsequently, ship-borne missiles. In 1959, Isayev's team was detached from NII-88 to become the independent special design bureau OKB-2, which later became the Chemical Machine Building Design Bureau (KB Khimmash), one of the country's leading firms in rocket and space engine construction.[20]

Neither I nor Isayev could foresee that our heart-to-heart conversation in the autumn of 1947 would be so fateful for cosmonautics. It has been said many times that history does not care what might have been. But if Isayev had remained to languish at NII-1, and if he had not accepted my offer, then who would have developed the maneuvering and braking propulsion system for the Vostok, Voskhod, and Soyuz spacecraft? Someone, of course, would have developed it, but I am not sure

---

19. TsAGI—*Tsentralnyy aerogidrodynamicheskiy institut.* TsAGI was (and still is) the leading Soviet/Russian R&D institution in the aviation sector. Additionally, the entire Soviet aviation design bureau system emerged from TsAGI in the 1930s.

20. This organization is today known as KB Khimmash imeni A. M. Isayeva (Design Bureau of Chemical Machine Building Named After A. M. Isayev).

that we would have launched a man into space on 12 April 1961.

IN 1948, I WAS DIRECTLY INVOLVED IN AUGMENTING NII-88 with yet another group from the aviation industry. In March 1948, Gonor gave me, as he put it, a delicate personnel assignment: "Yesterday I had a visit from Professor Karmanov, who works in an aircraft design bureau. The chief designer of that KB is Engineer Colonel Pavel Vladimirovich Tsybin.[21] They are located somewhere in Beskudnikovo. Supposedly, they no longer have any work and they are ready to conduct negotiations concerning collaboration. Your job is to see what kind of people they have there, get acquainted with the chief designer, and have a discussion with him. Don't make any promises until we have made arrangements here in our own ministry."

The next day I went to the northern edge of Moscow and barely managed to find barracks housing the design bureau that was headed by Pavel Vladimirovich Tsybin. Tall, well-built, and blue-eyed Engineer Colonel Tsybin received me cordially. However, when I alluded to Professor Karmanov's appeal to Gonor, he burst out laughing and shouted:

"Boris Ivanovich!"

The same Karmanov who had visited Gonor approached.

"This is Karmanov, but he is still a long way from being a professor. It just made it easier to gain access to your director."

When we turned to business, it became clear that we had many common acquaintances. Above all, it turned out that Tsybin was the chief designer of various gliders and before the war, he used to meet with Korolev.

"But in 1938, our association ended. Now I know why," he said.[22]

When he heard that I had been involved with the BI rocket-plane, he became very animated and said that now it was possible to explain the true cause for the death of pilot Grigoriy Bakhchivandzhi, thanks to the experience that had been gained using the flying laboratories that had been developed by Tsybin's KB and the flight research that was being conducted at the Flight-Research Institute (LII).[23] The work was now completed, but its future was unclear.

I announced that I was not leaving "this barracks" until Pavel Vladimirovich divulged the mystery of Bakhchivandzhi's death. The crux of the matter was that, having a great deal of experience in glider design, Tsybin had agreed to create gliders

---

21. KB—*Konstruktorskoye byuro* (Design Bureau).

22. Korolev spent six years in various Soviet prisons and labor camps after his arrest in 1938.

23. LII—*Letno-issledovatelskiy institut.* LII was one of the major Soviet testing facilities for high-performance aircraft. For the history of the BI rocket-plane, see Boris Chertok, *Rockets and People.* Vol. 1, ed., Asif A. Siddiqi (Washington, DC: NASA SP-2005-4110, 2005), pp. 193–200.

in the form of flying laboratories for aerodynamic studies at subsonic speeds.[24]

"In the wooden glider we reached near-sonic speeds, somewhat exceeding the speed of the towing aircraft," he noted.

The specially designed glider was equipped with a solid-propellant accelerator and was loaded not only with a pilot, but also with water ballast. Once it was lifted on the aircraft towline to a high altitude, the glider pilot released the craft from the towline and dove steeply toward the ground, switching on the accelerator during the dive. When the maximum allowable speed was reached, the pilot opened the water drainage valve, pulled out of the dive, and went in for a landing in the glider, which had dumped almost half its weight. The accelerators made it possible to reach a speed of almost 1,000 kilometers per hour during the dive. The glider's strong wing was attached to the fuselage on a dynamic suspension bracket, making it possible to determine the primary aerodynamic characteristics C_x, C_y and M_z and the distribution of pressure over the wing.[25]

For the first time in the USSR, snapshots were obtained in flight of sudden changes in compression, the nature of the airflow over the wings, areas where the flow was interrupted, and deterioration of the control surfaces' effectiveness.

"We conducted many dozens of flights and discovered very hazardous flight modes that were accompanied by losses of controllability. Something like that happened on the BI when Bakhchivandzhi reached maximum speed," explained Tsybin.

I could have listened to Tsybin for hours. But at that time, having painted an alluring landscape of our field, I convinced Pavel Vladimirovich to transfer to us at NII-88. A couple of days later, he met with Gonor and then with Vetoshkin. The orders of the two ministers were drawn up rather quickly. More than twenty of the thirty individuals who had worked at Tsybin's OKB transferred to NII-88. Tsybin himself was named chief of Department I for testing. His deputy was Leonid Voskresenskiy. "Professor" Karmanov was put in charge of aiming technology and worked at the launch pad.

LET'S RETURN TO 1946–47. The list of NII-88 SKB projects, taking into consideration modifications of all kinds, surpassed all the activities conducted at Peenemünde! And all of this was under the jurisdiction of a single chief, a mere artilleryman, Karl Ivanovich Tritko. Officially, Korolev was subordinate to him, as was another department chief, Kostin. When asked by Ustinov during the inspection of a V-2 in Germany in 1946, "Well, Pavel Ivanovich, can you make a missile like

---

24. While working at LII, Tsybin developed two advanced design rocket-planes, the LL-1 and the LL-3, to test various innovative wing designs at subsonic speeds. These vehicles flew about 30 and 100 test-flights respectively.

25. C_x denotes the drag-force coefficient, C_y the lift-force coefficient, and M_z, the pitching moment coefficient.

this?" Kostin's reply was "Of course, Dmitriy Fedorovich, if you give me about ten electricians."

"Well, I see you're a bold man," chuckled Ustinov. Tritko felt comfortable with SKB chief designers Sinilshchikov, Kostin, and Rashkov, former artillerymen themselves, and saw them as kindred spirits, more so than he did the initially enigmatic Korolev. And besides, Korolev had such a past. If pressure were brought to bear on him, then very likely no one would stand up for him. But, it turns out someone did stand up for him. Much, much later it came out that in 1946, before the order was issued appointing Korolev as chief designer of long-range ballistic missiles, Yevgeniy Sinilshchikov's more amenable and "clean" candidacy had been proposed.[26] Ustinov had been pressured to pick Sinilshchikov, and he began to waver. After all, why take someone from outside the fold if you have your own tried and true people? But here again Gaydukov played a role, and not for the last time. He was very familiar with the complex structure of the bureaucracy and the personal relationships that controlled job placement. He did everything in his power to prevent a fatal mistake, and the order appointing Korolev instead of Sinilshchikov was signed.

The second major structural unit at NII-88 was the block of scientific departments under the management of Chief Engineer Pobedonostsev. The primary departments were:

Department M, materials technology (Chief V. N. Iordanskiy);
Department P, strength (Chief V. M. Panferov);
Department A, aerodynamics and gas-dynamics (Chief Rakhmatulin);
Department I, testing (Chief P. V. Tsybin);
Department U, control systems (Chief B. Ye. Chertok);
Department T, rocket propellants.

Having received my own department, as well as being deputy chief engineer of the institute, I felt a certain degree of independence and on many issues went straight to Director Gonor, to Vetoshkin at the ministry, or to the office of Special Committee No. 2. As a result, by late 1947, it was possible to create within the department a well-equipped experimental shop staffed with skilled workers, a special instrument design bureau, and numerous specialized laboratories. The main problem was personnel, but the ministry was not stingy in sending us young specialists and encouraged the transfer of specialists from other enterprises.

In December 1947, after our return from Kapustin Yar, where the first A4 missile firings had taken place, Minister Ustinov ordered Gonor to assemble the Party and operations leaders from all of NII-88. More than a thousand persons gathered in the club of former Factory No. 88. After Gonor's brief report on the state of affairs at NII-88, Ustinov delivered scathing criticism of the leadership and especially of

---

26. In 1991, Korolev's deputy Vasiliy Pavlovich Mishin revealed that Sinilshchikov had been considered for Korolev's post in 1946. See B. Konovalov, "Iz Germanii—v Kapustin Yar" ("From Germany to Kapustin Yar"), *Izvestiya*, April 6, 1991, p. 3.

the experimental factory for the slow pace of its reconstruction, which threatened to disrupt the work schedule for the production of the first series of R-1 missiles.

During his speech he was handed a note, which he read aloud: "Comrade Ustinov, in your speech you praised Chertok for his organization of work on guidance systems. But Chertok owes his success to you. You have helped him more than others. Help the others and their projects will thrive." After reading the note, Ustinov replied: "It isn't signed. But it's not hard to guess its author. It's one of those individuals who has been criticized today. I am assisting Chertok only because I see that his projects are going somewhere and he is solving complex problems. I promise to help each one of you who organizes his work well. And if someone's project is still a mess, then why should I help him? He needs to be removed from the job." Ustinov's response to the anonymous note did not increase my fan club.

My whole Department U developed good relations with Korolev and his entire staff in SKB Department No. 3. We were united not only by our joint work in Germany, but, to an even greater degree, by the missile flight tests at the Kapustin Yar test range. There we were testing our characters as well as missiles.

Our relationships with the chief designers of surface-to-air missiles and SKB Chief Tritko shaped up quite differently. Tritko and Sinilshchikov were faithful to the artillery traditions. Surface-to-air guided missiles were shells to them. They considered that their main task was to produce good drawings so that these shells could be manufactured. It was just dandy that there was no need to design the "cannon" to shoot them. The missile would accelerate itself to a speed that surpassed that of an artillery shell!

Sinilshchikov used to love to say, "My designers in Department No. 4 draw better than the designers in the other departments." It seemed that the most important thing was the quality of the drawings, and whether an aircraft would be destroyed by a well-drawn missile, that was the guidance specialists' problem. Their guidance specialists worked at the NII-885 institute. There Ryazanskiy had set up the guidance department for surface-to-air missiles with Govyadinov in charge. He was a radio engineer coming into contact with this field for the first time. There was no qualified director of operations for the creation of the entire air defense missile system.

Sinilshchikov and Tritko complained to Gonor and to the Party committee that Department U, under Chertok's management, was only working on Korolev's projects—ballistic missiles—and was not devoting attention to surface-to-air missiles.

I ventured to tell Pobedonostsev that NII-88 was, in principle, not capable of fulfilling two programs: the long-range guided missile program and the air defense missile system program. He agreed with me, but who would dare report to the top that we were not capable of implementing a task that the decree signed by Stalin had entrusted to us?

Many of the conflicts between the NII-88 directors were temporarily eliminated by Ustinov's order dated 26 April 1950, which was prepared by Gonor with Korolev's involvement. According to that order, the SKB in NII-88 was abolished.

Using the facilities of the eliminated SKB, Specialized Design Bureau No. 1 (OKB-1) for the development of long-range missiles and OKB-2 for the development of surface-to-air guided missiles were established. Korolev was appointed head and chief designer of OKB-1. Tritko was assigned acting chief and chief designer of OKB-2. Korolev's appointment was logical and understandable. Everyone interpreted the second appointment, that is, Tritko's posting, as temporary. Gonor told me, "This is so we can catch our breath."

The most difficult situation of all was mastering the missile technology at the factory. The factory was the third, and to a great extent, the defining structural unit of NII-88. The factory personnel—management and workers—were steeped in the traditions of the artillery factory. We used to joke, "They're still using technology from the times of Peter the Great and Demidov."[27]

In 1941, the main part of artillery Factory No. 88 was evacuated.[28] In Podlipki, for the most part, the work was armament repair. By the end of the war the factory was partially restored and had been set up for the series production of automatic antiaircraft guns. The factory personnel would have to be retrained. The new technology required a systemic approach not only during the design process, but also during the organization of production. The entire missile production process, from concept through the factory production process to the firing range tests, had to proceed from principles of unity and interdependence in the work of the drafter, designer, process engineer, and tester with great external cooperation. Reprimands were heard from above and below about the factory's slow reconstruction. Officially, the factory had its own director and chief engineer. But Gonor was still considered to be responsible for everything. The chief designers complained that the factory was fulfilling their orders slowly and with poor quality.

During the first years working on rocket technology, virtually none of the institute directors criticizing the factory were able to specifically spell out what needed to be done to improve the standard of production and to determine the role of each shop chief, foreman, and worker. There were too many abstract decisions.

Ustinov's attitude was merciless toward shop chiefs and production chiefs when it came to filth and uncouth behavior. During factory visits, he started with the bathrooms. As a rule, in the shops, long before you reached the bathroom, a dis-

---

27. Nikita Demidovich Antufyev (1656–1725) (he later took the surname Demidov) was a blacksmith from Tula who accumulated great fortune by manufacturing weapons and building and operating an iron foundry in Tula after receiving land grants from Peter the Great.

28. Here, Chertok is referring to the massive evacuation of industrial institutions (mostly factories) that took place after the Nazi invasion of the Soviet Union in June 1941. Through the end of the year, hundreds of factories were literally packed up and moved to the eastern Soviet Union. Like many Soviet factory locations, the site occupied by the postwar Factory No. 88 had a confusing history. During the Nazi invasion, this site was occupied by Factory No. 8, whose equipment was evacuated to several different locations to the east in October 1941. A new plant, Factory No. 88, was then established on the evacuated site in December 1942.

tinctive "aroma" wafted toward you. In the bathrooms themselves, you had to walk through puddles. Ustinov would fly into a rage and thunder, "I can look at a john and see what the shop chief is like. Until your johns are a model of cleanliness, there won't be cleanliness in your shops."

Many years have passed since then. The problem of cleanliness in the public toilets at our factories and institutes, however, just as in the country as a whole, has yet to be solved. This has proved to be far more difficult than creating the most formidable nuclear missiles and fighting for world superiority in cosmonautics.

To this day, the blatant lack of manners and of a standard for general industrial cleanliness and hygiene is one of the reasons for the low quality of many domestic articles. During the war and in the ensuing years, concern about elementary comfort in the shops and the creation of a general atmosphere suitable for and attractive to workers was considered an excessive and impermissible luxury. Yet, in the end, expenditures on cleanliness, comfort, and elementary service were repaid with interest by increased productivity and quality.

Chapter 5
The Alliance with Science

Beginning February 1947, during the first three months of work at NII-88, it seemed to me that I was recreating something resembling the Institute RABE in the four-story building allocated for my department. The laboratories were being stocked with equipment and staffed with artillery specialists left over from cannon Factory No. 88, young specialists who had arrived voluntarily after their demobilization from the army, and those ordered by the ministry to join us after graduating from technical institutes, universities, and technical schools.

A similar process was underway in all the other departments of NII-88. Officially, my direct chief was NII-88 Chief Engineer Pobedonostsev. We had established good relations as early as 1944, when NII-3 was transformed into NII-1.[1] He relied completely on my experience and did not bother with managerial instructions. Most of Pobedonostsev's time was taken up with a multitude of routine organizational problems concerning the departments of materials technology, testing, engines, aerodynamics, and strength, plus the conflicts that arose in the design bureau.

The new SKB chief, artilleryman production worker and former blacksmith Karl Ivanovich Tritko, was not highly esteemed among the chief engineers subordinate to him. Each of them demanded independence, priority in production, and direct access to the director and the ministry.

Director Gonor's greatest concern was for factory reconstruction. All of us were anxious over both the preparation for flight tests of the V-2 missiles brought in from Germany and their production processes. On one of those hectic days in May, Gonor requested that Pobedonostsev and I report immediately to his office. When we had seated ourselves in the soft armchairs arranged before his enormous desk, he leisurely opened a pack of Kazbeks, lit one up, offered the pack to us, and looking at us slyly through half-closed eyes, he paused. Pobedonostsev was not a smoker, and instead of having a cigarette, without asking permission, I poured myself a glass of

---

1. The original Reactive Scientific-Research Institute (RNII), formed in 1933, went through several different incarnations. It was known as NII-3 in 1937–42 and then was united with a number of other different teams and factories to become NII-1 in 1944.

sparkling Borzhomi mineral water from the bottle standing on the director's desk. Gonor asked an unexpected question:

"Comrade missile specialists, what, in your opinion, should be the objective of higher science in the postwar period?"

I recalled a statement from a philosophy course and answered, "The objective of science is knowledge."

"I just received a call from Dmitriy Fedorovich [Ustinov]," said Gonor, switching to an official tone, "He is not pleased that we have not yet established close contacts with the Academy of Sciences. He insisted that we draw up proposals. To begin with he asked that we personally acquaint President of the USSR Academy of Sciences Sergey Ivanovich Vavilov with our problems. The president will visit us next week; probably with a large retinue of scientists. Get ready. Think about what to show them and what serious problems to pose before academic science. Keep in mind that Vavilov is not an armchair scientist, but a prominent physicist with a great deal of organizational experience. During the war he personally managed the development of the most complex optical instruments and mobilized the Academy of Sciences to help weapons production. According to my data, there are currently more than 20,000 persons working at the Academy. Comrade Stalin personally supports Vavilov."

"While you were there in Germany studying German technology, very crucial events were taking place here for Soviet science," continued Gonor. "In July 1945, there was a very festive celebration of the 220th anniversary of the Academy of Sciences. I was not present for the festivities, but I have been told that supposedly the old Academy president, botanist Komarov, said something wrong, but I don't think that was the issue.[2] Iosif Vissarionovich [Stalin] had understood for some time that the Academy needed a president who was younger and more energetic and who had a closer relationship to industry. A month after the anniversary celebration, the Academy's general assembly elected a new president, Academician Vavilov. Vavilov is a scientist, a physicist with a worldwide reputation. He has been reorganizing the Academy's work for two years, and now is precisely the time to get the Academy scientists interested in our work. You should have shown the initiative yourselves without waiting for ministry instructions."

I tried to defend myself by indicating that I had already established contact with the Academy's Institute of Automation and Remote Control, but Gonor showed us out, having instructed Pobedonostsev to have the appropriate explanatory talk with Korolev and Sinilshchikov.

After returning to my office, first of all, I assembled my "inner circle" to report to them about the talk with Gonor and to get their ideas about the upcoming crucial meeting with the Soviet Union's foremost man of science.

---

2. Vladimir Leontyevich Komarov (1869–1945) served as President of the USSR Academy of Sciences between 1936 and 1945.

During the first months since our return from Germany I began to understand that the situation in Soviet science had changed substantially compared with triumphant 1945. Party and State monitoring of the behavior and attitudes of scientists had intensified. Scientists recruited for work involving the most important defense programs were, however, more protected from charges of "servility to the West and capitalistic culture." The scientific community supported proposals of the Party Central Committee and ministries on their participation in missile technology projects for three reasons. First, the participation of a scientist in such projects was a sort of confirmation of his or her loyalty. Second, this new field of activity really was an extremely beneficial field for creative forces freed from routine industrial burdens. And third, participation in "Top Secret" operations strengthened the authority of scientific organizations at the regional, municipal, and *oblast* level for solving a multitude of economic problems.[3]

Aside from these practical factors, the field of missiles attracted true scientists with its romantic appeal. Maybe we really would actually achieve the dream of interplanetary flight in our lifetime! The war had shown that an inflexible armchair scientist stood little chance of achieving great scientific and technical advancement. But missile technology promised just such a chance!

THE DAY AFTER OUR TALK WITH GONOR, Pobedonostsev surprised me. He was significantly better informed than I of the general political situation.

"At our upcoming meeting with President Vavilov, keep in mind that he had an older brother Nikolay, who was also an academician with a worldwide reputation, a famous biologist and botanist. He was elected an academician back in the 1920s, while Sergey was elected in the early 1930s."

I answered that I had heard about the scientific feats of the botanist Nikolay Vavilov even when I was a schoolboy. "But why do you say 'was'?"

"Here's the deal. God forbid you should mention him. He was repressed. I think he may no longer be among the living."

"So how can the brother of an 'enemy of the people' be elected president of the Academy of Sciences?" I asked.

"It's a very complicated matter," answered Pobedonostsev. "Perhaps by supporting the candidacy of Sergey Vavilov to the high post of president, Stalin wanted to prove his objectivity or soften history's verdict for the death of Nikolay Vavilov."

"I see. If Stalin supported Sergey's candidacy, then the academicians supported him all the more so. They thereby expressed their solidarity with Vavilov."

Sergey Vavilov was elected president by secret ballot at a general assembly of the Academy of Sciences. This election proved a success for the Academy and for all of Soviet science at that time.

---

3. *Oblast* is the Russian word for domestic geographical units just below the national level.

The details of the tragic fate of Nikolay Vavilov were not discovered until the end of the twentieth century, when historians gained access to the top secret archives. Both brothers, Nikolay and Sergey, were scientists with worldwide reputations. Sergey Vavilov headed the State Optical Institute and the P. N. Lebedev Physics Institute of the USSR Academy of Sciences (FIAN)[4]. Before he became president, Vavilov directed and coordinated all of the primary research in the field of optics and participated directly in the establishment of the optical instrument industry, which, from the beginning of the war, the People's Commissariat of Armaments managed. For this reason, Ustinov had known Vavilov well beforehand and deferred to him, not only because he was president of the Academy of Sciences.

During the war and during the early postwar years, the leaders of the branches of the defense industry, and above all of the People's Commissariat of Armaments, sensed the power that was possible from the interplay between creative scientific thought and industry.

For Sergey Vavilov, rocket technology was a field that catalyzed a wide range of research to be conducted in a whole series of new scientific areas. It was said of Vavilov that he possessed great courage and perseverance; he especially considered the recently discovered potential in the combining of science with technology promising. As early as 1934, it was Vavilov who had been appointed Chairman of the Commission for the Study of the Stratosphere under the Presidium of the Academy of Sciences. At that time, this was a very important new field of research. Vavilov also organized the All-Union Conference on the Stratosphere, which was held in Leningrad in 1934. At that conference, Korolev, still an unknown engineer, gave a report on a rocket-powered stratospheric airplane.

One must assume that Vavilov's collaboration with the influential leaders of the military-industrial complex made it easier for him to defend the Academy and many scientists against the new wave of repressions in the postwar years.

AS IT TURNED OUT, THE WEEK AFTER OUR CONVERSATION WITH GONOR, Academy President Vavilov came to see us, not with a retinue of venerable academicians, but accompanied only by a woman. On the scheduled day of the meeting with our high-ranking guest, Gonor telephoned and asked me an unexpected question:

"What institute did you graduate from?"

"V. M. Molotov Moscow Power Engineering Institute."

"Who is the director there?"

"Valeriya Alekseyevna Golubtsova."

"You must know she's no longer Golubtsova, but Malenkova, the wife of Georgiy Maksimilianovich [Malenkov]."

"I know very well, Lev Robertovich, but what do you want from me?"

---

4. FIAN—*Fizicheskiy institut akademiy nauk.*

"You've got to tell her about the problems that her institute can help us with, and I am going to introduce you as a graduate of the Moscow Power Engineering Institute working here in a managerial position. Perhaps she really will be of some benefit. Keep in mind that she is coming with Sergey Ivanovich Vavilov."

Gonor clearly wanted the NII-88 institute to make an impression on Golubtsova. After visiting us, who knew what she might say to Malenkov himself, who was not only a Politburo member, but also the chairman of Special Committee No. 2! Such a visit could have important consequences. Each serious undertaking for the development of rocket technology required the government's support, but, ultimately, it was signed by Stalin. And Malenkov had to report to Stalin. I didn't explain that I knew Valeriya Golubtsova even before she became director of the Moscow Power Engineering Institute (MEI), and I had no doubt that the meeting would be beneficial for both NII-88 and MEI.[5]

Gonor designated Pobedonostsev, Korolev, Sinilshchikov, and me to attend the meeting with Vavilov and Golubtsova. At the appointed time we stood at the main entrance of administrative building No. 49. Vavilov and Golubtsova arrived in the same ZIS automobile.[6] Vavilov let Golubtsova pass through just slightly ahead. Gonor decided to introduce each person. When she saw me, Golubtsova smiled amiably, extended her hand, and—uncharacteristically for a woman—gave my hand a firm squeeze.

"Well, Chertok, so this is where you've ended up."

Then she turned to Vavilov, evidently continuing a conversation that they had been having en route to Podlipki. "Look, Sergey Ivanovich, MEI can already report that its graduates are making rockets."

Thus, I was honored with the attention of the president, whom I was seeing for the first time. Very likely I was embarrassed because I couldn't for the life of me remember whether Golubtsova and I were on a first name basis back in our student days of long ago, but I quickly calmed down and adopted a business-like, focused attitude.

Gonor was clearly pleased that his honored guest liked his institute's first "exhibit." We went up to the second floor. I noticed that Vavilov climbed the stairs with great effort. In Gonor's office, Vavilov asked that we brief him on the institute's tasks and structure and, if we were prepared, that we tell him in the most general terms about the problems that the Academy might be able to assist us with. "Actually," he added, "the Academy itself is interested in these projects. In particular, completely new opportunities are emerging for studying cosmic rays, the upper layers of the atmosphere, and various phenomena in the ionosphere. It would be possible to

---

5. MEI—*Moskovskiy energeticheskiy institut*. The full name of the institute was the Moscow Power Engineering Institute Named After V. M. Molotov.

6. ZIS—*Zavod imeni Stalina* (Stalin Factory). ZIS cars were the most common official vehicles during the era.

conduct very important joint work to study the passage of radio waves through the ionosphere if we succeeded in installing the appropriate equipment on rockets."

Korolev was enthusiastic about the ideas expressed by Vavilov. He proposed that we switch from general ideas to specific proposals concerning experiments as early as the autumn of that year. "To do that," declared Korolev, "we need not only wishes, but descriptions and drawings of instruments, connection diagrams, and specialists with whom we could work on specific layouts."

My "bag of tricks" contained proposals for research on the properties of the ionosphere to reduce errors in radio control systems and the development of problems for the radio monitoring of flight trajectories. Gonor listed several problems in the development of new materials. All in all, the list of tasks for the Academy of Sciences became quite large. Vavilov listened attentively and took notes in his notebook. Golubtsova listened attentively and also made notes in her notebook. The wife of the second ranking figure in the government of the USSR behaved very modestly. She was wearing an austere, but elegant, tailored suit and no jewelry. I had seen her for the first time in 1936. At that time, she was beginning her graduate studies at MEI. Now I would say she was forty-six years old, but she hadn't put on weight, as one might expect of the director of a major institute, a well-respected lady, and the mother of three children.

The conversation with Vavilov ended with him saying, "Well, there is really nothing to see here yet. It's too soon. You'll have a special conversation with Valeriya Alekseyevna." Golubtsova did not disrupt our conversation with the president, but at the end she suggested that I come to MEI.

"We will assemble a small group of the faculty, and, if there are no objections, Chertok will present a report on the main problems. After that we can reach some agreement on the joint work of NII-88 departments with our departments. If necessary, we are prepared to conclude a contract to conduct scientific-research work using the personnel of our departments, but," she added, and this comment betrayed the experience of an administrator, "we are not terribly interested in simple remuneration and compensation for expenditures. MEI is interested in creating specialized laboratories, and for this we need help with equipment and instruments."

At the conclusion of her proposals, Golubtsova accused the industry of being too protective: "The People's Commissariats hauled away everything that they possibly could from Germany, and now they don't want to share it with the Academy or with institutions of higher learning. Therefore, if you want science to help, be so kind as to help science too." In contrast to Vavilov's mild-mannered way, typical of the old school intelligentsia, Valeriya Alekseyevna was tough-talking and exacting. "If you want to have good young specialists, if you want our scientists to help you, if you want us to conduct serious work for you in our departments, then really help us, and not with vague wishes."

Golubtsova felt it was necessary to speak about the differences between MEI's focus and that of other institutions of higher learning such as the Moscow Aviation Institute, the N. E. Bauman Moscow Higher Technical School, the Leningrad

Military Mechanical Institute, and several others. "MEI is very involved in issues of the economy generally. The postwar restoration of the devastated power engineering infrastructure, electric transportation, the mastery of the technologies of modern electric machine building, the electric instrument industry, cable production, vacuum tubes, and electric drives for the entire machine building industry—these are the kind of problems that determine the specialization of MEI graduates and, respectively, the scientific concerns of the departments."

The gist of the MEI director's very emotional speech was that she had decided to put in their place the missile specialists, who had gone too far in their excessive appetites. It is very likely that Golubtsova had already conducted similar "educational" work with the atomic specialists and with others aspiring to privilege in postwar science. But it all ended peacefully. She repeated her proposal, "Let Chertok come see us. I hope he still remembers how to get to MEI; we'll work it out."

When the guests had departed, Korolev did not pass up his opportunity to ask loudly, so that everyone could hear, "Well, Boris, confess, how did you distinguish yourself so that a director like that still remembers you after all these years?"

Now I can write about that. Back then, I shrugged it off with a brief response that I met Golubtsova during our studies at MEI. From 1943 through 1952, Valeriya Golubtsova was the director of the Moscow Power Engineering Institute, one of the country's largest institutions of higher learning. This amazing woman was a talented, intelligent, and determined organizer. She fully deserved the title "first lady" of the state, and in terms of her civic qualities, she personified the grand scale of the state. Unfortunately, in the twentieth century the careers of female leaders ended before they could reach their full potential. I have already mentioned the tragic fate of one such woman, Olga Aleksandrovna Mitkevich, in the first book of this series.[7] The story of her life, utterly unusual for a woman, would make a captivating novel. But among professional writers and journalists, no interested parties have been found. In this regard, Golubtsova was a bit more fortunate.

On the occasion of what would have been her 100th birthday, the MEI publishing house issued a collection of remembrances about Valeriya Alekseyevna Golubtsova.[8] These recollections of her colleagues, former students, and daughter and sons paint a picture of a courageous woman with a generous heart, "an amazing director," and a loving mother who determined the fate of many of our country's scientists. The collection contains a chronology that enabled me to fine-tune my own recollections, which came out in the first volume of the publication *Rockets and People*.

I became a student at the Moscow Power Engineering Institute in the autumn

---

7. Boris Chertok, *Rockets and People*. Vol. 1, ed., Asif A. Siddiqi (Washington, DC: NASA SP-2005-4110, 2005), pp. 79–93.

8. *Valeriya Alekseyevna Golubtsova: Sbornik vospominaniy* [*Valeriya Alekseyevna Golubtsova: Collection of remembrances*] (Moscow: MEI, 2002).

of 1934. At that time the prefix "V. M. Molotov" was an obligatory part of the Institute's name. I really didn't want to quit my job at Factory No. 22, because the pay was pretty good, and financial support from my parents wouldn't have been enough to study full-time during the day. So I enrolled in night school without terminating my factory service.

An incoming class of students made savvier through industrial and life experience had been selected to enter the institute. Almost all of them had already advanced on the job to the level of foreman or technician, and their studies at the institute enriched them with knowledge not for successfully passing the next examination, but for use in their selected specialty.

They all had the same specialty of electrical engineering. The electromechanical department, where we were enrolled, had identical programs in all disciplines for the first three years for the entire incoming class. The night school class turned out to be very impressive. Many of my classmates later became chief engineers, chief designers, and directors of design departments. We even had a future academician in our midst. We were united not only by academic interests, but also by industrial interests. Gathering for lectures and seminars from our various enterprises without even having had a chance to cool off after the workday, we swapped news from our factories.

We were supposed to begin our narrow specializations during our fourth year. At that time our entire class was broken into three groups: electrical equipment for industrial enterprises; aircraft and automotive electrical equipment; and cable technology. The majority of us already had three to five years of industrial experience before entering the institute, and on average we were within two to three years of each other in age.

For me, the most difficult times were the end of the third academic year in the spring of 1937 and the beginning of the fourth in autumn 1937. This was the time of the famous transpolar flights. I was saddled with the responsibility of preparing the electrical and radio equipment, first for a squadron of TB-3 aircraft that landed the Papanin expedition on the North Pole and then for the N-209 aircraft in which Sigizmund Levanevskiy was supposed to have flown over the Pole to the United States.[9]

Due to the heavy work load at the factory, I earned an academic incomplete. I had no opportunity to take the last exam in the fundamentals of electrical engineering taught by Academy of Sciences Corresponding Member Professor Krug, nor the first exam in future Academician Trapeznikov's course on electric machines, nor had I completed the course project on the strength of materials.[10] The incompletes

---

9. See Chapter 7 of Chertok, *Rockets and People*. Vol. 1.

10. Karl Adolfovich Krug, who was elected Corresponding Member of the Academy of Sciences in 1933, founded the Soviet school of theoretical electrical engineering. Academician Vadim Aleksandrovich Trapeznikov (1905–), a pioneer in Soviet control theory, headed the Institute of Automation and Remote Control (from 1969, the Institute of Control Sciences) from 1951 to 1987. In 1998, the institute was renamed the V. A. Trapeznikov Institute of Control Sciences.

were still standing in the fall. But in September, when my fourth academic year had already begun and I was supposed to make up the incompletes within the first two weeks, work was continuing on the aircraft for the search expeditions that were being sent out to find Levanevskiy. I wasn't even able to show up at the institute for the beginning of classes.

I hoped for a "bail out" in the form of a letter written on the letterhead of the People's Commissariat of Heavy Industry (specifically its Main Directorate of Aviation Industry) signed by Andrey Nikolayevich Tupolev himself.[11] This letter, addressed to MEI director Dudkin, said that I had been very busy with crucial work on the preparation of transpolar flights and therefore the State Commission requested permission for me to take the exams in October or November 1937.[12]

Appearing for the first time at the general lectures two months late, I received an assortment of rebukes from my classmates and a warning from the dean's office that I should report immediately to the director for the decision as to my subsequent fate.

My plight might have been harder on my classmates than it was on me. The feeling of camaraderie and "one for all and all for one" among the night school crowd was very strong at that time. Lev Macheret, our oldest classmate, whose student nickname was Bambula, a sobriquet he had received for his solid, rotund physique, and who, incidentally, was to become the chief engineer of a cable factory, announced that he knew how to help me.

"Bambula is coming to Bumba's rescue," proclaimed Macheret. "And 'Sonny Boy' is going to help me." As revenge for the nickname "Bambula," he called me "Bumba." We called the very youngest among us long-in-the-tooth students "Sonny Boy." His real name was Germogen Pospelov, a technician at the Moscow Electric Factory. Sonny Boy was a brilliant student and many years later became an academician and a world-famous scientist in the field of artificial intelligence.[13]

At our next meeting Bambula and Sonny Boy told me that under no circumstances should I go to Director Dudkin. "Go to Golubtsova in the Party Committee. We explained everything to her."

Student Golubtsova didn't appear in our class until our third year. Naturally, at first we wondered why a woman, clearly five or six years older than our average age, needed to study with such blue-collar types. Outwardly very reserved, always modestly but elegantly attired; from the very beginning Golubtsova enjoyed deferential attention among us students. We concluded that a woman with such qualities was

---

11. Andrey Nikolayevich Tupolev (1888-1972) was the most influential and successful aviation designer of the Soviet era. His organization, OKB-156, produced several generations of bombers and civilian aircraft. Like many of his compatriots, Tupolev was arrested and thrown into prison during the height of Stalin's Great Terror in the late 1930s.

12. The State Commission was the ad hoc body composed of various industrial representatives responsible for the polar flights.

13. Germogen Sergeyevich Pospelov (1914-), a specialist in automatic control, was elected an Academician of the Academy of Sciences in 1984.

fully capable of playing the movie role of a factory director who exposes a saboteur, the factory's chief engineer. The all-knowing secretary of the dean's office hinted that she was an official in the Central Committee apparatus and that we shouldn't be up to any foolishness when she was around. But the outwardly severe Golubtsova was compelled more than once to turn to her classmates for help. We established good, comradely relations with her, including swapping course outlines, crib sheets, and the usual mutual assistance that goes on among students.

Suddenly, the most informed person in our class, Teodor Orlovich, who went by the nickname "Todya," and who would later become the chief designer of the cable industry Special Design Bureau, in strictest confidence informed a tight circle of comrades that Golubtsova was her maiden name and that she was actually Malenkova, the wife of that very same Malenkov, who... "you know."

We were filled with pride that such a distinguished woman shared our student ranks, but we soon became accustomed to it because she treated us as equals, rode public transportation after class in the evening, and got quite objective marks. We decided that we should be happy that Comrade Malenkov, well-known to the entire country, had such a good wife, who in the next three years would become a fine electrical engineer.

However, we were wrong about her intentions. Golubtsova had graduated from MEI back in 1934, and had been working as an engineer at the Dynamo factory, while dreaming of leaving to pursue science. She entered the MEI graduate school in 1936. Here it became clear that the accelerated four-year course for engineers from the ranks of so-called *Parttysyachniki* ("Party captains of thousands"), which she entered in 1930, was too condensed.[14] She would need to fill the gaps in the basic electro-technical disciplines, and after enrolling in graduate school, she became a part-time student of our night school.

Then it turned out that while I was working on transpolar flights and rescue missions, they had elected a new Party Committee at the institute and Golubtsova, a fellow student, became its secretary. At that time, the Party Committee secretary of an institution of higher learning could have as much influence as a director. In any event, it was impossible to expel a Party member from an institute without the approval of the Party Committee. On the other hand, the Party Committee could demand that a disagreeable student be expelled for any political sins. In that case, the director did not resist.

Following the advice of Bambula and Sonny Boy, I went to the Party Committee. Golubtsova received me like an old acquaintance. Her Party authority had not gone to her head in the least. As before, her outfit was modest, beautiful in its own

---

14. The *parttysyachniki* were a huge demographic granted preferential treatment for entrance into institutions of higher learning in the late 1920s and early 1930s. The Soviet government sought to train tens of thousands of younger workers and peasants for important industrial, military, and Communist Party positions that had previously been occupied by those educated under the Tsarist regime.

way, and tasteful. She stood up and with a kind and cheerful expression gave me a firm handshake. Golubtsova did not start moralizing, but simply asked me when I would be able to fulfill my incomplete work. And then, instead of a simple response, I handed her the letter signed by Tupolev.

Recalling that episode now, I think that at that time I wanted to attach more significance to my persona. Let the new Party Committee secretary know that I was not some lazy student. Tupolev himself would intercede on my behalf! But the effect was unexpected. The benevolent smile disappeared. Golubtsova frowned; she walked over to the safe standing in the corner, placed the letter inside like a secret document, and locked the safe. Turning to me, she said quietly, "Forget about Tupelov. He's been arrested. Don't even think about telling anybody about that letter, and if you don't pass your exams by December, you have only yourself to blame."[15]

After such a warning I bolted from work for several days in a row and hunkered down in the cozy reading room in the Park of Culture and Recreation. By November I had worked off my incompletes, while at the factory I had been reprimanded for failing to issue the next batch of documentation on time.

Soon rumors were in broad circulation about enemies of the people in the aviation industry and about the conspiracy that Tupolev himself had led. Comrades at the institute asked bluntly, "What was going on there with you all in the aviation industry?" My involvement with the transpolar flights was well known, and Bambula, who had a keen wit, reassured me, "If they didn't take you when they took Tupolev, then it was simply a matter of sloppy work. Now they're not about to correct their mistake, but you still better not fall behind, so it doesn't catch up with you."

Bambula and Todya organized a separate group specializing in cable technology and talked Golubtsova into switching to that group during her fifth and last year. That way, she graduated from the institute, as it were, for a second time, receiving a diploma in electrical engineering with a specialization in cable technology, although she had been considered a graduate assistant in the cable technology department since 1936.

During our fifth year we were supposed to have full-fledged daytime classes and take a leave of absence from the factory. I took my leave of the factory and once again met up with Golubtsova, this time in order to be placed on the Party roster at the institute. She had time to complain about the difficulties of combining Party leadership at the institute with her fifth year studies, and at the same time she asked me to delve into the affairs of the electromechanical department's Party organization.

"You've been a Party member since 1932, and now you could be in charge of organizing your department."

---

15. The Soviet secret police, the NKVD, arrested Tupolev on 31 October 1937. He remained incarcerated until July 1941.

I responded that I had a lot of gaps in my knowledge and would like to devote all my spare time to new problems in electroautomation. Nevertheless, when all was said and done, she had managed to persuade me to "bring order" to the Party organization of the electromechanical department.

According to some unwritten law, Party members, even leaders who belonged to the same Party organization, were on a first name basis. That is why seven years later at our meeting at NII-88, Golubtsova addressed me with familiarity, letting me know that she had not forgotten our Party association at MEI.

After defending my final thesis, I once again visited the Party committee office, this time to remove my name from the Party roster. Beforehand someone had warned me, "Don't forget to congratulate Valeriya Alekseyevna. She defended her dissertation." After we congratulated one another, Golubtsova recommended once again that I enroll at the institute, but this time as a graduate student without taking a leave of absence from the factory. When I wavered, she insisted, "You graduated with distinction, you have a great deal of factory experience; consider it arranged." As I was leaving, Golubtsova said, "And you have good friends." Bambula, Todya, and Sonny Boy really were good friends.

In the fall of 1940, I became a graduate student in the MEI department of aviation electrical equipment. Department head Professor Frolov even entrusted me with giving some of his lectures, because he had a heavy load at the Air Force Academy. The war interrupted my scientific career, which had begun at Golubtsova's suggestion.

In the autumn of 1941, like all Moscow institutes, MEI was to be evacuated to the east. Here is where Golubtsova's character and will emerged. She organized, to the extent possible, a normal evacuation, and then the continuation of the institute's academic activity at its new site. Bambula and Todya were mobilized for some particularly vital cable production projects and, having received exemptions from being drafted into the army, were working in Moscow like soldiers. Sonny Boy was called up for the army and fought to repel the Germans' attack on Moscow using an 1891-model rifle. Given his nearsightedness, this was terribly frustrating for him and he sent us desperate letters. Now, instead of Bambula and Todya, it was Lev Macheret and Teodor Orlovich, who appealed to Golubtsova in September 1941. They requested that distinguished MEI graduate Germogen Pospelov be relieved of his military duties and that he use his engineering knowledge for victory.

Golubtsova had not forgotten these men, whom she had referred to as my good friends. Pospelov was detached to the air force just twenty-four hours before the battle in which his rifle unit was completely wiped out. He finished the war at the rank of captain as an engineer working on special equipment for a major air force formation.

Having earned many combat decorations, Pospelov enrolled as a graduate student at the N. Ye. Zhukovskiy Air Force Academy. He became an instructor, a senior lecturer, a professor, and even a general. He developed the theory for and directed the creation of an experimental blind landing system for airplanes. In 1964, he was elected a corresponding member, and in 1984, a full member of the USSR Academy of Sciences. If not for the initiative of his classmates and the intervention of Golubtsova, Sonny Boy would have laid down his brilliant head on the bloody battlefields outside Moscow and Soviet science would not have had an Academician Pospelov.

In October 1941, the families of all the VKP(b) Central Committee and Politburo members were to be evacuated from Moscow. Malenkov and Stalin remained in Moscow, having shifted into a state of siege, and Golubtsova was forced to travel to Kuybyshev (now Samara) with her children and temporarily part with MEI. In Kuybyshev she was appointed instructor of the *oblast* Party committee for the defense industry.

In late 1942, the State Defense Committee decided to return a contingent of students and instructors to Moscow from evacuation. On 9 January 1943, yet another decree was issued. It defined a new developmental phase for MEI, in particular, substantially increasing the number of students and calling back professors and instructors from active duty in the army. It called for building and equipping laboratories with new equipment and providing students and instructors with housing and a food supply. It was astonishing that during the country's most difficult period of the war, when Hitler still believed in final victory, the high-ranking political leaders of the USSR made an unprecedented decision on mobilizing human resources for the country's future power in engineering systems! We still had the Battle of Kursk and more than two years of war ahead of us! Decrees similar to the MEI decree were also implemented for other principal Moscow institutes of higher learning and for the Academy of Sciences. In a country that was bleeding to death, the top political leaders made truly heroic efforts not only to preserve the scientific cadres, but also to ensure their numbers increased in the future.

In June 1943, Valeriya Golubtsova, instructor for the Party defense industry *oblast* committee, was called back from Kuybyshev and appointed director of the Moscow Power Engineering Institute. A tremendous responsibility lay on the shoulders of this forty-year-old woman. In Moscow, where air raids continued, she needed not only to restore the academic process, but also build new academic buildings and dormitories, acquire equipment for the laboratories, find food for the half-starved students, and, most important, begin scientific developments to restore the devastated power systems and for new radio electronic weapons systems.

Golubtsova was well-known in the upper echelons of state and Party organizations, and, in addition, they knew her as the wife of a Politburo and State Defense Committee member. This helped, of course. But her personal qualities were the primary and decisive factors behind MEI's success during the war years. So as not to be accused of having a subjective attitude toward Golubtsova, I will cite excerpts from the recollections of Petr Zhakovich Kriss, former MEI student and radio specialist, who collaborated closely with all the various transmutations of the Korolev collective. Of all the testimonials about Golubtsova published in the aforementioned collection, I have selected these because I have known Kriss for many years. No one has ever doubted his objectivity, honesty, and decency.

This is how he recalls his first meeting with the director in 1943:

"Before us stood a very interesting woman who seemed young even to us, young men twenty years of age. She was modestly, but elegantly and tastefully dressed, with a cheerful, kind expression… In today's terms, one might say that she radiated a powerful, positive energy, which each of us obviously sensed. She did not possess a drop of snobbishness, which often alienates young people. She won people over both by her simple, ingenuous manner of speaking, and by her kind, motherly look, and easy humor… I trusted her, and subsequently there was nothing that could disillusion me about a single feature of her blessed image."

She was just as Petr Kriss so aptly described her when she met me in her director's office in 1947. This was a week after the meeting described several pages ago at NII-88 attended by the president of the Academy of Sciences.

"In ten minutes our scientists and department heads will gather here. You tell them everything that you think is necessary and topical for MEI."

Out of all the institute scientists assembled there, the only one I recall is Vladimir Aleksandrovich Kotelnikov, dean of the radio department. Later I found out that senior lecturer Tkachev was also there. He was one of the pioneers in the development of inertial navigation systems. I became acquainted with him much later. His ideas at that time significantly surpassed the level of what we and the Germans had brewing in terms of autonomous control systems.

I told them briefly about our program of operations at NII-88 and the principles and problems of long-range missile flight control. I focused on the need to develop new multichannel telemetry systems and reliable radio monitoring of the flight trajectory along the entire flight path.

Within a short time, the results of this meeting exceeded our most optimistic expectations. Thirty-nine-year-old Professor Vladimir Kotelnikov was in charge of developing the ideas I had posed. But the efforts of a single radio engineering department were insufficient. What we needed was an institute-wide effort and pilot factory. Literally about ten days after my meeting with the MEI scientists, Golubtsova's office issued a governmental decree signed by Stalin on the creation of a special operations sector at MEI.[16] A year later, the collective that had rallied around Kotelnikov was already developing the *Indikator-D* system, which we used during the flight tests of the first R-1 domestic missiles in 1948. Beginning with this development, all subsequent missiles were equipped with MEI radio systems during test flights.

In 1951, the MEI collective entered a competition for the creation of telemetry systems, and the first R-7 intercontinental missile was equipped with its now legendary *Tral* system. Soon Kotelnikov acquired a young, energetic, and hard-charg-

---

16. This "experimental scientific-research profile" sector was officially created on 25 April 1947.

ing assistant, Aleksey Fedorovich Bogomolov.

In 1954, Kotelnikov became an academician and the director of the Academy of Sciences Institute of Radio Engineering and Electronics. Later Bogomolov was put in charge of operations at MEI. His hard work resulted in the creation of the Special Design Bureau (OKB MEI), a powerful organization that was fully involved in the production of complex radio electronic systems for the space industry. Kotelnikov and Bogomolov staffed their collective with MEI's most capable graduates.[17]

Unrestricted by previous projects in this field and by any rigid schedules imposed by ministries, OKB MEI became famous for many original and unique developments. Sometimes their ideas outstripped industry's technological capabilities, but they always remained a very strong stimulus for the developers of the space industry's radio electronic systems. Kotelnikov and Bogomolov became indispensable members of the Council of Chief Designers.

Academician Kotelnikov, who became vice president of the Academy of Sciences, vice president of the International Academy of Cosmonautics, and chairman of the Interkosmos council, always kept his activities associated with space.[18] We run into each other regularly at ceremonial sessions in honor of Cosmonautics Day and many other occasions. Vladimir Aleksandrovich never forgets to remind me, "You know, you were the one, Boris Yevseyevich, who got me into this business of cosmonautics some time ago." Now, that "some time ago" is more than fifty-five years behind us.

After the war, director Golubtsova showed exceptional dedication to the construction (of new academic buildings, the pilot factory, a Palace of Culture, dormitories, and housing for professors and instructors) and expansion of the research facilities. Thanks in large part to her energy, combined with her closeness to the country's higher authorities, an entire town sprouted in the area of Krasnokazarmennaya Street, consisting of the Moscow Power Engineering Institute and its OKB, which to this day are the foremost organizations in the field of rocket and space radio engineering.

Golubtsova, an engineer without any outstanding achievements in the fundamental or applied sciences, became the director of a major scientific technical institute. But in this case, the MEI was fortunate. God generously endowed her with organizational talent. Her natural feminine sensitivity helped her fuse the efforts of all the institute's scientists with a minimum of conflicts. At any rate, the very reputable MEI faculty supported the director in all of her deeds.

Over the ten-year period that Golubtsova was a member of the institute's governing body, her perseverance and day-to-day exactitude, and the close interaction of

---

17. OKB MEI was created in September 1958 by expanding the original experimental scientific-research profile sector at MEI.

18. Interkosmos was the international cooperative effort between the Soviet Union and other socialist countries, established to facilitate joint work on space research and applications satellites.

the institute's scientists with engineers from industry produced very tangible practical results. Golubtsova defended her Candidate of Science thesis in 1948 while
she was director. In 1952, she handed over her post as director to her deputy and
became a senior lecturer in the MEI department of general electrical engineering.
In 1953, she received the position of deputy director of the USSR Academy of
Sciences S. I. Vavilov Institute of Natural History and Technology, the very same
Vavilov with whom we met at NII-88 in 1947. In 1955, she defended her thesis for
a Doctor of Science degree in the history of electrical engineering.[19]

In 1957, after sacking a group of his former colleagues from Stalin's Politburo,
Khrushchev sent Malenkov into exile in Kazakhstan, appointing him first director
of the Ust-Kamenogorskaya and then of the Ekibastuzkaya thermoelectric plants.
Golubtsova could have stayed in Moscow, but she and her children followed her
husband. She did not return to Moscow until 1968. She completed her journey on
Earth in 1987, and was buried in Moscow at the Kuntsevskaya Cemetery.

President Vavilov passed away on 25 January 1951. Throughout his tenure in
that post he closely observed the participation of Academy scientists in our work.
Almost all of the firing range test launches were attended by FIAN scientists and
future academicians S. N. Vernov and A. Ye. Chudakov and by a group of young
up-and-coming scientists, who later served as the nucleus of the Institute of Space
Research (IKI).[20]

I PERMITTED MYSELF TO DESCRIBE IN SUCH DETAIL MY MEETING IN 1947 AT
NII-88 with Academy of Sciences president Vavilov and MEI director Golubtsova
because that event illustrates the quest to integrate three elements of our scientific
and industrial infrastructure—the Academy's fundamental research in various fields
of science, the scientific potential of the institutes of higher learning, and the most
leading-edge industrial technology—into a single, systemic statewide program. In
subsequent years, this unity was actually attained. In the early 1950s, Korolev managed to achieve relative independence, and in 1953 he was elected an Academy of
Sciences corresponding member. He took particular care to strengthen this triple
alliance of the sciences and zealously guarded it against destructive departmental
tendencies toward autonomy.

---

19. The Russian (and former Soviet) postgraduate educational system uses two academic degrees at
the doctorate level: *Kandidat nauk* (Candidate of Science) and *Doktor nauk* (Doctor of Science). The
former corresponds to the Ph.D. degree in the United States while the latter is equivalent to a higher,
second doctoral degree.

20. IKI—*Institut kosmicheskikh issledovaniy* (Institute for Space Research), founded in 1965, was
the Academy of Sciences' foremost institution to manage the development of payloads for Soviet
scientific and deep space missions.

Chapter 6

Department U

Unlike the other scientific departments in the NII-88 structure—approved by Ustinov and based on the concept developed by Gonor and Pobedonostsev in 1946—
the head of the guidance systems department (Department U) also served as the
institute's deputy chief engineer. I was indebted to Pobedonostsev for that. First,
he wanted to emphasize the importance of guidance systems for rocket technology
and, second, to grant me personally a position with greater authority and greater
independence. Moreover, Pobedonostsev told me in one of our first serious meetings that he personally did not want to bear the responsibility for too great a variety
of guidance system projects. Not only he, but Gonor and the ministry entrusted this
responsibility completely to me. "As for Korolev," Pobedonostsev added peevishly,
"Sergey always has his own opinion. He wants Department U to work entirely on
his projects. But now that's impossible. We are obliged to work on surface-to-air
guided missiles and to help Sinilshchikov and Rashkov." Skipping ahead a bit, I will
note that after I left this post in late 1950, no subsequent Department U chiefs were
appointed deputy chief engineers.

I was also pleasantly surprised that the department had been given what were for
those times quite nice accommodations. A separate five-story building was added
onto the old director's building. Before I arrived from Germany, the young control
surface actuator specialist Georgiy Stepan, whom I had sent from Bleicherode, and
radio engineer Dmitriy Sergeyev, who was appointed as my deputy, had been running the show rather successfully. I agreed to their draft of the office layout.

The fifth floor housed the radio laboratory; the fourth floor was home to the
design bureau; the third floor contained the instrument laboratory (after late 1947,
this included gyroscopic, stabilization, and astronavigation instruments); the second
floor was allocated for the integrated laboratory for general schematics and tests;
and the first floor, which was practically half underground, but the most spacious,
housed the experimental instrument shop. Many young and capable specialists came
to the department with an ardent desire to work, and there was not a whiner among
them, which was pleasantly surprising. Sergeyev deserves a great deal of credit for
setting up the department during the first one-and-a-half to two years. He was certainly a talented radio engineer. He immediately established contact with German

radio specialists and, for all intents and purposes, the development of proposals for the radio control system of the rocket developed by Gröttrup was conducted under his leadership. But he also established contacts with the new NII-885 (with Ryazanskiy, Boguslavskiy, and Borisenko), where they were developing radio control systems, and NII-20, where Degtyarenko was creating the *Brazilionit* telemetry system in place of the German *Messina* system. However, Sergeyev soon realized that we would not attain good missile flight control using those systems, and he created groups to develop our own system for monitoring the speed and coordinates of a missile in flight using standard radar systems.

The laboratory staff was quickly filled. Among its ranks were engineers who had been demobilized from the army. That is how radio engineer Aleksey Shananin came to the department. His frontline experience helped to quickly establish contact with comrades working in the laboratory and with a multitude of subcontracting firms. Later Shananin's capabilities were noticed and he was lured away to work at the Commission on Military-Industrial Issues (or Military-Industrial Commission, VPK) under the USSR Council of Ministers, where for a long time he was one of the leading and truly competent specialists.

In March 1947, recently demobilized radio engineer Oleg Ivanovskiy, who had worked at the neighboring Ministry of Defense's Central Scientific-Research Institute of Communications (TsNII Svyazi), caught Stepan's attention. Ivanovskiy's passion for the radio engineering field, organizational talent, and energy also did not go unnoticed. He went down in history forever as the "lead designer" of the Vostok that carried Yuriy Gagarin into space.[1] He also deserves credit for being the first specialist, rather than a professional journalist, who described the epic of the creation of Vostok and Gagarin's launch in his memoir *First Stages*.[2] The censorship office prohibited the book's publication under the author's real name, and so the book appeared under the name of a totally unknown Aleksey Ivanov. Subsequently Ivanovskiy also worked in the VPK offices in the Kremlin, and then transferred to the Lavochkin Factory. When Ivanovskiy left the Kremlin, radio engineer Aleksandr Ivanovich Tsarev from Department U was promoted to take his place in the VPK offices.

Mikhail Krayushkin, former artillery battery commander, proved to be a very serious theoretician in the field of radio wave propagation and antenna design. He would go on to brilliantly defend his Doctor of Science dissertation and organize a

---

1. "Lead designers" were different from "chief designers," in that the former were much more junior in the design bureau hierarchy. Whereas chief and deputy chief designers were deeply involved in the R&D stages of a program, lead designers were typically responsible for the production phase of weapons development cycle.

2. Aleksey Ivanov [Oleg Genrikhovich Ivanovskiy], *Pervyye stupeni (zapiski inzhenera)* [*First Stages (Notes of an Engineer)*] (Moscow: Molodaya gvardiya, 1970). Later variations of this book appeared under different names. See, for example, Aleksey Ivanov, *Vpervyye: zapiski vedushchego konstruktora* [*The First: Notes of a Leading Designer*] (Moscow: Moskovskiy rabochiy, 1982).

unique group of rocket-space antenna specialists.

Nadya Shcherbakova was a colorful figure among radio specialists. A nurse during the war, she graduated from a communications institute and attacked the problems of monitoring flight trajectory during firing range tests with vigor unusual for a woman. Her exacting nature, exceptional performance, and intolerance of everything that, from her point of view, stood in the way of our technology, occasionally led to conflicts that sometimes did not end in her favor. Nadezhda Pavlovna Shcherbakova enjoyed great prestige among missile specialists and later headed the radio department at TsNIIMash, which was formed at the NII-88 factory.

Shcherbakova's dedication was not the exception. During the very first year in the department other female engineers also began to work energetically on a par with men. I think it's fitting to mention Vera Frolova, Shcherbakova's right-hand woman for the organization of firing range tests, and Zoya Melnikova, the indisputable authority on telemetry sensors. Melnikova served as a sort of intermediary between the physical value being measured, its electrical analog, and the data transmission radio system. She had several other female engineers—sensor and telemetry specialists—subordinate to her. Because they were still supposed to show up at "hot spots," at the factory, with the Germans on Gorodomlya Island, and at the firing range, they used to joke about sharp-tongued Zoya Melnikova that she would be perfect commanding a female "death battalion," but all she got was a brigade of "bluestockings."[3] However, these "bluestockings" were by no means totally devoid of humanity. They fell in love, got married, and were happy not only when they were on the job, and their unhappy times were not only due to technical failures.

I should also mention the great role that engineer electrician Aleksandra Melikova played. Coming to us with experience from her work as an electrical engineer involved with relay automatics, she quickly mastered the problems of developing and testing the general electrical circuits of a missile and simply became an indispensable specialist in that field, especially when troubleshooting was required for an off-nominal situation in the behavior of relay electroautomatics that wasn't "by the book."

The development, testing, and series production of control surface actuators, on the other hand, ended up being strictly men's business. After returning from Germany, Stepan attracted several engineers to the department, including Ovchinnikov and Shumarov. Soon the strong-willed and broadly educated engineer Viktor Kalashnikov transferred from the Mytishchi tank KB. He became the supervisor of this entire direction of work and would go on to become one of the leading specialists in Korolev's design bureau.

At first, engineer and optics specialist Kabalkin headed the design bureau of our department, but soon Semyon Chizhikov replaced him. He spent most of

---

3. This is a reference to the famed Bluestocking societies of the late 19th century in England and France, which advocated further educational and social advances for women.

his entire life, practically side-by-side with me, beginning with Factory No. 22, Institute RABE and NII-88 and then continuing on through all the Korolev programs. The work he did at NII-88 and then at Korolev's OKB-1 created a rocket instrument design bureau that was unique in terms of its universality. When I write above of Chizhikov's "entire life," I can't help but recall the last hours of his life. I walked into his apartment when two emergency medical teams were already at work. Oxygen, artificial respiration, numerous injections, and electroshock were to no avail. A massive heart attack after a heavy work day ended his steady passion for energetic work.

Chizhikov founded a dynasty. His son Boris leads a staff of designers working on spacecraft docking assemblies, and his granddaughter, Marina, a mathematician, calculated the dynamics and strength of spacecraft docking mechanisms.

In the beginning, the biggest concerns were for the integrated laboratory. Its managers were engineer practitioners Valentin Filipov and Dmitriy Shilov, and its true ideologue was engineer and communications specialist Emil Brodskiy. The laboratory was supposed to develop a test rig that would serve as a simulator to test and verify the circuits of automatic launch systems and as a site to test and verify operational documentation for factory and flight tests. The subject was not the problem; daily tasks were so labor intensive that they swamped the staff's resources.

The experimental shop was quickly outfitted with captured precision machine tools. Using his connections at the factory, Troshin, the shop's first chief, selected the best all-round machinists, metal workers, and precision mechanics, that is, those who had the "magic touch." Thus, we were almost independent of the factory's production shops. We distributed the most unique mechanics among the laboratories.[4]

Despite establishing a strict secrecy regime at the institute, all the leading specialists understood the need for communications with scientific organizations that were not directly involved in our cooperative network and with scientists from institutes of higher learning. Thus, from the first months of 1947, there were joint projects with the Academy of Sciences' Institute of Automation and Remote Control. Future Academicians Vadim Trapeznikov and Boris Petrov and future Academy of Sciences Corresponding Member Vyacheslav Petrov cooperated in our project.[5] When the Presidium of the Russian Academy of Sciences awarded me an Academician B. N. Petrov Gold Medal in 1992, it was particularly gratifying because it reminded me of my joint work with the fine gentleman Boris Petrov back in the very nascence of

---

4. *Author's note:* This first instrument shop received the first administrative censure for failing to meet deadlines for the production of control surface actuators for the first R-1 series. But that's another story.

5. Academician Boris Nikolayevich Petrov (1913–80) was one of the leading Soviet scientists specializing in the theory of control systems. He contributed immensely to the early Soviet missile and space program, especially in the field of propulsion system control for ballistic missiles and spacecraft. From the 1960s, he was a prominent public spokesperson for the Soviet space program.

rocket technology.

Speaking of scientific connections, I would like to note the essential difference in the way our department formulated control theory problems—and, in particular, controlled systems stability theory—and the methods that were proposed at that time by pure theoreticians such as Nikolay Moiseyev, the N. Ye. Zhukovskiy Air Force Academy professor and author of the so-called "theory of technical stability." We preferred to conduct research using more empirical methods, without excessive use of profound and complex theoretical constructions that were not readily accessible to a practical engineer.

At that time, so-called frequency methods, based on the analysis of phase amplitude and frequency amplitude characteristics, were well developed at the Institute of Automation and Remote Control. Admittedly, they came to us from abroad, from the United States; we benefited from the famous research at the Massachusetts Institute of Technology (MIT). The work at MIT was the practical response of theoreticians to the real wartime problem of creating a radar system capable of automatic search and tracking of airborne targets.[6]

It turned out that if one didn't try to use pseudopatriotic and excessively complex approaches, then using these new methods [from the United States], one could successfully solve the problems of missile stability and control. The engineer should of course, have mastered classic oscillations theory. But that science was well developed and accessible in the works of our scientists Andronov, Bulgakov, Gorelik, and others.[7] Moreover, as a result of the development of our own radar stations we had also produced some interesting works on the theory of nonlinear systems. Therefore, our young theoreticians observed the fierce battles of the "founding luminaries" from the sidelines. They themselves did not meddle in the scuffle and laughed at the theatrics of the scientific and technical councils on these subjects.

Out of critical necessity, specialists at NII-885 studied stability. Here Pilyugin, an engineer with a great deal of experience, declared in a way typical of him, that you could believe theoreticians as long as you were dealing with paper, but "if I am responsible for selecting the parameters and adjusting the automatic stabilization system, then I need a simulator that I can put my hands on and feel everything, and the transitional processes must be visible on oscillograph tapes." We were in complete agreement about that. We had ordered the Institute of Automation and Remote Control to develop simulators instead of the primitive and scarce "Häuserman pendulums," but Pilyugin was bent on solving this problem on his own.

---

6. Here, Chertok is referring to the work of the Radiation Laboratory at MIT, which during World War II focused enormous resources on developing microwave radar systems.

7. Aleksandr Aleksandrovich Andronov (1901–52) was one of the pioneers of Russian control engineering. Boris Vladimirovich Bulgakov (1900–52) was a specialist in mechanics who contributed to the theory of oscillations and the theory of gyroscopes.

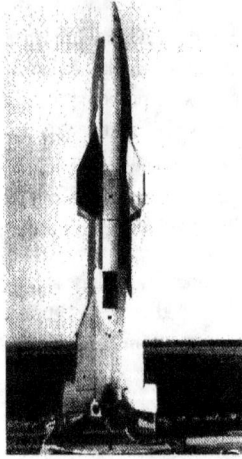

Dr. Matthias Uhl.

Many mistakenly misinterpret the May 1946 decree that founded the Soviet missile industry as being entirely focused on reproducing the German V-2 ballistic missile. In reality, the decree devoted comparable resources to both the V-2 and the Wasserfall cruise missile which, unlike the V-2, was never used in battle by the Nazis. The original German Wasserfall is shown here in June 1944 ready for launch.

As DIRECTOR OF THE GUIDANCE DEPARTMENT AND NII-88 DEPUTY CHIEF ENGINEER, I came under criticism from the chief designers of surface-to-air missiles. Sinilshchikov was particularly aggressive, because his Department No. 4, which was subordinate to Tritko, was marginalized. He argued that "Chertok's entire department is working only on Korolev's projects. Under these conditions we cannot create a missile like the *Wasserfall* because its guidance problems are a lot more complex than those of missiles like the A4. Either you switch Chertok to our project, or you create another similar department in NII-88."

At NII-885, departments were set up to develop guidance systems for surface-to-air guided missiles. Guidance specialists who had previously worked at the Institute Berlin worked there. Govyadinov was in charge of this sector under Ryazanskiy's authority. The surface-to-air missile guidance specialists also grumbled that the necessary conditions hadn't been provided for them at NII-885. Thus, the resentments of the NII-88 and NII-885 surface-to-air missile chief designers fused and by and large they were justified. The level of work on air defense missiles was clearly below even what the Germans had done in Peenemünde.

The successes achieved during the first three years of establishing the two fields of domestic missile technology—long-range ballistic missiles and surface-to-air guided missiles—differed very substantially. Admittedly, the initial starting conditions for long-range ballistic missile technology had greater advantages. This was very obvious when the two fields came together at the two institutes, NII-88 and NII-885, under the same ministers, directors, and chief engineers. Long-range ballistic missiles, even if they were a German model, had already begun to fly in the autumn of 1947. In 1948 and the first half of 1949, there were numerous firing tests, and whether they were good or bad, the missiles flew, new ones were designed for longer and much longer ranges, various projects were discussed, factories were loaded down with orders for series production, and the military had something to accept.

In that context, the surface-to-air teams working at the same two head institutes looked very colorless. Their work had not moved beyond drawings of missiles very similar to the *Wasserfall.* After all, the Germans had conducted experiments with a *Wasserfall* guidance system on an A4 as early as 1944. Before their evacuation from Peenemünde in 1945, they had already executed dozens of flight tests, which, while unsuccessful, provided them with invaluable experience.

One cannot say that Sinilshchikov and Rashkov had general working conditions that were worse than Korolev's. In terms of pay, all sorts of "privileges," budgetary appropriations, and other benefits, all the teams were in basically the same situation. Nevertheless, no urging from the top helped. Many years of experience showed that even the most well-provided-for groups entrusted with special projects to develop new systems, no matter what state-of-the-art equipment they had at their disposal and no matter how much financial support they received from the state budget,

From the author's archives.

Shown here are Sergey Korolev (left) and Sergey Vetoshkin (1905–91) during the R-1 test-launches at Kapustin Yar in 1948. In later years, Vetoshkin would become a very powerful administrator in the Soviet military–industrial complex.

were incapable of completing the project if the team as a whole did not have confidence in the management. Faith in the management on all levels is considerably more important for the success of the work than the pay scale, a comfortable work site, and the prospect of receiving housing.

The surface-to-air missile specialists lacked such faith. Instead, they were sure that sooner or later new management would appear and say, "Quit fussing, everything needs to be done differently." On many levels the time had come for reorganizing surface-to-air missile technology. This gave me at least the moral right to feign deafness to the criticism from Sinilshchikov; Tritko, who supported him; and the Party Committee and to devote all of Department U's working potential to Korolev's projects. Now I can admit that this happened with Pobedonostsev's silent approval. But Gonor warned me that if I didn't find a good explanation for why we had ignored surface-to-air missile projects, then during the next wave of anticosmopolitism I would be risking my neck.

For some reason I had a firm conviction that salvation would come from the outside! After all, there must be sensible people in our country who would understand that NII-88 was not going to save Moscow from the Americans' atomic bombs using R-1 missiles and future R-2 missiles.

Sergeyev and I discussed the situation and decided that the work of Sinilshchikov, Rashkov, and the other surface-to-air missile specialists had no future at our facility. It wouldn't be long before that would become clear to higher management.

They didn't have much time left at the institute, and it was no use for us to sink with them. Therefore, relying on Pobedonostsev's moral support and on Korolev's criticism that we were not putting out the volume of work that he demanded, we would pursue a strategy of quietly ignoring surface-to-air missile related projects.

The tragic death of Dmitriy Sergeyev in the Caucasus was a very heavy blow for me and for our entire team. He had dashed off to take part in what was for those times a complicated traverse of several peaks. After Sergeyev's tragic death, radio engineer Anatoliy Shustov, who had come to the institute after his demobilization, was designated chief of the radio laboratory and would later become chief of the radio department. From the very beginning of the establishment of OKB MEI he managed to set up a very productive collaboration with Kotelnikov's collective.

Despite the fact that at that time it was fashionable to contend that "no one is indispensable," I was convinced of the opposite. Each person who does things his own way is indispensable. We were all indispensable. I was convinced that if it hadn't been for the death of Roman Popov in 1944 and Dmitriy Sergeyev in 1948, much in our missile radio engineering would have turned out to be considerably more effective. Each of them had been capable of becoming a prominent scientist or leader.

However, as they say, history does not care what might have been, "What would have happened if…" In this regard the specific meetings and events that I have described characterize the general atmosphere in the country and in the industry in which we were working during those first postwar years. The events underscore the thesis that sometimes the natural progression of historical events can be changed, seemingly by chance. However, as a rule, these very chance incidents are the manifestation of the natural order of history.

I OCCUPIED THE POSITION OF NII-88 DEPUTY CHIEF ENGINEER AND CHIEF OF DEPARTMENT U FROM JANUARY 1947 UNTIL THE END OF 1950—just short of four years. Of these four years almost a year was spent at the State Central Test Range (GTsP) in Kapustin Yar. For three months, beginning in September 1950, I was tormented by a special ministry commission that single-mindedly studied my activity. The formal justification for this were letters to the Party Committee, Ministry, and even to the Party Central Committee. The letters, whose authors were the unsuccessful inventors of new missile guidance principles, accused me of sins, which I considered to be accomplishments. The main charge was that I had created an astronavigation laboratory, supposedly staffed with incompetent specialists, whom I selected based on our personal compatibility.[8]

Two other charges ensued from the main one. The first was that the diversion of manpower for the "stellar adventure" hampered the development of missile guid-

---

8. *Author's note:* My passion for stellar navigation was so serious that later I will devote a special chapter (Chapter 12) to it.

ance systems, which were more promising than those being developed at NII-885. And the second charge was that I colluded with Ryazanskiy, Pilyugin, and their colleagues to undermine work on new domestic systems and to prioritize the reproduction of German technology, which did not require intellectual exertion.

All the leading associates of Department U were drawn into the squabbles during the commission's work. They all supported me. My deputy Viktor Kalashnikov advised me, "Go on vacation, relax, and we'll deal with this mess without you." Gonor was no longer there. The new NII-88 director, Konstantin Rudnev, had nothing against my taking a vacation, but said that he wasn't going to get involved with the squabbles and the Commission. It wasn't on his list. Pobedonostsev had already transferred from NII-88 to the leadership of the Academy of the Industry. I obtained a voucher to vacation at the Kislovodsk sanatorium and departed with the hope that the curative powers of the Caucasus would not only strengthen my cardiovascular system, but would also purge my brain of unhealthy worries. Upon my return I was immediately summoned by First Deputy Minister Ivan Zubovich. He upbraided me for deciding to take a vacation at an inopportune time, saying that I should have defended myself against the charges. Now he couldn't help me at all. Based on the commission's findings, I would be removed from work as NII-88 deputy chief engineer and chief of the guidance department. I was free to select my future workplace, but he advised me to make an arrangement with Korolev and go under his supervision.

Reflecting on the events of a half a century ago, I have come to the conclusion that for the second time mysterious forces inflicted what appeared to be a shattering blow on my career. But in actual fact, these forces rescued me from potential genuine disaster.

The first rescue force came in 1933, when I was removed from supervisory *Komsomol* work at Factory No. 22 and was almost expelled from the Party. If this hadn't happened, I would have surely advanced to the next stages of a political career. For many of my comrades at that time—especially after the arrest of Olga Mitkevich—such political work ended their careers or their lives with the 1937–38 repressions. Seventeen years later, my removal from a high post rescued me from further enticing advancement up the administrative ladder in the ministerial system and made me one of Korolev's closest colleagues.[9]

WITH THE PERSPECTIVE OF A HALF CENTURY, what accomplishments do I take credit for during the first four years of the establishment of Soviet rocket technology?

I shall try to formulate them and, for simplicity's sake, list them point by point:

1. The organization of the country's first staff of specialists who initiated the integrated method for creating complex guidance systems for long-range missiles in

---

9. In effect, Chertok's move from being Chief Engineer of NII-88 to a senior designer in Korolev's OKB-1 led him to an entirely different career path.

all aspects: development, production process, factory and prelaunch tests, in-flight monitoring, processing of flight-test results, and introduction into series production.

2. For hundreds of engineers with the most varied backgrounds (in my department by early 1950 there were more than 500) there was perhaps an intuitive sense for a systemic approach. Each individual had to understand and feel that he or she was participating in a large system of operations, and, accordingly, understand the decisive importance of intersystem connections.

3. At my initiative, with the assistance of Director Gonor, the Ministry, and Special Committee No. 2, new creative organizations were established in academic institutes and affiliated with institutions of higher learning (OKB MEI, and departments at MGU and MVTU).[10]

4. We studied and mastered virtually all of the basic experience on German rocket technology guidance and measurement.

5. We established close creative and business ties with the staffs of all the chief designers, the military specialists at the missile test range involved with the armament acceptance process, and with central directorates.

6. We developed the theory and technology for electrohydraulic power drives for control surface actuators. We laid the foundation making it possible in later years to create a unique school of power drives for missile and later for space technology.

7. We invented and experimentally tested automatic stellar navigation methods. We set up an astronavigation laboratory, the staff of which, after being transferred to the Ministry of Aviation Industry, became the basis for the independent astronavigation design bureau that supported the world's first soft landing on the Moon.

8. We set up an antenna systems laboratory, which served as the starting point for the development of a scientific design school for onboard antenna feeder systems for missiles and later also for spacecraft.

9. For the first time domestically, we developed a method for bench testing a complex of electrical circuits and instruments and their interaction in a ground-to-air system.

10. Working jointly with testers, we developed instructions in the form of technical documents for factory and flight-design missile tests.

11. We introduced methods into experimental instrument production for testing the reliability of electrical and electromechanical instruments when exposed to external factors such as temperature, vibration, and electrical interference.

12. I consider as one of my special personal accomplishments bringing the teams of Chief Designers Isayev and Tsybin (in 1948) and Babakin (in 1949) into the NII-88 institute structure. Officially, their transfer was formalized by ministers' orders, but it was up to me to prove to Isayev, Tsybin, and Babakin that their transfers to NII-88 made sense and to persuade Gonor and Vetoshkin that it was an absolute necessity to take these design groups into NII-88.

---

10. MGU—*Moskovskiy gosudarstvennyy universitet* (Moscow State University); MVTU—*Moskovskoye vysshyeye tekhnicheskoye uchilishche* (Moscow Higher Technical School).

Chapter 7

Face to Face with the R-1 Missile

I have already written about the 1947 firing range tests on the A4 missiles assembled in Germany. After 1947, we stood face to face with the task of creating and launching R-1 missiles. These missiles were supposed to be precise copies of the German A4s.

To this day, among connoisseurs of the history of our rocket technology, there is still controversy over whether it was worth it in the years 1947–48 to begin broad-scale projects for the reproduction of German rocket technology. The war's results had shown the ineffectiveness of A4 missiles, even when fired against such obvious targets as London. It was clear that if the A4 missile had become obsolete as early as 1945, then its domestic analog, which could not appear in mass production before 1950, would be all the more hopelessly obsolete. We too had these same doubts back then.

This situation was perhaps harder on Korolev than anyone. He had been designated chief designer of a rocket whose actual developers had only yesterday been our mortal enemies. We had all experienced firsthand how difficult its field operation could be and its low degree of reliability during firing range tests in 1947. Besides that, at whom were we going to fire with a range of just 270 kilometers? This was a more difficult issue for the Soviet Union than for Germany in 1944. As if that wasn't enough, the frantic campaign against foreign influence was still brewing.

From the standpoint of today's understanding of history, one must admit that the decision to reproduce the A4, approved by Stalin, was correct. However, the initiative belonged to Minister of Armaments Ustinov. Despite the wavering of designers and many government officials, Ustinov, along with Ryabikov and Vetoshkin, insisted on this decision, consistently and strictly monitoring its implementation.

The following considerations influenced the decision to precisely reproduce the A4. First, large groups of engineers and workers needed to be brought together, trained, and taught to work. To do this, they needed a specific and clear task and not distant prospects. Second, the factories needed to be kept busy. If they were left idle, somebody else might borrow them. The nuclear community's paws were particularly dangerous. They were not only building factories, but taking away other people's factories using Beriya's protection. But to keep production going, we

119

needed verified, good-quality, working documentation. Where were we going to get it? Were we going to develop our own new documentation from scratch or rework the Germans'? The answer was obvious. The second option was two years shorter. Third, the military had already formed special units and they had actually created the State Central Test Range. We couldn't leave them with nothing to do! Fourth, we needed to get domestic industry involved in missile technology as soon as possible. Let them immediately start making engines, instruments, fittings, wire, and connectors, for which specifications already existed, and voila! our own drawings would appear.

When all this new cooperation got the kinks worked out and started working on a specific project—the series production of the R-1 missile—that's when we, having established a foundation, could allow ourselves to make a leap forward, switching to the development of our own missiles, which, at that point, the army would really need. These were the basic considerations that drove our planning, and I repeat, from today's standpoint they seem even more appropriate than they seemed back then.

The Americans immediately took a different route.[1] History has shown that during that segment of time we were the wiser, although it is sometimes more difficult to reproduce something "exactly" than it is to make it your own way.

We faced our biggest problems during the stage when design documentation was issued and during production. NII-88 SKB Department No. 3 headed by Korolev had the main role in preparing the technical documentation for production. Issuing the documentation, which had to meet the rigid artillery requirements of the customer, GAU, was a very painful experience. Korolev, Mishin, Budnik, Bushuyev, Okhapkin, myself, Chizhikov—who was the director of my Department U's design bureau—and many others wanted to bring aviation production approaches into NII-88. But we ran up against the harsh opposition of GAU officers and the NII artillery leadership, particularly SKB Chief Tritko. At first, compliance with the strict GAU specifications for technical documentation had seemed completely unnecessary to us. The so-called GAU "TU 4000" specification, which defined the drafting system, was very strict and rigid in terms of its formatting requirements.[2] These specifications had been formulated during the war, drawing on the experience of infantry and artillery armaments mass production. According to this system, the documentation that appeared in the shops of any factory, in any region of the

---

1. Instead of directly copying the V-2 rocket, in the postwar era, the U.S. Army fired dozens of them for research purposes. Later, companies such as North American Aviation used the experience of the V-2 design (such as propulsion) to develop the Navaho while the Aerojet and Martin companies developed completely indigenous rockets such as the Aerobee and Viking, respectively. By the time that more powerful American ICBMs and launch vehicles were available in the late 1950s, their lineage stretched back to both domestic and German origins.

2. TU—*Tekhnicheskiye usloviya*—literally "Technical Condition," but more appropriately "specifications."

country, had to enable production and product output without the assistance and involvement of the designers who had developed this documentation.

In the aviation industry, "on the spot" adjustments and slight deviations from the drawing that didn't affect the general tactical and technical requirements were considered normal, especially when running piping, cables, and so on. The artillerymen did not allow this. The situation required not only a new frame of mind from both sides, but also the judicious pursuit of compromises when daily job-related conflicts arose during the production process.

From the very beginning of work on the R-1 missile, in addition to these primarily formalistic conflicts, serious manufacturing problems also cropped up. The first of these was the problem of replacing all of the German materials with domestic equivalents. A problem affecting dozens of Soviet enterprises came crashing down on our materials technologists, who, it is apropos to mention, were not subordinate to Korolev at that time. In the production of A4 missiles, the Germans used eighty-six brands and gauges of steel. In 1947, Soviet industry was capable of replacing only thirty-two grades with steel that had analogous properties. The Germans used fifty-nine brands of nonferrous metals, and we could only find twenty-one of them domestically. It turned out that the most difficult materials were nonmetals: rubber, gaskets, seals, insulation, plastic, and so on. We needed eighty-seven types of nonmetals, and our factories and institutes were capable of providing only forty-eight!

Great difficulties arose during the process of mastering the manufacturing process for control surface actuators in the pilot-production shop of my Department U. We executed the drawings in precise compliance with the GAU requirements, but the first experimental control surface actuators assembled according to those drawings did not satisfy a single requirement for static and dynamic characteristics. Moreover, it turned out they weren't airtight. The oil that served as the working medium in these devices broke through the rubber seals when working pressure was generated and puddles formed under the test rigs.

One day Voskresenskiy dropped into our shop, and watching the tests on the first control surface actuators, he remarked, "You'll blow up the missile that way!" It was believed that the mixture of liquid oxygen that inevitably leaked during fueling and the oil from the control surface actuators was highly explosive. We immediately set up "explosion hazard" tests. Drop by drop we poured control surface actuator oil into a chamber containing scalding liquid oxygen. Nothing happened!

After this, the emboldened testers poured oil right out of the measuring glass. Again there was no explosion. Then they rigged up a device that mercilessly shook the explosive chamber, simulating the impact and vibration of a missile in flight. Still there was no explosion. Nevertheless, the fear of a possible explosion during missile launch preparation remained. Before beginning the oxygen fueling process, testers usually inspected the missile's tail section in the area where the control surface actuators were installed to make sure there were no traces of oil.

Late into the night, designers, factory process engineers, and metallurgists would labor over the control surface actuator gear pumps in the material technologists' laboratories. The primary parts of the pumps, which were made of special iron and steel, did not have the required cleanliness when treated. Sometimes the pumps fell apart. The relay/slide valve unit gave even more trouble. If the smallest speck got into the slide valve mechanism, it jammed. A "speck incident" would surely result in a loss of controllability and the inevitable crash of the missile.

But the greatest trouble of all awaited us when we began tests on control surface actuators cooled to temperatures below freezing. The oil thickened and caused such increased torque on the shaft of the electric motor turning the gear pump that it started to smoke from overload. The electric motor managed to burn out before it heated up and ignited the oil with its own energy.

We began a new search for hydraulic drive oils that wouldn't freeze. But they proved to be too liquid under summer temperatures at the test range, which had soared as high as +50°C (122°F). It was discovered that our factory, which had just mastered permanent mold casting of aluminum alloy actuator housings and had cheerfully reported this technological accomplishment, was not maintaining quality in casting. The actuator housings were porous. At high temperatures, the control surface actuators would "sweat" because oil seeped through the pores. Once again, talk started up about the explosion hazard posed by the control surface actuators. These flashes of memory highlight only a minute portion of the everyday problems that cropped up during the production process.

The USSR Academy of Sciences Institute of Automation and Remote Control decided to render us scientific assistance on all aspects of the actuator drive problem, and did so very enthusiastically, especially after Academy President Vavilov's visit to NII-88. Institute director Boris Nikolayevich Petrov, a young Doctor of Technical Sciences, had just taken over the directorship from Academician V. S. Kulebakin, who had been my guest in Bleicherode in September 1945. Petrov put his best personnel, under the leadership of future Academician Trapeznikov at our disposal. Academy scientists had a positive influence on raising our engineers' general theoretical level and cultivated a taste for strictness in technical reports and theoretical generalizations. But they could not suggest anything to prevent the mass failure of gear pumps or dirt jamming the slide valves.

If the scientific level was high, the general industrial culture was not up to the level of our tasks. Workers and process engineers needed a new mindset. This required a great deal more time than the plans and schedules had allotted.

A similar situation was developing in many other industrial sectors and among our numerous contractors. Unlike the Germans, we experienced no difficulties with graphite for the gas-jet control fins. Their fabrication was entrusted to the Elektrougli firm in Kudinovo. Fialkov, a specialist in carbon electrodes for primary-cell batteries, was in charge of this production. He was subordinate to the "chief electrician" of missile technology, Andronik Iosifyan. This facetious title invented by Korolev really flattered Andronik. When he heard that Korolev called me a "rusty electrician," he was very amused, and after that he loved to proclaim, "I am the 'chief electrician,' but my assignments come from a 'rusty electrician'."

The problem of the strength of the graphite control fins was so serious that Gonor ordered materials technology Department Chief Iordanskiy to create a special laboratory. Former GIRD engineer Fonarev was in charge of it. In his Department No. 3 Korolev gave the responsibility for the graphite control fins to young specialist Prudnikov. But the control fins remained a very unreliable part of the R-1 missile flight control system for a long time.

Kurchatov needed graphite for the moderator rods in nuclear reactors. He needed graphite with a particularly high purity, but mechanical strength was secondary in importance. We didn't need purity, but a high degree of strength was mandatory. How the Germans had achieved the strength of their graphite control fins, we did not know. Eventually, Prudnikov and the graphite production under his patronage run by Fialkov developed all the process secrets using their own wits. The control fins could be checked only on fire-testing rigs in the jet of a standard engine. NII-88 did not yet have such a rig. Glushko had the only such stand (in Khimki), but he had a heap of his own problems there. In Germany, it seemed that welding the large combustion chambers was not at all tricky. But in Khimki, welded seams were bumpy, burn-throughs abounded, and cracks occurred during tests.

All the engine specialists (or as we joked, the "trench people") in Glushko's entourage—Vitka, Artamonov, Shabranskiy, Sevruk, and List—had been through the *Sharashka* with him in Kazan and the fire-testing rigs in Lehesten.[3] They worked intensively. Here was yet another paradox. These were people to whom the existing regime had caused so much harm, against whom a scandalous injustice had been committed—seven years in prison, camps, or *Sharashki*—and they worked with self-denial and fanaticism that was rare even for those times. The tests on the gas-jet control fins interfered with their firing test program because they put additional stress and the consumption on the new engines. And there were never enough of them.

THE HEAVY BURDEN OF QUALITY CONTROL AND THE PRECISION REPRODUCTION OF THE GERMAN PROTOTYPES lay on the shoulders of military acceptance teams. Military engineers and our own engineers had gone through all the peripeteia of the Institutes RABE and Nordhausen. But while we had been workmates there, had enjoyed ourselves at the Villa Frank officers' club, and had helped each other in every way, now modest Colonel Engineer Trubachev, chief of military acceptance (regional engineer) could stop production with a single telephone call: "Friendship is friendship, but you better have a waiver anytime you deviate from documentation!"

I often used to recall a thought that Lavochkin expressed once when I met him in

---

3. *Sharashka* and *sharaga* were common slang words used to describe prison work camps for scientists that operated during the Soviet era.

Gonor's office.[4] "It'll take at least two or three years for us to get everything running smoothly." But we didn't have time for that. The R-1 missile series flight-design tests were scheduled to start in September 1948.

Operations on the R-1 had run full speed ahead since late 1947, but the decree for the development of long-range domestic missiles had not come out until 14 April 1948. The higher offices of government and Special Committee No. 2 were actively assisting us and our contractors to expand cooperation. This required process restructuring at many enterprises of other ministries. We were also aided by the spirit we still retained from wartime: "If the Motherland is in need, look for a solution, instead of an excuse if you fail."

Just to provide the entire gamut of new materials, the decree signed by Stalin stipulated that the following organizations be brought in for our projects: Central Scientific-Research Institute of Ferrous Metals (TsNIIChermet), the Academy of Sciences' Metallurgy Institute, the Scientific-Research Institute of Rubber Industry, the All-Union Institute of Aviation Materials, the Academy of Sciences' Institute of Physical Chemistry, the Central Institute of Aviation Fuels and Lubricants, the Serp i Molot (Hammer and Sickle) and Elektrostal factories, the Stupinskiy Light Alloy Industrial Complex, the Leningrad Rubber Technology Institute, and others.[5]

The decree made it incumbent upon the Ministry of Armaments to start building a rig to perform integrated firing tests on the missiles. In 1948, construction began at a very picturesque site fifteen kilometers north of Zagorsk. The rig was erected in a forest next to a deep ravine, which was to receive the rush of the engines' fiery jets.

This new facility under the codename Novostroyka (New Construction) was first declared the NII-88 institute's Branch No. 2, but then it attained its sovereignty and became the independent NII-229.[6] Nevertheless, this facility for missile firing rig testing continued to be called Novostroyka for thirty years. For a long time Gleb Tabakov was in charge of the facility. He was subsequently one of the deputies of the missile industry ministry.[7]

WHILE TESTING GERMAN MISSILES, INCLUDING THE A4, IN 1947, we found certain defects that could not be ignored, problems that led us to modify the A4 somewhat. In the design of the missile body, the tail and instrument compartments were

---

4. Semyon Aleksandrovich Lavochkin (1900–60) served as Chief Designer of OKB-301 in Khimki in 1945–60, during which time he directed the development of jet aircraft, supersonic cruise missiles, antiaircraft missiles, and drones.

5. TsNIIChermet—*Tsentralnyy nauchno-issledovatelskiy institut chernoy metallurgiy.*

6. On 14 August 1956, NII-88's Branch No. 2 separated to become the independent NII-229. The organization is known today as the Scientific-Research Institute of Machine Building (NII Khimmash)

7. Gleb Mikhailovich Tabakov (1921–95) served as director of NII-229 in 1958–63 and then as deputy minister of the Ministry of General Machine Building (MOM) in 1965–81.

reinforced. Hatches were incorporated into the tail compartment, enabling the control surface actuators to be changed out without removing the entire compartment. The nominal design range was increased from 250 to 270 kilometers. This required increasing the amount of alcohol fuel by 215 kilograms and making the appropriate ballistic recalculations, which were drawn up in the form of a range table. The work of the *Sparkasse* (Savings Bank) group at the Institute Nordhausen, which included Tyulin, Lavrov, Appazov, and German specialists, was used as the basis for the range tables.

The nose sections of the missiles in the first series were filled not with explosives, but with ballast, and were provided with an ampoule containing a smoke mixture, which facilitated the search in the area where it came down. As with the A4, the instrument compartment was located behind the nose section. It contained all the primary motion control hardware, which was now strictly domestically produced.

Three command gyroscopes controlled autonomous flight control: the GG-1 horizon gyro, the GV-1 vertical gyro, and the IG-1 longitudinal acceleration integrator. These instruments were substantially improved at the NII-10 institute after Viktor Kuznetsov and Zinoviy Tsetsior did a detailed study of the German models' defects. In particular, rather than pulses with a frequency of 45 Hz passing from the vibrator—which was inconsistent in its performance—to the horizon gyro program mechanism, they passed from a special collector mounted on a motor generator.

Other instruments installed in the instrument compartment had been developed with small changes by the staffs of Ryazanskiy and Pilyugin at NII-885. Based on A4 experience, filters were inserted in the layout and design of the *mischgerät* (amplifier-converter) for all three missile stabilization control channels. In 1947, Ustinov thanked Drs. Hoch and Magnus for introducing these filters. At the time, neither we nor the Germans knew that Ustinov had paid them a monetary reward in the amount of three months' pay with Stalin's consent.

The missile's general electrical system was identical to that of the A4 both in terms of the operating logic and the number elements and their function. The system's entire relay-control segment was concentrated in the main distributor. Time commands were issued by the "program current distributor" (PTR). That's the term we used to replace *Zeitschaltwerk*, the German phrase for timer. Lead batteries developed by Nikolay Lidorenko, and Andronik Iosifyan's motor generators supported the power supply system.

Instead of the four-channel *Messina-1* telemetry system, our domestic eight-channel *Brazilionit* system was installed in the instrument compartment. Degtyarenko, who was assigned this work back at the Institute RABE, developed this system at NII-20. Yevgeniy Boguslavskiy developed the fundamentally new *Don* telemetry system at NII-885. However, despite the developers' enthusiasm, this system was not ready to support missile flight tests in 1948. We used it only for the second series of R-1 missiles in 1949.

The alcohol and oxygen tanks were welded using an aluminum-magnesium alloy. The aviation industry supplied the material for the tanks. The welding pro-

125

cess, which was new for Factory No. 88, was mastered under the supervision of Leonid Mordvintsev. This was one of the key production problems for mastering missile production at Podlipki.

The R-1 propulsion system was developed under Glushko's supervision at OKB-456 in Khimki. They assigned it the designation RD-100. The greatest headaches during its reproduction and optimization had to do with the selection of nonmetallic materials for seals, various rubber and metallic parts, and problems with leaks at all the pneumatic and hydraulic interfaces.

As a rule, ignition was accompanied by a violent pop. Sometimes the engine never started. This defect was a problem that the engine specialists who had developed the engines worked on for a long time. The engines' components did not have an auto-ignition feature. Apropos that, one day when we were talking about our troubles, Isayev confessed that he had made a vow to develop engines with only auto-ignition components so that he wouldn't have to use the "horns and hooves" method that we had invented in Khimki in 1943, and he wouldn't have to live in perpetual fear of the ignition problem.

DURING THE FIRST DAYS OF SEPTEMBER 1948, the R-1 missiles designated for flight tests arrived at the State Central Test Range. The missiles to be tested were shipped in special enclosed freight cars under intensified guard in advance, so that the first ones had already been unloaded by the time we arrived. With NII-88 Director Gonor in charge, we followed the missiles out to Kapustin Yar to the State Central Test Range to take our first exam. Our work on the manufacture of a domestic long-range ballistic missile, the R-1, was about to be tested.

In today's terminology, missiles with a range up to 1,500 kilometers are classified as short and medium range. But at that time, 300 kilometers was a great distance. After all, we were only just developing the R-2 with a range of 600 kilometers, and the R-3 project with a range of 3,000 kilometers was in the distant future.

The launches were supposed to confirm that the R-1 was at least not inferior to the A4. We rode out on our special train, in the sleeping cars that would be our living quarters at the firing range. Vasiliy Ivanovich Voznyuk had not yet managed to build hotels.[8] There had been too many problems to take care of getting the firing range ready. The launch pad had been moved a little farther from the engineering facility and a thick concrete bunker built for launch control. Alongside the "steppe asphalt" road, a good concrete road was put down.

Construction workers also built sheds to house the three cinetheodolite stations. The launch site had a well-equipped shed that served as the facility for the Academy of Sciences' Physics Institute (FIAN), where the physicists who studied the intensity of cosmic rays during launches huddled together. This team of scientists included

---

8. Vasiliy Ivanovich Voznyuk (1907–76) served as Commander of the Kapustin Yar range in 1946–73.

two future academicians, Sergey Nikolayevich Vernov and Aleksandr Yevgeniyevich Chudakov. This was one of our country's first scientific space teams, which was just as interested as we were in penetrating into outer space. FIAN hardware was to be installed on at least two missiles.

The tests on the first domestic missile series were called factory tests. Ustinov approved the 1948 test program, and it was concurred with GAU. Vetoshkin was the State Commission chairman, his deputy was General Sokolov, and Voznyuk, Gonor, Korolev, Tretyakov, Yeremeyev, Vladimirskiy, and Muravyev were appointed commission members. Chief Designers Korolev, Glushko, Barmin, Ryazanskiy, Kuznetsov, Pilyugin, Likhnitskiy, and Degtyarenko were responsible for the technical management of the tests.

In all, twelve missiles were shipped for the tests. Of these, ten were equipped with the new domestic *Brazilionit* telemetry system instead of the German *Messina* system. Having retained the frequency multiplexing principles of the channels, the developers doubled the carrying capacity. It became possible to receive twice as much data as on the A4. The entire radio monitoring system of the missiles' flight and behavior was considerably enhanced. The number of radar stations had been increased, and their personnel had undergone preliminary training.

For the first time, the *Indikator-D* radar system responder was installed on several missiles. The *Indikator-D* was developed at OKB MEI by Kotelnikov as a result of our meeting with Valeriya Golubtsova.

We set aside a special railroad car for telemetry data processing. There, the members of the first serious measurement service, consisting almost entirely of young specialists who had graduated in 1946–47 from institutions of higher learning and were immediately "thrown into battle," began their careers. They all proved to be enthusiasts and soon held leading positions and commanded respect. Among them I need to name Nikolay Zhukov, Vadim Chernov, Arkadiy Ostashev, and Olga Nevskaya. Major Kerim Aliyevich Kerimov commanded the telemetry receiving-recording station. He would go on to become the permanent chairman of the State Commissions on manned launches.[9]

Radar tracking stations were manned by a staff military contingent. But NII-88 Department U radio engineers coordinated their work, developed the observation schedules, and processed the results. Nadezhda Shcherbakova and NII-4 radio engineer Grigoriy Levin supervised the operations.

The technical management included a group of ballistics specialists. These were men who would later become famous Soviet scientists and leaders in the space industry: Yuriy Aleksandrovich Mozzhorin, who in 1961 became director of the

---

9. Kerim Aliyevich Kerimov (1917–2003) served as the First Deputy Director of NII-88 (TsNIIMash) in 1974–91. From 1966 to 1991, he also served as Chairman of the State Commission for piloted space vehicles, that is, he oversaw flight operations for almost all Soviet human space missions including Soyuz, Salyut, and Mir.

main institute of the Ministry of General Machine Building; Svyatoslav Sergeyevich Lavrov, who in 1968 became a corresponding member of the USSR Academy of Sciences and in 1980, director of the Astronomical Institute; and Refat Appazov, a leading ballistics specialist in Korolev's OKB. Practically the entire ballistics *Kompashka*, as we called it, under Georgiy Tyulin's management, had worked well together at the *Sparkasse* of Institute Nordhausen back in Germany.

In early September, the Volga steppes, which had been scorched over the summer, were once again covered with vegetation. Ground squirrels scampered playfully over the roads. Steppe eagles stood watch on the telephone poles and transmission towers. Their life in the wild was continuously in danger. The ground squirrels, the eagles' main prey, were poisoned because it was believed that they were plague carriers. The missile specialists who now occupied this area had a penchant for the great eagle wings, a unique souvenir of the steppes, and this also became a reason for the extermination of those remarkable birds. People, not missiles, were destroying the unique animal world of that region.

In 1947, the electric firing, fueling, and other departments were staffed primarily with civilian specialists who had undergone training at the Institutes RABE and Nordhausen. German specialists attended the launches as consultants and prompters, but by 1948, there was not a single German specialist at the test range.

The launch crew was staffed with special purpose brigade (BON) officers and soldiers under the command of General Aleksandr Fedorovich Tveretskiy. An industrial specialist was paired up with each member of the military detail as a monitor. Despite the industrial specialists' obvious technical precedence, they quickly sorted things out with the military men and their joint work proceeded very amicably.

The military officers—launch team chief Major Yakov Isayevich Tregub, electric firing department chief Captain Nikolay Nikolayevich Smirnitskiy, his deputy Captain Viktor Ivanovich Menshikov, and stand-alone testing department chief Major Boris Alekseyevich Komissarov—all advanced to the posts of high-ranking generals, but they maintained their friendship with their missile launching comrades from the late 1940s. During those years, that is, during the period of operations at the Kapustin Yar test range in 1947–53, we all worked as a team.

Here, I'd like to give my due to Colonel (later General) Mrykin. In charge of the GAU missile directorate, he took upon himself the primary work of formulating the military's technical policies.[10] In the role of a strict and exacting taskmaster, Mrykin demonstrated exceptional steadfastness in dealing with Korolev and other chief designers, who were striving to rid themselves of the R-1 somewhat quicker and switch to forward-looking tasks. As a chief they thought he was harsh and too demanding. I have already mentioned that.

Mrykin's subordinates were somewhat fearful of him, but they respected him.

---

10. Mrykin's official title between 1946 and 1953 was First Deputy Commander of the 4th Directorate of the Main Artillery Directorate (GAU).

Industrial workers had mixed feelings. The chief designers who comprised the illustrious Council were obviously not overly fond of Mrykin because every time he brought an issue to them, they had to either respond to all of his demands or find some rational reason for declining. The chiefs' deputies and all the lower-ranking supervisors respected Mrykin. They saw and understood that his negative comments regarding technical imperfection, errors in calculations, or the need for solutions based on the analysis of the results from failed launches were essentially correct and required implementation.

Mrykin was not a military careerist. It wasn't easy for him to develop relationships with high-ranking leadership precisely because, being very devoted to his work and firm in his convictions and in the validity of his case, he fearlessly entered conflicts from which he did not always emerge the victor. His work had a great influence on raising the operational characteristics of all the long-range missiles of the first decade.

The unflinching natures of Korolev and Mrykin often clashed, and higher-ups had to resolve conflicts between them. The complexity of the relationships of these two men devoted to their work affected those surrounding them. More than once Korolev and Pilyugin reproached me for having good relations with Mrykin and making concessions to him in various joint documents. In 1965, Mrykin retired with the rank of lieutenant general. He went to work at TsNIIMash as deputy director of the institute and began to study problems of reliability, and, as a hobby, he collected and worked with materials on the history of aerospace technology.[11]

THE FLIGHT TESTS OF THE FIRST SERIES OF R-1 MISSILES HAD A TRAGIC BEGINNING. I must make a confession to the readers of the first volume of the Russian edition of *Rockets and People*.[12] In chapter 4, "Getting Started on Our Home Turf," on page 191, I write about the tragic death of one of the finest BON officers, Captain Kiselev. Over the eight years since the first volume was released, not a single reader pointed out my mistake to me. Captain Kiselev's death occurred not in October 1947, during preparation for the launch of the first missile, but during preparation for the launch of the first R-1 missile on 13 September 1948!

One of the negative comments of the military testers based on the experience of preparing the A4 missiles at the launch site was the inconvenient access to the instrument compartment, which was located directly under the warhead. Responding to the testers' wishes, Korolev had commanded his designers to develop a "cradle," a service platform that would be suspended from the nose section. After the missile was erected on the pad, the operator would be able to climb into this cradle from the upper part of the erector. When he had completed his work, the operator would

11. Mrykin served as First Deputy Director of NII-88 (TsNIIMash) between 1965 and 1972. He died in October 1972.
12. B. Ye. Chertok, *Rakety i lyudi* [Rockets and People] (Moscow: Mashinostroyeniye, 1994).

climb back onto the erector, open the locks of the metal belt secured around the warhead, and drag it onto the erector to descend to the ground.

The first missile was erected on the launch pad with the cradle placed on the nose section. Before work started, launch crew department chief Kiselev decided to personally check out the possibility of using the cradle for work in the instrument compartment. Standing in the cradle, without any safety precautions, he began to jump in the cradle to test the reliability of its fastenings. The catch clips did not withstand the dynamic load. The cradle holding Kiselev broke off and came crashing down from a height of twelve meters onto the concrete launch pad.

This happened right before the eyes of almost the entire launch crew. The ambulance on duty at the launch site delivered Kiselev to the test range military hospital. The hospital's chief surgeon reported to the State commission that the trauma that the captain experienced was fatal. Use of the cradle during missile tests was forbidden. All of this I have dragged out of my memory almost fifty-five years after the fact.

Everything that happened on that sunny September morning at the Kapustin Yar launch site I have tried to recreate in my memory by reading Korolev's letters published in the book ...*It Was a Time that Needed Korolev*.[13] The author who compiled that book, Larisa Aleksandrovna Filina, director of the Korolev Memorial House-Museum, has published the most vivid excerpts from Korolev's letters from the test range to his wife, Nina Ivanovna.

9–10 September 1948

"Our workday begins at six o'clock in the morning local time (5 o'clock Moscow time) and continues until late into the night.

Of course, I am very tired, being unaccustomed to it, but I think that I'll soon get into the routine. Meanwhile everything is going rather well."

30 September 1948

"On 13 September our dear friend and comrade-in-arms Pavel Yefimovich Kiselev died tragically. He was one of our main testers. The accident happened on the 13th, and he died without regaining consciousness on the 14th at 2:00 p.m.

A terrible chain of events led to his death. His personal courage and ardent love for the job entrusted to him tragically pushed him towards death. But given all of that, we designers and I, as their chief supervisor, bear a heavy responsibility for this incident. Officially they are saying that he is to blame, but I personally am taking this hard and cannot forgive myself that perhaps I might have overlooked something, and in any event, I should have watched more attentively. During the 24 hours that he was battling against death, all of us here lived with only one hope, that he would stay alive. On the

---

13. L. A. Filina, ed., "... *Byl veku nuzhen Korolev": Po stranitsam arkhiva Memorialnogo doma-muzeya akademika S. P. Koroleva* ["...It Was a Time that Needed Korolev": From the Pages of the Archives of the Memorial House-Museum of Academician S. P. Korolev] (Moscow: MDMA Koroleva, 2002).

14th at noon I couldn't help it, I fell asleep, and suddenly some force threw me from the sofa. I jumped up—it was exactly 2:00 p.m. At that moment he died ...

It seems so long since I had hoped for something so much, since I had prayed so hard to the powers that be or to some powerful God that he would stay alive—and now it's all over for once and for all!

This situation is all the more tragic because five days later his daughter was born.

We buried our dear Pavel and the following day we returned once again to our work; the work for which he gave up his life."

12 October 1948

"The distant and dear-to-my-heart hours and happy minutes of our life...

Now, here in this windy, sandy wilderness, during my few minutes of rest and calm I recall them. It is 1 a.m. Reveille is at 3:30. This is my second night without sleep, but I wanted to send you these few words of love and greeting...

I want you to be strong and steadfast in life. Know that I, too, am trying to be the same, and above all, for you.

No matter how hard our separation, it is necessary for the sake of this important work for our Fatherland, and we must be steadfast."

In 1948, Pavel Tsybin was the chief of testing at Department No. 12 at NII-88. He had transferred to us from the aviation industry. His deputy was Leonid Voskresenskiy. Both of them were present at the launch when Kiselev tested the ill-fated "cradle." After Kiselev's burial and the customary ceremonial funeral banquet, I listened to Voskresenskiy's ruminations; he had clearly been shaken by what had happened. He was tearing himself up that he had not demanded any documents from the designers clearing this cradle for operation, and there had been no preliminary factory tests whatsoever.

THE TESTS OF THE FIRST SERIES OF R-1 MISSILES SHOWED THAT WE HAD ADDED OUR OWN DEFECTS TO THOSE OF THE A4. Whereas the A4 had endured in-flight accidents, the R-1 stubbornly refused to lose contact with the launch pad. It took twenty-one instances of the engine failing to go into main stage ignition to get nine missile launches. To a certain extent, these failures came as a surprise to us. We had not expected the R-1 to be so reluctant to fly. The cause turned out to be powerful pops from micro-explosions of fuel entering the combustion chamber after the "ignition" command. The pyrotechnic igniter, located on a special device made of wooden strips in the combustion chamber, was supposed to burn a portion of the alcohol mixed with liquid oxygen vapor. After this, upon the issuance of the "preliminary" command, a significantly greater quantity of fuel was fed to the fire that had formed, and for seconds the steady, roaring jet of the preliminary stage was formed. Then, the "main" command occurred. The main oxygen and alcohol valves opened at full flow. The main stage jet was generated with the characteristic roar, thrust increased, and the missile took off from the pad. But this did not happen during the first attempt to launch the first R-1 missile.

After the "ignition" command, a powerful "pop" sound could be heard, a noise a lot louder than a rifle shot. The shock effect caused the message lights of the selected commands on the launch console to begin blinking and the circuit to reset. Transition to the intermediate stage was prevented, and power was cut off from the electric motors of the control surface actuators. For another launch attempt they had to reset the circuit to the initial state, cut off power to the onboard instruments, and replace the igniter, which required climbing almost inside the engine's already "wet" nozzle. These operations, including all the discussions and arguments, took one or two hours.

During the second launch attempt there was once again a powerful pop accompanied by the consequent resetting of the circuit. A significant evaporation of oxygen accompanied the launch delay. Ground support needed to bring up the oxygen filler again and top off the tank with oxygen. They preferred to drain the oxygen back into the filler and investigate the causes of the launch failure. After draining, the missile needed to be dried, for which they decided to bring in aircraft air heaters. They decided to drain both the alcohol and the hydrogen peroxide in order to completely repeat the electrical tests on a dry missile and find the cause. This took three days. They never did find the exact cause. Everything seemed to be working properly.

The ground crews fueled it up and began a third launch attempt. Using the idea of one of Pilyugin's testers, former seaman Nikolay Lakuzo, they decided to take the extra precaution of manually ensuring the launch's reliability. Even if a pop occurred, they would not allow the system to reset. To do this, Lakuzo crawled behind the launch console in the bunker, removed the back panel, and at the required moment, he manually retracted the armature of those relays that popped loose during the pop. That way, the system had to continue the automatic process of proceeding to the "main" stage.

This forced start mode really did make it possible to proceed to the main stage mode. But, evidently still reacting to the assault on its electrical system, and having reluctantly taken off, the missile immediately leaned forward and went into horizontal flight. All the observers jumped into the open trenches that had been dug beforehand. After flying approximately ten kilometers with its engine running, the missile went into a dive and crashed into the ground. But that's not all. Not just the missile, but also the heavy launch pad flew off and was hurled twenty meters from the launch site, and all that was left there was fused or swept away by the force of the fiery squall. Examining the mutilated launch pad, Glushko noted sarcastically, "I didn't think that my engine could make launch pads fly too."

All night we analyzed the system and finally realized that when Lakuzo was pulling in the relay armature, he didn't pull in the armature of the control surface actuators' power supply. The missile had flown without control surface actuators, that is, as an unguided projectile.

Forty-four years later, Dr. Vadim Chernov, who in 1948 had been at the test range as an MAI student, told me his version of the crash of the first Soviet R-1 long-range guided missile.

From the author's archives.

Missile launch control members during a break in the winter of 1948 while launching the R-1 ballistic missile. From left to right are B. Ye. Chertok, N. A. Pilyugin, L. A. Voskresenskiy, and N. N. Smirnitskiy.

"I was responsible for the first crash," declared Chernov. "At the launch site, Korolev saw me, called me over to the launch pad, and explained, 'This missile is Soviet, but the launch pad is still German. Do you see the onboard skid contact? It starts the timer at the moment of launch. Its rod rests in a corresponding niche on the launch pad. The pad needs to be fixed so that everything will be ready by morning.'" Chernov was devising and designing all evening. He woke up the metalworkers in the middle of the night and by morning in the workshop on the special train they had produced his version of the skid contact stop, or more correctly speaking, the liftoff contact. According to Chernov's version of the story, his student design did not withstand the powerful pop, and the contact broke after the "ignition" command rather than *after* the missile lifted off from the launch pad. The horizon gyro timer started ahead of time; a pitch command was sent to the control surfaces, tilting the missile immediately while it was still on the pad. As the missile was leaving the pad, the plume was pointed, not vertically, but at an angle, and it hurled the pad off into the steppe.

After this incident, Korolev instructed Chernov to calculate what gas dynamic forces were affecting the launch pad to the extent that it could fly so far. This was the first scientific-research project of the MAI professor, prominent specialist in the field of missile instrumentation, and member of the Russian Engineering Academy and Russian Academy of Cosmonautics.

The second missile proved to be even more obstinate. To begin with, the ground crews eliminated all the defects in the ground-based cable network. Next, during two launch attempts the engine did not start, despite the fact that the system did not reset. After long experiments on a missile standing on the pad, they discovered that the main oxygen valve had frozen. During subsequent attempts, resets were sometimes accompanied by the intervention of firefighters. Puddles of propellant components were burning under the launch pad.

Eventually they removed the oxygen valve from one of the missiles and checked its ability to freeze. They determined that the cause of the failure was the stiffening of the abundant amount of oil in its bellows assembly. The missile tests were discontinued. The main oxygen valves were removed from all the missiles and sent to the factory in Khimki for degreasing. This was a powerful blow to engine designer Glushko's self-esteem. Up until that point he had been bad mouthing "the resetting circuits of those guidance specialists and electricians."

It wasn't until 10 October that the missile reached the target area. But three days later, after three attempts, the next missile once again remained on the launch pad. In terms of psychological effect, the disruptions caused by the freezing of the oxygen valves surpassed the pops that had accompanied the launch attempts.

Soon after the flight tests had begun, Ustinov, Artillery Marshal Voronov, and former State Commission chairman Artillery Marshal Yakovlev flew in, not only as observers and enthusiasts but also as supervisors and threatening bosses. Their appearance coincided with the beginning of a whole series of failures and disappointments, and left all the testing participants in a state of constant stress. The high-ranking leaders had been fully convinced that we had not only studied and reproduced German technology, but had substantially increased the missiles' reliability. And now suddenly they discovered that the missiles, for various reasons, simply refused to fly.

According to the established traditions, we all were due strict reprimands. Supposedly this would be beneficial. In the conference car of the special train a State Commission meeting convened, including the chief designers and leading specialists. Glushko's deputy Dominik Sevruk reported on the causes of the pops. He just managed to explain the cause, but all he could offer in the way of solutions was: "Let the guidance specialists figure out why their system resets. Pops during launch are inevitable."

Pilyugin was offended and tried to argue that if you "hit all the relays with a sledge-hammer, you can't help but disturb the contacts and that's what causes the system to reset. The German missiles did not have this backfiring problem."

During the meeting, I was sitting in the far corner of the car between Smirnitskiy and Tregub. Before this we had rejected the idea of the relay contacts in the main distributor being disturbed. The main distributor was located far from the engine and the missile's entire structure should have damped the pop. I assumed that during powerful pops, the contacts were disturbed between the multiple-strand ground cable and the flange in the skid connector designed by the Prozhektor Factory.

I liked this idea so much that, despite the threatening mood of the visiting brass, I smiled and started to whisper this idea to Smirnitskiy. Infatuated with the hypothesis, I didn't notice that the meeting's arguing had died down. A strong shove in my side from Tregub stopped me. Ustinov's menacing, mocking voice rang through the silence. Turning to Voronov he said, "Nikolay Nikolayevich, have a look at Chertok. We've all been sitting here for days and they can't explain to us why the missiles won't leave the pad. We have to report to Iosif Vissarionovich [Stalin] that we have mastered production of the missiles, but it turns out, they refuse to fly. And through it all, Chertok is still smiling."

I immediately stopped smiling. But now, looking in my direction, Voronov smiled and kindly said, "So let's hear Chertok's explanation as to why the missiles would fly for the Germans, but not for us."

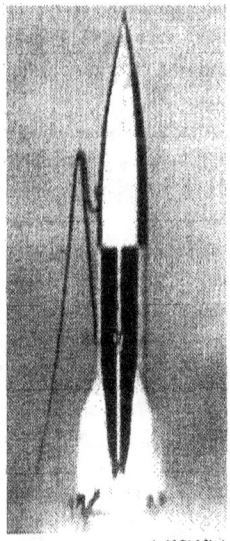

Despite the cramped quarters, I stood up and reported that as yet not everything was clear, but that tomorrow we would conduct an oscillographic analysis of the contacts' behavior, which would enable us to understand and correct whatever was causing the system to reset when powerful pops occurred. After the meeting my comrades attacked me. "What oscillograph? Where? What have you promised without consulting with anyone? Start drying bread crusts! Serov's agents have already memorized your smile."[14] Together with Boguslavskiy, whose artistic abilities for using an electronic oscillograph I had admired back at the Institute RABE, we worked out a system for monitoring the operation of the skid plug contacts. Right away, helpers and fans of the idea appeared out of nowhere. During the next pop we actually saw the blips of the tripping contacts on the oscillograph

Asif Siddiqi.

The R-1, shown here just after launch, was the Soviet version of the German A4 (V-2) missile. Its reproduction allowed Soviet industry to master a huge array of industrial practices necessary to develop much more complex missiles. The missile was eventually deployed for battle operations in 1950.

screen, which explained the system's reset logic. Goltsman, the chief designer of the ill-fated plug, devised an external spring that increasing the contact's reliability. The missiles began to fly!

I received no gratitude for the idea. But Voskresenskiy kept his head, and after the next successful launch, when the necessary quorum had gathered in one of the compartments of the special train, he explained, "Everyone drinks his own booze, but for *hors d'oeuvres* we'll have Chertok's dried bread crusts." We drank to

---

14. The comment about bread crusts was a typical contemporary euphemism to warn about the threat of arrest and imprisonment.

our success too soon. In addition to the skid connector, we also needed to increase the "pop resistance" of the skid contact. Chernov's work on that problem had proved insufficient.

Added to all our troubles was another incident that finally drove the visiting brass to distraction. The next missile launch scheduled for 1 November was postponed due to severe fog. During the night, the sentry guarding the launch site showed exceptional vigilance and for some unknown reason shouted, "Stop! Who goes there?" No response came out of the fog and he fired a warning shot. The guard raised by the alarm found nothing suspicious in the surrounding area.

Arriving at the site the next morning, the launch team immediately smelled the strong scent of alcohol. An inspection showed that the shot the night before had not been fired into the air, but rather into the filled alcohol tank. The missile's entire tail section was drenched with alcohol from the bullet hole. They removed the missile and shipped it to the factory in Podlipki for restoration and sent the sentry to the brig. Voznyuk was advised of the guards' utterly unsatisfactory training.

Factory tests of the R-1 at the firing range had begun during the marvelous days of September. They finished in cold and rainy November. Of the twelve missiles, nine were launched. Despite the very discouraging results of the flight tests, the findings of the State Commission were very generous:

"The first series of R-1 domestic missiles in terms of their flight characteristics, as demonstrated by the flight tests, were not inferior to the captured A4 missiles. Fundamental issues during the reproduction of R-1 missiles from domestic materials were correctly resolved ... The flight characteristics of the first series of R-1 missiles conform to the characteristics specified by the tactical and technical requirements, with the exception of range scatter."

Essentially, such an assessment was necessary to overcome the skeptical and even hostile or negative attitude toward missile armaments on the part of many military chiefs, who had gone through the war and emerged victorious using conventional armaments.

This calls to mind a statement by one of the combat generals who was invited to the firing range for familiarization with missile technology. After a modest banquet arranged in the special train in honor of completing testing on the first series, and being somewhat mellow from drinking our traditional "Blue Danube," that is, rocket fuel tinted with manganese crystals, turning to Pilyugin, Kuznetsov, and me, he said confidentially so that the marshals sitting nearby could not hear: "What are you doing? You pour over four metric tons of alcohol into a missile. And if you were to give that alcohol to my division, they could take any city easily. And your missile wouldn't even hit that city! Who needs it?"

We, of course, started to defend ourselves, and to argue that the first airplanes were not perfect either. But he proved to be not such a simpleton and crushed us with this simple argument: "The Germans manufactured and released thousands of missiles. But who could tell? In Berlin, I met both Brits and Americans. They told me frankly that they did not suffer any particular loss from the missiles. So they only affected morale. And the troops had no idea that the Germans had such a secret weapon. But what if, instead of thousands of their V-missiles, the Germans had hurled thousands of tanks or aircraft at the front! Now that's something we would have felt!"

I don't remember this general's surname. His tunic was decorated with an impressive abundance of sparkling medals. Having overheard our conversation, Vetoshkin smiling cunningly, proposed a toast, addressing it more to that general than to the rest of the gathering: "Don't look at what's on your chest, look at what's ahead!"

I must give credit to the sense of the future and the courage of the high-ranking military leaders Voronov, Nedelin, Yakovlev, and Marshal Zhukov himself, who at that time was deputy defense minister. In spite of all their accomplishments and highest authority, they risked more than we did. Ultimately, we were legally in the clear: there was a decree signed by Stalin that each of us was obligated to fulfill. But as regards this or that marshal, that same Beriya during his next meeting with Stalin could say that so-and-so supports missile projects demanding enormous resources and the ineffectiveness of such projects has been proven and was obvious already at the end of the war. And with that, the marshal's career, and perhaps his freedom, would have come to an end. After all, it was in 1952 that the highly upstanding Marshal Yakovlev was arrested on a much less serious charge!

Of course, Minister Ustinov, Vetoshkin, other ministers and the directors of *Gosplan* and Special Committee No. 2 were also taking a risk gambling on our obsession. By late 1948, operations had expanded so broadly that doubts and retreats would have been much more dangerous for everyone than their intense continuation. But we had to keep in mind to ensure a "significant increase in reliability, failure-free operation, and improvement of the operational parameters of all the assemblies and systems comprising the R-1 missile."

This is a citation from the State Commission's resolution. In 1949, we were faced with eliminating the defects discovered in the first series and once again traveled out to the firing range no later than September for joint factory tests on the R-1 second series. Before our departure from the firing range Korolev impressed on all the chiefs and persuaded Vetoshkin that the second series must comprise at least twenty missiles. No one objected to this proposal.

During work on the second series we were all, to a certain extent, freed from the obligations to precisely reproduce German technology. For that reason, a great deal of resources were devoted to experimental work, new ballistic calculations, compiling new range tables, reviewing all the factors determining accuracy, and, finally, developing new monitoring systems and instrumentation.

The year 1949 was also taken up with preparation for the production of the new 600-kilometer range R-2 missile, which was a departure from the German A4. The fabrication of the R-2E experimental missile was already underway at full speed. This missile was supposed to confirm the propriety of the primary design solutions adopted for the R-2. But who would support the prospects if we didn't vindicate ourselves with the new R-1 series?

At one of the unofficial meetings of the technical management upon returning from the firing range, Korolev clearly expressed his belief that the primary work to achieve failure-free launches "on the first attempt" must be conducted at NII-885 and OKB-456. As for NII-88, the main task would be bringing order and professionalism to the factory, increasing the reliability of the control surface actuators (this was addressed to me), and establishing a way to monitor what was going on in Pilyugin and Glushko's organizations.

Upon our return, Gonor very actively set about redesigning the factory and introducing new production processes. Traditionally, the factory had little trouble with machining processes. New processes that required nonferrous castings, a large volume of copper, riveting, and welding work were taken on very reluctantly.

By late 1948, Gonor had strengthened the management of the main shops, and after making arrangements with Lavochkin, sent more than fifty process engineers and factory foremen to Lavochkin's pilot-production factory for training in aircraft production.[15] Their primary objective was to study the processes for bending, forming, and welding aluminum alloys. They created an independent fittings shop with a closed production and testing cycle. Later, this shop became a high-capacity and very state-of-the-art engine fittings production facility.

While working on the guidance systems for the second series of R-1s, my comrades and I needed to concentrate on four primary areas: optimization of airtight (nonleaky) control surface actuators; perfection of factory electrical testing procedures and processes and, correspondingly, test equipment; mastery of the new *Don* telemetry system; and keeping track of what was going on at NII-885.

A serious technical achievement of 1947 was the creation of the new *Don* telemetry system, which was installed on all "series two" missiles instead of the *Brazilionit*. The *Don* was developed by Boguslavsky's small team at NII-885, and a short time later it went into series production.

The increase in the number of parameters measured on each missile, the development of new sensors, and the general electrical circuitry of the telemetry system required an increase in the number of telemetry specialists. The *Don* system ground-based receiving station was equipped with an electronic monitor that enabled real-time observation, satisfying the curiosity of the authorities in the event of an accident, without waiting for the film to be developed and dried. Instead of recording on paper using mirror-galvanometer oscillographs, for the first time measurement results were recorded onto motion-picture film using an electronic oscillograph. All the system ground tests were successful, and Boguslavskiy proposed that we also conduct aircraft tests before the firing range tests. They were conducted at LII. For the first time, the aircraft testers envied the missile specialists when they realized that the system would make it possible to understand flight phenomena, especially in

---

15. This was Factory No. 301 located in Khimki near Moscow.

critical situations, without waiting for the findings of the accident commission.

The year 1949 was the most stressful in terms of the number and variety of missile launches. During April and May, experimental launches of the R-1A were conducted. The primary objective of these launches was to optimize the separation principles of the nose section. But one could also not miss the opportunity for conducting a whole series of experiments necessary for the future during these launches.

The missile's nose section was equipped with a cowling that ensured its static resistance during entry into the atmosphere. A parachute system made it possible to recover the nose section, which held containers of scientific equipment that were designed to study the atmosphere up to an altitude of about 100 kilometers. Four missiles were launched to a range of 210 kilometers, and two were launched to an altitude of 100 kilometers.[16] At the same time, the capability was tested for radar tracking the missile body and the separated nose section separately. During the process of vertical launches, for the first time serious research work was conducted on the passage of SHF and UHF radio waves in the upper atmospheric layers. It turned out that the main interference for reliable radio communications with the missile was not in the ionosphere, but in the engine plume.

During vertical launches, patterns were very clearly recognized. While the engine was operating, information proceeded from the missile with transient errors. As soon as the engine was shut down, reliable communications were established, especially in the UHF band. Boris Konoplev at NII-20 had developed the equipment for experiments in that band. He also developed the radio control system for the future R-3 missile.

Konoplev was the staunchest supporter of combined control systems, that is, the combination of an autonomous inertial radio system and one that corrects its errors. I first became aware of his almost fanatical devotion to radio engineering and his ineradicable desire to solve any radio engineering problem his own new way after making his acquaintance back in 1937, during preparation for the transpolar flights. At that time he was working in the Main Directorate of the North Sea Route (*Glavsevmorput*) and, without having a degree in radio engineering, he was the most authoritative radio specialist. During the war he set up radio communications on the routes of northern sea convoys. Then he took a keen interest in radar. Finally, in 1947, he decided that his place was in missile technology, and he devoted all of his enthusiasm and talent to it.

During tests, radiating optimism, he would report the results of his research on the attenuation of radio waves in the engine's plume and countermeasures against this phenomenon to all whom he considered worthy of introducing to radio engineering. Pavel Tsybin, who at that time was NII-88 testing department chief,

---

16. Six launches of the R-1A took place on 7, 10, 15, 17, 24, and 28 May. The last two were "vertical" launches for reaching altitude rather than downrange.

devoted a witty satirical ode to Konoplev and to the problem of the engine jets' effect on radio waves. It was a sensational success among testers and radio specialists, who considered Konoplev a great radio ham, but a dilettante in radio physics.

In early 1950, Konoplev transferred to NII-885, where he was in charge of the entire field of radio engineering. Air defense problems were the exception. In this case his aspirations did not always match the technical view of Ryazanskiy and Pilyugin. However, these differences did not lead to the antagonism that often occurs in organizations when several outstanding talents are working on similar problems, dividing an entire team into warring camps.

Chapter 8

The R-1 Missile Goes Into Service

The second series of twenty-one R-1 missiles was broken into two batches: ten so-called ranging missiles and ten qualification missiles. One missile was designated for firing rig tests. The grim lessons of the first series were not lost on us. The second series, manufactured and tested at factories using newly revised documentation, showed substantial progress in reliability.

From September through October 1949, all the launches were conducted under incomparably calmer circumstances than in the past. It bears mentioning that the living conditions at the firing range had also been substantially improved. For the first time, we were living in hotels instead of the rather shabby special train. We now dined in normal dining halls rather than tents. All the roads were concrete-paved, and the old firing range song about "dust and fog" was receding farther and farther into the realm of folklore. Finally, missiles at the engineering facility for horizontal tests were given a significantly more comfortable assembly and testing building.

The new order at the firing range included brief periods of rest and relaxation. As a rule, we took advantage of them to go fishing. The Akhtuba River and its myriad tributaries were in the immediate vicinity of Kapustin Yar and our residential area. I don't consider myself much of an angler, but I genuinely enjoyed the fishing parties back then, both the actual fishing, and the subsequent socializing, when the main and only dish was marvelous fish soup cooked on the riverbank.

Flight tests on the second series of R-1 missiles were conducted in September through October 1949. As for the numerical ratings, the results looked pretty good. Of the 20 missiles, 16 landed in the 16- by 8-kilometer rectangle that the technical requirements specified. Only two missiles fell short of the target. One instance was caused by the "popping"; its shock triggered the premature uncaging of the integrator, which generated additional error because of gravitational forces. The other was caused by an error in the integrator tuning. Two missiles experienced mishaps in the launch area: one from prelaunch leaks in the fuel lines, resulting from the violent pops, and the other from an explosion of the oxygen tank during the fueling process, resulting from a faulty pressure release valve.

There was not a single failure during engine startup caused by system reset. Pilyugin and his people were very proud of this, although the pops still took a toll on the testers' nerves. In the first series, six of the ten missiles were removed from the launch pad because of launch failures. In the second series, not a single missile was removed.

After the launches were completed, they set up a review commission, of which I was a member. We worked from dawn to dusk, forcing our typing pool to collapse from exhaustion. They revised and retyped the conclusions, proposals, and findings dozens of times.

Mrykin believed that the missiles' shortcomings were still so serious that it was too soon to launch a large production run of them, much less propose putting them into service. Korolev was extremely disgruntled by this stance. He insisted on roughly the following wording: "begin series production, during which defects identified during flight tests shall be eliminated." These controversies would have to be decided in Moscow at the level of ministers and marshals.

Ustinov, Vetoshkin, Gonor, Korolev, and all of us R-1 developers believed that to give the new technology a sense of worth and to add to the respect of the entire field, the series production documentation needed to begin with the statement "accepted into service." After four years of unyielding work, the failure to hand over for production a missile that the Germans had been producing in series as early as 1944 would have been a blow to our prestige.

In 1949, THE PRIMARY TASKS FOR NII-88 AND ITS PILOT-PRODUCTION FACTORY were the manufacture of a second series of R-1 missiles and its preparation for flight tests, which were scheduled for September through October. The factory was not designed for large-scale series production and was working under tremendous stress. Factory horizontal tests in the control and test station of assembly shop No. 39 were conducted jointly by shop testers and Department U specialists, without whom it would have been impossible to figure out the intricacies of the electrical circuits and to master the new *Don* telemetry system. By brainstorming, we quickly learned to overcome all the difficulties we encountered during tests at the factory. But Pilyugin, our chief launch tester Voskresenskiy, and I were plagued by the memories of the "popping incidents" during the 1948 missile launches. We very much wanted assurance that the modifications to the electrical circuits, the strengthening of the cable connectors' fastenings, and the changes in the ignition devices were sufficient. How could we verify the reliability of all these modifications before flight tests began?

At first we hoped it would be possible to conduct firing tests on the missiles' launch process on the rig under construction near Zagorsk at NII-88 Branch No. 2. But in a meeting with Gonor on the readiness of the first firing test rig we were told that it would not be ready before December. The rig at the firing range was too far away for us to use. It was inconvenient and very expensive. It was unrealistic to ship a special missile and an expedition of hundreds of specialists to the firing range for the sake of two or three "pops."

"I've got it!" announced Leonid Voskresenskiy, who was already recognized by everyone as the highest authority in testing.

"I propose that we conduct experiments using a real missile in the woods near our airfield. We'll set the missile up on the launch pad. We'll fuel it and start the engine until it goes to the preliminary stage. As soon as it goes into the preliminary stage, we'll shut it down. We'll sort out the latest 'pop.' Glushko, Pilyugin, and Chertok, if they are capable, can devise modifications, and we can do a restart. And so we'll start up the engine as many times as we need to. Going to the preliminary stage isn't hazardous stuff."

"We'll install all the electrical starting equipment in the cab of a truck, and we'll ask Barmin and Goltsman to round up their assemblies from their factories."

At first the proposal seemed crazy.

"We're only fifteen kilometers from Moscow. What if the engine accidentally goes to main stage after preliminary? The missile will take off! Where is it going to crash?"

"The engine's not going to go to main stage! We'll switch off the circuits of the main stage valves, and that will be a full guarantee of safety."

Back in Bleicherode, we counted Voskresenskiy among the ranks of the "hussars."[1] This proposal underscored his hussar nature.

From the author's archives.

Leonid Voskresenskiy (1913–65) was one of Sergey Korolev's most well-liked and respected deputies, responsible for all field operations and flight-testing at the launch range. A man who had little time for the formalities of official hierarchies, he had a well-deserved reputation as a seat-of-the-pants adventurer.

Gonor placed Voskresenskiy in charge of the experiments. Everything was ready in a week. Before the standard experiments began, Voskresenskiy decided to conduct a general rehearsal on a quiet Sunday in April, so that all the high-ranking guests could be invited to the missile launches beginning Monday on the outskirts of Moscow.

The missile was erected in a forest glade on approximately the same site where the main building of world-famous TsNIIMash is located today. After a general rehearsal for the fueling of all the propellant components, right down to the hydrogen peroxide and sodium permanganate, having made sure that everything was going normally, the testers cast their fates to the wind and decided to take the rehearsal up to the point of the first ignition and first "pop."

On the first attempt the engine achieved nominal thrust without a pop. But

---

1. "Hussars" were members of various European light-cavalry units for advance scouting, modeled on the 15th century Hungarian light-horse corps. Here, the word is used in the Russian stereotype of hussars as good-natured, courageous, and "hard living."

despite all the commands from the control panel, the engine did not want to shut down and roared into the preliminary stage. The fiery squall thundering against the conical splitter of the launch pad spread out over the ground, igniting the grass, the bushes, and various and sundry trash. All attempts to shut down the engine failed. The launch pad and missile tail section were in danger of overheating. If the white-hot launch pad did not bear up, the missile would fall and an explosion would follow, shattering the windows in the nearby factory settlement of Finnish-style log homes.

Voskresenskiy asked the fire-fighting crew that rushed to the fire to direct the streams of water at the launch pad to cool it off as much as possible. The supply of propellant components kept the engine running continuously in preliminary mode for ten minutes. The propellant components were finally used up, but the launch pad and missile tail section were still steaming for a long time from the drenching with water.

On that sunny Sunday I was looking forward to strolling around Sokolniki with my sons. But a call from the NII-88 duty officer cut short my plans: "Lev Robertov-ich [Gonor] ordered that I find you immediately and bring you to the launch site. I have already sent a car." When I arrived, steam was still rising from the overheated launch pad and missile tail section.

Surrounded by individuals unknown to me, Gonor was explaining something to Deputy Minister Ivan Zubovich. Off to the side stood the small group of those responsible. Judging by their appearance, they had really "gotten theirs" and were waiting for further instructions. After the higher-ups had gone their separate ways, Gonor calmly explained to all of us that a state security representative had asked for a brief summary of why all of this was necessary and what had been the probability of the missile actually taking off.

"Chertok and Voskresenskiy shall write this at once. And tomorrow I will send Zubovich a copy of my order with severe reprimands for the gross violation of safety procedures, for unauthorized deviation from the program of experiments, and for lack of a backup means for shutting down the engine. But this is a precautionary measure. Ivan Gerasimovich [Zubovich] will try to keep us from being hounded by investigators." And so the attempt to fire up the R-1 missile engine at Podlipki ended in disgrace.

Novostroyka, NII-88's Branch No. 2 in the vicinity of Zagorsk, had been con-ceived to facilitate the integrated ground testing and verification of a missile with an operating engine. Firing tests on rig No. 1 at Novostroyka began after the tests on the second R-1 series in December 1949.

I wouldn't even have remembered them if it hadn't been for the 50th anniversary of NIIKhimmash. At the ceremonial anniversary meeting, NIIKhimmash direc-tor Aleksandr Makarov gave me a precious gift. It was a photocopy of the original report on the first firing tests on rig No. 1, dated 18 December 1949, article 1R No. 24, with propulsion system RD-100.

At the time, factory Director Makarov announced that this document was historical because it had been presented to the Moscow regional administration as material evidence allowing the anniversary events to be conducted. Moreover, over the fifty years of its activity, the former Novostroyka had been transformed from a small settlement into a flourishing modern scientific city. And it deserved its own name and location on a map. It had been proposed that it be given the name "Peresvet," the symbol of the ancient warrior hero, whom St. Sergius of Radonezh blessed for battle with the Tatars.[2]

The report, from which the secret stamp had been removed so that it could circulate freely among the administrative offices of the Moscow regional governor, was bound in a beautiful gold-embossed hard cover. The functionaries had not noticed a historical discrepancy: the gold embossing on the cover said "NIIKhimmash 1949." But in 1949 there was only an NII-88 branch, but no NIIKhimmash.[3] But let's not split hairs! For me the important thing was that I had received the right to refer in my memoirs to my direct involvement in the birth of a new city in the Moscow region.

The report on the first firing test on rig No. 1 of the NII-88 branch, now the city of Peresvet, had only three signatures: those of chief of Department No. 12 flight-test station No.2 (LIS-2) Voskresenskiy; chief of NII-88 Department No. 16 Chertok; and NII-88 Branch No. 2 chief engineer Tabakov.

THE CONSTRUCTION OF BRANCH NO. 2 BEGAN THE SUMMER OF 1948. By late 1949 Novostroyka was already a small, completely closed forest housing development. Its central and first installation was rig No. 1, closed off from the residential area. It also included service facilities supporting the missile's preliminary preparation for tests, the compressor unit, and yet-to-be-completed propellant component storage facilities, instrumentation building, and workshops. Rig No. 1 was built for missiles with cryogenic components. They had in mind the R-1 missile, its development, the R-2 missile, and subsequent modifications. A great deal of space remained for the construction of missiles using high-boiling components.

The initiator and primary enthusiast behind the creation of Novostroyka was Minister Ustinov. Its chief builder, and in the early years its actual manager, was Georgiy Sovkov, who had transferred to our ministry from the Academy of Sciences. Gleb Tabakov was the chief engineer overseeing the creation of the firing section and the technical ideologue. My Department U, or No. 16, was tasked with

---

2. The monk Aleksandr Peresvet fought for the Russians against the Tatars in the great Battle of Kulikovo in 1380 on the banks of the river Don. St. Sergius of Radonezh (1314–92) was a Russian Orthodox monk whose spiritual beliefs and social programs gained him fame as a highly respected spiritual leader. Zagorsk (now known as Sergeyev Posad) is the site of the monastery founded by St. Sergius.

3. At the time, NIIKhimmash was still known as NII-88 Branch No. 1. In 1956, the branch separated from NII-88 to become the independent NII-229. Today, the organization is known as NIIKhimmash.

developing and fabricating the electronic testing and launch equipment using the manpower of the experimental shop and also with developing and setting up instrumentation during firing tests.

With the consent of Gonor and Korolev, Pavel Tsybin made Voskresenskiy responsible for organizing and conducting tests in Department I, or No. 12. Together he and the rest of us verified the test rig's readiness for all the departments and the technical documentation for the first firing test. The factory was responsible for delivering the assembled missile, preparing it for firing tests, and installing it in the rig.

In 1949, Novostroyka did not yet have its own staff. For that reason, the primary functions were performed by mass expeditions from Podlipki to Zagorsk on the commuter train, and from the station to Novostroyka on service buses. Gonor ordered Voskresenskiy and me to arrive at Novostroyka a day before the launch, having warned us that this was a historical event and that Minister Ustinov, Marshal Yakovlev, and even Ivan Serbin from the Central Committee would be coming to the "performance."

"This launch is politically vital! It must go off without a hitch. And at precisely the time announced! I will say 6:00 p.m. OK?"

Gonor addressed those parting words and that question to us on 16 December. We agreed, and on the morning of 17 December we convened an operational meeting at Novostroyka and heard reports. The horizontal tests were completed successfully. The missile was installed in the rig. The instrumentation was inspected and calibrated, and stand-alone engine tests were performed.

"We've started general testing, but meanwhile we've discovered glitches in the circuit. We'll find out whether it's an onboard or ground problem," Emil Brodskiy reported to us. In effect, he was held responsible for the failings of the developers, testers, component manufacturers, and "everybody who concocted that missile."

The entire night of 17 to 18 December they worked to find the cause of the launch control system failure. When the "ground-to-air" command was issued, the circuit reset, and the engine's automatic startup process terminated.

"Perhaps this situation is similar to the demonstration of our 'foolproof' system that Marshal Rokosovskiy didn't trust." I reminded Brodskiy how we had demonstrated our smart automatic launch control system to almost all the marshals of the Soviet Union in 1947.[4]

As we were searching for the cause, Pilyugin's most scrupulous tester, Nikolay Lakuzo, proved that Pilyugin's onboard system was not the culprit. He asked us to look for errors in the ground-based control panels. And, of course, we found them. The great length of the cable network from the panel in the bunker to the rig caused voltage drops. The relay in the panel failed to pick up the commands coming from

---

4. See Chapter 4.

on board, and the failure to execute the necessary operation at the prescribed time caused the automatic circuit reset.

"Yes, that very same 'foolproof' feature."

We changed the relay and increased the entire system's power supply voltage to the maximum allowable, but we still didn't achieve stable operation.

Gonor arrived. When he found out that we, our eyes rimmed with dark circles from lack of sleep, had not yet granted permission for fueling the tanks, he demanded an immediate decision:

"If there is no hope of a launch, then tell me honestly. I will call Ustinov and request that the launch be postponed to another day."

"There's no need, Lev Robertovich. We have an idea and we're going to check it out. Give us thirty minutes."

The idea was fundamentally simple. We would take the tester who was smallest in size, but with a good head on his shoulders. We would shove him under the console where the monitoring and launching equipment was installed. The relay cabinet containing the unruly relay was located under the launch pad. The console operator controlling the launch would track the process by monitoring the illuminating display lights, and as soon as the process reached the point of the "ground-to-air" command, he would nudge or press down on the "under-the-pad operator" with his foot. The operator, in turn, having readied himself in front of the relay in question, from which the protective cover had been removed, would press on the armature with his finger, simulating its actuation.

Thirty minutes later, we had checked out the idea and the general tests proceeded "without incident."

Voskresenskiy and I reported to Gonor and he called Moscow and gave the green light for the high-ranking guests to make the trip out. We finally gave permission to fuel the tanks.

"During the launch we will have ten guests in the bunker. God forbid that anybody should notice that we have a man hidden behind the consoles, and that anybody should get the notion to speak about that afterward."

The fueling process proceeded with delays that were understandable because this was the first time. It had also turned bitterly cold, and even wearing gloves it was impossible to work on the rig exposed to the wind from all directions. The guests had no desire to freeze and the bunker was packed to capacity. Petya Vishnyakov, whom we had hidden behind the console before our guests arrived, was tired of being scrunched up like a snail in its shell and he started asking to be let out to stretch his legs.

Finally, at 8:50 p.m. Voskresenskiy announced momentary readiness. I nestled up close to the console, monitoring the displays, and all I could think about was the operator under the equipment.

The first fire up was successful. The engine ran for fifty-five seconds. Half of the high-ranking guests were watching the fiery squall for the first time. The spectators began to exchange hearty congratulations. Georgiy Sovkov proposed that all the guests and test directors come over to his Finnish log cabin for the post-test wrap-up. The "wrap-up" lasted late into the night, as we sat around a table with unlimited quantities of "rocket fuel," hot potatoes, pickles, and sauerkraut. Gonor was one of the first to get up to leave and suggested I ride with him.

"Afterward the driver will take you home," he said.

I sat next to the driver. Despite my own intoxication, I had just had time to realize that the official driver was drunk when I received a violent blow to the forehead. Later I found out that all the drivers had received large portions of missile alcohol as a reward. But not all of them could endure the long wait for their passengers. Our driver swerved off of the road and was unable to make out a stump hidden under the snow, and he hit it with the automobile's front axel. I flew forward and hit my head against the mounting of the windshield heater. I had a deep gash in my forehead and blood was streaming over my face.

Taking their prerogative as hosts, Sovkov and Tabakov gave me first aid. Voskresenskiy drove me to Podlipki with my head wrapped up in bandages.

"Alyona!" he shouted to his wife, as he dragged me, putting up a fight, into his apartment. "I have brought our friend, who was injured during the execution of his official duties. He can't go home looking like this."

Yelena Vladimirovna tried as best she could, and early in the morning, Voskresenskiy delivered me, cleaned up and rebandaged, to my home.

"The rig is poorly lit. Boris hit his head against the structure. But, according to the trauma specialists' findings, it's nothing serious." That's how he explained the situation to [my wife] Katya.

ALMOST ALL OF 1950 WAS FRITTERED AWAY AT OKB-456 for the experimental development of a new shock-free liquid ignition system to replace the pyrotechnic system to battle the pops. Kuznetsov reworked the integrator's impact resistance. Pilyugin tormented his subcontractors with his striving to increase the reliability of the relays and all of the contact couplings.

The production engineers and I had already announced for the umpteenth time our campaign for cleanliness and high standards during the production of control surface actuators. In this field I had a powerful assistant in Viktor Kalashnikov. In 1948, he transferred to us at NII-88 from the Mytishchi Machine Building Factory, along with designer Falunin and tester Kartashev. While I was coping with the dust at the firing ranges, my deputy Stepan got all three of them settled in to work on the subject of control surface actuators. Kalashnikov showed extraordinary organizational capabilities. By late 1949, he was already my deputy in Department No. 16 (Department U).

Kalashnikov managed the development, production, and testing of control surface actuators. Despite digressions into several other fields, Kalashnikov remained true to that subject until the end of his life. In 1951, Falunin left for Factory No. 586 in Dnepropetrovsk. There he later successfully headed the Ukrainian control field, which was more than a little irritating to his former boss Kalashnikov, who

148

believed that only he and I should define the technical policy for the development of the ideas and principles behind control surface actuators for all types of missiles.

Soon thereafter the talented designer Lev Vilnitskiy transferred to us from the SKB's Department No. 4, involved with air defense missiles. He proved to be a really remarkable designer and one who took a highly valued, unconventional approach to complex designs and mechanisms. His authority among designers and production engineers was indisputable. More than once, Vilnitskiy rescued Kalashnikov and me from seemingly hopeless design fiascos in terms of reliability, mechanism characteristics, and production deadlines. Later he was the one tasked with developing the most complex electromechanical assemblies for spacecraft docking.[5]

Possessing design and engineering talent like a gift from God, Vilnitskiy could not resign himself to the failure of his hip joint after a severe illness. Together with some surgeons he developed an artificial joint mechanism. He persuaded some doctors to perform an operation to replace his natural hip joint, and this restored his ability to walk using just a cane instead of crutches. Over the course of many years of joint work Kalashnikov and Vilnitskiy formed a duo that not only supervised the development of control surface actuators and mechanisms, but also established a discipline that became the leading one in our field.

AFTER LONG DISPUTES AND CONTROVERSY AT THE HIGHEST LEVELS, the decree on the acceptance of the R-1 missile into service nevertheless was issued in November 1950. The reconstructed Dnepropetrovsk automobile and tractor plant was transferred to the Ministry of Armaments for the series production of missiles. The factory was assigned the number 586 and it became yet another "post box."[6] A large group of specialists headed by Korolev's deputy Vasiliy Budnik was transferred from NII-88 to Dnepropetrovsk on a voluntary basis.

Korolev believed that the primary task of the designers sent from OKB-1 of NII-88 was to introduce the manufacturing process for the R-1 missile and then for the R-2 missile—and none of their own inventions. Budnik, departing from Podlipki, thought otherwise. When Korolev was no longer among the living, this is the story Budnik told:

"When I left the Moscow suburb of Kaliningrad in 1952, S. P. Korolev personally reviewed the list I had compiled of specialists to be transferred to the Ukraine and he crossed off all the designers, saying that there would be nothing for them to do there. At my request, Deputy Minister I. G. Zubovich put all the crossed off names back on the list.'"[7]

---

5. Vilnitskiy participated in the development of docking systems for the early generations of Soyuz spacecraft.

6. During the Soviet era, organizations conducting sensitive operations were assigned post box numbers which were used as both official designations and addresses.

7. B. I. Gubanov, *Triumf i tragediya 'Energii'*, t. 1 [*Triumph and Tragedy of 'Energiya'*, vol. 1] (Nizhniy Novgorod: Izdatelstvo NIER, 2000), p. 126.

Factory No. 586 was tasked with mastering not only missile assembly, but the manufacture of engines, control surface actuators, and all the fittings for the mechanical and hydraulic systems. The missile fabrication schedules were disrupted from the moment production began. The situation with mastering the manufacturing process of engines and control surface actuators was particularly critical. Ustinov appointed Leonid Smirnov as director of the new factory. Smirnov gave exclusive attention to the mastery of serial production at this major factory.

More than once I had to travel to Dnepropetrovsk to take part in setting up production of control surface actuators and integrated missile tests. The second half of 1951 and all of 1952 were especially critical for the factory. Kalashnikov, Andronik Iosifyan, and Nikolay Obolenskiy, director of the Moscow-based Mashinoapparat Factory, and I spent more than two months at that factory involved in organizing the large-scale series production of control surface actuators. Despite of all of his ministerial duties, Ustinov worked almost this entire time at the factory, substituting for the director and chief engineer. Availing himself of his authority in industry and contact with the local Party leadership, he staffed the factory with solid personnel. But they did not meet the missile rollout deadlines.

To help the factory master the manufacturing process for combustion chambers, Ustinov sent technical directorate chief Sergey Afanasyev on temporary assignment from the Ministry of Armaments. Afanasyev would later become minister of general machine building.[8] In 1998, I was part of a large group of veterans who celebrated the 80th birthday of our former minister. In his ceremonial speech at our convivial table he also recalled the difficult years of mastering missile production at the Dnepropetrovsk factory. There was one episode that characterizes the general circumstances of those times, and he shared his reminiscences about it in greater detail at other meetings later, as well.

I reconstruct Afanasyev's story from memory:

"The factory in Dnepropetrovsk was still under construction while workers were simultaneously mastering the production of the R-1 and R-2 there. There was a large brigade of specialists at the factory from institutes, design bureaus, and other factories of our industry. Minister of Armaments Dmitriy Fedorovich Ustinov was personally in charge of the brigade. Being chief of the ministry's technical directorate, as part of the brigade I was appointed chief of the Factory No. 586 combustion chamber shop. After the shop was up and running, I was appointed chief of the most complex engine shop at the factory.

There were very many difficulties. According to Valentin Petrovich Glushko's documentation, the missile engines were to be brought into large-scale series production first. At that time Lavrentiy Beriya, who headed Special Committee No. 1 for the atomic

---

8. Sergey Aleksandrovich Afanasyev (1918–2001) served as head of the Ministry of General Machine Building (MOM) in 1965–83. MOM supervised the development of nearly all Soviet missile and spacecraft during the late Soviet era.

problem, decided, supposedly on Stalin's instructions, to check how things were going in our ministry as far as getting missile technology into production and, in particular, at the Dnepropetrovsk factory. He made an official phone call to Ustinov almost daily and raked him over the coals. Almost a thousand kilometers away, Ustinov would stand at attention as he carried on his conversation with Beriya, and would ask us to leave the office.

One night Ustinov called me from the shop and ordered me to be ready in an hour to fly to Moscow. He said nothing as to why and for what reason. We flew into Moscow and drove straight from the airfield to the Kremlin for a meeting with Beriya. There we were given an earful over our inability start producing engines at the Dnepropetrovsk factory. Beriya asked Ustinov to give an explanation. But Ustinov replied that engine shop chief Afanasyev was in charge of engines, so let him give a report; in other words, he put the whole burden on me.

I started to report on the situation. I spoke about the difficulties we were having mastering the new materials, the test rigs, the construction of which was behind schedule, about our progress mastering parts and assemblies, the fabrication of rigging, and about the necessary remedies and deadlines. All of this I knew perfectly well and reported from memory without any notes.

"When will there be an operating engine and when will it go into series production?" asked Beriya.

I responded that, according to the approved production preparation schedule, the engine would be ready in eight months.

This angered Beriya. He started to shout and curse. Then he asked, "What will it take for the engine to be finished in two months?"

"Time," I answered.

"We'll have your head!" bellowed Beriya.

I was in a difficult situation. The meeting ended, everyone started to leave, and I left the room too. Beriya's secretary requested that I stay. Everyone passed by me, including Ustinov, with whom I had flown in. The last one to leave was Deputy Minister Ivan Gerasimovich Zubovich. He was the one responsible for missile technology in the ministry. He came up to me and said, "Let's go!"

"But the secretary forbade me to leave," I said, referring to Beriya's instructions.

Ivan Gerasimovich, clearly agitated, returned to Beriya's office. The door was slightly ajar, and I was able to hear the conversation. Beriya shouted, "I will throw you and your Afanasyev in prison."

Ivan Gerasimovich kept his cool. I could hear Beriya's cursing. Zubovich left the office, grabbed me by the arm and dragged me to the exit, after telling the secretary that it was all settled with Lavrentiy Pavlovich.

Ivan Gerasimovich and I fled onto Red Square through the Kremlin's Spasskiy Gate in a depressed state. Zubovich warned me that I should not stop off at home, nor at the ministry, but that I should wait there by St. Basil's Cathedral for the car to pick me up and it would take me straight to the factory in Dnepropetrovsk. I hadn't been home [in Moscow] for almost a year. I so wanted to drop in on them, if only for five minutes!

Twenty-four hours after my return [to Dnepropetrovsk], I found that the strictest confinement-to-barracks discipline had been established at the factory. All the specialists involved in putting the engines into production lived in the break rooms of the tool-and-die shop. They were not authorized to leave the factory premises. I had two KGB colonels assigned to me. They worked around the clock since they also wrote down every one of my oral or written instructions.

We slept no more than three or four hours. Just as I promised, the engines were up and running and went into series production eight months later.[9]

In 1953, the series production of missiles—which had replaced automobiles at the factory—was running smoothly. Incidentally, tractor production was set up as a parallel operation. The factory began to produce them even for export. Later on, everyone forgot about the factory's number and the world came to know it as the Southern Machine Building Factory (*Yuzhnyy mashinostroitelnyy zavod* or Yuzhmash).

The first ballistic missile, the R-1, was accepted into service in the Soviet Army along with a set of ground equipment almost five years after the Institute Nordhausen was established.[10] We were all very well aware that, were a war to break out in the very near future, this missile's acceptance into service would frighten neither a strong enemy nor a weak one. Moreover, it posed absolutely no threat to the NATO block. The strategic importance of the R-1 missile was not in its frontline combat qualities.

It served as good training material for many designers and scientific and testing centers, for organizing missile production, consolidating military and civilian specialists scattered throughout various agencies, and, in the final analysis, creating in the Soviet Union the foundation for a powerful missile infrastructure.

The R-1 missile was accepted into service with reservations. In order to ensure that all the identified defects had been eliminated, it was stipulated that the third and fourth production batches would undergo testing. Tests on the third production batch took place in January 1951. In particular, the missiles were checked out at an ambient temperature as low as minus 26°C (−14.8°F). The tests on the fourth production batch, which were called verification tests because they verified the Dnepropetrovsk factory's series production manufacturing process, also took place without significant negative findings. One hundred percent of the two batches reached their targets and landed within a 16- by 8-kilometer rectangle. The greatest deviations under a purely autonomous guidance system did not exceed 5.5 kilometers.

Despite the apparent success and the decorous wording in the test reports, one of the comments was not given proper attention. The matter concerned the breakup of

---

9. The incident that Chertok describes has been published in several different sources, including an essay by Afanasyev himself. See Yu. A. Mozzhorin et al., eds., *Dorogi v kosmos: I* [Roads to Space: 1] (Moscow: MAI, 1992), pp. 40–42.

10. The R-1 missile was officially declared operational on 28 November 1950.

missiles in the descent portion of the trajectory during entry into the dense layers of the atmosphere. This phenomenon had been one of the primary defects of the A4 missiles. Of course, not every missile armed with explosives blew up, but, without fail, one or two out of ten did. Despite the large number of experiments and the organization of special measurements in the impact area, for a long time we were unable to figure out the true cause for the aerial explosions.

It wasn't until 1954, when we were already working on a missile with a nuclear warhead, that we finally and unambiguously succeeded in figuring out the secret of the R-1 missiles' premature explosions. And the credit for solving the riddle belongs not to the designers, but to military engineers—GTsP Deputy Chief A. A. Vasilyev and the director of the firing range measurement service, A. L. Rodin. Ultimately, the cause of the trajectory explosions turned out to be the heating of the TNT charge. Its intense evaporation increased the pressure in the warhead's pressurized compartment, which led to the rupture of the metal housing. The resulting dynamic loads caused the graze percussion fuse to trip and naturally explode the entire charge. If you begin counting with the Germans, who never managed to figure out the true cause of the aerial explosions, then it took almost ten years to solve this puzzle!

Altogether, from the beginning of development until a relatively reliable (compared with other types of armaments) missile system was obtained, sixteen years passed! Of these sixteen years, Germany spent seven; two years can be considered a joint Russian-German project; and for seven years it was purely our project. In this regard, the R-1 with its entire set of ground equipment set a record for the length of time required for its total production cycle.

Evidently, the total number of missile launches conducted only for testing and verification also remains a record to this day (here, we will not take into account the Germans' combat launches, although they provided information that was extremely useful for developers). The total number of A4 and R-1 launches on Soviet territory for testing and verification was over 200.

One must not downplay the historic significance of the A4 and R-1. This was the first breakthrough into a completely new field of technology. Neither the Germans nor we had any practical experience or theory for creating large integrated technical systems combining many scientific disciplines and the most diverse technologies. Both in Germany and in the Soviet Union, in terms of the equipment and the top-level totalitarian state leadership, the optimal conditions were created for this work. In addition, both in Nazi Germany and later in the Soviet Union, the respective governments demanded a maximum reduction of the production cycle from all those involved in the work. And still it took sixteen years! All subsequent developments of much more complex and advanced missile weapons systems took no longer than six to eight years.

It is not the threatening dictates of governmental leaders, but the experience and knowledge of scientists, engineers, and all those involved in the development of large systems that determine the cycle of their creation. Those who are responsible for the security of the nation must concern themselves not only with material support for experimental and design work. They must make sure that in creating new and large systems, they find bright and forceful leaders with organizations devoted both to the leaders and their ideas.

Chapter 9
Managers and Colleagues

The Communist Party Central Committee defense department paid particular attention to selecting and appointing managers and administrators to implement large government-sponsored military science and industry programs.[1]

When Stalin was alive, it was impossible to appoint directors of leading organizations without his approval. After Stalin's death, a multistep procedure was gradually authorized to regulate the appointment of directors for the entire hierarchy. The Central Committee Secretariat passed decrees appointing missile industry directors and chief designers or removing them from office.[2] Only after such a step did governmental decrees and the corresponding orders of the industry minister appear.

In the one to three years since NII-88 was created, many new scientific-research, design, and production teams were created. Almost all of them faced the problem of finding qualified management. During the war, the office of the Central Committee, which monitored all scientific and design organizations and factories had appointed directors without burdening itself with concern over their rapport with the staff. The Committee favored strong-willed directors, who would spare neither themselves nor their subordinates to meet a military technology production deadline while obeying precisely the instructions from their superiors.

For the most part, this tradition persisted during the first postwar years. However, new scientific and technical problems required technical management that would play a much stronger role. The central figure became the chief designer, rather than the director-administrator bestowed with many government awards.

---

1. Technically, the Central Committee was the highest body within the Communist Party of the Soviet Union. In practice, because of its relatively huge membership and infrequent meetings, the whole Central Committee rarely acted as a deliberative body. Instead smaller, more powerful bodies within the Central Committee, such as the Politburo (also known as the Presidium) acted as the true centers of power during the Soviet era. The Central Committee also had several departments (such as the defense industries department) that effectively controlled almost every facet of Soviet political and economic life. The Council of Ministers was the most powerful deliberative body in the Soviet *government* (as opposed to the Party).

2. The Secretariat of the Central Committee was the cabinet-level body within the larger Central Committee composed of several "secretaries," each with a different portfolio.

That is what happened in the aviation industry, and the intention was that the nuclear field would work in the same way. At NII-88 in the rocket industry, something different took shape.

During the fall of 1949, Yuriy Pobedonostsev left his post as NII-88 chief engineer to become the rector of the Academy of the Industry. Karl Tritko, while remaining SKB chief, was temporarily assigned the post of chief engineer. Until April 1950, Korolev remained only chief of the SKB's Department No. 3, officially subordinate to Tritko.

At the firing range, Korolev was the ultimate ideological and technical leader, and every year his authority grew. Deputy ministers, industrial management central office chiefs, and chief designers from other ministries implicitly recognized Korolev as the missile program chief. Whenever he returned from the firing range, however, everything changed. At NII-88, Korolev was losing his status as top leader, unlike Glushko, Ryazanskiy, Barmin, Kuznetsov, and other chief designers who remained at the top of their organizations in a rapidly developing cooperative project. This depressed Korolev. His deputies, Vasiliy Mishin in particular, were also unable to come to terms with this mistreatment. Korolev began a struggle within NII-88 for greater autonomy. All his colleagues from the Institute Nordhausen and people arriving from the aviation industry supported him in this. Ustinov understood the absurdity of the NII-88 structure, but did not immediately opt for a serious reorganization. The paradox was that the minister himself could not resolve an issue such as reorganizing an institute subordinate to him or conferring greater rights on Korolev. The all-powerful office of the VKP(b) Central Committee defense department headed by Ivan Serbin was over Ustinov. Sometimes Serbin was aptly referred to as "Ivan the Terrible."[3]

It was mandatory that he approve all personnel reshuffling, firings, promotions, awards, and punishment of directors. Later, at various conferences attended by Serbin, I saw for myself that the ministers really were afraid of this man and never risked an argument with him.

Only in the nuclear sector, and later also in the air defense sector, was the appointment, promotion, and transfer of directors not coordinated with the office of the Central Committee. In the nuclear sector, where Lavrentiy Beriya was in charge, decisions on the appointment of program managers, directors of institutes and factories, and design bureau chief designers were made by Boris Vannikov, coordinated with Igor Kurchatov, and submitted for Beriya's approval. The small office of Special Committee No. 1 prepared draft resolutions on appointments, which Beriya presented to Stalin for signature. The power of "Ivan the Terrible" did not extend to the managers of the atomic program. In modern lingo, they had a more powerful *krysha*

---

3. Ivan Dmitriyevich Serbin (1910–81) served as Chief of the Central Committee's Defense Industries Department in 1958–81, during which time he was responsible for approving all high-level appointments in the defense industry on behalf of the Communist Party.

(literally "roof," or mafia protection).

AT NII-88 DIRECTOR LEV GONOR'S REQUEST, Ustinov agreed to give an audience to the obstinate Korolev. Under the pretext of discussing complex issues of the plan for the year 1950, Ustinov summoned Gonor, Korolev, Tritko, and me to his office on a Saturday night in May at 10 o'clock.

We took two cars; I rode with Korolev while Gonor rode with Tritko. On the way, Korolev said that he would not so much lay out plans for the minister as complaints about the structure and chaos at NII-88. He would demand that an independent OKB with its own pilot production and its own guidance, testing, and materials department be split off from the SKB. Disregarding the driver's presence (which was risky back then), rehearsing his speech, he turned to me and delivered an impassioned argument:

"You've got the entire guidance system with all of the cables. You pulled the guts out of the live body of the missile. I have to ask you before I change anything, as if I were asking for a favor. The testers obey me only at the firing range. It's a good thing that Chertok, Tsybin, and Voskresenskiy are people that I can trust, and we can always talk things over. But the material engineers could send me packing if they felt like it, if Tritko told them not to listen to that Korolev. And the factory, it doesn't consider my opinion at all! We can't continue our work without test rigs, without preliminary checkouts.

The factory has its own plan, and, plus, it does work for other projects. At the factory, they say that they are working on Korolev's project only now. But they don't want to listen when we say, we'd like you to do some testing using provisional sketches or something not included in the plan. Unbelievably stupid! I must have my own production! I have been to Glushko's facility in Khimki. There everyone hangs on his every word. But I have to kowtow to Gonor for every little thing. If Gonor were gone tomorrow and some empty suit were in his place, then everything would go to hell. It seems like that's what it's coming to.

Ryazanskiy complained to me that when Chertok come from NII-88 to NII-885, you go to the director and ask to change the schedule or do something over and above any plans, and they obey you at an institute that isn't even your own. And here at our own place, we aren't in charge. What's more, Gonor is afraid of spoiling relations with the Party committee. Utkin is there, and at least he's a decent man, but there are more than enough complainers.[4] He is also afraid of catching grief from the Central Committee if they receive letters complaining about Korolev."

But Korolev wasn't able to say all of that to the minister. First, Ustinov started the meeting not at 10 p.m., but an hour and a half later. Sitting in the waiting room chain smoking, half asleep, we wondered what time it would be when he dismissed

---

4. Ivan Ivanovich Utkin (1910–85) was the Communist Party secretary at NII-88 in 1947–50. Party Committees within organizations typically had significant influence in personnel appointments.

us. Second, he began the meeting by telling us about the prospects of a project involving surface-to-air guided missiles. Here, he alluded to the possibility of transferring that entire field to the aviation industry, bearing in mind that the brass were already reviewing such proposals for the reorganization of operations.[5] But for the time being, it was early to make such a decision. Therefore, he requested that we amicably and harmoniously work in the existing structure, taking into consideration the extreme complexity of the plans for 1950. Ustinov asked that we devote particular attention to the R-2 missile, saying that this was a test of our ability to work independently. He mentioned that Sinilshchikov still hadn't produced any good results in reproducing the *Wasserfall;* therefore, Korolev's work would become decisive for the fate of NII-88.

Korolev tried to cut in on the minister's lengthy lecture to express his views on how operations were being run, but Ustinov was not disposed to opening up the discussion. He glanced at his watch and said that we were all very tired, that it was already 1 o'clock in the morning. He wished us success, told us to enjoy our Sunday off, and dismissed us.

We left, terribly disappointed that not one of us had had the opportunity to speak his mind at the meeting. Tritko suddenly proposed that we go have supper. "The Moskva restaurant is open until 5 a.m.," he said. "While we were waiting, I called up and reserved a table. They're expecting us." Korolev and Gonor were not overjoyed, but agreed. Despite the late hour, the third-floor restaurant was crowded with people who, by the looks of things, had also just come from late-night meetings. The military-industrial elite caroused into Sunday. It turned out that Tritko was a regular here. The foreigners, the partying types, and the women, he explained, were kicking up their heels on the roof of the Moskva, so we could all talk frankly here. But in order to have a real "heart-to-heart" conversation, we all needed to toss down a drink "artillery style." Those who could drink "artillery style" would remain combat buddies forever. Such, according to Tritko, was the tradition of true frontline artillerymen. On his command, the experienced waitress quickly set our table with four half-liter bottles of vodka, four empty beer mugs, and two large pitchers of beer and filled soup bowls with steaming hot, fragrant *solyanka.*[6]

Gonor was the first to protest: "What were you thinking—ordering a half liter for each of us?! *Solyanka* and beer is enough for me." Korolev sat sullenly, waiting for our "heart-to-heart" conversation. But Tritko quickly filled the beer mugs to the brim with vodka and commanded: "You have to drink a half-liter of vodka in one breath without taking your lips off the mug! Then we drink beer and eat *solyanka.*" He demonstrated. I was the youngest in the group and felt compelled to show that

---

5. Ministry of Armaments' NII-88 supervised development of both surface-to-surface and surface-to-air missiles. By the late 1940s, Soviet industrial leaders believed that it might be more efficient to transfer the latter to the Ministry of Aviation Industry.

6. *Solyanka* is a savory, somewhat sour soup made with meat or fish, vegetables, and pickles.

artillery officers weren't the only ones capable of such a feat. After I had chugged a half-liter of vodka, chased it with a mug of beer, and finally started on my soup, I have no recollection of what happened. I didn't even see what Korolev and Gonor did. I regained consciousness the next morning when I woke up at home, completely fresh, and tried to remember how I had gotten home and in what condition.

My wife, Katya, said that I had showed up at 5 a.m., explaining that we had had a very tough meeting with the minister; I asked her not to wake me up in the morning and turned down any offers of food. In her sleepiness, she had not noticed anything abnormal in my behavior. On Monday, Tritko felt obliged to verify by telephone whether I had shown up for work. Finding that everything was in order, he said, "You're a real artilleryman!"

Gonor's chauffeur told me what had really happened: "We barely got Lev Robertovich [Gonor] and Korolev into the cars. Korolev wanted to scuffle, and you and Tritko separated them. I drove you home. You got out all right, but Gonor was in really bad shape. When we got to his place, he couldn't even get out of the car." Subsequently, neither Gonor nor Korolev ever mentioned this nocturnal "heart-to-heart" conversation.

IN 1950, THEY BEGAN RESHUFFLING PERSONNEL ANYWAY, upsetting the stability that Ustinov had asked us to maintain. In June, the director's office was empty for a while. Gonor flew off to Krasnoyarsk, without even being allowed time to say good-bye. In August, Konstantin Rudnev was appointed the new NII-88 director. He belonged to the young generation of military industry managers and was transferred to us from Tula. In Tula, Rudnev had been director of the famous ordnance factory. We began our acquaintance with the new director by peppering secretary Anna Grigoryevna with questions. She had been the permanent secretary under Gonor. As a rule, a new director would bring "his own" secretary with him, but from the very beginning Rudnev aimed to inspire a sense of trust and had no intention of reshuffling the staff, including the director's secretary.

Anna Grigoryevna could tell many stories about the managerial personnel. During her 56-year career as secretary or administrative assistant, she saw eight men pass through the director's office of NII-88, later renamed the Central Scientific-Research Institute of Machine Building (TsNIIMash). She reassured us all that the new director was very civil, had not yet manifested any despotic tendencies, and had an open door policy for anyone who requested an audience.

I considered myself an experienced manager by that time and decided that, before meeting the new director, I needed to find out about him in greater detail by taking advantage of my acquaintances in the ministry offices. Sergey Vetoshkin's secretary, Irina, whom he renamed Irene, was my neighbor on Korolenko Street. When asked what people in the central offices for industrial management were saying about our new director, she said that everyone considered him a very capable manager with a bright future. Acquaintances considered excessive leniency and civility his shortcoming. The people in Tula felt very sorry for him and thought that the missile specialists at NII-88 would "eat him alive," while the Soviet ordnance manufacturers were losing a good manager.

Gonor had worked as director in this hot spot since August 1946. He visited us in Germany. He knew us all inside and out. Over the course of four years he had gained an understanding of the technology and established good relations with all the subcontractors. Everyone respected him. Korolev often clashed and argued with him, but now he had to start everything from scratch. And why should the experience and knowledge that Gonor had gained go right out the window while he started making guns again? In meetings with other chief designers, Korolev grumbled, but he understood all too well that Gonor's departure was not Ustinov's whim, but Stalin and Beriya's policy, and it was better to keep quiet.

Rudnev actually turned out to be a polished, unobtrusive, and fairly modest manager. Of course, he was not familiar with our technology and, therefore, he was forced to study people in order to understand whom to lean on and whom to trust completely. When they got to know Rudnev, managers who had grown accustomed to stuffy conversations in the director's office were surprised by his inexhaustible good sense of humor. He didn't hide the fact that he favored people who understood a joke and preferred to "work rather than just follow instructions." Soon even Korolev announced that he could work with Rudnev. They found a mutual understanding, and the new director supported Korolev's proposal for reorganization.

Actually, already before that, Gonor had drawn up an order for the minister that stipulated a change in the NII-88 structure. The SKB was divided into two OKBs. Department No. 3 was converted into OKB-1, and Korolev was named its chief designer and head. Tritko was relieved of his post as SKB chief and was named head of OKB-2 in place of Sinilshchikov.

The post of institute chief engineer remained vacant. Here Rudnev, probably with someone's prompting, tried to explore the possibility of appointing me, especially since I was currently a deputy chief engineer. After his proposal was turned down, he attempted to bring back Pobedonostsev. The latter also was declined.

A rumor spread through the ministry offices that Korolev was laying claim to both posts, chief engineer *and* head of OKB-1. The ministry staff feared such a turn of events since Korolev's single-mindedness and character had always made the officials fearful that he would become uncontrollable and that all the projects at NII-88 would be completely under his influence.

Unexpectedly, Rudnev got Mikhail Ryazanskiy as "first deputy" and chief engineer.[7] It had not been very hard for Ustinov to persuade Ryazanskiy to abandon that same post at NII-885 and come to the rescue of NII-88 to help the young director straighten things up. Ryazanskiy felt uncomfortable about Korolev. He had been a subcontractor to Korolev as the chief designer of a guidance system, and now sud-

---

7. Russians use the term "first deputy" to denote someone who is "first among the deputies."

denly he outranked Korolev, having been appointed Rudnev's deputy. But Ryazan-skiy explained to me frankly that when the Central Committee was discussing the chief engineer vacancy, he had been warned that the Central Committee had many denunciations directed at me. They primarily had to do with the development of the automatic astronavigation system. But it wasn't only a matter of technology; rather it had to do with the fact that the current situation required a different personnel lineup and therefore I could no longer remain in the post of deputy chief engineer. I did not have the right ethnicity. If the fifth line of the personal history form had said "Russian" or even "Ukrainian," then it would have been a different story.[8]

With Ryazanskiy's arrival, power at NII-88 was actually shared by the triumvi-rate of Rudnev, Ryazanskiy, and Korolev. Korolev reorganized his Department No. 3 and began to form the full-fledged OKB-1, which soon was destined to become a historic organization, ensuring the Soviet Union's primacy in missile and space technology.

As I muse over the past many years later, it occurs to me that there were many blessings in disguise. More importantly, when the circumstances were bad, there were good people. In late 1950, new Deputy Minister Ivan Zubovich announced to me that he was very sorry, but by his order I was relieved of both posts—chief of Department U and deputy chief engineer of the institute—and was being sent to the disposal of the NII-88 personnel department. The directive was implemented, and I was thus removed from the *nomenklatura*.[9] Above all, this was a blow to my morale but I endured it relatively easily because Korolev and Rudnev had warned me in advance. Both sympathetically told me that with the minister's consent they would not dismiss me from NII-88.

The personnel department obeyed the director's command and transferred me to the post of deputy chief of Department No. 5 in the new OKB-1. This department, by Korolev's design, was the foundation of an integrated guidance system depart-ment, which was to be part of OKB-1 and subordinate to Korolev and not to the institute chief engineer.

Ryazanskiy supported this line. If I hadn't experienced such persecution, perhaps my subsequent fate would have taken a different turn. Now I was not just subordi-nate to Korolev in terms of subject matter, but also administratively.[10] My immedi-ate boss was Mikhail Kuzmich Yangel. Someone from the top levels of management clearly had his eye on him when, after working in the aviation industry and gradu-ating from a year-long course at the Academy of the Industry, it was recommended that Ustinov keep Yangel in the pool for further promotion. Korolev warned me

---

8. At the height of the anticosmopolitanism campaign, Chertok's Jewish ethnicity was a huge liability.

9. The *nomenkletura* was the list of Communist Party-approved individuals who could occupy important positions in industry.

10. In other words, Chertok reported to the NII-88 chief engineer in 1946–50, but after that he reported directly to Korolev, thus moving his career into a different direction.

that I would serve as guidance department chief Yangel's deputy on a temporary basis. Yangel was not a specialist in matters of guidance and automatics; therefore, Korolev would place the responsibility on me and I would answer to him.

The staff of the new department received both Yangel and me well since there were too many projects and technical problems. In addition, no one tried to shirk or shift the workload; on the contrary, everyone tried to take on a little more and take full responsibility. Such a work ethic was one of the conditions for our success during the first decade.

Yangel asked me to take on all the projects related to electrical circuitry, control surface actuators, and telemetry and radio systems. I was given free rein to make the decisions I deemed necessary without consulting him. But he retained the right to review them with my participation and prepare proposals for Korolev on matters of flight dynamics and the coordination of these matters with NII-885, that is, with Pilyugin's dynamics experts.

In 1951, the R-5 missile was already being designed. In terms of its dynamic characteristics, the R-5 required fundamentally new approaches to guidance system development. For that reason, we needed to be in constant contact with Pilyugin's theoreticians. Here, Yangel relied completely on my friendship with Pilyugin's team because conflicts arose from the very start.

Thus, Yangel and I came to terms and worked for almost a year in a very friendly atmosphere. A year later, Yangel was transferred to the post of deputy chief designer. Among other matters, Korolev assigned him design oversight over the series production of R-1 and R-2 missiles in Dnepropetrovsk.

In June 1952, NII-88 once again lost its chief engineer. Ryazanskiy moved to the ministry structure to head the main directorate—fortunately, not for long. Unable to endure the bureaucratic rigmarole, he soon returned to his old NII-885 organization.[11] Also in 1952, Rudnev was transferred to the ministry to the high post of deputy minister. To everyone's surprise, including Korolev, Yangel was appointed NII-88 director. Later the ministry officials said "in secret" that this had been the Central Committee's initiative. This appointment proved to be a difficult test of the good relationship between Yangel and Korolev.

Unfortunately, they did not pass the test of peaceful, amicable, ideological, and practical interactions. Both of them encouraged work contacts through their deputies and staff and met with one another only for meetings when summoned to the ministry or at other high levels. Our missile and space technology probably could have developed even further if these two managers had consolidated their efforts rather than being antagonists. Their relations were strained to the point that they tried to avoid one another and would not speak to each other. Korolev used me, Mishin, and his other deputies as go-betweens to communicate with the new director.

---

11. Ryazanskiy rejoined NII-885 in 1954.

At that time, we, that is, OKB-1 staff members subordinate to Korolev, blamed Yangel for the strained relations. Korolev's lust for power, ambition—which was rather understandable—and difficult personality irritated Yangel. Korolev's merits at the time—six years after beginning his work on the series production of domestic missiles—were great even by today's standards. Korolev and his organization worked selflessly and obsessively.

Like almost any new director who suddenly finds himself heading a powerful organization, Yangel decided to change its methods, goals, and structure to his liking. He made it his goal to "reform" Korolev so that OKB-1 would serve NII-88, but Korolev demanded that NII-88 projects be subordinated to OKB-1 tasks. At that time Korolev was certainly right. But Korolev's failure to accept Yangel's leadership threatened to destroy the institute structure, which was fragile at best. The ministry and Central Committee reached a compromise, and in late 1953, they removed Yangel to the post of institute chief engineer, thereby relieving him of the right to be in charge of Korolev. Having worked for less than a year in that role, frittering away his energies on workaday, routine administrative work, Yangel agreed to leave for Dnepropetrovsk, where he became chief designer of the Dnepropetrovsk OKB.[12] Here he gained the opportunity to begin to actually implement ideas for the production of missiles using high-boiling propellant components, rather than just talking about it. Yangel began with the development of the R-12 missile as a counterweight to Korolev's R-5M. After Gonor, Rudnev, and Yangel, the NII-88 director slot was filled by Aleksey Spiridonov, who until then had been chief engineer of the ministry's main directorate.

IN EARLY 1953, OKB-1 EMPLOYED OVER 1,000 PEOPLE and was an organization capable of leading both practical work and scientific research for missile technology development. The ministry, too, had finally realized that the missile industry needed a head institute like TsAGI, which had emerged as the lead institute in the aviation industry.

On 14 August 1956, the minister signed an order making OKB-1 an independent organization, that is, separate from NII-88. The structure of this new organization had already practically been worked out over the course of the last two years, and, therefore, no radical shakeups were required in the ranks of the main staffs. But it added many new concerns for Korolev personally and for his immediate retinue.

Under the new structure, the factory became subordinate to the OKB chief and also separated from NII-88. Production is fundamental for any design bureau. Without it, the most perfect ideas and designs will remain on paper. To have a truly state-of-the-art factory transferred to OKB-1, Korolev had to endure frequent battles at different levels. The factory's primary work was, after all, manufacturing

---

12. Yangel was appointed chief designer of OKB-586 at Dnepropetrovsk in July 1954.

missiles developed by OKB-1. The Dnepropetrovsk factory had already taken over series production of R-1, R-5, and R-5M missiles. Production of naval modification R-11M missiles had been transferred to the Urals.[13] Specialized instrument building design bureaus and factories were created in Kiev, Kharkov, and Sverdlovsk.

In 1955, the pilot production Factory No. 88 began manufacturing parts of the first R-7 intercontinental missile at full steam. The government decree for the development of this missile was issued on 20 May 1954. Before its final release, all the chief designers, their immediate deputies, ministry offices, and *Gosplan* had thoroughly reviewed the contents of the voluminous decree. The document devoted proper attention to production problems. It was logical to make Factory No. 88 subordinate to the OKB chief, in this case, the chief designer. But, despite the fact that it was officially part of the structure of OKB-1, at the ministry's insistence, the factory maintained a certain degree of independence. It had its own operating account at the bank, and its plans had to be coordinated with the ministry.

Roman Anisimovich Turkov was appointed factory director. At the same time he also acquired the status of Korolev's first deputy. Turkov had gone through the brutal school of war, when he served as chief engineer and then as director at a Krasnoyarsk artillery factory. He considered it perfectly natural to take on, in addition to production process problems, the burden of social problems—housing, public services, kindergartens, schools, hospitals, transportation, and a lot of other concerns that now would fall on Korolev.

Where, in what other country, must a scientist—a designer and director of a highly complex scientific and technical program—concern himself with relocating his staff out of ramshackle huts or with construction of roads and children's daycare centers? At that time, and even decades later, such was the hard lot of the director of a large-scale enterprise in the Soviet Union.

Sometimes people try to compare the creative achievements of von Braun and Korolev. They forget that while Korolev was developing the intercontinental missile and first spacecraft, he was going to great lengths to make sure that a municipal Palace of Culture was built, taking care of the orphanage, reviewing all the housing distribution lists, and making the rounds ordering food for the city.

Turkov was his invaluable assistant in this work. They understood each other well. Turkov had a knack for recognizing real craftsmen and exposing slackers on the production line and personally sorting out complex technological processes. He quickly gained the respect not only of the factory workers, but also the designers, with whom he loved to maintain contact, studying the drawings of the most complex assemblies.

Korolev kept Vasiliy Mishin as first deputy chief designer for drafting and design work. They had worked very well together back in Germany. In the process of form-

---

13. These two factories were Factory No. 586 (at Dnepropetrovsk) and Factory No. 385 (in the Urals).

ing a working team within the NII-88 system, Mishin was even more aggressive than Korolev, insisting on the unconditional subordination of institute scientific department projects to the front-burner problems of the KB.

Responsibilities were distributed among the other deputy chief designers approved by the ministry order as follows:

Konstantin Davydovich Bushuyev—drafting departments;

Sergey Osipovich Okhapkin—design departments and everything associated with technical documentation;

Leonid Aleksandrovich Voskresenskiy—firing rig and flight tests;

Anatoliy Petrovich Abramov—ground complex, including construction at the firing range.

Somewhat later Mikhail Vasilyevich Melnikov, who transferred from NII-1, became deputy chief designer for propulsion systems. So-called lead designers, on the other hand, were independently close to Korolev despite being lower in the hierarchy. By that time Dmitriy Ilich Kozlov, Viktor

Roman Turkov (1901–75) served as the director of the experimental pilot production facility at Korolev's OKB-1 located in the outskirts of Moscow. As such, early production runs of all of Korolev's missiles and spacecraft were directed by Turkov.

Petrovich Makeyev, and Mikhail Fedorovich Reshetnev had already distinguished themselves. I mention them first, because while Korolev was still alive they ventured outside OKB-1, first heading branches and then independent organizations.

The prestige of Viktor Makeyev—academician, two-time Hero of Socialist Labor, general designer of submarine missiles—was so great that in 1976 he was offered the post of minister of general machine building. However, Makeyev turned it down.[14]

In 1952, after Yangel's transfer from OKB-1 to the post of NII-88 director, I became department chief. But now it was no longer the NII-88 guidance department, but NII-88 OKB-1 Department No. 5. The administration of this department entailed problems of flight dynamics and guidance, telemetry and radio trajectory measurements, development of emergency engine shutdown systems, general onboard and ground electrical equipment, and a lot of other issues that, in one way or another, had to do with electricity and information transmission and processing.

---

14. Viktor Petrovich Makeyev (1924–85) was the Chief/General Designer of SKB-385 (KB Mashinostroyeniya) in 1955–85 and oversaw the development of several generations of submarine-launched ballistic missiles, which constituted one of the most important elements of Soviet strategic military power.

Over the course of three years (from late 1950 through late 1953), as the old song goes, the department's primary staff went through "fire, water, dust, and fog" on the expanses of the Kapustin Yar firing range and on the Novostroyka firing rigs near Zagorsk. The department workers loved their work. They were sympathetic to production difficulties, factory problems, and contractors and worked well as a team. Personnel from NII-88 and new young specialists joined the department, and its ranks grew rapidly.

In 1954, the department grew so much that we made arrangements with Mishin and proposed to Korolev that it be converted into a three-department "complex." Department No. 5 held onto the development of guidance systems, onboard and ground electrical equipment, antennas, supervision over all types of radio engineering, onboard telemetry, and measurement system sensors. Once again they set up design Department No. 18 and tasked it with the independent development of onboard and ground instrumentation and drafting. Design Department No. 4 was set up to develop all sorts of control surface actuators, drives, and other mechanisms. Korolev demanded a discussion over the structure of these three departments and the candidates for the posts of department chiefs and primary subdivision chiefs in each of the departments. If he personally did not know the individual very well or was not sure about him, it was impossible to convince him that this person be hired.

In early 1954, Igor Yevgenyevich Yurasov was appointed as my deputy. He had already gotten his feet wet in his research work at NII-88, but gladly broke away from his dead-end theorizing and immersed himself in our urgent and always eventful affairs. His involvement gave me the opportunity to shorten my stays at the firing range.

In 1954, Oleg Voropayev and Valentin Mukhanov arrived from Bauman Moscow Higher Technical Institute (MVTU). I sent Voropayev to Vetrov's dynamics sector. He did not object to design and theoretical work. Almost every young specialist dreamed of this. Soon he became the lead specialist in Korolev's OKB for missile dynamics and control systems. He distinguished himself with his ability to visualize outwardly complex phenomena and find their inner simplicity. Many years have passed since then. Two chief designers and two general designers have come and gone, but Voropayev served continuously as director of the dynamics department until his well-deserved retirement in 1992.

Mukhanov was very upset when I offered him work on control surface actuators in the design bureau rather than in the research laboratory. It got to the point that I gave my word to transfer him from the design department after six months if it became unbearable for him there. That request never came. Mukhanov became engrossed in the work of the control surface actuator design team and in optimizing their parameters. He was one of the leading specialists in this field. Young specialists reluctantly went into design work, and more than once I had to resort to that stratagem: to give my word that "if you don't like it, I'll transfer you in six months." As a rule, no one took advantage of my promise.

166

One such headstrong individual was Vladimir Syromyatnikov. Having reluctantly begun work on electrical drives and control surface actuators under the supervision of the indefatigable Lev Vilnitskiy, he found a successful combination of theory and practice in the development of complex mechanisms. Twenty years later, Candidate of Technical Sciences Syromyatnikov used his experience to develop androgynous docking assemblies for the famous Soviet-American Apollo-Soyuz Experimental Flight (EPAS).[15] Soon thereafter, Syromyatnikov defended his doctoral dissertation, received recognition abroad, and, in 1992, was selected as a member of the International Academy of Astronautics. Professor Syromyatnikov still heads the world's only team that designs docking assemblies.

We were also fortunate with our talented young theoreticians. Beginning in 1952, at my request, Mishin, who handled creative problems, sent graduates with a good university education to Department No. 5, which was responsible for control dynamics problems. He did this with the consent of Korolev, who understood that in the field of theoretical mechanics, universities provided greater basic preparation in higher mathematics and physics than the Bauman MVTU, MAI, or the Leningrad Military Mechanical Institute. Our design and testing departments were staffed with an abundance of their graduates.

University theoreticians quickly got involved in the solution of practical problems and were recognized as having a mathematical background superior to that of their bosses, who had a strictly engineering education. Over the course of many years of work I found that a talented theoretician acquires engineering experience more rapidly, while the theoretical heights of mathematics and mechanics that universities provide remain downright inaccessible to the engineer.

In 1953, three theoreticians arrived almost simultaneously in the dynamics section that Georgiy Stepanovich Vetrov supervised: Igor Fedorovich Rubaylo, a graduate of the Moscow State University (MGU) physics and technology department; Leonid Ivanovich Alekseyev, a graduate of Rostov University; and Yevgeniy Fedorovich Lebedev, a graduate of Gorky University.

The personnel department settled each newly arrived employee in a dormitory for "young specialists," and then, without giving much consideration to their individual goals, sent them to NII-88 department chiefs or to Korolev in OKB-1. The subsequent fate of each person, to a great extent, was determined during those first days of wandering from office to office.

We had a rule at OKB-1: each young specialist must go through Korolev's office. Lebedev recounted: "I'm sitting in Korolev's waiting room. I wait more than an hour. The secretary says that Ustinov is meeting with Korolev, and that I might not get in to see him. She took a chance and let me in to see Mishin. Once Mishin had an idea of my background, he sent me to Svyatoslav Lavrov, who at that time was

---

15. EPAS—*Eksperimentalnyy polet Apollon-Soyuz*—was the Russian name for the Apollo-Soyuz Test Project (ASTP).

filling in for drafting department chief Bushuyev on design and theoretical projects. Lavrov sent me to Vetrov, who was in charge of dynamics. Vetrov suggested I have a look at a NII-4 report in which they were studying ideas for launching a missile with a cluster configuration. Next, I was supposed to analyze the launch dynamics of a missile, which by that time already had a configuration and parameters close to those of the future R-7 intercontinental missile."[16]

In 1954, Lebedev was developing the dynamic procedure for launching the R-11FM ballistic missile from a rolling submarine. None of the "brass" had checked his calculations, but the engineers who had developed the special rig that simulated a submarine's roll believed the young specialist and were not surprised that everything turned out brilliantly. They weren't elated because they never expected any other outcome. It is difficult for today's engineers to understand how, without a single computer, specialists took on the responsibility for the critical solutions of control problems for missiles carrying nuclear warheads, launched from the ground, from a rolling submarine, from a submerged submarine, or from a silo.

Both the old and battle-seasoned personnel and the new young specialists worked at a very intense pace. I wouldn't say that the stress was generated from the top. The often unrealistic schedules themselves had inflexible deadlines; all kinds of criticism circulated among Party and business leaders, but these "signs of the times" did not have much of an effect on the mood in the organization.

Korolev, whose lead we took, made no concessions for youth. This was a good incentive for everyone who had come straight from college. Korolev loved to lecture time and again: "Youth is not the main shortcoming."

The overwhelming majority of engineers worked with genuine enthusiasm. The technical problems that needed to be solved "come hell or high water" also distracted them from their uncomfortable living conditions and the difficulties of daily life on the other side of the front gate. They went to work not just because they had to, but primarily because it was interesting. Despite poor conditions at the firing range, no one had to be persuaded to drive or fly in for a temporary assignment there.

OF THE FIRST MISSILE DECADE, the last three years were certainly the most interesting in terms of science and engineering. The people who joined the missile programs during 1954–56 to a great extent determined the subsequent development of our cosmonautics program. While these people were still relatively young, someone's quip caught their fancy: "According to personal history forms, our personnel fall into one of two categories: they are either Tsiolkovskiy's best students or individuals whose youth isn't their main shortcoming." "Tsiolkovskiy's best students" referred to

---

16. *Author's footnote:* I will write later about Rubaylo and Lebedev's contribution to the theory for the dynamic configuration of the R-7 missile—which to a great extent determined the missile's longevity—when I describe the history of the R-7 itself. (See Chapter 16.)

the chief designers and everyone who began working with them in 1946–47.[17]

Here I should put in a good word for Ivan Utkin, who was the first Central Committee Party organizer and arrived at NII-88 in 1947. After graduating from the physics department of Moscow State University, Utkin dreamed of a career as a scientist and entered graduate school. The Central Committee unexpectedly summoned him and informed him that as a Party member he would be going to the newly created missile institute, where he would head the Party organization. A good university education, good-natured temperament, and dreams of scientific work were hardly required of a Central Committee Party organizer. After devoting three years to administrative Party work, Utkin had not, however, earned the confidence of the Party's upper echelons that would have enabled him to rise in the ranks of the central offices.

As soon as OKB-1 was set up within NII-88 in 1950, Utkin begged Korolev to take him on. When I arrived in Yangel's department, Utkin had already set up the measurements laboratory. Soon, this laboratory developed into a department that managed to acquire a staff of capable, energetic radio engineers. With the formation of new missile design bureaus and factories, the problem of radio telemetry measurements became so acute that it exceeded the OKB's capabilities.

Korolev belonged to the category of managers who thought on the scale of the national interests as opposed to financial considerations of the day. When Utkin and I approached him with the idea of creating a specialized scientific-research institute of telemetry for the entire industry, he assessed the proposal's prospects right away and said that he would release Utkin and all his specialists to such an organization. And so, at the very entrance to the city of Kaliningrad (now the city of Korolev) from Yaroslavskyoe highway, a state-of-the-art scientific-research institute was founded, without which it would have been unthinkable to test a single modern missile.[18] Its first director was Ivan Utkin. He was replaced by Oleg Shishkin, who would become the last minister of general machine building. After Shishkin's departure, Oleg Sulimov became the institute's director, while Oleg Komissarov became chief engineer. They had begun their careers at the OKB-1 telemetry laboratory in 1950 under my supervision.

IN THE FALL OF 1953, at the GTsP there was a demonstration of missile technology for the managers of various ministries. Among the invitees were aviation industry general designers, including A. N. Tupolev. Ustinov and Nedelin were the gracious hosts. I had not seen Tupolev since 1937, when he came to the Air Force

---

17. Konstantin Eduardovich Tsiolkovskiy (1857–1935) was, of course, the "founding father" of Russian cosmonautics.

18. This was the NII Izmeritelnoy tekhniki (Scientific-Research Institute for Measurement Technology), which today is known as NPO Izmeritelnoy tekhniki (Scientific-Production Association for Measurement Technology).

NII Shchelkovsko airfield, where we were preparing Levanevskiy's N-209 aircraft for the transpolar flight to the U.S. At that time, as a government official and chairman of the State Commission for transpolar flights, he had painstakingly studied the aircraft's preparation.

Boris Konoplev had met with Tupolev both before and after the war to discuss aircraft radio issues. He diverted me over to the car where the rotund and weary Tupolev was sitting. Tupolev had already been shown the R-1, R-2, and R-11 missiles. Konoplev announced, leaving no room for objections, the "old man" would now see the R-5. The "old man" was then only 65 and would continue living and working until he was 84! Konoplev drove Tupolev up to the R-5 standing on the launch pad and, with his ever-present enthusiasm, began to explain the advantages of the radio control system. When Tupolev learned that the missile would cover a distance of 1,200 kilometers in 12 minutes, he smiled skeptically and said, "That's impossible."

The demonstration launches took place several hours later. The R-5 was also launched. During the launches I was at the *Don* telemetry system receiving station. Konoplev had stayed with Tupolev and later he told me that the "old man" was so amazed that he was ready "to throw in the towel on his airplanes and build missiles." Luckily, that didn't happen. Tupolev could be quite content with the work of his former graduate student Korolev, who was then only 47 years old. Neither of them yet knew that Korolev would attain posthumous worldwide fame as great as Tupolev's.

Back then, Tupolev believed he had a monopoly on nuclear bomb–carrying airplanes. After the launches, at a dinner for a very select group, when Ustinov and Nedelin in strictest confidence let it slip that Korolev was supposed to retrofit the R-5 to carry an atomic bomb, Tupolev said, "That's dreadful business. What if it falls on our own territory?" We, too, understood that this was dreadful business and developed interlock systems in case the missile went off course.

These episodes in autumn 1953 have also stuck in my memory because, after seeing me in the assembly and testing building, Ustinov rapidly approached me, gave me a firm handshake, and asked, "Is everything all right?" I assured him that everything was "quite all right." He wished me all the best and returned to the crowd of distinguished visitors. I understood that my two years of disgrace had ended. In part, the general thaw after Stalin's death and Beriya's arrest contributed to such a change.[19] A period was dawning when slanderers and careerists had their tails between their legs. The top-level leadership's ubiquitous suspicion and distrust of the managerial staffs was being replaced by sober assessments of business qualities, talents, and real achievements. Unfortunately, even during the Khrushchev thaw, it

---

19. Three months after Stalin's death in March 1953, the post-Stalin Politburo had Beriya arrested. Six months later, in December, Beriya was tried, sentenced to death, and shot on orders from the new leadership.

wasn't everywhere, or for very long, that scientists and designers could demonstrate their own will and work confidently without looking over their shoulders at the all-powerful state and Party apparatus.

In February 1956, NII-88 Party activists held a meeting on the results of the 20th session of the CPSU.[20] All present were surprised that Colonel General Serov, Lavrentiy Beriya's former deputy for counterintelligence, delivered the report on behalf of the Central Committee. This was the same Serov who had arranged for the German specialists to be sent from Germany to the USSR in 1946, and in 1947 had been a member of the State Commission on A4 missile launches. Serov's report depressed his audience. The people could not imagine that such horrible crimes had been committed in their country at the will of a man whom each in attendance believed to be great, infallible, all-powerful, wise, and gracious.

In March 1953, I was at the firing range in Kapustin Yar. We were getting ready for flight tests on the R-5 missile. Suddenly, in the assembly and testing building the Moscow call-sign resounded over the loud speakers with the official announcement of Stalin's death. Combat officers, men who had gone through the war, whom I had known from Germany and would never have suspected of sentimentality, broke down in tears! Unashamed of our tears, we turned to each other with the question that was being asked by millions back then: "What will happen now? How will we live?" Such was the hypnotic force that Stalin's name possessed. After all, we heard the announcement of Stalin's death while standing next to the missile we were developing on his instruction. Everything that had been created for missile technology both here at the firing range and in our country had been his will, aimed at protecting the country and each of us from the inevitable aggression of American imperialism. Those were our thoughts at that time in 1953.

Three years later much had changed. Stalin's name was no longer worshipped. But what Khrushchev reported to the 20th session and Serov was telling us soon after stunned us much more than the announcement of Stalin's death in 1953. After Serov finished his report, the dead silence of the auditorium was broken when a woman cried out in a loud voice:

"Ivan Aleksandrovich! Explain to us, where were you? Who were you, what were you doing? You probably shouted 'Hail Stalin!' loudest of all. What right do you have to talk about Beriya's evil deed, if you were his deputy?"

Everyone looked at the elderly woman standing in the middle of the auditorium. As I later heard, she was from the metal working shop. Serov remained silent

---

20. The Twentieth Party Congress in 1956 was one of the most famous sessions of the Soviet Communist Party. During the meeting of all the assembled delegates, Party Secretary Nikita Khrushchev publicly denounced the late Stalin and his cronies and enumerated a list of their unimaginable crimes, putting into motion a series of events that would result in the first stages of de-Stalinization of Soviet society and culture.

for a long time. The audience waited, also in silence. Finally, Serov stood up and replied:

"I am certainly guilty in many respects. But so are all of you sitting here. Didn't you praise Stalin at all of your meetings? And how many times did each of you stand up and applaud to exhaustion when you heard Stalin's name at your conferences and meetings? Now the Party wants to free itself from this cult. It is difficult for all of us, and let's not keep score with one another."

There was no discussion or debate. As the meeting broke up, we departed feeling ambivalent: depressed by the horrifying facts that had been exposed, but hopeful that now everyone could breathe easier. Perhaps this might even lead to the end of the Cold War.

Many years later I struck up a conversation with a quiet, modest, elderly colleague who had worked in our secret documentation department. I had heard that he was hired by OKB-1 at Serov's personal request. I asked him what he knew about Serov's activity as Beriya's deputy. It turned out that during the war he had been Serov's aide-de-camp. He related several episodes describing the exceptional fearlessness of Serov at the front during the most trying ordeals. Together they had gotten into such messes that it was a wonder they had emerged alive. Serov had no direct involvement with repressions, but, of course, he knew a lot. He was not afraid of Beriya, and it isn't clear why Beriya put up with him.[21]

AFTER MY FIRST BOOK, *ROCKETS AND PEOPLE*, CAME OUT IN RUSSIA IN 1994, I was justly criticized for not mentioning or for saying very little about the accomplishments not only of my immediate colleagues, but also upper management. I must agree with this criticism. My only excuse is that some of the people worthy of inclusion in this book are beyond the grasp of my memory and my literary and physical capabilities. Nevertheless, I shall try to insert additional information into this new publication of my memoirs, starting with the life of one of the leading managers of our rocket and space industry.

In 1959, Yuriy Filippovich Dukhovnov was sent to work with us. He had graduated from MVTU after serving in the army. He had a difficult relationship with his immediate boss. At that time, I was a deputy chief designer and chief of Branch No. 1 ("the second territory"), where Viktor Gladkiy's department was located. Gladkiy, who Pilyugin used to call "Mr. Rough," was in charge of problems of rigidity, elasticity, and the fundamentals of strength in missile hulls and large spacecraft designs.[22]

I don't remember exactly when, but in 1963, Military-Industrial Commission (VPK) Deputy Chairman Georgiy Nikolayevich Pashkov called me via the Kremlin

21. In the post–Cold War era, historians and scholars who have explored Serov's life have illuminated much of his callous disregard for the value of life. See for example, Michael Parrish, "The Last Relic: Army General I. A. Serov, 1905–90," *The Journal of Slavic Military Studies* 10 no. 3(1997):109–129.
22. "Gladkiy" in Russian means "smooth."

direct line.

"I could go directly to Korolev, but I am afraid he might do something rash. I have a personal request of you. My nephew Yuriy Dukhovnov is working at your facility. For some reason, he isn't getting on with his immediate boss for theoretical problems. I don't want him to lose his job with you all, especially since he is also enrolled at night school in the mechanical engineering and mathematics department at MGU."

I must admit that I personally knew nothing about Pashkov's relative, but I carried out his request. My conversation with Viktor Fedotovich Gladkiy was very difficult. He accused Dukhovnov of being unable and unwilling to work with an organization that had its own traditions.

I had three more conversations with Pashkov on this subject. Unfortunately, the time came when Gladkiy and I no longer had any administrative power, and under the pretext of staff reduction in 1992, Yuriy Dukhovnov, who was a member of the American Mathematical Society, was dismissed, supposedly "voluntarily."

Yuriy Dukhovnov lived for many years with Georgiy Pashkov's family, and, as I understand, was very close to him. Firsthand and with his approval, I have included new information that I have found out about the life and work of Georgiy Pashkov in this new edition of my memoirs.

Pashkov was born in 1909. He began his working career as a lathe operator. He graduated from the Leningrad Military Mechanical Institute, was a correspondence course graduate student of the same institute, and in 1948 defended his candidate's dissertation. In what were for our generation almost standard career phases, everything seemed normal. After the repressions of 1937–38, there was an acute shortage in the managerial ranks of the central state offices. During the period of 1939–48, Pashkov headed the second department of USSR *Gosplan*, that is, its armaments department. Actually he served as *Gosplan* Chairman Nikolay Voznesenskiy's deputy for defense industry planning and development of armaments.

I first became acquainted with Pashkov when he was in Bleicherode as part of a government commission during the summer of 1946. He was actively involved in distributing responsibilities for the future development of missile technology between ministries and in dividing the German missile stock of the Institutes Nordhausen and RABE.

Not only I, but even future Chief Designers Korolev, Glushko, and Ryazanskiy were a bit surprised at how attentively and virtually without objection Artillery Marshal Yakovlev, Minister Colonel-General Ustinov, Institute Nordhausen chief Major General Gaydukov, and other masters of our future destiny received the modest man dressed as a civilian.

During the war, Pashkov often made visits to Stalin's office with his boss Voznesenskiy, the *Gosplan* chief, and Beriya, who was in charge of domestic intelligence, including scientific and technical intelligence. But he also had several "face-to-face" meetings with Stalin. The first meeting took place at his personal request in late 1942. Pashkov asked Stalin to let him go to the front in order to avenge the death of his brother who had died during the battle of Kharkov. Voznesenskiy had refused Pashkov's request, and now only Stalin could reverse his decision. Stalin heard Pashkov out in silence and calmly said, "We will avenge your brother's death. But we need arms to reckon with the fascists. You and your comrades will provide them for the army."

The second face-to-face meeting took place in 1945. Pashkov was included in Artillery Marshal Yakovlev's government commission and not just with the broad-ranging rights of a *Gosplan* representative. The commission served as his "cover" for carrying out a secret assignment to meet with the now legendary Soviet agent Kim Philby.[23] Through Pashkov, Stalin gave Philby the assignment to copy documents from the file of von Braun, who surrendered to the Americans in May 1945. Thanks to Kim Philby's efforts, supposedly the mission was a success.

I am just surprised that none of the individuals who were actually interested in that file ever saw it! However, during the first years after the war, I had the opportunity to learn that the KGB leadership sometimes shelved scientific and technical materials that were obtained through the efforts of its foreign agents.

During their third meeting, Stalin gave Pashkov a very delicate assignment. As everybody knows, Joseph Kennedy, the older brother of future U.S. President John F. Kennedy, was a captain in the U.S. Army Air Force. At the end of the war, he was the commander of a sabotage group that was assigned to destroy an industrial site, which according to intelligence information was involved with the development of new types of aircraft technology. To carry out this assignment, Joseph Kennedy received an airplane packed with explosives. Several miles before reaching the target, the crew was supposed to switch the aircraft into automatic control mode on course to dive toward the target. The crew was supposed to bail out and parachute into an area where underground resistance members were waiting for them. Several miles before the bail out zone, however, the airplane exploded under mysterious circumstances. The crew perished, and the sabotage was not carried out.[24]

During this period, the Kennedy brothers' father, Joseph P. Kennedy, Sr., was the U.S. ambassador to Great Britain. He went to the USSR Ambassador to Great Britain Mayskiy and also to Churchill and Stalin requesting assistance in an investigation of the causes for the explosion. Stalin sent Pashkov to meet with Kim Philby and, if necessary, to call in additional agents to uncover the causes for the loss of the American sabotage group. Thanks to the efforts of the agents and Philby personally, the assignment was fulfilled. The individuals who had placed a detonator with a timing mechanism on the aircraft during its preparation were identified as intel-

---

23. Harold Adrian Russell (Kim) Philby (1921–88) served in the British Secret Intelligence Service while simultaneously being a spy for the Soviet KGB during the early Cold War. He was quite possibly one of the most successful spies in the history of espionage.

24. Joseph Patrick Kennedy, Jr. (1915–44) was killed on 12 August 1944 when his naval B-24 airplane exploded over the coast of England. He was involved in a mission to destroy V-1 and V-2 launch sites in German-occupied territory in France.

ligence agents of Nazi Germany.

In 1948, Stalin summoned Voznesenskiy, Malyshev, Zhukov, Vasilevskiy, and Ustinov to his office. He picked up a file on his desk and said: "These are the letters of Academician Petr Kapitsa. He writes that Beriya is a good organizer but doesn't have a very good grasp of physics. Recommend a man for the post of Beriya's deputy on these matters." No one said a word. Suddenly Ustinov uttered, "I know such a man. Georgiy Nikolayevich Pashkov."

Stalin smiled and said, "I approve of this candidate."

At that time Pashkov was carrying out some routine assignment for Stalin in Germany. Summoned back to Moscow, Pashkov flew in and set out to look for his car at the airfield. Suddenly he was approached by two sturdily-built young men in civilian clothes. They said, "Come with us. You need to get into this car." They pointed to a black ZIM.[25]

Pashkov understood that these people were from the KGB. All three sat down in the back seat—Pashkov in the middle, and the KGB guys on either side.

"Those devils," thought Pashkov, "They didn't even let me say goodbye to my family."

They drove past the *Gosplan* building straight to the KGB building on Lubyanka. However, rather than leading him into an interior prison cell, they took him directly into Beriya's office. Here, Beriya himself apprised Pashkov of the decree signed by Stalin transferring him from *Gosplan* to the KGB.

In 1954, to centralize the planning and management of all scientific-research and experimental design projects in the armaments and the defense industry sector, Khrushchev personally formed the managerial staff of the VPK, the Commission on Military-Industrial Affairs under the USSR Council of Ministers. Dmitriy Ustinov was appointed chairman. His first deputy was KGB representative Georgiy Pashkov.[26]

Thus, in 1948, Ustinov recommended Pashkov as Beriya's deputy, and, in 1954, enriched by his wealth of KGB experience, he brought him into the Kremlin, above all, to manage missile building. Pashkov spent over 15 years in that post.

At various large conferences Pashkov was always very reserved and all of his speeches were extremely brief. Many times I had to visit him in his Kremlin office. In face-to-face conversations or with other VPK colleagues he was far less laconic. But at the same time, not once did I hear hurtful insults directed at anyone.

Sometimes his reticence was frustrating. But chief and general designers, their

---

25. ZIM—*Zavod imeni Molotova* (Molotov Factory). ZIM refers to a model of car manufactured there.

26. Recent evidence suggests that the predecessor to the VPK, the Special Committee on Armaments for the Army and Navy, was established in April 1955, not in 1954. It was renamed the VPK in December 1957. In 1955–57, the Committee was headed not by Ustinov but by Vasiliy Mikhaylovich Ryabikov. Pashkov served as deputy chairman of the Special Committee in 1955–57 and then of the VPK in 1957–70.

deputies, ministers, and other figures of the defense-industrial elite knew they had to go through Pashkov in order to solve problems of funding, deadlines, and recruitment of subcontractors and to push through anything that required the help of the central authority.

His great wartime experience planning for the military industry, his joint work with "economic dictator" Voznesenskiy, and six years working in the KGB, which gave him access to all the secrets of immediate developments and a wealth of information about the state of military technology abroad, had made Pashkov a highly competent government administrator; he managed the most important state programs in the missile and space technology sector and in associated industries using the principles of mobilization economics.

He demanded that his staff and the managers of large programs develop a detailed general schedule from the conception of the idea until the product was put into service that overlooked nothing and clearly described everyone's responsibilities.

I have failed to fulfill my wish to write about "managers and colleagues" in a single chapter. I was intimately involved with too many interesting, talented, and enthusiastic colleagues and managers. The life and work of each of them has in one way or another remained as a contribution to world cosmonautics. But who has the capacity to recall and write about each of them?

Chapter 10
NII-885 and Other Institutes

In this chapter, I write about the organizations that designed and built guidance and control systems for Soviet missiles in the postwar period. The most important of these was NII-885. The same decree that gave rise to NII-88 also created NII-885, the head institute for long-range ballistic missile and air defense guided missile guidance systems. The Ministry of Defense factory on Aviamotornaya Street in Moscow was selected as the NII-885 facility. A large group of specialists in the field of relay technology and telephone and telegraph equipment worked at this factory. Many of them had been evacuated in the spring of 1942 from the Krasnaya Zarya Factory in besieged Leningrad. To begin work on missile technology instruments, the factory had absolutely state-of-the-art equipment and highly skilled workers, who had also been evacuated from Leningrad and had settled in Moscow. Two weeks after the 13 May 1946 decree was issued, many telephone and telegraph control engineering specialists were sent to Germany and turned up at the Institute RABE. Among them were Pilyugin's future deputy Georgiy Petrovich Glazkov, Abram Markovich Ginzburg, and Yakov Stepanovich Zhukov. Upon their return from Germany, they held key engineering posts under Pilyugin's supervision at NII-885.

While they were still in Germany, it was announced that NII-885, which was being established at the Ministry of Defense factory, was being transferred to the Ministry of the Communications Systems Industry (MPSS). This ministry assigned its own director to the new institute. Unlike our Director Gonor, this man had absolutely no knowledge of missile technology.[1]

At first he relied on Mikhail Ryazanskiy for everything. The latter had been appointed chief designer and deputy scientific director. For the first three years Pilyugin did not officially have the chief designer title and was referred to as deputy chief designer. Actually, management of development at NII-885 was neatly divided between Pilyugin and Ryazanskiy.

NII-885 received its share of German specialists from Germany. They had been separated from our German specialists who came to work in Podlipki as soon as

---

1. N. D. Maksimov headed NII-885 in 1946-49.

they arrived in the Soviet Union. They settled down in Monino, where a sanatorium building had been made available for them. It was too far and costly to transport the Germans to Aviamotornaya Street in Moscow. For that reason they worked in Monino and specialists from NII-885 gladly commuted to see them at the sanatorium.

The NII-88 and NII-885 organizations worked very closely together, in terms of both design and everyday routine work. For example, I was a member of the NII-885 scientific technical council. Ryazanskiy and Pilyugin were in Podlipki almost every week, participating in scientific-technical council sessions or meetings of the Council of Chief Designers. We consulted with them not only on technical issues, but also on organizational matters, including personnel problems.

From today's perspective and a common sense perspective, it is incomprehensible why the subsidiary branches of the NII-88 and NII-885 institutes and their German specialists were completely isolated from one another when their main staff people enjoyed such close contact. Back then this was attributed to "top priority governmental interests for the absolute preservation of state secrets." Presumably the security services demanded such arrangements.

The group of German specialists who worked at Glushko's OKB-456 was isolated in precisely the same way from our NII-88's German branch on the island of Gorodomlya; we were, in fact, totally unaware of the German group at NII-885 in the electrical industry. One can't blame only the security services for this arrangement. If the chief designers and three or four ministers had demanded that all the Germans be combined for the benefit of a project, the government absolutely would have accepted such a proposal.

It wasn't that being considerate was alien to our chief designers and the ministers standing over them. A German team that was too strong could generate serious competition for our own developments. And, above all, this group would have to be provided with an experimental and production base. But at whose expense? Of course, at the expense of Korolev, Ryazanskiy, Pilyugin, and Glushko, whose production capabilities were already limited.

Once, I had indiscreetly expressed the idea of a merger to Ryazanskiy and Pilyugin. I proposed that all the German guidance specialists be transferred from our branch in Gorodomlya to NII-885 so that they could develop a guidance system for the missile that Gröttrup was designing. They both pounced on me and demanded that I not dare come out with such an idea at NII-88, much less at the ministry. If that's the way friends reacted to this idea—and they really were my friends—then naturally I never broached that subject anywhere again.

While equipping my guidance systems laboratories, I, as NII-88 deputy chief engineer, consulted with Ryazanskiy and Pilyugin, who were remodeling the factory areas for laboratories. We even started a competition to see who could acquire the best laboratory setup.

Ryazanskiy complained that Minister Ustinov, a former artilleryman, was helping me more than their "switchboard operator" minister. However, soon thereafter

Sergey Mikhaylovich Vladimirskiy was appointed their deputy minister and the immediate supervisor of guidance systems operations in the Ministry of the Communications Systems Industry. He was a very energetic and knowledgeable man and, in every sense of the word, a decent man. He was appointed after NII-885's historic fire, which destroyed almost the entire new laboratory facility.

One early Sunday morning in the summer of 1948, I was awakened by a telephone call from Director Gonor.

"There's been a great disaster. Ryazanskiy's institute has burned down. I'm taking Korolev with me, and I'll be by to pick you up. Be ready soon."

When we drove up to the grounds of NII-885, we got a sense of the disaster by the dozens of fire trucks. The fire was already out, but steam mixed with smoke was rising up over the factory grounds. The administrative building was completely packed with managers of all ranks. Sooty and dirty, Ryazanskiy, Pilyugin, Boguslavskiy, and another five of their colleagues were busy writing explanatory memos listing the laboratory equipment that had been destroyed.

The administrative building housing the managerial offices was unharmed, but the factory had burned. The production shops and newly created laboratories had been hard hit; the collapse of the burning ceiling and roof had damaged them. The roof was held up by timber joists, and the ceilings had been coated with resin for waterproofing. It was the perfect fuel for a fire. When we struggled through the debris into the flooded shops and laboratories we sized up the scale of the disaster. They sent someone to find Gonor, Korolev, and myself and led us into an office where Zubovich, Vetoshkin, and the MPSS managerial staff were waiting.

"Lev Robertovich," said Zubovich, turning to Gonor, "Ryazanskiy and Pilyugin will be coming over to your place. They're going to look around Chertok's laboratories. I want you to give them everything they ask for without any arguments. Is that clear? We will assist them with machine tools and other factory equipment ourselves."

A week later we started to dismantle and transfer instruments and rigs from my departments to NII-885. A month later, the sparkling clean and orderly factory shops and restored laboratories of NII-885 were up and running as if no fire had ever occurred. They found that the culprit for the blaze was a soldering iron that was had been turned on in one of the shops over the weekend.

I visited NII-885 almost every week. I needed to coordinate layouts, check up on instruments being prepared for missiles, and let off some steam from the daily grind with Ryazanskiy, Pilyugin, and Boguslavskiy. We had some heated arguments about the integrated guidance system.

Even during those first years I advocated a purely autonomous inertial guidance system for long-range ballistic missiles. I started up arguments on this subject, having first studied the prospects in Viktor Kuznetsov's shop for increasing the precision of gyroscopic command instruments. Ryazanskiy argued that my proposals were unrealistic. Pilyugin tried to stay out of the arguments, but privately he agreed with me. At our next one-on-one meeting he said, "Don't get Mikhail started again.

179

It's still early, and there's no need to offend him. And don't get Sergey set against radio control. It's not time yet."

At the same time, Pilyugin was very jealous of my ideas and work on astronavigation, since this directly encroached on his prospects. Unlike Pilyugin, Kuznetsov was very interested in this work and promised any assistance from his production operations when needed.

By late 1948, the structure of NII-885 had already taken shape. Laboratories and shops for the development and production of onboard and ground equipment for missiles were up and running at the former telephone and telegraph equipment factory. Ryazanskiy, Pilyugin, and Boguslavskiy formed an alliance with other specialists, "buddies" from the Institute RABE. They had to endure many battles with ministry bureaucrats, defending their structural concept and placement of managerial personnel.

Right off the bat there was a split into two disciplines similar to what had occurred at NII-88: long-range ballistic missiles and air defense missiles. Ryazanskiy held on to his post as chief designer for long-range ballistic missile guidance. At the same time he served as first deputy director and chief engineer, and therefore he was also responsible for air defense missiles. Govyadinov was involved exclusively with air defense missiles as chief designer of the guidance system. Like Sinilshchikov in our operation at NII-88, Govyadinov thought that this field was being stifled. There were similar conflicts on this subject at both institutes.

NII-885 was directly subordinate to communications equipment industry Deputy Minister Sergey Vladimirskiy. At their respective ministries, he and Vetoshkin supported in every way a policy for forcing out air defense missile developments to the Ministry of Aviation Industry. Ultimately, this happened and both NII-885 and NII-88 halted their developments in the field of air defense missiles. In 1950, this project moved to KB-1 in the Third Main Directorate under the Council of Ministers.[2]

This certainly pleased Pilyugin because it freed up production facilities for his projects and, in addition, transferred talented specialists to him. Among them was Mikhail Samuilovich Khitrik, who subsequently became Pilyugin's deputy. He was one of our country's leading scientists in missile guidance systems.

From the very beginning, projects on long-range ballistic missiles were split into three areas at NII-885: inertial guidance systems, radio guidance systems, and radio telemetry systems. By mid-1948 Pilyugin's department had a staff of over 500 and had been converted into a complex of specialized laboratories and departments. Georgiy Glazkov became Pilyugin's first deputy. We had become accustomed to seeing him at the Institute RABE, where he constantly studied the operating prin-

2. The Third Main Directorate (TGU) of the Council of Ministers was a top secret body organized in February 1951 to manage development of the Moscow air defense system code-named Berkut. KB-1 was the primary systems integrator (and designer of missile control systems) within TGU.

ciples and layouts of general electrical diagrams. In Germany we had considered him the main Soviet specialist to have figured out all the fine points of the layout of relay automation of the A4 missile's "ground-to-air" integrated circuit.

Abram Ginzburg, who had also gone through the Institute RABE with Glazkov, supervised the integrated laboratory. He possessed a unique "circuit memory" and the gift for being able to react quickly to the unpredictable behavior of complex relay circuits. When he needed to send for or find Ginzburg, Pilyugin's memories took him back to the historic *bankobus* in the autumn of 1947, where we used to hold our meetings by the firing rig in Kapustin Yar. He repeated the words of General Serov: "Show me this Ginzburg."[3] Ginzburg really made his mark in the field of integrated design developments, and, in 1952, he was appointed chief designer of the Kommunar Factory in Kharkov.[4]

The Kommunar Factory traced its lineage from a labor commune of homeless children established by the renowned educator Makarenko. At this factory they mastered the production of FED cameras—replicas of the German Leica—and also electric drills.[5] All of this extremely necessary and useful production would be shut down or squeezed to the side in order to begin making equipment for the R-1 and many other missiles.

The Kommunar Factory became a series-production facility for the majority of Pilyugin's developments. But this was just the beginning of Kharkov's relationship with missile technology. The government of Ukraine, where the series production of R-1 missiles was already under way in Dnepropetrovsk, had future plans for R-2 production and wanted very much to have "its own" production, without having to rely on Russian contractors. Moscow encouraged such an initiative. Soon thereafter, a very powerful cluster of missile instrumentation OKBs and factories sprouted up in Kharkov. Chief Designer Ginzburg became a highly regarded figure in the town and in the ministry. But he didn't forget his first firing tests.

Many years later Ginzburg and I ran into each other in Kislovodsk at the entrance to the *Krasnyye kamni* (Red Stones) sanatorium. I pointed my camera to take a picture of him with the sanatorium in the background and joked that now I could show all my friends "this Ginzburg." He confessed that he still got horrible chills up his spine when he recalled Serov's voice saying, "Show me this Ginzburg." In the early 1990s I wanted to revise my memoirs with Ginzburg's reminiscences. He had a good sense of humor and at one time had promised to tell me a lot of interesting things about the establishment of the missile instrumentation industry in Kharkov. Alas! I was unable to do it. Abram Ginzburg had moved to the U.S.

---

3. See Chapter 2.
4. This plant was also known as Factory No. 897.
5. FED stood for Feliks Edmundovich Dzerzhinskiy, who was the founder of the predecessor to the KGB.

MAKUSHECHEV, ANOTHER SPECIALIST WHO WENT THROUGH THE INSTITUTE RABE IN BLEICHERODE, was appointed supervisor of the NII-885 laboratory for the coordination of operations with gyroscopes and control surface actuators. Actually, this was an external relations inspection laboratory, that is, its goal was to keep track of what Kuznetsov and Chertok were doing, so that they wouldn't come up with any "independent actions" that were harmful to NII-885. Was this intermediate control necessary? Makushechev's laboratory became a source of conflicts. Viktor Kuznetsov and I convinced Pilyugin that military acceptance was already monitoring us and one more supervisor would only interfere with our work. Ultimately, under our pressure Pilyugin adapted that laboratory for internal needs.

Pilyugin immediately established his own autocratic, totalitarian regime for autonomous systems. Not all of Pilyugin's associates liked his autocracy. There were obstinate individuals who had their own ideas on various technical problems that differed from the chief designer's way of thinking. One of these recalcitrant types was Nikolay Semikhatov, who managed the stabilization controller laboratory and was responsible for developing all sorts of transducer amplifiers. Pilyugin's differences of opinion with Semikhatov on technical issues affected their personal relationships, but benefited missile technology.

Pilyugin showed no enthusiasm for Korolev's ideas and projects on the development of submarine-launched missiles and disapproved of his young deputy Finogeyev's enthusiasm.[6] After Korolev handed over the development of naval missiles to Makeyev, he recommended that Pilyugin follow his example.[7]

Without any misgivings, Pilyugin agreed to Semikhatov's departure to a newly established firm for the development of naval missile guidance systems. Now the manager of a new independent organization, Semikhatov exhibited engineering talent and extraordinary organizational skills. Within a short period of time under his management Semikhatov set up an instrumentation institute in Sverdlovsk with its own pilot plant.[8] Enterprises in the Urals became the primary facilities producing submarine-launched missiles. Nikolay Semikhatov received every conceivable governmental award in this field and attained all the academic degrees and titles; he even became a full member of the Russian Academy of Sciences. He created his own scientific and technological school of long-range submarine-launched missile guidance. Even today, these missiles compose the primary armament of the submarine fleet.

Semikhatov devoted 45 years of his life to solving the most difficult problems of accuracy, reliability, safety, and the operation of flight control systems for mis-

---

6. Vladlen Petrovich Finogeyev (1928–) served as Pilyugin's deputy in 1957–70.

7. In 1955, Korolev distributed the development of submarine launched ballistic missiles to a branch in Miass under Makeyev's tutelage.

8. Nikolay Aleksandrovich Semikhatov (1918–) served as Chief Designer of SKB-626, the primary supplier of guidance systems for naval ballistic missiles.

From the author's archives.

Shown here at a meeting is Chief Designer Nikolay Pilyugin (1908–82), the founding patriarch of guidance systems for early Soviet ballistic missiles. He is flanked by two senior industrial managers of the Soviet space program, Vasiliy Ryabikov (left) and Sergey Vetoshkin (right).

siles launched from submarines cruising at full speed. By the late 1970s, the Soviet Union was on equal footing with the U.S. in terms of the strength of the submarine component of its strategic nuclear missile forces. All submarine missile guidance systems from the very first to the last were purely autonomous; that is, radio systems were not used to increase the target striking accuracy.

THE SITUATION WITH RADIO SYSTEMS TURNED OUT TO BE MORE COMPLICATED. Ryazanskiy entrusted the reproduction of the German *Viktoriya* lateral radio correction system to our new colleague Mikhail Borisenko. From the very beginning, this was a bone of contention between the two managers of radio developments at NII-885, Borisenko and Boguslavskiy.

Yevgeniy Boguslavskiy had begun to develop what was for those times the advanced *Don* radio telemetry system in place of the very low-capacity German *Messina* and its domestic modification, *Brazilionit*. The *Don* system was used widely after the R-1 firing range tests in 1949. It found a secure niche on all subsequent missiles up until the first intercontinental missile, when it was replaced by the *Tral* system developed by the OKB MEI. The *Tral* had an even greater information handling capacity.

The transfer of Boris Konoplev from NII-20 appreciably enhanced the radio engineering program at NII-885. During 1948–49, he served as chief designer of the R-3 missile guidance system. However, his arrival at NII-885 put quite a strain on rapport between the managers.

Konoplev thought that he was fully qualified on all problems of missile radio engineering and had no patience for Ryazanskiy's guidance. Soon he took over all the radio engineering projects in the institute. The strained situation was one of the reasons why Ryazanskiy accepted the offer of Ustinov and Rudnev and took the vacant position of NII-88 chief engineer after Yuriy Pobedonostsev left for a teaching position in 1950.

NII-885's frequently replaced directors had to spend a great deal of time resolving conflicts between Konoplev, who was striving for the "radiofication" of guidance, and Pilyugin, the system's actual manager. They usually dragged me, and then Korolev, into heated arguments over these problems. Korolev sympathized with all the warring factions, since all of them were top-notch, brilliant specialists, devoted to their work. But diverse viewpoints on the prospects for the development of guidance systems aggravated personal relationships. You couldn't accuse any of them of being dishonorable.

Not wanting to complicate his personal relations with Ryazanskiy, Pilyugin, Konoplev, or Boguslavskiy, Korolev resorted to a very wise tactic if a technical matter needed to be decided in favor of one of them. Once he had considered the situation and prepared proposals to resolve the problem, he assigned me or another one of his deputies to get involved in the conflict. If the conflict couldn't be resolved with our participation, all of us turned to Korolev with complaints against each other. He took on the role of arbiter. Here, much to the delight of the subcontractors, he pounced on his own people, who supposedly weren't reporting to him objectively or hadn't sorted it out. Everything usually ended with decisions that appeased everyone, while Korolev, slyly smiling with obvious pleasure, signed them.

One such conflict arose over the control surface actuators for the R-2. The first flight tests on R-2E missiles, which we had used for the experimental development of principles for the R-2 in 1949, showed a dynamic stability problem in the stabilization controller. For the first time, as an experiment, in place of the classic gyro horizon and gyro vertical we installed a gyro-stabilized platform that Kuznetsov and Tsetsior had developed.

Although the platform was very similar to the one that Kuznetsov had demonstrated to us in Berlin at the Kreiselgerät factory, Tsetsior assured us that his development was better. He had studied all the German achievements, found the weak points there, and reworked a lot of things. This was not a copy but indeed his own original development. Installing his platform in the R-2 pressurized instrument compartment caused the designers a lot of trouble. We ran a risk, because unlike the R-1 layout, the R-2 instrument compartment was located right by the engine, a source of vibration and powerful acoustical effects that could be transferred along the airframe.

The first R-2 missile with the new platform crashed. Our interpretations of the causes differed. Tsetsior himself attributed it to the platform's lack of vibration toler-

184

ance. Resonance phenomena had occurred in the platform's elements, and engine vibrations caused oscillations of its substructure. The new engine had been boosted appreciably compared with the RD-100 engine of the R-1 missile. Therefore, the intensity of the vibrations had also increased.

Despite Tsetsior's self-criticism, Pilyugin believed that the control surface actuators were to blame. In his opinion, the linear part of their performance was too small for the guidance principles underlying the stabilization controller developed by NII-885. I contended that no amount of linearity would help us as long as he, Pilyugin, was going to saturate the commands controlling the control surface actuators with interference. The vibrations caused such high-frequency oscillations of the sensors on the gyro-stabilization platform that interference blocked the legitimate signal, and the entire stabilization controller became a nonlinear system. Moreover, I chided Pilyugin for replacing the electronic tubes in the transducer amplifier with static amplifiers without having thoroughly studied the transient phenomena occurring in the electrical circuits containing iron-core windings. I maintained that doing away with the tubes was a progressive measure, but the static amplifiers could introduce much stronger nonlinearity than the control surface actuators.

It was a very heated argument. While preparing for the next R-2 launch, Pilyugin and I debated the issue so loudly right on the launch pad that the chief of the launch control team was forced to announce: "Since you are disturbing the peace and using inappropriate language during prelaunch tests, I request that you step away from the missile."

Pilyugin found Korolev and asked for his help. But Pilyugin had already explained his version to Korolev before our dispute, and Korolev had not given his approval to develop the new, more powerful control surface actuators; instead, he had advised that Pilyugin work it out with me. Now it was up to him to resolve the conflict. I explained that we were developing new control surface actuators strictly for future use, but their series production would be no easy task for our factory. It would take several months; the schedule for the R-2 flight tests would be disrupted.

In those days there were as yet no computers making it possible to conduct an experiment under laboratory conditions. The first simulator, Dr. Hoch's *Bahnmodel*, was not put into production after he left for KB-1. Pilyugin had just begun to develop his own electronic simulators. Korolev had to compensate for the shortcomings of research technology with his own intuition and will. He made a decision, which many years later served as a model to us for resolving seemingly dead-end situations. "Given the information at our disposal, no one is justified in asserting categorically the cause of the dynamic instability. Therefore, we are making the decision to hold all the suspects accountable."

Right there on the launch pad Korolev announced: "Boris, you're going to make new control surface actuators with our factory and you're going to coordinate their performance data with Nikolay. Nikolay, you will show us the performance data of your transducer amplifier using static amplifiers and if their performance is worse than the tubes, then take it in stride and redo it. As far as the gyro-stabilizing platforms are concerned, as regrettable as it might be, I have already come to an agreement with Kuznetsov without your input, to return to the gyro vertical and gyro horizon. Apparently we're not quite ready for [Ginzburg's] platform. But so that you won't all be offended, I have decided to replace the duraluminum tail with a steel one. Our strength experts think this will reduce the intensity of the vibrations in the instrument compartment."

As a result of these decisions, flight tests on the first series of R-2 missiles were broken into two phases and conducted from October 1950 through July 1951. In 1952, Pilyugin put his first electronic simulators into operation, substantially facilitating decision-making "under conditions of uncertainty."

THE ARRIVAL OF TALENTED YOUNG PEOPLE captivated by new problems played a major role. In the late 1940s, a group of engineers who had graduated from MAI came on board at NII-885. They included Georgiy Priss, Nina Zhernova, and Mariya Khazan. Soon Priss became the leading specialist and supervisor of integrated circuit development for the electroautomation of all the guidance systems for Pilyugin. Zhernova had a rare combination of feminine charm and an intuitive understanding of the dynamic processes of the stabilization controller. She brilliantly mastered research technology using the still faulty electronic simulators and possessed the ability to predict the behavior of a guidance system under different conditions.

I had to participate in the review of various accidents a number of times where Zhernova had been tasked with analyzing the behavior of the stabilization controller. She provided objective findings that did not always coincide with Pilyugin's views and were sometimes contrary to the departmental thinking of the entire firm. In such cases, Zhernova asked for time to repeat the research and simulation.

After numerous flight simulation sessions on a rig incorporating an electronic simulator, actual control surface actuators, and a transducer amplifier, Zhernova and Khazan spread out still-wet oscillograms on huge tables to prove that they were right.

When Mikhail Khitrik arrived in Pilyugin's collective, all research on motion control dynamics was transferred to him. He could combine profound theoretical research with practical recommendations for equipment under development. Khitrik established close contact with Korolev's dynamics specialists.

When I officially transferred from NII-88 to OKB-1, Georgiy Vetrov's group was already working there. It had been tasked with researching stability problems. This was supposed to have been integrated research in very close contact with the guidance system developers.

The chief developer of the guided missile did not have the right to categorically dictate his requirements to the guidance system developer. Success could be achieved only if the missile were designed as a single complex system. Problems of the structure, propulsion system, guidance system, and flight dynamics had to be studied in the closest cooperation with specialists from all the organizations responsible for this work. One of the most important services of the Council of Chief

Designers was the actual support of this activity. Korolev deliberately cracked down on any manifestations of egotism among colleagues who considered themselves "top dog." This won over specialists from contractor organizations.

This continuous working cooperation aimed at solving guidance dynamics problems did not occur overnight. Quite conscious of the need for a systemic approach, Korolev strove for direct contacts with the lead specialists of other organizations and, above all, with the NII-885 dynamics specialists.

Matters of ballistics, aerodynamics, structural loads, stability, controllability, accuracy, and inertia fell under the immediate purview of the chief designer. The guidance system chief designer also needed all of the baseline data on these problems. Therefore, Pilyugin and his people were not the end users, but active creative participants in the resolution of these problems.

During the development phase, integrated stands simulating the preparation and launch processes and in-flight operation of the entire intricate guidance complex were widely used. Integrated laboratories were developed for these simulators. The chief of the integrated laboratory was responsible for providing the entire system's operating technology, for knowing the special features of the missile itself, and for working in close contact with specialists from the entire institute and even closer contact with the head design bureau. Priss was the chief of such an integrated laboratory at NII-885 dealing with the R-2, R-5, and R-5M missiles.

Pilyugin appointed the young, talented, and very energetic engineer Vladlen Finogeyev as chief of the integrated laboratory for R-11and R-11M missiles and the R-11FM naval modification. While Korolev was absorbed with arming submarines with missiles, Finogeyev took advantage of his special situation. Soon Finogeyev became Pilyugin's deputy. He was awarded the Lenin Prize, and in 1961 he was named a Hero of Socialist Labor during a large award ceremony in honor of Gagarin's launch. But Finogeyev's brilliant persona somehow got under Pilyugin's skin. Having known Nikolay Alekseyevich for a long time and quite well, it was painful for me to observe that as the years passed he began to display jealousy toward his deputies who enjoyed great prestige outside his institute. Probably Khitrik was the only member of Pilyugin's inner circle who remained above suspicion until the latter's death. The spat that occurred through no fault of Finogeyev's resulted in his accepting the post of defense industry deputy minister. If Korolev had been alive, he would not have put up with that. It turned out that bureaucracy was not Finogeyev's calling. He returned to engineering work, but in a different field.[9]

Vladimir Lapygin and Boris Dorofeyev arrived from MAI to work for Pilyugin at the same time as Finogeyev. Lapygin was one of those individuals who were keen on gyroscopic platforms and who supervised their development when they were being produced under Pilyugin after the latter broke away from NII-885 in 1963

---

9. Finogeyev served as Deputy Minister of the Ministry of Defense Industry in 1970–81 and then Deputy Director of TsNII Avtomatiki i gidravliki (Central Scientific-Research Institute of Automation and Hydraulics) in 1981–2000.

and headed NIIAP.[10] After Pilyugin's death in 1982, Lapygin was appointed NIIAP director and chief designer.

Having passed through Pilyugin's school and having gained firing test experience at Novostroyka outside Zagorsk, Dorofeyev transferred to OKB-1 to work for Korolev. Soon thereafter he was appointed chief designer of the super-heavy N-1 launch vehicle for the lunar expedition. Dorofeyev shared the tragedy of this project to the full extent.[11]

THE FIRST FLIGHT TESTS OF THE R-5 MISSILE IN 1954 showed that the control fins started to vibrate in flight, and then the entire missile. This had not occurred at all in simulation during the design process on analog simulators of the "missile-stabilization controller" closed loop system. In those cases, engineers returned most meticulously to the analysis of previous launches of other missiles. Such excursions into the past very frequently showed that, from the standpoint of theory, irregularities in the behavior of standard control systems and oscillatory processes had also occurred earlier, but proper attention had not been paid to them if the flight ended without a crash. If the missile flew along the designated trajectory but intense vibrations developed throughout the entire hull around the center of mass, this posed a hazard because the missile structure experienced additional loads, especially if deviations during reentry into the atmosphere caused a high angle of attack. For future structural strength analyses, the load factors needed to be determined and standardized. Errors in load calculations meant unnecessary structural metal, a reduction of the payload mass, or reduction of flight range.

While still setting up Department No. 3 at the SKB, Korolev included the few load factor specialists he had in the design bureau and he pooled together strength analysts and designers. Viktor Gladkiy was one of the leading load factor theoreticians in Department No. 3 and later in Korolev's OKB-1 from the very beginning of NII-88's involvement in missile projects. He was supposed to calculate loads, taking into account acceleration forces, aerodynamics, tank pressurization, control deflection, and even vibrations. The results of calculations sometimes required that guidance specialists complicate the dynamic design to make control less rigid and more flexible in reducing loads. This really got on Pilyugin's nerves, and he would start to argue with Gladkiy, whose name means "smooth." After a typical quarrel, Pilyugin declared to Korolev, "Your Gladkiy is not the least bit smooth. He's rough."

The Germans who developed the guidance system for the A4, and after them

---

10. NIIAP—*Nauchno-issledovatelskiy institut avtomatiki i priborostroyeniya* (Scientific-Research Institute of Automation and Instrument Building).

11. Although relatively unknown in the West, Boris Arkadyevich Dorofeyev (1927–99) was one of the chief architects behind the famous N-1 Moon rocket. Between July 1972 and May 1974, that is, until the program was suspended, Dorofeyev served as chief designer of the N-1 program. Chertok will describe the N-1 program in Volume 4.

our specialists who developed the R-1 and R-2 guidance systems, viewed them as controllable objects possessing the properties of a "solid body," meaning that when exposed to loads, the missile hull would not deform at all. Such an assumption proved inapplicable for the R-5 missile, which was more than 20 meters long with hull diameter of 1.65 meters, like the R-1. The missile hull bent under the effect of loads from the control fins. The flexural elastic modes of the hull were transferred to the gyroscope bases. The gyroscopes responded naturally to these modes and sent commands to the guidance system, causing the control fins to shift. The loop closed and entered an unexpected self-oscillation mode.

In a joint effort, the OKB-1 and NII-885 dynamics specialists developed measures to limit the effect that this newly discovered phenomenon had on guidance. At one of the meetings that we had on this problem, I reminded Pilyugin about our materials resistance course at the institute. They had taught us that we could use a structure within the limits of its allowable elastic deformation. His comeback was, "We'll rock the missile with the control fins so much that your Mr. Rough will have to reinforce it with steel longerons." We introduced various filters into the system, but at NII-885 they continued to bad-mouth the "protection against Mr. Rough."

Another new curse for the guidance specialists was the effect caused by filling the missile with liquid. The control fins' vibrations not only bent the missile hull, but also disturbed the liquid oxygen and kerosene in the tanks. The fluctuations of the liquid surface caused additional perturbations. We needed to develop ways to counteract the effect of the filled tanks.

The effect of flexural vibrations and fueled tanks on stability proved very hazardous. The frequency of these vibrations fell within the guidance system's frequency band. Cooperative research was set up at OKB-1, NII-885, in the scientific departments of NII-88, and the military's NII-4 to study the new phenomena. Khitrik was in charge of this work at NII-885; Vetrov, Degtyarenko, and Gladkiy headed the project at OKB-1. At NII-4 Georgiy Narimanov made a special study of the effect of the liquid in the missile's tanks.

Through their combined efforts they developed a guidance theory allowing for the new phenomena. During 1955–56, guidance equipment was developed that was supposed to ensure stabilization over the entire dynamic structure. During this period the R-7 missile was designed, applying the experience derived from the R-5. To this day, missile and guidance system designers have to consider liquidity and elasticity as integrated factors from the very initial design stage.

THE BALLISTICS THEORETICIANS WERE IN A MUCH MORE ADVANTAGEOUS POSITION. Svyatoslav Lavrov and Refat Appazov, who had reconstructed the ballistics of the A4 at *Sparkasse* in Bleicherode with Dr. Wolf, the chief ballistics expert of the German arms firm Krupp, worked in Department No. 3, and later at OKB-1. Missile ballistics differs substantially from the ballistics concepts employed in artillery. Calculating the flight trajectory was an extremely labor-intensive business. For the users of the first Strela domestic computers and large electronic calculators (BESM), ballistics proved to be anything but simple.[12] Ballistics specialists came into play at the very beginning of missile design. They were also involved in the final phase of flight assignments for missile launch.

Range, accuracy, payload mass, aiming procedures, and adjustment of the automatic range control unit, consideration of the engine specifications, rate of propellant component consumption, and a myriad of other problems, including predicting the missile impact point in the event of possible crashes—all of this was part of the ballistics experts' job.

The first missile decade saw an unofficial interdepartmental association of ballistics experts from various organizations. Employees from the Academy of Sciences Department of Applied Mathematics headed by Dmitriy Okhotsimskiy, military theoreticians Georgiy Narimanov and Pavel Elyasberg under Tyulin's leadership at NII-4, the aforementioned Lavrov, Appazov, young Makarov, Karaulov, and Florianskiy in OKB-1, and the group of "guidance system" ballistics specialists headed by Nayshul set up in Khitrik's department formed a sort of ideological association. The military ballistics experts at the firing range were also part of the mix. They didn't simply follow the calculations of their industry colleagues, but actively intervened in drawing up the firing tables and flight profiles and tracking the flight trajectory.

One of the motives behind combining the ballistics experts from different sectors was their common interest in developing facilities for extra-trajectory measurements. It all began with the German cinetheodolites used to track launches in 1947. By late 1956, state-of-the-art radar tracking and data transmission systems had already been produced. They covered the entire flight path of the future intercontinental missiles. The combined ballistics experts initiated the creation of ballistics data processing centers. With the dawning of the space age, these centers and their tracking stations served as the basis for the first mission control centers and the entire Command and Measurement Complex (KIK).[13]

This example of interdepartmental ballistic solidarity is very illustrative. The various disparate teams coordinated their work not because of guidelines from the top, but because of a natural need to unite for a more effective solution to a common problem. The departmental quarrels that ensued among ministers, directors, and other managers did not break this professional solidarity. This unity of the first generation of the scientists and engineers of the first missile decade had an enormous significance for our work in the subsequent space age.

The joint work of the NII-88, OKB-1, NII-885, OPM, and NII-4 was not lim-

---

12. Russian language speakers typically refer to computers with the acronym BESM (*Bolshaya elektronnaya-schetnaya mashina*), which literally stands for Large Electronic Computation Machine.

13. KIK—*Komandno-izmeritelnyy kompleks*. The KIK was the official name of the Soviet ground communications network to support the missile and space program.

ited to interaction strictly on routine experimental design projects.[14] They also conducted joint scientific-research projects to predict missile technology development and to develop new ideas. They called in scientists from the Academy of Sciences, NII-4, and other organizations for these research projects, but invariably OKB-1 played the lead role in the NII-88 system. Korolev strove by all means, both practically and legally in guideline documents, to strengthen his role as chief designer and OKB-1 as the head organization. He did this very tactfully, with respect to everyone involved in the projects—with the exception of the NII-88 management. Until OKB-1 split off from NII-88 and became an independent organization, Glushko was also jealous of OKB-1's ever-increasing leading role and of Korolev personally.

Korolev strove to establish a completely independent organization. He wanted not only to escape from the guardianship of the NII-88 director, but also to separate completely from NII-88. It was not until 1956 that he finally succeeded. Inspired by this example, Pilyugin strove to acquire a great degree of independence within NII-885 and then to break away into an independent organization as well. But this did not happen until 1963. After Pilyugin's group departed, NII-885 became purely a radio engineering organization.[15] Mikhail Ryazanskiy was the technical director until the day he died in 1987.

NII-10 SHARED THE SAME NEIGHBORHOOD ON AVIAMOTORNAYA STREET WITH NII-885. This organization developed command gyroscopes for guidance systems as ordered by the same historic decree of 13 May 1946. A high concrete wall separated NII-10 from NII-885. But, in addition, a departmental wall also separated them. NII-10 was subordinate to the Ministry of the Shipbuilding Industry. In contrast to the new NII-88 and NII-885 organizations, in 1946 NII-10 was already a fully operational facility for the development of new instruments and was ready to fulfill "party and governmental" assignments. During the first years that NII-10 was involved with missile projects, its director was Valeriy Dmitriyevich Kalmykov, future USSR Minister of the Radio Electronics Industry. In addition to gyroscope technology for the navy, NII-10 developed naval radar, heat-sensitive radar, and naval radio navigational systems.

For us, the most important man at NII-10 was Viktor Ivanovich Kuznetsov. In Volume 1 of this series I wrote about my first encounter with Kuznetsov, in May 1945, at the Kreiselgerät Factory in Berlin.[16] Kuznetsov, whom we all simply called Vitya, was such a colorful figure that I will briefly describe his engineering background. Vitya Kuznetsov did his thesis project at the Elektropribor Factory in Leningrad. By the mid-1930s, a subdivision had formed at this factory for the

---

14. OPM—*Otdel prikladnoy matematiki* (Department of Applied Mathematics).

15. NII-885 is currently known as the Russian Scientific-Research Institute of Space Device Engineering (RNII KP).

16. Chertok, *Rockets and People.* Vol. 1, pp. 278–279.

development of gyroscopic instruments under the supervision of Vasiliy Nikitovich Tretyakov, future deputy minister of the shipbuilding industry, and distinguished scientist and engineer Nikolay Nikolayevich Ostryakov. Sergey Fedorovich Farmakovskiy was also developing a ship artillery fire control system at the factory.[17]

YOUNG ENGINEER KUZNETSOV VERY QUICKLY ENTERED THE INNER CIRCLE OF SELECT GYROSCOPE EXPERTS. In 1938, he received the first Stalin Prize for perfecting a fire-control system. In 1939, scientist/gyroscope experts Academician A. N. Krylov and Professors Ye. L. Nikolai and B. I. Kudrevich approached the government with a proposal to create a diversified scientific production center for gyroscopes to develop domestic instruments. By order of the People's Commissariat of the Shipbuilding Industry a center was created in May 1940 at NII-10 in Moscow. In 1940, Viktor Kuznetsov was appointed chief of the gyroscope section in the new institute, just two years after defending his final undergraduate thesis. At that time, Valeriy Dmitriyevich Kalmykov was NII-10 Chief Designer.

Physics and mathematics Candidate of Sciences Aleksandr Yulyevich Ishlinskiy, the future world-famous scientist and member of many academies and international science associations, was invited to Kuznetsov's department to provide theoretical reinforcement.

In October 1940, Kuznetsov departed on special assignment to Germany to take over equipment for naval vessels that had been ordered on contract. The Germans were preparing for war, but at the factories they hid nothing from him. The Gestapo was keeping a special dossier on Kuznetsov, which our security agency discovered after the war. The Germans highly appreciated his erudition and expertise.

On the morning of 22 June 1941, in Berlin, Kuznetsov was interned and taken to a camp where all Soviet citizens in Germany at that time were gathered. They offered him German citizenship. However, 22 days later the internees were transported through Austria, Yugoslavia, and Bulgaria to neutral Turkey. It was August before Kuznetsov and the others deported from Germany managed to reach Moscow. He immediately left for Leningrad to join his family and was evacuated with them to Birsk. Through the "Road of Life," the half-dead specialists of the Elektropribor Factory moved across Lake Ladoga and evacuated to Moscow.[18] Kuznetsov returned to Moscow and rebuilt the scientific-research gyroscopic section in the deserted NII-10 building.

In 1943, the NII-10 naval institute was reborn. Here, Kuznetsov and Ishlinskiy headed a section in a special gyroscope design bureau. They succeeded in devel-

---

17. *Author's footnote:* How our fates intertwine! I met Farmakovskiy in 1936 when I began visiting the Elektropribor Factory to place orders for the aircraft bomb sight for Chief Designer Viktor Bolkhovitinov's DB-A airplane.

18. The "Road of Life" *(Doroga zhizni)* was the transport route across the frozen Lake Lagoda that provided the single access to the besieged city of Leningrad during the infamous 900-day Siege of Leningrad in 1941–44 when German and Finnish forces cut off all land access into the city.

From the author's archives.

Chief Designer Viktor Kuznetsov (1913–91) was the main gyroscope specialist among the Council of Chief Designers. During his career, he worked at the NII-10 and NII-944 institutes.

oping a heavy-duty stabilizer for a tank gun. But Germany surrendered, and, in May 1945, wearing the uniform of a Red Army colonel, Kuznetsov flew to Berlin, where he had been interned in June 1941. This is where I had my first encounter with Kuznetsov.

The decree of 13 May 1946 switched the focus of Kuznetsov's work from ships and tanks to missiles. From Berlin he came to see me at the Institute RABE. Together we traveled by automobile to Peenemünde. He was actively involved in the reconstruction of German gyroscope technology and arranged for supervision of gyroscope production at the Zeiss factory in Jena and their simultaneous reproduction at NII-10 in Moscow. The gyroscopes *Gorizont, Vertikant,* and *Girointegrator* (there were just three to begin with) were guidance system components, and their official customers were Ryazanskiy and Pilyugin. However, Korolev, who had a knack for quickly appraising a person's intellectual potential, made Kuznetsov a full-fledged member of the Council of Chief Designers. Soon, Kuznetsov was appointed chief designer of gyroscopes both for the R-1 missile and for all the subsequent missiles for which Korolev was chief designer. In 1953, a special design bureau was created at NII-10 on the basis of Kuznetsov's section. In 1955, this design bureau was reorganized into a gyroscopic stabilization institute with Kuznetsov as its chief designer.[19] Korolev agreed with Kuznetsov's decision to decline the combined positions of institute director *and* chief designer. "Vitya is not cut out for administrative work. If he had consented, we would have lost a chief designer," he said.

Despite his seeming naiveté, kindness, and openness, Kuznetsov had a knack for selecting, drawing in, promoting, and protecting people; above all, he paid attention to engineering talent. He also had a certain innate sense of decency that protected his team against the intrusion of schemers and careerists.

During those hot days at the firing range, I often had to meet at various councils and meetings and have regular personal contact with Nikolay Khlybov, Zinoviy Tsetsior, Georgiy Geondzhan, Oskar Raykhman, Mark Effa, and Izrail Blyumin. They were all close co-workers of Kuznetsov whose engineering and human qualities commanded my respect. To this day, Kuznetsov's deputies Illariy Nikolayevich

---

19. The institute was known at the time as NII-944.

Sapozhnikov and Valentin Ivanovich Reshetnikov help me in my work in the academic section of the Council on Motion Control. On the occasion of Kuznetsov's 90th birthday, I had the honor, along with Sapozhnikov and Reshetnikov, to take part in a historical film dedicated to his memory.

In 1965, Kuznetsov's institute of gyroscopic technology became part of the newly created Ministry of General Machine Building. Minister Sergey Aleksandrovich Afanasyev loved to hold up Kuznetsov as an example. The latter had assured him that "there would be no squabbling and intrigue in our ranks." He created a unified, efficient, close-knit organization that combined designers and researchers with a highly disciplined production culture. Our domestic cosmonautics effort is particularly indebted to Viktor Kuznetsov. With the dawning of the space age, command instruments for spacecraft attitude control and navigation became one of the primary tasks of the Scientific-Research Institute of Applied Mechanics (NIIPM), the new name conferred on NII-944 beginning in 1965.[20]

"According to an ancient legend, the Earth rests on three whales. They ensure the Earth's proper orientation vis-à-vis the Sun, Moon, and Stars. Over the centuries humankind has realized that the whales themselves are well stabilized in space. In our nation, NII-885, NII-10, and NII-627 are the whales that stabilize and provide electricity for missiles and spacecraft. In their work, the first two whales mentioned [NII-885 and NII-10] must obey Maxwell's equations, theoretical, and celestial mechanics. As for the third one, NII-627, which is just twenty years old today, knowledge of Ohm's law is sufficient for its difficult work."

With these words I began my congratulatory speech on the occasion of the 20th anniversary of NII-627, which was renamed the All-Union Scientific-Research Institute of Electromechanics (VNIIEM) in 1959.[21] My first meeting with NII-627 director Andronik Gevondovich Iosifyan took place in 1946 in Germany. He had come to the Institute RABE for just one day to find out which mechanical converters—"Umformers" and electric motors for missiles—he would need to manufacture by order of the minister of the electrical industry.

I visited NII-627 during the first month after my return from Germany. I was told that the address was at "Khoromnyy Cul-de-sac" and that "the only access pass office [would be] there." Upon arriving, I found myself on palatial grounds paved with granite slabs. When I climbed the granite and marble steps of the main entryway, opened the heavy carved doors, and entered the main building, I continued to be amazed by the magnificent and deeply artistic beauty of the surroundings. The walls, ceiling, stairs to the second floor, and doors—everything drew you in and shouted "feast your eyes while you are here." An old friend of mine from our student

---

20. NIIPM—*Nauchno-issledovatelskiy institut prikladnoy mekhaniki*. Kuznetsov's team was originally part of NII-10 in 1946–55 but then separated into the independent NII-944 in 1955. *Author's note:* I plan to describe NIIPM's work separately.

21. VNIIEM—*Vsesoyuznyy nauchno-issledovatelskiy institut elektromekhaniki*.

days at MEI, Yevgeyiy Meyerovich, met me.

"How do you work here?" I asked him, pointing to the ceiling of the stateroom into which he had led me. The ceilings in this hall and the offices that I visited were colorfully ornamented with images of ideally beautiful nude female bodies. You could linger over the pictures admiring them as if you were in a museum.

"We're already used to them," explained Iosifyan, when I found myself in an office where the walls were adorned with French Gobelin tapestries. "We very quickly find a common language with everyone who comes here on electromechanical business. The fine art that surrounds us has a humanizing effect even on bureaucrats from the ministry when they come to do various and sundry inspections and audits."

I was pleasantly surprised when I discovered old acquaintances at NII-627 from my prewar work. Aleksandr Goldobenkov had been responsible for delivering electric generators and generator regulators for the N-209 aircraft before its transpolar flight in 1937. Teodor Gustavovich Soroker was the chief designer in the electric machine laboratory of the All-Union Electrical Engineering Institute (VEI).[22] He had designed all sorts of alternating current electrical machines that Factory No. 293 had ordered for my long-range "B" bomber project that was supposed to fly using alternating current. Nikolay Sheremetyevskiy was also an old prewar school chum at MEI. He was now involved with new, cutting-edge synchronous servos, which I had needed so badly in 1939; they were needed for both the remote control guns and machine guns on the exotic "I" fighter that Aleksey Isayev had designed and on the "B" bomber, for which I had designed the high-frequency alternating current system.

During my meeting with Teodor Soroker at NII-627, we reminisced about our joint prewar developments and came to the conclusion that we had done a lot of interesting things. "We were ahead of our time," said Soroker. "True, we haven't been bored here at *Krasnyye vorota* (Beautiful Gates) either."[23] Visiting VNIIEM, I fought for setting deadlines for mastering the production of several devices: domestic two-unit mechanical DC-to-AC converters to power combat missile guidance and telemetry systems and electric motors and polarized relays for control surface actuators and control fin drives. We needed these for our entire missile program, although they were not of particular interest to the NII-627 organization at the time.

Andronik Iosifyan became a celebrity among specialists back in 1930 when, as a student at Baku Polytechnical Institute, he invented the "helically cammed electric gun" and began to work in the VEI machine-hardware section under Academician

---

22. VEI—*Vsesoyuznyy elektrotekhnicheskiy institut.*

23. *Krasnyye vorota* is an area in northeast Moscow where the institute is located. A major Moscow metro station is located there.

K. I. Shefner.[24]   Our electrical engineering and automation superstars worked in this section. They supported my proposals to introduce alternating current systems into aircraft.

Academy of Sciences Corresponding Member A. N. Larionov was the supervisor of my diploma project. Academy of Sciences Corresponding Member K. A. Krug was the founder of VEI, and at MEI he directed the "fundamentals of electrical engineering" department.

In 1936, while trekking through the mountains of the Caucasus with my girlfriend, I carried along Krug's heavy fundamentals of electrical engineering textbook in my backpack in the hopes of making up the academic incomplete I owed to Krug in the fall. Professor G. N. Petrov was an authority in the field of transformers and electric machines. When I was a student at MEI, he was deputy science director. Among these men, the quite young Iosifyan was appointed director of the laboratory of electromechanical servo systems for antiaircraft artillery control.

To this day, various nations continue their attempts to create electric guns. Now they are not for firing conventional shells, but for launching antimissile projectiles armed with nuclear warheads or for boosting spacecraft. They are just as far from practical realization as they were in the early 1930s. Iosifyan assessed the scope of the problems and switched his inventive enthusiasm and incredible creative energy to the creation of noncontact synchronized transmissions (selsyns).[25] In this field, together with Svecharnik, he not only achieved real synchronous remote angular transmission, but he also patented the invention in many countries. In 1940, Iosifyan defended his doctoral dissertation, "The Theory and Practice of Non-contact Selsyns." Simultaneously, Iosifyan invented a centrifugal machine gun, an artillery fire control system, and an electric helicopter, and, jointly with TsAGI, he was even designing a twin-engine electric airplane.

In 1941, Iosifyan was already the chief of the VEI's OKB. When the Germans attacked Moscow, he and another inventor, future science-fiction writer A. P. Kazantsev, were inventing small electric self-propelled wire-controlled tanks.[26] According to the inventors' conception, these "land torpedoes" were supposed to spring out from the front gates of buildings and houses and blow up German tanks if they broke through into the city.

On 24 September 1941, an order of the people's commissar of electrical industry named Iosifyan director of Factory No. 627, which occupied the building of the former palace at the *Krasnyye vorota* in Khoromnyy Cul-de-sac. During the war, history was the last thing on people's minds; nevertheless, when the Germans

---

24. Academician Klavdiy Ippolitovich Shefner (1885–1946) was one of the pioneers of the Soviet electrical engineering.

25. Selsyns are systems consisting of a synchronized generator and motor.

26. Aleksandr Kazantsev (1906–) was a well-known Soviet-era science fiction writer known for his space-themed works. In 1946, Kazantsev advanced the theory that the mysterious Tunguska explosion may have been caused by an extraterrestrial spaceship.

were driven away from Moscow, the artistic monument protective services came to Iosifyan requesting that during all necessary defense work he coordinate potential reconstruction with them.

The palace, in which the factory's managerial offices and Iosifyan's office were located, was built in the 1880s by Sergey Pavlovich Derviz, son of the railroad magnate. In 1904, the building was sold to Lev Konstantinovich Zubanov, son of a petroleum industry millionaire who owned oil fields in Baku. Foreign master craftsmen and top-notch artists remodeled, decorated, and painted the main building and its interiors. After the October Revolution the Special Technical Bureau (Ostekhbyuro) of the All-Russian Council of the National Economy (VSNKh) was located in the mansion, and later NII-20.[27] All the while, the artistic splendors were unharmed. In September 1941, NII-20 was evacuated from Moscow and Iosifyan took possession of the magnificent estate. Thus, to this day, since 1941, the VNIIEM management has occupied the building, whose splendor the government protects.

Visiting NII-627 in the late 1940s and early 1950s, I felt like a "petty missile electrician." The list of missile technology orders for NII-627 was very small compared with the creative interests of both Iosifyan and the representatives of Soviet electrical engineering who had gathered around him.

At the factory, scientific-research projects for the development of new electrical engineering materials and new principles of electrical machine building achieved a broad scale. During the war the small "land torpedo" tanks were involved in combat operations, but were not broadly used. The main project was the development and production of electric power sources for combat radio stations, foot- and hand-driven electric generators called "soldier motors." Iosifyan was also the chief designer of power sources for radar of all types that had just emerged. These projects led to the development of a whole series of diesel and gasoline mobile electric generators that were the sole sources of power for the engineering troops.

Iosifyan received his first Order of Lenin for the development of selsyns for artillery fire guidance radar stations. The mass production of demolition dynamos for the guerrillas was also set up. Iosifyan himself was at one time fascinated with the idea of creating a "flying infantry." A special engine was supposed to help a man execute a flight lasting tens of meters. The future academician and Iosifyan's successor Nikolay Sheremetyevskiy invented electric grenades. Physicists who came to the institute developed thermoelectric tea kettles for the guerrillas that were electric power sources for radio stations.

---

27. *Osoboye tekhnicheskoy byuro* (Ostekhbyuro) was a special R&D organization operating in the interwar years whose mandate was to develop innovative armaments for the Russian Navy and other branches of the armed forces. The VSNKh—*Vserossiyskiy sovet narodnogo khozyaystva* (All-Russian Council of the National Economy)—was the top economic management body for Soviet industry in the interwar years.

On 1 May 1944, Factory No. 627 was reorganized into Scientific-Research Institute No. 627 (NII-627). In 1945, the most urgent task for the institute was the reproduction of electrical equipment and a fire control system for the Tu-4 aircraft, a copy of the American Boeing-29 Flying Fortress. In 1949, the first series of 28 bombers was put into service; project managers B. M. Kogan and N. N. Sherem-etyevskiy were awarded the Stalin Prize for this work.

A large project was also launched to produce new types of electric machines of various power ratings and mobile electric power plants and also to develop a single series of induction motors. These projects were economically crucial. Another major economic task was the development of regulating and switching equipment for a broad range of industrial applications.

I am citing this far-from-complete list in order to show that during the early years of missile technology development, rocket technology did not determine the main workload of NII-627 and the future VNIIEM. Nevertheless, Iosifyan was clearly offended that he wasn't officially in the first Council of Chief Designers. He accepted all of our new proposals, and, ultimately, in the early 1960s, he became a de facto member of the Council.

Purely inventive and design work did not slake the creative thirst of the inde-fatigable "nation's chief electrical engineer" in his pursuit of ultimate truth. Iosifyan was elected as a full member of the Armenian Soviet Socialist Republic Academy of Sciences and soon became its vice president. We were all certain that he would also be elected into the USSR Academy of Sciences. However, his innate enthusiasm betrayed him. Iosifyan dared to encroach upon what was for modern physicists the "holy of holies"—Einstein's general theory of relativity. It would have been fine if he had given lectures in his tight "electric" circle, but he dared to publish his own research in works for the Armenian Academy. Academic physicists could not forgive this. On the other hand, in the field of missile technology and cosmonautics, year after year VNIIEM seized all the new beachheads. But that is another story and another chapter.

Chapter 11
Air Defense Missiles

In Chapter Five of the first book of my memoirs in Russian, *Rakety i lyudi* (*Rockets and People*), I ventured to touch on the history of the development of air defense missile systems.[1] As a result, the creators of these systems sent me comments that justifiably argued that the scope of the operations for the creation of air defense missile systems in the Soviet Union, and later for antimissile defense, deserved a more precise and extensive description in my memoirs.

I hope that the creators of these unique air defense systems, and later the antimissile defense systems, will write a thorough work on the history of those developments in the USSR.[2] As far as my memoirs are concerned, for the new edition I have rewritten a very small part of this story, taking into consideration their criticism and the publications that have appeared after the first edition of my memoirs came out in Russia.

Karl Samuilovich Alperovich, one of the leading scientists and creators of the Moscow air defense system during the 1950s, rendered me essential help in revising the content and sequence of events on the history of the development of air defense systems. His memoir, *Years of Work on the Moscow Air Defense System 1950-1955*, is still the most complete account of the first guided surface-to-air missile system and how it was created.[3] I thought I might add my own reminiscences and musings on this topic. I regret that they were excluded from the first edition out of space con-

---

1. See chapters "Getting Rid of Surface-to-Air Missiles" and "A Call in the Night" in the original Russian version: B. Ye. Chertok, *Rakety i lyudi* [*Rockets and People*] (Moscow: Mashinostroyeniye, 1994), pp. 270–73, 293–98.

2. A number of well-researched works have appeared in Russian recently on the history of the early air defense project. Probably the most comprehensive is Mikhail Pervov, *Zenitnoye raketniye oruzhiye protivovozdushnoy oborony strany* [*Anti-Aircraft Rocket Armaments of the National Air Defense Forces*] (Moscow: Aviarus-XXI, 2001).

3. K. S. Alperovich, *Gody raboty nad sistemoy PVO Moskvy 1950-1955* [*Years of Work on the Moscow Air Defense System 1950-1955*] (Moscow: Art. Biznes-Tsentr, 2003). See also the earlier K. S. Alperovich, *Rakety vokrug Moskvy: zapiski o pervoy otechestvennoy sisteme zenitnogo upravlyayemogo raketnogo oruzhiya* [*Missiles Around Moscow: Notes on the First Domestic Anti-Aircraft Guided Missile Armament*] (Moscow: Voyenizdat, 1995).

siderations. I begin with two different accounts of the "prehistory" of strategic air defense, in particular, on the formation of the now famous SB-1 organization.

August 1947 was the hottest month of the summer for everyone who had a role in the upcoming first ballistic missile launches in the USSR. Horizontal tests on the missiles that had been brought out of Germany and assembled from German components at our factory in shop No. 39 were running round the clock. I would often spend the night at the factory. NII-88 Director Gonor, who checked daily on the missiles' preparation for shipment to the test range, warned that despite the all-hands work regime, the institute's Communist Party committee was on the verge of reviewing the status of operations on surface-to-air guided missiles any day now. Yevgeniy Sinilshchikov, Chief of Department No. 4, would deliver the main report. After all, he was the chief designer responsible for the reproduction of the German *Wasserfall* surface-to-air missile.

"I have the impression," said Gonor, "that when it comes to the *Wasserfall* missile, we are circling around the problem, not knowing which end of it to grab. Sinilshchikov will accuse you of ignoring his area of work. He has every reason to say that work on the complex is being held up because we are not fully clear on the principles of guidance."

"In that sense he's right," I responded to Gonor. "The issue isn't the drawings for the German *Wasserfall* missile. They'll have no problem drawing the body with the structural rings and longitudinal beams. But we still don't have our own idea of the entire missile complex. In my opinion, our NII-88 is not capable of bearing the responsibility for that, because unlike long-range ballistic missiles where the radar designers play a subordinate role, in this case, with the *Wasserfall*, the radar team should be the lead organization. If they confront me in the Party committee or even if they reprimand me, I will still say that NII-88 in its current form is not capable of being the lead institute for air defense complexes."

"I don't advise you to take that approach," said Gonor. "The same governmental decree from 1946 that assigned NII-88 to work on ballistic missiles also assigned us to work on the *Wasserfall*. When the decree was being drawn up, Vetoshkin and I told Ustinov to assign surface-to-air missiles to the aviation ministry, but the document that reached Stalin for his signature stated that our ministry, that is, NII-88, would be responsible for all types of missiles. For now, I promise to ask the Party committee to postpone its review of the matter until after our return from the test range."

Now, I can no longer remember all the arguments that I raised during our discussion of the problem of surface-to-air missiles and the reasons why we were lagging so much in that area.

One of the important aspects of such large-scale programs is attitude, specifically the enthusiasm of the team as a whole. This, in turn, depends on attention and exactitude from higher-ups. Whereas the Ministry of Defense had shown interest in ballistic missiles, we sensed no interest whatsoever in the surface-to-air missiles. The Main Artillery Directorate (GAU) was working closely with us only on long-range missiles. There was no real supervisor over surface-to-air missiles—not in a military department and not in Special Committee No. 2.

The Party committee meeting on surface-to-air issues was postponed until November, but this gave me no comfort. Sooner or later, they would ask us why the decree signed by Stalin was not being fulfilled. All the problems of surface-to-air guided missiles required radical solutions. The entire staff of Department U and I were deeply interested in this, not only out of general patriotic sensibilities, but also out of egotistical ones. We were expecting a miracle. And our salvation arrived! I will begin in chronological order.

On one of the last days of that hot August of 1947, Vetoshkin summoned me to the ministry. He was quite upset and warned me that the two of us would now go see the minister. There we would discuss a new, interesting proposal for a guided missile system. I had been invited as an expert and would have to give the findings as to whether this design could be implemented at NII-88 and to what extent my staff and I were capable of participating in its implementation. "Don't ask me any questions, Boris Yevseyevich. You'll figure it out when you get there, but keep in mind that rash, hasty responses could have serious consequences for you."

When we entered Ustinov's office, I saw the chief engineer of our NII-20 radar institute, Mikhail Sliozberg, and optical sight developers whom I had known back when I was working on aircraft. Ustinov seated us all along one side of a long conference table and announced, "We'll leave this side free. Soon some comrades will be coming who will report to you the gist of their proposals. Your job is to comment only on the scientific, research, and industrial facilities needed for implementation."

Two men entered: an armed forces engineer colonel and a communications troops major. Ustinov introduced them: "Sergey Lavrentyevich Beriya" and then with surname alone, "Colonel Kuksenko." Good God! How did I not recognize him? The famous Kuksenko, radio engineer and my idol from the ham radio days of my youth.[4] When I was first introduced to radio engineering as a schoolboy, Kuksenko was already teaching us a thing or two in our ham radio club on Nikolskaya Street and was often published in radio magazines. I read everything there was to read. But instead of the young, well-built radio engineer that we schoolboys had looked on as a radio demigod, now I saw a gray-haired, heavyset colonel, who evidently had a hard time standing. He bowed to those assembled and hurried to sit down.

Young Beriya began to hang posters. Everyone grasped right away that Lavrentiy Pavlovich [Beriya]'s son was standing before us, and we fell silent. We were surprised because the posters were of diploma-level quality. Then it became clear that that's

---

4. Pavel Nikolayevich Kuksenko (1896–1980) served as Chief Designer in 1947–53 of the top secret SB-1 (later named KB-1) design bureau that oversaw the development of the first Moscow air defense system.

what they were. Sergey Beriya was defending his dissertation for the second time in the office of the USSR minister of armaments. He did this not of his own volition, but at the instruction of his father, who had called Ustinov and "requested" that he assemble specialists and let them listen. But their task was not to evaluate the project, but to decide where to implement it! The project should be put before a panel of experts to see whether it merited implementation.

Sergey gave quite a respectable report. The subject was a naval guided cruise missile. The project consisted of two parts. The first part described the missile itself, which was ejected from an airplane and was equipped with an aircraft turbojet engine. That was an innovation at the time. The second part, judging by the posters and report, proposed a radar system to detect enemy ships and simultaneously radio-control the missile from the airplane on the detector beam. On the whole, despite evaluating the dissertation as excellent—or a five on a scale of one to five—an experienced expert would immediately have discovered naïve proposals and methods that had been rejected earlier.

For the sake of historical accuracy, I must say that for the first time in the USSR, and perhaps anywhere, Sergey Beriya and Pavel Kuksenko proposed an "air-to-sea" missile system that years later, in different modifications, went into service in the USSR and U.S. as the primary system for aviation warfare against surface ships. The report was followed by several questions, which Sergey asked Kuksenko to answer, having introduced him as the academic advisor. Kuksenko answered for Sergey, but it was already clear to everyone that this particular and still rough project itself was not the issue.

Ustinov proposed that we speak our minds as to whether the proposal was feasible and where it could best be implemented. I took the floor first. It seemed to me that I gave a very reasoned speech to the effect that a missile with a turbojet engine did not at all fall under the subject matter of NII-88. Moreover, we had virtually no radar specialists and, therefore, the realization of such a project would require the creation of a special organization, possibly using the facilities of an aviation industry enterprise. In contrast, Sliozberg argued that his institute had all the conditions for the realization of the radio engineering part of the project. Ustinov thanked everyone and adjourned the meeting. When I stopped in to see Vetoshkin, he was very pleased with my speech. "That will be the end of Sliozberg. Mark my words."

On 8 September 1947, Ustinov's decree was issued on the creation of Special Bureau-1 (SB-1) in the NII-20 system of the Ministry of Armaments for the development of guided air-to-sea armaments.[5] P. N. Kuksenko was appointed chief and chief designer of SB-1. His deputy was S. L. Beriya, who had graduated in the spring of 1947 from the radar department of the Military Academy of Communications in Leningrad.

---

5. SB—*Spetsialnoye byuro* (Special Bureau).

Many years later, through the stories of Pavel Tsybin and legendary test pilot Sergey Anokhin, I found out about the fate of Sergey Beriya's diploma project and the events that followed the creation of SB-1.[6] I deem it necessary to speak of this work because, first of all, it reflected the spirit of the time, and second, it was the prelude to the beginning of construction in 1950 at the small SB-1 facility of the enormous KB-1, which was in charge of developing unique air defense systems.[7] It was KB-1 that inherited all work from the NII-88 on the *Wasserfall* missile involving both Chief Designer Sinilshchikov and myself.

The duo of Kuksenko and Sergey Beriya, who remained at NII-20, cooperated with the Mikoyan KB to develop the Kometa, a cruise missile that was supposed to strike a sea target after it separated from the carrier aircraft.[8] The Kometa was launched from the Tu-4 carrier aircraft approximately 150 kilometers from the target and was supposed to enter the beam of the radar mounted on the carrier aircraft. The radar guided the Kometa to the sea target. When the Kometa was approximately thirty kilometers from the target, it was supposed to switch to homing mode and strike the target. This was the projected mission that Sergey Beriya and his academic advisor Pavel Kuksenko presented at the meeting in Ustinov's office in 1947. The radar portion of the system was to be developed at the small SB-1, which had been located at NII-20. The plan was agreed upon by Mikhail Sliozberg, who had announced at the aforementioned meeting that this work fell within his area of expertise.

NII-20 and the Ministry of Armaments did not know how to design and optimize winged vehicles. Without prolonged wrangling, the elder Beriya assigned the development of the winged vehicle to Mikoyan's aviation KB and the flight optimization of the Kometa to the LII. Beriya summoned Engineer Colonel Pavel Tsybin from NII-88 to direct the Kometa flight tests. With Tsybin's participation, LII converted the unmanned vehicle into a manned analog for the flight tests. To replace the volume and mass of the warhead, they set up a pilot's workstation with manual control and modified the landing gear. The cockpit was very cramped. The Kometa's turbojet engine would start up when it received a command from the carrier aircraft after its radar detected a target. After launch, the missile radio system received a signal reflected from the target at a range of around thirty kilometers,

---

6. Pavel Vladimirovich Tsybin (1905–92) worked briefly on the development of the Moscow air defense system in the early 1950s before serving in several aviation design bureaus. In his later life, he served as a deputy chief designer at OKB-1 (now RKK Energiya) under Sergey Korolev, helping oversee the development of many generations of space vehicles. Sergey Nikolayevich Anokhin (1910–86) served as one of the top test pilots at the Gromov Flight-Research Institute (LII) in 1943–64 before managing the training of civilian cosmonauts at OKB-1/Energiya in 1964–86.

7. In August 1950, SB-1 was reorganized into the much bigger KB-1.

8. The "Mikoyan KB" (or OKB-155) was headed in 1942–70 by Chief Designer Artem Ivanovich Mikoyan (1905–70), who designed some of the Soviet Union's best fighter aircraft, including several generations of MiG fighters.

thus activating the semiactive homing mode during the final flight segment.[9]

The pilot was "built in" to the unmanned control system to speed up the test phase and enable the single-use aircraft to be used multiple times. The pilot was supposed to monitor the system's transition to the homing mode. He was permitted to interfere in the control process only if the automatic system failed. To switch to manual control the pilot gave a command to pyrotechnic cartridges that shot off the mechanical linkage between the aerodynamic control fins and the automatic system's control surface actuators. At the very last moment during the normal target approach the pilot would take control and break away to a coastal airfield. Landing was the most hazardous part of a test flight. The light "bicycle landing gear" was supposed to withstand impact with the ground at a landing speed in excess of 380 kilometers per hour.

Flight tests were conducted on the Black Sea in the Crimea. As Tsybin recounted, after each flight they had to report to Lavrentiy Beriya. Decades later, Tsybin would mimic Beriya's voice with style. To the great pleasure of his audience, and nodding in the direction of Anokhin, Pavel Vladimirovich would say:

"All *he* [Anokhin] had to do was fly, report, and he was free. But I had to wait for the call from Moscow, sometimes not being able to leave for hours, and then stand at attention with the telephone receiver and answer questions that were not always easy."

"And why did you have to stand at attention?" asked someone from Tsybin's audience.

"That was the psychological situation, like it or not. When reporting to Stalin and Politburo members you stood up and gave the report standing at attention as if they could see you from Moscow."

Pavel Tsybin, a superb and witty storyteller, told us about top secret flights that the famous test pilots Sergey Anokhin and Amet-Khan Sultan performed under his supervision in the unmanned vehicle.[10] Anokhin and Sultan showed exceptional mastery; each flight involved great risks.

This naval project and all of its various updates could have been Kuksenko's first and last creation had it not been for Stalin's personal initiative.

All projects on the development of a Moscow air defense system invulnerable to atomic bombers had a considerably higher classification than our projects on long-range missiles. The most probable reason for the top-secret classification of these projects was the fact that they were under Lavrentiy Beriya's guardianship. As chairman of the Special Committee for the development of atomic weaponry, he introduced the strictest regime of secrecy and attached state security officers to the

---

9. In the West, Kometa was known as the AS-1 Kennel.

10. Amet-Khan Sultan (1920–71) was one of topmost Soviet test pilots. After a stellar career during World War II, he tested over 100 different Soviet aircraft in the postwar era. He was killed in an air crash.

scientists as overseers. There was nothing like that in our Ministry of Armaments system. Lavrentiy Beriya's son Sergey had already been working in his father's system before studying at the Academy of Communications. Which brings us to the second account of the "prehistory" of these systems. In an interview, Sergey Beriya gave his version of the beginning of operations with Kuksenko:

"Once, not long before I graduated from the Academy, my father introduced me to two prominent scientists, Aksel Ivanovich Berg, chairman of the radar committee, and Admiral Aleksandr Nikolayevich Shchukin, also a radio engineering specialist. They proposed to me that I work for a while with German, British, and American materials, select some subject, and do my dissertation on it.

…And so I did my diploma project, not without getting advice and help from experienced specialists. In my project, I proposed a cruise missile that followed a radar beam not 15 to 20 kilometers like the prototype, but around 150 kilometers, and during the final flight segment it switched to homing guidance. The diploma got a good grade and a recommendation that it be put into production. This recommendation might have remained just that, but one time Stalin was talking with Berg about some radar business and Aksel Ivanovich mentioned my diploma. Stalin asked that I be brought in and he started to inquire whether this could in fact be realized. More than anyone, it was Berg who convinced Stalin of the feasibility of this project, after which the decision was made to begin operations. This is roughly what Stalin said to me:

'No offense, but you are young and don't know anything yet. We need to put an old-timer over you who would manage everything.'

I requested Kuksenko. Stalin was not inspired by his candidacy.

'Kuksenko has already been in prison twice.' To which my only response was, 'That means he's truly an upstanding man, since they've released him twice.'

Stalin looked at me and said that I'd gotten those ideas from my father.

Right after graduation from the Academy in 1947, Stalin ordered my father and Georgiy Malenkov to set up the design bureau. At my father's suggestion, Dmitriy Fedorovich Ustinov, the Minister of Defense Industry, took the design bureau under his wing.

'Why not put it under the aviation industry?' asked Stalin. But my father thought that the people who developed the control system should manage the project, while the aviation folks could only be a sort of cab driver for this very guidance system.

Ustinov set up the design bureau on the premises of the so-called twentieth institute [NII-20], which was working on radar. Kuksenko was nevertheless appointed chief of the organization. He was also involved with management, in a constructive sense—"they failed to take this into account, they overlooked that, etc.."

We developed the cruise missile over a four-year period. The vehicle went into series production as the Kometa. It was jet-propelled with a triangular wing, that is, a winged missile. They sent in the obsolete cruiser Krasnyy kavkaz as a training target. They removed the crew from the ship, having first set the rudder so that the ship traveled in a circle with a 30-kilometer diameter. Several times we hit the shore with a blank shell. Ultimately, they asked us to show whether we could manage to sink a ship with a single projectile armed with an explosive charge. Not a single one of the admirals believed that *we could. But we sank the* Krasnyy kavkaz *on the first attempt. By the time we completed the program, the Korean War was in full swing. The* Kometa *had not yet officially been put into service, but a batch of 50 units had already been produced.*

Stalin assembled the designers at the Defense Council. He invited us and aircraft designer Mikoyan, who had developed the vehicle for the Kometa *system. Then Stalin asked whether we would be able to sink American aircraft carriers. At that time there were eight of them off the coast of Korea. We, of course, declared that we could. Stalin was thrilled. But my father and Marshal Vasilevskiy objected, saying that the projectiles could under no circumstances be used against the aircraft carriers: they would strike the target, but the Americans would respond with a nuclear strike against Moscow. Stalin was infuriated. 'You mean Moscow is not protected?'*

Right then and there we were told to leave the proceedings. We didn't have the rank to participate a discussion [about the protection of Moscow]. "[11]

Now, to the actual events involving the creation of the first Moscow-based air defense system. The history of the project has several versions. I am not about to judge which of them is the more authentic one, but it all boils down to the fact that the initiative for the creation of the organization that we know today as Almaz came personally from Stalin.[12]

Under great secrecy, in 1950, we passed on scraps of information to each other about the government decree promoting Pavel Kuksenko and Sergey Beriya to a higher level in the hierarchy of defense technology developers. The rumors were quickly confirmed. And how all of this took place actually managed to be bound into a single historic sequence many years later.

In 1950, the U.S. proclaimed "absolute supremacy" as its strategy in the Cold War. When the USSR obtained the atomic bomb in 1949, U.S. territory was still secure. For that reason, the military and political aspect of U.S. strategy revolved around using atomic weaponry against the USSR to inflict "preventive" strikes from the air.

After the R-1, we had already developed new medium-range missiles, and work had begun on super long-range bombers and fighter jets. But there was no system capable of reliably protecting Moscow and other important strategic centers against American [Boeing B-17] Flying Fortresses.

---

11. *Author's note:* The text is cited from an interview published in the newspaper *Sokol*. For the younger Beriya's memoirs, see S. L. Beriya, *Moy otets, Lavrentiy Beriya* [*My Father, Lavrentiy Beriya*] (Moscow: Sovremennik, 1994).

12. The Almaz Scientific-Production Association (NPO Almaz) is the direct descendent of the original SB-1 formed in 1947. During the Soviet era, it was one of the largest, most influential, and most secret defense industry organizations and produced various tactical and strategic weapons systems, including antiaircraft missile systems, antiballistic missile systems, and antisatellite systems. During its existence, it has been known variously as SB-1 (1947–50), KB-1 (1950–66), MKB Strela (1966–71), TsKB Almaz (1971–88), NPO Almaz (1988–95), AOOT Almaz (1995–96), TsKB Almaz (1996–2001), and NPO Almaz Named After Academician A. A. Raspletin (2001–present).

Stalin understood this and probably feared a repeat of 1941 with an atomic scenario. Apparently he consulted with Beriya as to who might be put in charge of the work on an air defense system. Lavrentiy Beriya had a competent adviser on this problem, his son Sergey. Probably he gave his father the idea, and the elder Beriya passed it on to Stalin that Kuksenko be consulted. Mind you, this is conjecture.

According to one of the latest publications, the air defense system got its start one summer night in 1950. It is common knowledge that Stalin liked to work at night and would send for people he needed as late as four in the morning. For this reason, the leading ministers, as a rule, did not go to bed before three or four in the morning.

Below I cite a version from the memoirs of Aleksandr Pavlovich Reutov.[13]

"One summer night in 1950, Kuksenko was summoned to the 'nearby dacha,' Stalin's Kuntsevo apartment. The apartment's host received Pavel Nikolayevich in his pajamas, browsing through a pile of papers on his couch. After a certain amount of time, tearing himself away from reading his documents, Stalin posed the question:

'Did you know that the last time an enemy airplane flew over Moscow was 10 July 1942? It was a solitary reconnaissance aircraft. And now, imagine that a solitary aircraft also were to appear in the Moscow sky, not a reconnaissance plane, but rather one carrying an atomic bomb. It appears that we need a completely new air defense system capable, even in the event of a mass attack, of preventing even a single airplane from reaching the site being defended.'

After this, according to Pavel Nikolayevich [Kuksenko], Stalin asked him a number of questions on a subject that was so alien for him. This was natural since radio-controlled missile weaponry was in an embryonic state, and for Stalin this was a new military technical field. Kuksenko emphasized that the complexity and the immensity of the problems here were on a par with developing atomic weapons

After listening to him, Stalin said:

'Comrade Kuksenko, one opinion has it that we must immediately begin creating a Moscow air defense system for repelling a massive attack of enemy aircraft from any direction.'"

In this regard, K. S. Alperovich noted in his memoirs that no one except Kuksenko himself could have said anything about the nighttime meeting with Stalin. Kuksenko, however, was a very private man. Two arrests and years of work in top secret NKVD *sharashki* had inured him to silence. Therefore, Alperovich believed that Kuksenko could not have told anyone about such a conversation with Stalin.

The only thing that one should consider credible is that Stalin personally entrusted Kuksenko with leadership over the program. His responsibility, however, was under the rigid control of Lavrentiy Beriya, whose son was officially listed in the new organization as a chief designer just like Chief Designer Kuksenko.

---

13. A. V. Minayev, ed., *Sovetskaya voyennaya moshch ot Stalina do Gorbacheva* [*Soviet Military Power from Stalin to Gorbachev*] (Moscow: Voyennyy parad, 1999), p. 493.

In those days, ideas voiced by Stalin were announced in the form of governmental decrees within several days. Stalin signed such a decree on the creation of the new KB-1 on 9 August 1950. As a result, on 12 August 1950, Minister Ustinov issued the order reorganizing SB-1 into the new KB-1. Initially, Ustinov's deputy minister K. M. Gerasimov was appointed chief of KB-1. Kuksenko and Sergey Beriya were named chief designers of the new system, which was given the name Berkut; wisecrackers claimed it to be derived from the first syllables of the chief designers' surnames.[14]

Vetoshkin's prophesy about the unfortunate fate of NII-20, which he had expressed in 1947 after the meeting that took place in Ustinov's office, came true. All of the facilities belonging to the Ministry of Armaments' NII-20 located at the fork of Leningrad and Volokolamsko Highways were, in fact, were transferred to the new KB-1. NII-20 itself was promptly moved to Kuntsevo. At the fork of the two highways, construction began on an enormous thirteen-story building for KB-1. Right away, without taking their views into consideration, the primary radio engineering scholars were taken from another organization, TsNII-108, the main radar scientific-research institute, headed at that time by Aksel Ivanovich Berg. The first leading specialist taken from Berg was Aleksandr Andreyevich Raspletin. At TsNII-108 he had directed the main developmental laboratory for radar systems. Kuksenko and Beriya agreed to the appointment of Raspletin as deputy chief designer on the Berkut system and as chief of the KB-1 radar department. This appointment had decisive importance for the fate of the Berkut system. It was Raspletin in particular who had the fundamental ideas that gave Berkut its unique technical characteristics, unparalleled in the world.[15]

KB-1 would not stay long in the Ministry of Armaments system. The elder Beriya decided to free his son from Ustinov's custody and to provide the Berkut operations with a scope comparable to that enjoyed by the atomic field in the First Main Directorate. Following that pattern, the Third Main Directorate (TGU) was created under the USSR Council of Ministers.

Vasiliy Ryabikov was appointed chief of the TGU, and his deputies were Valeriy Kalmykov and Sergey Vetoshkin. They were selected at Stalin's volition to create this new third field (after nuclear and rocket weapons).

The already approved budget did not provide funding for the new field of air defense during the first year. Therefore, Lavrentiy Beriya issued an order funding the rapid development of operations of the TGU at the expense of the first (atomic) one. The chief of the First Main Directorate, Boris Vannikov, was assigned not only to provide funding, but also to monitor the course of operations on the Berkut

---

14. The Russian word *berkut* also means "war eagle." Berkut was also said to be derived from "*Beriya*" and "*Kuksenko.*"

15. Academician Aleksandr Andreyevich Raspletin (1908–67) was one of the Soviet Union's leading designers of antiaircraft and antiballistic missile systems. He served as a chief designer at KB-1 in 1955–67. The KB-1 organization, known today as NPO Almaz, is now named after him.

system. This I saw for myself soon thereafter, not knowing at the time about the events of the top-secret story described above.

None of the individuals organizing operations on the Berkut system had any complaints against NII-88 or NII-885, which were simultaneously supposed to create an air defense system based on the *Wasserfall* missile. In fact, neither the NII-88 director, nor Chief Designer Sinilshchikov, were assigned to work on the Berkut surface-to-air missile. The government decree instead appointed Semyon Lavochkin, Factory No. 301 chief designer of the famous "La" fighters, as developer of the Berkut's surface-to-air missile. The aircraft construction OKB-301 in Khimki, in terms of its capacity and competency, of course, far exceeded Sinilshchikov's small Department No. 4 at the NII-88 SKB.

The development of the liquid-propellant rocket engine for Lavochkin's missile was entrusted to Isayev's OKB-2 at NII-88. Thus, Isayev remained NII-88's only link with the surface-to-air missile field. Vladimir Barmin, a member of Korolev's Council of Chief Designers, received the assignment to develop a launch system for Lavochkin's new surface-to-air missiles.

In early 1951, Amo Sergeyevich Yelyan was appointed as the new KB-1 chief in place of Gerasimov. Like Gonor, he was one of the first Heroes of Socialist Labor, having received the high award during the war for his success in the mass production of artillery armaments at the Gorkovsky Factory. The creation of the building where KB-1 was founded is linked with Yelyan's name. All of these events took place two years after I had been threatened with Party discipline for the failure of projects on surface-to-air missile control systems.

The designs for the Berkut system were so grandiose and radical that no one at NII-88 could even dream of them. But some talented people, including Georgiy Nikolayevich Babakin and his team, were assigned to the project.

IN LATE 1948, THE NII-88 SCIENTIFIC-TECHNICAL COUNCIL (NTS) was assembled under the chairmanship of Gonor.[16] There they heard the 35-year-old Babakin, the self-styled, as many believed, chief designer of surface-to-air guided missiles. Back then, it was customary for anyone who had anything to do with the field of missiles to conduct such work in closed organizations under the jurisdiction of the Soviet defense ministries. Babakin was from NII Avtomatiki, which was under the jurisdiction of the All-Union Council of Engineering Associations; that is, he was from a social organization.[17] Nevertheless, this organization managed, through a contract, to receive money from the Ministry of Defense, and, guided by Babakin, who was extremely gifted with engineering intuition, common sense, and organi-

---

16. Most Soviet research institutes operated a NTS—*Nauchno-tekhnicheskiy sovet* (Scientific-Technical Council)—to review proposals and projects undergoing at the organization. The NTS would be staffed by both in-house scientists and outside academics and scientists.

17. NII Avtomatiki—*Nauchno-issledovatel'skiy institut avtomatiki* (Scientific-Research Institute of Automatics).

zational talent, it was able to develop a perfectly competitive design for a surface-to-air guided missile and its guidance system, including the ground-based radar component.

As with any integrated design, the new ideas gave rise to a multitude of questions and critical comments. This was by no means a college paper. Babakin's team was made up of fully qualified specialists. According to my assessment, they had gained a better understanding of the flight control technology for a single surface-to-air missile than we had at NII-88 and than the specialists in Govyadinov's department at NII-885.

I liked Babakin right away. When I had heard the supplementary reports and responses of his colleagues to all sorts of questions, right then and there I got an idea. Babakin and his entire collective needed to be taken into NII-88. He was capable of relieving Department U of its obligations with surface-to-air guided missile control, and, who knows, maybe he could even be put in charge of the entire field. My first overture to Babakin proved unsuccessful. He was afraid of losing his independence. Knowing our structure, he absolutely refused to be under Tritko, much less under Sinilshchikov. Then I started working on Pobedonostsev, Gonor, and Vetoshkin. Finally, after long negotiations in the ministry and State Committee No. 2, a decree was issued on the basis of which, in December 1949, Babakin was transferred along with his design group to NII-88. Here he was put in charge of the new department of surface-to-air missile control, thus removing a considerable load from my Department U. Thus, officially NII-88, now strengthened by Babakin's team, continued its work designing surface-to-air guided missiles together with NII-885. Babakin and I not only found a common language for many technical problems, but our friendship soon developed to the point where we and our wives together went picking mushrooms in the forest.

The rumors about the decree calling for the creation of KB-1 reached us at NII-88 somewhat late and evoked conflicting reactions among us. I was ecstatic. "So, finally, the state's most vital problem is going to be solved in a radical way." Babakin and Sinilshchikov at NII-88 and Govyadinov at NII-885 would be left on the sidelines. However, as soon as it became clear that Lavochkin had been entrusted with the development of the missile for Berkut, Babakin began negotiations for a transfer to Khimki at Lavochkin's Factory No. 301.

By the end of 1950, under the management of the TGU, KB-1 had far surpassed the capabilities of NII-88 in terms of capacity, ideas, staffing, and production. It was senseless for Babakin to compete with that organization. At the same time, Lavochkin, who had been tasked with developing the actual missile, finally understood that he could not get by without good guidance specialists. Sinishchikov's work on the *Wasserfall* missile at NII-88 lost its significance. At NII-88, it was decided to shut down work on surface-to-air missiles all together. At the same time, this line of work was called off at NII-885.

After working with us for a year and a half, in 1951, Babakin transferred with his team to Lavochkin's facility and became his deputy for surface-to-air missile technology. In 1960, at the test range in the Balkhash region, Lavochkin literally died in Babakin's arms.[18] A few years after the death of Semyon Alekseyevich Lavochkin, Babakin was put in charge of his organization. While working on surface-to-air missile weaponry, the design bureau of Factory No. 301—under the management of Lavochkin, and then Babakin—was transformed from a narrowly specialized aviation enterprise into a scientific-production complex, which inherited Korolev's projects on robotic interplanetary spacecraft in 1965. In this field, Babakin's talent flourished. In 1970, Babakin was elected a corresponding member of the Academy of Sciences. And on 3 August 1971, just as suddenly as Lavochkin, Babakin died of a heart attack at the age of 57.

BUT LET'S RETURN TO THE 1950S. In early 1951, the development of surface-to-air guided missiles was under way at full speed at OKB-301 on a scale worthy of this highly complex problem. The missile itself, as a flying vehicle, was under the system of the Ministry of Aviation Industry. Having in due course rejected the problem of long-range ballistic missiles, aviation was forced to take on the role of manufacturer of the new flying vehicles designed to destroy airplanes.

My second meeting with Lavochkin was in connection with this new project. At this point, I cannot pinpoint the precise date, but it was in February or early March of 1951. Late on a Saturday night I was awakened by a telephone call from the ministry officer on duty.

"Boris Yevseyevich, a car will be waiting for you at your home in 15 minutes. Get ready quickly. This is an order from the minister." It was 2 a.m., so that meant there would be no Sunday for me. When I walked out of the building, the minister's ZIS was already parked out front. There was no one inside but the driver. I was certain that I had been summoned by the minister. He loved to do this late at night, and so I did not question the driver. But when the car tore off down Gorkiy Street past the ministry I asked, "Where are we going?"

"To Khimki."

What was going on in Khimki? Why had Ustinov sent his car for me? I didn't have to rack my brain for long. We flew out onto Leningrad Highway, careened across the canal over the same bridge that Isayev had wanted to defend with a guerrilla band against the Germans in 1941, and rolled onto the premises of the aviation factory. All I knew about the factory was that it had been transferred to Lavochkin several years before.

I was sent into the reception room, where I found Ryazanskiy and Pilyugin. Both of them were in a foul mood, but when they saw me they brightened up. Pilyugin was smoking a Kazbek and was spinning tales about what a wonderful dream he was having before he received the minister's telephone call.

---

18. Lavochkin died of a sudden heart attack.

Vetoshkin emerged from Lavochkin's office, and, having confirmed that all three of us were there, asked us to come in. In the spacious office at the end of a long table sat Boris Lvovich Vannikov. Gathered around the table were so many luminaries that we didn't know where to look first. Way at the back, seated separately at a small desk, was Lavochkin, evidently accompanied by two of his deputies whom I did not know. At the table seated closer to Vannikov were Sergey Beriya, Ryabikov, and Ustinov. Next were our radio engineering elite: Aleksandr Lvovich Mints and Aleksandr Nikolayevich Shchukin, who were already corresponding members of the Academy of Sciences; Valeriy Dmitriyevich Kalmykov; and many unfamiliar figures. Judging by the empty tea glasses, Borzhomi mineral water bottles, sandwich trays, and ashtrays overflowing with cigarette butts, they had been there for quite a while.

When we entered, Ustinov nodded in greeting, stood up, and announced, "Boris Lvovich, here come our specialists, whom I promised to call to help us figure out what's wrong with the pyrotechnic cartridges."

Vannikov turned to Mints and said, "Aleksandr Lvovich, this matter is your responsibility. Tell the comrades about it and report their suggestions to us in an hour." After receiving such a crucial assignment from the builder of the most powerful radio stations in the world, Mints accompanied us through hallways that were already familiar to him. We stepped into one of the design halls, where, despite the late hour, several men were working at drawing boards. We were met by chief designer of aircraft electrical equipment Fedoseyev, whom I knew from the aviation industry.[19] He explained everything to us.

Lavochkin was developing the surface-to-air missile. Sergey Beriya was in charge of the entire missile control complex.[20] He, Fedoseyev, had been sent here to help the few electricians of S. A. Lavochkin's design bureau develop the onboard electrical system of this missile. Isayev's engine was on the missile. Tanks of compressed nitrogen fed the propellant components under pressure into the engine. All the compressed nitrogen lines feeding oxidizer and fuel were shut off by pyrotechnic valves. Before startup, these valves needed to be opened in a particular sequence. These were one-time valves. In order to open the valve, an electrical pulse had to be fed to the built-in pyrotechnic cartridge. When it exploded, a path was opened up for the gas or propellant component. The first missiles that had passed all the electrical tests had already been fabricated. Before shipping the first experimental batch to the test range, some of the missiles had to undergo firing rig tests at Novostroyka near Zagorsk. But as soon as the pyrotechnic cartridges got into the act, inexplicable things began to occur. After the first detonation of a pyrotechnic cartridge, the rest failed and the valves did not open. Sometimes several more were tripped, but not in

---

19. Aleksey Frolovich Fedoseyev served as Chief Designer of the Ministry of Aviation Industry's Factory No. 25.

20. Lavochkin's missile for the Berkut system was known as the V-300.

the specified sequence. During the last attempt, oxidizer was fed into the chamber, but fuel wasn't. Experiments with the electrical system had already been going on for a week. One missile was knocked out of action, and they weren't able to start up the engine; but when Isayev had it on the rig, that engine started up and ran flawlessly. There were instances when the pyrotechnic cartridges detonated for no apparent reason when voltage was supplied on board.

Local electricians had developed the onboard electrical system and that of the ground-based console for the rig tests. Fedoseyev and his colleagues discovered many errors and recommended that the system be redone. But the dates for the rig tests had already been pushed back a week. After this, the dates for the first stage of firing range tests were postponed. It would take another two to three weeks to modify the system. But Lavochkin didn't have much time.

But now, Mikhail Ryazanskiy, the most experienced diplomat in our company, told Mints, "Aleksandr Lvovich! You can go relax for about an hour. During that time we will look into this problem, and we'll tell you right away what we think."

Mints gratefully withdrew, but Vetoshkin stayed behind to keep an eye on us. Pilyugin became livid and lashed out at Vetoshkin. "Was it worth it to bring us here for this? An entire week is going to go haywire now. Let them figure this out themselves."

But with his usual composure, Vetoshkin quickly cooled him off. "Iosif Vissarionovich [Stalin] is monitoring our work personally! Lavrentiy Pavlovich [Beriya] has assured him that range tests demonstrating the destruction of an American Flying Fortress would take place yesterday, and it looks like they won't even happen tomorrow. For that reason, Nikolay Alekseyevich, you'd better stop fuming and think how you can help so that you don't get stuck here for a month instead of a week."

I was absorbed in studying the electric circuit laid out on the table. After twenty minutes I understood that the circuit was fundamentally unsuitable for controlling pyrotechnical devices. It was a single-wire circuit in keeping with the principles of aircraft construction. All the positive wires ran directly to the pyrotechnic cartridges through single-contact relays without any redundancy. The body served as the negative wire.

That circuit had tormented me as far back as 1934 on the TB-3 when bomb racks activated by pyrotechnic cartridges had appeared for the first time. At that time this was considered a great achievement. The designer was Aleksandr Nadashkevich, Tupolev's deputy for aircraft armaments. For lack of experience then, he had not taken into account the unreliability of the single-wire circuit, and the mock-up bombs suspended in the electric bomb racks came raining down at their whim. The airfield at Factory No. 22 was jam-packed with enormous dark green four-engine bombers that the Air Force representatives had refused to accept for this reason. Olga Aleksandrovna Mitkevich, the factory director at that time, assembled specialists, including myself, and implored us, "Do something!" Together with the TsAGI electricians we partially redesigned the circuit, and two weeks later the bombers' acceptance process began. In honor of this occasion we threw a sumptuous banquet with *Trekhgornoye* beer in our dear old Fili factory kitchen with the crews of the Far East Air Force commanded by the famous Shestakov who had arrived to receive the bombers.[21]

Now a banquet wasn't in the cards. We could make the missile's single-wire circuit into a two-wire circuit on paper in two days if we got enough sleep beforehand. Next we needed to develop a wiring diagram for the electrical circuit and use it to develop documentation for the cables. The cables needed to be refabricated, and, considering that many of the instruments were also single-wire, we needed to reconsider everything down to the substructure. This would take at least 10 to 12 days. When everything would be manufactured, the rig layout would need to be assembled and we would have to hunt for errors. We would have to correct and modify the cables and instruments, and test them again and again—another ten days or so. Finally, we would have to assemble everything on the first flight-ready missile and conduct tests. In a word, when Fedoseyev and I made a quick estimate, it turned out that a radical surgical operation was absolutely necessary, but it would be at least a month before the missile with modified circuitry would be ready, if not a month-and-a-half!

I asked Fedoseyev why he had needed us to help him figure this out. He explained that he had understood it all completely, but they did not believe him and had forbidden him to even mention any suggestions on how to redesign the circuit. "Now they will start looking for someone to pin the blame on, but whose fault is it? Only the inexperience of the local electricians."

After hearing us out, Vetoshkin, obviously pleased, suggested doing the following: "We were invited as consultants, so we will report our recommendations to Mints. After that, it's his problem how he reports everything to Vannikov. But so that we don't put Lavochkin in a difficult situation, we need to explain everything to him without the top brass around. Since, of all those present, Chertok knows Lavochkin, let him do the explaining. After that, you three—Ryazanskiy, Pilyugin, and Chertok—have to disappear so that no one here will even remember you. And God forbid you should show any more initiative!"[22]

An hour later we returned to the meeting still in progress, and Mints reported to Vannikov: "Boris Lvovich! The specialists recommended by Dmitriy Fedorovich have provided us with some very valuable advice on the reliability problem in the electrical circuit with the pyrotechnic cartridges. I think that we will have to give them a thorough going over with Semyon Alekseyevich, draw up a schedule of operations for their possible completion, and after that we will give you a report."

"When?"

---

21. Semyon Aleksandrovich Shestakov was one of the pilots on the TB-1 "Land of the Soviets" airplane, which flew directly from Moscow to New York for the first time in 1929.

22. Here, Chertok is underscoring the NII-88 group's fear that it would be transferred full-time to work on the air defense project because of their expertise as consultants.

"Today by the end of the day."

"It is now 4:30 a.m. We still need to discuss a number of issues here, and then we do need to get a little rest. I agree that we should have all the modifications done today and tomorrow, and then on Monday we will hear what you all have to say."

My blood ran cold. I looked at Vetoshkin. He was making some kind of sign to Ustinov, but the latter was very pleased with Mints' kudos regarding "his specialists" and did not respond. Then Vetoshkin threw off his reserve. "Permit me, Boris Lvovich! Our specialists have looked into the situation rather well, and they have found a fundamental shortcoming in the circuit. But as we see it, to correct it will require a serious design revision. But Semyon Alekseyevich is the one who should do that. We are prepared to explain everything to him. But he should report the schedule and possible timeframe to you tomorrow."

Vannikov understood Vetoshkin perfectly, but he had to play out the performance according to all the rules. He turned to young Beriya and said, "Sergey Lavrentyevich, do you have any questions?"

"No."

"Then we all thank comrade Ustinov and his specialists for their help, but I request that *you*, Aleksandr Lvovich [Mints], follow up on this work and report everything to us tomorrow, if necessary, together with Ustinov's specialists."

The three of us and Vetoshkin quickly slipped out of the office. Vetoshkin ordered us, "It's a good thing Ryazanskiy and Pilyugin have their own car . . . so, scram! Chertok is now going to explain everything to Lavochkin, and I will wait for him. I won't leave without him."

They summoned Lavochkin with a note. I asked him to give me 15 minutes for the explanations. But we probably talked for an entire 40 minutes or more. He understood everything. He looked extremely tired, and therefore he perceived the extent of the calamity in a somewhat detached state. When we parted he thanked me and asked me to pass along his greetings to Gonor and Korolev, and suddenly, with a warm-hearted smile he added, "I don't know how it was for you, but I had it a lot easier during the war." I mentioned that Gonor was no longer with us.[23] Apparently Lavochkin was not aware of that and my statement depressed him.

Vetoshkin practically forced me out of the building, sat me down in the ministry car, and at around 8 a.m. we raced onto Leningrad Highway. Along the way he lectured me, "Don't get it into your head to call over here and ask whether your ideas are working. One word from Sergey, and if you catch his fancy or any of those academicians, and you might end up with them for a month or forever. That's why for the next week you need to disappear either on vacation or on a business trip."

After coming home, to the surprise of Katya and my sons, I drank half of a glass of vodka, and, after eating breakfast, to their great dissatisfaction, I flopped into bed.

---

23. Lev Robertovich Gonor was removed as director of NII-88 in June 1950.

On Monday I was sent to Leningrad for three days on a business trip—there was always some business going on there. It was a month later when I called Fedoseyev and asked how things were going. He reassured me that they had managed to avoid individual punishments and victims. They had modified the circuit, and tomorrow they would start up the rig tests again. There were a lot of other annoyances, but now everything was behind them: "We'll start flying soon."

After that episode, Babakin's entire team transferred from NII-88 to Lavochkin. The fears of both Fedoseyev and other "high-ranking" consultants from Ustinov's ministry that they might be moved to work under Lavochkin were allayed by the employment of many competent electricians at OKB-301.

IN APRIL 1951, ON MY LATEST TEMPORARY ASSIGNMENT TO OUR "DEAR OLD" TEST RANGE IN KAPUSTIN YAR, I found out that quite nearby there was a new test range for surface-to-air missiles. From Voznyuk they had taken the experienced tester and launch team director Lieutenant Colonel Yakov Isayevich Tregub. He was appointed chief engineer of the new test range and practically directed operations for the preparation and launch of surface-to-air missiles within the scope of all the test range services.

Later, in the spring of 1953, when we were working at the test range with R-5 missiles, Tregub gave us a friendly warning that Sergey Beriya, Ryabikov, Vetoshkin, Vannikov, and Shchukin were flying into their test range. "There will be an interesting salute," said Tregub. Indeed, when we got the call from Tregub's test range, we ran out of our building to watch the target aircraft, a gorgeous Tu-4, being turned into a formless mass of debris in the sky after the missile's direct hit. In all, during the spring of 1953, five target aircraft were fired on and brought down. Korolev, Pilyugin, and I, having come from the aviation industry, felt very sorry for the target airplanes as we observed the surface-to-air missile launches. But it was also something to admire. These missile launches supported development of KB-1's Berkut air defense system. They built two rings of surface-to-air missile defense around Moscow, where missile launchers were positioned. Later, in 1955, General Designer Raspletin's Berkut system went into service as the famous System 25 (S-25) in 1955.

My contribution to this technology was officially quantified as just one sleepless night, an event that Ryazanskiy, Pilyugin, and I often revisited over a few drinks. During these reminiscences I loved to tease Pilyugin, saying that although "we both received medals 'For the Defense of Moscow' back in 1944, I was the first one to express the idea of two-wire electrical circuits for the missiles. Therefore, I have a medal for real work. Yours is just for digging antitank trenches in 1941."

During my frequent flights in and out of Moscow, if the weather permitted and if my fellow passengers weren't in the way, I would press up against the window, trying not to miss the moment we crossed the two rings of the Moscow missile defense. This was a far cry from the antitank trenches for which Pilyugin and I had received government decorations. When the characteristic patterns of the surface-to-air missile complex emplacements came into my field of view, I experienced something akin to satisfaction with my own, albeit small, involvement in this project.

BESIDES THE SLEEPLESS NIGHT IN 1951 DESCRIBED ABOVE, I believe I can be credited with other air defense achievements. In 1944, working with Roman Popov and Abo Kadyshevich at Factory No. 293, we invented a radar guidance system to guide the rocket-propelled BI fighter toward an aircraft. Aksel Ivanovich Berg advised us. Factory No. 293, in whose formation I had a hand beginning in 1940, became the primary facility for the development of surface-to-air guided missiles 15 years later.

After the first S-25 system—which used V-300 missiles developed by Lavochkin at Factory No. 301—the development of surface-to-air guided missiles was moved to Factory No. 293, where the chief designer was future Academician Petr Dmitriyevich Grushin. Working in collaboration with Academician Boris Vasilyevich Bunkin, who had been brought in by Aleksandr Aleksandrovich Raspletin and who would later be in charge of the Almaz organization, they developed systems on a qualitatively new technical level to knock out the most varied means of air attack.[24]

On 1 May 1960, the S-75 system using Grushin's missile shot down the American U-2 reconnaissance aircraft.[25] I had the opportunity to brag to my close friends that a missile developed in Khimki at Factory No. 293, which I mention in all of my curricula vitae, had shot down the American, Francis Gary Powers.

On 27 October 1962, the same Bunkin-Grushin system developed at Factory No. 293 shot down yet another American U-2 reconnaissance aircraft, this time over Cuba during the Caribbean (Cuban Missile) Crisis.[26] During the Vietnam War the S-75 system shot down hundreds of American bombers. The Soviet government generously rewarded the creators of the air defense guided missile systems with honors. However, although the name of the American pilot shot down near Sverdlovsk on 1 May 1960 spread over the entire world, even 30 years later, the names of our missile designers still remained unknown to the world.

---

24. In November 1953, the new OKB-2 of the Ministry of Aviation Industry (located at Factory No. 293) under P. D. Grushin inherited development of guided surface-to-air missiles from S. A. Lavochkin's OKB-301. In subsequent years, the development of such missiles was supervised by a single systems integrator (the giant KB-1 or Almaz organization) led by A. A. Raspletin and then later by B. V. Bunkin. Actual missile development remained at Grushin's OKB-2 (which was later renamed Fakel). The Almaz and Fakel organizations jointly developed the S-75 system, which downed Francis Gary Power's famous U-2 spy-plane in 1960.

25. The S-25 and S-75 antiaircraft missile systems were known in the West as the SA-1 Guild and SA-2 Guideline respectively.

26. In Soviet/Russian vernacular, the Cuban Missile Crisis is commonly known as the Caribbean Crisis.

Chapter 12
Flying by the Stars

The first time I encountered an astronavigation problem was in 1937, while preparing our new four-engine DB-A bomber (assigned the polar designation N-209) for a flight over the North Pole to America. The airplane's commander, Sigismund Levanevskiy, was not concerned about the issue of astronavigation, but navigator Viktor Levchenko was another story. He demanded that I, the ground crew's lead engineer for electrical equipment (including navigational equipment), provide the aircraft with an astrodome and solar heading indicator.

We refined the solar heading indicator (SUK) as per Levchenko's instructions and, with his input, selected a place for the astrodome in the navigator's cockpit in the nose of the fuselage.[1] As for an astral sextant, Levchenko agreed to include one in the equipment, but noted that he would probably not have to use it. During the arctic day over the pole, the stars were virtually invisible, and you couldn't make them out at all through the dome (if it was dirty, fogged up, or iced over), even at night, until you reached the Arctic Circle.

These events came back to me 10 years later, in late 1947, when I was faced with the guidance problem for a cruise missile *(krylataya raketa)*, which at that time still did not exist.[2] In 1949, the R-1 missile with a range of just 270 kilometers had not yet been put into service. The R-2 missile with a range of 600 kilometers was still in the design process. But Korolev had already released the draft plan of the R-3 missile with a range of 3,000 kilometers. On this project he had already written: "The development of a cruise missile is an area with great potential in the evolution of long-range missiles. The realization of a cruise missile has a certain bearing on the successful development of long-range ballistic missiles…"[3] Missiles with this range

---

1. SUK—*Solnechnyy ukazatel kursa.*

2. In Russian the phrase *krylataya raketa* literally means "winged missile," but more generally means "cruise missile," that is, missiles that do not fly along ballistic trajectories but are guided by aerodynamic surfaces (such as wings) through most of their trajectories.

3. Here, Chertok is quoting from the introductory essay of the draft plan for the proposed R-3 missile, entitled "Principles and Methods of Designing a Long-Range Missile," which has been reproduced in M. V. Keldysh, ed., *Tvorcheskoye naslediye akademika S. P. Koroleva* [*The Creative Legacy of Academician S. P. Korolev*] (Moscow: Nauka, 1980), pp. 291–318.

still couldn't reach the U.S. from our territory, but all the American air bases that were home to the Boeing B-29 Super Fortresses in Europe and Asia would come within reach.

Which missile would be most efficient for the task: ballistic or cruise? We needed to analyze both. By the same token, alternative flight control systems also had to be examined. While discussing these problems, Ryazanskiy and Pilyugin announced that they were going to start developing ballistic missile control systems together with Kuznetsov or with the new NII-49 naval gyroscopic organization in Leningrad. The cruise missile needed to be controlled over its entire trajectory all the way to the target. This was a very difficult task, and NII-885 was not yet prepared to handle it. The principles that the Germans proposed in the A9 and A10 designs were inadequate. Radio control over enemy territory would be knocked out of action by jamming, while autonomous control still generated completely unacceptable errors.

In fact, the normal drift of the gyroscopic system (the Kreiselgerät firm's best gyroscopic platform at that time), one angular minute per minute of time, yielded a position error of 1 mile, that is, 1.8 kilometers. The best gyroscopic systems, even with air suspension, might have drift as high as one degree per hour. If a 3,000-kilometer flight lasted two hours, then the position error for a purely autonomous system might even exceed 200 kilometers. Who needs a missile like that?

But this evidence did not divert Korolev from his "cruising" ideas. Back in his SKB Department No. 3, he found individuals eager to grapple with potential layouts for the cruise missiles. One of them, Igor Moisheyev, sensibly reasoned that in two or three years we'd see proposals for a guidance system if we found solutions for the design of aerodynamic supersonic winged vehicles and energy-efficient sustainer engines. We continued to heatedly argue over the issue of cruise missile guidance.

That is when I remembered the N-209 aircraft's astrodome and navigator Levchenko's boasting that on a clear, starry night he could determine his geographical location with an error no greater than 10 kilometers using an astral sextant.

A navigator's work entailed searching the night sky for "navigational" stars that had been previously designated for the northern hemisphere, measuring the altitude of at least two stars using an astral sextant, determining the precise time of the measurement using a chronometer, and then determining his coordinates using special, rather complicated calculations and graphic plotting using a map. An experienced navigator, using specially prepared tables and spending 15 to 20 minutes per session, could determine his or her location with an accuracy of within five to seven kilometers. To verify that, I went over to the High Command's Air Force NII. Luckily, I still had a lot of acquaintances there, and they confirmed for me that there really were navigators, real astronavigation aces, who determined their location with an error of just three to five kilometers.

When I suggested in conversation with the Air Force NII specialists that we at NII-88 wanted to begin developing an automatic astronavigation system and dispense with a navigator, I was met with sneers.

"So you've flown as a navigator?"

"No."

"Look, study the difficult job that a navigator performs. We'll let you fly. And then you'll see for yourself how hopeless this venture is. You're just wasting your time."

But the aircraft navigators' skepticism did not change my mind. Instead of a human being, an automated mechanism—an automatic astronavigation system—could perform all the operations! By no means did it have to duplicate everything that a person does. The problem would be solved if we succeeded in developing this system and combined it with an autopilot. The system would issue heading control correction signals to the autopilot and, upon reaching the target's geographical area, it would switch the missile into a dive. Easy to say! Of course, you can't come up with everything single-handedly.

First we needed to set up a laboratory. The laboratory absolutely would have like-minded individuals. It would be best if these like-minded individuals were young and knew nothing about the professional problems of an aircraft navigator. More experienced people might doubt the feasibility of the task and would only be hampered by their own skepticism. It was a good thing that I did not accept the offer to fly as a navigator. It is very likely that after finding out how complicated astronavigation was, I would have abandoned my adventurous undertaking.

No, I didn't consider myself sentimental, especially after the war. But my memory of the N-209 crew haunted me. I shared some guilt there. If only they had had a real automatic navigation system! Now that we had the means and the need and could assign the research to an extensive staff, I couldn't pass up this opportunity.

I went to Korolev and announced, "I have an idea! I'm going to start developing a navigation system for the cruise missile, under the condition that you really are going to make this missile." Korolev immediately accepted the idea, but he said that I needed to get Yuriy Pobedonostsev's agreement to set up the new laboratory, and it would be best if I managed to do that on my own without his help.[4]

Back in those early years, Korolev still hadn't let go of his idea for the rocket-plane that he'd been working on at RNII before his arrest.[5] Now he had the opportunity, without any of the letters to Beriya or Stalin that he had written from prison, to invest in the realization of an idea that was much more daring than the design of the stratospheric airplane from a decade before. I understood him to mean "get to work and then we'll see." He did not reject the cruise missile idea. In fact, soon after, he proposed the so-called EKR (Experimental Cruise Missile).[6] At the time,

---

4. Pobedonostsev was Chief Engineer of NII-88 at the time.

5. In the 1930s, while at RNII, Korolev proposed the development of a rocket-plane known as the 218 (and later, the 318). A testbed for the vehicle with a different design, known as the 218-1 (and later, the 318-1) was developed and flown on a few test flights in 1940, while Korolev was already in prison.

6. EKR—*Eksperimentalnaya krylataya raketa*—literally stands for "Experimental Winged Missile" or more generally "Experimental Cruise Missile."

even the R-1 missile hadn't flown.

Korolev had a lot of complex problems in his relationships with the NII-88 managers. If he were to start demanding that they create one more laboratory working on his projects in my Department U, Yevgeniy Sinilshchikov would protest. He would obtain new proof that Chertok was gaining a hold on air defense projects in his department and that almost all the guidance specialists were working for Korolev. Sergey Pavlovich was right; we needed to be circumspect in this matter. He was aware of navigational problems and questioned whether I had specialists in my department to develop such an idea.

When putting together a team of people you hope will share your ideas, it is very important to formulate for each of them the specific task that will constitute their contribution to the solution of the problem as a whole. For a creative individual, this "whole problem" must be rather attractive. Wasting no time, they need to seize the beachhead. They need to find their own solutions without waiting until all the methods are invented and facts discovered to fully develop the system. In this matter an obvious problem needed to be solved, without expecting to know what the whole system would eventually become. The problem was searching for, identifying, and automatically tracking stars. To begin with, as our basis, we would take the procedure used by navigators at sea and in the air. After finding and identifying the necessary stars, we needed to solve at least two more problems. First, we had to determine the altitude of the stars above the horizon or the angle between the direction to the star and the direction of the vertical. Then we had to insert the measurement results into the prepared calculation procedure. But then we needed to come up with a computing instrument that, depending on the automatically measured angular distances between two stars, would calculate everything, process the navigation commands for the missile's autopilot in order to follow the optimal flight path, and issue the final command to dive toward the target.

And so the first task was to develop the automatic device that would track the stars from a stationary platform. To begin with, this would be the laboratory window. We needed to start with the simplest things. The first employee of the new laboratory, which was not yet officially entered on the organizational charts, was Larisa Pervova. She and I had developed an electric arc ignition during the war. Unlike many female engineers at that time who would only do exactly what their manager assigned, she showed initiative and a tendency toward independent action. In this case, while we in the new field were still sorting out the problem, it was a valuable quality, especially since I was only able to devote attention to this project by fits and starts.

Soon laboratory workers appeared. They procured and obtained optical measurement instruments, various photocells, and electronic multipliers and established contacts with the electronics and optical laboratories of other institutes. During the formative years of NII-88, we ran into no difficulties when it came to funding any new ventures. All you had to do was to show that the funds were needed for the future of missile technology. When it came to bringing new specialists on the job, difficulties arose during processing in the personnel department only if their background information wasn't "clean" enough.[7]

We needed inventors to work in the laboratory. The situation required them to design and then realize a fundamentally new system, the likes of which had never been seen before, not even abroad. A manager/administrator was not suitable to head such a laboratory. We needed a manager who possessed God-given talents—a creative source—who at the same time was a realist. In this case, the director of such a laboratory would also have a systematic mindset as well as an education in electrical engineering or at least in mechanical engineering and mathematics. Should this person turn out to be a capable administrator to boot, then so much the better. We needed someone with ideas.

Where could we find someone like this? I didn't see a suitable candidate at NII-88. What's more, even the first modest venture—creating a group to develop automatic star-tracking methods—had already raised objections among employees at Department U. We quickly found opponents who argued that the whole astronavigation project was a gamble.

There was also an individual who was one of those ideological government supporters. He made it clear to me that if I entrusted this promising work to him, he would overcome all the obstacles and work by the sweat of his brow. But, if not, he would publicly and by every means insist that we were trying to realize a foolhardy scheme in the laboratory. I didn't heed his warnings. But he kept his word, and over the course of three years, one commission after another tried to track down the foolhardy origins of the astronavigation ideas.

During one of my ministry visits I shared my problems with Vladimir Sergeyevich Semenikhin, a specialist who worked there on antiaircraft fire control instruments and all kinds of optics. He had recently been transferred from the Zagorsk optico-mechanical factory to head a department in the ministry. Suddenly Semenikhin announced that he was going to help me: "I have a candidate specifically for that job. He fits all your parameters, except for item five on the personnel background sheet.[8] But that's your problem. If you approve, I will be able to transfer him from Geofizika to Podlipki."[9] I had to go kowtow to the KGB colonel who was the institute's deputy director for personnel. And that is how Izrael Meyerovich Lisovich became director of operations for the automatic astronavigation system for many years.

As for Semenikhin, he turned out to be a man with ideas. He reached the rank of deputy minister for radio-electronics, but soon realized that a purely administrative career was not for him. Semenikhin headed a large scientific-research institute, was

---

7. Here, Chertok is referring to what at the time were considered dubious backgrounds (e.g., extended family abroad, older members of the family involved with pre-Revolutionary intelligentsia, former arrest records, member of persecuted minorities, etc.)

8. Item five referred to ethnicity.

9. Here "Geofizika" denotes TsKB-589, a Moscow-based design bureau that developed optical elements for guidance and attitude control systems.

selected as an active member of the USSR Academy of Sciences, and was awarded many medals and the title Hero of Socialist Labor.[10] Once when we met at a routine gathering at the Academy of Sciences, I reminded him of the good deed he had done in 1947. He could not recall and then asked, "What was the outcome of that whole endeavor?" For cruise missile navigation that endeavor lasted 15 whole years. For modern cosmonautics, astronavigation is a routine practice.

The laboratory was staffed, and soon thereafter Lisovich saw to it that gyroscopic problems were assigned to G. I. Vasilyev-Lyulin. He proved to be a talented engineer, both in theory and in design. In 1949, the three of us, Lisovich, Vasilyev-Lyulin, and I, were awarded an inventor's certificate classified "Top Secret." Essentially, we had developed all the main principles and tested them on mock-ups in 1948 and 1949. We proved the feasibility of automatic astronavigation using the domestic instrumentation available at that time.

It was still long before the time of transistors, microelectronics, and computers that made it possible to solve automatic control problems and complex calculations electronically, while ensuring reliability with multilevel fault tolerance. We were headed down a path of pure electromechanics, depending on the reliability of classic methods because of the simplicity of the ideas and design.

We tackled the development of a star tracking system first. The most complex problems here turned out to be those of light interference, such as from the general background illumination and the hazard of "fixing on the wrong star." To track two stars with a single telescope we came up with a device with a swiveling mirror. Gyroscopic stabilization made it possible to hold the direction to the star even if it hadn't been observed for some time. In the laboratory, such a mock-up using two collimators worked excellently on a dynamic platform and did not lose the artificial stars.[11]

After the star problem, the second problem involved inventing a vertical. An artificial vertical line was supposed to point to the center of the Earth. The angle between the direction to the star and the direction of the vertical line made it possible to determine the "altitude" of the star above the horizon and to construct the so-called "circle of equal altitudes."[12] If two circles of equal altitudes were constructed using two stars, then one of the points of intersection of these circles on a map would be the position of the aircraft, spacecraft, or missile. At that time, the

---

10. Academician Vladimir Sergeyevich Semenikhin (1918–90) served as director of NII-592 (also known as NII Avtomaticheskiy apparatury) in 1963–71 and 1974–90, during which period he oversaw the development of guidance systems for several Soviet ballistic missiles. He was elected an Academician of the Academy of Sciences in 1972.

11. Collimators are typically small telescopes attached to bigger ones used to establish precise line of sight.

12. In navigation lexicon, "altitude" refers to the angular distance above the horizon, that is, the arc of a vertical circle between the horizon and a point on the celestial sphere.

creation of a vertical was a brand-new problem.

The same Professor Shuler, who in 1945 visited us in Bleicherode to familiarize himself with the work of the Institute RABE, back in 1923 had discovered and published the principle of a pendulum device that maintained the direction of the vertical when exposed to accelerations. Moving along a geodesic line over the Earth's surface, this pendulum would have an oscillation period of 84.4 minutes! But a physical pendulum with such a period would have to have a suspension that was three times the Earth's radius in length. We needed to find other principles.

Vasilyev-Lyulin learned from reading the literature that back in 1932, Soviet engineer Ye. B. Levental had proposed a gyroscopic vertical with so-called integral correction. B. V. Bulgakov developed a theory for it in 1938.[13] But the practical realization of such a vertical with errors no greater than one to two angular minutes relative to the true vertical of a given location proved impossible. In our proposal (Vasilyev-Lyulin developed the ideas), the direction to the star replaced Levental's free gyroscope. Right away this eliminated the then-large error of the free gyroscope. In Academician A. Yu. Ishlinskiy's classic work, just the mathematical formulation of such a vertical takes up 14 pages.[14]

The third problem, developing a computing instrument that generated commands to the autopilot, was realized using a cam mechanism. It is curious that the conventional error of such a primitive instrument did not exceed one angular minute.

All the ideas and principles were tested on functional laboratory mock-ups. All of our research required thorough documentation, because a number of people were very actively opposed to it, stating that Chertok was spitefully diverting many specialists for a foolhardy project. The Communist Party committee and ministry received ever more incriminating letters, which were quite the fad in those days.

Despite the vociferous attacks of local crusaders, the first successes of our research and the creation of functional mock-ups of automatic astronavigation instruments inspired those who supported the cruise missile option in Korolev's team. One must give credit to his personal objectivity and conviction. He promoted scientific-research work on "Integrated Research and Determining the Primary Flight-Tacti-

---

13. Corresponding Member of the Academy Boris Vladimirovich Bulgakov (1900–52) was a major Soviet scientist in several fields, including gyroscope theory, oscillation theory, and navigation systems theory.

14. A. Yu. Ishlinskiy, *Orientatsiya, giroskopy i inertsialnaya navigatsiya* [Orientation, Gyroscopes and Inertial Navigation] (Moscow: Nauka, 1976). Aleksandr Yulevich Ishlinskiy (1913–2003) was a leading scientist behind the theory of inertial navigation for spacecraft. While serving as a senior scientist at NII-944, he worked closely with Korolev, Glushko, Pilyugin, and other major chief designers of the Soviet space program.

cal Characteristics of Multistage Long-range Cruise Missiles" in every way.[15] The individual directly responsible for these operations was Igor Moisheyev, with whom I met regularly. He contended that "the intercontinental concept can be achieved only with wings."

During this time of trouble, Korolev and his first deputy Mishin were both subjected to harsh criticism from a very orthodox segment of the Party committee. The two were accused of being conceited, filling the team with apolitical personnel who didn't belong to the Party, failure to exercise self-examination, and a host of other sins.

During one of his meetings with me, Korolev showed a surprising awareness of the state of affairs in Lisovich's laboratory. I asked how he knew all of that. Korolev answered that he was personally very interested in this work and therefore he had his own sources of information. "But keep in mind," he said, "in your time you've taken on some people without particular attention to their personalities. You have a lot of scumbags in your group, and in this climate you can't get rid of them. Once again the times are such that even a minister can't always intercede."

Korolev asked whether I knew that Vera Nikolayevna Frolova, who supervised work on ballistic missile gyroscopes, was going to marry Lisovich. I confessed that I did not meddle in the personal relationships of my staff. "That's a mistake. I know everything that's going on in your department. Let them get married. But you must without fail persuade Frolova to transfer to Lisovich's laboratory. We need to support him. Believe me, she will be able to."[16] Frolova really did show combative qualities in defense of the astronavigation concept. Lisovich gained a strong co-worker and lifelong friend. Once again I saw that Korolev had a knack for keeping an eye on the behavior and relationships between people, even when they were not his subordinates, and his skill for intervening—always appropriately—in the assignment of personnel.

Suddenly Korolev came out with a very welcome suggestion that temporarily disarmed opponents of astronavigation. In order to finally determine whether the proposed principles were valid or not, we needed to test the system on an airplane. This would enable us to experimentally confirm the validity of the principles and the precision promised in the calculations. Institute and ministry leaders accepted

---

15. The study cited by Chertok, "Integrated Research and Determining the Primary Flight-Tactical Chara.cteristics of Multistage Long-Range Cruise Missiles," was part of a broader research project entitled "Research into the Prospects of Creating a Missile Capable of Long-Range Flight with the Goal of Obtaining Its Primary Design and Flight-Tactical Characteristics" initiated in December 1950. The goal of the broader research project was to study future configurations of intercontinental (ballistic and cruise) missiles. The study on the cruise missiles referenced here was presented to an advisory group in January 1952 at NII-88, part of which is reproduced in Keldysh, *Tvorcheskoye naslediya...*, pp. 328–41.

16. *Author's note:* In this conversation I was not referring to radio engineer Vera Aleksandrovna Frolova, whom I mentioned earlier, but another Vera Frolova who was a gyroscope specialist.

Korolev's proposal. In 1950, we began to develop and fabricate a functional mock-up of an automatic night astronavigation system.

There was one more episode that determined the eventual fate of this whole project, but it happened before the events described above. In early January 1949, NII-88 Director Gonor called me to his office. Lately, when he called me in, he was always trying to prevent something. He warned me about the latest letters to the Central Committee or ministry and about the preparation of commissions to investigate the work my department was doing. Gonor himself treated me very kindly, but always hinted that "if something happened," he wouldn't be able to defend me. One time he even insinuated that it would be better if I replaced Lisovich in this whole "star" brouhaha with somebody else who had a "cleaner" background, because they would also flatly accuse Gonor of having a particular tendency in selecting and placing personnel.[17] "Keep in mind, this isn't coming from Vetoshkin or from Ustinov. There are forces that even they can't resist."

When I entered Gonor's office, having prepared myself psychologically for another unpleasant warning, I saw that he was not alone. Sitting in the armchair next to Gonor's enormous desk was a major general. I immediately determined that he was from the aviation industry and he had a very familiar face. When I walked in, he stood, gave me a firm handshake, and smiled as he introduced himself, "Lavochkin." So that's why his face seemed so familiar! More than once I had seen his picture in the newspaper; the celebrated general designer of the LaGG, La-5, and La-7 fighters that were famous during the war. Lavochkin was tall and slightly stoop-shouldered. His general's uniform with its Hero of Socialist Labor star went very well with his host's general uniform with the same gold star. Only the shoulder boards were different. Lavochkin had aviation shoulder boards, while Gonor had artillery.

As usual, Gonor smoked his favorite Kazbek cigarettes, and evidently continuing his report on the structure and projects of NII-88, he turned to me and said, "Semyon Alekseyevich [Lavochkin] is visiting us for the first time. I just learned that he knows Korolev quite well and even put engines that he and Glushko made while they were still in Kazan in his planes.[18] I have briefed Semyon Alekseyevich on the structure and subject matter of our institute. And he wanted to hear first-hand from you about your projects."

I was surprised that such a famous fighter aircraft designer had suddenly taken an interest in ballistic missile guidance systems. As best I could, I briefed him on the structure of Department U and the main projects that we were involved with, avoiding the mention of astronavigation out of caution. But Gonor, having noticed

---

17. Gonor, Lisovich, and Chertok were all Jewish, and hence more vulnerable to being under suspicion in the early 1950s at the height of the anti-cosmopolitanism campaign.

18. During Korolev and Glushko's wartime stint working at Factory No. 16 in Kazan, they developed propulsion systems for various Lavochkin fighter aircraft.

my circumspection, interrupted and added: "Boris Yevseyevich failed to mention a very interesting project, an astronavigation system for cruise missile guidance."

Lavochkin gave a start and listened very attentively. When I mentioned the accuracy figures—from 5 to 10 kilometers regardless of the flight time and range—he smirked.

"You're giving me a sales pitch, of course, but it's still very interesting. If you conduct flight tests, I will definitely request that you let me know the results."

Then the conversation returned to the subject that the two generals were apparently discussing before my arrival. Lavochkin was lecturing Gonor (I will give the gist because I can't reproduce it word for word from memory): "It's very important to select talented people. You need to give them the freedom to discover their capabilities and to adjust to one another. Your organization is young, and your people are still learning to work together on a single general project. There will be a lot of squabbles, believe me. It will take another two or three years before you work it all out, especially with such differences in subject matter and interests."

Lavochkin was right. More than three years passed before everything more or less fell into place and until Korolev finally occupied his well-deserved place as Chief Designer at OKB-1. Lavochkin continued:

"I was telling Lev Robertovich that I worked with Korolev when we were both quite young. In Krasnaya Presnya a Frenchman named Richard had a design bureau.[19] At that time Korolev was very fascinated with gliders. There were a lot of talented young people there. Next he became absorbed with reactive motion. Just before the war I found out about his misfortune.[20] And quite recently we tried to introduce the engines that Glushko and Korolev developed in Kazan. They flew. But they're not worth it now. We realized that liquid-propellant rocket engines are not for airplanes."

At that time Lavochkin had every reason to talk like that. Fighter aircraft had already broken the sound barrier. And this revolutionary leap for aviation had been made thanks to airplanes featuring turbojet engines rather than liquid-propellant rocket engines. In the race for the most advanced jet fighter, Lavochkin had at first lagged behind Mikoyan and Yakovlev, but once he had created a serious scientific-technical facility at the new factory in Khimki, he not only began to work on a more advanced fighter than his competitors, but he also agreed to develop and produce air defense missiles.[21] Lavrentiy Beriya's son Sergey acted as the official administrative guru for the control systems for them.

---

19. Paul E. Richard was a French engineer who briefly worked in the Soviet Union in the late 1920s and early 1930s as an aviation designer at the Experimental Section-4 (OPO-4) also known as the Naval Experimental Aircraft Building All-Union Aviation Enterprise (MOS VAO). Korolev briefly worked for Richard during his schooling years.

20. Lavochkin was referring to Korolev's arrest and incarceration in 1938–44.

21. Aleksandr Sergeyevich Yakovlev (1906–89) designed a diverse array of Soviet military and civilian aircraft during his tenure as head of OKB-115.

Then Lavochkin complained that he had invited Korolev to come talk with him. He himself had fished for an invitation to go talk to Korolev, but "it never got any further than telephone conversations." I proposed that Semyon Alekseyevich come with me and have a look at the laboratory and a missile launch simulation rig. He thanked me, saying that he wouldn't want to make such a tour without his specialists. They might feel slighted.

When Lavochkin left, I asked Gonor straight out why he had not invited Korolev to the meeting. Gonor explained. In the first place, Gonor *did* invite Korolev, but Korolev immediately said that he had to leave on some urgent business. Second, Lavochkin himself telephoned and gave advance notice that he wanted to speak with the director.

Lavochkin's visit to NII-88 had important consequences for astronavigation. The functional mock-up of the system for tests on an airplane was manufactured by the staff from the laboratory and our experimental instrument shop over a period of a year and a half and was ready to be installed on an Il-12 aircraft by early 1952. The pilot was supposed to fly the airplane so that, to the extent possible, the indicator needle remained in the zero position. This meant that the airplane was flying along the route indicated by the astronavigation system. During target approach a red indicator light would light up on the navigator's and pilot's consoles. The navigator's duty was to determine the aircraft's actual position according to ground reference points. It was a good thing the flights were conducted only on clear nights. By determining the actual position along the flight path when the "target" signal appeared, it was possible to determine the system's error.

Nine flights were completed on the Moscow-Daugavpils (Latvia) route, a distance of approximately 700 kilometers. The tests were conducted jointly with the High Command's Air Force NII throughout the second half of 1952 and the first half of 1953. All the laboratory's leading specialists headed by Lisovich participated in these tests, which would decide the system's fate. The flight tests brilliantly confirmed the validity of the fundamental design. During the entire time there was not a single failure, and the navigational error did not exceed seven kilometers. Subsequent calculations showed that if the gyroscopes and other elements of the system had been manufactured with the degree of accuracy obtainable by the technology of the 1970s, then the error would not have exceeded one kilometer! And if the calculations had been executed using the onboard computer technology of the 1980s rather than an electromechanical calculator, the error would have been 10 to 20 meters! I was not involved in these tests, but felt for my comrades, with whom I had begun this project in 1947.

By the time of the tests, Gonor had already been removed from his post as NII-88 director. The management of Department U had been dissolved; I had been relieved of my duties as the institute's deputy chief engineer, and, based on the highly partisan conclusions of the ministry's special commission, they threatened to fire me. But Korolev came to my aid in the nick of time. As soon as he learned of the commission's conclusions, he invited me in for a tête-à-tête. "You have yourself to blame for what happened to you. You need to use your head when you select people, and get rid of all the rotten apples promptly. Look how things are going with me. My people support each other. Nobody writes denunciations. Attacks only come from the outside. But it's different with your department. That's why I have already made arrangements where I need to. You are transferring over to me in OKB-1 with a demotion in rank. I'm appointing you deputy chief of Department No. 5. We're going to set up our own guidance department, and we'll be independent of the NII. Your boss will be Mikhail Kuzmich Yangel. You don't know him. I don't know him either. I don't think he'll be with us long. I don't think he understands what you're doing and he doesn't have our experience. But, judging by everything, he is a decent man. Ustinov sent Yangel to us. I took advantage of that and arranged with Ustinov to have you transferred to me. By the way, I think he was happy with the proposal because he hinted that he couldn't do anything more. He asked me to pass that along to you so that you wouldn't be offended."

"Outside" forces brought Professor Petr Krasnushkin to NII-88 to manage Department U, my creation. He was a specialist in the propagation of very long radio waves. This area of radio engineering had virtually no relation to our project. But once Krasnushkin had discovered our renegade astronavigation laboratory, he immediately announced that he was developing a high-precision navigation system for cruise missiles using very long radio waves.

In January 1952, Korolev spoke at a presidium session of the institute's scientific-technical and academic council devoted to summing up the scientific-research work on the subject "Integrated Research and Determining the Primary Flight-Tactical Characteristics of Multistage Long-range Cruise Missiles."[22] This subject was a component of the larger scientific-research project (NIR) entitled "Prospects for the Development of Long-range Missiles."[23] At this session, NII-1 Director Academician M. V. Keldysh and TsAGI chief theoretician Academician S. A. Khristianovich delivered reports on problems of aerodynamics, engines, and multistage missile configurations.

This meeting decided the fate of the future carrier of intercontinental weapons. In the early 1950s, the Cold War was in full swing, stimulating the development and production of many promising weapons. We already possessed the atomic bomb, but the successes of the nuclear specialists were to a certain extent one-sided. The U.S. remained our primary opponent in a possible third world war. An ocean away, it remained untouchable.

The experience of developing new air defense missiles and jet fighters had showed

---

22. For an edited version of the report, see Keldysh, *Tvorcheskoye naslediye . . .*, pp. 328–41.

23. NIR—*Nauchno-issledovatelskaya rabota*—literally stands for "Scientific-Research Work" but represented a specific phase of initial R&D in the Soviet engineering sector. The complete title of the NIR was "Research into the Prospects of Creating a Missile Capable of Long-Range Flight with the Goal of Obtaining Its Primary Design and Flight-Tactical Characteristics."

that if our new bombers even managed to fly across the ocean or over the pole carrying a nuclear payload, there was very little chance of them dropping it or hitting the target. The flight crews participating in a possible attack on the U.S. would be doomed. It is clear that the loss of a hundred or so of our airmen didn't alarm Stalin very much. What disturbed him was that, basically, Soviet military technology could not inflict any damage on U.S. territory, while all the Soviet Union's vital centers were within reach of the American B-29 Super Fortresses, and according to intelligence information, they were especially vulnerable to the new long-range heavy jet bombers.

Intelligence turned up one more piece of news. The U.S. had supposedly started developing the Navaho long-range unmanned drone aircraft. The scanty information about this program confirmed that the Navaho was a cruise missile with a range of around 4,000 to 5,000 kilometers.[24] Consequently, if there were several hundred of these Navahos, the Americans were capable of striking almost the entire territory of the Soviet Union with atomic bombs from their European and Asian military bases surrounding it, without risking the lives of their own airmen. We discussed this at all of our meetings with tremendous concern. In those days the possibility of a new war seemed quite real.

That is why the scientific-research project "Prospects for the Development of Long-range Missiles" was particularly significant. Korolev was not yet ready to give preference to one of the two possible options: ballistic or cruise. Our affiliation with the Ministry of Armaments required us to develop a ballistic missile. At the same time, Ustinov, who had made this type of missile his own in 1945, was not opposed to transferring the development and production of air defense guided missiles to the aviation industry. The burden on the ministry and the degree of responsibility might be too great if he held onto both technologies. He even acquiesced to the transfer of his first deputy Ryabikov to work in Special Committee No. 3, which had been entrusted with managing all issues of radar defense and air defense missile control complex.

If cruise missiles proved to be a promising field for intercontinental ranges, then the development and production of such missiles would also be transferred to the aviation industry. And then, what would the Ministry of Armaments be left with after expending so much effort on the development of the missile industry? Who would need these R-1 and R-2 missiles that carried their 800 kilograms of TNT 300 and 600 kilometers? Actually, Korolev had already begun working on a missile with a 1,000- to 1,500-kilometer range. But that, according to the strategists in our ministry and our main customer, the GAU, was not the correct approach.

Korolev understood all of this perfectly. Obviously he would not be able to manage two fields, and if the aviation industry felt like taking the cruise missile

---

24. The projected range of the U.S. Air Force's Navaho (XSM-64 version) was about 5,600 kilometers. A later version, the XSM-64A, was designed to fly about 10,000 kilometers.

away from us, let them take it, but let it end up in good hands. After all, this missile had a lot of purely aviation-related problems. Keldysh, Khristianovich, and all the TsAGI elite understood this perfectly. Although, like me, Korolev's first deputy Vasiliy Mishin had gotten his start in the aviation industry, he hadn't become a fan of the new cruise missile field. In general, in Korolev's team, the cruise missile fan club had a very small membership. And that was understandable. The absolute majority of specialists were up to their ears in routine work on ballistic missiles.

But Korolev would not have been the greatest chief designer if he had allowed himself to make a superficial report on the cruise missile problem. Therefore he gave it serious consideration. He proposed a two-stage cruise missile with a range of 8,000 kilometers and a launch weight of around 90 to 120 metric tons. The first stage had a powerful liquid-propellant rocket engine to execute a vertical liftoff, accelerate, and gain altitude before separation from the second stage. By that time, a vertical liftoff had been put to the practical test on ballistic missiles many times and did not require complex launch facilities.

The second stage of the missile was winged and acted as an engine that was supposed to operate over the entire flight path. He proposed a supersonic ramjet engine (SPVRD).[25] Mikhail Makarovich Bondaryuk, the developer of this engine, had achieved significant success.[26] But large ground rigs that were not yet available were needed to perfect the engine. Nevertheless, calculations showed that at a cruising altitude of 20 kilometers, the required range could be achieved at a velocity of up to Mach 3.

Korolev had thoroughly analyzed two alternative navigation options: an astronavigation option and a radio option. He wrote:

"The primary advantages of the astronavigation method is that flight range and duration have no bearing on guidance accuracy and no need for communication with ground stations... The research conducted in this field shows the absolute feasibility in the near future of creating such a system, operating for the time being under nighttime or twilight conditions. So far, the primary drawback of this proposed option is the uncertainty about how to solve the problem of guidance under full daylight conditions at altitudes up to 20 kilometers...

The primary difficulty in creating the elements of an automatic astronavigation system is, above all, the very high degree of precision required of them...

Later this year, mock-ups of the primary fundamental units of the astronavigation system will be tested on an airplane. These tests should provide answers to many extremely important questions, and above all, should confirm the feasibility of achieving the necessary accuracy."[27]

---

25. SPVRD—*Sverkhzvukovoy pryamotochnyy vozdushno-reaktivnyy dvigatel.*
26. Mikhail Makarovich Bondaryuk (1908–69) served as chief designer of OKB-670, the leading Soviet organization developing ramjets.
27. Keldysh, *Tvorcheskoy naslediye . . .*, pp. 340–41.

Asif Siddiqi.

The Burya missile (with the internal designation '350') was the first Soviet intercontinental cruise missile. Developed as an alternative to the R-7 intercontinental ballistic missile, the Burya was launched nearly 20 times between 1957 and 1960 before being cancelled.

And then Korolev cited very convincing arguments against the system Krasnushkin proposed. The successful flight tests completed in 1953 removed all doubt as to the astronavigation system's performance. By that time we had also received the encouraging results from Bondaryuk's experiments with the SPVRD. The time had come to make a decision on the subsequent fate, not just of astronavigation, but of the whole cruise missile project. After much agonizing discussion, deliberation, and reflection, Korolev capitulated. Having come to terms with Keldysh, he decided to stop his work on the cruise missile project and transfer the entire thing to the Ministry of Aviation Industry (MAP).

A government decree on the development of cruise missiles placed responsibility on NII-1 to develop an astronavigation system to ensure the required precision of the main stage in its flight toward the designated target. Mstislav Vsevolodovich Keldysh was appointed scientific director for the development of cruise missiles. Now, as chief of NII-1, he was responsible for developing an astronavigation system in a branch of NII-1 specially created for this purpose. Lisovich's entire laboratory was transferred from NII-88 to MAP's NII-1 branch. Lisovich was finally named chief designer of the automatic astronavigation system. The resources he had to work with were considerably more extensive than in Podlipki. In 1955 the work force in his design bureau exceeded 500.

In 1954–55, flight tests were conducted on the newly fabricated aircraft mock-ups. This time they used the Tu-16 aircraft. In four flights with a range of 4,000 kilometers, at an altitude of 10,000 to 11,000 meters, with an average speed of 800 kilometers per hour, over a flight time of five to six hours, the system had errors ranging from 3.3 to 6.6 kilometers.

The decree on the development of intercontinental cruise missiles armed with nuclear warheads was issued in 1954. It called for work to be conducted simultaneously on two missiles: the lighter Burya (Storm), which was assigned to Semyon Alekseyevich Lavochkin, and the heavy Buran (Blizzard), assigned to Vladimir Mikhaylovich Myasishchev. Academician Keldysh was appointed scientific director of both these projects.[28] By this time, Keldysh was a member of Korolev's Council of Chief Designers. Thus, he was the most knowledgeable scientist on all the crucial scientific and technical problems that needed to be solved in order to create an intercontinental nuclear warhead carrier. Lavochkin's deputy, doctor of technical sciences Naum Semyonovich Chernyakov, was named chief designer of Burya.

As Korolev had proposed, the Burya multistage missile had a first stage with a liquid-propellant rocket engine. Isayev produced it out of a cooperative effort that had formed in Lavochkin's team. Bondaryuk, who had been working under Keldysh, developed and delivered the SPVRD sustainer engine for the second winged stage.[29] The total launch mass of the Burya exceeded 90 metric tons. Here, the full mass of the actual cruise missile was more than 33 metric tons. The system was designed for a range of 8,000 kilometers at a speed of Mach 3.1. The cruising altitude was 17,500 meters. During target approach the missile made an evasive maneuver, climbed to an altitude of 25,000 meters, and executed a nose dive. It was already assumed that the missile would carry an atomic bomb. Based on the astronavigation system's flight test results, the maximum error relative to the target center should not exceed 10 kilometers. Flight at supersonic speeds caused a significant increase in the temperature of the missile hull. For that reason, unlike an aircraft astronavigation system, this system needed to be mounted under a transparent, but heat resistant astrodome.

At Lavochkin's suggestion, the Dal (Distance) antiaircraft guided missile system developed at OKB-301 was used for supplemental astronavigation system reliability tests.[30] In the five flight tests performed, the system successfully "picked out" the

---

28. The decree (no. 957-409) was issued on 20 May 1954 and ordered OKB-301 (Lavochkin) to develop the "350" Burya and OKB-23 (Myasishchev) to develop the "40" Buran.

29. Isayev developed the S2.1100 engine (later replaced by the S2.1150) while Bondaryuk produced the RD-012U ramjet engine for the Burya.

30. Lavochkin's Dal was an advanced antiaircraft defense system designed to protect Leningrad. Despite extensive testing, it was never operationally deployed.

designated stars, during nighttime *and* daytime, and kept them in its field of vision for several minutes.

There were still many problems. But they were all overcome, and the flight equipment sets were delivered without delaying the beginning of the Burya flight tests. The first flight test vehicles went into production at the same time as the prototype at aviation Factory No. 1 in Kuybyshev, which produced 19 missiles. The flight tests did not begin until 1957.[31]

The Burya cruise missile obeyed its stellar navigator on its first flight. But in one flight after another it wasn't possible to determine the terminal range and target precision. The SPVRD engine was operating stably, but actual fuel consumption exceeded all ground calculations. The complex gas-dynamic processes that took place in that wily tube, as we referred to it, hadn't been studied sufficiently. Not a single missile reached the target. The fuel was used up well in advance.

For the nation's high-ranking military and Party leadership these tests were a good reason for terminating the project. By this time Korolev's intercontinental *Semyorka* and its updated R-7A had already been put into service.[32]

The development of the Buran cruise missile began in Vladimir Mikhaylovich Myasishchev's OKB-23 sometime after the Burya. When the Burya began to fly, the Buran had only just come off the drawing board and gone into production at aviation Factory No. 23, now known as the M. V. Khrunichev Factory. The Buran was supposed to be a significantly more powerful missile. Glushko was developing the liquid-propellant booster engines of the first stage. In all, four engines were installed, each with 57 metric tons of thrust on the ground. As in the Burya, the sustainer engine was Bondaryuk's SPVRD.[33] With a launch mass in excess of 152 metric tons, its flight range with a 3,400-kilogram payload was calculated to be 9,150 kilometers. The cruising altitude was 18.2 kilometers. At that altitude the sustainer engine was supposed to have a specific impulse of 1,690 kg/kgf. At that time, unlike a liquid-propellant rocket engine, this value could not be confirmed for a ramjet engine on the ground. The experience of the Burya showed that the specific impulse was lower than promised.

Therefore, in 1958 when the decision was made to terminate the Buran project, Myasishchev's organization took it in stride. They developed the new super long-range and super high-speed M-56 bomber, which outstripped the later famous Boeing B-52 in all parameters. But another danger was hanging over Myasishchev's design bureau. With Nikita Khrushchev's help, Vladimir Nikolayevich Chelomey had decided to incorporate Myasishchev's organization to work for his own missile

---

31. The first attempted Burya launch took place on 1 August 1957 but the flight never took place due to a pad abort. The vehicle successfully lifted off for the first time on 1 September 1957.

32. The *Semyorka* ("old number seven") or R-7 ICBM was formally declared operational on 20 January 1960. The improved R-7A reached operational status on 12 September 1960.

33. Glushko produced the RD-213 for the Buran while Bondrayuk developed the RD-018A ramjet engine.

projects. But that is an entirely different dramatic page in Soviet aviation history.[34]

Nikita Khrushchev decimated the aviation industry, asserting that now the sector was not at all necessary. Missiles could do everything. The Strategic Rocket Forces had already been established, and they had no need for cruise missiles. At that time, Lavochkin and his new first deputy Georgiy Babakin were absorbed in the problems of air defense and missile defense missiles. Lavochkin spent most of his time at the firing range in the Balkhash region. He put up no fight in Burya's defense, and soon a Politburo decision terminated this project.

During this period, a group of chief designers appealed to Khrushchev in a letter requesting that he grant permission to continue the project. The Burya and Buran project scientific directors Academician Keldysh and Minister of Defense Malinovskiy supported this request. Khrushchev announced that it was a useless project. He instructed CPSU Central Committee Secretary Frol Kozlov, who was second after Khrushchev in the Party hierarchy, to have a meeting of all the interested parties and explain to them the error of their views.

At this meeting, Lavochkin's deputy Chernyakov attempted to report on the launch results. Kozlov interrupted him. "Here you are boasting that your missiles reached a speed of 3,700 kilometers per hour. Our missiles already have a speed in excess of 20,000 kilometers per hour." Chernyakov understood that technical arguments were futile. When Malinovskiy appeared, Kozlov scathingly admonished him for defending the request to continue the projects: "After all, Nikita Sergeyevich said that this was useless." The Defense Minister could find no better defense than the phrase: "The designers misled me." And so you see, the fate of intercontinental cruise missiles was decided at such a high governmental level and at such a low scientific and military technical level.

FROM 30 MARCH THROUGH 3 APRIL 1992, a scientific conference devoted to the International Space Year was held in Moscow. The initiative for this conference came from the Machine Building, Mechanics, and Control Processes Department of the Russian Academy of Sciences, many academic institutes, the Gagarin Committee, the Interkosmos Council, the Yu. A. Gagarin Cosmonaut Training Center, the Central Scientific-Research Institute of Machine-Building (the former NII-88, now TsNIIMash), our NPO Energiya, and a number of other social and scientific organizations.

By decision of the Academy of Sciences presidium, Academicians V. S. Avduyevskiy and B. V. Rauschenbach were named co-chairmen of the organizational committee of this very imposing conference. Rauschenbach, who was highly experienced in such matters, appointed me and the recently retired TsNIIMash director Yuriy Aleksandrovich Mozzhorin as organizational committee deputy co-chairmen.

---

34. Myasishchev's OKB-23 ceased to be an independent entity in October 1960 and was attached to Chelomey's OKB-52 as a branch.

As usual, the main organizational concerns for conducting the five-day conference fell on the Russian Academy of Sciences S. I. Vavilov Institute of Natural History and Technology. Despite this academic institute's miserable economic situation, its limited scientific staff coped with the difficult assignment, including the most difficult aspect, accommodating foreign guests with a minimum expenditure of hard currency and a shortage of automobiles and gasoline.

At first this conference seemed like an unavoidable waste of time for me, tearing me away from work and from writing these memoirs. But as the program developed, I could see that it would be exceptional. The program of two plenary sessions and sections on the "History of Missile and Space Technology" called for reports on subjects that two or three years earlier would have been completely inconceivable. The reports, particularly those scheduled in the historical section, were sensational because the subject matter of some of them had been "top secret" until very recently, and of course, none of them could have been presented to an auditorium full of visiting American scientists and foreign press correspondents.

The first report in the historical section had three co-authors: I. M. Lisovich, Academician A. Yu. Ishlinskiy, and myself. The report was about the history of the development of the astronavigation systems and a description of the first of those systems tested in the USSR on the Burya intercontinental cruise missiles. The whole world knew that the first intercontinental ballistic missile—Korolev's R-7—had appeared in the USSR back in 1957. This missile became the first factor of the real nuclear threat for the U.S. After a series of modifications, the world knew the R-7 as a spacecraft launch vehicle. From 1957 through 1969, that is, until American astronauts landed on the Moon, by all parameters the R-7 remained the most reliable rocket in the world for manned flights.

This was the missile still in service in 1962, our only intercontinental missile that was almost launched against the U.S. during the Caribbean (Cuban missile) crisis. This is particularly worth recalling, as I was an unwitting witness to those events. The R-7 missile, which proved to be a champion in terms of the number of official world records in space, was also a record holder in terms of longevity. It had more than 45 years of operation in various modifications! (Originally, people had predicted operation for another 10 to 15 years.) In our age of technical revolutions, this is certainly a remarkable record.

However, not only abroad, but even among our missile specialists with clearances for top-secret projects, only a very narrow circle knew that the famous *Semyorka* had a strong competitor for the delivery of a nuclear warhead, the Burya multistage intercontinental cruise missile. Burya flight tests began before the R-7 flew, but they were halted in 1960.

Almost none of our missile and space specialists knew that the world-famous modern Buran winged space vehicle had a top-secret namesake that was also a cruise missile. It did not manage to make a single flight, but was discontinued in production after the first successful flight of the ballistic *Semyorka*. The biographies of the now famous creators of aircraft and missile technology Korolev, Keldysh, Lavochkin, and Myasishchev made no mention of Burya or Buran. This gap needs to be filled.

BURYA AND BURAN WERE CRUISE MISSILES THAT LITERALLY FLEW BY THE STARS. These programs have long since been forgotten. But flying by the stars went from being a series of unique experiments to being an everyday occurrence in the world of rocket-space technology. Navigation "by the stars" has found broad application in combat missiles and peaceful space technology. After breaking through a layer of water and clouds, ballistic missiles launched from submarines find their bearings by the stars to ensure accuracy. The guidance system for the Ye-6 spacecraft, designed for a soft landing on the Moon, used the stars Sirius and Canopus for astronavigation.[35] Star trackers have become an essential piece of equipment on the most diverse types of spacecraft. The names of those who developed the first astronavigation systems will be forgotten, but as long as cosmonautics exists, flying by the stars will continue.

---

35. The Ye-6 is better known as the second-generation Luna probes that were launched between 1963 and 1966. A Ye-6 probe *(Luna-9)* made the world's first survivable landing on the surface of the Moon in 1966.

Chapter 13
Missiles of the Cold War's First Decade

   The vast majority of Soviet historians and political writers who wrote about missile and space technology tried to skip over the decade of 1946 to 1956 as quickly as possible and break out into the realms of space. There are several reasons for this focus. The first, as I see it, was that during that period in the Soviet Union work was being conducted with the utmost intensity to develop the first combat missile complexes, that is, new types of armaments. The country's best scientific and technical forces had been mobilized for these top-secret projects. Only in the 1980s did the opportunity arise to write openly about this period. The second reason was more mundane: anyone who knows the history of that period either could not or did not want to write about it. Unfortunately, the majority of those who started the missile era are no longer alive. The third reason had only to do with the professionals, that is, the writers and journalists. For them the history of this period did not hold the sensations and mind-boggling wealth of achievements that showered down like manna from heaven beginning in 1957 after the launch of *Sputnik,* the first artificial Earth satellite.

   With few exceptions, historians and biographers of Korolev also speak very sparingly about his work during that period, and, evidently due to a lack of "baseline data," they make virtually no mention of the work of other scientists, engineers, scientific leaders, and promoters of the development of new technologies in industry. During the first postwar decade, only two nations, the USSR and the U.S., worked in the field of missile technology. When the Cold War started, our totalitarian state was cut off from contact with American science by the Iron Curtain. We were forced to develop and produce much of what could be routinely bought in the West. And we learned to do it no worse, and sometimes even better. The vigorous development of cosmonautics in subsequent decades depended upon the foundation that was created during this period.

   By the end of the first postwar decade, hundreds of thousands of people had already been drawn into the creation of this foundation. For some, work in this field did not require an abrupt restructuring of their lives; for others, work in missile industry "p.o. boxes," design bureaus, or far away in the swirling winds of the missile test range was just the beginning.

There were no millions killed on battlefields during the Cold War. But at our workplaces—the design bureaus, laboratories, classified shops, and test ranges—there was no less heroism than that shown by the people who created weapons for the front during the war. Not only my generation, but the entire populace of the former Soviet Union has a right to be proud of achievements both before and after the war, and the collapse of the Soviet Union should not justify a devaluation of history.

In spite of the inconsistency, lack of logic, and also at times the criminality of the Stalinist leadership's policies, the bulk of the intelligentsia sincerely believed in the absolute necessity of developing new military technologies, including weapons of mass destruction. No matter how difficult it might be for us, our military technology must not be inferior, but to whom? During the war it was clear: our military technology had to be superior to that of the enemy, Nazi Germany. But now, after the war? After Churchill's speech in Fulton, Missouri, and not without "Uncle Joe's" help, a new type of enemy took shape, the enemy in the Cold War."[1]

The politics of confrontation between the USSR on the one hand, and the countries of Western Europe and the U.S. on the other—a course set on intensifying a standoff, a policy on the verge of unleashing World War III—resulted from the actions of the most belligerent and expansionist factions of both Stalin and his entourage and the Western establishment. For Stalin, the Cold War was a convenient pretext for cracking down on any dissidence in the Communist Party and government.

In response to hard-line voices in the West, Stalin and his entourage implemented a policy of real support for militarized science, and with cost as no object, promoted the development of large-scale advanced weapons systems.

The Cold War was waged at a real "hot" pace in laboratories, at test ranges, and at classified factories. And it wasn't just a thin stratum of the scientific and technical intelligentsia that was aware of this. The implementation of fundamentally new ideas for the development of nuclear weaponry, missile technology, and radar equipment required the involvement of millions. The bulk of the workers, especially those who had experienced the superhuman stress and deprivation of four years of war on the home front, worked together and did not see themselves as separate from so-called "designers."

Creative activity and production during the postwar years in the rapidly expanding new branches of military industry were not without conflict. There were intense clashes, struggles between different scientific and technical concepts, and struggles for the supremacy of one trend or another. These were unavoidable phenomena, and perhaps even necessary ones. Conflicts accompanying the rapid development of

---

1. British Prime Minister Sir Winston Churchill introduced the term "iron curtain" in his famous speech on 5 March 1946, at Westminster College in Fulton, Missouri, where he received an honorary degree. He noted specifically that, "From Stettin in the Baltic to Trieste in the Adriatic an iron curtain has descended across the Continent."

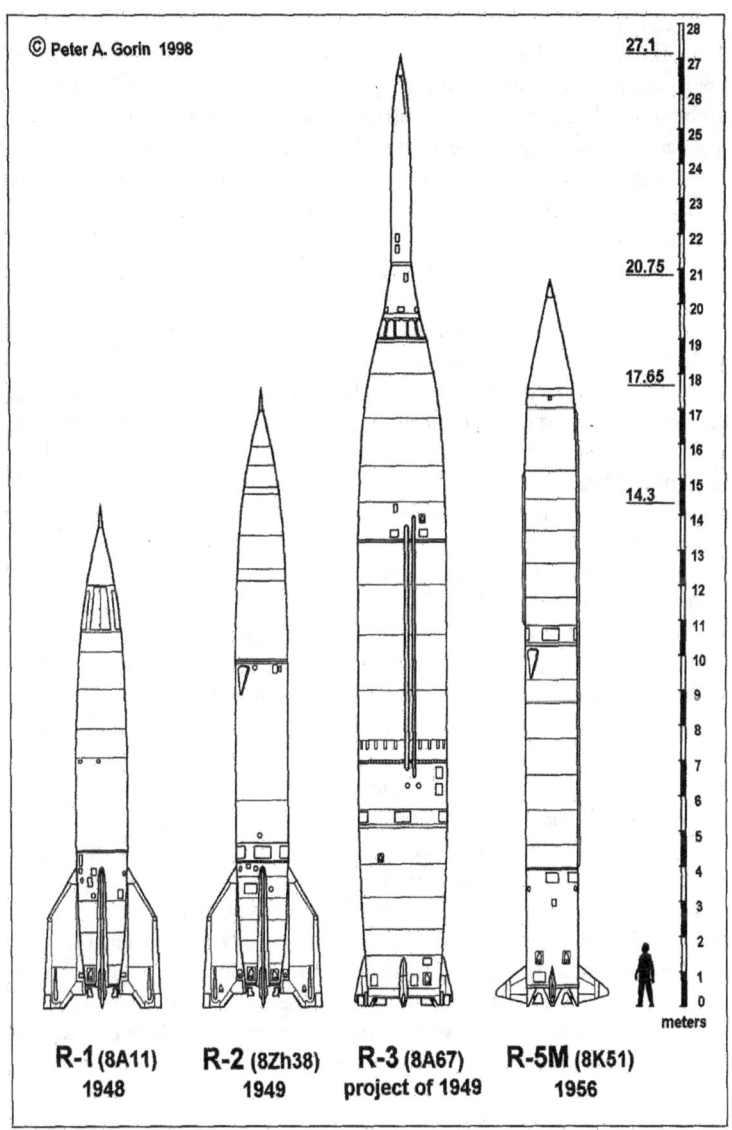

© Peter A. Gorin 1998

	meters
R-1 (8A11)	**R-2** (8Zh38)	**R-3** (8A67)	**R-5M** (8K51)
1948	1949	project of 1949	1956

27.1
20.75
17.65
14.3

Peter Gorin.

The drawing shows the gradual progression of Soviet ballistic missiles from the R-1, a copy of the German V-2, to the R-2, to the R-5M, the first Soviet ballistic missile capable of carrying a nuclear warhead. The R-3 missile was an ambitious long-range project that was abandoned in 1951 although it allowed Soviet engineers to adopt and abandon certain key technological paths.

241

new technology exist regardless of the social structure of society.

The history of NII-88 at that time is very revealing. The staff of that institute, along with its factory, and the entire cooperative missile effort over a 10-year period performed work that served as the basis for subsequent missile and space triumphs. In order to form the most general idea of the scale and scope of the operations that were carried out in our country for the birth of a new mighty and deadly force, nuclear missile weapons, one should examine the list of projects conducted at this leading missile enterprise and, above all, the projects conducted by Korolev's team, who were part of the institute until August 1956.

WORK AT FULL CAPACITY ON THE FIRST DOMESTIC R-1 MISSILE BEGAN IN 1948. Already by autumn of that year the first series of these missiles had undergone flight testing. Flight tests on the second and third series took place during 1949–50, and, in 1950, the first domestic missile complex using the R-1 missile was put into service.[2] The launch mass of the R-1 missile was 13.4 metric tons. It had a flight range of 270 kilometers and a conventional explosive warhead with a mass of 785 kilograms.[3] The R-1 missile engine was an exact copy of the A4 engine. For the first domestic missile we required a striking accuracy in the dispersal rectangle of 20 kilometers for range and 8 kilometers laterally.

A year after the R-1 missile was put into service, we completed flight-testing on the R-2 missile complex. It was put into service with the following specifications: launch mass—20,000 kilograms; maximum flight range—600 kilometers; warhead mass—1,008 kilograms.[4] The R-2 missile was equipped with radio-controlled course correction to improve lateral precision. Therefore, in spite of its longer range, it was just as accurate as the R-1. The R-2 missile's engine thrust was increased by boosting the R-1 engine. In addition to range, there were substantial differences between the R-2 and the R-1 missile: the implementation of warhead separation concept, incorporation of a load-bearing tank into the body design, and the transfer of the instrument compartment to the lower part of the body.

In 1955, testing was completed on the R-5 missile complex and it was put into service.[5] The R-5 had a launch mass of 29 metric tons, a maximum flight range of 1,200 kilometers, and a warhead mass of around 1,000 kilograms, but it was possible to carry another two or four suspended warheads for launches with ranges of 600 to 820 kilometers. The missile's accuracy was increased thanks to the use of an integrated (preprogrammed and radio-controlled) guidance system.

---

2. In the absence of original Soviet designations for missiles, Western agencies such as the U.S. Department of Defense (DoD) or the North Atlantic Treaty Organization (NATO) would assign their own names in classified documents. The R-1, for example, had the DoD designation SS-1 and the NATO designation "Scunner."

3. The Russian phrase *vzryvchatoye veshchestvo* (VV) literally means "explosive matter."

4. The R-2 missile was known as the SS-2 and Sibling by the DoD and NATO, respectively.

5. The R-5 missile (and its nuclear variant, the R-5M) were known as the SS-3 and Shyster by the DoD and NATO respectively.

The R-5M complex was a fundamental upgrade of the R-5 missile complex. The R-5M missile was the first missile in the history of world military technology to carry a nuclear warhead. The R-5M missile had a launch mass of 28.6 metric tons and a flight range of 1,200 kilometers. It had the same degree of accuracy as the R-5.

The R-1, R-2, R-5, and R-5M combat missiles were single-stage, liquid-propellant missiles. Their propellant components were liquid oxygen and ethyl alcohol. Korolev was the chief designer of all four missile models, and Glushko was the chief designer of their liquid-propellant rocket engines.

In 1953, at NII-88, we began to develop missiles using high-boiling propellants such as nitric acid and kerosene. Isayev was the chief designer of the engines for these missiles. Two models of missiles using high-boiling components were put into service, the R-11 and the R-11M.

The R-11 had a range of 270 kilometers, with a launch mass of just 5.4 metric tons. It was armed with a conventional explosive warhead with a mass of 535 kilograms and was put into service in 1955.[6] The R-11M missile was already the second missile in our history to carry a nuclear warhead. In modern terminology, this was an operational-tactical nuclear missile. In contrast to all preceding missiles, the R-11M was placed on a mobile tracked vehicle-mounted launcher. Because of a more up-to-date preprogrammed guidance system the missile had strike accuracy within an 8- by 8-kilometer square. It was put into service in 1956.[7]

The last combat missile of this historical period was the first submarine-launched missile, the R-11FM. In terms of basic specifications it was analogous to the R-11, but with a fundamentally altered guidance system and adapted for launch from submarine launchers. Thus, from 1948 through 1956, seven missile complexes were developed and put into service, including two nuclear missile complexes and one naval complex. In order to do all of this, it was necessary to produce experimental missiles and conduct preliminary flight tests on them. For example, to optimize the principle of warhead separation, the R-1 missile was used as a basis to develop the R-1A missile, which was launched several times. Before the R-2, the R-2E missile was created, and before the R-5, the experimental R-2R was created. When the development of an intercontinental missile was already under way, its many systems had first been tested on M5RD and R-5R series missiles.

Korolev did not forget about his meeting with Sergey Vavilov in 1947.[8] Korolev initiated an extensive research program on space, the upper atmospheric layers, and the behavior of living organisms during high-altitude rocket launches. Thus, we

---

6. The R-11 missile was better known by the U.S. DoD and NATO as the SS-1a and Scud, respectively.

7. The R-11M missile was known by the U.S. DoD and NATO as the SS-1b and Scud-A, respectively.

8. Academician Sergey Ivanovich Vavilov (1891–1951) served as President of the USSR Academy of Sciences between 1945 and 1951.

developed the R-1V, R-1D, R-1Ye, R-2A, R-5A, and R-11A missiles for various payloads. For design bureaus, production, testers, and test-range services, these were independent developments that were sometimes more labor intensive than combat missiles. Institutes under the Academy of Sciences developed instruments that were installed in the missile payload containers that could be recovered with the aid of parachutes. With the launch of these missiles, for the first time we obtained data on the makeup of primary cosmic radiation and its interaction with materials, determined the physical and chemical makeup of the air at different altitudes, the spectral makeup of solar radiation, the absorptive capacity of ozone, and so on.

Long before Vostok launch vehicles, dogs and smaller creatures had already flown in combat payloads. In contrast to the later famous Layka, who perished in space, the "missile" dogs successfully landed with the aid of parachutes, but without any publicity in the mass media.[9]

Thus, in the Soviet Union over a period of 10 years (through 1956, inclusively), at NII-88 alone, Chief Designer Korolev, with the direct involvement of chief designers from cooperating enterprises Glushko, Pilyugin, Ryazanskiy, Barmin, and Kuznetsov—members of the old Council of Chief Designers—and new Chief Designers Isayev and Isanin, developed and tested 16 types of liquid-propellant guided ballistic missiles with flight ranges up to 1,200 kilometers and flight altitudes in excess of 200 kilometers.

Except for the naval R-11FM, all the missiles were launched from launch sites at the Kapustin Yar State Central Test Range. The total number of launches over this period exceeded 150. Of these launches, 30 missiles were armed with conventional explosive warheads and one missile had a real nuclear warhead. Back then we were much too daring. Under present-day conditions the launch of such a missile would be impossible not only for the obvious political reasons. We, the missile specialists, never found out the environmental consequences of that launch in 1956.

In 1953, there were two more experimental launches that left us with an unpleasant aftertaste. The R-2 missile had already been put into service under the code name "article 8Zh38" and was considered more or less reliable.[10] However, its combat effectiveness, which in terms of its impact was only marginally superior to a one-ton aerial bomb, was the weakest point of all missiles of that time. There was no nuclear warhead for missiles until 1956. Neither then nor many years later did we found out the true authors of the experiments conducted to modify the R-2 missiles under the code names *Geran* (Geranium) and *Generator*.

It all started in the crowded conference room at our hotel at the test range,

---

9. Dogs were killed during flight or on landing on 4 (out of the 29) suborbital launches carried out under the biological program carried out in 1951–58.

10. Typically, all Soviet armament systems were given a secret code in the form of a "number-letter-number" designation such as 8Zh38 ("Zh" is a letter of the Russian alphabet). These designations were assigned by the Main Artillery Directorate (GAU) in the development phase and used in all production documentation to disguise the true nature of the weapon systems to both subcontractors and people outside the program.

when they showed us the film *Serebristaya pyl (Silver Dust)*. This was one of the first semifictional films dealing with the horrors of a future war. The "silver dust" represented radioactive powder that was dispersed over a large area through high-altitude detonation of specially developed aerial bombs. The radiation dose was lethal for everything living in the zone affected by the silver dust. No protective clothing or gas masks saved the population. The contaminated Earth was deadly throughout the entire "half-life" period.

The film was created in consultation with specialists who had studied the effect of nuclear explosions. They were intending to show that it was not at all necessary to drop atomic bombs from airplanes, an idea that anticipated the idea of the neutron bomb with which every nonliving thing would remain intact and unharmed, while the people would die, and after a certain amount of time, without having to fight, the victor could occupy the territory with all its resources preserved. There is an old adage about a "dream come true." This was a "film come true."

According to the designers' conception, the warhead container mounted on the R-2 *Geran* missile contained a radioactive liquid. When detonated at a high altitude this liquid was supposed to disperse, settling in the form of lethal radioactive rain. *Generator* differed from *Geran* in that the same radioactive liquid was also placed in the warhead container but in a large number of small containers rather than a single chamber. Each of them was to burst above the Earth independently.

During launch preparation of the first *Geran*, a turbid liquid trickled out of the top of the missile standing on the launch pad. Evidently, the chamber containing the lethal liquid had sprung a leak. The entire launch team hurried to get away from the missile. But what were they to do with it? Voskresenskiy, never at a loss in critical situations during a launch, ambled up to the missile. With the launch crew watching where it had taken refuge a hundred meters away, Voskresenskiy climbed up the erector to the height of the tail section so that everyone could see him. He gracefully stretched out his hand and ran his finger down the side of the missile through the liquid trickling down. Then, turning to the dumbfounded spectators, he stuck out his tongue and placed his "radioactive" finger on it. After climbing down, Voskresenskiy sauntered over and said, "Guys, let's get to work! It tastes like crap, but it's harmless."

He was convinced that the liquid just simulated the atomization process, and he was not mistaken. Nevertheless, that evening at the hotel he availed himself of an additional shot of alcohol to "neutralize the substance and to allay the terror" that he had endured. *Geran* and *Generator* were not continued.

THE FIRST MISSILES WITH NUCLEAR WARHEADS, the R-5Ms, were already in series production at the Dnepropetrovsk Factory (later Yuzhmash and then Factory No. 586) and were set up for combat duty in the Far East and in the Baltic region. Thus, the creation of the notorious nuclear missile shield began in 1956 with the R-5M missile, which was referred to in production and drawing documentation and also in unclassified documents as "article 8K51."

In 1956, OKB-1 Chief Designer Korolev left NII-88 and obtained full independence. Within 10 years he had a monopoly in the development of long-range ballistic missiles. In the next decade of the Cold War, new leading organizations would emerge. They would be directed by Mikhail Yangel, Vladimir Chelomey, Aleksandr Nadiradze, and Viktor Makeyev.

Chapter 14
On the First Missile Submarine

Recounting the story of German missile technology in Bleicherode in 1946, Helmut Gröttrup mentioned a project he called ridiculous and unrealistic. As soon as Germany's top-ranking military leaders determined that the V-2 (A4) missiles had just barely developed a knack for flying and decided to begin firing on England, some enthusiasts proposed developing a system for the underwater delivery of missiles to the shores of America to fire on the U.S. mainland.

A small group was created in Peenemünde to design the system with the participation of submarine specialists. The system consisted of submarines towing V-2 missiles in containers. The submarine containers would somehow be stabilized or placed on the bottom at a depth of no more than 100 meters and about 100 to 150 kilometers from the U.S. coastline. The submarine was supposed to have a crew that prepared and launched the missiles.

A simple calculation done by the Peenemünde specialists showed that this idea was absolutely impractical. In addition to the container holding the missile, the submarine needed to tow liquid oxygen tanks across the ocean under the water. These tanks were so large that the cost of the entire system was several dozen times greater than the damage that the missile warhead was capable of inflicting, even if it reached the center of New York City. Dornberger and Wernher von Braun opposed this reckless project, and it was laid to rest.

After flight tests on the first series of the R-1 missile, that is, the domestic copy of the German V-2, our Navy command approached Korolev with the proposal to develop a version of the missile to be installed in specially adapted submarines. According to Vasiliy Mishin's story many years later, he quickly convinced Korolev that the plan was unrealistic, and he was absolutely right. It was completely unfeasible to adapt an R-1 or some similar missile that required storing liquid oxygen for many days or even weeks during a submarine voyage.

The situation changed radically with the development of the R-11 missile. To this day veterans in the field of missile technology argue about who first came up with the idea of using the R-11 missile as a weapon on a submarine. Without getting caught up in these arguments, I can confirm that, despite the large number of skeptics, including Pilyugin and Ryazanskiy, Korolev consistently supported this idea and perhaps was even the first to propose it. Korolev approached the development of a missile for the Navy with genuine enthusiasm, pushing aside other seemingly urgent business, including a variety of other projects and responsibilities, in favor of it.

At a scientific and technical conference held in St. Petersburg in February 1991, in honor of the 35th anniversary of the world's first launch of a ballistic missile from a submarine, some said that the idea of arming submarines with ballistic missiles had been put forth by naval officers as far back as 1952. In particular, the initiative of a group of enthusiasts headed by Engineer Rear Admiral N. A. Sulimovskiy and Engineer Vice Admiral L. A. Korshunov was mentioned. The naval officers' proposal, unfortunately, could not be implemented using the older R-1 or R-2 missiles. It became feasible to develop a modified long-range ballistic missile launched from a submarine only after the development of the R-11, which used high-boiling propellants and was designed for mobile launch. Unlike land-based infantry commanders, naval officers were very enthusiastic about this new type of weapon. I have already written about the degree of skepticism that many combat generals expressed when comparing the effectiveness of conventional armaments and missiles. Naval officers were considerably more far-sighted and proposed creating a new class of ships, missile-carrying submarines possessing unique capabilities. A submarine armed with torpedoes was designed only to strike enemy ships, while a submarine armed with ballistic missiles would be capable of striking, from the sea, land-based targets located hundreds of kilometers away, and in the future even thousands of kilometers, while remaining invulnerable.

The first successful launch of a land-based R-11 missile took place on 21 May 1953. Flight tests had identified the need for a large number of modifications, but Korolev did not want to lose time and simultaneously coordinated the design specifications for the naval version of this missile with the naval officers and ship builders.

While Stalin was alive, he alone made decisions as to what new military technology should be developed, when, and by whom. After his death, there was a brief hiatus in the issuance of decrees for the development of new types of armaments. However, on 24 January 1954, the USSR Council of Ministers issued the decree "On Carrying Out Experimental-Design Work to Arm Submarines with Long-range Ballistic Missiles and the Development of a Technical Design for a Large Submarine Armed with Missiles Based Upon This Work." The decree assigned the work to:

- Central Design Bureau-16 (TsKB-16) (later renamed Malakhit) headed by Chief Designer N. N. Isanin—for all work on the submarine[1];
- NII-88 OKB-1 headed by Chief Designer S. P. Korolev—for the missile armament.

---

1. TsKB—*Tsentralnoye konstruktorskoye byuro.*

Korolev had a passion for developing new ideas and demanded the same degree of passion from his close associates. But, in such an unusual undertaking, he needed new strong allies among the *sudaki*, the shipbuilders.[2] Chief Designer of TsKB-16, Nikolay Nikitovich Isanin, proved to be Korolev's ally. He was an experienced shipwright who had begun to study submarines after being schooled in the construction of heavy cruisers and liners. During the war he worked on the most popular type of vessel at that time, torpedo boats. Isanin became the chief designer of diesel submarines just two years before he met Korolev. He boldly took on the task of adapting the V-611 submarine as a carrier for R-11 missiles, which had been assigned the designation R-11FM in its naval version.

When they initialed the draft of the government decree, both chief designers Korolev and Isanin took on a great responsibility. Indeed, the R-11 missile had not even been fully developed in its land-based version. And how was a crew supposed to launch it at sea from a submarine, where there was no escaping to a bunker? What if, as happened with us on land, the missile did not lift off from the rocking launch pad and the engine spewed hot propellant components into the submarine? But both chief designers were bold optimists.

Outwardly, nothing in particular set Isanin apart from other naval engineers with whom we were beginning to meet, but right off the bat he was quite prepossessing. Despite his self-effacing, natural modesty, he had a composed and solid confidence when making crucial decisions. After my first encounters with him in 1953, I had the feeling that we had known each other for a long time. He quickly studied Korolev's character and developed very amicable relations with him, although Isanin used to good-naturedly poke fun at Korolev's trademark short temper. I must admit that these jokes endeared Isanin even more to those of us who were close associates of Korolev.

After the submarine project was handed over to Viktor Makeyev in 1956, I had no business contact with Isanin for many years, but at Academy of Sciences meetings we always found many subjects in common to discuss. In conversation he was just as amiable and ironic. During boring, formulaic reports, we would often slip out of the conference hall and reminisce about days gone by. All the while he smoked like a chimney and poked fun at the new procedures just as good-naturedly as before.

According to our calculations, modifying the R-11 missile for a naval version, plus developing new naval testing and launching equipment instead of the old ground-based equipment, would take three to four years. Korolev did not even want to hear about a timeframe like that.

Being opposed to the Navy mania, Nikolay Pilyugin entrusted the solution of all guidance problems to Vladilen Finogeyev. We were won over by Pilyugin's tall,

---

2. The word *sudak* is a slang expression referring to shipbuilders. It is a play on words using the word *sudak*, which means pike perch, and the prefix *sudo*, which means ship.

fair-haired, very dapper and energetic young deputy immediately. We dealt with him on guidance issues without addressing Pilyugin. Viktor Kuznetsov, whom we considered an old "sea wolf," understood better than the others which new and difficult problems would need to be solved using gyroscopes. There were disputes and debates, but with Kuznetsov's approval, the entire problem of command instrumentation development was assigned to the Leningrad-based NII-49, where by that time the chief designer was young Vyacheslav Arefyev.

Of the remaining members of the Council of Chief Designers—after Korolev, Pilyugin, and Kuznetsov—Vladimir Barmin had not much to do with the submarines. It was clear that the shipbuilders themselves could work on their own launch systems. Also, because the missile didn't have any radio flight control system, Chief Designer Ryazanskiy only had to provide the telemetry system, and even then, only if we didn't select the system of OKB MEI headed by Academician Kotelnikov. The chief designer of the R-11 missile engine was Isayev rather than Korolev's traditional associate Glushko. Thus, except for Korolev himself, from his Council of Chief Designers, only Pilyugin was involved with the development of the submarine missile. And still, without the traditional Council's support, Korolev fearlessly and without hesitation plunged under water.

But it wasn't just enthusiasm and desire that drove Korolev. From conversations that he had with his closest compatriots, one could conclude that in addition to everything else, it was the idea of a safety net in case we really blew it and missed our deadline for the development of the first intercontinental missile. In fact, if the intercontinental missile was not ready within the 1956–57 timeframe, then we would remain defenseless against the real threat of nuclear attack.

To begin with, we could mount missiles with conventional warheads on diesel submarines, and then ... after all, we had already modified the same R-11 missile for the atomic scenario. So what if its range was just 200 kilometers? If tens of submarines carrying missiles approached within 100 to 150 kilometers of the U.S. coast, then that would somehow compensate for our failure to develop the intercontinental missile that we were obliged to do in 1956. But the jury was still out on this missile, although we already had complete confidence in the R-11; we had a real submarine, but the main thing was that we had outstanding sailors, shipbuilders, and military men.

The most important thing that we first had to resolve was how we would launch the naval R-11, from the surface or from underwater. Isanin convinced Korolev that we needed to break up our development process into two stages. In the first stage, we would modify existing submarines or those already under construction for surface launch. To do this, we would retrofit at least two vertical tubes equipped with special elevating devices into the solid hull of a submarine. The tubes would be covered with hatches that would be opened before launch. The missiles in a primed state would be in the dry tubes, ensuring their failure-free operation after a prolonged underwater cruise that could last more than a month. This problem of long-term storage was not a simple one for Isayev either. He had always been apprehensive about leaks in the lines and the corrosive power of the propellant components he had selected for the propulsion unit.

The submarine selected for modification as the first naval ballistic missile carrier belonged to the V-611 series of diesel submarines. The subs in this series were built to operate on ocean routes and in the areas of remote enemy naval bases. The preproduction submarines of this design went into service in 1953 and were built until 1958. In all, 26 V-611 submarines were built.[3] The primary weapons of the submarines were 533-mm torpedoes housed in the bow and stern sections. The submarine was equipped with radar to detect and identify surface and air targets and sonar systems to detect foreign submarines and surface ships. Its maximum diving depth was 200 meters. In its missile version, the main modification of the V-611 design involved cutting two vertical tubes into the solid hull in the centerline plane behind the conning tower.

Korolev hadn't forgotten about safety. He understood that a haywire missile on a submarine could have catastrophic consequences for the entire crew. He proposed providing the capability to jettison the missile overboard or to flood the tubes in the event the missile didn't launch normally.

Another problem was purely naval in nature. When the submarine surfaced, it would inevitably be subjected to rolling. Launching missiles from a rolling pad rather than from a launch pad standing firmly on the ground did not immediately fit into our land-based ways of thinking about targeting technology and the subsequent behavior of the missile. The behavior of the liquid contained in the tanks was also cause for concern. Finally, how would the missile hit the assigned target area if we didn't put the exact coordinates of the launch site into previously prepared ballistic calculations and tables? An entire team of surveyors worked at the firing range on the steppes. They precisely fixed the launch site to geographical coordinates, determined the line of fire, and after the launch, reported the deviation of the point of impact from the calculated point with accuracy to within meters. But how could we do this on a stormy sea?

All of this proved relatively easy for the sailors. The institutes of the *sudaki* handling these projects patiently explained how to adapt naval navigational technology to our joint tasks. Arefyev proposed principles of prelaunch orientation and the idea of integrating the missile guidance system with the submarine's navigational complex. The axes of the onboard gyroscopes were aligned with the axes of the ship's navigational complex before launch. The missile would launch from the surface, having a momentary angle of inclination and angular velocity determined by the sway. After launch, the gyroscopes, having recorded the prelaunch setting, would correct the missile first vertically and then "lay" it on the programmed flight path in the firing plane. The implementation of these principles required developing a

---

3. These Soviet diesel electric submarines were known in the West as Zulu class submarines.

special shipboard coordinate converter that connected the submarine's navigation instrumentation complex and its motion control system with the missile's onboard control system. Two special institutes of the shipbuilding industry performed this work.

At the sailors' suggestion, the project was named *Volna* (Wave). The all-hands job to develop the first R-7 intercontinental missile, the first to carry a thermonuclear warhead, began in 1954. OKB-1 associates close to Korolev, surprised at his intense interest in the naval project, wisecracked that the "little R-11 heaved up such a mighty 'Wave' that it might wipe out the big R-7."

There were many skeptics. Perhaps that was precisely what inspired Korolev and Isanin, who, rather than conducting the experimental-design work authorized by the decree, decided to develop a submarine with real missiles and conduct real launches within what was by today's standards an unimaginably short time. The naval officers supported them. That was the only way to pave the road for a new type of weaponry in the navy.

Despite Korolev's attitude, the *sudaki* postponed the second phase, that is, launching from underwater. They argued that in order to do that, they would have to come up with a new submarine design; it would really take at least three or four years before they could successfully perform the first launch from an underwater position. The naval command agreed with this strategy and ordered all the services involved in the first phase to go "full steam ahead."

The creation of a rig to test and develop the missile launch equipment under rolling conditions became an urgent task. The naval TsKB-34 was assigned the task of developing and creating a dynamic rig. NII-49 and NII-303 developed the control equipment for the dynamic rig. However, TsKB-34 Chief Designer Ye. G. Rudyak announced that the deadline specified by Korolev for the creation of the rig was unrealistic. Then Korolev decided to build the first somewhat simplified model of the rig using his *own* resources. Korolev gave our chief ground-segment specialist Anatoliy Abramov the super-urgent assignment of developing a similar rig as soon as possible. Isanin decided to help, and TsKB-16 fabricated a launch table for the rig with mechanisms to raise and turn it. The rig designer, Abramov's deputy P. V. Novozhilov, pulled off the assignment (no one knows when he slept), and in April 1955, they conducted the first three test launches at Kapustin Yar.

Despite the exceptionally complex situation in preparation for the R-7 missile range tests, Korolev sent his main tester Leonid Voskresenskiy to Kapustin Yar to direct launches of the R-11FM from the rig. The rig simulated the procedure of raising the missile into the upper section of the tube for a surface launch. At the firing range they nicknamed the rather dynamically complex structure the "elbows and knees" contraption. Three launches of missiles with the R-11 guidance system were sufficient to determine that the submarine mock-up would withstand the engine's fiery plume. Later, a significantly more advanced dynamic rig came along, also Rudyak's creation. The electric drives could simulate rolling action equivalent to that generated by a storm measuring four points on the Douglas scale.[4] Here, the deviation amplitude was as high as ±22 degrees.

Eleven missiles were launched under rolling conditions from Rudyak's rig with the naval control system, and surprisingly all of them were successful. By this time, the first submarine had already been retrofitted with two tubes rigged with "knees and elbows." The naval officers were quite actively involved in all the operations on the White Sea. Experienced submarine officer, Hero of the Soviet Union, Captain 2nd Class Khvorostyanov commanded the open sea firing range that had been set up and the special unit that had been organized. The time had come to go out to sea.

Severodvinsk was the base for the first missile submarine. In the early 1950s, it was still named Molotovsk. This coastal city had all the necessities: a shipbuilding factory, a base for ground-based missile storage and testing, a submarine crew base, and, most importantly, an atmosphere that was "maximally conducive" to our work.

For the first sea tests, seven missiles equipped with the new naval control system were set up. The testing and launching ignition systems in the naval *Saturn* and *Dolomit* versions were developed jointly with us and NII-885 and by the naval institutes MNII-1 and NII-10. Telemetry receiving stations were erected on the shore to monitor the flight while special ships conducted observation and communications. Onboard telemetry transmission and orbital control equipment was installed in the nonseparating missile payload and operated on a slot antenna.

The first R-11FM missile launch from a Northern Fleet B-67 submarine built according to the V-611 design was carried out in the White Sea on 16 September 1955, at 5:32 p.m. Korolev and Isanin personally directed these tests.

N. N. Isanin served as chairman of the State Commission on the first submarine missile launches, and S. P. Korolev was his deputy and also technical director. According to the missile tradition that had developed under Korolev's influence, the naval launching team included V. P. Finogeyev and V. P. Arefyev, by dint of their status as chief designers; Novozhilov, as the primary designer of the launch contraption; and OKB-1 lead designer I. V. Popkov. In addition to this civilian contingent on the team, Navy Commander-in-Chief Admiral L. A. Vladimirskiy attended the first launch. The commander of the first missile-carrying submarine was Captain 1st class F. I. Kozlov. A. A. Zapolskiy was in charge of the launch team.

Seven launches on the White Sea were successful, three of which were launched after prolonged storage. The launches were performed when the submarine was stationary and when it was moving at a speed up to 10 knots with swells measuring 2 to 3 on the Douglas scale.

---

4. The Douglas scale is an international scale of sea disturbance and swell ranging from 0 to 9. It is named after Sir Henry Douglas (1876–1939), former director of the British Naval Meteorological Service.

The R–11FM was the world's first submarine-launched ballistic missile. This photo shows the missile launched during a test in 1955.

Deputy Commander-in-Chief of the Navy Admiral Vladimirskiy, Marshal Nedelin, and commanders of fleets and flotillas were invited to the final launch in September 1955. The process of the submarine's surfacing, opening the tube hatches, raising the missile using the contraption, and, finally, the effective launch at the precisely specified time evoked thunderous applause from the guests on board the destroyer. And that marked the beginning of the arming of our fleet with long-range ballistic missiles.

In November 1955, despite the happy ending of the first naval tests, Korolev announced to Finogeyev, Abramov, and myself that he was going to take us with him to Severodvinsk. He explained that he wanted one more chance to study the preparation and launch process in detail. In his words, we needed to gain a better understanding without all the extraneous hoopla. We gladly accepted his command, especially since we wouldn't be flying, but rather taking the fast train from Moscow to Severodvinsk via Arkhangelsk.

It was the first time I had been to the naval firing range. Korolev and Finogeyev had been there several times and had been on the submarine. They considered me a greenhorn. The night I spent on that train is well preserved in my memory. We occupied two sleeper compartments. The four of us gathered in one of the compartments over a bottle of cognac. Afraid to interrupt Korolev, we listened as he dispassionately related stories about his journey to Kolyma and back. This was the first time I had heard him personally speak about this part of his life. Usually, he very much disliked recalling and talking about this painful period. I don't know what came over him during that nighttime journey. After his death, I had the occasion to hear and read different variations, oral and written, of the stories he told us that night. I don't want to retell what I heard, so as to avoid getting into arguments with the authors of numerous publications and films about Korolev. I suppose it's not particularly important now, since for the most part the facts line up, and each author paints the details in his own colors. It would be nice if there was a conscientious historian who, after studying all the stories, publications, and documents, would do a special study of the period in Korolev's life from 1937 through 1945, without introducing his own personal speculation.[5]

The hospitable naval command met us in Severodvinsk. Representatives of the Northern Fleet showed the expertise of their crews, who independently conducted the missiles' electrical tests, fueling them, and loading them into the tubes of the submarines standing at the pier of the Severodvinsk shipbuilding factory. I admired the efficient and well-coordinated work of the naval sailors. It was just a little over a year since the base and submarine crew had begun to master the missile business, but they worked with much greater confidence than their land-based colleagues.

Only when the submarine was ready to put out to sea did Korolev announce that he had made an arrangement with the command. They were permitting him, Finogeyev, and me to join the cruise. Participation in the submarine cruise was essential for Korolev. He steadfastly held to the principle "it is better to see something once than to hear about it 100 times." Besides us, lead designer of the R-11FM missile Ivan Vasilyevich Popkov was required to take part in such cruises. He and several industry representatives were part of the crew as official submariners. Finogeyev and I hypothesized that Korolev was not reporting to the fleet high command, but had received permission only from the local command. Therefore, our participation in the cruise was unknown until a couple of hours before we shoved off.

When I saw my first missile submarine up close it didn't seem like a large ship. Until we made our way into its compartments, I couldn't understand how they crammed all the complex missile launching equipment and additional missile team in there.

The submarine departed from the pier early in the morning, and soon thereafter the command to dive was given. I, of course, was interested in everything because my conception of the goings on inside a submarine when it dove and cruised underwater came only from literature. Korolev was in his element on the submarine.

---

5. For one of the best accounts of Korolev's time as a prisoner, see Yaroslav Golovanov's magisterial *Korolev: fakty i mify* [*Korolev: Facts and Myths*] (Moscow: Nauka, 1994). Korolev moved through many different locations during his incarceration in 1938–44. From June 1938 to August 1939, he spent time in various prisons in and around Moscow. From August to December 1939, he was at labor camp in Kolyma in northeastern Siberia. From March to September 1940, he was imprisoned in Moscow before being transferred to Tupolev's *sharaga* prison design bureau (TsKB-29), where he remained until November 1942. Finally, from November 1942 to July 1944, he was incarcerated at Factory No. 16 in Kazan as part of another *sharaga*. He was officially released from prison in July 1944, although he was not formally rehabilitated until 1957.

He went straight to the conning tower, where he studied the submarine's control engineering and looked through the periscope. He didn't forget to warn us, "If you walk around the ship, don't crack your head open." Despite the warning, I repeatedly bumped into various and sundry mechanism parts protruding where I didn't expect them and cursed the designers for the narrow bulkheads that separated the compartments from one another.

All the equipment to set up launch control was located in the special "missile" compartment. It was very tightly packed with consoles and racks holding naval electronics. Before a launch, this compartment was supposed to accommodate six men at their battle stations. When the submarine surfaced and the tube hatches were opened, only the metal of those tubes would separate the men from the cold sea. It was impossible to move to other compartments after the sounding of battle quarters. All the hatches were battened down. The missile compartment combat team was in charge of the entire preparation, and they conducted the launch from the submarine's central station.

Four hours into the cruise, when it began to feel that in the submarine's cramped quarters we were bothering everyone and they were fed up with our questions, the command was given to surface. Korolev, who tracked down Finogeyev and me in the torpedo compartment, told us that now all three of us were supposed to be by the tube from which they were going to raise and launch the missile. Why did he require such a demonstration of bravery? Should something happen to the missile, whether in the tube or even in the upper section, we would be goners. To this day, I do not understand why the submarine commander permitted Korolev to sit by the tube during launch. If a catastrophe occurred, it wasn't the commander's head that would be knocked off. Admittedly, afterward, a submariner said, "If something had happened, there would have been no one left to accuse."

At the 30-minute alert, the commander's "battle quarters" command was sent to all the submarine compartments and, for good measure, the klaxon was sounded. I was reminded of my youth in the *Komsomol*.[6] In 1932, on the battleship *Marat* in Kronstadt, the ship's combat training was demonstrated to us, the young "captains." We had heard the same klaxon and the same "battle quarters" command.

Exchanging clipped sentences, the three of us sat uncomfortably, jammed up against the cold metal of the tube. Korolev clearly wanted to "showcase" himself and his technology, as if to say, "Look how much trust we have in our missiles' reliability." In the tube, you could hear scraping and rumbling as the "knees and elbows" mechanism elevated the missile. We grew tense in anticipation of the engine's startup. I expected that the roar of the engine and plume of flame shooting into the tube would be terrifying. However, the launch was surprisingly quiet.

---

6. *Komsomol* was the informal name for the All-Union Lenin Communist Union Youth (*Vsesoyuznyy Leninskiy kommunisticheskiy soyuz molodezhi*, VLKSM), a large Communist youth organization formed in 1918 to foster social and youthful activities that celebrated Communist rule. The *Komsomol* was dissolved in 1991.

Everything turned out fine! The hatches opened and the joyful commander emerged, congratulating us on the successful launch. Reports had already come from the impact point. Now they were verifying the coordinates. The telemetry stations were receiving, and according to preliminary data the flight had proceeded normally. This was the eighth or ninth R-11FM launch from this first missile submarine.

After the launch, our stress dissipated immediately. Finogeyev, who had participated in launches from this submarine before, smiling broadly, asked me: "So, feel better now?" "Yes," I answered, "that's no concrete bunker launch for you." Indeed, the psychological stress at a naval launch can in no way be compared to the launch of a ground missile.

The submarine returned to base, and we were all invited to a submarine dinner party. Our mood and that of the officers in the cramped crew's quarters over dinner was excellent. Korolev whole-heartedly praised the herring and navy-style macaroni. The commander joked that he himself could not remember tasting such delicious navy food. Korolev promised that in three years it would not be necessary to surface. It would be possible to launch missiles from a submerged submarine.

This all happened long ago; many of those conversations and jokes can no longer be retrieved from memory. But I still remember very well that rare sensation of bliss. And it wasn't personal, but something shared, bringing together and rallying completely diverse people. Perhaps this originated from Korolev, whom I couldn't ever imagine as happy. Here on the submarine, sitting at the dinner table after the launch he radiated that feeling. Neither before that time, nor after it, do I remember him being that happy.

From Severodvinsk we set out for Arkhangelsk on the factory motor boat. It was storming and the three of us got seasick. The ship's rocking didn't affect Korolev, and, teasing us, he offered us a swig from his bottle, but we couldn't manage to swallow. Finally we made it to the airport, where our airplane awaited. The crew was pleased that we would be able to get some rest at the hotel. The weather prevented us from flying out of Arkhangelsk, and we couldn't fly into Moscow due to fog. Despite our departure being justifiably forbidden, Korolev could not calm down. It was absolutely out of the question that we would spend the night in the airport or stroll around Arkhangelsk (which we, having never seen it, would not have been averse to doing). He set out to find the airport director, got through to the Air Force Commanding General via high-frequency communications, and somehow convinced him that we should be granted permission to depart. An hour later, we took off, and after landing at our own airfield in Podlipki (now the head institute TsNIIMash is located on the former airfield there), we found that there was no fog.

SOON THEREAFTER MANY OF US EXPERIENCED THE BITTER FEELING OF PARTING WITH OUR ROMANTIC NAVAL PROJECT. After realistically weighing the possibilities and delving even deeper into the problems of intercontinental missiles, Korolev came out with the proposal to create a special design bureau for naval missiles. He recommended Chief Designer Viktor Makeyev for the job. The Central Committee of the Communist Party and Council of Ministers accepted the proposal. Ten years later in the small town of Miass in the Ural Mountains, a mighty machine-building KB headed by General Designer Viktor Petrovich Makeyev was up and running, spawning urban development.

Korolev's decision to hand over the responsibilities of chief designer of submarine missiles to Makeyev took some time and was a very difficult one.

Makeyev was the "lead designer" for the land-based versions of the R-11 and R-11M missiles. A government decree assigned their series production to machine-building Factory No. 385 in the town of Zlatoust in the Urals. Korolev's first step was to propose that a branch of OKB-1 be set up at Factory No. 385. The chief of this branch would also serve as Korolev's deputy chief designer. He offered this post to Makeyev, the young, energetic former director of the Soviet Union's delegation at the 1952 Olympic Games in Helsinki and former NII-88 *Komsomol* committee secretary, who had been highly successful as the lead designer of all the R-11 and R-11M missile modifications.

Makeyev was already 31 years old. He was no longer *Komsomol* age. However, they remembered him well in the Central Committee of the All-Union Lenin Communist Youth Union (VLKSM) and he had come to the attention of the Communist Party Central Committee. Probably, in the long run, he could make a career in the Central Committee office or in the Council of Ministers.

Korolev failed to assess Makeyev's potential ambition. Abandon Moscow for small-town of Zlatoust in the Ural Mountains in order to be Korolev's deputy? Constantly squabble there with the local factory director and grovel before municipal bureaucrats, all the while listening to Korolev's rebukes via high-frequency communication that he had once again missed a production deadline and the military acceptance team was dissatisfied with the quality? To be asked, "What on Earth is my deputy doing out there—and he better be on his toes"? "I didn't send you there to sip tea with jam!" This was one of Korolev's standard scoldings. Such prospects had no allure for Makeyev. In response to Korolev's offer, he took the risk of declaring: "I'll go to the Urals only as a chief designer. I won't go as a deputy."

The government had already made its decision to transfer production of R-11FM missiles to Zlatoust. If another chief designer showed up there, then all subsequent modifications of this missile and the development of new ones would be his responsibility. The naval project would inevitably be transferred to him.

Korolev wavered. The State Committee of Defense Technology that Khrushchev had created to replace the Ministry of Defense Industries had already granted him approval to organize a branch in Zlatoust.[7] If he wished, in place of Makeyev he could find another, less ambitious candidate for the post of deputy chief designer for

---

7. In 1957–58, Khrushchev enacted decentralization reforms in Soviet industry. As part of this reform, several major ministries were transformed into "State Committees." The Ministry of Defense Industries which oversaw the missile program was reformed into the State Committee of Defense Technology.

On the left is Viktor Makeyev, the Chief Designer of all Soviet strategic submarine–launched ballistic missiles during the Cold War. On the right is Chief Designer Aleksey Isayev, who produced a number of sustainer rocket engines for Makeyev's missiles. Isayev's organization also developed engines for Soviet air defense missiles, launch vehicle upper stages, and spacecraft.

naval affairs. But that wasn't really the issue. Would naval missiles be an OKB-1 project or not?

Korolev's closest member on the Council of Chief Designers, Nikolay Pilyugin, had already announced that he was completely turning over development of naval missile guidance systems to his former colleague Nikolay Semikhatov in the Urals. And all of the high-ranking government officials concurred and promised to assist the new chief designer of guidance systems.

When I found out that Makeyev had been appointed as an independent chief designer, I asked Mishin, "Did S.P. consult with you?"

"No, he didn't ask me. I don't know who he talked with."

Mishin was miffed that in making such an important decision Korolev had not even informed his immediate first deputy. When, after assembling all of his deputies, Korolev announced his decision approved by the State Committee and the Communist Party Central Committee, we remained sullenly silent. We understood that this decision had not been easy for him.

Emerging from Korolev's office, we found Makeyev in the reception room. Korolev had invited him for one last conversation. Leonid Voskresenskiy immediately brightened: "Well, Viktor, now you are going to be the one 'sea soul' for all of us, and we are off to choke on the dust out on the steppe."

It was a prophetic wish. Makeyev's heart really was with the sea. Nineteen years after that event, in 1974 at the regular meeting of our academic department in the Academy of Sciences, Academy Corresponding Member Viktor Makeyev accepted warm congratulations from his Academy colleagues on the occasion of his 50th birthday in October. Smiling furtively, Pilyugin revealed the secret of Korolev's position at that time:

"Back in 1955, Sergey called me up, and I talked him into giving in and naming you chief designer and turning over all the naval projects to you."

THE FIRST U.S. EXPERIMENTAL NUCLEAR SUBMARINE *NAUTILUS* was developed in

1954. The first Soviet nuclear submarine, *Leninskiy Komsomol,* was commissioned in 1958. These submarines did not yet carry ballistic missiles. In 1959, Polaris A-1 missiles were installed on the American nuclear submarine *George Washington.* Our 667A design nuclear submarine carrying 16 missiles with a range of 2,500 kilometers reached operational status only in 1967! But by the time Makeyev celebrated his 50th birthday in 1974, we had bridged the gap in the total number of nuclear submarines armed with nuclear warhead missiles! We had 120 atomic missile submarines against the Americans' 90. This was a tremendous achievement for our shipbuilding industry.

Reminiscing about his talk with Korolev in late 1955 about Makeyev, in a conversation over tea and cookies in the guest room at the institute on Griboyedov Street, Pilyugin noted that he was a diabetic and that Corresponding Member of the Academy Makeyev and Academician Isanin both had heart trouble; in view of the fact that this locale didn't serve cognac, we should toast Korolev's wise memory with tea and cookies! During this improvised academic tea party, Academician Nikolay Isanin said that there were only two among us who had participated with Korolev in the submerged, cruising, and first submarine-launched missile firings: "Chertok and I."

Since that memorable cruise in November 1955, I have never again had the occasion to be on a missile submarine. From the stories of acquaintances who have participated in cruises on modern nuclear submarines, the living conditions are anything but easy. Back then we were on a very short cruise on a diesel submarine. For the first time I got the sense of how much easier it was to work as a ground-based missile specialist. No matter how difficult it was on the ground and in all sorts of bunkers and silos, the living conditions on a submarine were a hundred times more difficult.

Our OKB-1 together with TsKB-16 and a small group of naval officers in the 1950s laid the foundation for a completely new type of strategic naval force. In present-day strategy and policy, missile-carrying nuclear submarines play as important a role as the land-based Strategic Rocket Forces. The creators of the submarine fleet and naval missiles and the naval submariners deserve the highest respect, and we should remember their heroic service not only when we hear the sensational accounts of submarine accidents.

The B-67 submarine was the first missile-carrying sub. It may not have been the first launch, but in 1955 I had the good fortune, yes, indeed, the good fortune, to make, albeit brief, an underwater cruise on that submarine and to participate in a missile launch along with Korolev and Isanin. After our cruise in November 1955, the B-67 submarine was commissioned into the naval fleet as an experimental training vessel. The experience gained made it possible to retrofit submarines with missile launchers and to build another six submarines of this series, assigning it the designation AV-611. The first submarine missile complex was commissioned under the designation D-1. This was the first and last project that Korolev's OKB-1 performed for the Navy. The decree for the development of the new D-2 complex with the R-13 missile and its missile-carrying submarines was issued during the second half of 1955. Design series 629 diesel submarines developed by TsKB-16 and design series 658 nuclear submarines developed by TsKB-18 were armed with the D-2 missile complex.[8]

Viktor Makeyev was already the chief designer and would soon also be the general designer of the D-2 project. Nikolay Semikhatov was the chief designer of the onboard and ship-based guidance systems for all the submarine ballistic missile complexes. However, despite the qualitative advantages of the D-2 over the groundbreaking D-1, the D-2 did not solve the primary problem: just like our first R-11M missiles, the R-13 missiles were launched from a surface position.

A year after the first historic surface launch, the government issued a decree stipulating the conduct of experimental operations to develop the capability for *underwater* missile launch. NII-88's OKB-10, headed by Chief Designer Yevgeniy Vladimirovich Charnko, was designated as the head organization. Our OKB-1 was involved in projects to modernize the R-11FM missile. Korolev's deputy Anatoliy Abramov and Isanin's designers retrofitted the tubes and got rid of the "knees and elbows," and Semikhatov redeveloped the entire guidance system. The first underwater launch of the modernized R-11FM missile from a B-67 submarine took place on 10 September 1960. During that time Korolev's entire "band of warriors" was up to its ears in work preparing for the first manned flights and the new R-9 combat intercontinental ballistic missile.

I was in Mishin's office when a smiling Korolev came in and began to congratulate us. Accepting his congratulations, Mishin nevertheless couldn't restrain himself and said, "Sergey Pavlovich! This really isn't our work anymore. Sure, Abramov helped them, but we don't do underwater launches."

Such an attitude on the part of his deputy offended Korolev.

"How can you not understand?! If it hadn't been for our R-11 and our surface launches, this underwater launch wouldn't have happened. We share the success. It's just too bad that we no longer work with the Navy guys, but Makeyev, after all, isn't a stranger either; he's one of us. You two mark my words; he's going to go far."

I don't want to quote Korolev from memory; I'm paraphrasing what he said. He spoke with emotion and great conviction. Then, for the first time, he revealed that after the successful testing of the first domestic nuclear submarine *Leninskiy Komsomol,* the government decided to build a nuclear submarine fleet armed with missiles and thermonuclear warheads.

---

8. Typically the Soviets designated the entire submarine-launched missile system in the form "D-number," with the actual missile retaining the old "R-number" designation for designers and the "number-letter-number" designation for the military and production personnel. In the absence of knowledge about Soviet designations, in the West, the D-2 missile complex was codenamed SS-N-4 (by the U.S. Department of Defense—DoD) and Sark by the North Atlantic Treaty Organization (NATO). The D-1 system (with the R-11FM) was never assigned a designation by either the DoD or NATO.

"Makeyev took a stab at a range up to 3,000 kilometers," said Korolev, "More power to him!"

Five years later, already a Hero of Socialist Labor, Makeyev flew in for Korolev's funeral. At a meeting of a close circle of Korolev's deputies, Sergey Kryukov and I tried to persuade him to become *our* chief designer and return to Moscow.

He asked us directly to give up any such notions.

"I've been saddled with the entire D-5 complex.[9] You can't imagine what that entails. For the first time we and the *sudaki* are going to put 16 missiles with submerged engines on a nuclear submarine! We have more problems than you and those cosmonauts all put together! We still don't know when we are going to fly, but they are already after me to put the intercontinental missiles on the submarines.

"No, you are all nuts! Get this! Sixteen missiles, well, OK, maybe 12 intercontinental missiles on a single submarine. So the sub doesn't need to stray far from her own shores. She can deliver a massive blow with a salvo right from her base so that rocket troops with all their silos and bunkers won't be needed any more!"

The D-5 nuclear submarine complex that Makeyev told us about began to enter service in the fleet as early as 1967. Seven years later, 34 such vessels had been built. They received the official title "strategic missile submarine cruiser." The general designer of these submarine cruisers was Sergey Nikitich Kovalev.[10]

Two general designers, Kovalev and Makeyev, headed the "Naval Council of General Designers." It included guidance complex designer Semikhatov, rocket engine designer Isayev, chief designer of nuclear power plants Fedor Mikhaylovich Mitenkov, the top scientific director for the development of nuclear submarines and future President of the USSR Academy of Sciences Anatoliy Petrovich Aleksandrov, and many more chief designers of dozens of systems.[11]

The pilot 667A submarine with tubes for storing, preparing, and executing the launch of RSM-25 missiles was built in 1967. The first missile submarine cruisers carrying RSM-40 intercontinental missiles were produced in 1974.[12]

In 1968, Makeyev and I were elected corresponding members of the USSR Academy of Sciences in the department of mechanics and control processes. Members of this same department included Nikolay Pilyugin, Viktor Kuznetsov, and Vasiliy Mishin. Department meetings were usually held twice per year and, as a rule, on Griboyedov Street at the Institute of Machine Science. Since neither Pilyugin nor Kuznetsov had yet been involved in the development of submarine missile guidance systems, the three of us interviewed Makeyev; during these formal conversations we could not conceal our engineering curiosity about his work.

---

9. The D-5 was known in the West as the SS-N-6 or Serb. Makeyev began development of the system in 1964, and the missile was accepted into service less than four years later.

10. Academician Sergey Nikitich Kovalev (1919–) was the leading designer of strategic missile-carrying nuclear submarines in the Soviet fleet.

11. Fedor Mikhaylovich Mitenkov (1924–) oversaw the development of nuclear energy units

12. The RSM-25 and RSM-40 submarine-launched ballistic missile systems were known in the West as SS-N-6 (Serb) and SS-N-8 (Sawfly) respectively.

According to the rules of etiquette at that time, we did not marvel at Makeyev's achievements. But when Makeyev told us that, after emerging from under the water and breaking through the clouds, the first naval RSM-40 intercontinental missile determined its position according to the stars and hurtled toward the target with the help of an onboard computer, I thought, sure, we had star-guided flights 20 years ago, but, of course, not from underwater.

The developer of the system's electro-optical unit was the Geofizika Factory's design bureau in Moscow on Stromynka. Beginning in 1959, together with Rauschenbach, we collaborated closely with Geofizika to develop spacecraft guidance systems. I attempted to explain astronavigation in the chapter "Flying by the Stars" (Chapter 12). But all of our spacecraft astronavigation problems now seemed trifling to me compared with submarine missiles' "flight-by-the-stars" problems.

Pilyugin and Kuznetsov were up to their ears in problems developing ground-based combat complexes for the Strategic Rocket Forces. Nevertheless, you could sense jealousy in their responses in conversations with Makeyev, who was praising Semikhatov. Pilyugin considered Semikhatov his pupil. Now Semikhatov had surpassed his alma mater in terms of using computer technology and increasing the accuracy of his intercontinental missiles. However, the achievements of Makeyev and Semikhatov did not so much evoke a feeling of jealousy as admiration of the fact that now, in terms of the number of intercontinental missiles, the naval component would be a reliable backup and reserve for the ground-based strategic nuclear forces.

Submarine missiles, including the engines developed by Isayev, were produced in large numbers at the Krasnoyarsk Machine Building Factory. The manufacturing processes were new for Siberian factories, and this required the highest production standards and strict quality control. Through the fault of the factories, serious mishaps occurred during flight tests and missile firing exercises.

During evening strolls with Isayev along the "Walkway of Heroes," he once remarked that we could make a splendid feature film about missile submarines that would have all the intense emotion of the film *Ukroshcheniye ognya (Taming the Fire)*, a film for which we were both consultants.[13]

He described one of the incidents when, in his words, "I felt like blowing out my brains and the brains of those Krasnoyarsk slobs." During one of the RSM-40 missile development launches—in the midst of the prelaunch preparation on the submerged submarine—they overfueled a propellant tank and it blew up. It's a good thing the submarine commander kept his cool and ordered an emergency surfacing. It turned out that at the factory they had failed to remove the temporary plug from the inlet of the tube to the tank's pressure sensor.

At the ministerial board meeting Minister Sergey Afanasyev indignantly described

---

13. *Ukroshcheniye ogonya (Taming the Fire)* was a fictional film loosely based on the inside history of the Soviet space program that was released in 1972. The film was directed by Daniyl Khrabrovitskiy who, in conceiving the film, consulted with many "secret" designers, including Chertok.

another edifying case. During the attempt of a routine missile launch, everything was proceeding normally until they started the engine. The missile had not managed to emerge from the tube when the engine died. The tube flooded, the submarine surfaced, and they delivered the missile to the base. There, while examining the propellant tank they discovered rags that had stopped up the throat of the outlet pipe. An investigation of this affair led to the culprit, who explained that during assembly the process instructions called for special cloths to be used to clean the tank's interior surfaces and to thoroughly wipe everything. He had followed the instructions to the letter. When asked him why he had left the cloths in the tank, his calm response was that the instructions said nothing about removing the wipes.

By the 1970s Krasnoyarsk Factory began to move out from the danger zone. They started to arm submarine cruisers with missiles that had multiple warheads. Each warhead was a unit with its own nuclear charge and its own individual self-contained guidance system for a specific target.

I got together with Makeyev for the last time on 25 October 1985 in the Gorbunov Palace of Culture back in my hometown of Fili. He died the same day—on his 61st birthday. When you stand for five minutes in a graveside honor guard, for some reason you remember things that the everyday hustle and bustle has made you forget and will make you forget again. The brief obituary published by the central press said nothing specific about the actual work of two-time Hero of Socialist Labor, member of the Communist Party Central Committee, deputy of the Supreme Soviet, Academician V. P. Makeyev. In 1966, the great Korolev's work was declassified two days after his death. The feats of Academician Makeyev, a "Korolev school" alumnus, however, remained unknown. At the end of 1985, the Soviet Navy had 200 nuclear submarine cruisers armed with General Designer Makeyev's ballistic missiles. It is up to historians to assess the scientific, technical, and political outcome of the "submarine race" between the USSR and the U.S. in the 20th century.[14]

---

14. For a recent overview in English, see Norman Polmar and Kenneth J. Moore, *Cold War Submarines: The Design and Construction of U.S. and Soviet Submarines* (Washington, DC: Brassey's, 2004).

Chapter 15

Prologue to Nuclear Strategy

As we set about developing new missile complexes, we took great pains not to waste our R-1 experience. But as we worked on new tasks, new ideas emerged that sometimes went counter to the principles we had used in the R-1. More often than not, our experience led us to the conclusion that we'd been doing things all wrong. As we made the transition to new projects, when failures occurred, we no longer had the excuse that the Germans had conceived it that way and we had been forced to reproduce it. Now, we were required to know precisely who was responsible for reliability and safety.

A missile complex is the product of creative teamwork. For this reason it would be wrong to say, for example, that a missile developed at OKB-1 had crashed due to failure of an engine developed at OKB-456. In those days, when we experienced success we never played up our leading role, and when we suffered failures through the fault of our subcontractors, we did not use them as scapegoats. But we did demand reciprocity. If it was your fault, then own up to it, find the cause of your system's failure, but don't try to make excuses, shifting the responsibility to make yourself look good across the board. The Council of Chiefs worked in this style from the very beginning.

We did not achieve a qualitative leap in reliability during the development of the first domestic R-2 missile. Despite the wealth of experience we had gained during the production and launching of the R-1, reliability problems were solved intuitively. Much later, dozens of guidelines, hundreds of regulations, and all sorts of standards would emerge, regulating the process for developing all missile technology hardware from the initial technical proposals to the procedures for acceptance into service.

For modern-day launch vehicles, the degree of reliability estimated using statistical methods based on many launches is 90% to 95%. This means that on average, 5 to 10 launches out of a hundred may fail. One must pay a very high price to achieve such reliability, and, of course, it is based on the priceless experience of the past.

Before we made the transition to launching the first preproduction R-2 missile series, we tested the reliability of our new ideas on experimental R-2E missiles. Six of them were manufactured, and five were launched in 1949. Of those five launches, only two could be considered successful. But we gained experience that enabled us to launch 30 R-2 missiles during 1950–51. Of these 30 launches, 24 were successful by the standards of that time. All the failure cases were analyzed, and appropriate measures were taken to improve reliability. Nevertheless, during launches of the mass-produced R-2 missiles in 1952, 2 out of 14 missiles failed to reach the target. The R-2 missile was accepted into service even though an objective assessment showed that its reliability was no higher than 86%.

In 1955, the first R-11 tactical missile using high-boiling propellant components was put into service. This was a worthy replacement for the R-1. Unlike the R-1 and R-2 missiles, the R-11 did not carry the "birthmarks" of the German A4. It was purely a domestic development. Given its mobility (it had a mobile launch), the R-11 to some extent also replaced the R-2. Thirty-five launches were conducted before it was accepted into service. Of those, six could be considered failures. Thus, in 1955, the military put into service a missile that had a reliability of 83%.

At that time, the R-5 missile was the record holder in terms of range for a single-stage missile. Its conceptual design was completed in 1951. During flight tests conducted in 1953, 15 missiles were launched in two phases. Of these, only two failed to reach the target. Its reliability finally began to slowly approach the 90% level. And this happened despite the fact that many ideas incorporated in the R-5 missiles were new.

There are numerous works on the theory of reliability, and often they include a classification of failures according to their causes: structural, production-related, operational, and "miscellaneous." Under our conditions the numerous entries in the "miscellaneous" category included "not-in-your-wildest-dreams" failures.

In this regard, a prime example was the explosion of R-1 and R-2 missile warheads when they entered the atmosphere.[1] But there were two indirect causes for "not-in-your-wildest-dreams" failures: poorly developed telemetry technology and the ones we considered "unpremeditated sloppiness." I will provide an example of the former. During R-11 missile flight tests there were two failures generalized as "stabilization controller failures." But our modest telemetry capabilities could not reveal what, where, and why it had failed. We saw only that commands had proceeded from the gyroscopes, and the control surface actuators had started to do something inexplicable. Luckily, the first of these failures—and we had already chalked up quite a few—did not cause any casualties.

In April 1953, at the Kapustin Yar firing range on the Volga steppes, resplendent with the flowers and sweet smells of springtime, we began flight tests on the first phase of the R-1l. Nedelin and his retinue of high-ranking military officers flew in for the first tests of the new tactical missile that used high-boiling propellant components.[2]

---

1. See Chapter 8 of this volume.

2. Marshal Mitrofan Ivanovich Nedelin (1900–60) served in various senior military posts through the 1950s, during which time he effectively directed strategic missile procurement for the Red Army. In 1959, he was named as the first Commander-in-Chief of the Strategic Rocket Forces, the new branch of the armed services established to operate strategic missile divisions.

The launches were conducted from a launch pad mounted directly on the ground. They placed two panel trucks containing the *Don* telemetry system receiving equipment next to the FIAN shed one kilometer from the launch site in the direction opposite the trajectory of flight. They gave this observation post the sonorous name IP-1, Tracking Station 1.[3] All the automobiles bringing guests and the review team to the launch gathered at this point. Just in case, firing range chief Voznyuk ordered that several special trenches be dug in front of the station.

At the R-11 launches I was no longer responsible for communications from the bunker and collecting readiness reports via the field telephone. After completion of the prelaunch tests I gladly stationed myself in the IP in anticipation of the upcoming spectacle. It never occurred to anyone that the missile might fly not only forward along the route toward the target but also in the opposite direction. For that reason, the trenches were unoccupied. Everyone preferred to enjoy the sunny day on the lush steppe not yet scorched by the sun.

The missile lifted off at precisely the appointed time. Thrusting out a reddish cloud and resting on a glaring fiery plume, it lunged vertically upward. But about four seconds later it changed its mind, pulled a sort of airplane barrel-roll maneuver, and went into a dive headed, it would seem, straight for our fearless retinue. Standing straight up Nedelin loudly shouted, "Lie down!" All around him everyone dropped to the ground. I considered it humiliating to lie down before such a small missile (it was just five metric tons), and leapt behind the shed. I took cover just in time. An explosion rang out. Clods of earth knocked against the shed and the vehicles. That's when I really got scared. What had happened to those who hadn't taken cover? To make matters worse, now everyone might be covered with a red cloud of nitric acid vapor. But no one was injured. They got up from the ground, crawled out from under cars, shook themselves off, and looked with amazement at the toxic cloud being carried away by the wind in the direction of the launch. The missile had fallen just 30 meters short of the crowd. Analysis of the telemetry recordings did not enable us to unequivocally determine the cause of the mishap, and it was blamed on a stabilization controller failure.

There were 10 first-phase missiles launched in the spring of 1953, of which 3 were failures. The oxidizer tank on one of the missiles disintegrated, and another missile burned up due to a leak in the propellant lines. But the primary shortcoming of this series of missiles was the engine's low specific thrust versus its design thrust. For this reason, it didn't reach maximum range when fired; there were launches that fell short by 50 kilometers. Because of this fundamental shortcoming, we sometimes did not devote proper attention to other problems, with the excuse that "this is experimental development, after all."

The second phase of testing was conducted in the spring of 1954. By this time, Isayev had modified the entire engine system. Of the 10 missiles launched at a range

---

3. IP—*Izmeritelnyy punkt*—or literally "Measurement Point," but more commonly "Tracking Station."

of 270 kilometers, 9 reached the target, and 1 pulled a stunt very similar to the incident I described above. Admittedly, this time the missile took off to the left for 12 kilometers. We could no longer be content with the findings generally stated as "stabilization controller failure."

At the next session of the accident investigation commission I reminded folks of the jokes where doctors say that only a coroner can determine the actual cause of death. Control surface actuators had solid cast housings, and when a missile landed on soft ground they might remain intact. If we searched for them, then we might be able to prove that at least the control surface actuators were not the cause of the "stabilization controller failure."

We found the point of impact and, despite the lingering strong scent of toxic oxidizer, we removed the missile's well-preserved remains. Outwardly, the control surface actuators actually looked quite presentable. We put them on a test bench in the laboratory at the firing range. Two worked normally, and two didn't feel like obeying commands. When we opened them up, we discovered that in both of the nonfunctioning control surface actuators, there was a break in the steel wire that acted as a linkage connecting the armature of the electromagnetic relay with the control valve of the hydraulic system. After replacing the linkage wire, both control surface actuators were fully operational. Why and when had the linkage wires broken? My colleagues Kalashnikov and Vilnitskiy unequivocally stated that it was caused by the shock of impact. Well, if that's it, let's do a direct experiment. We arranged to drop control surface actuators from an airplane without parachutes. When we finally found them, we brought them to the laboratory, cleaned them off, and tested them. As the military controller reported, they were "completely normal." In other words, impact was not the cause of the breakage.

I hypothesized that the breakage was due to vibration. On the R-1 and R-2, these same linkage wires in the control surface actuators did not break because the vibration in Glushko's oxygen engines was probably not as strong as in Isayev's engine.

Isayev was outraged and said that was impossible since his engine had a thrust of just 9 metric tons, while the R-2 had 35! A more powerful engine shakes more. After hashing it over we put the actuators on the vibration stand. But at the firing range we couldn't get a vibration frequency to exceed 100 Hz. The actuators withstood the maximum intensity the rig was capable of producing. Then I sent a radiogram to Podlipki: "Urgently need to conduct vibration resistance test on control surface actuators in range up to 500 Hz." A day later we received an unexpected response: "The actuators fail at a frequency close to 300 Hz." The cause was the natural frequency of the string that we call the linkage, which according to calculations is close to 300 Hz. If an external effect has that frequency, then resonance sets in and the string breaks.

That's it! And there we were, not realizing during vibration testing that we had to expose the control surface actuators to prolonged vibration at that frequency. We took our complaints to Isayev, "So when you conduct firing tests, do you measure the vibration frequency and intensity?" Of course not. He didn't have the right equipment for that.

The control surface actuators were removed from all the missiles and returned to the factory. But to what frequency should the control mechanism be "tuned"? We did not know the true inflight vibration frequencies and intensities, and telemetry at that time couldn't give us an answer. After reflection, conjecture, and consultation with the engine specialists, we did some redesigning to ratchet the natural frequency to above 800 Hz. After that, there were no more accidents due to stabilization controller failure.

This "resonance" resulted in a three-month delay in testing. But this harsh lesson was not in vain. We sat down right away to develop a procedure and equipment for measuring vibration. For onboard equipment, we introduced the requirement that each instrument and assembly be tested to see whether there was a possibility of resonance-induced failures or deviations from the norm within a very broad range. Domestic industry had not yet produced vibration stands for frequencies higher than 500 Hz. The events described gave us the opportunity to obtain funds to import test stands reaching frequencies as high as 5,000 Hz.

Proceeding from the principle that "God helps those who help themselves," we checked the possibility of a similar "resonance" failure of the R-1 and R-2 control surface actuators. It turned out that they, too, could be disabled when exposed to a vibration frequency close to 300 Hz. Without panicking, we decided to modify as soon as possible the control surface actuators during the series production process and to replace them on all manufactured missiles. When several mysterious failures of years past were reinvestigated, one could assume that they had the same cause but we simply hadn't known that at the time.

We drew one more conclusion for future investigators into failure and accident causes. If someone expressed a hypothesis concerning a probable cause for a failure, we required that this failure be simulated on the ground. For example, that is how we dealt with oxygen valves when we surmised that they failed to open due to freezing of a lubricant that they didn't need. It was worse if, during missile preparation on the ground, the failure self-corrected. A self-correcting failure does not recur during repeated checks and during all sorts of tests intended to induce it. In such cases, we repeated a cycle of horizontal and vertical tests many times and thought that "there are no glitches, the testers imagined something." If, after this, the missile was launched anyway, then more often than not this defect manifested itself in flight and caused an accident.

Having learned the law of the "universal vagaries of missiles" (that is how the witty folk of those romantic times explained the occurrence of certain failures), we lived by the rule: if you cannot precisely determine the cause of a self-corrected failure during preparation of the missile at the firing range, then at least replace all the suspect instruments and even the cables and repeat the tests. This was by no means always possible.

Manufacturing defects most often resulted in failures with disastrous consequences. A break in the soldered joint of a wire at the point where it attached to a connector plug caused a command transmission failure, which meant the loss of stabilization or a failed command to shut down an engine. In the best case, a radio control instrument might fail in flight, which wouldn't affect the actual progress of the flight.

A classic example of a failure caused by "flagrant sloppiness" was a case that entered the annals of missile folklore. According to schedule, the launch of a combat R-2 missile was coming up. The warhead had been armed not with an inert payload and smoke mixture but with a real TNT charge. State Commission Chairman General Sokolov told Korolev that he wanted to observe the launch from a trench and invited along several other men.[4] That was a safety rule violation. Trenches had been dug not far from the launch pad to provide shelter for the launch team in case it didn't have time to take cover in the bunker. When a missile was launched with a warhead, the entire launch team was supposed to take cover in the bunker.

I was in the bunker in communication with all the firing range and radio control services, verifying their readiness. That time, special safety measures consisted only in a stepped-up security guard, which had driven all curious loiterers a little further away from the launch. Voskresenskiy and Menshikov were standing at the periscopes. Voskresenskiy loudly barked the commands: "Ignition! Preliminary! Main! Liftoff!" The roar of the engine filled the bunker, but stopped short followed by an unusual premature silence. "The missile is falling…" A few seconds of silence … "Fire on the launch pad!"

Suddenly Korolev, who had been standing next to Voskresenskiy, dashed for the exit, grabbed the fire extinguisher in the passageway, and ran up the steep steps leading out of the bunker.

"Sergey, get back!" yelled Voskresenskiy. Korolev did not stop, and Voskresenskiy darted off to catch up with him. Up there in the roaring flames of the gigantic bonfire fed by the mixture of alcohol and oxygen lay the missile payload containing a metric ton of TNT. Despite the dangers, some force compelled Menshikov, the chief of the launch team, and I to exit the bunker.

When we ran out, Korolev stopped. The hot wind prevented him from moving further. Voskresenskiy was trying to take the fire extinguisher away from him. He managed to do so and banged the fire extinguisher on the ground. A white stream squirted out, but it was impossible to get closer to the fire because of the unbearable heat. Voskresenskiy threw the fire extinguisher aside, grabbed Korolev by the arm and started to drag him toward the bunker. When he saw us he shouted, "What are you guys doing here? Everybody in the bunker! It's going to blow!" Breathing heavily, Korolev and Voskresenskiy were the last to return to the bunker. An oppressive silence fell over us. We waited for the explosion and wondered what had become

---

4. Andrey Illarionovich Sokolov (1910–76) served as director of NII-4, the leading R&D institute within the Ministry of Defense responsible for defining requirements and projections for strategic missile development.

of General Sokolov and all those whom he had enticed to observe the launch from the surface. Among them were Barmin and Goltsman.[5] About 10 minutes later the observer at the periscope reported, "Fire trucks are coming."

Three fire trucks rolled up and streams of water rushed at the burning missile. The warhead did not explode.

Goltsman related that during the launch, he, Barmin, and several other brave souls were standing next to General Sokolov about 50 meters from the launch site. When the missile toppled over and the fire started, Sokolov gave the command, "Everybody follow me!" They ran to the trenches, tumbled down into them, and lay in anticipation of the explosion until they had determined that the fire brigade was busily going about its business among the remains of the missile.

When the fire had been extinguished and the ground had cooled down, General Sovolov clambered out of the trench and ordered that a guard be stationed and for everyone to leave. Korolev, Voskresenskiy, and I were allowed, as members of the accident investigation commission, to inspect the accident site. About 15 minutes after we began our inspection, we determined the cause of the accident without any analysis of the telemetry recordings. Voskresenskiy discovered a tank that was filled with a sodium permanganate catalyst to break down hydrogen peroxide. The filler hole on the tank was open! The plug, which required many turns to screw in, was missing. Thus, after the tank was filled the plug was not screwed in. The required pressure could not be generated in the open chamber. Sodium permanganate was not fed to the gas generator.

The turbopump assembly, which is set into rotation by the hot steam/gas mixture formed by the decomposition of the hydrogen peroxide, received this fluid only to start the engine's operation, and then it stopped. The engine died, and the missile collapsed on the launch site. Sloppiness or sabotage? Of course, a security service representative was involved in the inspection. And it just had to happen that he was the one who found a wrench in the missile's remains. Picking it up, he asked, "Is this tool supposed to fly too?"

As I recall, the State Commission hushed up this scandal. In any event, no one was repressed.[6] Their punishments were limited to administrative reprimands. Explosion experts questioned at commission meetings explained that there was not supposed to be an explosion. The detonating fuse was set to respond only to an electrical command to shutdown the engine. That is why there was no explosion when the missile toppled over; in the end, the firefighters had time to cool off the warhead with water and everything turned out all right.

---

5. Aleksandr Mikhaylovich Goltsman served as Chief Designer of OKB-686 (later GOKB Prozhektor), which developed power sources for several generations of Soviet ballistic missiles.

6. During the Soviet era, it was common to use the word "repressed" to describe a person's arrest, incarceration, or execution by the Soviet security services (such as the NKVD or KGB).

WE OFTEN RECALLED THAT INCIDENT, when in 1953 we first began to meet with the developers of the atomic and then the hydrogen bombs. Korolev and Mishin received an invitation to the atomic bomb tests at the firing range in the Semipalatinsk area of Kazakhstan. They returned completely shaken. Mishin told us that if you don't see the results of the explosion with your own eyes, it is simply impossible to imagine.

At this point we were tasked with the problem of making a qualitative leap in the reliability and safety of this formidable warhead's delivery vehicle. Recalling the incident when Korolev dashed up to the smoldering missile with his useless fire extinguisher, Voskresenskiy, half joking and half seriously, suggested that to begin with we should move the launch site about 20 kilometers away from the bunker, and that the launch of a missile carrying an atomic warhead should be radio-controlled, "to teach Sergey not to run off with the fire extinguisher again." The science fiction–like idea about radio-controlled launches was realized 35 years later but out of quite different considerations.

This incident with the fire took place three years before the government issued its decree on the development of the R-5M missile, the nuclear warhead carrier. The R-5M missile was designed on the basis of the R-5, which we were supposed to revamp so that it could be a reliable atomic bomb carrier.

Based on our own many years of experience, as well as that of the Germans, we knew that no orders and entreaties would guarantee the reliability of all the electrical equipment, the onboard cable network, and control instruments, since any single failure such as a broken wire, loss of contact in a plug and socket connector, or random short circuit would cause a missile to crash. Furthermore, the single-stage R-5 was a statically unstable flying vehicle. Unlike the R-1 and R-2, it had no stabilizers. Only after a thorough analysis and study of the behavior of this long missile in flight did we begin to understand the hazard of disregarding the elastic vibrations of the entire structure and the effect of liquid-fueled tanks. The guidance system also needed to have a significantly greater margin of resistance and controllability in terms of its dynamic characteristics than its predecessors.

The development of a multichannel telemetry system was a new and powerful means for optimizing reliability. Constant vigilance was required of the telemetry monitoring service and its specialists, even if outwardly the flight had ended quite successfully. The "film report" procedure became an indispensable feature of the launch preparation process and of the analysis of launch results. Sometimes a careful examination of the films performed by the trained eyes of telemetry experts after a launch revealed glitches, over which, like it or not, the chief designers would have to rack their brains in pursuit of explanations.

Nikolay Golunskiy and Olga Nevskaya, who later became husband and wife, were virtuosos at hunting down difficult-to-explain fluctuations in the readings of various sensors recorded on motion picture film. Nevskaya had a service record that dated back to the *Brazilionit* era.[7] We were accustomed to Lelya Nevskaya's calm

---

7. *Brazilionit* was the Soviet modification of the German *Messina* radio control system.

reports, which Vadim Chernov then tried to interpret from a theoretical standpoint. Arkadiy Ostashev provided a practical explanation for these processes. Kolya Golunskiy rapidly advanced up the career ladder. He alone had the authority to interact with the telemetry developers and with firing range service personnel and also the sole right to report to management on these issues.

The presence of such constant, vigilant monitoring was crucial for the entire process of increasing the missiles' reliability. After a successful launch, the systems' developers were always very optimistic. Their satisfaction, public recognition, and accolades for the missile's good behavior in flight were sometimes ruined by the telemetry specialists' subsequent reports, which showed that it was a miracle that the missile hit the target. In such cases, if the comments had to do with Pilyugin's or Glushko's systems, they would usually go into a rage, demand reverification, and declare to Boguslavskiy, "your *Don* [telemetry system] is lying again. The missile flies just fine, but the telemetry is recording who knows what."

But the union between the telemetry developers and telemetry recording analysts was rarely proved wrong. Even after the successful outcome of an R-5 flight, when the telemetry recordings detected inflight vibrations that were inexplicable in amplitude and frequency, Pilyugin accused the measurement system, offering the hypothesis that it had been affected by electromagnetic blasts that had nothing to do with the guidance system. After a thorough analysis they determined that the measurement system had not been in error. Yevgeniy Boguslavskiy, who had been working with Golunskiy's team for several days analyzing the last launch and all the preceding ones, triumphantly announced, "Nikolay [Pilyugin] is my friend, but the truth is more precious. The telemetry recordings correspond to the behavior of the missile and the guidance system."

After numerous debates in the Council of Chiefs and various and sundry other echelons, reliability policies were developed that altered traditions born out of seven years of our missile work. The primary move was to introduce redundancy into the guidance system. From the gyros down to the control surface actuators, all of the electrical circuits had a backup. In the gyros the potentiometers were modified so that any single break at any point would not deprive the system of controllability in any of the channels. Redundancy was introduced into the amplifier-converter so that two loops would be in parallel operation for each of the three stabilization channels.

The failure of any loop would change the system parameters, but these changes would remain within a range that ensures stability. Instead of four control surface actuators, we installed six. The relay windings in the actuators were redundant, and each of them had its own path to the amplifier-converter. According to our model, the failure of one control surface actuator would not cause a loss of controllability. The model, however, generated many disputes. Skeptics believed that despite the positive results of simulation in the laboratory, if a control surface actuator failed in actual flight, a crash would nevertheless be inevitable.

Pilyugin and I proposed launching one missile in the schedule of upcoming flight

273

tests with a control surface actuator that was deliberately disconnected. Mrykin supported us, but spitefully asked, "You're not going to insist that this missile have an atomic warhead on it, I hope?" We promised we wouldn't insist. We allowed ourselves to joke around like this until we started dealing with the legendary nuclear experts.

NOW THAT WE ARE FACED WITH THE FACT THAT A NUCLEAR MISSILE WAR CAN NOT ONLY DESTROY A GOVERNMENT, but could also annihilate life on Earth, it is instructive to recall the history behind the emergence of the term "nuclear missile weaponry." A nuclear weapon was used for the first time by the Americans in 1945. The R-1 and R-2 missiles were put into service in 1950 and 1951, respectively. And it wasn't until 1953 that fully practicable ideas appeared for combining the two types of armaments, which earlier had been developed completely independently. After these two achievements of the human mind and modern technology were combined, all the previously existing principles of war developed by many theoreticians were only of historical interest.

The practical beginning of the development of the R-5M missile was the first step toward turning a missile into a weapon of mass destruction. In August 1953, Malenkov, the Chairman of the Council of Ministers, delivered a report at a session of the USSR Supreme Soviet. His report contained many new statements on foreign and domestic policy. At the end of his speech, he said that the USSR had everything for its defense; it had its own hydrogen bomb.

We had already had our first contact with the nuclear experts, as we had started designing the R-5M missile, our atomic bomb carrier, but we had not yet heard anything from them about this new weapon, the hydrogen bomb. And it was not our custom to ask questions that a person was forbidden to answer. Tests on the first hydrogen bomb in the USSR were conducted on 12 August 1953. It was impossible to hide this from the world. Physicists had already learned to record each test nuclear explosion no matter where it took place.

But we could not help but ask ourselves, and one must assume that we were not alone in asking, how would this bomb be delivered to its target? In 1953, air defense missiles were being developed fairly successfully. From our firing range in Kapustin Yar we had the opportunity to observe the effectiveness of Lavochkin's new surface-to-air guided missiles, which were being tested at the air defense firing range about 30 kilometers from us. As targets, they used airplanes controlled by autopilot after the crew had bailed out.

Once, we saw a Tu-4 flying at high altitude. It was a reproduction of the American Boeing B-29, the last model of the Super Fortress. These were the aircraft that dropped the atomic bombs on Hiroshima and Nagasaki. We had been warned in advance about the testing of Lavochkin's new surface-to-air missiles. In the bright rays of the morning sun, I could not make out the missile's hurtling flight. But when, against the backdrop of the clear blue sky, instead of the distinct contour of an airplane a formless gray cloud appeared, from which some sort of shimmering debris showered down, I felt sorry for the airplane. This hydrogen bomb carrier could not pose a threat for the U.S., our potential enemy.

At the very end of 1953, there was a meeting of the Central Committee Presidium, at which Vyacheslav Aleksandrovich Malyshev, the new head of the atomic agency (the Ministry of Medium Machine Building) and simultaneously deputy chairman of the USSR Council of Ministers, made a statement about the latest achievements in nuclear weaponry development.[8]

Two resolutions were passed at this meeting. The first had to do with the development and testing of a new thermonuclear bomb. Unlike the hydrogen bomb that was detonated on 12 August, this one was to be suitable for transport. Andrey Sakharov proposed the idea for this new "article."[9] The second resolution committed our ministry (at that time it was called the Ministry of Defense Industry) to develop an intercontinental ballistic missile for the thermonuclear warhead and tasked the Ministry of Aviation Industry with developing an intercontinental cruise missile. But until there was an intercontinental missile, this same Council of Ministers resolution proposed that we develop the R-5M missile, identical to the R-5 missile, but with a nuclear warhead.

Andrey Sakharov wrote about these resolutions in his *Memoirs:*

In essence, this meant that the weight of the thermonuclear charge, as well as the dimensions of the missile, had been fixed on the basis of my report. The program for an enormous organization was set in this manner for many years to come. The rocket designed for that program launched the first artificial satellite into orbit in 1957, and also the spacecraft with [Yuriy] Gagarin aboard in 1961. The thermonuclear charge that provided the original rationale for all this, however, fizzled out, and was replaced by something quite different.[10]

What constituted an intercontinental missile was still not very clear at that time. By that time we had conducted very meticulous, but still only exploratory work. First and foremost I should mention the design of the R-3 missile. The N-3 project and its subsequent refinement in the T-1 project were a continuation of the quest for ways of achieving intercontinental ranges. The T-1 project entailed the study of various layouts, making it possible to develop a two-stage ballistic missile with a

---

8. In 1953, the First, Second, and Third Main Directorates of the USSR Council of Ministers, which managed the nuclear weapons, uranium procurement, and air defense programs, respectively, were consolidated into the "super" Ministry of Medium Machine Building headed by V. A. Malyshev. Between 1953 and 1955, this ministry managed all strategic weapons development in the Soviet Union.

9. Andrey Dmitriyevich Sakharov (1921–89) was a Soviet nuclear physicist and "father" of the Soviet hydrogen bomb who would later become an outspoken advocate for human rights and reform in the Soviet Union. He was awarded the Nobel Peace Prize in 1975.

10. Andrey Sakharov, *Memoirs* (New York: Alfred A. Knopf, 1990), p. 181.

range of 7,000 to 8,000 kilometers.[11]

The R-3 missile design was never realized. Perhaps that is for the best. It would have taken a great deal of manpower, and the 3,000-kilometer range that was envisioned for it would not have made it substantially better than the actual R-5 missile and its nuclear modification, the R-5M.

Work on the N-3 project was officially finished in 1951. In the findings, Korolev wrote that "the most reliable path to achieving a flight range of 7,000 to 8,000 kilometers is to create a two-stage ballistic missile ..." However, the thermonuclear warhead that Sakharov proposed in 1953 could not be delivered to a range of 8,000 kilometers by a two-stage missile that had a launch mass of 170 metric tons. I am not able to judge to what extent Andrey Sakharov personally determined the design and weight of the warhead intended for the first intercontinental missile. But certainly, Sakharov's actions required the development of the missile we designed under the code number R-7. And so, Sakharov's name must also be mentioned in the history of cosmonautics!

WE WERE FIRST EXPOSED TO NUCLEAR SECRETS IN 1953. Korolev formed an especially restricted group to work on the first nuclear missile payload. Officially this group, which was headed by Viktor Sadovyy, was part of the design department subordinate to Konstantin Bushuyev.[12] Correspondence with the nuclear experts was classified at least "top secret." But, in addition, papers also appeared stamped "Critical." But documents weren't the only sources of "critical" government secrets.

Nuclear weaponry was developed in closed cities where not only simple mortals but even we who had access to top-secret projects were denied entry and passage without orders. These cities did not appear on a single geographical map. It wasn't until the 1990s, from a plethora of sensational publications, that the public was able to piece together an idea about the work conditions of the country's best physicists, scientists of other specialties, and finally, of the workers, servicemen, and their family members in those cities.

Our first personal contacts with nuclear experts began with their visit to NII-

---

11. Here, Chertok is referring to a set of research projects that preceded the development of the final ICBM. In 1948–53, various Soviet organizations engaged in three major R&D projects designed to study a single-stage missile (the R-3) with a range of 3,000 kilometers, a missile using storable propellants, and exploratory work on an intercontinental ballistic missile. These study projects were known as N-1, N-2, and N-3, respectively. On termination of these R&D themes, two new R&D studies were performed in 1953–55, T-1 and T-2, focused on future intercontinental ballistic and cruise missiles, respectively. All five of these studies were carried out simultaneously and in coordination by several leading missile development organizations, including NII-88, NII-4, and institutes of the Academy of Sciences. The research led directly to the eventual creation of the first Soviet ICBM, the famous R-7.

12. Konstantin Davydovich Bushuyev (1914–78) served as deputy chief designer in OKB-1 in 1954–72, during which period he managed several important human spaceflight projects. In 1972–75, he served as Soviet chief of the Apollo-Soyuz Test Project (ASTP).

88. I remember a meeting in late 1953 with Samvel Grigoryevich Kocheryants and General Nikolay Leonidovich Dukhov. Kocheryants worked in the now famous, but then top-secret city of Arzamas-16, where he was directly involved in the design of the atomic bomb.[13] Dukhov had received the title Hero of Socialist Labor during the war. He was the chief designer of heavy tanks, including the IS tank.[14] In Moscow he was assigned to head the design bureau and factory that were developing and manufacturing all of the electroautomatic controls for the atomic bomb and later also for the hydrogen bomb. Viktor Zuyevskiy, the leading specialist on atomic electroautomatic controls, was in charge of developing the general electrical system and integrating it with the missile's system. For that reason, my dealings were first and foremost with him.

In Arzamas-16 they were developing the warhead itself and its mechanical framework. The later famous physicists Yuliy Khariton, Yakov Zeldovich, Andrey Sakharov, Kirill Shchelkin, Samvel Kocheryants, and many others lived and worked in Arzamas-16 itself. We couldn't exactly understand the division of responsibilities among them at that time. But we clearly understood that there was a division between great theoreticians, who were removed from the pedestrian problems of reliability, and engineers/unskilled workers, who were responsible for the construction down to the last nut and bolt.

In his memoirs, Andrey Sakharov described who was who among the physicists, with descriptions of their essentially human qualities. Evidently, he had little contact with the designers and those who actually manufactured, assembled, and tested the "article" with their own hands. At that time, for secrecy's sake everything was called an "article" *(izdeliye)*. That's what we called our missile, and that's what the nuclear experts called their atomic and hydrogen bombs.

Besides the simple "article" concept, there was also the more complex "article in its entirety." It turned out that Dukhov was responsible for the "article in its entirety" since the "article" equipped with the nuclear explosive could only be actuated by the second part, a case stuffed with all sorts of automatic electronic control devices. The entire "article in its entirety" needed to be contained in the R-5M payload section. And to do this required the joint work of designers from Arzamas-16 and our group headed by Sadovyy.

At our institute, Sadovyy's group was treated like a delegation from a foreign country. It had special rooms closed off from other work rooms and had its own top-secret records management system so that documents containing nuclear secrets would not make their way around every "Department No. 1" (information security departments) and dozens of administrators.

---

13. Arzamas-16 was the closed city where one of the Soviet Union's two major nuclear weapons laboratories was founded.

14. The IS, named after Joseph [Iosif] Stalin, was a series of wartime heavy tanks such as the IS-1, IS-2, and IS-3.

We were faced with developing a process for the joint testing of two "articles in their entirety" after their integration, along with a whole multistage engineering operations plan at the launch site. Korolev delegated this work to Voskresenskiy's young deputy, Yevgeniy Shabarov. Why not to Voskresenskiy himself? Here, for the umpteenth time, I witnessed Korolev's knack for selecting the right people for the job. Voskresenskiy was a top-notch tester, endowed with exceptional intuition. Someone aptly encapsulated his personality, remarking that if he had been a pilot, he would have taken risks like Valeriy Chkalov.[15] Guerrilla operations like that of Voskresenskiy were, however, absolutely inadmissible in relations with the nuclear experts. Besides the basic operations, the process also needed to be formalized concisely and methodically.

What would happen if a missile containing an atomic bomb toppled over at the launch site during preparation because of something akin to the sloppiness mentioned earlier when they failed to seal the tank containing sodium permanganate? The nuclear experts' work procedure called for a triple check of all assembly and testing operations. The head of assembly or testing would hold the instructions and listen as the tester read aloud the steps of an operation. For example: "Unscrew five bolts securing such-and-such a cover." The performer of the operation would unscrew them. A third participant in the operations would report: "Five such-and-such bolts have been unscrewed." The controller, a military acceptance representative, would report that he accepted the operation's execution. A notation to that effect was made in the appropriate document. Only after this could the entire team move on to the next operation. Work went slowly and scrupulously, with the mandatory reading aloud and mandatory reporting aloud about the execution of an operation and a notation to that effect in a special process logbook.

We did not usually have these strict formalities in the missile industry. When Shabarov told Korolev about this whole procedure, the latter decided that since we were going to be working together, we needed "to show them that we were just as good." As far as our own work was concerned, for the R-5M missile we needed to revisit all the instructions on the preparation procedure at the engineering and launch sites and also implement a triple check procedure. The primary operator was to be from the military (an officer or a soldier); he would be monitored by another officer from the appropriate firing range division and, always, an industry representative.

There were to be two phases of testing on the R-5M, which had been assigned the military designation 8K51: flight-development tests and qualification tests. During the flight-development tests, the intent was to optimize the reliability of

---

15. Valeriy Pavlovich Chkalov (1904–38) was one of the greatest Soviet pilots of the interwar years, who gained fame during the 1930s when the Soviet government used worldwide aviation exploits to legitimize its various claims to greatness on an international and domestic stage. He completed the first nonstop Moscow-to-U.S. flight over the North Pole in a single-engine ANT-25 in June 1937.

the launch vehicle and all of its onboard and ground-based systems and to check out the documentation ensuring reliable operation. We began flight-development tests in the spring of 1955, just a year and a half after the R-5 missile flight tests ended. Fourteen R-5M missiles were presented for the first phase. In addition to redundancy in the guidance system, other measures were implemented in this series to enhance its reliability. The engine underwent numerous firing rig launch tests in extreme modes that substantially exceeded the nominal. The onboard instruments were first shaken and "fried and steamed," eliminating anything that might have stirred doubts during laboratory and factory tests.

We also developed a new emergency Automatic Missile Destruction (APR) system.[16] If, as a result of inflight malfunctions, the missile were to stray sharply away from the target or threaten to strike our own territory instead of the enemy's, it needed to be destroyed in flight. But! The question was how to destroy it without scattering the radioactive fallout where it wasn't supposed to go. I was personally responsible for developing the APR system. Nikolay Dukhov, chief designer of the atomic "article in its entirety," reassured me that "all you have to do is give us an electrical signal indicating there's a problem and that the missile needs to be destroyed. We'll take care of the rest."

The atomic bomb contained a rather powerful charge of conventional explosive, which was used as the detonator for the atomic explosion. How could we actuate this detonator without destroying the atomic warhead? Korolev wanted me to provide the answer to this question. I asked Korolev to request an explanation from the inventors, confessing that the nuclear experts had not explained this secret to me.

Korolev scolded me for not adhering to his philosophy; he added, in that case he would raise an objection to using an APR system. Anything could set off this dangerous system; and then we would be guilty of causing a nuclear explosion! Since all first-phase flight tests were conducted without a nuclear warhead, the APR system could fly without a fuss and have its reliability verified in telemetry mode.

The development of separating payloads for the R-2 and R-5, which were designed to carry a conventional TNT warhead of 800 to 1,200 kilograms, was not particularly complicated. Ivan Prudnikov was the chief developer of payload sections. His direct boss was Korolev's deputy, Konstantin Bushuyev. Our factory's job was to manufacture the conical steel housing, apply the thermal-protective coating, and run the cable from the control system to the percussion fuse, which was installed only at the launch site.

To rig the payload sections with explosives, engineers sent them to "powder" factories, where this operation did not pose a complex technical problem. All of the performers understood everything right down to the size of the crater that should be produced in the impact zone by a warhead rigged with conventional explosives.

---

16. APR—*Avtomaticheskiy podryv rakety.*

A fundamental difference between a missile warhead and an aerial bomb was the requirement to destroy the missile in flight if it decided to fly somewhere "out of bounds" rather than into the target area. The aforementioned APR system was developed for this very event. Aleksandra Melikova and Aleksandr Pronin developed the logic and electrical circuitry for this "scary" system. Semyon Chizhikov was directly responsible for developing the automatic control instruments. The APR system was considered particularly "scary" because its false, or as the instructions stated, "off-nominal," actuation could lead to the detonation of the warhead on the missile's normal flight trajectory in the best case and, in the worst case, at the launch site during launch preparation. After heated arguments and analysis of failed launches during flight tests and of the many criteria for determining accident rates, only two were selected: premature shutdown of engine operation and missile deviation by more than seven angular degrees from the specified value of any of three angles monitored by the gyroscopes.

Multistage inhibitors were removed as each launch preparation phase was completed; over the course of the flight they provided protection against a false command. The first inhibitor was called "arming APR system." This meant that right before the launch, power was fed to the system's main buses, allowing it to execute its task if this were required. In principle, until the missile lost contact with the launch system, the APR command could not be transmitted. The second inhibitor was provided by the "liftoff contact" signal. The third protective inhibitor was time. Regardless of the nature of the failure, an explosion was not supposed to be possible until a specific second of the flight. There were very many disputes as to precisely what moment that would be. For the R-5 missile we set an inhibitor for the 40th second.

During flight-development tests of missiles not equipped with a warhead, the APR execution command was replaced with the Emergency Engine Shutdown (AVD) command in the event that the missile with its engine running deviated significantly from the specified impact zone.[17]

After they had familiarized themselves with our achievements in safety engineering and test procedures, the nuclear experts announced that, when a nuclear warhead was present, they would take over the development of all types of emergency and nominal detonation and all inhibitors, right down to the detonating fuses. Nuclear warheads were complex systems, and we weren't allowed to come near them. As for the payload section, if they had had the appropriate manufacturing facility at their disposal, they would have completely taken over the manufacture of the entire payload section. In any event, that is what they did with airborne atomic

17. AVD—*Avariynoye vyklyucheniye dvigatelya.*

bombs. In our case, they could not take on responsibility for normal separation from the missile and integrity of the payload section along the descent path until it hit the Earth's surface.

During their first meetings with Korolev, Khariton, Kocheryants, and Dukhov announced that the conditions that would affect their "article"—vibrations, loads, temperatures, and atmospheric pressure—needed to be cleared with them and we must guarantee the proper conditions, not only in flight, but during all instances of ground preparation. We learned for the first time that the "article" had problems with low temperature. It turned out that on days when the temperature dropped below freezing it was necessary to put an insulating cover on the payload section and maintain a specific temperature. Even the payload's interior layout was the object of joint study.

"Our automatic electronic controls are a lot more complex than your APR," said Viktor Zuyevskiy after his first encounter with our system. "We're not going to connect with your onboard power sources. We'll have our own independent sources. But we have to obtain information from the control system so that the safety inhibitors can be removed stage by stage and our system can be prepared for activation."

We imagined in the most general terms the form of an airborne atomic bomb from the secret lectures that a general from the engineering aviation service, specially invited by Korolev, presented to us in 1953 at the firing range. He spent a great deal of time trying to explain the principle of the explosive chain reaction. But when it came to describing the specific technology of combining two or more pieces of uranium-235 of noncritical mass into one supercritical mass, he proved "incompetent."

Bushuyev and the immediate designers of the R-5M payload section, Prudnikov and Vorontsov, and Sadovyy, who maintained a direct communications link with Arzamas-16, coordinated the dimensions and fixtures of the individual payload parts on paper down to hundredths of millimeters, but no one explained to them what would be inside. When it came to interfacing the electrical connections and our missile systems, I dealt with Dukhov, Zuyevskiy, and their colleagues at what had once been the aviation industry's Factory No. 25.

The creators of the atomic bomb, including Yu. B. Khariton, at first underestimated the problem of creating automation and all the electrical instrumentation for nuclear warheads. A search for facilities to develop these systems resulted in the selection of aviation industry instrument Factory No. 25. The chief designer at this factory was A. F. Fedoseyev, whom I already knew through our work developing the electrical circuitry for the Lavochkin surface-to-air missiles, and the factory director was the first Hero of the Soviet Union, polar pilot A. V. Lyapidevskiy.

Yu. B. Khariton considered it unacceptable to develop, in a ministry other than his own, the automation controlling a nuclear explosion. The all-powerful Beriya was no longer around, but, nevertheless, Khariton, with Malenkov's aid, managed to have Factory No. 25 transferred from the Ministry of Aviation Industry to the Ministry of Medium Machine Building. The factory was converted into KB-11's Branch No. 1.[18] Nikolay Dukhov became the director and chief designer of Branch No. 1 in 1954. When in 1954 I first visited the site that I had earlier known as Factory No. 25, Fedoseyev and Lyapidevskiy were no longer there. Former chief designer of super-heavy tanks, three-time Hero of Socialist Labor Nikolay Dukhov mastered what was for him a completely new field of technology extremely quickly.

Despite all the prohibitions, the nuclear experts were forced to reveal their basic "secrets" to us. It turned out that there was no uranium in the atomic bomb! It was the plutonium that exploded! The plutonium was produced from uranium in nuclear reactors. The process for obtaining plutonium was complex and labor intensive, resulting in a large number of fatalities from radiation exposure. Even many of the leading specialists who first developed the chemical process for extracting plutonium were exposed to life-threatening doses of radiation. And all of this was taking place, not at Arzamas-16, but in the Urals in another closed city.[19]

To make a warhead, they formed a sphere from a noncritical mass of plutonium. They then surrounded the sphere with a solid spherical mass of TNT or a mixture of TNT and another conventional explosive. The surface was finished with great precision and contained a large number of fuses that were supposed to be actuated synchronously with a time scatter of microseconds. The explosion of the conventional explosive was directed so as to form a converging spherical blast wave that squeezed the globular mass of fissionable plutonium and converted it into a supercritical state. However, this did not guarantee a chain reaction. In order to start a full-fledged chain reaction, one more detonator was needed. This was the neutron detonator, which "sprayed" neutrons inside the collapsing sphere.

The complex electrical device needed for the neutron triggering of the blast was the "neutron gun." High voltage, up to 20,000 volts, was used to trigger all the detonators. The weight of the entire nuclear warhead was determined not by the weight of the active plutonium but of the heavy steel hull, the interior walls of which acted as neutron "deflectors." The automatic controls, safety inhibitor electrical elements, and communications with the missile guidance system were contained in a separate unit inside the payload and were connected with the missile's electrical circuitry via a pressurized plug-and-socket connector on the bottom of the payload container.

During flight tests the nuclear "article in its entirety" was first tested without plutonium. Coordinating the test conditions and analyzing the results inevitably

---

18. KB-11, formed in 1946, was the leading design bureau assigned to develop the first Soviet atomic and thermonuclear weapons. It was located at Arzamas-16 near the industrial city of Gorky and headed by Yu. B. Khariton.

19. The Soviet government set up a competitor organization in competition with the original KB-11 in 1954. The new organization, NII-1011, was established at a closed facility in Chelyabinsk-70, and competed through the 1960s with KB-11 for contracts to develop new generations of thermonuclear weapons.

required not only official interactions, but also good old-fashioned, face-to-face shop talk. Now the nuclear experts were forced to share their professional secrets with us.

The preparation and launch of history's first ballistic missile carrying a nuclear warhead in February 1956 provided good training in interaction between missile specialists and the developers of nuclear warheads. This association marked the beginning of the era of nuclear-tipped missiles, the hottest weapon of the coldest war.

We introduced the concept of "combat readiness" into the R-5M preparation procedure, and for each of them we developed a process-oriented plan of actions for the missile teams. The preparation process called for conducting all sorts of tests before the warhead was attached at the launch site. The missile was hauled out to the launching pad like a "headless horseman." They assembled and prepared the warhead for integration with the missile in a special building with particularly high security. The new secret building for the firing range, which we had made our "home," was erected three kilometers from the R-5M launch site. A tall fence surrounded the area around the building. This first nuclear facility at the missile firing range was guarded by special troop subunits of the KGB that were formed during Lavrentiy Beriya's tenure as the head of the entire nuclear project. The nuclear personnel stayed in a separate hotel constructed for them. For that reason, aside from our on-the-job contact, we almost never ran into the nuclear experts. Even when it came to the motor transport service that the firing range services provided for us in full, they had their own.

The missile's payload section was equipped with various sensors and a telemetry system to determine the conditions that the warhead would be exposed to in flight. Once and for all, the prepared payload section containing the nuclear warhead was placed inside a special thermally insulated vehicle that delivered it to the launching pad. Integration with the missile took place right at the launch site.

The missile arrived at the launch site in a special transport assembly along with the firing table. This assembly placed the missile and payload in a vertical position. After performing the laying operations and installing the onboard batteries, they checked out the "ground-to-missile" power switchover and, just in case, the "abort launch" system. Next came the fueling operations, and then the standby-for-launch command was issued. One had to be able to perform all of these operations reliably not just during the day, but also at night using portable lights.

Flight-development tests were conducted from January through July 1955. Of the 17 missiles launched, 15 missiles reached the target. Two missiles deviated by more than the seven degrees permitted and the engine was shut down by the APR system.

Five missiles were submitted for qualification tests. The payload sections of four were equipped with functioning mock-ups of a nuclear warhead. Essentially these were not mock-ups since they were equipped with everything that was required for a nuclear explosion except for the products initiating the chain reaction. Ground personnel checked out the integration with the missile's systems, the preparation process, and the inflight operating reliability of all the automatic controls. Launches began in the cold January of 1956. Four launches proceeded normally. The last launch, the fifth, was the worst of the worst. Korolev was on edge because of delays with the missile preparation. By no means did he want to let Nikolay Pavlov, who had supervised the preparation of the payload section and warhead, report to State Commission Chairman Nedelin that the warhead was prepared for rollout, but that the missile specialists were causing a launch delay.

As deputy technical director, I was responsible for missile preparation at the engineering facility. There we had conducted stand-alone tests on all the systems and integrated horizontal tests on the entire missile with the electrical equivalent of the payload section. Leonid Voskresenskiy had the same title and was responsible at the launch site for preparatory operations and launch execution.

Korolev delegated Shabarov to maintain contact with the nuclear experts' facility and to observe them throughout the preparation of the entire payload section. Shabarov was admitted to that "Mecca" only after warhead deputy chief designer Yevgeniy Negin arrived at the firing range. That night, I reported to Korolev about a glitch that occurred during tests on the stabilization controller. I recommended replacing the amplifier-converter and repeating horizontal tests, but that would require another three to four hours. He answered, "Take your time. Their neutron gun failed, too." My knowledge of nuclear technology wasn't sufficient to grasp what that meant and how much time we had gained.

I reported to Korolev that all the glitches had been eliminated and asked, "How are they doing with their gun?" He said, "Drop by and I'll explain." Despite the late hour, Voskresenskiy and Shabarov were sitting in Korolev's hotel room, which also served as his office. They had already reported for the umpteenth time on readiness and the operational procedure in effect at the launch site after the payload section containing the nuclear warhead arrived there in the "specially guarded vehicle."

Korolev had received very scanty information from Negin as to what a "neutron gun" was and why it was capable of holding up a launch. He understood only that the nuclear experts were reassuring themselves and were checking out all their automatic controls again and again.

On 2 February 1956, for the first time in history the R-5M carried a missile armed with a nuclear warhead through space. After flying the prescribed 1,200 kilometers without breaking up, the warhead reached the Earth in the Kara-Kum desert near the Aral Sea. The impact fuse went off and the surface nuclear explosion marked the beginning of the nuclear missile era in human history. No publicity followed this historic event. American technology did not have the means to detect missile launches. Therefore, they recorded the nuclear explosion as a routine nuclear weapon ground test.

We congratulated one another and wiped out the entire champagne supply, which until then had been zealously protected in the pantry of the executive dining room. Later, after we had returned from the firing range and had once again readjusted to the problems of the R-7 intercontinental missile, at a small gathering, Korolev remarked "with great secrecy," "Do you know what they told me? The yield of the blast was greater than Hiroshima."

Ryazanskiy joked grimly, "And you're not afraid that someday they'll try us as war criminals?"

The R-5M missile was put into service in March 1956. Many years later I met up with Lieutenant General Academician Yevgeyiy Arkadiyevich Negin. He had been elected to the Academy of Sciences in our department of mechanics and control problems. All of his nuclear science and technology colleagues were in the physics and power engineering departments. He usually didn't associate with any of us. Nevertheless, I managed to get him to talk with me about the events of 1956. Regarding the yield of the warhead carried by the R-5M missile launched on 2 February 1956, he said that the missile head carried a warhead with a yield of less than 3 kilotons, whereas they only put warheads of 80 kilotons or more on missiles that went into service. Several years later, instead of nuclear warheads, they started to come out with thermonuclear warheads with equivalent yields up to one megaton for the R-5M missiles that had already been put into service and were on duty in the Baltics, Crimea, and Far East.

Soon after the first successful launch of an R-5M missile with a real nuclear warhead, Korolev and Mishin were awarded the title Hero of Socialist Labor. Another 20 NII-88 employees, including me, received the Order of Lenin. The enthusiasm that our entire team had for our work was strengthened by the government decree awarding an Order of Lenin to NII-88.

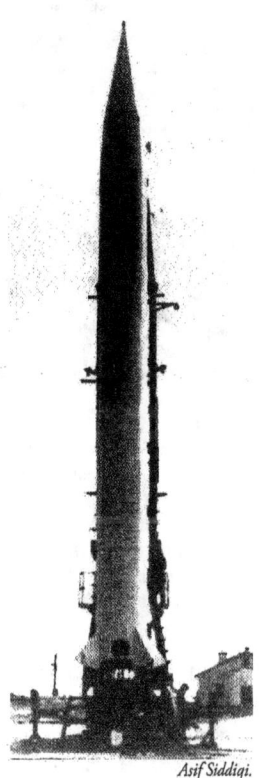

Asif Siddiqi.

The R-5M was the first Soviet ballistic missile capable of carrying a nuclear warhead. In February 1956, the R-5M was launched with a live nuclear charge as a test of its reliability.

We had a real celebration on our street during this time, when work on the development of the first R-7 intercontinental missile was in full swing. Glushko, Barmin, Ryazanskiy, Pilyugin, and Kuznetsov received the gold stars of Heroes of Socialist Labor. A large number of individuals who were involved in projects in almost all of the subcontracting organizations were generously awarded orders and medals.

During work on the R-5M and R-7, Korolev often arranged social gatherings with leading and *crème de la crème* of the nuclear experts. We joked that he invited an "exclusive group of narrow-minded people" to a party with famous scientists. Mishin, Bushuyev, Prudnikov, and Sadovyy from our institute were also usually at

285

This famous picture, known as the "Three K's", has been prominently reproduced in many Russian books for it shows the three scientific giants of the Soviet military-industrial complex: (from the right) Sergey Korolev, Igor Kurchatov (the "father" of the Soviet A-bomb), and Mstislav Keldysh (later President of the USSR Academy of Sciences). On the extreme left is Korolev's first deputy ("first among the deputies") Vasiliy Mishin. In most reproductions of the photograph, Mishin's image was typically cut out, especially during Mishin's "banishment" from public view prior to the late 1980s. The image dates from July 1959 when Korolev, Mishin, and Keldysh visited Kurchatov's institute.

the gatherings that I often attended there. Several times Yuliy Borisovich Khariton, Kirill Ivanovich Shchelkin, and Nikolay Leonidovich Dukhov attended. They were always accompanied by "secretaries," that is, KGB officers, whose lives were on the line to keep their charges safe and out of trouble.

Not once during our joint work did I have the opportunity to meet with Sakharov. In his memoirs, Sakharov writes that he had been at our facility and met with Korolev:

> ... *after we returned to the Installation, Malyshev organied a series of 'excursions' for us, including a trip through a ballistic missile plant where I met [Sergey] Korolev, the chief designer, for the first time. We had always thought our own work was conducted on a grand scale, but this was something of a different order. I was struck by the level of technical culture: hundreds of highly skilled professionals coordinated their work on the fantastic objects they were producing, all in a quite matter-of-fact, efficient manner. Korolev explained things and showed us some films.*[20]

On 23 October 1953, Korolev and Glushko were elected corresponding members of the USSR Academy of Sciences. For Korolev, who was at that time known

---

20. Sakharov, *Memoirs*, p. 177.

to a very narrow segment of the academic community, this was a victory signifying that they believed in him, that they were placing their bets on him. Back then, it was impossible to be elected to the Academy of Sciences against the will of the Central Committee. Moreover, if the Central Committee felt that someone absolutely must be elected, then the academicians were persuaded to implement such decisions.

At this same general assembly of the Academy, the scientific director of Arzamas-16, essentially the most important chief designer of nuclear warheads, Yuliy Borisovich Khariton was elected an Academy member. Khariton's deputy, Kirill Ivanovich Shchelkin, and Nikolay Leonidovich Dukhov were elected corresponding members. The general assembly also elected doctor of physical and mathematical sciences, 32-year-old Andrey Dmitriyevich Sakharov, straight into the ranks of full academicians without first passing through the traditional corresponding member ranks.

All of the nuclear experts at the Academy meeting had been decorated with one or two of the gold stars worn by Heroes of Socialist Labor. Before 1953, when the opportunity arose, enemies of Korolev and Glushko used to reproach them that all their work was nothing more than the reproduction of German technology. Their inclusion in such a glittering constellation was to a certain extent an advance payment. No one else from among the Council of Chiefs members was elected to the Academy of Sciences in 1953. Of the scientists who had collaborated with us, only Vadim Trapeznikov and Boris Petrov were elected corresponding members. The 1953 elections marked the beginning of the establishment within the Academy of Sciences of a powerful coalition of rocket scientists who had worked in the military-industrial complex.

From the standpoint of "fundamental" Big Science, our work differed from that of the nuclear experts. We began to develop our own school of missiles, relying on technology, production engineering, and pure engineering science. As soon as we delved into projects involving the intercontinental missile, we ran up against problems requiring fundamental research, which in academic circles they liked to call Big Science.

At the beginning of their careers, almost all of the nuclear experts were theoreticians, servants of pure science, or experimental physicists. They worshipped science for science's sake, above all because mankind had to know why the world was arranged this way and not any other way; they wished to discover its building blocks. And then when they had figured out that the conversion of a substance into energy—which could be theoretically explained on paper—could be realized in practice, they had to recruit engineers and throw themselves into the problems of technology.

Chapter 16

The Seven Problems of the R-7 Missile

The creation of an intercontinental missile carrying a thermonuclear warhead required large capital investments for constructing new production facilities and test rigs and for searching for a new firing range.[1]

All the work on the new missile stemmed from a Communist Party and government decree issued in 1954. Specialists from all of the affiliated ministries pored over the draft of the new resolution before its submission to the Central Committee and government. As always, they wanted to anticipate all eventualities, maximize the return on the investment, and overlook nothing and no one. However, many years of experience had shown that no matter how thoroughly such resolutions were prepared, several days after they were issued, it seemed that something had always been left out. That's when you heard the comforting words, "Wait for the next one. You are not the only ones who are forgetful."

The Council of Ministers and Central Committee resolution on the development of the R-7 intercontinental missile was issued on 20 May 1954. The first launch of the first missile took place on 15 May 1957. What a lot we had to accomplish over those three years! In May 1954, we didn't even have a draft plan! Now it is difficult for me to imagine how we managed to do a job like that. After all, we were working on the R-11, R-11FM, R-5, and R-5M missiles at the same time.

By early 1956, we had not yet performed the first test of a missile carrying a nuclear warhead, and just a year later, in 1957, we were already taking a stab at a missile carrying a thermonuclear warhead!

Beginning in 1954, we were confronted with one difficult scientific, technical, or organizational problem after another. We hadn't, however, identified or even recognized all of these problems during the time we were drawing up the R-7 missile's draft plan. The design was issued in 1954, in record time.[2] We acknowledged the need for many new modifications on the already developed design of the missile, as well as during subsequent experimental research.

---

1. *Author's note:* The mass of the warhead along with its payload container was 5.5 tons.
2. The R-7 draft plan was officially signed on 24 July 1954.

I will allow myself to list just a few of the solutions that were fundamentally new for missile technology of that time. They are also illustrative in that they completely refute statements expressed in some Peenemünde veterans' memoirs and some foreign publications to the effect that supposedly the Russians got the first artificial Earth satellite because of a launch vehicle developed with the assistance of German scientists. In fact, the R-7 missile is noteworthy in that developing it we were negating to a great extent our past achievements that had used German ideas.

I shall not list the problems in order of their importance. To one extent or another, they *all* required heroic work, inventiveness, collective brainstorming, and tremendous organizational efforts.

Problem number one. After researching and designing alternative layouts for a two-stage missile, a clustered version was selected. The first stage consisted of four strapon boosters surrounding a central sustainer, which was also the second stage.[3] We had no experience firing a powerful liquid propellant rocket engine in space. Glushko could not guarantee that it would fire reliably somewhere out there far away, under unknown conditions. As a result, we decided to fire all five engines simultaneously under ground control. But then the central second stage would be operating for over 250 seconds, twice as long as the graphite jet vanes could withstand. But even if they had been made of something more fire resistant than graphite reinforced with tungsten, there still would have been two arguments against the jet vanes. First, their use led to a loss of range due to resistance generated at the engine nozzle exhaust outlet. And second, the precision of the velocity measurement affected errors in the projected range. When the terminal velocity design value was achieved, the control system issued a command to shut down the second stage engine. But it turned out that no matter how remarkable the control system was, after it had executed its command to shut down the engine, an uncontrolled residual fuel burn-off occurred, causing the so-called aftereffect burn.

Based on R-5 experience and bench tests, the scatter of aftereffect burn values was so great that it surpassed the scatter of control system–generated errors by several factors of 10. For that reason alone, range errors for the intercontinental missile could exceed 50 kilometers.

There were many suggestions made in this regard, the majority of which amounted to propulsion system modifications, which Glushko rejected. We eventually found a solution that killed two birds with one stone. Instead of using jet vanes for control, we decided to use special control engines. These same engines could serve as the last stage's vernier thrusters. After the shutdown of the second stage main engine, a

---

3. Typically, in the West such a configuration, that is, a central sustainer with a number of strapons, would be called a "one-and-a-half" stage vehicle, especially if at launch, both the central sustainer and the strapons fired simultaneously (as they did on the R-7). The Soviets (and later, Russians), on the other hand, refer to such designs as "two-stage designs," with the strapons being the first stage and the central sustainer being the second.

precise velocity measurement would be performed with only the control engines in operation. Upon reaching the specified velocity, they would shut down with virtually no aftereffect burn. Glushko, however, refused to produce control engines. He had enough to worry about with the main engines, and he was already in danger of missing the deadline for completing them. At Vasiliy Mishin's initiative, we invited Mikhail Melnikov, Ivan Raykov, and Boris Sokolov to join OKB-1 to develop control engines. The three of them had gotten stuck at NII-1 with Keldysh after Isayev had left with his engine specialists.[4]

Our factory was already producing engines, but only Isayev's high-boiling component liquid-propellant engines for antiaircraft missiles and the R-11. We needed once again to set up the production of low-thrust liquid-oxygen engines and create a ground station for all sorts of tests, including firing tests. I should note that back in Bleicherode, Pilyugin and I had dreamed of a system without jet vanes.

Mishin was all for this idea and took it further. If you can do away with jet vanes in the central sustainer, then why keep them on first stage strapon boosters? We came to a revolutionary decision: there would be absolutely no graphite jet vanes on the missile. The missile's entire ascent would be guided by the control engines, which would use the same propellant components as the main engines and would receive power from the same turbopump assemblies. Glushko had developed essentially a single engine with four combustion chambers for the first and second stages. Now to this engine on the second stage, we added four more small vernier thrusters, and on the first stage, two small chambers on each engine of the strapon boosters. The draft plan had called for three jet vanes and one aerodynamic fin on each strapon booster for control. We decided to replace both the jet vanes and aerodynamic fins on the strapon boosters (with control engines) only *after* defense of the draft plan. At that time, it was a revolutionary design.

Instead of a single combustion chamber, which we were accustomed to dealing with on any missile, all of a sudden there were 32! This design is almost 40 years old. But not only is it not aging, it is now experiencing a third youth. Thirty-two chambers required systems to control the preparation of turbopump assemblies for startup, the opening of dozens of valves in the required sequence, and the simultaneous ignition and subsequent operation in all modes.

Our responsibility for the coordination of operations in the OKB-1—OKB-456—NII-885 triangle increased dramatically. OKB-1 developed the general hydraulic system; NII-885, the general electrical system; and OKB-456, the engines' layout and thrust sequence. It wasn't easy for Glushko to agree to have another 12 oscillating chambers hooked up to his propulsion systems! But Mishin's uncompromising stand plus the enthusiasm of the Melnikov-Raykov-Sokolov team showed an unusual way out of a hopeless predicament. It also cleared the way for many

---

4. Isayev's engine group moved from NII-1 to NII-88 in July 1948.

subsequent missile and spacecraft control systems.

Jointly with control surface actuator specialists Kalashnikov, Vilnitskiy, and Stepan, my group was supposed to develop new control surface actuators that would have greater reserves in terms of dynamic parameters and the capacity to overcome friction in the assemblies feeding oxygen and kerosene to the oscillating engines. The whole kit and caboodle—Glushko's engines, Melnikov's control chambers, and our control surface actuators—after being developed separately, needed to undergo developmental testing during joint firing! They had to tested first on the OKB-456 rigs in Khimki and then at the Novostroyka at NII-88 Branch No. 2 in Zagorsk.

Problem number 2. No matter how hard the engine specialists tried to produce absolutely identical engines, they would have manufacturing tolerances for specific and absolute thrust values and consequently discrepancies in propellant component consumption. As a result, over the same period of time, each of the strapon boosters would consume a different amount of oxygen and kerosene. When we calculated all the figures, we were horrified. By the time the first stage shut down, the residual propellant discrepancy would reach tens of metric tons, endangering the missile structure and controls with asymmetrical loads and outright loss of range. It turned out that even when the engines were painstakingly matched up in sets with identical characteristics, we failed to use tens of tons of precious propellant components. Missile specialists had never before experienced such problems. We guidance designers came to the aid of the engine specialists and claimed that we could synchronize the consumption of the propellant components from all of the strapon boosters if we were allowed control over the total consumption and the ratio of kerosene-oxygen consumption in each engine. Such a system proved to be essential.

Once again the adage, "no good deed goes unpunished" held true. We were not only granted permission, but also obligated to develop a system to regulate the propellant component consumption ratio and to synchronize the consumption between all the strapon boosters. And for good measure, this proposal was reinforced, as was the custom at that time, by a resolution of the Central Committee and Council of Ministers.

OKB-12, headed by Aleksey Sergeyevich Abramov, developed the electronics for this system. This was the same Scientific Institute of Aircraft Equipment (NISO) that I had worked with during the war.[5] I flew to Germany in 1945 for the first time with General Petrov, who was then NISO chief. Gleb Maslov, an experienced aircraft instrumentation specialist, was involved with the system's theory and the design of its electronics. He was adept at critically interpreting a problem and embodied the qualities of a theoretician, a designer, and a tester. In Maslov we acquired yet another reliable subcontractor and good comrade, with whom we always found common

---

5. NISO—*Nauchnyy institut samoletnogo oborudovaniya*. In 1964, NISO and OKB-12 merged to become the new NII Priborostroyeniya (Scientific-Research Institute of Instrument Building).

ground in the future during extremely critical situations during flight tests.

The new system was called SOBIS, the Tank Depletion and Synchronization System.[6] Within the Academy of Sciences, the Institute of Automation and Remote Control was assigned to develop a theory for regulating the engines for operation. There, the young scientist Yuriy Portnov-Sokolov studied these problems.

Much of our manpower was devoted to the study, design development, and testing of sensors measuring the levels of liquid oxygen and kerosene in the tanks. Konstantin Marx [Marks], famous for his inventiveness, was responsible for this development. He had an excellent grasp of the theoretical bases of electrical engineering and was famous for the engineering art of transforming his ideas into actual instruments.

After numerous experiments to select the principle of measurement, we settled on capacitance discrete sensors. However, it turned out that the task of positioning points for discretely recording the propellant levels was anything but cut and dried, partly due to the special features of the tank design and flight program. When something was out of whack when they were fitting the level sensors for each tank, wisecrackers never passed up an opportunity to joke, "Even Karl Marx doesn't know the answer."

The development of a fundamentally new system, both in terms of tasks and execution, always involves a tremendous amount of trouble, but the SOBIS became one of those systems that were incorporated and essential to missile technology. It would have been impossible to imagine the R-7 missile without automatic controls that both optimized the propellant component consumption ratio and thrust of the engines and also synchronized consumption among the strapon boosters.

Having mentioned Maslov and Portnov-Sokolov, I would like to note that our acquaintance was not limited strictly to the job. Maslov's wife, an artist, painted several portraits of Korolev, which after his death adorned the interiors of our institute and other firms. Portnov-Sokolov and I also shared a hobby, a passion for kayaking. Now that we no longer have the physical strength for such voyages, we are forced to limit ourselves to the fond memories of our paddling expeditions.

<u>Problem number three.</u> Not one of the cluster layouts proposed in the draft plan proved reliable when integrated with the proposed launch facilities. Beginning with the A4 (R-1), we were accustomed to a free-standing missile launched from a pad. But how was one to erect a cluster of five missiles on the pad without it falling apart? In this configuration, the load on the aft section was so great that the structural reinforcement required to ensure support exceeded reasonable limits. According to our calculations, given a wind speed of up to 15 meters per second, the cluster's tremendous "sail effect" (the width of the cluster at the aft section was 10 meters) generated loads that threatened to knock the missile off of the pad. Korolev asked Barmin to

---

6. SOBIS—*Sistema oporozheniya bakov i sinkhronizatsii.*

From the author's archives.

Vladimir Barmin (1909-93), shown here in a picture from the 1980s, was the Chief Designer responsible for the majority of Soviet missile and space launch complexes. The organization that he founded in 1946 is now known as the Design Bureau of General Machine Building (KB OM).

design a wall around the launch area to protect against the wind. Barmin firmly refused this job saying, "Building the Great Wall of China around the launch pad isn't in my job description."

At Barmin's design bureau they were working at full steam to design a "carriage" that would haul the assembled missile out from the Assembly and Testing Building (MIK) in a vertical position and erect it on four pads, one for each strapon booster.[7] Few were inspired by this idea. It was complicated and expensive. Additionally, when the designers calculated the possible overturning moment due to variation in the strapon boosters' absolute thrust and added it to the possible wind-generated loads, they were convinced that they couldn't possibly manage without the Great Wall of China to surround the whole vehicle. At the same time the very idea of a wall provoked so much valid opposition that the general consensus boiled down to the short phrase, "It can't go on much longer like this." The situation was critical.

The cluster's load-bearing system had been selected so that in flight, the stress from the thrust of the strapon boosters would be transferred to the central sustainer through the upper load-bearing connecting points. The side boosters would drag the entire cluster as they rested on the "waist" of the central second stage. This configuration proved to be optimal for flight conditions. The principle of connecting the strapon boosters with the central sustainer, the transfer of stress to it, and the subsequent nonimpact separation procedure, that is, the breakup of the cluster so that the central sustainer could easily proceed without any danger of collision, all of that was cleverly and creatively invented and developed. Pavel Yermolayev headed the design group that concocted all of this in the department that first Konstantin Bushuyev and then Sergey Kryukov directed.

Korolev kept a watchful and hypercritical eye on the proposals for the cluster assembly process and the separation system. Mishin, who even as a student had

---

7. MIK—*Montazhno-ispytatelnyy korpus.* The MIK was the Soviet equivalent of the American Vehicle Assembly Building (VAB).

loved offbeat proposals in practical mechanics, devoted a great deal of attention to this problem. Along with such innovations as control chambers instead of control surfaces, these problems began to spill over into heated discussions among the Councils of Chiefs.

Someone came up with the idea of doing away with launch pads and creating conditions close to flight conditions for the missile on the ground. Instead of erecting the missile on a pad, it would be *suspended* in a launch assembly resting on its trusses in the same location to which the strapons' stresses were transferred, that is, at the point on the waist of the central sustainer where the tapered points of the strapon boosters met. If the historic repercussions of these decisions had occurred to anybody back then, and if inventor's certificates had been issued in the names of the design's creators, then Mishin, Yermolayev, and Kryukov would have had to be mentioned at the top of the list. Their proposal could also overturn the developments on which Barmin had already expended so much effort. The ground crews continued to defend their position, that is, resting the aft compartments of the strapon boosters on the launch assembly.

Korolev instructed Mishin to report his and his associates' new revolutionary ideas to the Council of Chiefs and to Rudnev, who at that time was a deputy minister of the defense industry under Ustinov and responsible for fulfilling the resolution on the development of intercontinental missiles. NII-88 was once again subordinate to Rudnev.[8] With his participation, the Council of Chiefs reviewed the new and unconventional proposal for the R-7 launch system.

Mishin's report was passionate. He proposed assembling the cluster not vertically, but horizontally in the Assembly and Testing Building. The assembled missile would be transported to the launch site in its horizontal state and then raised, and rather than being mounted on a pad, the entire cluster would be suspended in the launch system by the load-bearing mechanisms on the strapon boosters where they would be attached to the central sustainer. Using this approach, it was proposed that the bottom portion of the missile be lowered, since the launch pads had been eliminated. Now the launch system trusses would bear the wind loads, but the missile structure would not have to be strengthened; only flight loads were taken into consideration. In this conception, Barmin would develop a simpler integrated transporter-erector assembly. The Great Wall of China would not be needed.

The creators of the launch facility have every right to be very proud of the unique engineering originality of the system which they created in 1955. The strapon boosters on the launcher were suspended on the support booms by their nose cones, while the central booster rested on four points on the spherical heads of the strapon boosters' nose sections. The design prevented the radial crushing forces from being transferred to the missile. During missile launch, the support booms would track

---

8. Rudnev had headed NII-88 in 1950–52, before moving to his ministerial position under Ustinov.

the movement of the missile. After the support boom heads emerged from the special support recess in the nose sections of the strapons, the support booms and trusses would be jettisoned, swiveling on the support axes and freeing the way for the missile to lift off. During launch, the missile and the launch facility formed a single dynamic system. The missile's movement could not be analyzed independent of the launch facility. The dynamics of the moving parts of the launch system, in turn, could not be studied without analyzing the missile's behavior.

In terms of the distribution of responsibilities in Korolev's OKB-1, missile motion control problems fell under the jurisdiction of Department No. 5, the guidance department. As I have already written, Mikhail Yangel was the first organizer and chief of the department, but when he was promoted to a higher post, I became department chief. While still working with Yangel, we began to put together a dynamics sector, which according to our plan would be used to study the theory of motion control problems during all phases, give recommendations to our design departments, and skillfully draw up the specification for the chief designer of the control system.

Georgiy Vetrov was appointed sector director. Years later he would make a name for himself through his historic research of Korolev's work.[9] We sent young specialists to work in Vetrov's section, and when possible, chose specialists with the broader educational background granted by a university. In 1952, Igor Rubaylo joined the sector. He had graduated from the physics and technology department of Moscow State University. This was the first graduating class of the department, which, at the initiative of the Academy of Sciences, was converted into the Moscow Physics and Technology Institute. To this day I have the honor of being a professor of that institute.

In 1955, Rubaylo headed a group of university-educated theoreticians. The group included Nellya Polonskaya, who had graduated from the Moscow State University mechanical engineering and mathematics department, Leonid Alekseyev from Rostov University, and Yevgeniy Lebedev, a graduate of Gorky University. By today's standards, these specialists were still quite young and "green." But back in the 1950s their youth gave them no special advantages.

Rubaylo's group was tasked to perform a theoretical analysis of the dynamics of the missile's behavior in conjunction with the launch assembly. Just a year and a

---

9. After his retirement as an engineer, Georgiy Stepanovich Vetrov became a historian of the Soviet space program. Because of his unprecedented access to formerly secret archives, his publications opened the way for future generations studying the topic. He also helped compile two remarkable collections of formerly classified documents on Korolev's legacy in: M. V. Keldysh, ed., *Tvorcheskoye naslediye Akademika Sergeya Pavlovicha Koroleva: izbrannyye trudy i dokumenty* [*The Creative Legacy of Academician Sergey Pavlovich Korolev: Selected Works and Documents*] (Moscow: Nauka, 1980); and B. V. Raushenbakh and G. S. Vetrov, eds., *S. P. Korolev i ego delo: svet i teni v istorii kosmonavtiki: izbrannyye trudy i dokumenty* [*S. P. Korolev and His Affairs: Light and Shadow in the History of Cosmonautics*] (Moscow: Nauka, 1998).

half after graduating from the university, Yevgeniy Lebedev had developed a simple, very effective, and original procedure for calculating the movement of the R-11FM missile relative to the launch assembly, that is, the "elbows-and-knees" contraption used when launching from a rolling submarine. He was the only "theoretician" who had been involved in the testing of a submarine launch system during the developmental testing of the design on the special rig at Kapustin Yar. This work had been a sort of proof for the planners and designers that design intuition and many years of experience were good, but if they were bolstered by calculation based on theoretical mechanics, then it was even better.

Among the theoreticians at NII-88, and then at Korolev's OKB-1, the ballistics specialists enjoyed the greatest prestige. And it is easy to understand why. The OKB-1 ballistics experts, headed by Svyatoslav Lavrov and Refat Appazov, had started their careers in missile technology in 1945, when the *Sparkasse* (Savings Bank) theoretical design group was organized at the Institute RABE.

It was self-evident that without the ballistics experts, not a single missile would fly. Dynamics specialists were still a long way from being recognized as being similarly essential, especially since, according to the distribution of duties among the chief designers for the missile's inflight stability, Pilyugin was supposed to be responsible for the dynamics of motion relative to its center of mass.

Things were simpler when missiles were launched from a simple pad. The complex launch systems of the first R-11FM naval missile and the first R-7 intercontinental missile, however, required joint design and experimental analysis of the missile's dynamics in conjunction with its launch assemblies.

Pilyugin declared with good reason that this was not his concern but rather Barmin's. Officially, however, Barmin was not responsible for the missile's behavior. If it had not been for Korolev's sense of responsibility, a "no man's land" might have developed. Without reservation, Korolev decided that OKB-1 would take responsibility for the end-to-end solution of the problem. Korolev's deputy for all "ground" issues, Anatoliy Abramov, together with our dynamics experts, drew up the specifications for Barmin and participated in formulating the launch system.

Theoretical analyses and deliberations over the drawings could not provide complete confidence that the selected layout was reliable and free of design errors and that all the dynamic parameters had been correctly selected. In those days, we were not confident about the effectiveness of computer simulations. Besides, specialists needed confidence in the design before beginning the enormous task of assembling the entire launch facility. We needed a direct experiment before the assembly of the launch system began at the firing range. The question was, where and how could we simulate an R-7 missile launch? Our search led to the Leningrad Metal Works (LMZ), then still named for I.V. Stalin.[10] In the enormous building where they

---

10. LMZ—*Leningradskiy metallicheskiy zavod.*

assembled major caliber gun turrets for naval ships we found suitable height, the proper depth, and all the requisite hoisting cranes.

In June 1956, the launch system, instead of being sent to the firing range, arrived at LMZ, where it was assembled and optimized under the leadership of Barmin with the participation of the factory workers. Specifically for this experiment, our pilot factory fabricated full-scale mock-ups of all the boosters of the missile, which were assembled at LMZ. In Leningrad, the missile met up with its launch assembly for the first time. After the assembly was erected, test "launches" began. Instead of engines, the factory crane lifted the missile.

Korolev sent Anatoliy Abramov and Yevgeniy Shabarov to Leningrad to help Barmin set up the labor-intensive experiment. A month before the missile liftoffs at LMZ, Igor Rubaylo was sent to make sure that sufficient measurements were taken and to analyze the launch dynamics.

The experimental work at LMZ in the winter of 1956 had surprising outcomes. In the 1994 publication of my memoirs, at Yevgeniy Lebedev's prompting, I described the experiment of 40 years before.[11] Later, in 2001, Igor Rubaylo, who was Lebedev's boss in 1956, supplied a different version. In the late 1950s, I visited LMZ only once and was not able to recall whether I had seen Rubaylo or Lebedev there. Abramov, Shabarov, Barmin, and other living witnesses were gone. What is a memoirist or historian to do in such a situation?

For this new edition, I have decided to cite the description of events from late February to March 1956, according to Rubaylo's version. This version seems more credible to me than the one in the first edition of my book. I cite the description of operations at LMZ based on the text that was kindly passed on to me by Candidate of Technical Sciences, senior scientific associate, and distinguished RKK Energiya specialist, Igor Rubaylo.[12]

"Two liftoffs were performed. The first one was conducted on the unfueled missile (mass = 26 metric tons) without simulating the combined effects on the missile of the disturbance forces, moment generated by wind, and the moments resulting from the thrust differentials between the various Strapon Propulsion Units (BDU).[13] The second liftoff of the missile filled with water (mass = 260 metric tons) was performed with lateral force exerted on the missile from a weighted cable connected through a pulley.

Already as they were lifting the unfueled missile, the support booms came out of the missile by themselves (after each boom pulled out, the total vertical reaction on the missile decreased, it hovered, then after several tenths of a second, the next boom pulled out, etc.).

---

11. B. Ye. Chertok, *Rakety i lyudi* [*Rockets and People*] (Moscow: Mashinostroyeniye, 1994), p. 400.

12. RKK Energiya—*Raketno-kosmicheskaya korporatsiya 'Energiya'* (Rocket-Space Corporation Energiya or RSC Energiya). Energiya is the current name and incarnation of Korolev's old OKB-1.

13. BDU—*Bokovaya dvigatelnaya ustanovka.*

The technical documentation on the launch assembly noted that the hydraulic mechanisms connecting the support booms and trusses should ensure a strictly simultaneous separation of all four trusses (the specific value of the possible allowable time difference in the pullout from the missile of the various support trusses was not indicated), regardless of disturbance forces and moments; only this would ensure launch safety. Therefore, all those involved with launch safety were seriously concerned, even panicked, over the actual scene that played out during the first 'liftoff.'

After the first liftoff, it became clear to me that the buckling of the missile and the launch assembly during the fueling process—and the BDU thrust achieving the first intermediate level—might seriously affect the launch dynamics. At my request, before the second liftoff, they arranged to measure the displacement of the missile's lower section from the moment they started filling the missile with water to when the last support boom pulled out. These measurements later made it possible to take into account the buckling of the missile and launch system during launch and to properly adjust the elements connecting the missile and the launch assembly in the lower load-bearing zone, and to correctly identify weak points in plug connectors and other types of connectors when preparing the missile for launch

In order to attempt to ensure that all the trusses of the upper load-bearing zone released simultaneously as stated in the launch assembly blueprints, before the second liftoff, we increased the pressure in the hydraulic mechanisms connecting the support booms and trusses. However, due to limitations in the allowable vertical response of the support booms to the missile, this pressure increase could not be significant, and only around 10%. As a result, during the second liftoff we observed the same scene as during the first. We were now faced with an extraordinary situation.

Because of the unexpected test results during the first liftoff, S. P. Korolev, Deputy Minister G. R. Udarov, A. P. Abramov, Ye. F. Lebedev, G. S. Vetrov, specialists from V. P. Barmin's design bureau, and rather high-ranking military acceptance officials had come to the second liftoff at LMZ.[14] At the beginning of the meeting to discuss the liftoff results, V. P. Barmin took the floor. There was a sense of bewilderment in his talk that reflected the general feeling among those responsible for launch assembly development.

In his first speech, Barmin said nothing about possibly modifying the launch assembly and then testing it. He said only that according to information he had, our nation was approximately a year ahead of the U.S. in the development of an intercontinental missile. If even several of the first launches of the missile were unsuccessful, but at the same time the launch assembly was not harmed, then we would lose little ground in the race with America. If, however, the missile got caught on the launch assembly, fell over, and exploded at the launch site, then our time advantage over the Americans would disap-

---

14. Grigoriy Rafailovoch Udarov (1904–91) served as deputy minister of machine building in 1956–57, the ministry that supervised Barmin's design bureau. Later, he was a deputy minister at the Ministry of General Machine Building (MOM), which oversaw the Soviet ballistic missile and space programs.

pear. He asked all those attending the meeting to keep that in mind.

A. P. Abramov, who spoke after Barmin, attempted to postpone discussion and a decision on the matter in order to have time to calmly think through the situation and find a solution. He said that in view of the limited scope of the measurements and their poor quality, it was impossible to analyze the results within a short period of time. However, the measurement specialists' team leader Lyudmila Georgiyevna (I don't recall her last name) quite decisively rejected his claims against the measurements. She announced that not only had all the measurements written in the program been performed and interpreted, but also additional measurements had been inserted in the program between the first and second liftoffs at the request of OKB-1 representative Rubaylo.

Not one of the attending specialists who had worked on the launch assembly design and calculated the missile's motion dynamics in relation to the launch assembly elements had any specific proposals. Then S. P. Korolev took the initiative. He consulted briefly with his specialists and asked if it might be possible within a certain period of time working jointly with Barmin's representatives to find some solution. We could only answer, "We can probably come up with something." Then Korolev asked Barmin.

'If we complete the launch assembly tests at LMZ now, and make whatever modifications are required during the couple of months while they're getting the firing range ready to begin assembling the launch system, then would we be able to carry out the same launch simulation at the firing range that we did at LMZ?'

After consulting with his specialists, Barmin replied that, if the missile needed to be modified, simulation could be performed using the erector that transports the missile from the Assembly and Testing Building, mounts it on and removes it from the launch assembly. At that point, Korolev proposed writing down the following in a protocol:

- complete phase of tests on article liftoff from launch assembly at LMZ; disassemble launch booster assembled at LMZ and deliver to firing range;

- perform system modification by March [1957]; eliminate operational deviations;

- at firing range begin final phase of construction and equipment assembly at launch site and, after assembly of the launch booster, conduct additional launch simulations at the firing range using an erector and perform all necessary measurements.

The meeting participants affiliated with Barmin and OKB-1 agreed with this proposal, but then their senior military representative, Colonel Yuriy Fedorovich Us, took the floor. Holding in his hands the design draft of the launch assembly, he quoted: 'Under exposure to wind effect and perturbing factors, launch safety is ensured when the hydraulic mechanisms in the upper load-bearing zone [of the strapon boosters] ensure a strictly simultaneous separation of the support trusses.' He said that he could not sign the protocol on the completion of tests at LMZ since it was unknown whether a way would be found to ensure that the trusses would separate simultaneously, thereby guaranteeing launch safety.

Korolev, however, knew how to convince the colonel that if he refused to sign the protocol, then he would lose his prestige with army and national leaders, since "the brass" would sooner agree with Korolev, Barmin, Udarov, and their specialists. Colonel Us gave in and signed the protocol. The protocol was signed and approved. It wasn't long before *the launch system was disassembled and shipped off to the firing range. However, no matter how much Barmin's and our specialists racked our brains, we couldn't come up with something radical that would guarantee the simultaneous separation of the trusses stipulated in the design documentation. The launch assembly had already been fabricated, and the parameters of the missile and the propulsion systems had been nailed down."*

Problem number three was granted the right to a definitive solution at the firing range.

Problem number four. Varying thrust from the strapon boosters during the buildup could lead to very great destabilizing moment values. Because the strapon boosters were not rigidly attached longitudinally to the central sustainer, it was possible for any strapon booster to come off the cluster at launch if the thrust of its engine was less than the others. This meant an inevitable disaster and launch pad destruction.

Korolev demanded that Glushko synchronize the thrust of all the engines during their build-up. Glushko categorically refused. Indeed, our synchronization system was designed to regulate thrust in flight at steady-state output. The engine specialists were unable to control the buildup transient. Neither full start, nor slow build-up, which causes a hang-fire liftoff, solved the problem. The principles of the selected launch system had to be changed. One idea was to conduct a "forced" launch. To do this they would have to somehow hold the central sustainer by the "tail" until it was certain that all of the strapon boosters had achieved buildup. When the total thrust substantially exceeded the weight of the cluster, the command would be transmitted to the locks, unlocking the center's "restraint," and the missile would suddenly take off.

A second alternative proposal was to use a special automatically generated launch sequence. First only the strapon boosters would fire. They would be allowed to build up sustained thrust to an intermediate stage that was less than the weight of the entire cluster. Here, the destabilizing moment resulting from the thrust variation at the intermediate stage would be counteracted by the reactions of the launch system supports. The central sustainer would be permitted to build up to full thrust after the stable operation of all the strapon booster engines had been electrically monitored. As the central sustainer engine gained thrust, the missile would begin to lift off and it would safely separate from the launch system. In flight, the strapon booster engines would build up to full thrust nominal mode. This second proposal was thoroughly calculated and analyzed. But Glushko's consent was required to introduce the special intermediate stage and delay during the central sustainer engine's buildup to full thrust.

There was no unanimous opinion on the selection of the alternative version. In Department No. 5, all those interested in making the final decision for Korolev's subsequent approval assembled in my office: Kryukov, Bushuyev, Voskresenskiy, Abramov, Shulgin, Yermolayev, Vetrov, Rubaylo, and Lebedev. The very makeup of the meeting underscored the importance of the issue under discussion. All the participants, with the exception of Kryukov, favored the simpler version, which to us also seemed the more reliable one: the forced retention of the missile by the "tail" of the central sustainer and the opening of the locks by electrical command after the steady buildup of the side units. Kryukov said that Korolev might have another opinion.

When we reported to Korolev, it turned out that Kryukov was right. Korolev rejected the version calling for the missile to be held by the tail, arguing that it was unreliable and required the development of a complex electromechanical device. It turned out that he and Glushko had already talked via the "Kremlin hot line" and Glushko had given his consent to introduce the new intermediate stage. Soon we heard Korolev's routine command, "full steam ahead!" The true author of the new dynamic launch sequence was still unknown.

I am writing about this at such great length, knowing that I risk wearing the reader out with technical details. I am trying to show that when a large group of people is involved in intense creative work—during the course of which lots of problems arise requiring innovative and unconventional thinking—the names of the actual authors, the ones who were the first to express the idea that saved the day, are usually lost. In such situations, only the shamelessly immodest and particularly ambitious filled out inventor's certificates, and as a rule they would invite their immediate supervisor to be co-author.

Later, plans, that is, target figures, were sent out to the departments for inventor's certificates. So as not to be among those lagging behind, the departments strove to lay a claim to any sort of claptrap before the All-Union Committee on Inventions.[15] But during those gung-ho Korolev years of the birth of the R-7, such activity was viewed as a distraction from our primary work and was not encouraged.

It was up to us guidance specialists to develop the launch automatics for the new system. The process required a "cautious" launch sequence for all the engines, beginning with the purging operations, ignition, build-up, and escape from the launch system. The entire sequence of operations, which was rather complex for those times, was to be executed by the control system, with many protective interlocks.

The engineering teams of Korolev, Barmin, Glushko, and Pilyugin worked extremely closely. Despite constant quibbling on hairsplitting issues, a general atmosphere of truly creative enthusiasm prevailed. Staying late into the night, in Podlipki, Khimki, or on Aviamotornaya Street, we discussed a wide range of processes that culminated in the decisive moments of the launch. The gas dynamic processes in 32 engines needed to be tied into a single monitored sequence with the missile's motion dynamics and the launch system mechanisms.

We had the feeling that we were working on the creation of some kind of ani-

---

15. During the Soviet era, the All-Union Committee in the Sphere of Science and Inventions, offered services similar to a patent registering body in the West. The committee (which held many different names at different times) was subordinate to the USSR Council of Ministers.

mate, anthropomorphized system, and not on a purely electromechanical structure. Thus, the commands now familiar to millions of television viewers "Key to ignition" and "Launch" were born long ago during that inspired technical outburst.

Before the very end of 1956, work at LMZ continued nearly around the clock to optimize the missile and launch system. In the process, the number of drawing and design errors, as well as all kinds of glitches found in the operational documentation exceeded several hundred.

At LMZ, hundreds of designers, assembly personnel, design engineers, and servicemen swarmed about filing, welding, reassembling, and writing, debating, and deliberating. They had completed almost six months of work in Leningrad. The launch equipment had been disassembled and shipped to the new firing range for final assembly. There, on the firing range launch system they would have to run and test hundreds of electric cables and pneumatic and hydraulic lines connecting the missile's systems with the ground testing equipment during the preparation process.

The first firing tests of the individual boosters on the rig that had been reconstructed at NII-229 showed how difficult it was to anticipate everything without testing. The beginning of firing tests in 1956 on only individual boosters immediately revealed many defects. Preparing for the general firing tests of the whole cluster of boosters, scheduled for early 1957, was like preparing for a final exam.

Problem number five: Production. For a single R-7 missile launch, it was necessary to manufacture five boosters, each of which surpassed the former single-stage missiles in terms of labor intensity. Each booster was tested independently. Next we assembled the cluster and conducted many days of horizontal tests on it in the new assembly building. Assembly and testing shop No. 39 became the factory's most popular shop, and assembly foreman Vasiliy Mikhaylovich Ivanov became the most esteemed shop foreman.

Guidance specialists were the most essential specialists there. Without the presence of control system design engineers, the electrical tests at the factory's controlled-testing station did not go well at first. The testers and systems developers were merged into integrated brigades and jointly worked out the test process that later had to be transferred to the firing range. Officers from the new missile unit from the new firing range also participated in this work. During World War II, fold-away beds would be placed right in the shop locker rooms in factories so that workers could take short rest breaks. In shop No. 39 they recalled this, and added a peacetime upgrade in comfort. They furnished bedrooms for the testers so that those living far away could sleep right there at the factory.

The first missile cluster for tests on the firing rig at NII-229 and the second standard cluster for the first launch were released in December 1956. The factory manufactured a full-scale test cluster before that and shipped it to Leningrad and then to the firing range.

Many new engineering processes were adapted at the factory for the sake of the R-7. A new instrumentation production building was built and equipped. They created a separate clean room that was pristine by the standards of that time.

Instead of the usual 4 control surface actuators on a missile, each R-7 required 16! And they were all structurally new, were more powerful, and had redundant electrical systems. The new control surface actuators, new SOBIS instruments, APR, and measurement systems required the development of new test consoles, instructions, and assembly guidelines. Transistor circuits were also used for the first time. We sent a stream of new drawings to the factory. Another stream of comments would come back, saying "that won't work." Hundreds of change notifications were issued, causing deadlines to be missed. I was torn between the factory and my departments and subcontractors. I'm not about to hide the fact that no one had to twist my arm to make me fly to Kapustin Yar, where the R-5M tests were completed in 1956. And we still had to test the M-5RD and R-5R.

Problem number six. According to the most optimistic calculations, a two-stage missile consisting of five smaller missiles had to be five times less reliable than a single missile! Except for the R-5M, a single failure in the control system usually led to one accident or another on all of our missiles. Consequently, if the reliability of each booster was even brought to 0.9 (90%), then according to the probability theory, the reliability of the entire cluster would be $0.9 \times 0.9 \times 0.9 \times 0.9 \times 0.9 = 0.59$, or 59%! But that result needed to be multiplied at least two more times by a factor of 0.9, taking into consideration the reliability of the interunit mechanical, electrical, and dynamic linkages in the cluster itself and the reliability of the launch system, which was a very complicated mechanical complex with hundreds of electrical and hydraulic lines. Given all that, we arrive at the absurd value of 0.425 or 42.5%. Thus, according to optimistic calculations, using fundamental concepts of probability theory, we calculated that out of every 10 missiles, at least 5 would strike the wrong target.

Those systems that were in any way related to electricity were in the most vulnerable position for failure. We started to provide backup everywhere we could. Here, for the first time, besides simple redundancy we used "voting" principles at the most critical points. Such systems are broadly used today; they are called majority voting. For example, we installed three longitudinal acceleration integrators. An integrator transmitted the engine shutdown command only after receiving two confirmation signals. The failure of one of the three instruments was allowed. It was rather simple to use the "two out of three" principle in relay contact circuits. It substantially increased reliability but complicated preparation and testing. We had to make sure that we were sending into flight a missile that had all three voting instruments or systems that were fit as fiddles. In those places where voting didn't work, we were limited to redundancy. Each chief designer of each system was required to strictly follow the principle that any single failure anywhere in any instrument must not lead to a system failure. This was so much easier said than done. It was even more difficult to verify that, given any failure—such as a breakage or a short circuit—there really would be no system failure.

On the first test equipment sets—still not yet the flight models—we began to experience failures that we called "foreign particle" failures. The designer was forgiven for them, but production would hear about them. This mysterious "foreign particle" had a knack for shorting out two contacts in close proximity to one another in an instrument. This led to the most surprising effects. A particle would get into the slide valves of control surface actuators, causing them to deflect the control engines to the maximum extent without receiving a command. A "foreign particle" even found its way under a valve seat, causing the valve to continue to "let off" high pressure when it wasn't supposed to. These particles could explain more than 50% of all the glitches that we had chalked up on the ground during testing and preparation. We waged a relentless struggle for cleanliness and production discipline. Alas, the strictest orders in this regard could not have changed the situation in one or two years.

All of the teams were really governed by a genuine, sincere desire to do good and conscientious work. Although there were no pep talks for workers, they were very enthusiastic in their work, all of them united in tackling a problem that just might decide the fate of humankind. And for all that, minutiae like "foreign particles," dirty contacts, and loose connectors were capable of nullifying the work of thousands of people and wasting unknown billions of rubles during the last phase.

The engine specialists were in the worst situation. After all, it was impossible to provide redundancy for an engine and its hydraulic fittings. But even if you could imagine that you might sometime succeed (and later on the N-1 rocket we really did succeed, and this is also done on spacecraft), then a different hazard would appear. For example, for inexplicable reasons the engines had the habit of switching from a normal vibration mode into a high-frequency mode. Usually, the high-frequency pulsations led to the explosion of the combustion chamber and an engulfing fire. Here, increased redundancy did not lead to increased reliability! Then we understood the need for—and began to demand from ourselves and our subcontractors—the most thorough, multistage, and comprehensive ground optimization.

In addition to ground optimization, we carried out experimental missile launches. One such experimental missile was the M-5RD, essentially an R-5 missile on which we tested out both the principle and equipment for regulating the engines for the R-7 and new inertial navigation equipment. The R-5 missile used a new automatic stabilization control system that used a system for correcting the missile's center of mass position based on information from off-range and lateral acceleration transducers. To optimize the trajectory and increase range accuracy we tested an Apparent Velocity Regulation (RKS) system on the M-5RD. This system's sensors acted via amplifiers on the drive regulating engine thrust. On this same missile we checked out the operating principles of the tank depletion control system, the fuel and oxidizer tanks' liquid level damping system, and the slosh amplitude measuring system. In all, five M-5RD missiles were manufactured and launched. The launches took place at GTsP from July through September 1956.

August through September was considered the "mild" season for the Kapustin Yar region. The heat subsided; magnificent Astrakhan tomatoes and the season's first watermelons began to appear everywhere. On top of all that, the fishing was superb and the living conditions were quite tolerable. It was no wonder that the number of people who wanted to improve the reliability of the future R-7 by participating in experimental M-5RD launches was always more than necessary.

The second experimental rocket, the R-5R, was developed on the basis of the phase three R-5 missile as a result of the special government decree issued on 20 May 1954 and was designed to test the principles of radio control. Four missiles were prepared. I spent from May through June 1956—glorious spring and summer days in Kapustin Yar—at the launches of this experimental missile in the company of Boris Konoplev, who represented the interests of NII-885, and Yevgeniy Panchenko, who was then an engineer captain and representative of the Main Directorate of Reactive Armaments and would later become a general.[16] The primary goal of the tests was to check out the principle of the radio measurement of missile velocity in pulse operating mode in the super high-frequency (centimeter) band and to determine the effect of the engine exhaust jet on the operation of the SHF interrogator and responder links. The program consisted of three launches. I dissuaded Konoplev from insisting on a fourth. Korolev supported me.

Problem number seven: The firing range. The selection of a firing range for the testing of intercontinental missiles proved to be anything but simple. The R-7 draft plan specifically called for a radio-control system. At the request of Ryazanskiy, Borisenko, and Guskov—the primary system developers—two Radio-Control Ground Stations (RUP) had to be placed symmetrically along both sides of the launch area at a distance of from 150 to 250 kilometers.[17] One of these two stations was the main base station and the other was the relay station. For accurate range control, a third station was needed, situated 300 to 500 kilometers from the launch site. This station would take precise missile velocity measurements using the Doppler effect and would issue engine shutdown commands when the design values were reached.

Thus, as we used to say, the launch site had a "radio moustache" and a "radio tail." Immediately after launch, there had to be a direct line of sight between the radio-control station antennas and the onboard antennas mounted on the second stage. For that reason, the use of mountainous terrain was ruled out. The second condition was the need to expropriate the land in the areas where the first stages might land. The flight path had to pass without encroaching on large populated areas so that, in the event of an emergency shutdown of the engines, the missile

---

16. Boris Mikhaylovich Konoplev (1921–60) was one of the leading guidance systems chief designers in the Soviet missile program and worked in several different institutions such as NII-20, NII-885, NII-695, and OKB-692. He was killed in the so-called Nedelin Disaster in 1960. The Main Directorate of Reactive Armaments (GURVO) was the procurement agency for missiles within the Red Army's Main Artillery Directorate (GAU).

17. RUP—*Radioupravlyayushchiy punkt*—literally stands for "Radio-Control Point," but more generally denotes "Radio-Control Ground Station."

impact would not cause harm. But the most important requirement was that there must be, at the very least, 7,000 kilometers between the launch site and the warhead impact site.

The selection of the flight path and firing range area was traditionally the military's business. But Korolev could not to come to grips with the fact that this was to be done without his or his deputies' input. He assigned Voskresenskiy to participate in this operation, and authorized me to handle conflict resolution, if any should arise over the placement of RUP sites.

It was natural that Voskresenskiy, after poring over maps, was inclined to put the beginning of the flight path at good old Kapustin Yar and the end at Kamchatka.[18] That provided a range of 8,000 kilometers but put the impact fields for the first stage strapon boosters over population centers; additionally, one of the RUPs would have to be located on the Caspian Sea, and then in Iran. We shifted over the map to the Stavropol Territory.[19] After determining that the first stage impact fields fell on the Caspian Sea, we dreamed that our future work at the firing range might take place under resort conditions. Now I bitterly recall with what derision the irate team of radio specialists rejected our proposal. Ryazanskiy telephoned Korolev and sniped that he dreamed of conducting launches from the mineral water spas of the Caucasus just as much as Voskresenskiy and Chertok, but his radio link couldn't get through "all those mountains."[20]

In a fit of agitation, Korolev informed us that a reconnaissance commission had been set up to select the firing range site. GTsP Chief Vasiliy Ivanovich Voznyuk would be in charge, "So stop fantasizing." Korolev delegated Voskresenskiy to establish contact with the commission and, to the extent possible, to influence its work so that we wouldn't be driven into the Arctic. Having lost hope for the Stavropol option, Voskresenskiy and I halted our initiative.

Voznyuk's commission studied four options:

- in the Mari ASSR[21];
- in the Dagestani ASSR;
- east of the city of Kharabali in the Astrakhan region; and
- in the semi-desert of Kazakhstan by the Tyura-Tam station in the Kzyl-Orda region on the bank of the Syr-Darya River.

---

18. Kamchatka is a huge peninsula about the size of Japan on the very eastern end of the Russian landmass. Still sparsely populated, the peninsula contains Russia's largest volcanic belt.

19. The Stavropol Territory is located in the northern Caucasus between the Black and Caspian Seas and borders the nation of Georgia on the south.

20. This is a reference to a poem by Mikhail Yurevich Lermontov (1814–41), the leading Russian romantic poet and author of the famous *Geroy nashego vremeni* (Hero of Our Time), published in 1840.

21. ASSR—*Avtonomnaya sovetskaya sotsialisticheskaya respublika* (Autonomous Soviet Socialist Republic)—was the subordinate geographical and political unit within the larger Soviet republics such as Russia or Ukraine.

After heated arguments, reconnaissance flights, and trips to the sites, the fourth option, the Kazakhstan option, was adopted. In our opinion, they should have taken the option in the Astrakhan region that Voznyuk originally proposed. The proximity of GTsP, the already familiar climate, and the Volga delta removed a whole series of problems that would occur when setting up a firing range at a new site.

The fourth option was the most complicated in all respects due to the extremely difficult climatic conditions: temperatures as high as 50°C (122°F) in the shade and dust storms in the summer, and winds with temperatures as low as minus 25°C (–13°F) in the winter. Not only was it desert terrain, but according to health service data, it was a breeding ground for the plague, transmitted by millions of ground squirrels. In no way could we imagine conditions for a "civilized" life. The closest regional centers, Kazalinsk in the west and Dzhusaly in the east, were more than a hundred kilometers away from the potential new housing construction site. The first two to three years at Kapustin Yar, GTsP military specialists and officer staff were accommodated along with their families under very difficult conditions in the cottages of local residents. And still, at the very least, there was someplace to lay one's head, prepare the food, and bathe the children. There was no shortage of fresh fish, black caviar, and watermelons; there was plenty of meat, milk, and vegetables in the collective farm market. And for the provision of the entire garrison, Stalingrad was just 70 kilometers away. But at the newly selected site in Kazakhstan there was nothing, absolutely nothing.

We were supposed to start the R-7 tests in 1957. According to the most conservative estimates, in all, more than 1,000 military and civilian specialists were supposed to participate in them. To the numbers of servicemen, you needed to also add their family members, and over and above that, all the public amenities, medical, cultural, and transportation services. Then we had to figure out how many construction workers were needed for all the aforementioned individuals to have living quarters, roads, production buildings, workshops, and communications systems. Even before site selection, plans for the future included the construction of an oxygen plant, its own powerplant for a reliable power supply, a hospital, bakery, radio stations, tracking and radio measurement stations, etc., etc.

Based on the results of Voznyuk's commission, on 12 February 1955, the USSR Council of Ministers passed a decree approving the site and measures for the construction of Ministry of Defense Scientific-Research and Test Firing Range No. 5 (NIIP-5).[22] This name is long forgotten. The firing range is known to the world today as the Baykonur Cosmodrome.

The name "Baykonur" was created after 1961, when the press needed to refer to a launch site in official communiqués on the latest space triumphs. There really is a Baykonur city located 400 kilometers northeast of the Baykonur Cosmodrome.

---

22. NIIP—*Nauchno-issledovatelskiy ispytatelnyy poligon.*

This renaming was done to "confuse" enemy intelligence services and to keep secret the true location of the intercontinental missile launch site.[23] When, before the next scheduled TASS report, someone proposed designating the new site as Baykonur instead of its true geographical name, not only did Korolev, Keldysh, and the entire Council of Chiefs have no objections, they even supported this sham.[24]

In April 1955, Lieutenant General Aleksey Ivanovich Nesterenko was appointed the first chief of NIIP-5. Before this, Nesterenko had already worked in the missile field as chief of the Academy of Artillery Sciences NII-4. This was the first scientific-research institute within the armed forces dedicated to studying missile armaments. Later he was head of the faculty of reactive propulsion at the Artillery Engineering Academy.

From the author's archives.

Mikhail Melnikov (1919–96) was the leading engine designer in Korolev's design bureau and was responsible for producing a series of reliable engines that were later installed on important upper stages. These include engines for the Blok L on the Molniya booster and the Blok D on both the N-1 and the Proton launch vehicles.

I had been acquainted with General Nesterenko during his stint as NII-4 chief in Bolshevo, which was close to our NII-88.[25] A massive number of multicolored combat ribbons decorated his chest, enabling him in peacetime to rest on his laurels, to bask in the tranquility of a general's dacha and an easy staff job somewhere. It turned out that he belonged to the category of the obsessed; there were quite a few of them in the military.

The assistance of artillery Marshal Nedelin, who at that time was deputy defense minister, as well as Nesterenko's personal connections, contributed to the fact that by late 1956, the NIIP-5 garrison had been staffed with a very good cast of military specialists.[26] I knew some of them from the GTsP; soon I would have to get to know many of them quite well, and then over the course of many years share in the hard-

23. Ironically, U.S. intelligence services had a very good idea about the location of the new firing range by 1957.

24. TASS—*Telegrafnoye agentsvo Sovetskogo Soyuza* (Telegraph Agency of the Soviet Union)—was the official (and only) media agency during Soviet times.

25. Bolshevo is a suburb very close to Podlipki (later Kaliningrad, now Korolev) where both NII-88 and OKB-1 were located.

26. Nedelin served as deputy minister of defense (for reactive armaments) in 1955–59.

ships of work, the joys of our first missile conquests and space triumphs, and the tragic accidents. It seems to me that until recently, our media, which are dominated by the work of professional writers, journalists, and screenplay authors, have failed to appreciate the self-sacrificing work and heroism of the military engineers.

In most narratives on missile technology and cosmonautics—whether nonfiction, memoirs, or fiction—cosmonauts, chief designers and their associates, and flight directors seen in the lush interiors of mission control centers are the ones that stand out. Rarely does an officer standing in a bunker at a periscope or as an extra pressing the buttons on some obscure control panel flash onto the movie or television screen.

Jumping ahead, I will mention that in 1970 Deputy Commander-in-Chief of the Strategic Rocket Forces Colonel General Mikhail Grigoryevich Grigoryev and I agreed to act as consultants for the film *Ukroshcheniye ogonya (Taming the Fire)*, written and directed by Daniyl Khrabrovitskiy.[27] The screenplay included numerous missile launches with their various aftereffects. I insisted that everything should be authentic: the officers and soldiers should be dressed "by the book." Khrabrovitskiy responded, "That's out of the question." To my astonishment, Grigoryev agreed with him. Those who saw this film might have been pleasantly surprised to see the launch site personnel in gorgeous, multicolored costumes that look more like Olympic team outfits. The few initiated mercilessly reviled me, "How could you agree to such a sacrilege?!" But this desecration struck the fancy of Ustinov—Communist Party Central Committee secretary at that time—and the staff of the Central Committee defense industries department, which endorsed the film and permitted its release.[28] Khrabrovitskiy spelled it out for me in no uncertain terms that "if we had portrayed everything as it really was, the film never would have appeared on the screen." I realized that he was right and waved it off; at least, the movie tried to come close to a true portrayal of the technology. In that regard, Khrabrovitskiy certainly succeeded with the frames depicting the failed R-7 launches and blastoffs. I would like to note that for all the plot shortcomings in *Taming the Fire*, to this day it remains the only feature film in which a director tried to portray the creative process of developing a missile in all its dramatic variety.

For us "civilians," our stay at the new firing range with all its hardships and difficulties was a temporary assignment. To a certain extent, it was even a romantic and exotic experience. We knew that in a month or so we would return to the civilized world, where we would find a familiar climate and our dear old central Russia landscapes. If we felt like it, we could go to the Sandunov Baths and on Sunday g

---

27. Daniil Yakovlevich Khrabrovitskiy (1923–80) was a famous Soviet Jewish writer and direc who contributed to a number of films in the 1960s and 1970s. Mikhail Grigoryevich Grigor (1917–81) served as first deputy commander-in-chief of the Strategic Rocket Forces in 1968- Before that he had served as the first chief of the Mirnyy (Plesetsk) firing range in 1957–62.

28. Ustinov served as Secretary of the Central Committee for space and defense industrie 1965–76, that is, as the effective head of the Soviet missile and space program during that period

skiing or kayaking, depending on the season.[29] Military personnel were deprived of these normal pleasures of life. After agreeing to go to the new firing range, the officers and their families had to abandon populated areas and work for several years under extreme conditions and develop virgin land in Kazakhstan that was much wilder than what Kazakhstan's own grain farmers faced.[30]

Upon arrival at their new post, the first officers lived in the old railroad cars of the same special train produced for the military by order of the Institute Nordhausen that had been so useful in 1947 in Kapustin Yar; in 1955 these special trains again came to the rescue, this time in Tyura-Tam. The rank-and-file and noncommissioned officers were housed in tents. During the day, the railroad cars and tents warmed up to +45°C (113°F). Nearby, they built dugouts where personnel could escape the heat during the day. Troop trains were constantly arriving carrying construction materials, military construction brigades, and new officers. Also beginning to arrive were the families, who had no idea of the living conditions. They were all housed in the old railroad cars and hastily constructed dugouts. A lucky few got the first prefabricated huts.

Nearby was the Syr-Darya River, still deep for the time being. Its murky water was, however, not potable. The problem of pure fresh water was one of the most acute. Artesian wells provided brackish water and it needed to be specially treated. Even now, many years after the beginning of firing range construction, the water supply problem for the population and for production needs has not been completely resolved. There have been instances when, staying at the most comfortable cosmodrome hotel, washing up meant sparingly using bottled Borzhomi or Narzan mineral water obtained at the snack bar.

Was it necessary to set up a firing range in such a hell-hole only because originally, according to the map, it was convenient to position three radio-control stations that proved to be superfluous just five years after construction began? I am certain that in 1954, and even in early 1955, if we had had a better feel for the prospects for developing inertial navigation systems, Voznyuk's commission would have selected the Astrakhan region. Now the heroic firing range construction campaigns are the stuff of legends. Under other conditions, without a doubt, everything would have been considerably easier. But every cloud has a silver lining.

The town of Leninsk—recently declassified—sprouted up on the banks of the Syr-Darya.[31] The Baykonur Cosmodrome, the missile firing ranges surrounding it,

29. The *Sandonavskiye bani* (Sandunov Baths) were famous and ostentatiously decorated baths built in the early 19th century on the Neglinnaya River that were very popular with Russian nobility in the imperial era. The baths are still open today.

30. The reference to "virgin lands" alludes to a massive national program initiated in the mid-1950s to plow and irrigate huge portions of Soviet central Asia for grain cultivation. The Virgin Lands project, sponsored and supported by Nikita Khrushchev, ultimately proved to have negative consequences to both the land and its inhabitants.

31. The original settlement known as Zarya was renamed Leninsk on 28 January 1958.

and the combat missile positions spread out many hundreds of kilometers over the vast steppes. Missiles of different designs are launched independently from dozens of launch sites. Sovereign Kazakhstan became the proprietor of an absolutely unique autonomous region called the Baykonur Cosmodrome. After the collapse of the USSR, Baykonur began an uncontrollable process of self-destruction. On the eve of the 21st century, the future of Baykonur, the first space port in human history, seemed uncertain.

However, the example of the northern firing range in Plesetsk shows that problems can also be solved piecemeal.[32] Instead of a single grandiose firing range, larger in area than a country like the Netherlands, it would be better to have several specialized, smaller firing ranges that don't require the expropriation of so much of the Earth's surface and would be less expensive. But what's done is done. Veterans of Tyura-Tam have a right to be proud of their contribution to the transformation of the desert that, in Korolev's vivid words, was the "edge of the universe."

---

32. The original firing range at Mirnyy close to the town of Plesetsk was founded in January 1957 as the Scientific-Research and Test Firing Range No. 53 (NIIP-53). Later, in August 1963, a portion of the range was converted into a space launch center. Since 1966, the majority of the world's satellites have been launched from the Plesetsk site.

Chapter 17

The Birth of a Firing Range

At the time that the decree for the creation of NIIP-5 was issued in February 1955, to the best of my recollection, the chief designers, headed by Korolev and their primary deputies, were at the well-settled Kapustin Yar State Central Firing Range. On 21 January, we successfully began factory flight tests on the R-5M missile. Despite glitches in the flutter of the small and, in my opinion, unnecessary control fins, our mood was optimistic. We modified the fins, increased the rigidity of the drive, and executed two more launches, introducing malfunctions right down to the shutdown of one of the control surface actuators. The missiles reached their targets as if they hadn't noticed the malfunctions we had deliberately inflicted. We felt optimistic. Despite the harsh winter, we lived and worked under what were for those times comfortable conditions.

One evening after a successful launch, we gathered in the hotel's cozy dining room "for managerial staff." Smirking, Korolev raised his champagne glass, announcing that a decree calling for a new firing range in the Kara-Kum Desert had been issued, and that we here were relaxing in Vasiliy Ivanovich Voznyuk's domain perhaps for the last time. Voskresenskiy could not pass up an opportunity to needle Ryazanskiy:

"Mikhail Sergeyevich, this is all because of your demands that we place the radio-control stations hundreds of kilometers from the launch site. We'd just gotten ourselves set up almost like Europeans, and now once again, without a decent night's sleep, we're going to have to shoot out of our tents in the morning and expose our naked butts to the icy wind!"

The next day in the midst of our routine cares we forgot about the new firing range. We didn't know that one of these days in the Kyzyl-Kum Desert, at the heretofore completely unknown whistle-stop of Tyura-Tam, 2,500 kilometers from Moscow and 1,000 kilometers from Tashkent, Lieutenant Igor Nikolayevich Denezhkin had disembarked with a platoon of soldiers from a train that stopped for three minutes. They were the first to ask the railroad employees for lodging and announced that in a day or so hundreds of railroad cars would begin to arrive with cargo and many, many soldiers.

The secret decree from 1955 declared construction of the new missile firing range—the future cosmodrome—to be of paramount importance, ranking with national tasks for the postwar reconstruction of cities and villages destroyed by the war. The governmental decree tasked the Ministry of Defense with construction of the firing range. In prior times, tens of thousands of Gulag prisoners had built closed cities, factories, and nuclear industry silos. But after the elimination of the all-powerful chief caretaker of the nuclear industry, Lavrentiy Beriya, the Gulag empire had crumbled. After the war, the Ministry of Defense not only retained its military construction potential, but substantially increased it. The minister of defense had a deputy who was involved solely with construction. At that time, the deputy was Aleksandr Nikolayevich Komarovskiy, a talented engineer, splendid organizer, professor, and Doctor of Technical Sciences. He was a sort of commander-in-chief of an enormous army of military builders. Almost all the officers of this army were war veterans.

Another Ministry of Defense deputy, Mitrofan Ivanovich Nedelin, acted as the construction "client." During the war he had commanded the artillery of the 3rd Ukrainian Front. In 1953, he took command of all the artillery forces from Chief Marshal of Artillery Nikolay Nikolayevich Voronov and had already anticipated the great future of missile armaments. Colonel Engineer (later Colonel General) Georgiy Maksimovich Shubnikov was tasked with direct supervision of construction in the desert.[1] He was faced with moving from comfortable Tashkent to the desert, where he would take charge of and bear full responsibility for meeting deadlines and ensuring the quality of a great construction project, the importance of which for the future of humankind no one could really imagine at that time.

A few landmarks stood at the tiny Tyura-Tam stop: a small brick building proudly bearing the sign *Vokzal*, a modest water tower for steam engines, a dozen or so trees barely clinging to life, several cottages for the station personnel, and five mud huts with local Kazakhs living on who knows what.[2] Endless desert was all around.

IN THE SPRING OF 1955, an avalanche of freight and people descended on the tiny station. Lieutenant Colonel Ilya Matveyevich Gurovich, who arrived in this desert in April 1955, devoted 20 years of his life to the construction of the future Baykonur cosmodrome. In 1966, with the participation of the Council of Veteran Baykonur Builders, Gurovich's daughter succeeded in publishing the book *Before the First Launch*, at her father's behest.[3]

After retiring from his post as chief builder of Baykonur due to illness in 1975, General Gurovich, a highly cultured man, left behind valuable and vivid reminis-

---

1. Georgiy Maksimovich Shubnikov (1903–65) officially served as chief of the 130th directorate for engineering work of the Ministry of Defense.

2. In Russian, a small train station in an outlying community is a *stantsiya* while a major station serving a city is a *vokzal*.

3. I. M. Gurovich, *Do pervogo starta* [*Before the First Launch*] (Moscow: A.D.V., 1997).

cences about the first three truly heroic formative years of the "future shore of the Universe." While re-editing this chapter, I made use of Gurovich's memoirs in order to travel back in his "time machine" to the bleak steppe where he had been required to establish the first space port in the world within a two-year period.

Thermometer readings in the sun during the day in this desert were off the scale at over 60°C (140°F). Glacial gales blew in the winter at wind speeds up to 30 to 40 meters per second. In the early spring the desert blossomed and you felt like taking deep breaths of air. But as soon as the sun heated up the earth, dust storms started. Dust filled the folds of your clothing, your eyes, ears, and lungs. Dust was everywhere.

In the early spring, tiny yellow tulips on the thick clay crust covering the surface were a delight to the eyes. It seemed easy to race over this vast expanse in an automobile in any direction. In Kapustin Yar we really did race around in Jeeps and "Gaziks" over the "steppe asphalt."[4] Here, after trucks rolled over the "steppe asphalt," it was pulverized. Deep ruts filled with dust formed and the trucks sank in up to their bellies. The dusty roads of the battered desert got to be two to three kilometers wide.

In two years it was necessary to create conditions for people to lead a normal life. Thousands of specialists were supposed to live there permanently with their families. However, first and foremost, there was a demand that the builders fulfill the "Primary Objective," building the engineering facility and launching site for the first R-7 intercontinental missile.[5]

A special institute planned and developed documentation for the builders to erect the launching site at Site No. 1 and the engineering facility at Site No. 2. The project chief engineer, Aleksey Alekseyevich Nitochkin and a team of design engineers moved from Moscow to the desert to quickly resolve the issues.[6]

A town sprouted up from prefabricated barracks and wooden cabins. The two-year construction plan included apartments, a general data processing center, headquarters building, officers' residences, a department store, a chain of other stores, a bread factory, a hospital and outpatient clinics, hotels, a heating and power plant, a complex water supply system from the muddy Syr-Darya River, and a sewage system (with particularly strict sanitation regulations).

The future town was named *Desyataya ploshchadka* ("tenth site" or Site No.10). It would become a real oasis in the desert; hundreds of thousands of trees would be planted along the streets. Parks, boulevards, and riverside recreational areas would be built. For the plants to survive, water had to be brought in to each of them.

---

4. Gaziks (*Gaziki*) were vehicles produced by the Gorkiy Automobile Factory (*Gorkovskiy avtomobilnyy zavod* or GAZ).

5. The phrase *Tekhnicheskaya pozitsiya* (TP) literally means "Technical Position," but in the context of the missile industry more typically means "engineering facility."

6. Nitochkin officially served as a senior engineer in TsPI-31 (Central Planning Institute No. 31) in Moscow.

The engineering facility and launching site were located more than 30 kilometers from the town. At the facility, the construction work was the most intense. Construction workers from organizations specializing in "special steel structures," electrical networks, wire and radio communications, and assembling radio receiving and transmitting centers could not begin work until the main construction workers had completed their project. But the builders weren't finished with their work until a railroad line was extended from the town's railroad station, choked as it was with railcars packed with equipment, and until an ordinary concrete road was paved from the concrete factories and construction warehouses to the firing range's missile pads and all its necessary sites.

Construction continued around the clock! Tens of millions of cubic meters of land were redeveloped, hundreds of thousands of cubic meters of concrete were poured, tens of millions of bricks were delivered to various sites, and hundreds of kilometers of pipe of various diameters, thousands of reels of all sorts of cable, and lots and lots of things unmentioned in historical literature were brought in.

On 14 August 1956, OKB-1 split off from NII-88 and became an independent enterprise. Soon after, in late August, OKB-1 Chief Designer Korolev flew to the firing range for the first time, one and a half years after construction had begun. There still was no airfield near the firing range. For the first two and perhaps even three years, we flew as far as the regional center of Dzhusaly on Il-14 or Li-2 airplanes.[7] From Dzhusaly we had to rattle over the dusty off-road for another three hours to Tyura-Tam in Gaziks or in the best case in Pobedas.[8]

An "OKB-1 expedition" was already working at Site No. 2 in the summer of 1956. This group was supposed to receive, lay out, and set up communications and transport for all the missile specialists. Korolev took on this responsibility. At Site No. 2, five structures were assembled from prefabricated wooden units. These were five barracks, which the builders considered to be comfortable hotels, each for 50 individuals equipped with public lavatories with cast iron pots to answer nature's call. These were designed for soldiers who were allotted three minutes to sit on the throne. According to Gurovich's memoirs, Shubnikov told the disgruntled Korolev, "our officers live worse." And this was true. Korolev responded, "I don't care how your officers live; my staff isn't going to live like that. Fix these hotels up a bit better. My people are golden." Shubnikov promised to comply, but replied, "With regards to people, Sergey Pavlovich, if your people are golden, then my builders are surely steel." On his return several days later, Korolev arranged for the air shipment of toilets, lampshades, linoleum, paint, and a lot of other things for Site No. 2.

Soon, for the first time, Korolev and Shubnikov inspected both the Assem-

---

7. "Il" represents "Ilyushin," while "Li" denotes "Lisunov." The Li-2 was the Soviet version of the American Douglas DC-3.
8. The Pobeda (Victory) was a luxury car produced by GAZ.

Without proper paved roads, transportation at Tyuratam in the early days was a mix of horses, trains, and automobiles. Shown here is a soldier waiting on a horse–driven carriage outside of the housing area at Site No. 10 in 1957.

bly and Testing Building that the builders had already handed over to be fitted out, and the on-going rush jobs at the launch site. When Korolev first visited the sites with Shubnikov, an obscure engineer from the Spetselektromontazh concern named Bakin and a young radio and telephone communications specialist named Pervyshin were working there.[9] Years passed. Boris Vladimirovich Bakin became the USSR Minister of Assembly and Special Construction Work while Erlen Kirikovich Pervyshin became the USSR Minister of Communications Systems.[10]

In our view and according to all guidelines, 1957 was supposed to be the year of the birth of the first R-7 intercontinental missile. The R-7 designation did not appear in technical documentation. None of the unclassified drawings, correspondence, or even the numerous secret documents referred to it as a missile, but rather an "article" with the designation 8K71. Only CPSU Central Committee and Council of Minister decrees, resolutions of the Commission on Military-Industrial Issues, and the ministers' orders issued in furtherance of these decrees and resolu-

---

9. Spetselektromontazh—*Spetsialnoye elektricheskoye montazh* (Special Electrical Assembly)—was one of many specialized organizations within the government tasked with construction operations for civilian and military industry.

10. Boris Vladimirovich Bakin (1913–92) served as Minister of Assembly and Special Construction Work in 1975–89. Erlen Kirikovich Pervyshin (1932–2004) served as Minister of Communications Systems in 1974-89 and then Minister of Communications in 1989–91.

tions mentioned the R-7 intercontinental missile by its real name. However, in our internal secret documentation, in accordance with the standards for maintaining technical documentation, most often the numbers and letters were reversed: not R-7, but 7R. This was also the case for all preceding "articles." Systems that were part of the whole missile complex were also assigned designations authorized for use in technical documentation and unclassified correspondence.

Such "triplicate bookkeeping" for the names of missiles and the dozens of systems comprising them required either a good memory or reference books, that is, exactly the type of notebooks that security regulations forbade. We used to joke that, "If we can't figure it out, then how in the world are those poor CIA station chiefs going to?" Incidentally, the names *Semyorka* and *Pyatyorka* became rather firmly affixed to articles 8K71 and 8K51, respectively, and were widely used in verbal communication.[11]

In 1957 the *Semyorka* took up all of our official work time. But during our brief rest periods and even at home, our heads were also crammed with the problems this missile posed. Watching the tests on the launch system at the Leningrad Metal Works, the firing tests of individual boosters on rigs, and, finally, the rig tests of the entire cluster in Novostroyka near Zagorsk that shook us with an avalanche of fire, I felt as I had never felt before with any of our previous creations. It was a feeling of respect for this unique technical creation, pride in being directly involved in its development, and finally, dread for its future fate.[12] Since 1947, we missile specialists had grown accustomed to the spectacles of missile launches gone awry. It was painful and scary, being in the immediate vicinity of the launch, to watch missiles burning and tumbling in flight. It was frightening to imagine that something similar could happen with the *Semyorka*. How many hopes were tied up with its subsequent fate! How much work had been invested in its creation! We also felt a tremendous responsibility. We viewed the *Semyorka* with its nuclear warhead, whose yield was still unknown to us, as a beautiful goddess that protected and sheltered our country from a dreadful transoceanic enemy.

Nuclear weaponry, both "ordinary" and hydrogen, had already been created. For the first time its fantastic power was combined with target-striking speed in our R-5M missile. But the U.S. still remained outside the range of our *Pyatyorka*. The *Semyorka* was supposed to strip the U.S. of its invincibility.

Assigning responsibility between his deputies, Korolev came to an understanding with Voskresenskiy and me about the upcoming work at the firing range in preparation for the first *Semyorka* launch. He proposed that I supervise missile preparation

---

11. The Russian word *Semyorka* can be translated as "ol' number seven," that is, an affectionate reference to "number seven." It is derived from the Russian word for seven (*sem*). Similarly, the R-5 (or 8K51) was informally known as the *Pyatorka* ("ol' number five"), derived from the Russian word for five (*pyat*).

12. Novostroyka was the informal name for NII-88 Branch No. 2, later NII-229, the rocket engine firing test facility.

and tests at the engineering facility, including the preparation of all the testing equipment. Voskresenskiy was supposed to concentrate on the as yet neglected, but also most crucial area, the preparation of everything needed for launch. Abramov, who supervised Barmin's work on the construction of the launch system, was assigned to step up all assembly and construction operations to put the unconventional launch facility into operation.

Before we got there, Yevgeniy Vasilyevich Shabarov, who was then the aide to the chief designer for testing, had been at the new firing range for a long time. After returning from a temporary assignment, he gave us a detailed rundown of the state of affairs in a meeting in Korolev's office. Thus, we were apprised of the procedures at our new habitat.

I would like to mention that in such ticklish matters as the distribution of responsibilities and the best placement of specialists along the entire frontline of operations, Korolev never held to the principle of having only his "own" people everywhere. If he noticed among the subcontractors an outstanding specialist whose human qualities caught his eye, he would press to arrange for that person to be entrusted with a critical portion of the work.

In February 1957, we, Korolev's deputies, gathered for the first time, not in well-settled Kapustin Yar, but in the desert of Kazakhstan. We flew from Vnukovo airport early in the morning on an Il-14 cargo-and-personnel aircraft. It would be a long flight, with an intermediate stop over for refueling in Uralsk. The firing range airfield was not yet prepared to receive Ilyushin transport planes, so we would have to make our final landing in the regional center of Dzhusaly. Its airport served the Moscow-Tashkent line.

After four tedious hours of flight we gladly deplaned to stretch our legs and take a stroll in Uralsk. To our surprise, in the drab, barrack-style airport building, we discovered a small cafeteria with a superb assortment of hot dishes. Voskresenskiy, who in our circles was considered not only an authority on fine wines, but also a sophisticated connoisseur of food, announced that he could not remember having such splendid tongue with mashed potatoes and such thick sour cream in ages. I proposed that we not pass up the opportunity to stopover in Uralsk again on the return trip to which he replied, "But will there ever be a return trip?"

We flew with stopovers in Uralsk until our flight detachment got Il-18 and An-12 aircraft. It became part of our steady tradition to have breakfasts of tongue with a side of mashed potatoes and a glass of very thick, cold sour cream. Someone joked that such incredibly delicious sour cream could only be made from camel's milk!

There was nothing like the Uralsk service in Dzhusaly. I can't remember how many hours we hung around there before we finally settled ourselves on the Tashkent-to-Moscow train. We got off at the former whistle-stop, which was now the lively Tyura-Tam station.

Our first impression was one of sorrow and melancholy from the sight of the dilapidated mud huts and dirty back streets of the nearby village. But just beyond this first unsightly landscape, a panorama opened up with the typical signs of a great construction project. It was early morning. The sun's warmth was spring-like, although it was still February. Mikhail Vavilovich Sukhopalko, who was responsible for taking care of all the new arrivals, met us. His job description covered everything from the procurement of foodstuffs, to transport, housing allocation, food service, and construction of cottages for the chief designers and barracks for everybody else at our Site No. 2.

To begin with, we drove out to the future town, which was then officially named Site No. 10. In general, the builders, who back then were the real bosses here, called every facility, "site number such-and-such."

Thus, the launch site was called "Site No. 1." Located one and a half kilometers from the launch site, the engineering facility, correspondingly, was called "Site No. 2." In the future, this second site would become a well-furnished hotel community for all the specialists involved in testing. With the birth of the firing range, people also quickly developed a unique range slang; they shortened and simplified certain standard expressions that were used frequently in everyday discourse. Thus, instead of *Desyataya ploshchadka* ("tenth site" or Site No. 10), the majority of us to this day say *Desyatka* ("the Ten"), and instead of *Vtoraya ploshchadka* ("second site" or Site No. 2, we say *Dvoyka* ("the Two"). Instead of saying *Tekhnicheskaya pozitsiya* (engineering facility), we say *Tekhnichka* (the Tech) or simply TP. Over time, in official correspondence the term *Pozitsiya* (the Facility) was replaced by *Kompleks* (the Complex), and so now they use TK instead of TP. But no one said "SP" for *Startovaya positsiya* (launch facility); the majority used the word *Pozitsiya*, while the abbreviation "S.P." referred only to Sergey Pavlovich Korolev and never the launch facility. Sometimes, in keeping with the established pattern, we called it *Yedinichka* (the One).

The military specialists, who were already longtime veterans at the firing range, lived at Site No. 10, the future town of Leninsk on the bank of the Syr-Darya River. The distance between "the Ten" and "the Two," from main office to main office, was 35 kilometers. Subsequently, military design engineers and builders adhered to the principle of placing the launch sites and engineering facilities approximately one to two kilometers apart. Hotels, cottages for civilian subcontractor specialists, bachelor officers' quarters, and barracks for the soldiers of the troop unit attached to the missile complex were built 500 meters from the MIK. Meanwhile, for our *Semyorka* and for the future missiles of Chelomey and Yangel, the rule they followed was to withdraw further from the future town of Leninsk and the Moscow-to-Tashkent rail line.

The principle in effect was "God helps those who help themselves." Over the past 35 years, thousands of launches of various caliber missiles from the firing range's numerous launch sites have never posed a hazard to the town's residents.

The main office of the firing range, the manual computing facility, rear services for the various troop units, and construction administration were located at Site No. 10. In the late 1950s, everything was housed in barracks-style buildings. But construction was under way at full speed on a multistory military hospital, modern

From the author's archives.

Shown here is the construction of the original Assembly-Testing Building (MIK) at Site No. 2 at Tyuratam in 1956-57. All R-7 ICBMs were assembled within this facility, one of the largest at the launch range.

buildings for the future headquarters and all of its services, a three-story department store, and numerous two-story brick residential buildings.

From the station, we headed over to see Lieutenant General A. I. Nesterenko, head of the firing range. He received us with open arms and introduced his deputies, with whom we were very well acquainted from Kapustin Yar: Engineer Colonel A. I. Nosov, his deputy for experimental-testing work, and Engineer Colonel A. A. Vasilyev, his deputy for scientific-research operations. We were also introduced to two already quite "dusty," as they put it, graduates of the F. E. Dzerzhinskiy Artillery Academy, Engineer Colonel Ye. I. Ostashev, the older brother of our telemetry specialist and tester Arkadiy Ostashev, and Engineer Major A. S. Kirillov. Yevgeniy Ostashev had been named chief of the first directorate, which was in charge of our project, and Anatoliy Kirillov was chief of the department for testing and preparing missiles.

Both Ostashev and Kirillov had graduated from the military academy after four years at war. Kirillov had commanded an artillery battery until the end of the war in Europe and then had participated in the war in the Far East in the defeat of the Japanese Guandong army.[13] The service ribbons on the chests of Nosov, Ostashev, and Kirrillov spoke for themselves. Even Voskresenskiy, who had a tendency to behave

---

13. The Guandong Army (or Kwantung Army) were an elite (and vicious) unit of the Imperial Japanese Army formed in the early 20th century, whose battles included encounters with the Red Army in 1938–39.

in a patronizing chip-on-the-shoulder way toward those in the military, spoke in a deferential and tactful manner.

Nesterenko complained that the builders were behind schedule in handing over the MIK for the installation of equipment. But the main hall was ready to receive the missile. The most precious acquisition for the MIK was a crane manufactured by special order. No other domestically produced crane had such a precise, fine-tuned degree of play; now the missile could be assembled with millimeter precision. "The rest you'll see for yourselves. For the time being, our lives are difficult. At Site No. 2, however, there's a whole passenger train with all the conveniences for the chief designers and their main personnel; it has everything, except, I beg your pardon, toilets. Take it or leave it, but you have to go outdoors. No more than a month from now, individual cottages will be ready for the chiefs and hotel-barracks for the rest."

We drove over to "the Two." On the left they were putting in the concrete road to Sites Nos. 1 and 2. We overtook dump trucks that had fresh mortar dripping from their sides, trucks carrying all manner of boxes and building materials, and vans carrying military construction workers. It reminded me of the military roads in the rear areas of armies, the same work-weary drone of hundreds of trucks, each hurrying along with its cargo. Here there were no rumbling tanks and guns, but soldiers were sitting at the steering wheels of all the vehicles and in the cabs.

Unlike the nuclear cities, our NII-229 near Zagorsk, or many other secret facilities, there were no prisoner construction workers here. The army did the construction work. And, we soon realized, the military builders had the knowledge and skills to do it all.

We, Korolev's deputies, having arrived for the first time at the new firing range where intercontinental missile launches were to begin in three months, had to deal with issues for which we bore no direct responsibility. But the universal sense of responsibility for everything that in one way or another affected our projects, unrestricted by any bureaucratic directives, made us take an interest in problems from the most diverse branches.

The firing range was not subordinate to the minister of the defense industry, much less to Korolev. Firing range Chief Lieutenant General Nesterenko was immediately subordinate to Deputy Minister of Defense and Chief Marshal of the Artillery Nedelin. The army of builders, who were actually creating the largest scientific and testing missile center in the world in this desert, were subordinate to another deputy minister of defense. For that reason, the chief of firing range construction was not officially subordinate to the firing range chief. Tracking missile flights over virtually the nation's entire territory required precise and reliable communications work. The chief of the signal corps, who was also a deputy minister of defense, was responsible for setting up the communications system at the firing range and outside its boundaries. In order for the airport at the firing range to finally begin operating, it was necessary to approach yet *another* deputy minister of defense, the

Air Force commander-in-chief.[14]

The railroad was the only avenue available for the delivery of missile boosters, propellant components for fueling, and thousands of tons of freight for the construction, and crucial activity of the ever-increasing number of sites; it was also used to transport people 20 kilometers to work from Site No. 10 in town every day. The Ministry of Railways and the Ministry of Defense railway troops were responsible for building railroads from the Tyura-Tam station in many new directions.

Kazakhenergo was supposed to provide electric power to the firing range.[15] To do this, it was necessary to install poles and run hundreds of kilometers of high-tension power transmission lines. Until that had been accomplished, special railway mobile power plants supplied electric power. From the very beginning of construction, power and water supply were critical and acute problems.

Before we could conduct flight tests a whole list of projects had to be completed. The builders had to finish the launch pad at Site No. 1 (for the time being they called it *Stadion*, "the stadium"). The Assembly and Testing Building at Site No. 2 needed to be suitable for work along with all the auxiliary services including hotels, dining halls, first aid station, and even a store. They needed to build a good concrete road connecting the airfield with the town and all the sites, a wide railroad track to transport the future missile cluster from the MIK to the launch site, and much more. We learned to write telegrams and letters to the VPK, which coordinated each and every project, about jobs that were the most urgent and pressing, but which, as a rule, couldn't be performed by the desired deadlines. In early 1957, Vasiliy Mikhaylovich Ryabikov headed the VPK. He had known us since our Bleicherode days, and we did not pass up the opportunity to notify him when deadlines for putting projects into operation and for making necessary deliveries were missed. In order that the reaction would be rapid and the information we were "ratting out" wouldn't be put on the back burner, we had to use a specific phrase, which became a classic: "And despite our repeated appeals, the delivery dates (or dates for putting into operation or completing construction) continue to be disregarded, which threatens to disrupt the fulfillment of CPSU Central Committee and Council of Ministers decree number such-and-such, dated such-and-such."

When a situation really did reach the point of "threatening to disrupt," Korolev could come down very hard on the alleged guilty party at meetings, especially if the inquiry had taken place in his presence. He really disliked signing reprimands with such wording to the higher echelons. However, if he thought that there was no other recourse, he would first call up and warn the offender: "Keep in mind that I will be forced to go to so-and-so or so-and-so." Often, after one of these conversations

---

14. The Soviet position of minister of defense combined the positions analogous to the American secretary of defense and chairman of the joint chiefs of staff. Several deputy ministers of defense served the "head" minister, each one heading a service such as the Air Force, Navy, or Ground Forces.

15. Kazakhenergo (Kazakh Energy) was the energy producing authority in Kazakhstan.

The first commander of the Scientific-Research and Testing Firing Range No. 5 (NIIP-5) centered at Tyuratam was Maj.-General Aleksei Nesterenko (1908-95), shown here in the 1950s. Earlier, Nesterenko had served as director of NII-4, the military R&D institute that defined operational parameters of all Soviet ballistic missiles.

the need for letter writing to higher-ups fell away. This working style at OKB-1, instilled from the top, fostered in his managers a sense of involvement and responsibility not only for their own specific work sector, but for the whole enormous front in creating our missile power. The specific nature of our style drove me to many discussions and meetings involving the firing range builders.

Soon I determined that there were only three true bosses who could resolve almost any issue at the firing range: the Council of Chief Designers, which had trusted Korolev to defend the interests of each of them, firing range chief Nesterenko, and construction chief Shubnikov. In 1957, Georgiy Maksimovich Shubnikov was still a colonel. Tall and dapper, with a forthright, frank look, he always spoke very calmly and responded to the carping and fault-finding of high-ranking superiors with a sense of inherent dignity. His unconventional nature was charming. Whenever my immediate boss Korolev came down hard on Shubnikov for what seemed to be a trifle, I always felt that it wasn't appropriate.

It was Korolev's way sometimes to put routine demands in a very harsh form. Even those who had worked with him for a long time and knew how uncompromising he was toward all kinds of technical slovenliness and irresponsibility, could not always endure the tone of his tongue lashings calmly. Sometimes when he was dealing with a new individual whom he intuitively guessed to be a strong personality, you could observe his desire to test the latter's tenacity. If this new individual did not hold up, if he gave in and confessed that he was guilty of everything, Korolev lost interest in him. If he rebuffed Korolev, harshly saying something like, "Sergey Pavlovich, what are you doing giving orders around here? This is none of your business," and so on in a similar tone, their relationship was ruined for a long time to come.

But with Shubnikov this did not happen. Shubnikov understood that he was working on an assignment of special national importance and the final stage of its implementation had been entrusted to Korolev. The chief builder of the firing range did not argue and did not clash with the chief designer. Ultimately they became allies. Behind their backs, in private conversations with us about construction work at the firing range, Korolev particularly cursed high-ranking managers for the trying conditions in which they had placed the builders. But he always spoke respectfully of Shubnikov and his deputy Ilya Matveyevich Gurovich.

Once Ryazanskiy complained to Korolev in my presence that there was a lot of substandard construction work at what we called "the third elevation," where they were erecting IP-3—the orbital radio tracking station—and the AVD-APR (Emergency Engine Shutdown and radio-controlled Emergency Missile Destruction) command radio-link station.[16] Officially, we were supposed to go to the firing range chief, and Ryazanskiy requested that Korolev call up Nesterenko. But Korolev telephoned around until he found Shubnikov. He related Ryazanskiy's complaint to him and, after listening to his response, thanked him kindly.

"Here's the thing, Misha [Ryazanskiy]," said Korolev, "You need to deal directly with the construction workers about all construction issues, rather than going around the barn to get to the front door. I have excellent relations with Shubnikov. He'll do everything necessary, but now at the "third elevation" there's a water storage reservoir and they're starting to build an oxygen plant. You've got a very difficult situation on your hill. So don't waste any time. Go meet with Shubnikov yourself; he'll send out all the necessary directives, and if you like, I'll call his deputy, Gurovich, too. His name is Ilya Matveyevich, and he understands everything perfectly. But don't go quibbling uselessly. Believe me, they have it even harder than we do." Ryazanskiy already regretted that he'd gone to Korolev. Now he really would have to meet with Shubnikov or Gurovich.

I had to attend meetings where Shubnikov or Gurovich were reporting. Even the generals outranking them—and Marshal Nedelin himself—grumbled, but they did not raise their voices at the builders. It was evident that there, until the missiles took off from the firing range, the builders were the true bosses. During those first years of setting up the new firing range, their very difficult work determined the future outlook of our work. It seems to me that only there at the firing range did I really begin to understand and appreciate the military builders and their difficult work.

A quarter century after the true heroes of missile and space achievements were no longer kept secret and could share the cosmonauts' celebrity, the builders were still left out. Celebrating their truly outstanding achievements in astronautics, the Americans, too, failed to praise those who had built the remarkable installations at Cape Canaveral. For some reason, the builders' plight galls me. Evidently, it wasn't just in the Soviet Union that builders experienced this fate.

Site No. 10, the future town of Leninsk, and the future Baykonur are very indebted to General Shubnikov and the entire army of builders. Shubnikov died in July 1965, having lived and worked in Kazakhstan for only 10 years. I recall

---

16. AVD-APR—*Avariynoye vyklyucheniye dvigatelya ili avariynyy podryv rakety.*

that Korolev was shaken by this news. This was also the last year of his life. But he not only grieved, he also instructed his deputy and factory director Turkov, "If Shubnikov's family wants to live in Kaliningrad, do whatever you need to, but find an apartment for them and set them up in it with a certificate of domicile and all the rest."[17] I don't know the details, but Shubnikov's family lives in Kaliningrad. In Leninsk now there is a school, a park, and a street named after him.

In October 1992, we celebrated the 35th anniversary of the launch of the world's first satellite. I was in Berlin at the time and visited the memorial of the Soviet soldier in Treptov Park for the first time.[18] To my surprise, I saw here numerous quotations from Stalin's speeches inscribed in pristine gold letters on the polished granite slabs. Descending from the hill where the victorious soldier nestled the child he had saved against his stone chest, at the exit from the memorial plaza, I saw the names of those who had built this architectural structure engraved in a red granite frame in tiny black letters. The name "Shubnikov G. M." was on the very first line. I recalled that during those postwar years when we were working in Germany, Shubnikov had restored the demolished bridges there and then built the unique architectural ensemble in Treptov Park in Berlin. He was involved in the construction of many vital military installations and, shortly before building Baykonur, he built the airport in Tashkent. So as the link to the past will not be broken, there should also be a memorial plaque listing the names of the builders at the now legendary "stadium" of Site No. 1 at the Baykonur Cosmodrome.

AT SITE NO. 2, AS NESTERENKO PROMISED, we settled into the two-berth compartments in the sleeping cars. We hadn't even had time to have our own traditional arrival celebration before we received an invitation to visit the dining car. Dinner was plentiful and delicious. The dining-car's waitresses and imposing director were perfectly courteous and affable. Their starched, snow-white uniforms were completely incongruous with the circumstances surrounding this train. Very impressed with this unexpected service, Lenya Voskresenskiy decided to indulge me. Employing many epithets, he introduced me to the restaurant director and asked her to be sure to remember that soon it would be 1 March, comrade Chertok's (45th) birthday. She promised not to forget, and, indeed, we were able to celebrate the date with a dinner that would have done credit to a good big-city restaurant. Saiga antelope was the meal's main delicacy, artfully prepared and exceptionally tender and delicious to eat.

Hunting for saiga antelope was forbidden. But what did prohibitions mean when they came from the far removed Republic authorities?! Herds of saiga antelope, at

---

17. Roman Anisimovich Turkov served as the director of the Experimental Machine Building Factory (EMZ), the pilot plant in Kaliningrad attached to OKB-1.

18. The Treptov Park memorial was dedicated to Soviet soldiers but based on the exploits of Nikolai Maslov (1923–2002), a Soviet soldier who saved a little girl from certain death during the Berlin siege. The statue shows a soldier with a rescued girl in one hand and a sword in the other and was sculpted by Soviet artist Yevgeniy Vuchetich.

that time numbering tens of thousands, roamed freely over the forbidden territory of the firing range, oblivious that the missiles that would kill them would do so, long before they would destroy their real targets. The hunting of saiga antelope became quite popular as soon as construction of the firing range began. Hundreds of the antelope fell victim to our first nuclear missile tests. Radio operators who had set up the radio-control stations near Kazalinsk told us that they had seen many saiga antelope skeletons in the Aral Kara-Kum Desert. Local residents reported that in February 1956, they were all moved out along with their livestock. But there was no justice for the saiga antelope. They perished during the first nuclear missile explosion.

Each morning we dispersed to our various sites. At the engineering facility we had already begun to assemble the test equipment for numerous systems. Teams from our factory were working to prepare the first two missile clusters for unloading and acceptance. The NII-885 and the Prozhektor Factory team had installed the test consoles, and with the help of soldiers was laying the cables to the work stations and to the power sources (motor generators). Other teams were testing and verifying the battery charging station and preparing a special telemetry film developing room. Each day on the spur lines by the MIK they were unloading railcars carrying new equipment.

At the "stadium," that is, the launch complex, the builders had poured more than a million cubic meters of concrete. Two hundred meters from the launch facility a bowl was excavated where they would build the concrete control bunker. After they filled it in and a concrete-encased hill was built upon it, specialists told us that you could calmly sit and drink tea in such a bunker and take a direct missile strike. Nedelin, who had witnessed the tests of the first atomic and then hydrogen bombs, remarked that in such a situation it would be better to drink tea about 50 kilometers down the road.

On our first visit, Voskresenskiy and Abramov spent a lot of time at the "stadium." The amount of installation and adjustment work was enormous. The whole time there was always something that they needed, someone who was late, something that wouldn't hook up to something else. I also visited the launch complex often, and Voskresenskiy often visited me at the engineering facility. We had to discuss and resolve many issues.

The color image of launch complex at Site No. 1, or the "Gagarin complex," has become just as familiar to today's television audiences as the Mosfilm movie trademark depicting the famous Vera Mukhina sculpture *Rabochiy i kolkhoznitsa* ("The Worker and the Collective Farm Girl").[19] But in March 1957, when I first saw the

---

19. Vera Mukhina (1889–1953) was the Soviet Union's most famous sculptor and worked in many different styles, including Socialist Realism, Cubism, and Futurism. The sculpture "The Worker and the Collective Farm Girl" was probably her most famous work. It was first unveiled in 1937 and now stands very close to the Exhibition of Achievements of the National Economy (VDNKh) pavilion in Moscow.

launch facilities, I was anything but awestruck. I was both depressed and surprised by what had become of the fundamentally new and beautiful idea for a lightweight launch facility, the joint invention of Barmin's and our designers, in which four open-work trusses were to hug the waist of a cluster of five missiles.

The general configuration of the launch system with the missile built into it was redrawn repeatedly until finally Korolev and Barmin approved it. These drawings served as guidelines for the development of the construction design documentation. But in addition, so many new specifications were issued to the builders and design engineers that the harmony of the missile's contour with the retracting open-work trusses had really been squelched by millions of cubic meters of concrete. The builders poured concrete in the bowl that took the fiery shock of the rocket engines' plume; they also poured concrete at "ground zero," where the diesel engine carrying the transporter-erector, as well as the tankers and railcars with unknown cargo, were located.

In countless photographs and in televised reports to this day, the missile, surrounded by its steel crown, sits atop an empty expanse of concrete resembling an overhang. If you descend the steep stairs about 30 meters downward from the overhang, a panorama of the vast concrete-covered surface opens up, reminiscent of a hydroelectric power plant dam. The overhang with those very same retractable trusses that looked so beautiful in the design drafts jutted out over this concrete wall, which merged into the bowl.

The gas dynamics specialists had played it safe. They figured that the rocket exhaust's fiery squall at a temperature as high as 3,000°C (5,432°F) would destroy any substructure if its surface did not match the theoretically designed profile. The bowl was lined with a three-meter-thick concrete venting chute, the profile of which was maintained to the centimeter. Four rectangular concrete pylons were erected at the corners of the foundation slab of the overhang. Balconies with auxiliary space for gear and all sorts of equipment rested on the pylons. Ninety-meter-long lightning rods soared above the whole vast concrete and steel structure. They had little to do with the launch process, but were forever a fixture on the 20th-century missile landscape.

In March 1957, workers finished installing the equipment that had arrived from Leningrad after undergoing tests. The launch complex was crawling with assembly personnel running hundreds of meters of every type of cable and pipeline imaginable, or welding something, chiseling concrete, or performing leak checks on various tanks accompanied by the hissing of compressed air.

On one of my tours around the launch complex with Voskresenskiy and Abramov we were joined by Yevgeniy Ostashev and a construction officer whose name I don't recall. The builder said that during and after the war he had been involved in the demolition, restoration, and construction of so many facilities that he considered himself capable of building "palaces for Satan himself." But what he had to experience and endure here proved to be the "ultimate education" for him. While we were standing in the bowl below the overhang, the officer requested that we observe a minute of silence:

"Three soldiers died here," he said.

We removed our hats.

"A year ago," he said, "was the most difficult time in the construction of this 'stadium.'"

"The excavation required considerably more time than we had calculated. Dozens of powerful motor vehicles and excavators were mobilized for the rush job. When one of the heavily loaded dump trucks drove out of the bowl to the surface, at the very top of the slope, the bolts of the rear axle shaft sheared off and it tumbled backward. The driver was startled, and instead of braking, he jumped out of the cab. The dump truck rolled down the incline into the bowl. The din of the excavators drowned out the shouts of the bystanders. The dump truck smashed into a group of men completely absorbed in their work. Three soldiers died."

We remained silent for quite a while. Voskresenskiy was the first to ask the question that was tormenting all of us and answered it himself: "If, God forbid, we are the first to fire from this site at the Americans, then there won't be a second launch. Not just three men, but all of us at all the sites will be able to admire the illumination [of a mushroom cloud from a retaliatory strike] in the next second. Although the bunker you have built for us is excellent."

Back in March 1957, none of us could foresee that this intercontinental launch Site No. 1 had a great future. The bowl where we had observed a moment of silence in memory of the three fallen soldier/builders would withstand the fiery squall of the first satellite's launch vehicle and after that, many hundreds more launches in this new realm of human endeavor. We couldn't even imagine that this super-secret scrap of desert would be a bright and glorious spot in human history rather than a dark one.

MANY TELEVISION VIEWERS ON THE PLANET HAVE ADMIRED THE FIERY LAUNCHES OF THE R-7, no longer an intercontinental missile but an interplanetary one, now called the Soyuz. But probably few have seen any television footage of the bunker from which the rocket launch commands were issued.

Unlike the one-room bunker of Kapustin Yar, the new bunker was a spacious five-room suite. The prelaunch test and launch consoles were installed in the largest hall, equipped with two naval periscopes. Everything on them was new and different from the primitive consoles of the early years of rocketry except for the firing key. I remember when we were just looking over the R-7 electrical launch circuits, I said to Pilyugin that it ought to be time to do away with this traditional key that we had borrowed so long ago from the German A4 consoles. He agreed with me and gave instructions to develop a special switch instead of the firing key. To his surprise, the military vehemently protested this idea. Missile units had already been formed and military console operators were accustomed to beginning the launch operation with the command "Turn key to fire!"

When the matter reached the chief of the Main Directorate of Reactive Armaments, his deputy Colonel Mrykin felt compelled to call up Korolev and request that the conventional firing key design be kept in the R-7 rocket launch consoles.[20] Korolev asked Pilyugin, and the latter referred to my initiative. To my surprise, S.P. did not make a snap decision, but invited me in for a discussion. I explained that I had operated not so much from technical considerations as from considerations of prestige. The *Semyorka* shouldn't have birthmarks. It was something new and strictly *our* page in the history of rocket technology. After thinking about that, S.P. said, "When the *Semyorka* starts to fly, no one will remember those birthmarks. The military has asked that we leave the firing key. This is, after all, also our history."

The command "Turn key to fire!" also remained. Among the various souvenirs that I keep is a firing key that was given to me in 1962 by the military testers. When this modest, but in my view, precious gift was presented to me on my 50th birthday, the firing range envoy promised that 50 years hence he would present a firing key of the very same design that would be used to send an expedition to Jupiter.

The bunker's second large room was for "guests." This room was intended for State Commission members, for high-ranking guests, and those chief designers who would be in the way in the console room. Two other rooms were filled with fueling control and firing mechanism control instrumentation and with gear for the radio tracking systems. There were also hallways and auxiliary rooms for communications specialists and security personnel. A lot of space in one of these rooms was taken up with multichannel recording equipment. This system served as a partial backup for telemetry while the missile was still at the launch site. In addition, it recorded the behavior of the launch system itself during the launch process.

Only four persons could view the launch from the bunker. There were two periscopes in the console room and two in the guest room. If the missile successfully left the launch pad, everyone else had to manage to jump out of the bunker in order to admire its flight. This required charging up about 60 steep steps and running another five to seven meters once you'd reached the top.

The Assembly and Testing Building was the main structure of the engineering facility at Site No. 2. That is where we had to conduct all the operations to prepare the missiles before they were transported to the launch site. Diesel engines pushing railcars carrying missile boosters rolled freely into the large high-bay of the MIK. Here in the high-bay the boosters were unloaded and placed on handling trailers for testing and then the missile cluster was assembled from the individually tested boosters.

---

20. Anatoliy Ivanovich Semenov (1908–73) served as chief of the Main Directorate of Reactive Armaments (GURVO) in 1954–64. GURVO was the main missile procurement and acceptance agency within the Soviet armed forces in the 1950s.

Three stories of laboratory/service rooms were immediately adjacent to the high-bay where assembly operations were performed. Back in Moscow when they were planning out the rooms, people fought over each of them. Along with Nosov, Osta-shev, and Kirillov, I had to make the final decision as to where which system went and where to run the power and communications lines. There were a lot of labora-tories for all sorts of systems.

During this period, Nina Zhernova and Mariya Khazan arrived at the firing range and settled in on the train. Pilyugin assigned them to participate in the assem-bly and adjustment of the integrated stand for electronic analog simulation. They explained that Nikolay Alekseyevich wanted to be able to conduct all the necessary research with the actual stabilization controller equipment *here* rather than flying to Moscow to ask the institute about each glitch. These were good intentions, and from his reserves, Kirillov set aside a large room on the upper floor in the MIK for them known as the "Personal Laboratory of Nina Zhernova on behalf of comrade Pilyugin."

Each system had its own chief designer, who in no uncertain terms demanded a "separate individual suite." It didn't matter if it was just one room and cramped, as long as it was private, with no unauthorized intruders. Thus, space was allocated for the control surface actuators, for both the Tank Emptying System (SOB) and Tank Emptying and Synchronization System (SOBIS) and for gyroscope instruments.[21] To be on the safe side, we pressure tested each of the fittings for each system.

The assembly and adjustment of radio systems gave us the most trouble. The onboard radio control system equipment required such an abundance of all sorts of racks crammed with test units that they set aside the most spacious rooms for it on the second floor.

There were also disputes with the young rivals of Ryazanskiy's radio electronics monopoly. Back during the tests on the R-1 and R-2 missiles in 1950–53, we used the *Indikator-T* radiotelemetry system and the *Indikator-D* trajectory measurement systems developed by young MEI graduates under the supervision of Academician V. A. Kotelnikov. The young, dynamic, and enthusiastic team, having gained its first missile firing range test experience, decided to begin developing the next generation of radio engineering devices. Their work blatantly and brashly intruded into the work of Ryazanskiy, Boguslavskiy, Borisenko, Konoplev, and of the special organiza-tion SKB-567 under the supervision of Yevgeniy Gubenko, recently created in the State Committee on Radio Electronics.

Back then many theoretical and practical radio electronics issues still lacked clar-ity. Scientists and engineers continued to debate over the attenuation of radio waves in the ionosphere, the influence of the engine plume plasma, antenna design, and sites for their installation. Unreliable radio tubes and the first semiconductor ele-

---

21. SOB—*Sistema oporozhneniya bakov*; SOBIS—*Sistema oporozhneniya bakov i sinkhronizatsii.*

ments presented the biggest headaches to equipment designers. Their production technology simply was not ready for our strict requirements.

MEI responded to the 1954 decree for the development of the intercontinental missile with great enthusiasm. Just a year later, they developed experimental models of the onboard equipment and ground stations. Teams under the leadership of Aleksey Fedorovich Bogomolov—who succeeded Kotelnikov—developed these systems.

Korolev gladly agreed with my proposal to support Bogomolov and encourage competition between Bogomolov and radio industry organizations. Minister Kalmykov and his deputy Shokin did not approve of our initiative.[22] However, whenever the opportunity arose we worked items into Central Committee and Council of Ministers decrees that obligated the Ministry of Higher Education to create all necessary conditions for the development of radio equipment for the R-7 at MEI.

The government announced no official competition for the development of radio-telemetry equipment for the R-7. Nevertheless, designers scrambled for a place on board the program. Our obvious support of Bogomolov irritated Ryazanskiy. State Committees had not taken Bogomolov's OKB MEI seriously, and when the occasion arose they poked fun at our patronage of this "orphanage" and as a countermove supported the development of Gubenko's telemetry system in every way. Nevertheless we succeeded in setting up an expert commission, which decided to conduct comparative aircraft tests. The expert commission findings were—in a rare instance—fully unanimous: they recommended that the *Tral* (Trawl) system developed by the OKB MEI be used for the R-7 rocket. It was no accident that the *Tral* won the competition. The young, talented engineers used the most cutting-edge electronics achievements, which were considered to be premature in indigenous technology. *Tral*'s 48 measurement channels enabled us to make a comprehensive study of the missile in flight.

But having lost the competition, Gubenko, Bogomolov's main competitor for the radio-telemetry system, was not left without work. The shortcoming of Bogomolov's *Tral* at that time was its inability to record rapidly changing parameters such as vibrations and pressure pulsations in the combustion chambers. By 1956, Gubenko had developed a new telemetry system to record these phenomena, the "rapid telemetry" RTS-5. We developed vibration sensors for it and the system was also installed on the first R-7 rockets.

During the period 1954–56, series production of stationary and mobile onboard equipment and ground stations was set up at radio engineering factories. In just two years, 1956 and 1957, more than 50 sets of ground-based units were produced.

---

22. Valeriy Dmitriyevich Kalmykov (1908–74) served as minister of the radio-technical industry in 1954–74, during which time the ministry was known under several different names. Aleksandr Ivanovich Shokin (1909–88) served as his deputy in 1954–61.

The firing range and all the tracking stations from Tyura-Tam to Kamchatka were equipped with them.

We installed three autonomous *Tral* sets on the R-7 rocket:

• in the nose section;

• in the second stage, the central Block A booster; and

• on the strapon Block D booster to monitor the parameters of all four boosters of the first stage.[23]

We called our first missiles measuring missiles; the total number of parameters measured exceeded 700. The mass of the entire instrumentation complex was so great that it reduced the missiles' range from 8,000 to 6,314 kilometers. There was one more reason for reducing the range: at full range the nose section reached the Pacific Ocean, and we did not yet have any tracking facilities available there.

Kamchatka was the maximum range that we could achieve while leaving our tracks on terra firma. Therefore, we set up ground tracking station NIP-6 in the area of Yelizov on Kamchatka. This station on the edge of the Soviet territory was supposed to measure the parameters of the nose cones and receive the telemetry data emitted by the *Tral* transmitters. Soon thereafter, a second tracking station, NIP-7, was also opened there on Kamchatka in the area of Klyuchi.

The Bogomolov team's "assertive" actions did not end there. In "strict confidence" Bogomolov related that he had made an arrangement with the leading radar factory in Kuntsevo for the joint development of a trajectory radio-monitoring system. *Gosplan* department manager Georgiy Pashkov actively supported him in this undertaking. This "secret" conversation took place in 1955. Korolev too, after a "confidential" meeting with Bogomolov, ordered that the *Rubin* (Ruby) transponder be installed on the R-7 at once. This innovation determined the missile's current range. After the measurement results had been processed, the ballistics specialists were able to determine the nose cone's points of impact with a high degree of precision.

Kama ground stations, which worked with the *Rubin* onboard transponder, were a modification of air defense radar systems. Their series production had been set up long before, which worked to the advantage of Bogomolov's proposal, as opposed to using systems based on the highly complex and expensive RUPs. Within the MIK, the telemetry equipment was located in separate rooms removed from the other radio-emitting systems to avoid electromagnetic interference. I very much enjoyed the contact I had with the boys from OKB MEI who worked enthusiastically assembling and checking out their stations. Mikhail Novikov, who supervised the operations, spoke about the principles and layout of the systems with such pride that you couldn't help but want to help him in any way possible. Our telemetry

---

23. Each booster in the R-7 cluster was called a *Blok* (Block). The center was Block A while the four strapons were known as Block B, Block V, Block G, and Block D, reflecting the first five letters of the Cyrillic alphabet.

specialists headed by Nikolay Golunskiy and Vladimir Vorshev very quickly came to an understanding with the OKB MEI engineers so that subsequently they were all considered part of the same "gang."

The first flight-ready R-7 missile arrived at the firing range engineering facility on 3 March 1957, with its full complement of five boosters. It carried factory number M1-5, and in conversation we referred to it as number five or simply *Pyataya* (the Fifth). They began off-loading the boosters and placing them on the handling trailers. On 8 March a large group of designers headed by deputy "lead designer" Aleksandr Kasho flew in. They brought with them a long list of modifications that needed to be introduced based on the results of the firing rig tests.

The operations on the heatshield of the aft compartments promised to be the most labor-intensive. During the firing rig tests the aluminum alloy skin of the aft structure burned through in many places. Even the feedback potentiometers of the control chambers and cables burned. They would have to sheathe the exterior of the aft compartments with thin sheets of chrome-plated steel and wrap the interior with a layer of asbestos to protect all the vulnerable parts.

I had spent almost a month at the firing range. In late March I was given the opportunity to briefly abandon the hotel in the hospitable train and return to Moscow while the first cycle of modifications was under way.

I flew into the firing range for the second time with Korolev in April 1957. For the first time we landed at the new Lastochka airfield, the future Baykonur airport, destined many years later to attain international fame. Many of our colleagues were flying on this airplane. Korolev felt that he needed to send as many of his employees as possible through the firing range school so that they could get a sense that "We're not here to sip tea with jam."

As we were getting into the vehicles, Korolev seated me in his Gazik. For the first time, I rode with Korolev over the steppe of the new firing range. The roads were already producing clouds of dust and I couldn't pass up the opportunity to remind Korolev that if it hadn't been for Ryazanskiy's requirements for the placement of the radio-control stations we wouldn't have gotten ourselves into this semidesert. Sergey Pavlovich surprised me with a very effusive response: "Good grief, Boris, Boris! You totally irreparable and rusty electrician! Take a look and feast your eyes on the limitless space that surrounds us! Where else can you find such a perfect playground? We are going to do great things here. Believe me and stop your belly-aching."

He told me this, turning around from the front seat. His usually preoccupied or even stern expression glowed with youthful and delighted animation, unusual for Korolev. It is precisely this atypical image of his face beaming with delight that is etched in my memory. Almost a half century after that memorable trip with Korolev, while editing this chapter it dawned on me for the umpteenth time that Korolev had a knack for foreseeing the future better than all his compatriots. We were dogmatists—enthusiasts—we created the first intercontinental carrier missile for a hydrogen bomb. And we really were convinced that if it were used for its express purpose, there would be no next launch. But he, Korolev, talked enthusiastically about the future great projects that we would be conducting in what was no longer a desert, but a verdant steppe. That is what should distinguish a true leader. He sees further than everyone around him.

Four individual cabins had already been built for chief designers and were waiting for their new tenants. In the future, two of these cabins would be awarded memorial plaques. Yuriy Gagarin spent the last night before his flight in cabin No. 1, and for eight years cabin No. 2 became Korolev's second residence after Moscow.

Considering the extremely difficult living conditions, Korolev arranged to have three cabins temporarily occupied on a "communal-democratic basis." The new hotel was not yet ready, and in Korolev's opinion, life in the barracks could diminish his deputies' authority. All the cabins had three rooms, and therefore, three men were settled in each of the three cabins. Cabin No. 1 was left vacant in case the State Commission chairman or Marshal Nedelin wanted to take a rest or stay a while at *Dvoyka*. That's why until Gagarin's stay there it was called *Marshalskiy* ("the Marshal's"). Korolev allotted a room each to Mishin and me in his cabin. Barmin, Kuznetsov, and Voskresenskiy settled in the third cabin. Glushko, Ryazanskiy, and Pilyugin occupied the fourth cabin. Thus, in terms of firing range privileges, Korolev placed his three deputies, Mishin, Voskresenskiy, and me, on the same footing as the chief designers.

Returning to the firing range two weeks later, I saw the almost green steppe of springtime for the first time. You felt like strolling over it instead of driving. Here and there were scrubby, multicolored tulips and delicate, downy dandelions that we weren't used to seeing in Moscow, in what's called the central zone of Russia. They tenaciously withstood the wind without scattering their fluff.

The concrete roadway had been completed. Only tracked vehicles and very heavy-duty trucks traveled over the steppe. Almost the entire population of the train had moved from the hot, cramped quarters of the railroad cars to the multiple rooms of the barracks, which were divided, respectively, into men's and women's quarters, and rooms assigned by department. Distribution of rooms was carried out spontaneously on departmental, system, and group bases. As a result, there were rooms for telemetry specialists, ballistics specialists, engine specialists, land-segment specialists, assembly workers, and so on.

The barracks-like administrative building for the as yet modest administrative detachment completed the rectangular perimeter of the barracks housing complex. There was already a telephone for high-frequency communications installed in it and a large room for meetings and occasional film showings.

We rapidly cultivated a unique firing range lifestyle, filled not only with work, but also evening strolls along the concrete roadway, picking tulips, and all sorts of practical jokes. There was an atmosphere of optimistic expectations; good-natured humor lightened the hard work and difficult living conditions.

Soon we were pleased to learn that Vasiliy Mikhaylovich Ryabikov had been named chairman of the State Commission on R-7 missile tests. This was especially good news to those of us who had received him at the Villa Frank in Bleicherode in 1945. The commission included Marshal Nedelin (deputy chairman), technical director of testing Korolev, members assigned as deputies to Korolev, (Glushko, Pilyugin, Ryazanskiy, Barmin, and Kuznetsov), and, finally, regular members Peresypkin (deputy minister of communications), Mrykin, Vladimirskiy, Udarov, Nesterenko, and Pashkov.[24]

---

24. The R-7 State Commission was officially formed by decree on 31 August 1956. Such bodies as state commissions were temporary bodies formed to oversee testing of particular weapons systems. They were typically headed by a civilian (in this case Ryabikov, whose actual job in the government was Chairman of the powerful State Committee for Reactive Armaments for the Army and Navy) but staffed with a combination of designers (such as Korolev), scientists (such as Keldysh), military leaders (such as Nedelin), and industry representatives (ministers or deputy ministers such as Ustinov). A State Commission would oversee the entire testing phase and then certify the weapon as ready for operation in the Soviet armed forces. Once certification was done, the State Commission would be dissolved. Note that State *Commissions* and State *Committees* were entirely different administrative organizations. The former were usually temporary and fluid in membership while the latter were permanent and more rigid.

Chapter 18

15 May 1957

On 30 March 1957, the last rig firing tests on the R-7 flight article were conducted at what used to be a branch of NII-88, but today is called NII-229, outside Zagorsk. The tests revealed many new glitches that needed to be taken into account during modifications on the very first R-7 (missile No. 5), which was at the firing range. A difficult lot fell to the factory brigade. They had to perform operations in the MIK bay that, back at the factory, would have been performed by specialized shops. The operations that they hadn't managed to do at the factory were completed by a brigade from shop No. 39 under Tsyganov's supervision. They worked together smoothly and amicably, bringing with them all the materials, tools, and alcohol in excess of any and all norms for "flushing and cleaning."

Reinforcing the heatshield of the aft units of the booster was the most troublesome operation of all. This work had begun before I flew back to Moscow. The piping of the oxygen lines was replaced in order to eliminate stagnant areas where liquid oxygen would heat up, come to a boil, and cause shaking known as a "water hammer." A fire prevention process of purging the aft compartments with nitrogen was also introduced. The ballistics experts, who had already made use of the first computer, recalculated the trajectory. As a result, at the last moment the final stage shutdown time of the control chambers' thrust had to be changed in the timing units.

The list of modifications was long. The chief designers of the systems attacked missile lead designer Kasho, announcing that according to the latest results of the factory tests they needed to replace this or that instrument. While the missile was being transported from Podlipki to the firing range, unloaded, and prepared for testing, some defect was discovered in each system at the very last moment of the final factory tests. At the factory, these units could be replaced quickly, without any formalities. But here at the firing range engineering facility you could "jump onto the last car of a departing train," only after I gave the order. After that, the lead designer would have to explain the change. The final decision was up to Korolev. Before approving a document calling for a routine replacement, the system's chief designer or his deputy would collect as many authorizations as possible. After this, they would personally appeal to Korolev, who demanded strong arguments in favor of the replacement or modification. Finally, it would be announced that further replacements and modifications would be permitted in the event of failures or serious glitches, but only based on test results.

Glitches occurred on an hourly basis during horizontal electrical tests. It wasn't easy to report to Korolev about each glitch and have to explain causes, to boot. And to make matters more complicated, he would demand that he be called about any glitch, even at night. Voskresenskiy was more decisive and persuaded Kasho and me to buck this system; otherwise, later at the launch site it would be impossible to get anything done.

Late one night after the failure of yet another instrument (most likely it was the *Tral* or one of the radio control instruments), I decided to replace it immediately and reported this to Korolev after waking him up with a phone call. A half hour later, referring to my decision, Kasho repeated the same report over the phone. Another half hour later Voskresenskiy woke up Korolev with a third phone call and said that he was very troubled by these failures and by the instrument replacement that Chertok was performing.

In the morning, when he appeared at the MIK after his sleepless night, Korolev summoned us and said: "I know that you arranged to teach me a lesson. To hell with you. Let's set up this procedure to make a detailed entry about all glitches in the logbook. Every morning when I arrive, Kasho will call in whomever necessary if he can't explain it himself, and I will sign the logbook after you."

Radio control system instruments caused the largest number of glitches. Ryazanskiy grew haggard from having to frequently explain the situation to Korolev.

Throughout the entire cycle of horizontal tests, after introducing new procedures, we racked up such a number of instrument replacements, modifications, and glitches that we were down in the dumps. The deadline for launch before the May Day holidays became completely unrealistic.[1] After consulting, we decided to propose to the technical management that we conduct a second finishing cycle of tests, but without any freedom to make changes. Korolev agreed with us and took this proposal to the meeting of the chief designers. Everyone accepted it amicably, resigned to the fact that we would be celebrating 1 May at the firing range. Alas, no one would be able to use the complementary tickets to see the military parade from the viewing stands on Red Square.

At the meeting of the technical management, Korolev announced that all modifications would be halted completely during the finishing cycle of tests and that, as director of testing at the TP, I was absolutely forbidden to discuss any new proposals without reporting to him personally.

However, all of my problems during the horizontal tests of the missile's first

---

1. Although not celebrated in the U.S., May Day is a holiday in many countries of the world, recognizing the contributions of the labor movement. During the Soviet era (and continuing to the present day), May Day (May 1) was one of the most important national holidays in the year.

flight article in the Assembly and Testing Building seemed trivial compared with the experiment to "yank" an engineering mock-up of the missile out of the launch complex. This experiment was scheduled after the failed experiments with missile "liftoffs" using a crane at the Leningrad Metal Works, where we had been unable to demonstrate that the upcoming "real" launches would be safe, since the launch system support booms had not simultaneously released from the recesses on the missile's load-bearing ring.

The technical conditions for the launch assembly stated that the hydraulic mechanisms connecting the support booms and trusses should simultaneously separate all four trusses despite disturbance forces and moments. This was one of the conditions that ensured launch safety. In Chapter 16, I wrote that, as a result of the negative results of the experiment at the Leningrad Metal Works, at Korolev's recommendation, engineers decided to conduct another simulation at the firing range after all the modifications ensuring launch safety had been implemented. A large number of unknown elastic deformations of the launch system and missile elements, which had been impossible to calculate, had caused the original failure of all four trusses to separate simultaneously and release the missile's support ring at the waist. At Barmin's direction, the missile erector was specially modified to conduct the missile liftoff experiment at the firing range.

In late April, State Commission Chairman Ryabikov and Marshal Nedelin flew to the firing range just in time for the new tests.

"We didn't manage to put on our circus act without the brass," complained Barmin. "This is all because of your boys who came up with all these fancy measurements to take."

Indeed, our dynamics specialists Vetrov, Rubaylo, and Lebedev initiated a process of monitoring the experiment, such that installing and tuning the instrumentation required a great deal of time. On the day the decisive experiment was conducted, Ryabikov, Nedelin, and Korolev arrived at the launch site. Barmin supervised the experiment. Voskresenskiy and I had arranged to drive to the launch site together, but at the last minute I heard Kasho's voice coming over the phone from the MIK:

"There's smoke coming out of the instrument compartment of the central booster. We've aborted the tests. We need a decision immediately. I haven't reported to S.P. yet."

"Smoke" was a tester's most terrifying word. Voskresenskiy dropped me off at the MIK, and then he drove to the launch site.

"I'm not going to tell Korolev anything. When you've figured it out, you can give him the report."

This "smoke," the source of which in fact turned out to be ground engineering electrical connectors rather than the missile, kept me from attending the launch site "circus."

Many years later, Igor Rubaylo agreed to fill the gap in my memory. In March 1957, he participated in the experiments at the Leningrad Metal Works, where Korolev personally had sent him to supervise the installation of the launch assembly and its testing and to check Barmin's documentation with regard to dynamic parameters. He remembered:

"And so the day came when we were supposed to wrap up our many years of work on the design, manufacture, assembly, preliminary tests, and modifications of the launch assembly. At the very last moment before missile 'liftoff,' when the service trusses and assembly workers' ladders had already been taken away, someone noticed from below that one of the temporary mounting blocks that had been needed for the preparatory operations, but that was supposed to be removed before liftoff, hadn't been removed. (All such elements were painted red or had a red flag). And right in front of the top brass and a large number of specialists, a serviceman—a captain, still wearing his overcoat and boots and without any safety equipment—climbed up the support trusses to the upper load-bearing flange, removed the mounting block and climbed back down. This operation exacerbated the tense atmosphere as the attendees awaited the experiment results.

And then finally, the erector operator received the command to begin missile 'liftoff.' Soon, from the entire configuration consisting of the missile, the four trusses of the load-bearing launch system ring surrounding the missile's 'waist,' and the erector boom, one of the trusses broke free and the rest began to shift around horizontally. As the missile was 'lifted off,' each of the trusses pulled out individually and randomly instead of simultaneously.

All present were clearly dismayed. Marshal Nedelin simply waved his hand and said, 'That's it. The missile is totally screwed,' and he departed 'ground zero' and headed for his car. However, Korolev and Barmin kept their cool. Before calling a meeting, they tasked the group of specialists to conduct a thorough analysis of the telemetry results in order to check the coincidence of the actual truss separation speeds with the predetermined values and prepare their remarks. There were four men in this group: Vetrov and Rubaylo from OKB-1, Barmin's senior engineer Zuyev, and the senior lieutenant from military acceptance. No middle management from our team, Barmin's team, or military acceptance were there.

Despite the fact that the processing and deciphering of the measurement results were conducted as a matter of urgency, our group's analysis of the measurement results was not completed until two days after the tests had been conducted. We were working at Site No. 10. One of those evenings Marshal Nedelin dropped in on us, asked how things were coming along, and said, 'Boys, we're pinning all our hopes on you!'

We compared the relationships of the launch system elements' motion parameters from the experiment with the design data and found some surprising results. In particular, the measurement results revealed that one of the elements of the upper load-bearing ring had not been properly secured, but this could not affect the overall liftoff picture. I was, just as my colleagues likely were, feverishly looking for a way out of this very convoluted mess. We had been analyzing the measurement results for two or three days before it occurred to me that in a real launch, due to the tremendous speed of the missile's liftoff, in all probability, the likelihood that the support trusses would fail to release simultaneously was negligible. It was not possible to perform specific numerical calculations to evaluate these time intervals for an actual launch either under firing range conditions (especially within a short period of time), or in the calm atmosphere within the OKB walls, because for this we needed to take into consideration the system's elastic properties under axial and transverse force loads and the effect of vibration on the operating engines

Therefore, the launch safety findings that our group prepared, stated simply that at the speed that would be generated during the actual launch, the degree of asynchrony in the release of the support trusses would be negligible and would not be capable of appreciably disturbing the missile's motion relative to its center of mass. OKB-1 management and Barmin, let alone military acceptance, did not demand analytical documentation supporting this assertion only because so little time remained before the launch. If there had been another six months until the launch, as was the case after the mockup liftoffs at Leningrad Metal Works, these claims, which were not corroborated by the appropriate analytical documentation, would hardly have satisfied military acceptance. And even if such documentation had been prepared, there would have been any number of opportunities for it to be challenged.

Indeed, it didn't seem possible to evaluate how the transverse disturbance forces and moments would affect the missile's position relative to the load-bearing trusses supporting it when the strapon thrusters were building up to the intermediate stage, precisely when a large portion of the launch assembly deformation was occurring. Consequently, vibrations could also occur in the whole system at the time. After all, tests at Leningrad Metal Works and at the firing range showed that the 'lock' in the upper load-bearing flange would fall apart long before the first truss would pull out.

Despite the analysis group's signing of the launch safety findings, many still had qualms about the safety of the launch. Representatives from the middle echelon of military acceptance approached me several times and tried to find out how confident I was in the safety of the launch. I told them 200%. V. P. Barmin invited me to his room and, holding our findings out before him, asked, 'Comrade Rubaylo, were the tests performed sufficient to be confident in the safety of the launch? Or, perhaps, do we need to pull the article out of the launch assembly a couple more times? Then you will return to your young wife another two or three weeks later.' (Somehow he found out that I had gotten married six weeks before leaving for the firing range.)

I answered that I was 100% confident in the safety of the launch and that additional experiments would, in fact, yield nothing new. My confidence was based on the fact that during all three 'liftoffs' of the missile from the launch assembly, after the release of the first support truss there were no hitches between the missile elements and the launch assembly, despite the tremendous asymmetry of the missile relative to the launch assembly.

Despite the fact that the launch safety findings had been drawn up and signed by the analysis group and then signed by the entire management, there was not 100% confidence in the safety of the launch. And only actual launches could dispel these apprehensions.

Marshal Nedelin was not mistaken when he said, "Boys, we're pinning all our hopes on you."

For more than 40 years all sorts of versions of the *Semyorka* have lifted off more than 2,000 times! But not once was there a failed launch due to the support trusses falling away "out of synch." In December 1957, 28-year-old Igor Rubaylo was awarded the "Badge of Honor," defended his candidate's dissertation, and later received the title "Distinguished RKK Energiya specialist." Now he lives almost year-round in our gardening cooperative, Pirogovo, on the shore of the bay, at the same time performing the duty of watchman guarding our garden plots. Alas! It has proved impossible to protect our dachas against robbers with the same degree of reliability that he provided for launch safety. I learned that first-hand.

THE SECOND FINISHING CYCLE OF HORIZONTAL TESTS ON THE INDIVIDUAL BOOSTERS WAS COMPLETED ON 30 APRIL. After arriving at the firing range, Ryabikov announced that we would have the day off on 1 May, but first he gathered as many people as he could fit into the conference room and made a report. The report was surprising. Ryabikov told us about the crackdown in Moscow on the "anti-Party group" of Molotov, Malenkov, Kaganovich, and others.[2] This announcement left us with a bad taste in our mouths. After Stalin's death, the liquidation of Beriya, and after Khrushchev's grim speech at the Twentieth Party Congress, it seemed that a wise, just, and unified authority had finally gained a foothold at the very top. As we deliberated, we interpreted this as a clear victory for Khrushchev's line. But now it meant that once again there were enemies in the Communist Party, and once again we would have to fight, expose, and exclude. There were already supporters of this anti-Party group, but Ryabikov calmed us, saying that the Central Committee had completely and unanimously approved the exclusion of the former Politburo members from the Party and that the unity in the Central Committee was unshakable.

How many times had we heard this and applauded the complete unity in the Central Committee, in the Party as a whole, and the unity of the Party and the people? For the country and many of the peoples of the Soviet Union these were now largely hackneyed abstract slogans. It was another story here at the firing range in Kazakhstan. Indeed, for the sake of our common goal we were a unified, tight-knit, and gung-ho team—people from different departments, military and civilian, workers, engineers, scientists, rank-and-file employees, and high-ranking superiors.

We had made up our minds not to work on 1 May. At last we could sleep in and relax. We could revel in the not yet scorching sunshine or even take a trip to Syr-Darya river! But the break wasn't without incident. The telemetry service team had received a substantial amount of alcohol "to flush out the developing machines and dry the photographic film." That's how it was worded on the requisition. I'm

---

2. The "Anti-Party group" comprised leading pro-Stalinist members of the Presidium (or Politburo) such as Vyacheslav Molotov, Lazar Kaganovich, and Georgiy Malenkov, who, unhappy with Nikita Khrushchev's de-Stalinization policies, attempted to isolate and depose Khrushchev in the summer of 1957. Although they enjoyed a majority vote in the Presidium, they were unable to canvas sufficient votes in the Central Committee to oust Khrushchev. All of them were eventually forced out of political life by Khrushchev.

culpable, having approved the phony requisition signed by Nikolay Golunskiy and somebody from the military. What to do? To get alcohol in those days, people wrote requisition invoices "for flushing optical axes" and "antenna directional diagrams." Strict prohibition of alcohol was in effect at the firing range. Vodka was not for sale. But to reward those who particularly distinguished themselves on the job, they allowed alcohol to be dispensed free of charge from the auxiliary supplies.

After celebrating with the requisition all night, at six in the morning on 1 May, the telemetry specialists decided that it was time to join all those relaxing at Site No. 2 who were commemorating the international holiday of proletarian solidarity. The enterprising group together with Nikolay Golunskiy armed themselves with a red banner, a decanter of alcohol, a table glass, and a single lemon. One by one they visited the rooms of all the barracks and woke up the sleeping occupants. One of the revelers got up on a stool that he had been carrying with him. He pronounced a salutation on the occasion of the 1 May holiday, and extolled the solidarity of the workers and the success of our project. Then they let them have a sip of the alcohol and the single lemon, and proceeded onward under general guffaws of laughter or the swearing of their sleep-deprived comrades. Everyone chuckled good-naturedly about this demonstration, but in this amateur performance the firing range political office saw somewhat of a parody of the official way of conducting May Day festivities and made a statement to Korolev regarding the disorderly conduct of his employees in a high-security area.

Golunskiy and his comrades were spared exile from the firing range for want of specialists to replace them on the eve of a crucial launch. Korolev could do little more than threaten them and sternly warn that if there were any more warnings about their behavior he would "send this whole gang to Moscow on rails."

The threat of sending people "to Moscow on rails," for whatever infringement, was Korolev's way of expressing extreme dissatisfaction. But sometimes he exploded even more violently: "Get over to the typing pool and type up an order firing you without severance pay and bring it here for my signature!" If the guilty party returned and held out the typed order for Korolev, he yanked it away from him and yelled so loudly that everyone trembled. "What? You want to go home and sip tea with jam? Get back to work immediately!" Then he treated the guilty party as if nothing had ever happened. Bystanders, chuckling over the star of the latest incident, were afraid that now he wouldn't make it to Moscow any time soon on rails or any other form of transportation.

Indeed, it was considerably more difficult to fly *out* of the firing range than to fly in. Korolev introduced a procedure whereby the expedition chief was supposed to show him a list of passengers for every departing aircraft. If someone ended up on one of these lists without his knowledge, he ruthlessly crossed him off and demanded an additional report.

Once when Korolev was absent from the firing range, I saw lead designer Kasho; his face was all distorted. He had a huge dental abscess and an excruciating toothache. The local dentist said that Kasho needed an operation that he could not undertake.

Then I sent Kasho to Moscow on the condition that he would return on the next airplane right after the operation. Kasho returned the day before Korolev's arrival, but someone had already managed to snitch to Korolev that "Chertok let Kasho go to Moscow without reporting to you."

When Korolev showed up in the MIK an hour after his arrival at Site No. 2, he demanded a report from Kasho. To his great surprise, Kasho appeared and was prepared to report on the status of the missile modification work. My explanations came next. I told him honestly what had happened, and the incident ended there.

RIGHT AFTER OUR EXUBERANT DAY OF REST ON MAY DAY, all the firing range services continued their intense preparation for the first launch. Horizontal tests were finally completed in the MIK, and the assembly of the five-booster cluster began. This operation, which was being conducted here for the first time, drew a lot of spectators. Senior Lieutenant Sinekolodetskiy and our factory assembly foreman Lomakin supervised the integration.

The slight, thin, very nimble Sinekolodetskiy, who had changed into slippers, balanced like a performer on the surface of the missile boosters, giving orders to the crane operator. One after another the special lifting mechanism picked up the strapon boosters and smoothly raised them up from the ground supports, and together with the officer in charge of moving them, floated over to the central booster. The entire missile cluster was placed on a handling trailer and would then be transferred to the erector platform. The last electrical checkouts of the cluster were not completed until 5 May.

At 7 a.m. on 6 May, in keeping with a tradition religiously observed to this day, a diesel engine rolled the erector platform through the wide MIK gates. Carrying the booster, it crept along the new rail line to the launch site. The missile was out in front of the diesel engine, with all 32 of its nozzles facing the steel trusses of the launch assembly waiting to take them into their embrace.

On that day a tradition was established: the State Commission chairman, the chief designers, the firing range chief of control, and anybody who wanted to, would come to the solemn ceremony to see the latest missile hauled out of the MIK. This first time we all followed the very cautiously moving diesel engine on foot along the rails all the way to the launch site. Subsequently, those entitled to use vehicles would skip the hikes to the launch site.

A large number of fans attended the first installation of the R-7 missile on the launch pad. Everyone sensed that the most crucial phase of our work was beginning; a phase that would determine the fate of many for years to come. It wasn't until the end of the day that Barmin, having personally supervised the entire missile installation process, announced that he had completed his task and said, "now put it to the test!"

And the long—by today's standards—cycle of prelaunch tests began. For the time being all authority was transferred to Leonid Voskresenskiy and Yevgeniy Ostashev. The net "machine" time of all the electrical tests on the first R-7 mis sile No. 5 at the launch site was 110 hours. We tried not to work at night, but the launch site tests on the missile took seven days of round-the-clock work. including the analysis of all the glitches, review of the telemetry films, reports, and heaps of all kinds of procedures due to our lack of experience and sometimes also due to mistakes.

During the tests, while Voskrenskiy and I were standing on the concrete of the launch site clearing up a matter with Kasho about the modification and replacement of a valve in the vernier thrusters' feed lines, Barmin approached our group. After listening to our argument for a while, he said:

"You're going to make a lot more missiles, but this is the only launch pad. If your 'structure' doesn't take off, and falls on my launch assembly, then keep in mind that it will be at least two years before the next launch!"

What else could we do? We assured him it would take off.

"If you release our beautiful missile at the proper time, Vladimir Pavlovich; but, who knows, if your trusses don't pull away, then our girl will show you something."

From the author's archives.

This photograph shows the original R–7 (or 8K71) ICBM at the single launch pad, Site 1, at Tyuratam (now Baykonur) in 1957. This basic version had an unusual conical nosecone that was eventually abandoned during flight-testing in 1957-58. Note also the "tulip" launch structure around the base of the rocket.

Until the missile fueling process began, it wasn't hazardous to be near the missile. Here and there groups of people gathered, arguing and discussing the process of the electrical tests and reports coming in from the console operators in the bunker.

On the morning of 14 May, diesel engines began to bring steaming tanks of liquid oxygen up to the launch site. Ryabikov, who had been at the site, complained, "This is the second time we've left the country with no oxygen." Why the second time? It turns out that at the meeting of the State Commission in Moscow the Central Committee, that is, Khrushchev, announced the requirement to perform the first launch before 1 May as a gift in honor of the holiday. Nesterenko vehemently protested, showing rather convincingly that it would not be possible to prepare the firing range, launch complex, and the missile itself in the 20 days remaining before the holiday.

"Well, if you don't manage, we'll report to the Central Committee and explain why," said Nedelin apologetically.

Nesterenko asked that the order to ship liquid oxygen to the firing range be cancelled: "We can store anything, but we don't know how to store oxygen—it evaporates."

Indeed, in order to fill the missile with oxygen, three times the required amount needed to be sent to Kazakhstan from Russia. Railroad tank cars were not designed for the long-term storage of cryogenic liquids. There was a very high rate of evaporation. The oxygen plant and the storage facility at the firing range had not yet been built. We really did leave our industry, especially metallurgy, without oxygen.

Nesterenko's arguments had no effect. The instructions to ship oxygen to the firing range with delivery before 25 April were fulfilled. After 1 May, all the tank cars returned for a second filling, having enriched the steppe atmosphere with pure oxygen from the first shipment. But the second time, no one had any doubts that the oxygen would be used. By the end of the day all the glitches had been analyzed, the films had been reviewed, and the flight profile had been signed and reported to the State Commission. All the services—all the way from Tyuratam to Kamchatka— reported that they were ready for 15 May.[3] Along the way there were four tracking stations: Sary-Shagan, Yeniseysk, Ussuriysk, and Yelizovo. This is not counting the two local stations. The remote radio-control stations, the universal time services, and the firing range telemetry stations in KUNGs (large vans) were ready.[4]

We had all studied plans for evacuating all the services and residents of Site No. 2, the evacuation of the actual launch team, and the list of persons who would be in the bunker during the launch.

On the last day before the launch, no one managed to rest or get a good night's sleep. All of our time went to analyzing glitches by studying the *Tral* telemetry system films of the last repeated general tests. We needed not only to understand any upswing or downswing of a telemetry parameter on the film, but also to explain them to the State Commission. Finally, after all the readiness reports, the decision was made to begin fueling.

They announced strict procedures at the launch site that defined who should be where, when they should be ready to evacuate, and to where. A large portion of people who were not needed after T-minus one hour were sent to a reinforced area on a hill three kilometers from the launch site. The best place to observe and to receive immediate information in real time was IP-1, the first tracking station one kilometer from the launch site. Three KUNGs containing *Tral* equipment were set up there. The telemetry specialists' cabin had direct communication with the bunker. Just in case, trenches had been dug, and there was an awning to protect high-ranking guests from the rain and sun.

---

3. Although the original Kazakh rendering of the location includes a hyphen ("Tyura-Tam"), by the early 1960s, Soviet engineers simplified the name to simply "Tyuratam."

4. KUNG—*Kuzov universalnyy normalnykh gabaritov*—(All-purpose Standard Clearance Body).

When the list was being drawn up, many jockeyed to get a spot at IP-1, but Korolev and Nosov crossed names off without a thought, reasoning that, first, it was very close to the launch site, and second, extraneous people would interfere with the telemetry specialists' work. I was on the list of people to be in the bunker and thought that my recent stair-climbing speed drill up the steep flight of concrete steps might come in handy.

Launch day was here—15 May. It wasn't until that morning, before we drove out to the launch site, that I recalled that this was the 15th anniversary of the first flight of our BI-1 rocket-plane on 15 May 1942 at the Koltsovo airfield on the outskirts of Sverdlovsk.[5] With whom would I share such a revelation? Of those who participated in that historic event, Mishin, Melnikov, and Raykov were also here at the firing range. When I reminded them, they excitedly responded; after the launch we would have to celebrate two events.

Look at all that had happened over those 15 years! From the primitive plywood BI-1 with an engine that had one metric ton of thrust to the *Semyorka* of the missile era with engines that had more than 400 metric tons of thrust! And in the future the *Semyorka* would carry a warhead capable of destroying any city. But there was no time to indulge in reminiscences and philosophizing.

Launch day dragged on for an incredibly long time. The first fueling was stop and go. Korolev, Barmin, Voskresenskiy, Nosov, Yevgeniy Ostashev, and the officers and soldiers conducting the fueling process emerged and disappeared in the thick clouds formed by the hovering oxygen. I went down into the bunker. There, Pilyugin and Nikolay Lakuzo had joined the Novostroyka officers and console operators from Zagorsk behind a console, trying not to bother them.

Not everyone had gathered yet in the guest room. Glushko was sitting silently looking cool and calm. Kuznetsov was questioning his gyroscope specialist one more time about the setting of the integrator that was supposed to shut down the second stage engine when the missile reached the designated terminal velocity.

In the radio room Ryazanskiy was conducting a routine roll call of all his remote radio-control stations and IP-3 at "the third elevation," where the transmitter to issue the emergency missile destruction command was installed. There was no explosive device on this missile. Therefore, if the destruction command was issued, it would shut down the engines. We had electrically inhibited this command so that it could not proceed on board to shut down the engines before the 12th second of flight. This was sufficient time for the missile to get a bit farther away from the launch site and in the event of an emergency shutdown, it would not destroy it. At the same time, after 12 seconds of engine operation, regardless of what might have happened to the guidance system, the failed missile would not possibly be within striking range of any populated area.

---

5. See Chertok, *Rockets and People: Vol. 1*, Chapter 13.

At one of the last meetings of the State Commission, after yet another thorough review and all sorts of ballistic calculations conducted with the military (the firing range calculation bureau), Korolev reported that the estimated range would be 6,314 kilometers. We considered the primary objectives of the launch to include a launch procedure drill and testing the following: the first stage flight control dynamics, the process of the separation of the stages, the effectiveness of the radio-control system, second stage flight dynamics, the payload separation process, and trajectory of the nose cone before impact on the ground. The total thrust of the engines during launch would be 410 metric tons. The strapon boosters of the first stage would operate for 104 seconds and the central booster for 285 seconds. The estimated launch mass was 283 metric tons. The primary fire-prevention measure during launch was intensive nitrogen purging of the aft compartments of all five boosters.

To ensure the safety of populated areas along the missile's flight route, an integrated emergency engine shutdown system was put in place. If the missile began to spin hard about its center of mass, then once it reached angles of deviation greater than seven degrees, the emergency contacts on the gyroscopes would close, sending commands for the subsequent shutdown of the engines. There was a chance that the missile might begin a smooth departure from the design trajectory due to the zero drift of the gyroscopes themselves. In that case, very great deviations from the route with unpredictable results were possible. We began monitoring using optical observations from the ground and transmitting a radio command for such an eventuality. The responsibility for making the decision to issue such a command was huge. Out of fright, a person could wreck a good missile and disrupt the flight tests. Therefore, a group of the most highly qualified and responsible specialists consisting of Appazov, Lavrov, and Mozzhorin was singled out to supervise. They would be located directly in the shooting plane and observe the missile's behavior using a theodolite. Based on a three-way decision, they would relay by telephone to the bunker a password known only to them and the two launch directors, Nosov and Voskresenskiy. After receiving the emergency password in the bunker, they would press two buttons in succession. This served as a command to the radio-control station 15 kilometers away to broadcast the emergency signal using the directional antenna. An omnidirectional antenna was installed on the missile's central booster to receive this signal. Even if the missile was spinning at that time, it would receive the signal. Highly accountable officers and industrial representatives were at the radio-control station. I bore personal responsibility for the onboard autonomous unit of the system; Ryazanskiy, for the radio link; and the firing range communications chief for the reliability of the telephone and signal communications.

During the very last days before the launch spent at the launch site, we collectively decided to convert the APR system into the AVD (Emergency Engine Shutdown) system. The APR system used an explosive charge in each booster of the missile in order to destroy them before they hit the ground, whereas the AVD merely shut down the engines.

348

It took a lot of time to remove these explosive devices at the launch site. The soldier who dismantled the electrical assembly controlling the detonator reported that while he was performing that operation he dropped a washer that was securing the instrument inside the missile. Finding this washer was more difficult than finding a needle in a haystack. While hunting for it, they raked a heap of all sorts of trash out of the missile, but there was no washer. Finally, in order to end the hopeless search, it dawned on someone to get an identical washer from the "detonator" institute's representative. They furtively attached it to a wire probe with a magnet, and then began to "search" for the lost washer in the missile compartment. Finally, they triumphantly announced: "Found the washer!" They even showed Nedelin the catch that had been hauled in with the magnetic fishing lure. The detonator specialist and the soldier responsible for the violation confirmed that that was indeed the missing washer.

Wow, that emergency system gave my comrades, the developers of the missile's electrical circuitry—Melikova, Shashin, and Pronin, and myself—so much more trouble. Specialists wrote just as much material substantiating its reliability and safety alone as they did about the entire primary control system.

Ryazanskiy was supposed to think up a highly classified password that no more than six persons could know. After a long creative search process, he printed the word "Ivanhoe" in large letters on small strips of paper ripped out of a note pad and inserted them into special envelopes.[6] And so the hero of a novel from the days of knighthood entered the history of Soviet rocket technology.

Soon we realized that for emergency inflight shutdown without radio it was possible to limit our options to just the autonomous portion: a seven-degree contact on the gyroscopes, emergency monitoring of the turbopump assembly's rpms, and the pressure in the engines' combustion chambers. These parameters proved sufficient to cover various emergency situations.

As for "Ivanhoe," 20 years later in meetings with Ryazanskiy and me, OKB-1 chief ballistics specialist Refat Appazov, TsNIIMash (formerly NII-88) Director, Professor and General Yuriy Mozzhorin, and USSR Academy of Sciences Corresponding Member and Director of the Institute of Theoretical Astronomy Svyatoslav Lavrov delighted in poking fun at our general naiveté. Mozzhorin confessed that it was no picnic standing many hours on the desolate steppe in a whirlwind, waiting for the launch on the flight path, knowing that the missile might crash somewhere nearby despite "Ivanhoe."

"How young we were, how much faith we had in ourselves!" This faith in ourselves soon helped us decide to keep the radio-command emergency engine shutdown only in the event of an emergency situation at the actual launch site. We thought up a terrifying scenario: the engines started up, but did not generate the required

---

6. "Ivanhoe," was a reference to the famous historical novel *Ivanhoe* (1819) by Sir Walter Scott (1771–1832).

thrust; the missile remained in the embrace of the launch assembly; flames engulf the missile, the cables are damaged, the bunker's communication with the missile is now lost, and it is impossible to send the emergency command from the console to shut down the engines. In this case, the launch control officer would press the two buttons consecutively, and the saving command to shut down the engines would be broadcast from the radio station at "third elevation" to the blazing missile.

Nevertheless, on 15 May, after the State Commission gave its decision to launch, when it seemed that everything had been thought through, provided for, and reported on, uneasiness and anxiety still plagued me. I'd forgotten something.

That's it! I stopped Ryazanskiy at the launch site as he was rushing to the telephone as always. "Mikhail, this is urgent." At first he waved me away and ran to the field telephone that had just been set up. He repeated some instructions to his radio tower and then prepared himself to listen. I had made up my mind to tell no one but Ryazanskiy what had occurred to me. Others wouldn't understand or would laugh.

"You know, it seems to me that we all subconsciously are experiencing the feelings that overcame Pygmalion. He toiled long and with inspiration, carving the beautiful Galatea out of marble, and fell in love with her. We are all Pygmalions.[7] Here she is, our beautiful creation hanging in the embrace of the steel trusses. And today, if the gods are willing, she should come to life if we have thought everything through and anticipated everything. But if we have forgotten something, then the gods will punish us and either they will not bring her to life or we ourselves will kill her with our emergency commands."

In these surroundings Ryazanskiy didn't grasp right away why I was alluding to Pygmalion. But, after a moment of reflection he told me that my analogy was worthy of the pen of a small-town hack rather than Korolev's deputy.

"Hey, what say we put a smile on Lenya Voskresenskiy's face?"

And he walked right up to Voskresenskiy, who was ever present at the launch site, and grinned as he began to relate my analogy to him. Voskresenskiy remained true to form, and without missing a beat, responded: "If you and Boris are so inclined, then after the launch you won't have much trouble tracking down some live Galateas. As for this one, she's got a lot in store for us yet! We'll be sorry we got mixed up with her."

My romantic diversions ended on that note, but Voskresenskiy's words proved to be prophetic.

Finally, I went down to the console room in the bunker at T-minus 30 minutes. All the places there were already taken. Yevgeniy Ostashev was acting as chief console operator. Next to him was launch control officer Boris Chekunov and on either side the testers from Zagorsk who had conducted the firing rig trials. Nikolay Pilyu-

7. This is a reference to ancient Roman poet Ovid's play *Metamorphoses* about a sculptor, Pygmalion, who fell in love with his creation.

gin and his associates Georgiy Priss and Nikolay Lakuzo were on the left.

A chair had been left vacant for Korolev. The "brain bureau" was in the other rooms: circuitry consultants, electricians, and engine specialists to act in case something went wrong when setting up the launch circuit. Quick prompting was necessary. Inna Rostokina sorted out the most complex electrical circuits faster than all the men. She was the only woman who was permitted to be in the bunker during those hours. Worn out, bulky albums of all the systems' electrical circuits were laid out all over the radio room and "filling room." Nedelin, Keldysh, Kuznetsov, Ishlinskiy, Glushko, and Mrykin were in the guest room. The halls and passageways were already full of missile crewmembers who had completed their jobs at the launch site "ground zero."

At T-minus 15 minutes, Korolev, Nosov, Voskresenskiy, and Barmin went down into the bunker. Nosov and Voskresenskiy took their places at the periscopes. Dorofeyev was in communication with the first IP, where Golunskiy and Vorshev were supposed to comment on the events displayed in the form of shimmering green columns of parameters on the electronic screens of the *Tral* ground station.

T-minus 1 minute. Total silence now. Habit more than memory focuses on what have now become standard commands: "Broach! Key to ignition! Purge! Key to vent! Launch!"

I noted how Chekunov pressed the red button with particular zeal upon hearing the command "Launch!" Gazing at the console, Yevgeniy Ostashev commented:

"The 'ground-to-board' command has passed."

Voskresenskiy was glued to the periscope:

"The gantry has pulled away… Ignition … Preliminary… Main!"

A report came from the console:

"Lift-off contact."

Voskresenskiy exclaimed:

"Liftoff! The missile lifted off!"

The roar of five engines penetrated into the bunker.

Yevgeniy Ostashev informed us, "The console has reset."

There was nothing more to do in the console room. Pushing, I fought my way upward, oblivious to the steep climb, just annoyed at how slowly the crowd of people ahead of me was climbing. Where did they all come from? Finally I sprang out. It was dark; after all, it was 9 p.m. local time!

I made out Nedelin's imposing form next to me. The rapidly dimming exhaust plume blazed brilliantly against the dark sky. But what was this?! It became sort of lopsided. In addition to the main plume, another one had formed. The missile broke out of the Earth's shadow and it glistened, illuminated by the sun, which was invisible to us. It was an otherworldly, unforgettable spectacle. Now we would see the separation! But suddenly against the black sky the lights went out. A small flicker was still shining and moving away from the spot where it had just been blazing so brightly.

Trying not to knock anybody down, we descended into the console room. There would be a report from the telemetry specialists there. Only they could explain why our star had faded away ahead of time. There was a tremendous rush of activity in the bunker. Having cast aside his customary self-control, Mrykin was congratulating and embracing the still stunned Korolev. Voskresenskiy was on the phone interrogating Golunskiy. Everyone was exchanging theories, but no one could explain anything. Barmin was already calling up the bunker from "ground zero" to report that on first inspection no external damage was detected on the launch pad.

Finally, Voskresenskiy tore himself away from the telephone and loudly reported: "Telemetry has visually detected the passage of an emergency shutdown command at somewhere around 100 seconds. They will not say anything more precise. They are taking the film cartridges over to MIK for development.

Korolev couldn't stand it:

"Ask when they will be ready."

"Sergey Pavlovich, let's at least give them a night. By morning they will all be deciphered. It would be useless for us to conjecture who's to blame.

After debates as to what time we should gather for the report on the films, we nevertheless talked Korolev into leaving to have dinner, get some sleep, and after an early breakfast, to hear what the telemetry specialists had to say at 9 a.m.

When Voskresenskiy saw me, he said, "Boris, let's go to my place."

Korolev happened to pick up on this, and in a disgruntled, but rather loud fashion grumbled, "You'd better find out where that command came from and then sort out what's wrong. Boris, your AVD is probably the culprit."

Voskresenskiy's housemates in cabin No. 3 were Barmin and Kuznetsov. Despite our fatigue, we settled into Kuznetsov's most spacious room and over a bottle of cognac we discussed the events, scenarios, and repercussions for another couple of hours. Barmin was very satisfied that the launch system had passed the test. Yes, that alone was already a very big success.

But not just that; after all, the booster cluster had flown for 100 seconds. That meant the cluster dynamics also checked out and it was controllable. It didn't keel over during the very first seconds of flight. That was something we could drink to. At 1 a.m. I was getting ready to go over to the adjacent cabin and get some sleep, but Voskresenskiy got a call from Golunskiy, who reported the results of the film analysis. "Fire in the aft of Block D. The temperature sensors started to go off the scale and went out of order. It was outside the parameters. The temperatures began to rise during launch. Controlled flight lasted for 98 seconds. Then, by all appearances, the fire started and got so big that the thrust of the engine in Block D dropped abruptly and the booster separated without receiving a command. The remaining four engines were running and the control system was trying to restrain the missile. The control surfaces could not cope with the disturbance. They were at their limit and at 103 seconds the AVD command passed validly."

Voskresenskiy asked, "Did you call Sergey Pavlovich?"

"Yes, I gave him the report. He demanded that we find the source of the fire. Now let's examine all the other parameters.

352

"Look here, Boris," said Kuznetsov, "now you and I must drink another drink. It was my gyroscopes that gave the command and your emergency engine shutdown unit that did just that for the first time on the very first missile. Your control surface actuators put up a good fight for the missile's life."

These were sufficiently convincing arguments for us to polish off the bottle.

Chapter 19

No Time for a Breather

In the morning everyone knew about the fire. But what had caused it? Trumped up "authentic" versions were already circulating. The State Commission, technical management, and everyone who could elbow his or her way in gathered in the small meeting hall.

Voskresenskiy and Nosov reported what they had observed of the launch from the periscopes. They noticed an intense flame that rose up all the way to the support cones as the engines built up to the preliminary stage. The exteriors of the first stage boosters were engulfed in flame over their entire height, but as the engines were building up to the main stage, evidently a stream of air put out the flame and the missile blasted off completely clean. They noticed no fire during liftoff. Nevertheless, the source of the fire was clearly found in Block D. A kerosene pressure sensor downstream from the pump first showed a normal buildup and then the pressure began to fall and reached zero. This indicated a leak in the line feeding kerosene to the engine. The turbopump assembly of Block D operated normally, and kerosene under high pressure gushed through some hole. A fire started in the aft compartment right on the launch pad.

It was simply amazing that the missile was able to fly for another 100 seconds! It had fought heroically. It had been so close to separation! There were no glitches in the core booster. If it had held out for another 5 to 10 seconds, the command for separation would have passed, and then the second stage, having gained its freedom, could have continued the flight.

How frustrating that this occurred during the first flight! Such a routine defect should have been detected on the ground during testing at the engineering facility. Heated debates before and after the meetings confirmed that the leak might have developed during the lengthy, jostling transport by rail. Such cases had occurred even on the R-1. In 1950, a mandatory requirement was introduced for the R-1 and R-2, pneumatic testing after railroad transport. As the missile traveled thousands of kilometers, it bumped against the rail joints, loosening numerous flanged and fitting connections in the fuel lines. And during resonance there had even been instances of breaks in loosely laid pipelines. All missiles underwent pneumatic tests at the engineering facility, but they forgot about them for the R-7!

Asif Siddiqi.

Valentin Glushko (1908–89), the giant of Soviet rocket engine design, shown here around 1958, probably at the time of his induction as a full member of the USSR Academy of Sciences.

Although the leak occurred in lines that were under Glushko's jurisdiction, we all felt our share of guilt. Voskresenskiy, who rightfully considered himself a specialist on hydraulic systems and their testing, blamed not only Glushko's deputy Kurbatov and our designers Voltsifer and Raykov—who had supervised the propulsion systems—but also himself. Korolev did not feel at fault this time. He felt that Glushko had been rightly punished for his self-confidence and hubris before the launch.

The guidance specialists felt like golden boys. The behavior of the stabilization controllers, all the instruments, and control surface actuators had matched up almost completely with the oscillograms that Zhernova had obtained on the electronic simulator. We had thoroughly analyzed the oscillograms and compared various segments of the actual flight that had been recorded on the *Tral* films. Zhernova commented, "And for some reason I had been afraid for your control surface actuators. Look how well they responded to all the commands and how tenaciously they fought for the missile's life." Thus ended the life of the first *Semyorka* No. 5. The gods did not pass up the opportunity to punish us for letting our vigilance slip.

The State Commission decided to urgently prepare the next missile, No. 6, or factory designation M1-6. All involved in the project were forbidden to leave for Moscow. Commission members flew out only with the chairman's permission while industry employees could do the same only with Korolev's permission.

Engineers quickly formulated leak tests that were introduced for booster cluster No. 6. And that was a good thing. We found so many potential fire sources that it was a wonder that the Block D was the only one to catch fire on the previous *Semyorka*. After hearing the stories about the flames that had engulfed the entire lower part of the missile before liftoff, it was decided to place additional thermal shielding on all the onboard cables.

Meanwhile, Konstantin Nikolayevich Rudnev replaced Ryabikov as chairman of the State Commission. Korolev flew to Moscow to prepare a design and speed up plans for satellites. This was something he didn't like to discuss, I assume out of fear of jinxing it. He did have that streak in his personality. We knew this, but we didn't

356

say anything bad about it.

Unbearable heat set in at the firing range. The multicolored tulips were gone. The steppe dried out and began to scorch and take on a uniform reddish-gray color. The MIK gradually heated up, and it was pleasant to work at the engineering facility only in the evening and at night, when we opened up the wide doors to let the cool night air blow through. How many times over the course of a day did I step out of my cabin and tread over the path to the MIK that I had studied down to the last pebble! At that time they had just completed the military barracks on the hill, a fire station on the left-hand side of the concrete road, and a large dining hall on the way to the MIK. This was the road I followed from the residence area to the TP all the succeeding years, in unbearable heat, and in icy wind, inhaling deep breaths of the intoxicating steppe air in the spring.

During the first year of life at the firing range, the concrete road ran from Site No. 10 to our Site No. 2, and then over the bare steppes on to the MIK and the launch site at Site No. 1. If you were traveling from the Tyuratam station, just to the left there was a solitary railroad spur over which trains carried officers to work in the morning and took them home in the evening.

Gradually the steppe became more developed. Two years after the firing range was opened, you could walk over pedestrian sidewalks that had been laid alongside the concrete road to the MIK. We got respite from the heat from the shade of poplar trees planted along the roads and from the fine spray of the irrigation sprinklers that saved the first plantings from impending death.

In April 1991, at festivities marking the 30th anniversary of Gagarin's flight, I walked this route as a guest, veteran, and tourist, carrying my camera. I strolled along the same route that I had followed 34 years before, but this was a road where "everything was the same and nothing was the same." The steppe, that same insufferably hot and scathingly cold, dusty steppe of Kazakhstan, abloom with tulips, was simply not visible anymore. One could only admire the numerous service buildings, the official cottages, and the distant panorama of colossal buildings built for programs like the N-1, Energiya, and Buran.[1] Only the fire station to the left, the barracks to the right on the hill, and the smokestack of the first boiler house on the low-lying land by the railroad tracks remained untouched and took me back to that long-ago, difficult but wonderful time.

During those first years when I walked to the MIK or returned weary to have a turn at "horizontal tests," as we referred to a brief rest, everyone I encountered along the way was a comrade, a friend, or, at least, a like-minded individual. I was confident that I had no enemies here. There was nothing and no one to fear except the odd "monkey wrenches" in the works that might throw off the next missile. And this wasn't fear, it was the nature of our work. We all derived genuine pleasure from

---

1. The N-1 and Energiya were the Soviet Union's two "superbooster" projects. Buran was the piloted space shuttle designed for launch on Energiya.

searching for and discovering our own mistakes. When we were preparing new missiles for launch, we always found new and unexpected quirks, but these idiosyncrasies didn't get us down. We knew that the next quirk would not be the last.

During our first years of operations at the firing range a certain sense of community united us, individuals of various ranks: marshals and soldiers, ministers, chief designers and young engineers. We were all working under strict secrecy. The newspapers had still not written anything about us, and Levitan's voice had not yet broadcast our successes to the entire world over the radio.[2] But you can't hide a missile blasting off from thousands of eyes. Each person who had seen its exhaust felt connected to something that united him or her with all the others who were here, regardless of what their role was.

But missiles did not take our feelings into consideration. The second one launched—*Semyorka* No. 6—simply did not wish to take off. We had created the R-7 missile as a weapon. One of the most crucial parameters for a missile, even an intercontinental missile, is the time required to achieve readiness, that is, the length of the preparation cycle from the time it is delivered to the launch site until launch. For the first launch we spent almost 10 days at the launch site. Everyone understood very well that we would no longer have the luxury of such a prolonged preparation cycle. Therefore, in addition to all the other tasks, we decided to develop prelaunch tests, strictly standardizing the time spent on all operations.

Missile No. 6 was delivered to the launch site on 5 June, 20 days after the first launch. At that time, such an interval seemed reasonable considering the large number of modifications and additional pneumatic tests that had been conducted at the engineering facility. Preparation and testing at the launch site went considerably faster, and five days later the missile had already been fueled and was ready for launch. The entire launch schedule was repeated. While the first launch was fraught with much anxiety and various prognoses, everyone was much more optimistic about the second launch. After all, the previous launch had proceeded almost all the way to the most thrilling and enigmatic moment, the moment of separation.

The second launch attempt took place on 10 June 1957. According to the flickering displays, everything went normally up to the moment the "Launch" button was pushed. Ignition also occurred. Suddenly, shutdown! No fire engulfing the missile. One could barely hear the clicking of relays. The lights on the display console died out, and the message "Circuit reset" appeared.

This meant that the electric monitor in the form of end contacts and relays had detected that some valve had failed to open or that damage had occurred in the circuit. The albums of circuit diagrams stockpiled earlier in the bunker immediately

---

2. Yuriy Borisovich Levitan (1914–83) was one of the most famous radio announcers of the Soviet era. Few who lived through Soviet times have forgotten his momentous announcements of the great events of the era, including daily bulletins during World War II and the final defeat of the Nazis in 1945.

came into play. All the circuitry specialists dove back into the already tattered pages and, using their experience and intuition, tried to figure out what had happened.

While they were searching feverishly for the cause of the malfunction, Korolev, Voskresenskiy, Nosov, and Glushko decided to attempt another launch. To do this, some members of the launch crew had to run to the missile and change the igniters and others had to set the launch system in the initial state, rig the gantry, connect the connectors that were thrown off, and feed power to the missile from the ground source.

We couldn't dawdle; everything had to be done very quickly. Rapid evaporation was a weakness of oxygen-fueled missiles, and it also meant a decrease in the oxygen supply. Delays at the launch site might require that tank cars holding liquid oxygen, already withdrawn to a safe distance, would have to travel back over the railway spur to refuel. We estimated and calculated and decided to go for another attempt without refueling.

A little more than 2 hours later everything was ready for another launch attempt. We had ignition and then the circuit reset again. Now it became clear that until the cause was understood, it was pointless to reattempt a launch. It was getting on toward evening, and this workday at the launch site had started at 7 a.m. Someone gave the order, and something akin to a buffet was brought in. Even if it wasn't exactly a full lunch, you could at least drink some mineral water. The telemetry specialists once again came to the rescue. After the first circuit reset, they managed to send the film to be developed. After the second reset—when everyone felt nearly exhausted, the telemetry specialists joyfully reported, "The *Tral* recorded an indication from the KD—the contact sensor—that the main oxygen valve in Block V failed to open."[3] Again, a breakdown caused by Glushko's system. The circuitry specialists feverishly analyzed, discussed, and issued their findings: "As it should be, everything is OK."[4] All discussions were going on right then in the bunker. Luckily it was still the coolest place at the launch site.

Once again everything was reset to the initial state, and the igniters were replaced. Korolev asked Glushko, "What's your decision?" The latter gave it some thought. Voskresenskiy suggested, "Let's direct hot air from the air heater onto the valve. The valve probably froze up from moisture. Let's warm it up and try again."

What could we do? There were no other suggestions. We'd lost a lot of time, and even more would be required: we needed to give the command for the oxygen tank car to return and fill up the missile. This meant that we were returning to T-minus 4 hours. The entire multitude of services all the way to Kamchatka were given a 4-hour delay. Everyone, except those of us at the launch site, could relax.

There was no smoking in the bunker. Pilyugin—who still smoked back then—

---

3. KD—*Kontaktnyy datchik.*

4. In the original, Chertok uses the acronym "TDB" for *tak i dolzhno byt* which loosely translates as "as it should be."

Voskresenskiy, and I climbed up to the surface and sat down in the "smoking room," not far from the entrance to the bunker. They had already switched on the flood-lights at the launch site. The first stars were lighting up in the darkened sky. Pil-yugin was the first to cave in under the prolonged uncertainty. He demanded an answer from Voskresenskiy and me, asking, "What will happen on our third launch attempt?" I answered that the missile would take off, and then it would be time for us guidance specialists to pull a rabbit out of a hat. Voskresenskiy quoted: "Night is already falling and still no Herman. You won't be involved yet." He continued, "I have a feeling that Valentin [Glushko] hasn't exhausted his entire stock of 'monkey wrenches' yet. We won't blast off today."

And once again he was right. On the third attempt the ill-fated valve opened. The missile built up to the preliminary stage and … stalled there. There was no transition to main stage at the designated time. We guidance specialists had pro-vided a time lock in the automatics circuit for such an event. If, considering all allowable variations, the engines did not make the transition from the preliminary to the main stage at the designated time, there was a general emergency shutdown. The missile was soon engulfed in bright flames lapping in the darkness, and then … suddenly it all quickly died out.

This happened at midnight between 10 and 11 June. Now the discussion in the bunker did not come down to the question: "Why did this happen?" An urgent decision was required as to what to do with the missile. Glushko answered unequiv-ocally: "There can't be another launch attempt. As soon as the "Preliminary" com-mand is given, kerosene entered all the combustion chambers. We need to dry them out completely and maybe even replace them."

After the official report to Rudnev, Korolev announced the decision of the tech-nical management. "Drain the fuel and oxidizer, remove the missile, and return it to the engineering facility. Create a commission headed by Voskresenskiy to determine the causes of all of today's incidents." Our truly heroic battle with the headstrong missile ended so disgracefully.

We took our time with the third M1-7 missile, nicknamed the *Sedmaya semy-orka* ("seventh seven" or "seventh *Semyorka*"), and spent a month preparing it at the engineering facility.

The morning after the previous night's defeat at the launch site, I made a great effort to try to speed up the missile's preparations. Having examined the systems' readiness and the test results, I reported to Korolev that we would be ready to haul out the missile no sooner than 6 or 7 July. Considering that we would spend five or six days at the launch site, the next launch could be scheduled for 12 July.

Korolev agreed in principle, but requested, if Voskresenskiy's commission didn't add a lot of work, that we nevertheless shorten the preparation cycle at the engineer-ing facility by about 10 days. He himself had to leave and granted permission for all the chiefs to go home for a visit. The State Commission would fly out too. After clarifying the causes for the failures, Voskresenskiy would also be allowed a short respite, but I would be left to get the seventh *Semyorka* ready.

To make it easier for me to endure such a prolonged stay at the firing range, Sergey Pavlovich led me to his large refrigerator. "I want to sweeten your solitary stay in this cabin." He opened the refrigerator and pointed to an enormous chocolate cake. The cake was magnificent. "Nina Ivanovna sent it to me quite recently as a surprise. I give you permission to enjoy it, but not with too many people, and leave a little for when I return." Many years later I confessed to Nina Ivanovna that her cake really was a hit for several evenings in cabin No. 2 at *Dvoyka* (Site No. 2).

It wasn't easy, but the commission investigating the causes for the failed launch uncovered the truth. It was right out of the "not-in-your-wildest-dreams" department. During the factory assembly of the central booster's onboard hydraulic system, the engine's prelaunch nitrogen purging valve was installed backward. Although the valve had an arrow engraved on it indicating the flow direction, the fittings on the inlet and outlet had identical threading, which caused the error. The assembly worker simply turned the valve however he saw fit, because it wasn't his job to know which way the arrow was supposed to point. He would have had to study the hydraulic system to know that. Where were the inspectors and military representatives looking? The debacle did not slow down our investigations; hot on the trail, we also discovered precisely the same error on the next missile that we had just begun to prepare.

As a result of this mistake, the purging of nitrogen didn't terminate before the launch. Gaseous nitrogen entered the oxygen chambers of the main engine's and vernier thrusters' combustion chambers. The kerosene didn't ignite in the oxygen/nitrogen atmosphere, there was absolutely no engine build-up, and without waiting until the designated time of zero pressure in the combustion chambers, the control system automatics issued the command to shut down all of the cluster's engines. That is when we recalled the stringency of triple control that we had heard so much about in the past year while preparing the R-5M missile carrying an atomic warhead.

Fool-proofing is one of the most difficult problems, and not just in complex technical devices. An American driver's education handbook put it this way: "When you get the notion to drive out onto the roadway, remember that you are not the only idiot sitting behind the wheel at that particular moment."

This time we actually got off cheap. The missile was intact, and after undergoing a checkout procedure, it could be prepared for another launch attempt. The launch pad had suffered no damage either. Only the oxygen, once again taken from industry, was wasted. Each launch preparation cycle was good training for the officers and soldiers of the launch control team. And it also became clear to the industry employees that it was still too early to let things get to your head.

The engineering facility kept its promise regarding the deadline, and on 7 July the missile was hauled out of the MIK to the launch site for the third time. Everyone had flown back to the firing range by this day. This third departure from the MIK was just as festive as the first. The diesel locomotive slowly pushed the erector and missile ahead of it. "Cannons travel backwards into battle"—in these lines by

Tvardovskiy from *Vasiliy Terkin*, you might as well replace the word "cannons" with "missiles," at least when talking about the R-7.[5]

We were considerably more organized in preparing the seventh *Semyorka* at the launch site. Despite temperatures that shot up to 45°C (113°F) in the shade, we completed work on it without all-hands rush jobs at night or extra stress. The missile was ready in five days.

The same body of people once again gathered in the bunker, and the same battle-tested team sat at the consoles. This time I talked Korolev into letting me go to the first tracking station (IP-1) at T-minus 30 minutes. At last I would be able to take in the missile's liftoff from its first second rather that running out of the bunker 60 seconds into its flight! So for the first time I saw the launch of the R-7 missile on our third attempt on 12 July 1957.

This is how I remember that day: after the flash of ignition, a tumultuous dance of fire appears under the entire missile. A second later flame engulfs the missile over the strapon boosters from top to bottom. One begins to fear for it. It seems the tanks will explode now, destroying the launch complex and burning it down. But in an instant the engines build up and a stream of air pulls the whirling flame downward into the enormous unseen concrete escarpment. The trusses resting upon the missile's waist glide apart. Five engines merging together, the blinding, triumphantly roaring plume cautiously lifts the cluster's 300-metric ton body. Even from a kilometer's distance, the roar of the engines, incomparable to any other sounds, is deafening.

Taking its time, the missile rises upward. In the flame billowing about the aft section I can clearly see the supersonic fronts. Before I know it, the missile is on course and leaving the launch site. The first 5 to 7 seconds are terrible; what if a false command were suddenly to pass and shut down just one engine! Then the missile would fall apart and shower down on the launch site and perhaps even this tracking station.

I am overcome with the sense of being a part of this creation, dreadful, powerful, and yet familiar and dear. I want to imbue the fire with my own will, my aspirations, and my entire being. Come on, fly! Now I no longer fear for myself but for her, for the missile. Will she make it this time? But I didn't have to ponder for long. The time count was coming over the loudspeaker, and somewhere after 35 seconds, the missile made a smooth, triumphant liftoff and departed into the dark blue evening sky. The missile spun about its longitudinal axis, and suddenly the strapon boosters flew off of the core! The missile was destroyed!

The five hot, smoking boosters are now coasting along, but gradually they descend, and somersaulting, they fly over the hill to the horizon. A tragedy, this

---

5. Aleksandr Trifonovich Tvardovskiy (1910–71) was a famous Soviet poet who gained fame while working as a journalist on the war front during World War II. His poem "Vasiliy Terkin" about a resourceful Soviet soldier later became part of postwar Soviet folklore.

staggering spectacle as yet another *Semyorka* dies. What sort of curse hangs over this missile? When earlier missiles had crashed, it was sometimes horrible, sometimes intriguing, and always vexing. This time I experienced pain. As if before my eyes a close and dear individual was dying. And all of us left behind on the ground were powerless to help.

Jarred back to reality, I got into the KUNG that carried the *Tral*. Others had already packed up the film canisters and were getting ready to drive to the MIK to develop them. What had we seen? Golunskiy, Vorshev, and the rest said in unison, "At 38 seconds there was a big command for rotation. Everything started spinning! And you saw what happened after that." If a great disturbing torque or a false command for rotation occurred, then it's not surprising that the strapon boosters fell off. It seems that this time the gods were not angry with the engine specialists but with us guidance specialists.

Practically all night, Pilyugin, his team, the telemetry analysts, and I sat in the screening room poring over the films that had been brought to us still wet. The picture was clearer in the morning after thorough analysis. Back during preparation at the engineering facility it was a complete surprise for me that, beginning with this vehicle, an instrument had been introduced into the stabilization controller integrating the rotation channel signal for the strapon boosters. I couldn't understand why stabilizing the moment for rotation over and above that already flight-tested was needed. No one could explain this properly. The false command in the rotation channel apparently came from this instrument, the IR-FI (integrator for the rotation angle). I was incredulous, albeit post-factum. "Whatever possessed you, Nikolay, to insert that instrument? To begin with, you should have at least checked it out in telemetry mode." Crushed by his obvious guilt, Pilyugin could find no excuse.

It surprised me that his staff, who had always been "stauncher monarchists than the king himself" when something went wrong in their kingdom, did not stand up for their boss this time. They also felt guilty for the unnecessary innovation. When they began to sort out the situation in greater detail, they still did not find the clear causes for this intense command for rotation that occurred in flight. Even a failure of individual elements in the new instrument could not cause such wild behavior.

After running through many scenarios, one remained: a short-circuit in the controlling circuit inside the instrument. Only in this case could there be a signal comparable to the one that occurred in flight. For verification they opened up a spare instrument. A visual inspection did not suggest any ground faults. If one had occurred, then why did it occur 38 seconds into the launch and not sooner? But once we had decided to do away with this instrument in the future, we moved on to the scenario that always bailed us out in baffling situations: "foreign particle." This current-conducting malefactor had been hiding in the instrument from the very beginning. It had been left there as a result of an inadequate control process. During inflight vibrations and under the effect of g-loading, it began to move and managed to connect one of the exposed command circuit pins with a nearby cable shield.

Appropriately, for the next, now the fourth launch attempt, the following prophylactic measures were established: the rotation channel on all the strapons was disconnected from the integrator of the stabilization controller, and all plug and socket connectors were flushed with alcohol before their final mating and then wrapped in adhesive tape to prevent "foreign particles" from getting in.

When passions had subsided, I met with Zhernova and asked: "Nina, did you really simulate the processes of the stabilization controller with this IR-FI? The first time the missile got almost to the point of separation without it. Then you and I did a detailed analysis of the launch, and you even praised our control surface actuators. Why was this improvement necessary?"

Zhernova replied that she had been against this change, but she could not convince Nikolay Alekseyevich. He insisted, and the circuit was modified beginning with this missile. "Just please don't tell Nikolay Alekseyevich that we had this conversation. I feel very sorry for him now. He had so looked forward to this launch and had been so confident in it. Now it turns out that the crash is his fault."

My next hardship came in a conversation with State Commission Chairman Rudnev. Smiling, he began by joking that not just the engine specialists but also the guidance specialists had learned to kill powerful missiles with the aid of "foreign particles." And so that this would not happen in the future, he asked me not to leave for Moscow (though Korolev had given permission), but to stay to prepare the next missile. "I guarantee you that, regardless of the outcome of the next launch, we will let you go immediately either home or on vacation."

I was completely floored when he said that he and Mrykin had decided not to leave, but to stay at the firing range until the launch. "Here, of course, it's very hot, but in Moscow, if you show up, they give you such heat that you immediately regret that you went there."

At first I protested, "I haven't budged from this place in almost four months!" But Rudnev begged me and recommended that I go on a fishing trip. I gave in.

Before his departure, Korolev said that he was to have some very serious meetings with nuclear physicists in Moscow. They were proposing a new warhead for the *Semyorka* with a slightly reduced yield but almost two times lighter than the existing one. This would instantly increase our missile's range by about 4,000 kilometers. "Twelve thousand kilometers! We will be able to reach the Americans from any spot on our territory!" said S.P. with animation. "Just for the time being this shouldn't get around. Nedelin said that he made an agreement with Khrushchev about building operational launch sites near Arkhangelsk and we'll build one more here, a backup." I was surprised that Korolev wasn't at all dispirited by the demise of the last missile.

Rudnev gave me to understand that duties for the immediate future had already been assigned. Nedelin, Keldysh, and Korolev were giving presentations in Moscow before the Central Committee. There would be a meeting with Minister of Defense Malinovskiy and perhaps even with Khrushchev himself.[6] Despite the first setbacks,

---

6. Marshal Rodion Yakovlevich Malinovskiy (1898–1967), a famous wartime veteran, served as USSR minister of defense in 1957–67.

Korolev intended to insist on allocating two rockets for the orbital insertion of artificial Earth satellites. The Americans had announced that they were preparing such a sensation to commemorate the International Geophysical Year (IGY).[7] If they got the jump on us, this would be a severe blow to our prestige. While they were conducting an active defense in Moscow, we here must at all costs conduct a failsafe preparation of the M1-8 missile. A successful launch needed to be executed no later than early August. Otherwise, everyone could expect all hell to break loose.

At that time Rudnev was a deputy minister of the defense industry, a person who undoubtedly had assessed the situation critically and realistically. Taking advantage of the situation, I asked, "Well, what can they do with us? After all, these days it's considered bad form to imprison us or send us to Kolyma."[8]

"Yes, indeed," answered Rudnev, "No one is going to put us in prison. But they can assign our missile, or more accurately, your missile, to others. You shouldn't forget that Khrushchev is supporting Chelomey's proposals. Yangel also has proposals for a new missile."

It was difficult to say what capabilities Chelomey had, but we had trained people at the Dnepropetrovsk factory and Yangel's design bureau, and they were a powerful force. They had already mastered our R-5M, had developed their own first missile, the R-12, a rival of the R-5M, and now they were working on a new intercontinental design.[9] Yangel did not conceal his negative attitude toward liquid oxygen missiles. The military, too, was waffling. Of course, oxygen and kerosene was inert and safe. Nitrogen tetroxide and dimethylhydrazine were toxic components. Bluntly put, it was horrible using them, although we did just that with the R-11 and other nitrogen-fueled naval missiles and it was fine. We even got used to them on submarines.

Rudnev continued, "I had a frank conversation with Sergey Pavlovich. He has a lot of interesting proposals and far-ranging plans. But one or two more mishaps with the *Semyorka* and all of this might be transferred to other people. Keep in mind that even Nedelin might waffle. And, after all, he's the only one among all the marshals who understands our technology. We can't count on Malinovskiy's support. He doesn't even see beyond his own past experience as a combined arms division or even army commander. He tolerates us just because Khrushchev needs a missile. Nikita Sergeyevich has faith in us for the time being." Rudnev hadn't told me any-

---

7. The International Council of Scientific Unions (ICSU) designated the period between July 1957 and December 1958 as the International Geophysical Year (IGY), a time when scientists from all over the world would jointly study geophysical phenomena in remote areas of the Earth and the upper atmosphere. In October 1954, the ICSU decided to include satellite launches as part of the IGY program.

8. One of the Gulag's most notorious forced labor camps was located in Kolyma in northeastern Siberia.

9. This new intercontinental missile was known as the R-16 (later known in the West as the SS-7 Saddler).

thing fundamentally new, because we ourselves had learned to assess the political situation throughout these years, and based on the catchwords at numerous meetings attended by the highest officials, we had a sense of "who was who."

EACH MORNING, BEFORE THE UNBEARABLE DRY HEAT SET IN, I walked to the MIK. On 20 July we unloaded all the boosters of missile No. 7 and arranged them at the work stations. Each booster had been transported by rail in special closed four-axle boxcars. The central booster was so long that it had been constructed in two sections. Each of them was transported in a separate boxcar.

In the MIK we would assemble the core booster and mate a large number of electrical connectors, pneumatic lines, and hydraulic lines. The most crucial operation was connecting the large-diameter pipe using flexible corrugated tubing through which liquid oxygen was fed from the upper tank to the propulsion system through a tunnel passing through the lower kerosene tank.

Mikhail Lomakin, a very experienced machinist foreman from our factory, supervised the assembly of the core booster. When the missile boosters had been prepared for assembly, I asked him to complete those operations as quickly as possible, since we could not begin the electrical tests until all the machine assembly operations had been completed. When electric power was being fed on board, no one who might disrupt the progress of the electrical tests with their movements was supposed to be inside the missile boosters.

By mid-day the temperature in the main assembly and testing hall of the MIK began to exceed the temperature outside. Back then air conditioning was unheard of. Fans only pushed the hot air around, and besides, it was forbidden to switch them on. They raised so much dust that you couldn't work, and it was impossible to guarantee the reliability of the instruments and assemblies that weren't protected against the all-pervasive gritty dust.

Yevgeyiy Ostashev and I agreed to begin the electrical tests "when it's cold," that is, after the sun went down. I was about to rest a bit after lunch when suddenly the secretary of the State Commission called up and alerted me that despite it being 50°C (122°F) in the shade, Rudnev wanted to come meet with me and the military representatives and look over the missile preparation schedule.

In town at Site No. 10, the so-called "ground zero quarters" had been erected for particularly high-ranking administrators such as marshals, generals, State Commission chairmen, and their adjutants or secretaries. These quarters, comprising two hotel buildings, afforded maximum possible comfort under those conditions at that time. The area adjacent to the hotels was landscaped like a garden or small park descending right down to the Syr-Darya River. It was probably more psychological than actual, but the proximity to the water helped us endure "the mid-day heat in the desert of Kazakhstan." That's how the local wits reworded the lines from Lermontov's "The Dream."[10]

---

10. "The Dream" was one of Lermontov's last poems, the first line of which is "In noon's heat, in a dale of Dagestan."

366

It was not more than 30 minutes by car from the ground zero quarters to our *Dvoyka* (Site No. 2). For exercise I set out for the MIK on foot. After making my way over the hot expanse, when I entered the suffocating bay I was absolutely dripping with sweat. To my surprise the core booster had not yet been mated. One of the workers explained that Lomakin himself had been inside it for more than two hours now, and they were afraid that something might have happened to him. I started talking to Lomakin through the hatch. He promised to come out soon. While we were mulling over how he could work in such heat inside a missile in a space so cramped that any movement was restricted, Rudnev drove up. We went into the console room with him. There Yevgeniy Ostashev had prepared the schedule of operations. Soon thereafter, red as a lobster, Lomakin stopped by and asked if he could have a word with me. He explained that while installing the corrugated tubing connecting the two parts of the duct, he had lost one of the six 10-mm bolts along with the nut. He had connected all the flanges but he didn't know what to do next. My blood ran cold, "Are you sure that that bolt didn't accidentally end up in the duct?"

"Yes," answered Lomakin, "I can vouch for that. I removed all the bolts from the flange before the mating operation and placed them in a recess there. When I connected the flange and began the assembly, instead of six bolts, only five turned up. I felt and looked all around and they were nowhere to be found."

"Relax," I recommended, "Think about it, try to recall the whole situation, have a cold drink of water, and get back in there and have another look. Until we find it, there will be no work on the core booster. We must have an absolute guarantee that the bolt is not in the duct. If it's there, an accident is inevitable. It'll get caught in the oxygen pump and then . . . well, you get the picture."

When I returned to the console room, Rudnev wanted to know what had happened. I wasn't about to hide it and explained. He said that he wasn't going to leave the MIK until we found the lost bolt. Before once again setting out on his quest, Lomakin turned all the pockets of his coveralls inside out in the presence of the military rep and the controller to prove that he wasn't taking a spare bolt inside with him. An hour passed, then another. Word spread throughout the MIK about what had happened. Someone came up with the idea of "rolling" the booster and determining where the bolt was by listening for the sound of its rattling around. But we awaited Lomakin's return. After a little more than two hours, he emerged beaming from the missile and triumphantly lifted his find over his head for all to see. We all congratulated Lomakin, and he crawled back in to put the sixth and last bolt in place. Rudnev, who, it seems, was more pleased than we with the happy ending, proposed issuing a directive granting Lomakin a monetary reward for integrity and dedication in the fulfillment of his assignment. I wrote a directive to the expedition chief calling for a monetary reward of 250 rubles. Rudnev stamped it "Approved." When the completely haggard Lomakin climbed out of the core booster, having completed all of his work, I solemnly handed him this document.

The next day via high-frequency communication from Podlipki, Korolev called me and asked for an update on how things were going. I reported everything in detail, saving the incident with the bolt from the day before until the end. S.P.'s calm conversational tone instantly changed. Even over this voice-distorting communication line I sensed that he was choking with agitation. "You need to punish people for such doings, not reward them! You let everybody slack off, and you're still handing out prizes! Revoke that reward immediately and issue a reprimand! Mr. Nice Guy." When I told Rudnev about my conversation with Korolev, he was amused. "I'm the only one who can revoke it because it requires my permission. I am not going to revoke anything. Sergey Pavlovich will forgive us. When he gets here he'll have too much else to do."

Chapter 20
Mysterious Illness

Several days after the incident with the IR-FI, horizontal electrical tests began at full speed. I sent radiograms to Moscow calling all the specialists who had been dismissed for home leave back to the firing range. According to the schedules that Yevgeniy Ostashev, Anatoliy Kirillov, and I had developed, preparation of the missile at the engineering facility would be completed on 12 August—if there were no incidents. Considering the heat and any possible unforeseen circumstances, we decided to add three days and declare 15 August the date the missile would be hauled out to the launch site. There were 20 days remaining until this date. If you figured in another 5 days at the launch site, the launch could take place on 20 August.

It was already late evening when, having noted down all the key dates for the preparation process, I set out from the MIK to Korolev's cabin. Along the way I mulled over the conversation I was to have with him the next day via high-frequency communication. My task was to convince him to accept our proposal and at the same time not to recount all the mess-ups we'd racked up at the beginning of the tests. It would be simpler to explain everything to him when he arrived.

Striding along the well-traveled road from the MIK that I had treaded down so many times, a strange queasiness came over me. When I arrived at the cabin, despite the heat, I decided to take a hot shower to see if that would make me feel better. Each cabin had water heaters installed for baths and showers. The water heater was fueled by ordinary firewood. Firewood was scarce in this treeless area, but Lena, an expedition worker who took care of our household chores, always managed to get it for the chief designers' cabins. She managed to maintain exemplary cleanliness in all the cabins and kept us stocked with mineral water. I lit the water heater and took a hot shower. I gratefully remembered that the attentive Lena had left a bottle of Borzhomi mineral water in the refrigerator. But with my very first sip I began to shake with the chills. I crawled under a blanket trying to get warm. The chills didn't go away. Unbelieveable! The thermometer in the room read 30°C (86°F), and yet I felt cold. I went into Mishin's room and pilfered the blanket from his bed. Covered with two warm woolen blankets, I decided to go to sleep.

In the morning when Lena came to clean and discovered that the blanket was missing in the empty room, she suspected something was wrong. When her knock went unanswered, she entered my room, and as she later told me, got quite a fright. I was lying with my eyes wide open and did not respond to her questions.

She dashed over to expedition chief Sukhopalko. Having intercepted the local nurse en route, he came to my cabin. I remember that when I regained consciousness, I recognized him and asked what had happened. The nurse touched my forehead and in a panic showed Sukhopalko the thermometer that she had managed to insert in my armpit. When they removed the thermometer, it turned out I had a fever of just over 40°C (104°F). Sukhopalko guessed he should call up ground zero quarters. He asked for Mrykin so that he could rely on the garrison hospital chief to request that a doctor be sent immediately. Until the doctor arrived from Site No. 10, the nurse and Lena fed me hot tea with raspberry jam that they'd gotten who knows where. An hour later a lieutenant colonel from the medical service appeared. He brought with him a lab technician, who drew blood to analyze then and there. The doctor could make no diagnosis before receiving the results of the analysis, but he gave me some tablets to reduce my fever and some antibiotics. Fearing that I had been infected with some variety of the plague, the doctor forbade anyone to visit me, asked the nurse to stay, and, if I got worse, to call him immediately. He promised to return as soon as he got the results of my blood work.

Indeed, the doctor did return that evening, but to my surprise Mrykin was with him. From the doctor's lengthy explanation I understood that the results of my blood work had frightened the medical staff. According to all reference books, the results of my blood tests indicated radiation sickness. The results did not fit any other diagnosis.

I already felt better than I had that morning. I tried to stand up, but I wobbled. Mrykin announced that he had already made arrangements with Moscow: I would be admitted to Burdenko Hospital. He had booked a plane for tomorrow, and I had to be ready to depart. He had called Moscow to arrange for someone to meet me at the airport.

My mind was still racing with thoughts of the testing process, hoping that the launch would at last be successful . . . so this hit me like a bolt out of the blue. The surprising thing is that the next morning I felt almost completely well. My temperature was near normal, but Sukhopalko, who accompanied me to the airfield and handed over the paperwork for my trip, warned me against any foolishness: "Straight to the plane." We exchanged warm goodbyes and I promised to return in a week. The week stretched into six months. On the airplane, my tongue swelled up inexplicably. It filled up my mouth so that I dared not deplane in Uralsk to enjoy the traditional calf's tongue and sour cream praised by all of our travelers.

Despite Mrykin's strict instructions, I went home rather than to the hospital in the car that met me at the airport. My sudden appearance did not surprise Katya, but she came unglued when I started talking like a faulty loudspeaker. We decided that I would go to the hospital the following day once I had rested and the gift of gab returned to me. Indeed, the next day Katya, who accompanied me all the way to the receiving room, was sure that I was once again speaking with my "real" voice.

They put me in a building that dated from the times of Catherine the Great. Why did they build hospitals so extravagantly back then? Thick, fortress-like walls, large windows, and incredibly high ceilings. My neighbors turned out to be two sociable colonels. Both of them were heart attack patients. When they heard that I was suspected of having radiation sickness, they concluded that I was from one of the outfits involved with atomic weaponry. My attending physician Dr. Kostoglot and various consulting physicians persistently questioned me as to when and where I could have been exposed to radiation. I stubbornly rejected this possibility. Indeed, if it had been radiation, then why was I alone exposed and where and when? No, that was impossible. Katya visited me almost every day and passed on greetings from my comrades. She said that almost all my friends and acquaintances were back on temporary assignment (at the firing range).

On one of those typical days at the hospital, they had drawn blood for routine analysis, I had breakfast, and dozed off. Suddenly my neighbor the colonel woke me up. "Put on your headphones!" Following his instructions, I heard the second half of a TASS report about the development in the USSR of an intercontinental ballistic missile and its successful testing.[1] Wow! Finally victory! I could imagine what joy and what a celebration they were experiencing there at the firing range. The *Semyorka* had broken through to the target on the fourth attempt.[2] After the TASS report, now the whole world would be talking about it. And I was stuck here because of some unknown disease!

They allowed me to go and stroll a bit in the hospital yard. When I telephoned Kalashnikov all I learned was that everything was great. There was general euphoria at the OKB. Internal dissenters and pessimists had been put to shame, and external ones, that is to say, the Americans, well, let them tremble. Yurasov had replaced me at the firing range. New people were arriving at the OKB now. New projects were beginning. In a word, I needed to get well quickly.

Having lost all faith in conventional drugs and even the newest ones, my attending physician told Katya that on her next visit she should bring no more than 200 grams of cognac. He recommended that, without my neighbors noticing, I could take around 50 grams in the morning and in the evening for two days. I followed his instructions with pleasure. To tell the truth, on the second day I did not follow the regimen and consumed the entire remaining 100 grams in my first dosage after breakfast. It is amazing, but about two days later Kostoglot announced that my blood had significantly improved. To be on the safe side, he invited the most famous professor of hematology at that time Iosif Abramovich Kassirskiy to have a consultation with me.

The professor did indeed visit me. He studied my case history and interviewed

---

1. The TASS announcement was issued on 27 August 1957.
2. Including launch aborts, the four attempts were on 15 May (inflight failure), 11 June (launch abort), July 12 (inflight failure), and 21 August (success). There were also two additional attempts on 10 June, both of which were aborted just before ignition.

me in detail as to when and where I had experienced my first symptoms. When I said that I had become ill in Kazakhstan, Kassirskiy's expression brightened. "I don't think," he said, "that this is radiation sickness. Your blood has an extraordinarily high eosinophil index. Most likely you have an eosinophilic disease, which is rare, but it does occur in Soviet Central Asia. This is the body's reaction to parasitic microorganisms that have infected your liver. These parasites are common in that part of the world." He promised to give it some thought and have another look at me.

Once, during "rest time," the duty nurse woke me up and told me to go out into the lobby. There, I was surprised to see a large and cheerful group. Disregarding hospital rules for quiet, my comrades greeted, hugged, and congratulated me. They included Korolev, Voskresenskiy, Mishin, Yurasov, Kalashnikov, Bushuyev, and Okhapkin. From the random friendly chitchat that is normal in such situations, I gleaned that not everything had gone as smoothly as the TASS report had trumpeted to the entire world.

Korolev excused himself. He needed to get over to a meeting with Nedelin and Keldysh. He took Bushuyev, and before leaving, tossed off a parting remark: "Boris, fake it, but not for long."

My remaining visitors told me that during that victorious launch, colleagues hadn't found the nose cone in Kamchatka. They searched and searched, but found no traces of impact. By all appearances, the nose cone had burned up and dispersed in the dense atmospheric layers quite close to Earth. Telemetry communications were lost 15 to 20 seconds before the calculated time of impact with the Earth's surface. That is why Korolev and Bushuyev were now hurrying over to meet with Keldysh. He was arranging a consultation with specialists from TsAGI and other gas dynamics specialists. Nedelin also wanted to participate in the discussion.

Mishin expressed more concern than the others. In his view, it wasn't so easy to select a new configuration for the nose cone. Quite a bit of time would be required for wind tunnel tests and fabrication. What were they supposed to do now? Stop testing? There were already reports from America. They didn't believe TASS and considered this to be a hoax. To be honest, we indeed did have a missile, but we did not yet have a hydrogen bomb carrier. Who would entrust such a payload to us, if the payload container disintegrated and burned up long before it hit the ground.

"What's more," added Yurasov, "Right after nose cone separation, they determined that it collided with the body of the core booster."

"So that's the story," said Voskresenskiy. "Everybody's congratulating us, but we're the only ones who know the truth."

There was one missile left, No. 9. Preparation was under way at the engineering facility, but it was as yet undecided what measures engineers would have to take. Most likely, Korolev was pulling out all the stops to persuade Nedelin and Keldysh to launch the next missile and nose cone without modifications in order to accumulate some more data and then halt testing to perform the most important modifications. While the modifications were being performed, we would be busy with satellite launches. For the time being this would distract Khrushchev's attention from combat technology. This was more or less the tactic that my visitors laid out.

We were mistaken in our prognoses apropos "distracting attention." The first artificial Earth satellite did not "distract the attention" of our high-ranking leadership, but rather, having wedged itself into the flight-design testing program of the intercontinental combat missile, it caused a sensation all over the world and a real panic on the shores of the Potomac. In parting, Voskresenskiy did not fail to pass on greetings from Katya and slipped me a package; I immediately figured out that it was a bottle. "This is the best three-star medicine for you. Let's go, guys, before they catch us." On the whole, my friends left me gloomy news: four launches and still no absolute intercontinental weapon.

The next day in the hospital yard I had an unexpected meeting with Germogen Pospelov. We hadn't seen each other for a long time. He was already a general and Air Force Academy professor. Germogen was in the hospital for acute rheumatic heart disease. He knew where I worked and immediately congratulated me on our great success. But I couldn't even tell my old friend Germogen the truth. All I could do was change the subject to our adventures as students in Koktebel before the war. Germogen and I shared pleasant memories about our swim to the *Zolotyye vorota* (Golden Gate) and about the Karadag cliffs. We recalled how out of stupidity I had hung over a precipice that dropped straight down to the sea and Germogen had tied two towels together and thrown them to me. Using them I'd managed to climb to a safe place and afterward developed a special respect for rock climbers who get by without any towels.[3] A nurse called out, "Comrade general, it's time for your procedure," and interrupted our further reminiscing. With difficulty, Germogen stood up and, leaning on his cane, limped into the building.

In early September, Bushuyev, Yurasov, and Voskresenskiy visited me. Yurasov had just flown in from the firing range and was full of impressions. He seemed very excited; one moment, he would rail against someone there, then the next, he would be delighted with someone else. On the whole, however, he was upset.

On 7 September they launched the last of the missiles that had been prepared, No. 9. The primary action on it had been to increase the time between the shutdown of the second stage engine and the issuance of the nose cone separation command from the 6 to 10 seconds. To ensure the reliability of communications, they switched the external telemetry slot antenna from the head to the bottom antennas before entering the dense atmospheric layers.

The development of antennas for the radio telemetry systems of missile nose cones was a very complex problem both in theory and in practice. Once the separation pulse was received from the push rod on the missile body, the head could spin. Therefore, the antenna's radiation pattern must be, to the extent possible, circular. But uniform radiation in space in all directions reduces the energy reaching the

---

3. See Chertok, *Rockets and People.* Vol. 1, p. 153.

From the author's archives.

Chertok on vacation with his sons Valentin (right) and Mikhail (middle) in 1957.

antennas of the ground-receiving stations compared with the energy that directional radiation antennas can concentrate. During entry into the atmosphere, when the payload container equipped with a special stabilizing "skirt" stops erratically somersaulting and swoops toward the Earth, a layer of hot plasma forms around it due to the high temperature from braking in the atmosphere. This layer absorbs the energy radiated by the nose cone antenna to such an extent that for the last 30 seconds before it reaches the ground, almost no telemetry data get through. It was very important to place the antenna in a location on the structure where the concentration of electrons in the plasma was minimal and where there was still the hope of breaking through to the Earth. Antenna laboratory director Mikhail Krayushkin occupied himself with all these problems for us. He had a well-developed theory of antenna design for missiles and a practical method for simulating their characteristics. We didn't yet have a way to simulate the behavior of an antenna in plasma at that time.

During the last launch, despite delaying the separation command after engine shutdown by as much as 10 seconds, the body once again collided with the separated nose cone. This collision might have damaged its heatshield. Once again the payload container disintegrated in the atmosphere. But, nevertheless, fragments reached the ground and parts of them were found. From these it was determined that there was a target overshoot of just three kilometers and a deviation to the right of one kilometer. Telemetry reception halted 30 seconds before impact. In addition, a failure of the tank pressurization system was recorded in flight, evidently due to damage in the liquid nitrogen line.

In the semidarkness of the hospital lobby the four of us discussed at length the latest situation and speculated on possible solutions. Yurasov noted, "The euphoria at the OKB has now been replaced by a certain level of bewilderment. But S.P. has refocused his energy on satellites. So, of course, it's less stressful. A satellite doesn't have to enter the atmosphere. But, no matter what, we need to solve the problem of the nose cone reaching the ground without disintegrating." Bushuyev added, "There is the danger that the nuclear specialists will lose faith in the *Semyorka*'s reliability and will switch over, along with their payload, to work with Chelomey and Yangel."

According to intelligence from our "fifth columns," Yangel had been working intensively on the R-16 missile, which operated on nitrogen tetroxide and unsymmetrical dimethyl hydrazine propellants. Within the military, many strongly opposed our reliance solely on oxygen. They would actively support Yangel. According to information from "our people" who worked in Dnepropetrovsk, the R-16 could be ready in about three years. A draft resolution had even been prepared calling for the beginning of construction at our firing range of a separate engineering facility and launch site for Yangel. It specified a completion date of the first quarter of 1960. Chelomey, of course, could not produce an intercontinental missile within that period of time, but he could in about four years. Both Yangel and Chelomey had already received Glushko's reassurance that he would produce engines using those propellant components.

Bushuyev believed that if Glushko entered an alliance with Yangel and Chelomey, it would have an unavoidable impact on his relationship with Korolev, and consequently on our plans as well. We needed to hurry, but with what and where, that was the big question. S.P. had a lot of plans and areas of interest, many of which had not yet sparked enthusiasm among the military. We weren't going to gain a lot of active support from them now! Voskresenskiy lamented that under these circumstances Mishin was conducting himself improperly. He was not seeking a compromise with Glushko, and he was straining their relations over any trifle.

Bushuyev also had a lot of interesting things to say about meetings at various levels. He had participated in them with Korolev and sometimes by himself at Korolev's instructions. According to Bushuyev, Keldysh showed the greatest initiative when it came to cultivating a positive attitude toward the satellite program in the higher echelons of power. He was instrumental in persuading Academy of Sciences President Nesmeyanov, Academician Blagonravov, and many more scientists, on the future of satellites.[4] Thanks to our missile, they all dreamed of penetrating into space before the Americans, thereby proving the superiority of Soviet science. But we found ourselves in a complicated situation. We had been working for almost a year on Object D with the academicians, but the further along we were, the clearer it became that the work needed another year.[5] The equipment alone weighed more than 300 kilograms. Here, Voskresenskiy could not pass up the opportunity to tease Bushuyev. "An awful lot of interesting female scientists are buzzing around Kostya. Each one is trying to charm him into pushing their little instrument on board."

Yurasov complained that Konstantin Davydovich [Bushuyev] had gotten some very good electricians. He gave them to Ryazanov, and the latter was developing the satellite's onboard system. Though they really were capable kids, they were inexperienced; eventually, Korolev would make *us* figure it out. Bushuyev was not offended, but said that we would deal with both the women and with the electricians. However, in his opinion, the situation with the deadlines was hopeless.

Bushuyev continued, "Right after the launch, S.P. assembled our whole team and proposed that we temporarily halt work on Object D, and for the remaining month everyone should develop the simplest—"if only makeshift,"—satellite.[6] We had already estimated with the ballistics specialists that we could haul about 80 kilograms into an orbit with an apogee of 1,000 kilometers. S.P. thinks that this would be a great sensation. We need to manage to make not only that soccer ball, but also a fairing and special separation system for it. Krayushkin is fussing with the antennas there. We still haven't decided how to reliably deploy them. S.P. is terrorizing all of us with reports that someone is tossing at him or that he has thought up himself. Supposedly the Americans have announced that they will launch their own Vanguard satellite in October.[7] Keldysh doesn't think that they are capable of putting up a satellite greater than 10 to 15 kilograms, but they'll create quite a stir."

In parting, my comrades confessed that they had vented their feelings here in the hospital. Tomorrow morning they would again be wrapped up in such hectic activities that there would be no time to think straight.

---

4. Academician Aleksandr Nikolayevich Nesmeyanov (1899–1980) served as president of the USSR Academy of Sciences in 1951–61. Academician Anatoliy Arkadyevich Blagonravov (1894–1975) served as the academic secretary of the Academy's Department of Technical Sciences in 1957–63 and was closely involved with the development of a program for scientific research for the early Soviet space program.

5. The so-called Object D was the first Soviet satellite project approved for development. The Soviet government formally approved its implementation in January 1956 in time for a launch in the 1957–58 period. The satellite, which would carry a suite of scientific experiments into space, was named "Object D" because it was the fifth payload for the R-7 ICBM, after Objects A, B, V, and G, which were all nuclear warheads.

6. This satellite was also known as PS (*Prosteyshiy sputnik* or simplest satellite).

7. Vanguard was the name of the first U.S. "civilian" satellite program, formally approved by the Eisenhower Administration in 1955. Although Vanguard was touted as a completely civilian project, it used hardware, personnel, and funding from both the military and intelligence communities. The program was officially run out of the Naval Research Laboratory (NRL).

From the author's archives.

Chertok shown here in 1957 while on vacation at the Tetkovo sanatorium.

During his second visit, Professor Kassirskiy recommended that I be discharged from the hospital, switch to a regimen of certified sick leave home rest, and report to him at the clinic for special treatments at least three times per week. The military hospital, however, did not let go of me so easily. First they transferred me to clinical hospital No. 6, which specialized in the treatment of persons exposed to radiation. It was truly frightening to see actual radiation sickness patients. The regimen in this hospital was strict. Before I could be admitted, I had to present a certificate showing that I really was cleared for top-secret work. Visits with my wife, who had no clearance for secret work, were out of the question. You had to spend a day filling out paperwork to meet with colleagues from work. Packages were checked. There was no private telephone. Our food was excellent, but the quasi-prison regimen and isolation from the outside world forced me to feign an appearance of excellent well-being.

Despite my lousy blood tests, the "atomic" doctors considered me an alien who had accidentally turned up among real radiation sickness patients. Two weeks later I was driven out of this top-secret medical institution as someone who had landed there by mistake. Kassirskiy had a laugh and sentenced me to unpleasant treatments that entailed purging my internal "hydraulic lines" with pure oxygen. He assigned his graduate student as my attending physician. She confessed that I was a lucky find for her. Eosinophilic disease was the subject of her dissertation. Unfortunately, persons suffering from this rare disease were as scarce as hens' teeth in Moscow, and suddenly such a lucky break! Oxygen purging was the professor's idea, but there were as yet no statistics. Each time I showed up for a purging, she performed a quick

377

blood test and with a look of satisfaction announced that she observed a "slight tendency toward improvement." The regimen of home rest enabled me to keep up with events. Despite my fluctuating temperature and unusual feelings of weakness, once a week I visited the OKB.

Chapter 21
Breakthrough into Space

Analyzing the past with the knowledge of heretofore top-secret events rather than in terms of official history, one can't help but realize that sometimes what seemed to be colossal failures contributed to subsequent triumphs. I shall try to prove this paradox using the example of the first artificial Earth satellite in human history.

The idea of producing a satellite came up in an engineering memo written by Mikhail Tikhonravov in 1954.[1] He was working at NII-4 when it occurred to him that the intercontinental missile that Sergey Korolev, his former chief at GIRD in Moscow, had been assigned to develop was capable of carrying not only a thermonuclear warhead but also a satellite.[2] Tikhonravov's idea prompted Korolev to take immediate action. After working with Korolev for 20 years, I knew that it was against his very nature to lay new ideas aside or forget about them, no matter who had come up with them. On 16 March 1954, USSR Academician Mstislav Keldysh held a meeting where Tikhonravov gave a general presentation of his proposals. He understood that the overly general thoughts needed to be made comprehensible for the defense industry leaders. On 27 May 1954, Korolev sent a letter to Minister Ustinov to which he attached a memorandum "Concerning an Artificial Earth Satellite." The memorandum was written by Tikhonravov.[3]

---

1. Mikhail Klavdiyevich Tikhonravov (1900–74) was one of the founders of Soviet rocketry and spaceflight. As an engineer in GIRD, he designed the "09," the first Soviet rocket to use liquid propellants. In his later life, first as an engineer at NII-4 and then later at OKB-1, Tikhonravov played critical roles in the development of the first intercontinental ballistic missile (the R-7), the first satellite *(Sputnik),* the first human-rated spaceship (Vostok), and the first robotic lunar probes (Luna).

2. GIRD—*Gruppa izucheniyu reaktivnogo dvizheniya* (Group for the Study of Reactive Motion) was the first Soviet amateur group dedicated to developing rockets. Between 1931 and 1933, GIRD engineers, who included Korolev and Tikhonravov, produced the first Soviet rocket that used liquid propellants, the famous "09."

3. The memo cited by Chertok has been reproduced in B. V. Raushenbakh, ed., *Materialy po istorii kosmicheskogo korablya 'Vostok': k 30-letiyu pervogo poleta cheloveka v kosmicheskoye prostranstvo* [*Materials on the History of the 'Vostok' Space Ship: On the 30th Anniversary of the First Flight of a Man into Cosmic Space*] (Moscow: Nauka, 1991), pp. 5–15. Korolev's cover letter is reproduced in M. V. Keldysh, ed., *Tvorcheskoye naslediye Akademika Sergeya Pavlovicha Koroleva: izbrannyye trudy i dokumenty* [*The Creative Legacy of Academician Sergey Pavlovich Korolev: Selected Works and Documents*] (Moscow: Nauka, 1980), p. 343.

It would seem that at a time when the production of an intercontinental nuclear delivery vehicle was a "life or death matter for the Soviet Union," the minister's response to Korolev should have been: "Now is not the time. Produce the missile!" But Ustinov was not an ordinary minister. After consulting with Keldysh, he decided to legitimize the idea of a satellite with a governmental resolution.

In August 1954, the USSR Council of Ministers approved the proposals of V. A. Malyshev, B. L. Vannikov, M. V. Khrunichev, and K. N. Rudnev on the study of scientific and technical issues associated with space flight.[4]

On 30 August 1955, V. M. Ryabikov, chairman of the Military-Industrial Commission, convened a private meeting during which Korolev reported about the potential for using an intercontinental missile for space flight. Ministry of Defense representative Aleksandr Mrykin expressed strong concern that the deadlines for the development of the R-7 missile would not be met if "we get carried away with satellites." He proposed putting off the matter until the R-7 missile flight tests had been completed.

B.A. Smirnov

Mikhail Tikhonravov (1900–74) was one of the pioneers of Soviet rocketry and space exploration. He helped to design the first Soviet rocket that used liquid propellants (the '09'), proposed the concept eventually used on the R-7 ICBM, and led the teams that designed the Sputnik, Vostok, and Luna spacecraft. This photo dates from around 1970 during the making of a secret documentary about him. The star on his left lapel is the "Hero of Socialist Labor," the highest civilian honor given to Soviet citizens, which he received in 1961.

But nevertheless . . . . On 30 January 1956, a governmental resolution was issued calling for the production of an unoriented satellite (Object D), weighing 1,000 to 1,400 kilograms and containing scientific-research equipment weighing 200 to 300

---

4. All of these men were powerful administrators in the Soviet defense industry. At the time, Malyshev, Vannikov, and Khrunichev served in the Ministry of Medium Machine Building, the "superministry" that oversaw all Soviet strategic weapons programs in 1953–55. Rudnev was a deputy minister in the Ministry of the Defense Industry.

kilograms, to be inserted into orbit in 1957 or 1958 by the R-7 rocket. The Academy of Sciences was entrusted with general scientific supervision and production of the instruments for space research. The Ministry of the Defense Industry was tasked with producing the actual satellite. The Ministry of the Radio Industry was assigned to develop the telemetry system and command radio link. The item in the resolution calling for the creation of 15 stations on the territory of the USSR to track the satellite and receive telemetry information proved very important for the future. Now I can confirm with full authority that in terms of planning, we outflanked the Americans with this resolution by at least three years.[5]

In July 1956, the draft plan for Object D was completed and new space-related subdivisions started developing the actual structure. However, in the last days of 1956, based on the results of the firing tests on the R-7 rocket engines, it came to light that the specific impulse ("specific thrust" in old terminology) was 304 seconds, rather than the 310 seconds that Glushko had promised.[6] This had no particular significance for achieving the specified range of 8,000 kilometers carrying a nuclear warhead, but it did not guarantee the insertion of Object D into space. Moreover, it turned out that the deadlines for the development of the scientific equipment could not be met. Glushko promised to bring the specific impulse of the engines up to the design value by spring 1958. As a result, the VPK set a new deadline for Object D, 1958. It would seem that OKB-1 Chief Designer Korolev would have to simmer down now that he had been given more than an additional year. We already had enough to worry about.

I don't recall now where I read that Napoleon once said to his chief of staff Marshal Louis-Alexandre Berthier: "You are a superb chief of staff, but you will never become a real commander."[7] Korolev was not only a superb organizer and strong-willed chief designer. He possessed the innate qualities of a commander: faith in himself, in his own intuition, and in the fact that he was the only one who would make the decision that would result in success.

In the Council of Chief Designers in January 1957, Korolev reported that as a result of the low specific thrust values, they could only guarantee the on-orbit insertion of an artificial Earth satellite weighing up to 100 kilograms. And, for a margin of safety, it was proposed that the rocket be lightened as much as possible, that is, remove all the radio-control system equipment and provide for a one-step engine shutdown that depends on the integrator or upon receipt of an Emergency Turbine

---

5. The first American satellite project, Vanguard, was approved in August 1955. Earlier, in March 1955, the U.S. Air Force issued system requirements for a reconnaissance satellite system known as WS 117L. In June 1956, the Air Force chose Lockheed's Missile Systems Division to design and build military observation satellites under the WS 117L program.

6. The "specific impulse" is a measure to evaluate the efficiency of a rocket engine. It is equal to units of thrust per unit mass of propellant consumed per unit time and is expressed in seconds.

7. French Marshal Louis Alexander Berthier (1753–1815) served as Napolean's chief of staff in 1796–1815.

Contact (AKT) command (triggered when one of the propellant components is depleted).[8] There were also other, less innovative proposals for reducing the mass of the rocket itself. Korolev's colleagues in the Council of Chiefs showed little enthusiasm for his proposals. Ryazanskiy objected and Glushko was silent. Pilyugin took a neutral position. Keldysh, whom Korolev had prepared in advance, supported the proposal.

After a heated discussion, the decision was made to draw up the necessary governmental resolution. Here, the rationale was that the satellite needed to be launched before the beginning of the International Geophysical Year (July 1957).

The Council of Ministers' resolution, which amended preceding resolutions, was issued on 15 February 1957. It stipulated the orbital insertion of the "simplest satellite," observation of its on-orbit behavior, and study of the passage of radio signals through the ionosphere. It was proposed that two R-7 rockets from those prepared for the flight-design testing program be used for the launch. However, the launch of the "simplest satellite" would not be permitted until after one or two successful R-7 rocket launches.

I must now confess that, like the majority of the other participants in the development of the R-7 rocket, I was not at all excited by all these conversations and resolutions about satellites. We put up with Korolev's infatuation. The optimization of the R-7 missile together with—and especially with—the warhead and its deadly innards, that was the most important thing! But an utterly simplistic satellite …?! Ultimately, even if it were launched, this would in no way help solve the most important problem, the safe passage of the warhead through the dense atmospheric layers and how to achieve as much accuracy as possible. Incidentally, it wasn't just the missile elite who showed no particular enthusiasm for this satellite mania.

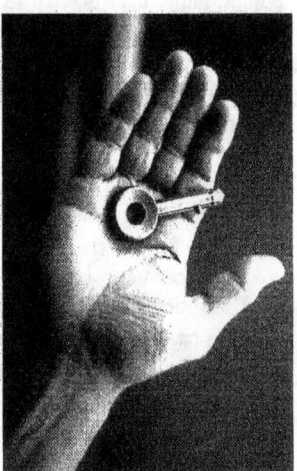

From the author's archives.
The original key used on the control panel that launched the first R-7 ICBM.

On 17 September 1957, in the House of Scientists' Hall of Columns a ceremonial meeting was held in honor of the hundredth anniversary of the birth of K. E. Tsiolkovskiy. Academy of Sciences Corresponding Member S. P. Korolev, who was not known to the public, delivered a report in which, among other things, he

---

8. AKT—*Avariynyy kontakt turbina.*

said, "In a very short while the first test launches of artificial Earth satellites will be conducted for scientific purposes in the USSR and the U.S."[9] One would think this would have caused a sensation! But no. There was no buzz in this regard either in the USSR or abroad.

THE NET RESULT OF 1957: The flight-design testing program of the first series of R-7 missiles showed that the structure and heat shield of the nose cone disintegrated during entry into the atmosphere. It seemed illogical to continue the launches until a new nose cone had been developed. However, Korolev insisted on launching with the clearly unsuitable nose cone, using the rationale that "we need to optimize the launch vehicle, not just the nose cone." Realizing that the launch vehicle had carried the unfit nose cone as far as Kamchatka and that in the best-case scenario the newly developed nose cone would be ready in six months, he insisted on using the remaining "headless" missiles to launch the satellites. TASS's August report about the production of an intercontinental missile was a bluff in the sense that the missile had no warhead. But aside from the very few of us who were privy to the secret results of the flight tests, no one knew.

I could stand it no longer and went to my good old OKB-1 one beautiful September day despite the doctor's firm orders to the contrary and the fact that my temperature continued to jump for no apparent reason from 36° to 38°C (96.8° to 100.4°F). None of my department chiefs were in their offices. They were all on the production floor or away at the firing range. I walked over to shop No. 39, which in those days was not only the site for final assembly, but also a laboratory for the optimization of the world's first (or so we hoped) artificial Earth satellite.

The factory was in a round-the-clock, all-hands work mode for the fabrication of the polished sphere with four long tails that were antennas. The radio operators had coordinated the "input resistances" for the transmitter with Krayushkin. Depending on those values, the antennas were first lengthened and then shortened again. At Korolev's request, Ryazanskiy personally developed and then listened to the coded signals on the special receiver. In the coming weeks the sound of this beep was destined to shake the entire world. But at that time no such thought had dawned on anybody at the factory or at the design bureau. Okhapkin and his designers were stuck at the factory 24 hours a day, racing to fabricate a special fairing to protect this beautiful sphere.

When they began working on the layout of the thermonuclear warhead for installation on the *Semyorka* in our department at the OKB, as I studied the dimensional installation drawings and electrical diagrams, I developed an anxious respect for

---

9. A slightly edited version of the speech was published in *Pravda* the same day. See S. P. Korolev, "Osnovopolozhnik raketnoy tekhniki: k 100-letiyu so dnya rozhdeniya K. E. Tsiolkovskogo" [The Founder of Rocket Technology: On the 100th Birthday of K. E. Tsiolkovskiy], *Pravda*, September 17, 1957. This was the last article published under Korolev's own name during his life.

this creation of human genius, which we modestly referred to as the "payload." And suddenly, instead of a multiton "payload," a sphere barely larger than a soccer ball and weighing just 80 kilograms was going to be placed on the *Semyorka*. Its internal electrical circuit was so basic that it would be a snap for any group of young hobby technicians to reproduce it.

IN LATE SEPTEMBER, THE OKB WAS EMPTY. All those involved and those called in for support flew to the firing range along with the "sphere," the accessories, and fairing. Fans who remained behind followed the preparation process via high-frequency communications and promised to alert me 24 hours before the launch.

On 4 October, I went to the OKB and joined the group of around 30 duty officers who had filled the reception area and Korolev's office, where the high-frequency communications phone was located. On Korolev's orders, on the other end of the line in the barracks at Site No. 2, sat our commentator, who transmitted information to us as he received it from the bunker. It wasn't until 10:30 p.m. that we heard the excited report that liftoff had proceeded normally. An hour and a half later, someone already quite hoarse was shouting, "Everything is OK. It's beeping. The sphere is flying." We went our separate ways from Podlipki late that night still unaware that from that moment humankind had entered the space age.

This was the sixth *Semyorka* launch. Of the five preceding ones, only two missiles had passed more or less normally through the powered flight phase, two had crashed, and one had failed to lift off at all. The world had no knowledge of all this background history when it heard Levitan's voice saying, "All the radio stations of the Soviet Union are operating. We are transmitting a TASS report ...."

On 5 October, the morning newspapers managed to run this report. It wasn't until 9 October that *Pravda* published a detailed description of the satellite, its orbit, the radio signals, and methods for observing it. It published a schedule of when the satellite passed over Soviet cities and many world capitals. For the first time, on a clear dark night against a background of motionless stars, it was possible to observe a single, fast-moving one. This was mind-boggling.

So much has been said and written about this historic event that it is very difficult to report anything new. What is well known to historians and has become banal for them is a revelation for today's youth. As such, I will take the liberty of citing excerpts from my material published in the book *Kosmonavtika SSSR (USSR Cosmonautics):*

"Although the satellite was referred to as rudimentary, it was an original, without any analogs in technology. There had been only one specification, a weight restriction (no greater than 100 kilograms). The designers rather quickly came to the conclusion that it would be advantageous to make it in the shape of a ball. The spherical shape made it possible to more fully use the interior space while having less body surface.

They decided to place two radio transmitters with a radiated frequency of 20,005 and 40,002 MHz on board...

The satellite was designed rapidly and the parts were fabricated as the drawings were issued...

384

A satellite 'twin' was mated with and separated from the missile body many times until we were convinced that all the circuits operated reliably: the pneumatic locks activated, the nosecone fairing separated, the antenna spike released from the stowed position, and the push-rod directed the satellite forward...

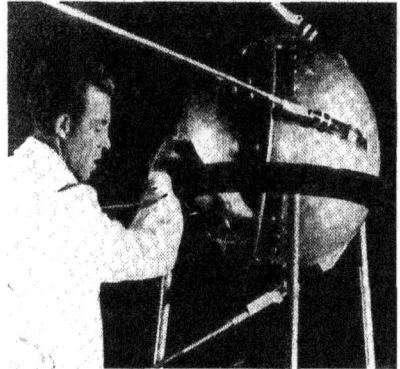

The Sputnik 1 (PS-1) satellite is shown here on a rigging truck in the assembly shop in the fall of 1957 as a technician puts finishing touches on it.

The satellite's radio transmitter was supposed to have radiated power of 1 W. This enabled its signals to be received at significant distances by a wide audience of amateur radio operators in the shortwave and ultra-shortwave ranges and also by ground tracking stations...

The satellite's signals were in the form of telegraph pulses with a duration of approximately 0.3 seconds. When one of the transmitters was operating, the other was in pause mode. The estimated continuous operating time was at least 14 days...

Electrochemical current sources (silver-zinc batteries) designed to operate for a minimum of two to three weeks provided the power for the satellite's onboard equipment...

On 4 October 1957 at 10:28 p.m. Moscow time a violent flash of light illuminated the night over the steppe and the rocket lifted off with a roar. Its flame gradually diminished and soon became indistinguishable against the background of heavenly bodies.

Newton calculated the first cosmic velocity, and now three centuries later a creation of the human mind and hands had achieved it for the first time...[10]

After the satellite separated from the last stage of the rocket the transmitters began to operate and the celebrated signals "beep, beep, beep," flew over the airwaves. Observations during the first orbital passes showed that the satellite had been inserted in an orbit with an inclination of 65°6', an altitude of 228 kilometers at its perigee, and a maximum distance from the Earth's surface of 947 kilometers. It took 96 minutes 10.2 seconds to complete each orbital pass around the Earth.

The Russian word 'sputnik' immediately entered the languages of all the peoples of the world.[11] *The headlines on the front pages of foreign newspapers during those historic October days of 1957 were full of admiration for our nation's achievement...*

When news of the satellite launch reached Washington, it was as if a bomb had exploded. It wasn't the scientific significance of the satellite's flight that shook the Penta-

---

10. Russians refer to the velocity required to reach Earth orbit as "the first cosmic velocity."

11. *Sputnik* means "fellow traveler" in Russian, although since 1957 the word has most commonly been used to denote artificial satellites of the Earth and other heavenly bodies.

gon specialists who had fought for a brink-of-war policy; it was the fact, now obvious to everyone, that the Soviet Union had produced a multi-stage intercontinental missile against which air defense was powerless.

A number of U.S. leaders declared that the Russians had thrown down the gauntlet in the fields of science, industry, and military might...

The first American satellite was launched four months later and weighed just 8.3 kilograms... The Americans could not help but feel disappointed and exasperated."[12]

A few qualifying comments should be made. The generally accepted notion at that time that at night one could visually observe the satellite illuminated by the sun without any special optical devices was incorrect. The satellite's reflective surface was too small for visual observation. In actual fact, we were observing the second stage, or core booster of the rocket, which had been inserted in the very same orbit as the satellite. This mistake was repeated again and again in the mass media.

During the launch of the rocket—assigned the designation M1-1SP—a delay was observed in its buildup to the first intermediate stage and to the main stage of the Block D main engine. This delay could have caused an automatic system reset. But it "squeaked by," and during the last fractions of a second of the Block D time check, it completed buildup. Sixteen seconds into the flight the SOB system failed, causing an increase in kerosene consumption.

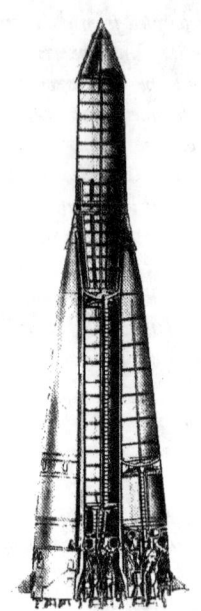

M. V. Keldysh, ed., Tvorcheskoye naslediye Akademika Sergeya Pavlovicha Koroleva: izbrannyye trudy i dokumenty (Moscow: Nauka, 1980).

This cutaway of the Sputnik launch vehicle (the 8K71PS) clearly shows the engines of the core and the strapon boosters. Note also the "hammerhead" shape of the core. The booster developed a thrust of about 398 tons at liftoff and stood 29.17 meters tall.

As a result, there was not enough kerosene in the tank to last until the designated time at which the integrator had been set, 296.4 seconds. The engine was shut down a second earlier by the emergency turbine contact signal. Freed from the load of the kerosene pump, the turbine began racing and the emergency contact controlling the rpms shut down the engine. At the very end of the powered flight phase, 1 second of engine operation substantially affects the orbit. The rocket and satellite were

---

12. Yu. A. Mozzhorin et al, eds., *Kosmonavtika SSSR* [*USSR Cosmonautics*] (Moscow: Mashinostroyeniye, 1986), p. 41.

Asif Siddiqi.

Sputnik, the world's first artificial satellite was launched near midnight local time at Tyuratam (now Baykonur). Because of the darkness, the existing photos of the launch are all, unfortunately, of poor quality.

inserted on orbit with an apogee approximately 80 to 90 kilometers lower than the calculated orbit. No subsequent public descriptions or reports contained information about these glitches.

No one in the OKB organization or among our subcontractors had expected such worldwide publicity. We were intoxicated with our sudden triumphant success. Lists of individuals to receive awards were drawn up; subcontractors were called up to determine what awards would be given, to whom, and how many. Suddenly all this activity came to a halt. Khrushchev called in Korolev, Keldysh, and Rudnev and hinted that a cosmic gift was needed in honor of the 40th anniversary of the Great October Socialist Revolution. Korolev protested that it was less than a month away. It made no sense to repeat the very same launch, and it was simply impossible to develop and fabricate another satellite. Privately, Korolev was justifiably apprehensive. This preholiday gift might end with another crash. Then the victory we had gained with such difficulty would be quickly forgotten. But Khrushchev was implacable. The political success that we had brought him—and could bring him again with another sensational space launch—was for him more important than refining the intercontinental nuclear missile. As a result, the second stage of the missile was converted into a space laboratory. The research subject was a dog. For health reasons I did not attend the meeting of the Council of Chief Designers where they decided to fabricate and launch a second satellite. Bushuyev managed to tell me that during this Council meeting, which convened immediately after the

387

conversation with Khrushchev, Korolev introduced the proposal about launching a dog. In this regard, he said that it was impossible to fabricate any other instruments to perform space research within the available timeframe. According to Bushuyev, Korolev had hoped that the Council members would resist Khrushchev's unrealistic proposal and ask to rethink his demands. But everyone embraced his idea for the immediate launch of a second satellite with a gambler's enthusiasm.

On 12 October, the decision was officially made to launch a second satellite in honor of the 40th anniversary of the Great October Revolution. The decision was a death sentence for one of the mutts as yet to be selected. About 10 days before the launch, military physician Vladimir Yazdovskiy picked Layka, who would go down in history.

We already had experience with high-altitude rocket launches of dogs. But before, it was a matter of pressurized compartment laboratories supporting 1 or 2 hours of vital activity.[13] Now we were required, without any preliminary experimental development, to create an experimental space laboratory making it possible to study a dog that would not be returned to the Earth. Everything that would happen in space could be tracked only via telemetry.

The second simple satellite was produced without any preliminary draft design or other plan. All the rules that had been in effect for the development of missile technology were abandoned. The draftsmen and designers moved into the shops. Almost all the parts were manufactured using sketches. Assembly wasn't conducted so much according to documents as according to the designers' instructions and on-the-spot fitting. The total weight of the satellite—508.3 kilograms—was already a qualitative leap by itself. An unexpected decision, but one of necessity, was the decision not to separate the satellite from the core booster. Indeed, if the rocket itself were inserted on the satellite's orbit and no orientation were required, then why not use the *Tral* already installed on the launch vehicle to transmit parameters? Thus, the second satellite was the entire second stage, that is, the *Semyorka's* core booster.

The launch dedicated to the 40th anniversary of the October Revolution took place on 3 November 1957. The electric power sources installed on the rocket's body to track the satellite were sufficient for six days. When the electric power supply was depleted, Layka's life was also over. Incidentally, biomedical specialists believed that Layka died much earlier from excessive heat. It was virtually impossible to create a reliable life support and thermal control system within such a short period of time.

It was a complete triumph. None of us doubted that the Americans had been put to shame. Only the British Society for the Prevention of Cruelty to Animals protested Layka's martyrdom. In response to this, our tobacco industry promptly issued

---

13. In 1951–60, OKB-1 launched over two dozen "vertical" shots carrying dogs and other animals up to altitudes between 100 and 500 kilometers. The design bureau used converted civilian versions of the R-1, R-2, and R-5 missiles for these experiments.

the Layka cigarette with a picture of this cute little dog on the pack. The launch of the second satellite was the last one in 1957. Finally, all attention was focused on finishing the nuclear missile.

The government made it worth our while after the two satellite successes. In December 1957, we were showered with governmental awards, including the Lenin Prize, which was reinstated after Stalin's death. At that time, the Lenin Prize was very highly esteemed. It was just as honorable as the title Hero of Socialist Labor. But if, as the old song said, "anyone can become a hero," then the Lenin Prize was given for especially outstanding achievements in the fields of science, literature, and art. According to the policy on Lenin Prizes, they had to be awarded in honor of Lenin's birthday, 22 April. But they made an exception for us. At Korolev's OKB, Mishin, Tikhonravov, Kryukov, and I received the title of Lenin laureate. All the members of the Council of Chief Designers who had received the Hero of Socialist Labor title in 1956 became Lenin laureates in 1957. Bushuyev, Voskresenskiy, and Okhapkin received the Hero of Socialist Labor title.[14] Individuals involved in the project in all the subcontractor organizations also received their share of the awards.

ALL THOSE WHO PARTICIPATED IN HISTORY'S FIRST BREAKTHROUGH INTO SPACE WERE PREPARING TO USHER IN THE NEW YEAR, 1958, with the awareness that we were entering a new field of endeavor. Before these first two rudimentary satellites, we white-collar missile men had looked down at our first cosmic draftsmen, but now we understood that a "cosmic weight" was being placed on us all.

I was already quite fed up with the oxygen purging procedures that were accompanied by excruciating bile sample extractions. The amount of eosinophils and leukocytes in my blood was dropping slowly, but they would not release me for real work. Taking advantage of my Lenin Prize laureate title, I obtained a voucher for Katya and me to visit the Valday health resort, which was part of the Ministry of Health Fourth Main Directorate system. This directorate took care of the health of high-ranking Party and government officials and individuals "considered equivalent to them."

In late January 1958, having left our two sons in the care of Kseniya Timofeyevna, their maternal grandmother, for the first time, Katya and I visited a government health resort. It was located within a vast restricted area on the shore of Lake Balday, almost midway between Moscow and Leningrad. This marvelous site was selected

---

14. Typically, the Hero of Socialist Labor award was the most prestigious civilian honor. The leading *Sputnik* designers (Korolev, Glushko, Pilyugin, Ryazanskiy, Kuznetsov, Barmin, and Mishin) were among those who received the award in 1956 for the development of the R-5M nuclear-tipped strategic missile. As a result, in 1957, the previous awardees were not given a *second* Hero of Socialist Labor (which was an extremely rare honor) but instead given the Lenin Prize. Three of Korolev's leading deputies (Bushuyev, Okhapkin, and Voskresenskiy) were the only ones awarded the Hero of Socialist Labor for *Sputnik*.

before the war as the residence of Zhdanov and Stalin.[15] According to the stories of old Central Committee functionaries whom I met at this resort, Zhdanov had proposed construction at this site, with Stalin's approval. It was assumed that the two of them would settle down in this secluded spot to collaborate over a great treatise, a new history of the revolutionary movement, a Party history, and a theoretical justification for building a communist society.

A building with two "luxury" suites, one intended for Zhdanov and the other for Stalin, and many rooms with all the conveniences for their closest aides, formed the central part of this country estate. There was a magnificent library, rooms for quiet relaxation, billiards, auditoriums for music and movies, and a large dining room. There was no room for me in this elite building. Instead, they placed us in a building that had been converted for vacationers from a former battalion barracks that had housed Stalin's security service. Local managers told me that the battalion and the numerous support services had served for almost an entire year before the war. Zhdanov used to come, but Stalin never showed up. The barbed wire that enclosed the vast pine forest bordering the lake and the "restricted zone" signs were reminders of this resort's previous function.

Despite my malaise, I decided to try a therapy that I had devised on my own. Right after breakfast I went cross-country skiing until I was on the brink of

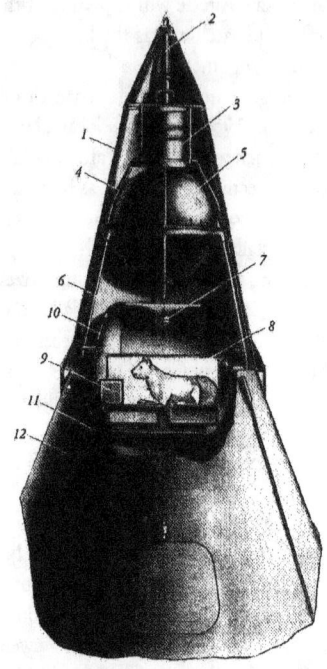

M. V. Keldysh, ed., Tvorcheskoye naslediye Akademika Sergeya Pavlovicha Koroleva: izbrannyye trudy i dokumenty (Moscow: Nauka, 1980).

Sputnik-2 carried Laika, the world's first living being to enter orbit around the Earth. The basic configuration of the payload was similar to the first Sputnik, but included a separate container to carry the dog. This container was itself derived from capsules developed in the mid-1950s to carry dogs on "vertical" trajectories into the upper atmosphere. Legend: 1) detachable protective fairing 2) mechanism to separate the fairing 3) instrument to study solar spectra in the short-wave range 4) framework for instruments 5) spherical container with radio-transmitter 6) thermally regulated cabin for experimental animal 7) air regulator 8) air regeneration system 9) food container 10) light 11) antenna 12) intermediate compartment.

---

15. Andrey Aleksandrovich Zhdanov (1896–1948) was a member of Stalin's Politburo (since 1939) and a major Communist Party leader in the Leningrad area. Zhdanov was the ideological instigator of the Soviet Union's postwar turn to extreme nationalism and stricter political control over intellectual and cultural life.

exhaustion. When I got back, I went to the shower room, lay down on the wooden grating, and took a steam bath under the surging streams until I reached a state of perfect bliss. After a short nap, I had lunch followed by the routine after-lunch rest hour. Then I was back on my skis, but this time joined by Katya and new acquaintances. The second skiing session was less grueling.

After two weeks of this regimen I felt completely healthy when I returned to Moscow. I reported to Professor Kassirskiy. After examining the quick blood test that had just been taken, he asked me, "So, tell me, who cured you so quickly? Your test is completely normal!" I told him everything as if I were at confession. He didn't have much faith in the stability of my new condition and requested that I come see him regularly. That was the end of the mysterious disease that had torn me away from work for more than six months.

Since then, along with the measles, scarlet fever, and an appendectomy—as well as a more recent case of influenza—I have to remember to list eosinophilic disease in my prior ailments whenever I check into a polyclinic or health resort. The only consolation in the ordeal was Kassirskiy's grad student's successful defense of her candidate's dissertation. Although I even received an invitation to speak at the medical board of academics, I deemed it best not to appear, so as to avoid undeserved celebrity, and confined myself to sending my best wishes by telephone.

The launch of the world's first artificial Earth satellite immediately removed all doubts that the Soviet Union had an intercontinental missile. Our sudden success shook the world. This had happened not only because Korolev had showed the qualities of a commander and an uncommon chief designer in a complex and rapidly changing situation. He had persuaded and captivated the Council of Chief Designers and the Academy of Sciences with his ideas and obtained the approval of the nation's leaders. As a result, in history, 1957 will forever remain the year that humankind broke through into space.

Chapter 22

Flight-Development Tests Continue

The launches of our first two satellites stunned those in charge of U.S. nuclear strategy much more than the August report about the creation of an intercontinental missile. Prominent publicist Professor Bernard Brodie of the RAND scientific and research corporation, which worked on defense issues, wrote that, "The Soviet satellites have dealt a blow to the Americans' complacency, having demonstrated for the first time that the Russians are capable of jumping ahead of us in technical achievements of great military importance."[1]

We had access to this sort of thinking and commentary by prominent American military men and scientists as classified information stamped "for managerial personnel only." This "managerial personnel" took great pleasure in familiarizing themselves with reports from across the ocean and at the same time realized that if, God forbid, the Cold War were to turn into a "hot" one, then we would be "big talkers, sham artists, and knights undeserving of our orders"—that is what Okhapkin, Voskresenskiy, and I called Bushuyev when we had the chance. As a design engineer he was officially responsible for the nose cones that disintegrated when they entered the atmosphere. Now Korolev was assigning him all the developments for new space projects.

Commander Korolev was so spellbound by the prospects of space that he wasn't even apprised of what precisely was going on with the nose cone that would contain a thermonuclear warhead. But anyone who grumbled about Korolev's swerve toward space-related subjects valued his foresight and his ability to handle resources available to him and rapidly enable very broad cooperation to solve new problems.

THE POSITIVE REVIEWS and praise in the global media and praise for our success,

---

1. B. Brodi [B. Brodie], *Strategiya v vek raketnogo oruzhiya [Strategy in the Age of Missiles]* (Moscow: Voyenizdat, 1961) p. 261. The original English version was published as Bernard Brodie, *Strategy in the Missile Age* (Princeton, NJ: Princeton University Press, 1959). RAND was formed in 1946, originally as part of the Douglas Aircraft Company, to conduct research on a variety of defense-related topics. Later, as a semi-independent research institution, it produced many ground-breaking works on the military, war, strategy, and foreign and domestic affairs.

which surprised Western society, sometimes frustrated us. The "unknown" chief designers felt deeply insulted. They, the council members, had put so much effort into developing the intercontinental missile, and now look what they got: complete anonymity.

But what was it like for Korolev to read the translations of the enthusiastic foreign press reviews and to hear the speeches of Soviet statesmen about our scientists' great achievements? On his desk like a red flag waved in front of a bull, I saw a translation of the magazine *Quick* devoted entirely to the "Red satellite" carrying pictures and comments from prominent scientists about the "artificial moon." These included Walter Riedel, a specialist in liquid-propellant engines who had worked in America with Wernher von Braun; Werner Schultz, a mathematician from the Federal Republic of Germany who had spent seven years in the USSR working on the island of Gorodomlya; and a man "who sees into the future"—astrophysicist Dr. Van Fried Petri from Munich. They all saluted the Russians' achievements. But who were these Russians?

This same magazine published photographs of the "father of the Red rocket," President of the Soviet Academy of Artillery Science A. A. Blagonravov, and the "father of the Red moon," Academician L. I. Sedov. The satellite's launch coincided with Blagonravov's attendance of a meeting for the International Geophysical Year in Washington, D.C. and Sedov's presence at the annual session of the International Astronautical Federation in Barcelona. These two Soviet scientists received the largest number of congratulations. They were photographed from various angles, and these portraits made the rounds in the international press. Having no direct involvement in the creation of the "Red rocket" and "Red moon," they nonetheless did not disavow the titles of "paternity" conferred upon them and accepted the congratulations and accolades. They knew full well the truth and the names of the actual creators of the rocket and satellite. Each of them could have been accused of immodesty, but what were they to do if they had no right to tell the truth?

Pilyugin was particularly miffed. He and Sedov had been at odds over questions of priority in inertial navigation. Pilyugin loved practical jokes, and at the Council of Chiefs he didn't pass up the opportunity to announce that "It turns out that it was Sedov and Blagonravov who launched the rocket and not us. I move that we induct them into our Council."

Korolev and Glushko, who were both fairly ambitious and who already had academic titles, were very touchy about these jokes and the misplaced praise by the global press. Unfortunately there was no one to complain to about it. Keldysh once mentioned that during his next meeting with Khrushchev he would ask permission so that our real missile specialists—instead of stand-ins—could participate in international forums. But as far as I know, this initiative on Keldysh's part never found any support, right up to the very death of Korolev.

All we could do was find consolation in such catchy foreign press headlines as "First Satellite Speaks Russian," "What's Keeping the Americans?," "Eisenhower Knew About Russian Rockets," "Man-made Moon Orbits Earth." All of this was

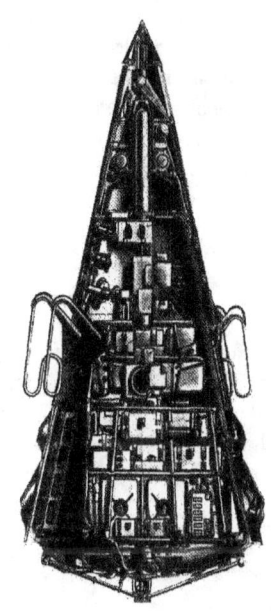

*M. V. Keldysh, ed., Tvorcheskoye naslediye Aka-
demika Sergeya Pavlovicha Koroleva: izbrannyye
trudy i dokumenty (Moscow: Nauka, 1980).*

**A cutout of the Sputnik-3 gives a
sense of the relatively complexity of
this scientific satellite designed to
study a range of natural phenomena
during the International Geophysical
Year. Known internally as the Object
D, the 1.3 ton satellite was originally
designed to be the first Soviet artificial
satellite. When its development was
delayed, Korolev opted to launch
a simpler satellite, later known as
Sputnik.**

accompanied by diagrams, fantastic pictures
of future satellites, predictions, and portraits
of specialists, not one of whom actually participated in the creation of our R-7 rocket and
our satellites. The Germans who had worked
on the island of Gorodomlya were unassum-
ing. They did not claim the laurels of those
who had participated in the creation of the
first "artificial moon." Judging by those publications available to us, they spoke ambigu-
ously about who had actually developed it all.

Incidentally, the younger generation of
specialists that followed us weren't yet seeking
fame. If anything, the atmosphere of secrecy
and protection that surrounded our work flattered their vanity, satisfying a patriotic sense
of personal involvement with great historic
events.

As I HAVE I ALREADY NOTED, IN JUNE 1957,
due to a series of defects, R-7 vehicle No. 6
was removed from the launch site. The rocket
engines underwent checkout procedures—all
sorts of tests—and once again was delivered
to the launch site. On 12 March 1958, during
a launch attempt, an emergency shutdown of
the engines occurred after buildup to the first
intermediate stage. Once more the culprit was
the main oxygen valve of the Block G booster,
which had opened prematurely due to the
failure of the break bolt. The testers quipped,
"How can one not but become supersti-
tious? This rocket is cursed, and it needs to
be removed from the firing range so that it
doesn't ruin the others."

Beginning in early 1958 the scope of projects at OKB-1 continued to grow
dramatically. The sudden success of the first two rudimentary satellites during the
enormous workflow for the development of the R-7 rocket was achieved relatively
easily. However, these successes exacerbated two new problems.

First, we recognized that we needed a more profound and serious attitude toward
space vehicles. Second, we needed to rethink the failures during R-7 launches, along
with many problems associated with this rocket. We accounted for all the unpleasantness that this rocket caused during various stages—its launch preparations,

during the launch itself, and during the powered flight segment—rather quickly. We were able to explain these problems using the engineering knowledge and experience that had been gained by that time. But it turned out that the rocket's very first successful full-range flight and the first space triumphs gave rise to problems that required not only new fundamental research, but also organizational restructuring. The original "old council" of chief designers didn't have solutions to both new problems, that is, of new research and organizational changes. As the council head, Korolev needed new compatriots and allies from outside the bodies governed by the council members.

AFTER MY ILLNESS, I returned to the heated rhythms of daily work revolving around my immediate problems of guidance, electrical, and radio engineering. Having done so, I sensed that our enthusiasm associated with the first space triumphs in 1957 would quickly die out and be replaced by expectations of a "miracle" that could save the missile's warhead. We had already learned to put spacecraft and even a dog into space, but what about returning them to the ground? If we weren't capable of preserving the warhead in the dense atmospheric layers, then it was much too soon to consider human flight. A human being is not a dog named Layka. In the future, how could we hope to return a human being to Earth alive if we could not protect a warhead from reentry?

Actually, having recognized how acute the problem was and without making a show, Korolev mobilized not only his own specialists, but recruited scientific forces from the outside to solve the problem of warhead reentry. In this endeavor it wasn't members of the council of chiefs who were very helpful, but rather Academy of Sciences Presidium member Mstislav Keldysh.

Since 1946, Keldysh had been head of NII-1, from which I was transferred to NII-88 in 1946. Due to all sorts of name changes, RNII, NII-3, and NII-1 finally became the Scientific-Research Institute of Thermal Processes. Keldysh indeed brought together first-class specialists in gas dynamics, thermal physics, and energy conversion in this very first rocket center in the Soviet Union. Georgiy Ivanovich Petrov, Vitaliy Mikhaylovich Iyevlev, and Aleksandr Pavlovich Vanichev headed fundamental research into various shapes of nose cones and future spacecraft during entry into the atmosphere. Jumping ahead, I will note that G. I. Petrov's work in this field contributed to his becoming an academician in 1958, while Vanichev and Iyevlev became corresponding members in 1962 and 1964, respectively. However, the work of the physical gas dynamics laboratory at the Academy of Sciences Leningrad Physical-Technical Institute (FTI) provided the greatest assistance to our nose cone developers (and not just Korolev's people, but all those who followed).[2]

As early as 1954, a decree was issued assigning Professor Yuriy Aleksandrovich

---

2. FTI—*Fiziko-tekhnicheskiy institut.*

Dunayev, who headed this laboratory, to develop a system for protecting the nuclear warhead of our R-5M missile "against the effects of the external air stream." Many FTI (today it's called the A. F. Ioffe Physical-Technical Institute) projects involved nuclear physics research, and therefore the level of security surrounding our subject matter there was even tighter than at OKB-1, where the FTI proposals were implemented. Korolev, our aero-gas dynamics specialists Viktor Fedorovich Roshchin and Andrey Georgiyevich Reshetin, materials specialist Aleksey Anatoliyevich Severov, and design specialist Ivan Saveliyevich Prudnikov dealt personally with Dunayev's laboratory.

After the 27 August 1957, TASS report about the successful flight of an intercontinental missile and secret reports from Kamchatka that nothing was found there, Korolev bitterly rebuked the military, saying, "You've got sabotage there on Kamchatka. Your officers and soldiers don't want to conduct a real search. I'm going to [have to] send my own people." And indeed, before the second launch toward Kamchatka, Korolev sent Andrey Reshetin there.[3] In order to satisfy their ambitions, the military sent their own scientists, including Engineer Colonels Narimanov and Elyasberg from NII-4. Flying once again into the Klyuchi area of Kamchatka, the scientific team found that hunting down the fragments of wreckage strewn over the undergrowth of the taiga was much more difficult than running into the Kamchatkan bears guarding their forest domains. Nevertheless, during the next launch they managed to visually determine the area where the fragments of the nose cone that disintegrated in the atmosphere must have fallen.

Decades later, chuckling, Andrey Reshetin related how after many days of hunting he reported to Korolev from Kamchatka about each new fragment found there. Dunayev had little interest in the Kamchatka fragments. According to one story, he was the first "theoretician" who proposed, developed, and, with the assistance of rocket materials specialists, introduced a new physical heatshield mechanism. The innovation consisted not in increasing the thickness and mass of the shield but in removing mass or "ablation." The "golden rain" of governmental awards for space triumphs did not bypass Dunayev. In 1961 he became one of the Lenin Prize laureates. A little over 30 years later he published a scientific treatise on the topic entitled "Development of a Technology for High-Temperature Coatings for the First Domestic Manned Spacecraft." Of course, even in the top-secret decree giving these awards, there was not a word about warheads.

In April 2004, Nobel Prize Laureate Zhores Ivanovich Alferov invited me to read a lecture on the history of cosmonautics at the A. F. Ioffe Physical-Technical Institute. The gift I received was a great surprise: it was a copy of the magazine *Neva*, issue No. 5 from the year 2003. The issue carried the memoirs of one Tatyana Vladimirovna Sokolova entitled "Terrestrial Anecdotes on 'Space Affairs.'" It was

---

3. *Author's note:* Professor Andrey Georgiyevich Reshetin is now a fellow department head at the Moscow Physical-Technical Institute.

the first time that our space literature published "nonscientific writings with lyrical digressions and everyday details" about how the nose cone heatshield of the first intercontinental missiles was developed. Before then, unfortunately, there had been no open publications on this subject.

In addition to a new heatshield, the nose cone also got a new shape. Instead of the pointed nose piece, it became blunt and spherical. For a detailed study of the phenomena that occurred during entry into the atmosphere, Bogomolov augmented a second *Tral*-G2 system with rod antennas. It was installed under the sheathing of the heatshield. The next significant step was to enhance the separation system in order to avoid collision with the main missile hull. After imparting a one-metric ton thrust to the payload container, the Block A core booster turned away to the side.

An R-7 missile with all the modifications, vehicle number M1-11, was delivered to the firing range around New Year's 1958. A month later it was prepared, and on 30 January 1958 the launch took place. Some evil fate continued to haunt the combat versions of this missile. The flight proceeded normally just until the strapon boosters began to separate. A defect in the mechanisms of the discharge nozzles of strapon boosters Blocks V and G damaged the tank pressurization line. The final stage was not able to generate the design thrust. The turbine raced and for some reason the emergency shutdown didn't kick in. Apparently the turbopump assembly (TNA) exploded.[4] The control pressure line was destroyed, and the cable network was damaged. The nose cone did not separate from Block A, and they entered the atmosphere together. Nevertheless, the new nose cone reached the ground for the first time, although it overshot the calculated point of impact by more than 80 kilometers.

Again we set about making modifications. Instead of a single separation push-rod, we installed three, each with one metric ton of force. A fundamental innovation was the installation in the nose cone of a "black box"—an automatic recorder with heavy-duty armor protection. This was the first serious project of Ivan Utkin's new organization, which had split off from our old NII-88 with a group of capable and enterprising radio engineers.[5]

Finally, on 29 March a missile with lucky number 10 (or M1-10) lifted off quite smoothly. This was launch number eight; counting the two satellite launches, it was the sixth launch of the R-7 intercontinental missile program. Kamchatka reported that, judging by the crater, the nose cone hit the ground without disintegrating. It had overshot the target by 7.5 kilometers and deviated 1.1 kilometers to the right. Telemetry received during the 8 seconds before impact with the ground confirmed that the nose cone did not disintegrate in the atmosphere. Nevertheless, after processing the information, typically the telemetry specialists would spoil the mood of

---

4. TNA—*Turbonasosnyy agregat.*
5. Utkin's Complex No. 5 separated from NII-88 in July 1966 to become the independent Scientific-Research Institute of Measurement Technology.

somebody or other among the chiefs. This time the tank emptying system (SOB) operated unstably, and false commands were issued. The radio-control ground stations failed to process the pitch tracking program. There was a dispute among the radio specialists—the ground station specialists versus the onboard systems specialists—as to who was more at fault.

Every new launch brought some new failure! Nevertheless, the missiles streamed in to the engineering facility. Three days after the first basically normal launch, another "lucky one" was moved to the launch site, missile number M1-12. The launch took place on 4 April. In the first report, Kamchatka scared us by stating that once again it was receiving no nose cone telemetry information and a new crater was not found in the impact square. However, the next day they cheered up: there was indeed a crater, but the nose cone had overshot by 68 kilometers, with a deviation of 18.2 kilometers to the right. Again, telemetry analysis provided an explanation: 142 seconds into the flight, tracking using the radio-control antenna ceased; apparently the programmed tracking mechanism malfunctioned.

Despite such a number of serious glitches, flight-development testing (LKI) of combat missiles was once again interrupted by space launches.[6] Object D's turn had come—it was fated to become a full-fledged Earth satellite. Unlike the first two satellites, the third one was prepared without resorting to all-hands rush mode, with the participation of many scientists who had been recruited for this program as early as 1956. Keldysh devoted particular attention to the preparation of this satellite. He held many meetings and conferences and reconciled conflicts that were flaring up between our "missile" interests and the aspirations of "pure" scientists. Factions passionately struggled over the volume and mass appropriated for scientific equipment.

In April, during the launch of Object D, slated to become the third Soviet satellite, our R-7 once again decided to show its stubborn side. It delivered the payload with all its precious scientific instruments "over the hill."[7] Keldysh and all of the young scientific space community were in mourning. But Korolev did not give up.

Assembly of a backup satellite was under way at the factory. S.P. assembled all of his close associates and announced that, despite the setback, each of them would be paid a substantial bonus if everyone would remain at the firing range and prepare the next launch vehicle. The launch needed to be conducted in mid-May. He and Keldysh would fly to Moscow to speed up the preparation of a new third satellite. This decision had not been easy, but there was no alternative. Commitments for the launch of a "scientific space laboratory" had already been made to Khrushchev.

Events surrounding the third satellite bear recounting at some length. On 15 May 1957, as we congratulated one another on the first R-7 launch, we consoled

---

6. LKI—*Letno-konstruktorskiye ispytaniya.*

7. The first attempt to launch the Object D satellite into orbit ended in failure on 27 April 1958 when the R-7 launch vehicle disintegrated about 96 seconds after launch.

Asif Siddiqi.

This photo shows Sputnik-3 (Object D) on the launch pad prior to launch in 1958. This satellite used a one-off variant of the R-7 ICBM known as the 8A91. Note the unusual tip of the payload fairing.

ourselves that it was as it should be, that "the first pancake is always lumpy."[8] The day of 15 May 1958 compensated to some degree for that "first pancake." R-7 vehicle number B1-1 inserted the third Soviet artificial Earth satellite on orbit. The satellite's imposing mass of 1,327 kilograms, of which 968 kilograms constituted the scientific equipment and instrumentation once again generated glowing press reports.

This was actually the first automatic spacecraft. It carried 12 scientific instruments, Bogomolov's *Tral* telemetry system with a recording device, and the *Rubin* transponder for orbital monitoring. This was also the first spacecraft equipped with the command radio-link that our new subcontractor NII-648 had developed.[9] In 1956, the very energetic and enterprising radio engineer Armen Sergeyevich Mnatsakanyan headed the institute. Under his leadership, command radio-links (KRL) were developed for our new spacecraft, and later Mnatsakanyan's organization began to develop space search and rendezvous radio systems for the Soyuz spaceships.[10]

---

8. This Russian idiom for "practice makes perfect."

9. *Author's note:* Now this institute is called the Scientific-Research Institute of Precision Instruments.

10. KRL—*Komandnaya radioliniya.* These search and rendezvous systems included the *Igla* (Needle) and *Kurs* (Course) radar systems.

The third satellite was a spacecraft that had required the development of a complex electrical power supply, program, and command control system for the individual science equipment. These developments were entrusted to two young engineers who had only recently been sent to OKB-1 upon graduation from the Taganrog Radio Engineering Institute. Yuriy Karpov and Vladimir Shevelev belonged to a group of young specialists who came on the scene at the very birth of the idea of space electrical engineering and automatics. When our work on space systems expanded, these two "highest guys at OKB-1" generated the ideas and principles for the development of onboard complex control systems (SUBK) for spacecraft.[11] For them the third satellite was their first really serious engineering task. In subsequent years, my close association with Yuriy Karpov and his team was always interesting, not only in a professional and engineering sense, but also on a personal, human level. I've had the opportunity to work first-hand with many engineers on a daily basis over the past decades, but I felt particularly warmly toward Yuriy Karpov and his circle of circuitry specialists. They created a sense of community among—as Korolev used to say—the "rusty electricians." On the job and in their lives they adhered to the principle of "all for one and one for all."

One of the sensational results obtained with the aid of the third satellite's scientific instruments was the discovery of a high concentration of electrons at high altitudes beyond the limits of the already known ionosphere. Sergey Nikolayevich Vernov, the MGU professor and primary investigator of this research, attributed this phenomenon to secondary electronic emission, that is, to the dislodging of electrons from the satellite's metal during collision with high-energy particles such as protons and electrons. I recall his elated report about this at a meeting in Keldysh's office, where scientists gave their accounts of the results of the scientific research on the third satellite.

However, two years later the American physicist James Van Allen proved that what the third satellite's instruments had actually measured was not from secondary emission, but rather from primary particles of the Earth's previously unknown radiation belts.[12] That is why the Americans named these radiation belts the "Van Allen Belts." In Vernov's defense it must be said that he erred due to the failure of the satellite's telemetry recording device. Vernov was not able to receive measurements of the radiation activity over the satellite's *entire* orbital pass, but he received measurements only in direct reception mode when the satellite was flying over the territory of the USSR. Van Allen made his discovery using the results measured by

---

11. SUBK—*Sistema upravleniya bortovymi kompleksami.*

12. James Alfred Van Allen (1914–) is a pioneering astrophysicist best known for his work in magnetospheric physics. Besides his fame in identifying the belt of charged particle radiation that is trapped by the Earth's magnetic field (the Van Allen Radiation Belts), he is also known as one of the main instigators in organizing the International Geophysical Year in 1957–58. He remains a prominent public commentator on the role of science in the exploration of space.

an American satellite.[13] He showed that there was a region in near-Earth space in which the Earth's magnetic field holds in charged particles (protons, electrons, and α-particles) that possess a great deal of kinetic energy. These particles remain in near-Earth space, held in what is referred to as the magnetic trap.

This discovery was a great scientific sensation and had important practical significance for cosmonautics. Spacecraft, whose orbits passed through radiation belts, were exposed to significant levels of radiation that damaged, in particular, the structure of their solar array sensors. For crewed spacecraft, a prolonged stay in these belts is not acceptable at all and can be very dangerous.

After the publication of Van Allen's discoveries, we decided, albeit belatedly, to correct the mistake committed through the failure of the recording device on the third satellite. In our literature, however, they started to refer to the radiation belts as the "Van Allen-Vernov belts."

This episode was a good lesson for scientists since it demonstrated how essential it was that direct measurement instruments and onboard service systems reliably operate when obtaining, storing, and transmitting the data obtained by them back to Earth. Unfortunately, equipment reliability for scientific research remained a weak spot in our cosmonautics for years to come. With the goal of "rehabilitating" Soviet science, on assignment from the Academy of Sciences, we urgently developed and launched four new spacecraft: *Elektron-1, -2, -3,* and *-4*. But they were not launched until 1964. These Elektrons made it possible over a long period of time to obtain comprehensive data about the Earth's radiation belts and magnetic fields.

AFTER 15 MAY 1958, a historic date for rocket technology, we once again returned to our regular flight-development testing program and suffered two disasters in a row. On 24 May, R-7 vehicle number B1-3, prepared at the launch site in a record short period of time—21 hours—lifted off normally. However, Kamchatka reported that it fell short of the target by almost 45 kilometers, with a slight lateral deviation. Once again telemetry helped to determine the cause. In the final phase of the second stage, the blow-off valve of the oxidizer tank failed. Without pressurization, the oxygen entering the pump contained "bubbles." The turbopump assembly broke down, damaging the adjacent lines. The nose cone entered the atmosphere along with the entire core booster.

How many hopes were tied to the last launch of this long-suffering R-7 first series! But our Galatea did not give in. Voskresenskiy reminded me with gentle derision that the Galatea of the ancient Greeks brought to life by the gods was probably more compliant.[14] "Just think—we have so many men and have worked already for

---

13. Van Allen used data from the *Explorer 1* and *Explorer 3* satellites to conjecture the presence of these radiation belts.

14. This is a reference to the Pygmalion myth from Greek mythology; Galatea was the name of a statue created by Pygmalion and brought to life by Aphrodite.

over a year and yet we haven't been able to make friends with the rocket that we brought into this world."

We tried to launch the last rocket, vehicle number B1-4, on 10 July. I write "tried" because the rocket was removed from the launch site due to the failure of the Block D strapon and the latest failure of the break bolt on the main oxygen valve. By then, of the 10 missiles not used as satellite launch vehicles, only seven had lifted off. Of those seven, only two more or less tolerably carried their payload equivalent to the target.

The State Commission was in a very difficult situation. They quibbled with the wording, retyping the conclusions and comments dozens of times. Ultimately, they wrote that the "experimental data on dispersion did not allow a full evaluation as to whether the design specifications had been met. But, according to preliminary data, in principle, the dispersion would not exceed the predetermined value." The report went on to cite a short list of systems that had demonstrated their effectiveness and a long list of all the defects and measures that should be implemented before.... Before what? The next phase was supposed to be joint Ministry of Defense and industry tests, the results of which were to decide the missile's fate. There was no way to retreat. After many days of meetings and many hours of discussions, the State Commission recommended going on to the next phase, that is, the joint tests.

Here I should make one more digression of no small significance. The general conviction that we would "bring the *Semyorka* up to speed" still outweighed the skepticism of the cautious and the fierce attacks of enemies of the liquid-oxygen engine approach that our OKB was using. For the next two years no other intercontinental missile project could compete with the *Semyorka* in terms of readiness. A strong production base needed to be prepared in advance for the mass production of the R-7 missiles, engines, and instrumentation. It was also necessary to build another two or three launch sites. It was quite evident that at the same time the "joint" launch program for the R-7 was under way, space launches would also be in the cards.

THE POLITICAL HUBBUB SURROUNDING THE DAWNING OF THE SPACE AGE reached such a pitch that in the launch plans for the next few years, considerably more launches might be required than those that would simply intimidate the Americans with the fact that we had an intercontinental thermonuclear bomb carrier. The R-7 was all that the USSR had for both tasks, and according to the most optimistic plans, there would be no other prospects before 1961. After producing the R-7 missile, our large network of cooperation headed by our OKB-1 carried double accountability. We were now responsible for both the military use of the missile and for using it to develop space technology. For the next few years, only the R-7 rocket would be able to slog down the road to space, which began on the territory of the USSR.

Depending on the results of the joint tests, the decision would have to be made whether to recommend putting the R-7 with a thermonuclear warhead into service. The military had a vital interest in a positive outcome. At Nedelin's initiative a draft

403

decree of the Council of Ministers was prepared on the creation of a new independent branch of the armed forces: the Strategic Rocket Forces (RVSN).[15] If such a decree were issued, the rocket forces would be equal to the conventional branches of the armed forces—the air force, navy, ground forces, and air defense forces. Each of these branches had its own commander-in-chief, headquarters, uniforms, military institutes, academies, and much more.

But such a decision could not be made until intercontinental strategic missiles were put into service. Up to then, troop formations that had Korolev's R-1, R-2, R-11, R-11M, and R-5M missiles, as well as Yangel's very new R-12 missile in service, had been called the Supreme Command Reserve (RVGK) engineer brigades.[16] Heavy artillery—supreme command reserve artillery brigades—also had similar status in wartime.

Sixteen R-7 missiles were manufactured for the joint tests—eight at the Progress Factory and eight at our pilot plant "where Comrade Turkov is director" (this was what the press wrote and what was said at conferences to avoid mentioning the number and location of a secret facility). The Progress Factory, new in our network of cooperation, had become part of the nascent rocket empire having been forced out of the aviation industry during Khrushchev's so-called campaign of "cannibalization" of that industry. All of the series production aviation factories were subordinated to regional Councils of National Economy *(Sovnarkhoz).*[17] A Council of Ministers decision delegated the organization of the series production of the R-7 missile to the Kuybyshev *Sovnarkhoz*, which proposed allocating the task to the aviation Factory No. 1, which was renamed the Progress Factory.

This factory had an illustrious history. Even before World War I, one of the first factories to build airplanes in Russia was the Moscow Duks bicycle factory. After the Revolution, the Duks factory switched over completely to the manufacture of airplanes to create the Red Air Force and was renamed State Aviation Factory No. 1. The factory specialized in the production of fighter planes and light reconnaissance planes and was located in Petrovskiy Park on the border of Khodynka Field. Later Khodynka became the airfield for Factory No. 1 and by 1925 was called the M. V. Frunze Central Airfield of the Republic. The entire area adjacent to Factory No. 1 and Khodynka Field, which was later renamed October Field, gradually turned into a military-industrial aircraft area. The design bureaus and the pilot plants of Polikarpov, Ilyushin, Mikoyan, and Yakovlev were located along the former Petrovs-

---

15. RVSN—*Raketnyye voyska strategicheskogo naznacheniya.*

16. RVGK—*Reserv verkhovnogo glavnokomandovaniya.*

17. In 1957, Khrushchev instituted nation-wide industrial reforms that decentralized much of the Soviet defense industry. As a result, defense factories (such as the Progress Factory) were subordinated to local councils instead of a central command in Moscow. These local authorities were called Councils of the National Economy (*Sovet narodnogo khozyaystva* or *Sovnarkhoz*).

kiy Park, now Leningradskiy Prospekt.[18] Here, a palace from the days of Catherine the Great stands out. For many years it has also been the main building of the N. Ye. Zhukovskiy Air Force Engineering Academy. It was also the founding historical site for the Air Force Scientific-Research Institute.

One of the first directors of aviation Factory No. 1 was Petr Dementyev, who would later become minister of aviation industry.[19] In 1941, at the beginning of World War II, the factory was commissioned to produce Il-2 fighter bombers. After their evacuation to Kuybyshev, the factory's employees accomplished an extraordinary feat of labor. At the new site, under the most difficult conditions, the half-starved people produced 12,000 of the famous Ilyushin fighter bombers. Shooting upward on an arched steel above the banks of the Volga in Samara, an Il-2 serves as a monument to the heroic labor of the war years. After the war, as one of the biggest and best factories in the aviation industry, this factory switched over to the production of MiG-9 and MiG-15 jet aircraft and Il-28 bombers.

The factory underwent a major overhaul for the production of mis-

From the author's archives.

In 1959, Sergey Korolev tasked Dmitriy Kozlov (1919-) with production oversight over the R-7A ICBM. Kozlov, shown here around 1970, eventually took over leadership in developing numerous launch vehicles (the Molniya, Soyuz, etc.) derived from the original R-7 missile. In later years, as Chief Designer of the independent Central Specialized Design Bureau (TsSKB), Kozlov supervised the development of the majority of Soviet optical photo-reconnaissance satellites. He retired only in 2003.

siles. In my first encounters and subsequent close acquaintance with factory director Viktor Yakovlevich Litvinov, he impressed me as a very gentle and sensitive individual, quite unlike a director. Nevertheless, he enjoyed indisputable authority in his organization. His instructions were carried out without him having to pound

---

18. Sergey Vladimirovich Ilyushin (1894–1977), Artem Ivanovich Mikoyan (1905–70), Nikolay Nikolayevich Polikarpov (1892–1944), and Aleksandr Sergeyevich Yakovlev (1906–89)—all famous aviation designers—headed some of the largest aviation design bureaus during the Soviet era.

19. Petr Vasilyevich Dementyev (1907–77) served as minister of aviation industry in 1953–77.

his fist on the table and without shouting and strong language. When he was tasked with mastering a completely new technology, he joked: "During the war Stalin threatened me with court martial if we failed to meet the deadline for the delivery of fighter bombers. After the war, one month before the Tushino air parade, we were ordered to produce a squadron of jet fighters. Now we have a new order: do away with the fighters and bombers at the factory and make Korolev's missiles. But we had just mastered the new bombers and were dreaming of working happily without rush jobs, if only for a couple of years.... So I wanted to send Korolev a hundred or so workers, engineers and technicians, for training to master the new technology. But they lost their tempers and complained that 'Korolev handed over missiles that fly on oxygen and good ethyl alcohol to the Dnepropetrovsk plant for series production, but here in starving Kuybyshev, we get a missile that runs on kerosene. If it ran on alcohol, we wouldn't argue.'" Litvinov loved jokes that took the edge off of difficult situations.

New shops and test benches were rapidly built, and cooperation was established between OKB-1 and the Progress Factory. In 1959, Progress confidently began the series production of R-7 missiles, and soon OKB-1 Branch No. 3 was created there. R-7 missile lead designer Dmitriy Ilich Kozlov was appointed chief of this branch; he expanded and reorganized the Kuybyshev branch into the independent Central Specialized Design Bureau (TsSKB).[20] Subsequently the TsSKB assumed all the responsibilities for the modification and production of the R-7, although the primary products of the TsSKB in subsequent years were spy satellites. Later Kozlov twice became a Hero of Socialist Labor.[21] He was elected a corresponding member of the Academy of Sciences and was awarded Lenin and state prizes.

After Khrushchev was overthrown, one of the first serious measures that the Communist Party leadership headed by Brezhnev took was to eliminate the Councils of National Economy *(Sovnarkhozi)* and restore the old ministry system. The Ministry of General Machine Building (MOM) was created to manage all rocket and space technology.[22] The Progress Factory and all series production missile factories, including the Dnepropetrovsk-based Yuzhnoye Machine Building Factory and our OKB-1, became part of the new ministry.

Progress Factory Director Litvinov was pulled out of Kuybyshev and appointed deputy minister of general machine building.[23] Certainly not every director of a

---

20. TsSKB—*Tsentralnoye spetsializirovannoye konstruktorskoye byuro*. TsSKB subsequently became one the primary developer of Soviet optical reconnaissance satellites. Today, it continues to develop new versions of R-7-based launch vehicles, as well as military reconnaissance, remote sensing, and microgravity spacecraft.

21. He was bestowed the award in 1961 and 1979.

22. The Ministry of General Machine Building (*Ministerstvo obshchego mashinostroyeniya* or MOM) was created in March 1965 to oversee all strategic missile and spaceflight programs.

23. Viktor Yakovlevich Litvinov (1910–83) served as deputy minister of general machine building in 1965–73.

large enterprise is pleased when promoted to such a high and, it would seem, honorable post. I had known many powerful managers; they were talented production organizers who had passed through all the levels from worker, foreman, and shop chief to chief engineer and director. The majority of them felt very uncomfortable when they found themselves in positions of authority in the central political apparatus. Litvinov did not conceal his own dissatisfaction with this promotion; this was, however, a decision of the CPSU Central Committee Secretariat, and Party discipline was sacred. You could grumble, but you were obliged to fall in line, part company with your dear organization, and plunge into the bureaucratic paper chase of the hierarchical central power *apparat* (bureaucracy).

During the arduous days of all-hands rush jobs when we began mastering the Soyuz manned spacecraft, I often dealt with Litvinov when he came to our facility. He frankly confided that he envied us, because no privileges granted to high-ranking officials of the central *apparat* could replace the genuine satisfaction that the manager of an organization experiences when working to produce new and complex technology.

At another Kuybyshev plant, machine building Factory No. 24 "where Comrade Chechenya is director," personnel were mastering the production of engines for the R-7 rocket. So as not to ruin the oldest aviation engine building factory, industry leaders persuaded Khrushchev not to devote the plant solely to the production of liquid-propellant rocket engines. They volunteered to arrange production of rocket engines while maintaining production of turbojet engines.

Other heavy machine building enterprises were drawn into cooperation with Barmin in order to create launch complexes at five new sites, one at the Tyuratam firing range and four in Plesetsk near Arkhangelsk.

Our nation's new missile technology was also a powerful stimulus for the development of the instrument making and electronics industry. While the best aviation factories could be restructured for the series production of missiles, thereby inflicting tremendous damage on our aviation technology, there was no one to take factories from to produce instruments; this branch of industry had to be created virtually from scratch.

Only the gyroscope production sector could benefit from the experience and facilities of the mighty shipbuilding industry. Enjoying great prestige in naval instrument building circles, Viktor Kuznetsov managed to set up series production of gyroscopes at his institute's factory, at the Saratov instrument building factory, and at a new factory that was under construction in Chelyabinsk. High-capacity, very well-equipped production of command gyroscopes was also set up in Leningrad at NII-49 under the management of talented engineer and great gyroscope technology enthusiast Vyacheslav Pavlovich Arefyev.

Nor was Gorodomlya Island on Lake Seliger forgotten. The sylvan island abandoned by the Germans caught Kuznetsov's fancy. He managed to convert NII-88 Branch No. 1 on the island into a branch of his own gyroscopic institute, and citing its exceptionally clean environment as a rationale, created a plant there that produced precision gyroscopic instruments using the most state-of-the-art principles. This new factory proved to be virtually the only one in the USSR where the toxic process of casting and machining parts made of super-light beryllium alloys was mastered. Thus, when the Germans departed in 1953, not only did the island not drop its "cover," it became even more secret.

THE ENTIRE SECOND HALF OF 1957 AND BEGINNING OF 1958, I was involved in very few important technical discussions of future projects and Council of Chiefs meetings. To begin with, I was constantly at the firing range, and then my illness also kept me away from work.

I regularly received information about the most important events occurring at the OKB and in the "higher spheres" surrounding it and also about attitudes and considerations in that regard from Ryazanskiy, Yurasov, Voskresenskiy, Bushuyev, and Kalashnikov. Nevertheless, when I finally appeared at work in spring 1958, I once again realized how swiftly events unfold. We, who were in charge of OKB-1, were the tip of a growing iceberg. Beneath us a thoroughly hush-hush mighty empire was being developed. Our iceberg was not the only one in a vast ocean of problems. A new missile giant had already come into view on the horizon; the Dnepropetrovsk-based Factory No. 586 switched from being a collaborator to a competitor after Yangel showed up there in 1954 as chief designer.

Comrades told me the details of a series of important discussions that took place in my absence. It started with a discussion of proposals for a prospective program at a meeting of the chief designers in June 1957. "Nonchiefs" in attendance from OKB-1 were Yurasov, Mishin, Voskresenskiy, Karpov, Bushuyev, Okhapkin, Lavrov, and Raykov. There were also some other deputy chiefs there. In the opinion of Bushuyev and Yurasov, the degree of consensus that had existed before was already lacking among the chiefs. And this was above all due to a rift in the relationship between Korolev and Glushko. The latter felt that it was necessary to use dimethyl hydrazine as a fuel along with kerosene. He also harked back to his previous proposals for the R-8 missile, contrasting it to the R-7.[24] His position was understandable; he had made engines that used high-boiling components for Yangel, and, therefore, he considered it proper and expedient to develop yet another heavy missile design along the same lines. In his opinion, liquid-oxygen missiles needed to be backed up with missiles using high-boiling propellant components. For the R-16 missile that Yangel had begun to design, a new guidance system chief designer had been found, Boris Konoplev. Initially, Konoplev went to Kharkov to set up operations for the radio-control systems, but then he took on the control complex for the R-16 in its entirety. Thus, Pilyugin and Ryazanskiy no longer held a monopoly.

---

24. In 1956, Glushko proposed a new ICBM (the R-8) as a successor to the R-7. The new missile would use ten 100-ton thrust engines working on storable propellants (such as unsymmetrical dimethyl hydrazine).

Meanwhile, Glushko remained the only monopolist in his field of rocket engines. Even Kuznetsov had already ceased to be the one and only developer of onboard gyroscopic instruments. NII-49 in Leningrad specialized in gyroscopic technology for submarine-launched missiles, but was also ready to develop other command instruments.

VPK chairman Ryabikov, who presided over the discussion, spoke out clearly in favor of optimizing the R-7. There was no place for wavering here. But the R-7's 8,000-kilometer range was insufficient. We needed to begin designing liquid-oxygen engines with greater range.

I heard quite unexpected news from Kalashnikov. In late January 1958, Fedor Falunin, our former lead designer for control surface actuators, came to us on temporary assignment from Dnepropetrovsk. Now he was working at Yangel's KB as chief of the control surface actuators department. Falunin told us about Yangel's sensational speech at the meeting of the expert commission on the R-16 missile's conceptual design. All of the numerous staff who had transferred from Podlipki to Dnepropetrovsk believed that they also had a stake in OKB-1. They were glad for our successes. Yangel's very tactless speech before this commission chaired by Keldysh astonished and offended them all the more. Instead of defending the R-16 design as such, Yangel lambasted OKB-1's technical policy, which in his words was leading our nation into a dead end. In Yangel's opinion, liquid-oxygen missiles were useless. Instead of these missiles, we needed to produce state-of-the-art and mission-capable missiles using high-boiling propellant components. Yangel's speech was so tactless that Keldysh had to interrupt him and ask that he stick to his presentation in defense of the R-16.

Why Yangel needed to expose his personal dislike for Korolev in this way at an official technical gathering attended by many people, I cannot explain. Having studied both their personalities well, now that neither of them is around, I believe that Yangel was primarily at fault for their falling-out. More than once I observed that he could not contain his emotions. With respect to Korolev, Yangel's emotions sometimes prevailed over his reason.

Later I had the occasion to meet with Yangel many times in Moscow, at Dnepropetrovsk, and at the firing range. Despite the fact that I was Korolev's deputy, we maintained good personal relations. Moreover, our lead specialists, who visited Yangel's KB on business many times, were always cordially received. There was no antagonism between our organizations, but the staff were unable to influence their managers and to have them achieve normal relations between themselves.

You have to give Pilyugin credit. He had good relations with both Korolev and Yangel. More than once, as he used to tell me, in one-on-one conversation he convinced each of them that they needed to reconcile to work out a unified missile policy in the interests of the cause. Let them both even agree to a healthy competition, a contest between liquid-oxygen and high-boiling component missiles. After all it was obvious that both types had a right to exist for the time being. Later life would show to whom the future belonged. But neither Korolev nor Yangel took the first step toward reconciliation. Subsequently, when the fire of enmity died down, Glushko threw oil on it. And later, Chelomey joined into this controversy. He didn't form an alliance with Yangel against Korolev. He pursued his own technical policy, competing against both of them.

Enemies of liquid-oxygen missiles had very solid arguments. Losses of oxygen to evaporation during transportation and storage were two to three times the fueling requirements. Korolev and, perhaps even to a greater degree, Mishin decided to study this shortcoming of oxygen in earnest. Together with specialists recruited for this problem they soon realized that the oxygen industry was not interested in developing a technology and methods to reduce the loss. Seeing that Mishin had thoroughly investigated the problems of oxygen economics, Korolev made him responsible for drawing up new proposals, relieving him of other responsibilities for the time being.

It was Vasiliy Mishin's nature to become utterly absorbed in any new idea. During such periods he devoted himself completely to the development of the new idea, trying not to waste time on other routine matters that had nothing to do with his current fancy. Korolev knew how to use this character trait of Mishin's to great advantage for the common cause. When Korolev noticed that Mishin was immersed in working out a problem that Korolev endorsed, he stayed out of his way. If I needed to meet and consult with Mishin on some matter that had no direct relation to his latest passion, regardless of the urgency of the matter I was coming to him about, he would tell me about the latest accomplishments, thoughts, and problems that completely engrossed him. Such was also the case with the oxygen storage problem that obsessed Mishin in the late 1950s and early 1960s. Mishin's intransigence, which at many meetings ended up in vehement confrontations, was rooted not in his personal attitudes toward one individual or another, but rather in his conviction in the rectitude of his ideas and proposals. Even a comrade and friend who at a given moment did not share his engineering idea could become an enemy for a while.

We needed to be able in the nearest future to transport and store liquid oxygen without losses. If this problem was not solved throughout the entire cryogenic industry it would be impossible to issue the proposals for the development of the new R-9 intercontinental missile on which we had already begun to work. If we did not stand up for the liquid-oxygen missile design at a range up to 12,000 to 14,000 kilometers, then after the R-7, the military would have no choice but to accept Yangel's new proposals, the R-16 missile using the "most toxic" components, nitrogen tetroxide and unsymmetrical dimethyl hydrazine.

In the struggle over these components for super long-range missiles, much less for space tasks, Mishin was a "greater monarchist than the king himself." He succeeded in firing up not only our OKB-1 specialists with his enthusiasm, but also many on the outside. Besides enthusiasm, we also, of course, needed the direct assistance of industry. For this, Korolev had to appeal to Khrushchev and Ustinov—who succeeded Ryabikov as chairman of the VPK. The majority of the measures proposed were realized—not in a year as proposed—but in three years. By 1961, new heat-shielding principles and materials, receptacle designs, and new pumps to service high-vacuum systems had been developed. I did not participate directly in solving the oxygen problems, but operations at OKB-1 assumed such a scale that it was simply impossible at that time to stand on the sidelines if there was an opportunity to help. The next time Mishin argued with his inherent ardor how important it was to achieve and maintain a high-vacuum for the vacuum shield thermal insulation, it reminded me of our meetings with Academician Vekshinskiy.

In 1944, working with Roman Popov and Abo Kadyshevich at NII-1 on the Aircraft Coordinate Radio Locator (ROKS), we invented a powerful new tube, a microwave-range radio wave pulse generator. Through our youth and inexperience we imagined that we had discovered principles that would cause a radio engineering revolution. Aksel Ivanovich Berg, who was then the leader of all radar engineers, advised us to turn to Sergey Arkadiyevich Vekshinskiy for consultation. Vekshinskiy was a famous scientist in the field of electronic tubes. He listened to us attentively and then led us to a laboratory and showed us the mock-up of the tube whose concept we had just presented to him. "America has already been discovered and settled," he joked, quoting from an old school song. We departed terribly disappointed.

Now, 15 years later, I was accompanying Korolev and Mishin to call on top-notch Soviet scientist and electro-vacuum technology specialist Academician Vekshinskiy. The Electro-vacuum Institute had grown up at the site of the modest laboratory, which was enormous even by our missile standards. The demands of nuclear and radar science accounted for its rapid development and opulent facilities. Institute Director Vekshinskiy, who cracked a plaintive smile when I reminded him of our meeting in 1944, said that back then, despite the war, work was easier and more light-hearted. After studying the oxygen problem, he promised to help. Vekshinskiy kept his promise. His institute developed a very economical system for maintaining a high vacuum in the thermal insulated chambers of the liquid oxygen storage tanks.

The oxygen problem had a significance that went far beyond the boundaries of missile technology interests. The problem of storing oxygen for combat launches of the R-9 missile was solved by late 1962, thanks to the fundamental work that Korolev and Mishin directed—not because of departmental affiliation—but because of their understanding of its importance for the state. Losses due to evaporation during the storage and transportation of oxygen were reduced by a factor of 500!

IN JUNE 1958, a general assembly of the Academy of Sciences took place. Despite the total secrecy of our missiles, the learned academic community understood that the developers of intercontinental missiles and satellites deserved the highest academic degrees and titles. At this meeting Glushko and Korolev were elected academicians, while Barmin, Kuznetsov, Pilyugin, Ryazanskiy, and Mishin were made corresponding members. At this same meeting, former *zek* Aleksandr Lvovich Mints was also elected as an active member in the USSR Academy of Sciences, joining former *zeki* Glushko and Korolev.[25] Nor did they pass over the developers of the first air defense missile systems. Comparatively young radio engineers Kisunko, Rasple-tin, and very belatedly, general designer of fighter aircraft and air defense missiles Semyon Lavochkin were also elected as corresponding members.

According to the academic rules, the last names and scientific achievements of the newly elected members needed to be published, if but briefly, in the press. Glushko was briefly described as "specialist in the field of thermal technology" while Korolev, Barmin, Mishin, and Kuznetsov were referred to as "specialists in the field of mechanics." Pilyugin's description, "specialist in the field of automatics and telemechanics," provided slightly more insight. Ryazanskiy, Kisunko, and Raspletin meanwhile were "specialists in the field of radio engineering." And then Lavoch-kin, already world-famous, received the straightforward label of "aviation designer." Chelomey, who had already gained strength, was elected a corresponding member. He, too, fell under the heading "specialist in the field of mechanics."

The results of the elections to the Academy gave the Council of Chiefs a substan-tial boost in prestige not only at the highest levels, but also among engineers. The managers of many subcontractor organizations received a very palpable incentive to step up their work in rocket-space technology. As later experience confirmed, many talented scientists were attracted to our projects in the hope that their achievements in solving scientific problems for rocket technology and space research would give them a chance to be elected to the Academy.

Another pleasant event was the Moscow Municipal Council of Peoples' Depu-ties *(Mossovet)* decision to provide more than a hundred apartments in Moscow to particularly distinguished specialists and individuals involved in the development of the first satellites. In particular, three sections were set aside for our organization in new apartment buildings along 3rd Ostankinskaya Street, which today is named for Academician Korolev. In building No. 5, there was a housewarming party for Korolev's deputies Bushuyev, Voskresenskiy, Okhapkin, Melnikov, and myself. The Chizhikov family, who had been part of our tight-knit group at the Villa Frank in Bleicherode, became our neighbors by the staircase landing. To this day, Mikhail Tikhonravov's family lives on the other side of our apartment wall. We occupied only two entryways out of ten in the enormous building, but the entire building eventually came to be called *Korolevskiy* (or "Korolevian").

By special governmental decree, Korolev and the other five chief designers obtained the right to build dachas at government expense. Barmin, Kuznetsov, Pil-yugin, and Ryazanskiy took advantage of this right and received large tracts of land, and cottages with all the conveniences in Barvikha, one of the most elite suburban Moscow areas. Korolev did not want to build outside Moscow and obtained per-mission to build a two-story cottage next to the Exhibition of National Economic

---

25. *Zek* is the slang for prisoner, the plural of which is *zeki*.

Achievements (VDNKh).[26] And this despite the fact that we, his closest associates, took the initiative and picked out an absolutely gorgeous spot for him for a dacha in the forested water conservation district on the high bank of the picturesque Pyalovsk reservoir. He didn't even explain why instead of two homes—a nice apartment in Moscow and a large cottage in the country—he chose to have one, a cottage right in the city. Eight years later, grief-stricken, we realized that back then, our S.P. was picking out a site to which the "public walkway wouldn't get overgrown." Now there is a memorial museum in his house.[27] Right next to it, thrusting upward into the Moscow sky is an obelisk in honor of the conquerors of space.

Once built up with suburban Moscow dachas, 3rd Ostankino Street, now Academician Korolev Street, begins at the space obelisk and ends at the Ostankino television broadcast center and the famous television tower. When the sun descends toward the west, the obelisk honoring the conquerors of space stands out quite distinctly against the background of the Kosmos Hotel. The Avenue of Heroes proceeds from the obelisk, at the base of which sits a stone Tsiolkovskiy. Memorials to Keldysh and Korolev stand at the end of the Avenue. Behind their backs, the neon lights of the Kosmos movie theater on Zvezdnyy (Star) Boulevard illuminate the evening. Tsander Street leads from Korolev's house to Zvezdnyy Boulevard.[28] Kondratyuk Street connects Tsander Street with Mir (Peace) Prospekt .[29] If you go down this street and cross the prospect, you end up on the broad Kosmonavt (Cosmonauts) Street. From Kosmonavt Street, if you turn right and go down Konstantinov Street, you will reach Raketnyy (Rocket) Boulevard. Another "rocket" street, Kibalchich Street, runs parallel to Kosmonavt Street.[30] Finally, in the mid-1980s, not far from the museum that was formerly Korolev's house, a vast neighborhood of cottages for cosmonauts closed off from pedestrians and street traffic sprouted up. And it all started with Korolev's cottage and our three sections on the former 3rd Ostankinskaya Street.

---

26. VDNKh—*Vystavka dostizheniy narodnogo khozyaystva*. The VDNKh traced its origins back to the All-Union Agricultural Exhibition (VSKhV), which opened in 1939 in Moscow. In 1954, the original complex was expanded to 80 pavilions spread over nearly 600 acres to highlight *all* Soviet economic achievements. One of the most notable pavilions at the VDNKh was the Kosmos Pavilion that showcased models and replicas of various Soviet spacecraft.

27. The S. P. Korolev Memorial Home Museum was opened to the public in 1975 as a branch of the nearby Memorial Museum of Cosmonautics.

28. The street was named after Fridrikh Arturovich Tsander (1887–1933), one of the pioneers of Soviet rocketry who founded the Group for the Study of Reactive Motion (GIRD), the earliest Soviet organization dedicated to the development of liquid-propellant rockets, in 1931.

29. The street was named after Yuriy Vasilyevich Kondratyuk (1897–1942). Kondratyuk, whose real name was Aleksandr Ignatyevich Shargey, was one of three major Soviet theorists of space exploration. In 1929, he published a book, *Zavoyevaniye mezhplanetnykh prostranstv* (The Conquest of Interplanetary Space), an innovative and in-depth exegesis that mathematically explored many different aspects of space exploration.

30. Nikolay Ivanovich Kibalchich (1853–81) was a member of the Russian underground revolutionary and terrorist group *Narodnaya volya* (People's Will). For his participation in the bombing death of Tsar Aleksander II on 1 March 1881, he was imprisoned and subsequently executed. While in prison awaiting execution he wrote up an idea for a crewed rocket-propelled flying vehicle.

Chapter 23
The R-7 Goes into Service

Of all the rockets developed early in the space age, the R-7 rocket has proved to have record-setting longevity. Having begun its triumphant journey in 1957 as the world's first potential carrier of a hydrogen bomb, the R-7 was upgraded in various modifications and continues to staunchly serve cosmonautics. According to all predictions, it will complete its service no earlier than the second decade of the 21st century.[1] The unaltered first two stages serve as the foundation to which the third and fourth stages are added. The history of this rocket has been described as an uninterrupted string of victories from one space triumph to another. Typically, the mass media presented each of these triumphs under the headline "World's First."

In the history of our aerospace technology during the Cold War, although each new success was enthusiastically recorded often even with technical details, the names of the actual commanders and rank-and-file soldiers on the scientific and technical front were never mentioned. In the era of human spaceflight, the yoke of celebrity fell mostly on Soviet cosmonauts and American astronauts. But even in democratic America, just as in our country, behind the visible trees stood an invisible forest of the unknown (and classified) names of those who actually built the shining monuments of modern day cosmonautics.

In the scheme of history, the R-7 was, more so than other rockets, the means for solving many military, strategic, political, scientific, ideological, and economic problems. The Soviet Union's top political leaders never missed an opportunity to play their "space" trump card in the foreign affairs game and to remind the people that only the leadership of the Communist Party and its Central Committee could produce achievements, demonstrating the clear superiority of the socialist system.

It was during Khrushchev's term in office that the R-7's life cycle began leading to the first space triumphs. He was, perhaps, the first to understand the unlimited possibilities available to those government leaders who enjoyed supremacy in the

---

1. The latest modifications of the original R-7 booster include the three-stage Soyuz-FG and the Soyuz-2 launch vehicles. The former continues to launch cosmonauts to the International Space Station.

field of rockets and spaceflight.

In September 1959, Khrushchev visited the United States at the invitation of President Dwight D. Eisenhower. During this period, R-7 joint tests were continuing and the rocket had not yet been put into service. This did not prevent Khrushchev from making a strong impression on the Americans, who lacked reliable information at that time. Khrushchev used the following words in his speech at a reception: "Our people have rallied around their government. People are burning with enthusiasm. They are striving to do their duty to the best of their ability and thereby strengthen their socialist regime even more. We developed the intercontinental ballistic missile before you. To this day you don't really have one. But, after all, the intercontinental ballistic missile is truly the crux of human creative thought."[2] If we take Khrushchev's words about "our people burning with enthusiasm and striving to do their duty to the best of their ability" to refer to us, the creators of the R-7 rocket, then Khrushchev was right. We really were enthusiasts and spared no effort to promote the R-7 rocket in military and space spheres. For the sake of historic fairness, one must admit that, regardless of the later accusations against him, Khrushchev's enthusiasm and intense activity certainly contributed to the accelerated development of rocket and space work in the USSR.

ENTHUSIASM IS ENTHUSIASM, but the real circumstances that had developed in late 1958, by the beginning of the joint tests of the R-7, were extremely difficult. The failed launch of the last rocket in the flight-development test series, along with three lunar launches, brought the number of failed launches of the R-7 to four in a row.[3]

With no time to recover, at the Ministry of Defense's urgent demand, we switched over to joint tests without a break. In order to somewhat improve the utterly unsatisfactory reliability numbers, by mutual agreement with the military we excluded the three first Moon launches of 1958 from the number used to calculate the reliability rating. At the same time, however, it was agreed that the results of subsequent Moon shots for the first two stages would be counted when summing up the results of the joint tests and making decisions about the fate of the R-7.

This was fair. The R-7 rocket faced service on two fronts. The two stage combat version had to wait in stand-by mode for a command that would mark the beginning of a nuclear missile war, while the space version, which had third and fourth stages, would fulfill humankind's striving for knowledge of the universe and maintain the

---

2. *Author's note:* From a speech at a dinner held by the New York economics club in honor of Nikita Sergeyevich Khrushchev on 17 September 1959.

3. *Author's note:* Between flight-development tests and the "joint tests" (or qualification tests) of the R-7 combat missile, planners wedged three launches of the 8K72 rocket into the schedule. This was a three-stage version of the R-7, modified for firing probes to impact on the Moon. The "fleeting rocket fire on the Moon" in 1958 did not bring us success. I will write about that in greater detail in Chapter 25.

prestige of a great power. R-7A flight tests were also scheduled to begin at the end of the year. This missile with the designation 8K74 had a range of at least 12,000 kilometers. Thus, taking into account the planned assault on the Moon, for the entire upcoming year we would have to perform no less than 22 to 24 launches.

The general preparation cycle of the R-7 rocket at the firing range, from the beginning of tests at the engineering facility in the Assembly and Testing Building until the launch, took 15 days on average. In 1957 and 1958 the chief designers and the entire "Korolev throng" spent a great part of their time at the firing range. Work on the numerous new space projects and the new intercontinental missiles required the presence of managers at their OKBs and at factories, their participation in scientific and technical councils, and hundreds of conferences at all levels.

Glushko was the first of the chiefs to rebel against the requirement to attend each launch. Kuznetsov supported him and then Pilyugin joined him. They showed that even if one were to abandon all other business, it would still be impossible to attend all the launches. Understanding that there were limits to what they could do in space and time, the chiefs agreed to the maximum extent possible to delegate responsibility and routine management of the flight tests to the military contingent at the firing range and to their most reliable deputies in charge of testing. Each of these deputies received all the authority to resolve issues in the subject matter of their organization and represented the chief designer at the State Commission. Thus, the Council of Chiefs' interdepartmental "shadow test cabinet" was formed. Over the course of 1959, its members spent, on the average, seven to eight months each at the firing range, participating in each launch of the combat R-7.

Korolev immediately entrusted this work to two of his deputies, Voskresenskiy, as the official deputy for testing, and Kozlov, to represent both the chief designer and the Kuybyshev branch of OKB-1. Pilyugin transferred his authority to Vladlen Finogeyev. Gleb Maslov kept track of all the propellant feed and synchronization systems. Bogomolov entrusted work on the *Tral* telemetry system to Mikhail Novikov. Glushko's first deputy, Vladimir Kurbatov, represented his interests. Barmin placed Boris Khlebnikov in charge of the ground complex. Vyacheslav Lappo oversaw Ryazanskiy's radio systems. On the whole, we assessed this staff of testers as "quite professional" and fully competent. Gradually all the minor everyday problems faded into the background. The people got into the intense rhythm of testing.

It bears mentioning that the band of individuals listed above formed very businesslike and congenial relations with the firing range military command—with its chief, General Konstantin Gerchik; and with the immediate operations managers Colonel Aleksandr Nosov, Colonel Yevgeniy Ostashev, and Major Anatoliy Kirillov; and with the entire officer staff of military testers.

OVER THE COURSE OF A YEAR, 16 missiles were launched under the joint testing program, in addition to four for the lunar program and two for the 8K74 program; one missile was removed after a failed launch attempt. The first launch under the joint testing program took place on 24 December 1958, and proceeded in keeping with "the first pancake is always lumpy" rule. Due to the faulty setting of the hydrogen peroxide pressure control valve, the Block V strapon prematurely consumed its propellant and separated from the missile 3 seconds ahead of schedule. The missile began to spin, and the AVD command shut down all the engines. State Commission chairman Rudnev and his deputy representing the military, Myrkin, correctly attributed this failure to sloppiness on the part of the military squad during preparation and to a lack of competent supervision from Glushko's representatives.

All 16 missiles presented for testing were launched. Four missiles reached the Kamchatka region, with large deviations due to errors in the tuning of the radio-controlled ground stations or defects in the onboard systems. Eight missiles flew normally. Their nose cones carrying a treasure trove of instrumentation reached the target with a circular error no greater than six kilometers.

The last launch on 27 November 1959 worthily concluded the whole series of joint tests. The missile completed all flight segments without a glitch. The nose cone reached Kamchatka with a deviation from the "peg" (*kolyshka*)—the calculated point of impact—of 1.75 kilometers in range and 0.77 kilometers laterally. These were dazzling results for the R-7. The nose cone did not contain a nuclear warhead, but everything needed to put one into action was installed and monitored by the nose cone's telemetry. The reports about the flight-test results of the nuclear warhead control system were so hush-hush that none of us saw them. At the State Commission it was only reported that the "results were satisfactory."

There were four missile crashes. Of these, two were due to engine problems; one was the fault of radio control; and one was due to a flaw in the missile construction. Thus, reliability was 75%. Compared with 45% for the flight-development tests, this was substantial progress.[4] The missile was put into service on 20 January 1960 by a special USSR Council of Ministers decree. Completion of R-7 joint flight-development tests contributed to the decision to establish the independent Strategic Rocket Forces. On 17 December 1959, Khrushchev signed the USSR Council of Ministers decree establishing the post of commander-in-chief of the Strategic Rocket Forces as part of the USSR Armed Forces. The "top-secret, special importance" decree stated that the commander-in-chief of the Strategic Rocket Forces—also a deputy minister of defense—would bear full responsibility for their status; for their combat use, combat and mobilization readiness, and material and technical support; for the development of missile armaments; for supervision of the construction and operation of weapons systems and special facilities; for troop discipline and morale of personnel; and also for coordinating issues concerning the creation, development, and introduction of special weapons and rocketry in all the branches of the Armed Forces. Even in this document of "special importance," the

---

4. OKB-1 carried out R-7 launches on 24 December; 17, 25, and 31 March; 9 and 31 May; 9 June; 18 and 30 July; 14 August; 18 September; 22 and 25 October; and 2, 21, and 27 November. The launch abort was on 21 February 1959.

code words "special armaments" were used in place of nuclear weapon. Chief Marshal of Artillery Mitrofan Ivanovich Nedelin was named the first commander-in-chief of the Strategic Rocket Forces.

Despite the decree's top-secret classification, news about it quickly circulated through all the OKBs directly involved with the production of strategic combat missiles. Our community of engineers and designers received the decree with great pleasure.

Nedelin's appointment surprised no one. Anyone who knew him believed that Soviet missile technology was very fortunate. In this regard, I was reminded of Korolev's story about his meeting with Chief Marshal of Artillery Nikolay Niko-layevich Voronov in 1950. When Voronov came to NII-88 he was still commander of all artillery, including missile technology. During firing range tests of missiles in 1947 and 1948, Voronov participated in the work of the State Commission and made a good impression on all of us with his amiable nature and his officer's decorum, which was certainly not the rule among high-ranking military officers.

Korolev did not hide his liking for Voronov. He valued a visit from Voronov highly and would talk about such a meeting as if it were a very important event. According to Korolev, Voronov introduced him to his chief of staff Colonel General Nedelin, whom Voronov had tasked to study and develop prospects for missile weaponry. In the 10 years from 1950—when Nedelin was effectively introduced to missile technology—he accomplished a great deal. After his appointment as commander-in-chief of the Strategic Rocket Forces in 1959, Nedelin had less than a year to live. But even over that short time, we saw for ourselves the inherent breadth, independence, and unconventional nature of his thinking.

These qualities were particularly essential for a deputy minister of defense who, by virtue of the system that had developed in our country, had the capability to directly affect the development of cosmonautics. Unfortunately, after Nedelin's death, the Soviet combined-arms marshals and World War II heroes that replaced him in that high post did not possess such qualities.

THE FIRST ARTIFICIAL EARTH SATELLITE LAUNCH WAS THE BEGINNING OF THE PROCESS CONVERTING THE R-7 INTERCONTINENTAL MISSILE FROM A THERMONUCLEAR WARHEAD CARRIER INTO A LAUNCH VEHICLE FOR THE MOST VARIED TYPES OF SPACECRAFT. A launch vehicle based on the two stage R-7 continues to be perfected even now, more than 40 years after its first flight. During Korolev's life alone, more than five modifications of the *Semyorka* were produced.[5] Each new modification

---

5. At least seven modifications of the basic R-7 (8K71) were used for the space program during Korolev's lifetime, that is, before 1966. These included the 8K71PS (for the first two satellites), 8A91 (for the third satellite), 8K72 (for the early lunar probes), 8K72K (for Vostok and Zenit-2), 8A92 (for Zenit-2), 11A57 (for Voskhod and Zenit-4), 11A510 (for US-A satellites), and 11A59 (for IS satellites).

was intended for a specific type of spacecraft, typically with the first two stages remaining unchanged. The primary modifications and enhancements of the rocket came about in order to increase the payload mass, making it usable for the on-orbit insertion of automatic interplanetary stations and crewed space vehicles. Since the 1950s, the guidance system has undergone the greatest enhancement. Currently, in the early 2000s, the rocket's motion control system is a completely autonomous inertial system requiring no radio correction.

Since 1957, the R-7 rocket has undergone 12 updates and modifications. In open publications it is referred to as Sputnik, Vostok, Molniya, or Soyuz, depending on its purpose. For us veterans it will remain the *Semyorka*.

While Korolev was still alive, the title of chief designer of the *Semyorka* was gradually transferred to Kuybyshev to Dmitriy Kozlov. Kozlov himself, who in the late 1970s became TsSKB general designer, devoted his primary attention to spy satellites. The most troublesome duties of the *Semyorka* chief designer were shifted to Kozlov's deputy, Aleksandr Soldatenkov. Without his summary reports, not a single State Commission would have been held to make decisions on crewed and other vital launches. Today Samara has a monopoly on the production of the most reliable launch vehicle in the world. As before, the production of the rockets themselves is concentrated at the Progress Factory, while engine production takes place at the M. V. Frunze Factory, formerly aviation engine Factory No. 24. After the collapse of the Soviet Union, a very difficult situation developed for guidance system production. As fate would have it, the Kharkov instrument factories ended up outside Russia, in a neighboring country.

By the early 1990s, the number of launches of *Semyorka* modifications had passed the 2,000 mark. Disruptions and problems that were not always technical began to crop up in the smooth running manufacturing process. The Baykonur Cosmodrome along with all of its services and the city of Leninsk, all of which were outside Russian borders, became the weak link that could undermine the reliability of the former Soviet rocket and space complex as a whole. In other words, the politics of sovereignty was an important factor reducing the working capacity of the Baykonur Cosmodrome, one of domestic cosmonautics' most advanced creations of the second half of the 20th century.

After much deliberation and long negotiations, the European Space Agency decided in 2003 to build a launch complex at the European Space Port at Kourou in French Guiana (in South America) for a specially modified R-7 rocket. Thus after its latest rejuvenation, the trusty old *Semyorka* is traveling abroad in its entirety not as a missile to drop a hydrogen bomb on America but as a launch vehicle fo spaceflights. If in the late 1950s you could have found a joker who predicted that rather than firing our *Semyorka* on America, we would be launching it from a American continent, in the best case scenario he would have been offered a cours of treatment at a psychiatric hospital.

Chapter 24

From Tyuratam to the Hawaiian Islands and Beyond

The maximum flight range of the R-7 rocket, which we had finally put into service, was determined by its separable nose cone that carried a thermonuclear warhead. In 1955, this warhead required the creation of a nose cone with a total mass greater than 5.5 metric tons. Carrying such a payload, there was no way the missile could cover a range greater than 8,000 kilometers. When firing from the launch pads of the Tyuratam firing range, this range was clearly insufficient. The notion that the U.S. had lost its advantage of nuclear invulnerability—a claim actively promoted by our propaganda—was terrifying for the Americans. In reality, the R-7 missile was not capable of reaching many strategic centers in the U.S. In order for the R-7 missile to become a real intercontinental weapon capable of reaching any point on the entire U.S. territory, its range would have to be increased to 12,000 to 14,000 kilometers, that is, by more than 1.5 times.

Work to upgrade the R-7 had already begun in 1957, long before flight-development tests were completed. In our internal communications we referred to the prospective updated missile simply as "No. 74" in contrast to the standard R-7 we had put into service, which was called "No. 71." This numerical slang was the abbreviated unclassified title that the military assigned to secret articles. In technical documentation the R-7 missile was called "article 8K71." Correspondingly, the R-7A was called "article 8K74." However, even in secret documents, often for the sake of security, the terminology "article 8K71" or "article 8K74" was used.[1] An administrator would insert "R-7 rocket" or "R-7A missile" into the typewritten top-secret text by hand; the idea was to conceal state secrets from the state's own classified pool of typists.

We needed to significantly increase the flight range without making substantial changes in the missile's design and without disrupting its series manufacturing process. That being the case, the only real way to achieve an additional 4,000 to 5,000 kilometers in range was to reduce the payload mass. As early as late 1957, after

---

1. *Author's note:* Later, the designations 8K72, 8K75, and 8K78 were assigned for all sorts of modifications for the space program but I'll discuss those later.

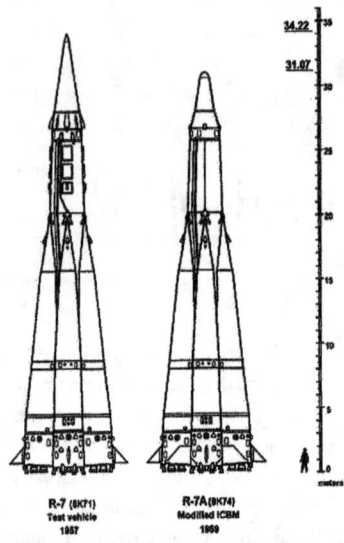

R-7 (8K71)
Test vehicle
1957

R-7A (8K74)
Modified ICBM
1959

Peter Gorin.

Shown here are the two basic variants of the R-7 ICBM, the original developmental version and the modified R-7A. The two were declared operational in January and September 1960 respectively. Both versions were known as the SS-6 by the U.S. Department of Defense.

the latest meeting with Academicians Khariton and Sakharov, Korolev said that they had confidently promised to cut the weight of their "doodad" *(tsatski),* as they referred to it, in half.

S.P.'s demeanor usually changed somewhat when it came to the subject of the nuclear warhead for which we were making the missile. He would lower his voice and by his entire appearance try to produce a sense of awe and reverence in his audience for this greatest state secret, as well as respect for this terrible force that was to be concentrated in our payload. I think there was more to this than just the top-secret atmosphere that surrounded everything directly associated with the development of nuclear warheads.

Out of necessity, we had studied the physical processes that took place in all the systems of our brainchild, the R-7, but when it came to the subject of what this warhead would contain, we all felt timid and fell silent. We had heard lectures, read the popular literature on nuclear physics, and had direct contact with nuclear specialists while working to integrate the warhead with the missile. Yet, the very essence of the titanic destructive force hidden behind the dry phrases of agreement protocols, dimensional installation drawings, and circuit diagrams remained in some ways opposite to our engineers' way of thinking.

It wasn't that we didn't understand anything at all. Of course, they explained to us that the hydrogen bomb consisted of a thermonuclear warhead containing no uranium-235 or plutonium-239. In and of itself, the thermonuclear warhead was harmless. It turns out that in order to compress and ignite the fuel for thermonuclear fusion, you first needed to detonate a "simple" atomic bomb. The explosion of this nuclear detonator produced the x-ray radiation, temperature, and pressure capable of generating an instantaneous thermonuclear reaction, that is, the detonation of the hydrogen bomb. The atomic bomb itself required a detonator in the form of a conventional explosive charge. This explosive, in turn, was detonated by detonating fuses, which gave us the greatest trouble of all when coming up with the nose cone layout. For everything to be reliable and secure, we were not required to delve into nuclear physics beyond what I have described above.

But often there were all kinds of problems associated with layout, fastening, thermal protection, vibration protection, accelerations, electrical connections, and inhibitors; as a result, we were forced to interact closely with the lead specialists of Arzamas-16 and the Moscow OKB headed by Nikolay Leonidovich Dukhov, former wartime chief designer of heavy tanks.[2] Viktor Zuyevskiy, with whom I dealt directly when coordinating technical issues, explained in layman's terms that in a hydrogen bomb the hydrogen isotope deuterium is converted into helium. This is fusion, which produces a relative magnitude of energy many times greater than that released during the explosion of an atomic bomb, which uses fission. During our contact with the nuclear people, I realized that nuclear physicists and nuclear designers worked with as much enthusiasm as we did on the missile to reduce the mass and dimensions of the hydrogen bomb, or in modern terminology, the thermonuclear warhead. They developed original designs for all the structural parts of the warhead—including the automatic detonation and neutron initiation devices—and developed new compact safety and control instruments.

Reliability was ensured using the principle whereby the malfunction or failure of any instrument would not cause a system failure or premature triggering of the warhead. For flight tests the nuclear specialists developed their own "atomic" telemetry and indestructible indicators, making it possible to record the operation of the automatic controls activating the non-nuclear detonation and neutron initiation of the warhead.

The lightweight thermonuclear warhead was intended not only for our R-7A missile. In Kapustin Yar in 1958, Yangel conducted flight tests of his "high-boiling" R-12 missile, a competitor of our R-5M missile. The range of the R-12 missile was 2,500 kilometers. In contrast to the R-5M missile, its separating nose cone carried not a "simple" atomic warhead with an 80-kiloton yield, but one with a one-megaton TNT equivalent thermonuclear warhead.

We imagined the difference between the aftereffects of the explosion of an R-5M 80-kiloton warhead and an R-12 one-megaton warhead in very abstract terms. Nevertheless, when the subject turned to the low yield of the R-5M, we immediately revamped its nose cone so that the warhead's yield was as high as that of the R-12. S.P. openly commented that he couldn't understand why this risky race for megatons on our missiles was necessary. We, too, believed that it was better to have a missile with a nuclear warhead of "just" one megaton with a range of 14,000 to 15,000 kilometers than one that could barely reach 8,000 kilometers with a warhead that was three times more powerful.

In 1957, a reference book on nuclear weaponry came out in the U.S., enabling anyone who so desired to calculate the effect of a nuclear explosion in terms of a

---

2. Nikolay Leonidovich Dukhov (1904–64), a corresponding member of the Academy of Sciences, was one of the leading Soviet designers of heavy tanks.

TNT equivalent.[3] Thus, the secrets carefully guarded by our nuclear specialists had become available for all missile specialists. According to this reference book one could expect a one-megaton warhead to be quite sufficient, if it hit the center of Washington, to completely wipe out the U.S. capital. We, of course, were outraged: "Why the hell would you need anything bigger?! Go ahead and put the R-12's warheads on the *Semyorka* and hit any range." But the top brass in the Ministry of Defense had other considerations; operations to perfect our thermonuclear warheads were under way with even greater intensity than our missile projects.

Actually, before long, the promises that Khariton and Sakharov made to Korolev took on the form of engineering designs, which enabled us to reduce the mass of the nosecone by 2.5 tons. At the same time, they promised that the yield of the new nuclear warhead would be at least what it had been with the standard *Semyorka*.

After the designers received directives to reduce the mass of the warhead by more than a metric ton, their calculations immediately increased the range by 3,500 kilometers. They picked up another 500 to 700 kilometers by simplifying and reducing the weight of the radio-control system, increasing the oxygen and kerosene load, and enhancing the precision of the propellant level control systems and the tank depletion synchronization systems of all the boosters of the R-7 in order to cut down the propellant safety margin. There were a lot of other miscellaneous minor design changes aimed at reducing the missile's weight and raising the propellant margins by increasing the volume of the core booster's tanks.

For No. 74 we made every effort to eliminate the danger of resonance phenomena occurring in the missile's elastic contour, specifically pressure pulsations in the engines' combustion chambers, which led to dramatic situations during the first lunar launches. Tests on No. 74 missiles completely confirmed the effectiveness of the damping system developed by the united forces of OKB-1, NII-1, and OKB-456. Thus, all the combined measures made it possible, without production stoppages, to switch over to rolling out missiles with a range up to 13,000 kilometers. To be on the safe side, the Council of Chief Designers decided to declare a maximum range of 12,000 kilometers. They left 1,000 in the "Chief Designer's reserve."

S.P. had a large globe in his office on which, using a special protractor, you could very graphically measure the distance between any points on earth's surface. Once, at a meeting after summarizing the results of all the teams for No. 74, the problem of flight-tests of the missile at full range was under discussion. Walking up to the globe, Korolev showed that, when firing at the Pacific Ocean, points of impact fall in the region of the Hawaiian Islands. Georgiy Tyulin, who was present at the meeting and was director of NII-88 at the time, couldn't pass up the opportunity to employ some glib battlefield jargon: "We can deliver the Americans such a jolt that

---

3. Here, Chertok is probably referring to the first volume of the following series: National Academy of Sciences-National Research Council, *Nuclear Theory Reference Book (Nuclear Data Project)* (Washington, DC: U.S. GPO, 1957/58–).

they'll remember Pearl Harbor as the good old days."

The 8K74 flight-tests began successfully in late 1959. In all, eight missiles were to be launched, of which at least three were to be fired to maximum range. The first launches aimed at the "Kama" region (the name used for the sake of secrecy for the nose cone impact area on Kamchatka) were successful. They confirmed the structural reliability of the new nose cone, whose weight had almost been cut in half, and the effectiveness of the measures to enhance the precision of the autonomous control system.

The main test for No. 74 was, however, the test at maximum range while firing at the Pacific Ocean. It was the Ministry of Defense's job to determine the coordinates of the nose cones' points of impact during intercontinental missile launches at full range; Korolev and the Council of Chief Designers had virtually no input into the solution of this problem. The Ministry of Defense's NII-4, the creator of the Command and Measurement Complex (KIK), was responsible for determining the points of impact on land, and consequently this organization also solved this problem for the ocean. For the 8K74 flight-development tests at full range it was necessary not only to manufacture missiles and equip the nose cones with automatic control systems for the detonation of the nuclear warhead, but we had to create marine tracking stations in addition to ground stations.

LONG BEFORE THE BEGINNING OF 8K74 FLIGHTS we recognized the need to use floating tracking stations. The 8,000-kilometer range of the standard 8K71 *Semyorka* was already beyond the border of Kamchatka. NII-4 began exploratory operations to create floating facilities as early as 1956. The special *Akvatoriya* project was created to accomplish this task. Georgiy Tyulin, my former compatriot on our long-term project in Germany, who was at that time NII-4 deputy chief for scientific operations, was in charge of the project. I remind the reader that Tyulin graduated from the Moscow State University mechanical mathematics department. During the war he was chief of staff of a *Katyusha* troop unit that General Tveretskiy commanded. Tyulin was one of the first combat "scientist-colonels" who occupied leading posts for the production of missile weaponry, first in the military and then in industry. In 1959, Tyulin was appointed director of NII-88, then later served as the first deputy minister of general machine building, that is, the "space industry" minister in 1965–76.

NII-4 Director Andrey Sokolov and Georgiy Tyulin personally convinced Minister of the Shipbuilding Industry Boris Butoma of the need to retrofit already built dry-cargo ships as missile tracking ships. The NII-4 scientific staff, who already had practical experience developing ground tracking stations and the Command and Measurement Complex, determined the makeup of the radio engineering, optical, and sonar equipment to receive nose cone telemetry information and determine the coordinates of the impact points in the ocean.

In early 1959, the Council of Ministers issued a decree calling for the Ministry of the Shipbuilding Industry to retrofit steamship coal ore–carriers to create a Floating

Measurement Complex (PIK-1) as per NII-4 design specifications.[4] At the Baltic and Kronshtadt Factories in Leningrad, within an unusually short timeframe for shipbuilders, the three ships *Sibir* (Siberia), *Sakhalin,* and *Suchan* were equipped with telemetry and orbital monitoring equipment, while a fourth ship, the *Chukotka,* was equipped with communications and relay systems to transmit data to the continent.

Work on retrofitting and manning the coal ore–carriers with missile specialists was completed in July 1959. The ships entered the naval fleet under cover as the Fourth Pacific Ocean Hydrographic Expedition (TOGE-4).[5] The ships flew the hydrography flag of the Soviet naval fleet and departed Leningrad for the Pacific Ocean via the North Sea route. Captain (later Admiral) Yuriy Ivanovich Maksyuta was named commander of the first floating complex. On 30 August, all four ships arrived in the base port of Petropavlovsk-Kamchatskiy, and on 15 September they were under way on their first cruise to southern latitudes to carry out their primary mission. The crews, who had spent a month in polar ice, would now be working indefinitely in tropical latitudes.

The TOGE-4 ships' cruising range was 10,000 miles; their cruising capacity was 90 days. Each ship had a crew of 200, including measuring systems specialists. When they received notification of an impending launch, three ships arranged themselves in a right triangle so that the nose cone's estimated point of impact fell in the middle of the hypotenuse. During the descent segment of the nose cone's flight trajectory before it entered the dense atmospheric layers, information was received through the *Tral* telemetry stations and the SK-2 telemetry system specially developed to monitor the warhead's "well-being." The Kama station monitored the trajectory. Aleksey Bogomolov had every reason to be proud of the fact that the OKB MEI systems had been put into service for the navy. Special photo-recorders captured the luminescence of the plasma in the dense atmospheric layers. The warhead had a detonator fuse that tripped the instant it hit the surface of the water. When the warhead was submerged several meters, it exploded and sent up a column of water. In addition, the explosion discharged a special dye that formed a colored spot on the water's surface. The ships' radar located the column of water from the explosion. The sound of the explosion was picked up by the sonar equipment. The ships were "armed" with a Ka-15 helicopter, which took off and hovered over the colored spot, providing an additional option of determining the coordinates of the point of impact.

The *Chukotka* communications ship received information from the continent and relayed information to the tracking ships about missile preparation and launch and estimated time. After nose cone splashdown the communications ship received information from the tracking ships and relayed it to the firing range. The TOGE-4

---

4. PIK—*Plavuchiy izmeritelnyy kompleks.*
5. TOGE—*Tikhookeanskaya gidrograficheskaya ekspeditsiya.*

ships were the first squadron of the Soviet Union's future naval space fleet. Its history is inseparable from the history of domestic cosmonautics. I shall try to return to this subject later in my memoirs.

The TOGE-4 ships were low-speed vessels; their cruising speed was just 11 knots. It took them almost five days to reach the impact area in the south of the Hawaiian islands. They also needed to have a "just-in-case" margin. In fact, the ships arrived in the impact area several days early and messed around in the ocean awaiting the readiness command, which was in the form of a coded notification about the launch date and precise time. If a launch delay or postponement occurred, we could provide no clear explanation to the TOGE-4 crews over the radio. Security agency specialists believed that radio exchanges with the fleet ships might contain information that, if intercepted, might enable U.S. intelligence to determine the launch targets and missions as well as the purpose of the TOGE-4 ships.

On more than one occasion I found that people who served in the state security agencies that guarded our missile technology were quote sensible. However, for some reason the security services thought that if we announced that TOGE-4 vessels were messing around in the vast expanses of the Pacific Ocean for many weeks on scientific missions, then the American special services would actually believe it. In other words, the high level of secrecy over the communications was only for internal consumption.

After Krushchev's visit to America, his meeting with President Eisenhower, and the speech at the UN outlining a program of global disarmament—all in September 1959—there was a distinct thaw in relations between the USSR and the U.S. Such an about-face in the usually confrontational international situation clearly went against the grain of Cold War apologists, especially since Khrushchev had invited Eisenhower to the USSR and proposed a meeting on the shore of Lake Baykal in the spring or summer of 1960. Who knows, they might actually agree to end the arms race and disarm. The new harmonious relations between the two countries was rudely interrupted by the TASS report about upcoming test missiles coming down in an area of the Pacific Ocean where the sole proprietor was the U.S. Without wanting to, we had given Cold War hawks an opportunity to accuse the Soviet leadership of cunning and of posing a real threat to the security of the U.S.[6]

To coordinate actions during the launches into the Pacific, the fleet staff sent its own representative to be a member of the State Commission. At one of the sessions he described how TOGE-4 ships got into tricky situations in the potential impact area declared by the TASS report as dangerous for shipping during the launches. American, British, and French naval ships had our ships under continuous watch. The American ships were extremely bold. They came right up to our unarmed ves-

---

6. The TASS announcement was widely reported in the United States. See Max Frankel, "Shipping Warned: Russian Space Vehicle to Land in Sea East of Marshall Islands," *New York Times*, January 8, 1960, pp. 1–2.

sels, barely avoiding collision. Submarine periscopes would suddenly appear about 20 meters from the board of our ships. Neptune all-weather reconnaissance planes were particularly brash.[7]

Sailors once reported the following incident. In overcast weather a helicopter lifted off the deck of one of the TOGE-4 vessels for a training exercise. At this moment a Neptune tumbled out of the low clouds and buzzed the ship to take photographs. It looked like a collision with the helicopter was unavoidable. The helicopter shot upward and managed to get away so that the Neptune passed between it and the ship. On its next pass the sailors shook their fists at the Neptune. Roaring with laughter, the aircraft's navigator returned the gesture. The next time the Neptune swooped in, the ship's commander decided to blind its camera using a searchlight. The aircraft's navigator once again shook his fist. But as soon as the TOGE-4 ships received radio notification of T-minus 4 hours and began to disperse to their designated areas—to the vertices of the triangle—all the naval vessels surrounding them moved 10 to 15 miles out of harm's way. To the surprise of our sailors, this evacuation of the foreign ships sometimes began before the TOGE-4 command had received the readiness notification. The Americans had some channels of their own for receiving reliable information about the actual situation back at our launch site.

Beginning with the launches in 1957, and specifically as of 27 August when the TASS report about the development of the Soviet intercontinental missile was first issued, the Council of Chief Designers composed similar communiqués. And that's the way it was—no matter what happened in missile technology or cosmonautics, the chiefs had to compose the first draft of all communiqués; the "partocrats" entrusted this work to the technocrats.

The first composition coordinated with the State Commission at the firing range was immediately transmitted to Moscow, reviewed and corrected in the Central Committee's Defense Department, and passed on to TASS for reports in the press and over the radio.[8] Usually the "chiefest of the chiefs" shirked this thankless job. At Keldysh's initiative Ishlinskiy was usually tasked with composing the first draft of the text. Keldysh loved to dignify him with the quasi-Latin title "el professoro." Aleksandr Yulyevich never took offense. Most often he recruited Okhotsimskiy and some other intellectual for the job. When the text was ready for discussion, Korolev and Keldysh assembled all the Council and State Commission members who were not involved in pressing business. They began proofreading and rewriting, which sometimes dragged on for several hours. Meanwhile Moscow was getting the jitter and putting the pressure on. As a rule, during the first years of the missile and spac

---

7. Chertok is probably referring to the Lockheed P2V-5 Neptune reconnaissance aircraft, whi first flew in 1950.

8. The Defense Department of the Central Committee supervised all ideological and personi issues of the Soviet defense industry (which included the missile and space sector).

programs, only Yuriy Levitan was trusted to read the TASS reports on the radio.

Before the R-7 launches into the Pacific planned for early 1960, it turned out that we had not publicly announced that the TOGE-4 ships were headed toward the Hawaiian islands—albeit in neutral waters. A communiqué needed to be released legitimizing their presence in the area where the nose cones and second stage debris would come down; this was the duty of the Ministries of Defense and Foreign Affairs.

At a meeting of the Council of Chief Designers on 30 December 1959—when among other issues they were also discussing the preparations under way for the first launch of No. 74 into the Pacific—Ryazanskiy took the initiative and reminded Korolev that they needed to promptly draw up the communiqué about the upcoming launches to legitimize the presence of TOGE-4 in the Pacific Ocean. Korolev flew into a rage and said that he would not be handling that and that Ryazanskiy should mind his own business. S.P. was clearly miffed and lashed out so harshly that Mikhail blushed, started to sulk, and had nothing more to say. At that moment a minister telephoned via the "Kremlin hot line." He said that the communiqué had been written and asked that Korolev listen to it and sign off on it. S.P. announced that Ryazanskiy was already attending to that, handed the phone to him, and turning back to us, grinned, "Look how Mikhail has been punished for his initiative. Let him sign off on it now." The communiqué came out the next day and caused an incredible stir in the world press and in all the radio broadcasts. Cold War hawks screamed that Khrushchev's appeals for global disarmament, peace, and friendship were pure propaganda, and that these new nuclear missiles were a real threat to the U.S. Either way, Eisenhower's visit to the USSR, which Khrushchev had arranged in September, proved doubtful.

About 10 years later, Katya and I received an invitation to vacation several days in one of the two cottages that had been built on the shore of Lake Baykal at Khrushchev's instruction. Both cottages and the entire interior decor were absolutely identical. One had been intended for President Eisenhower and his immediate staff; the second was for Khrushchev. The large entourage and press were supposed to be housed in the adjacent buildings of the "Baikal" sanatorium. The 8K74 missile tests scuttled the meeting of the two leaders, but the cottages remained. I never found out whether it was Khrushchev's or Eisenhower's cottage in which Katya and I spent several lovely days.

THE FIRST LAUNCH AT MAXIMUM RANGE WAS SET FOR 19 JANUARY 1960. However, according to the Pacific Ocean flotilla commander, there was such heavy fog in the impact area that they couldn't jeopardize a helicopter to search for the spot on the water.

The launch took place the next day, on 20 January. TOGE-4 commander Captain Maksyuta reported, "All okay." For security considerations, the coordinates of the impact point were not communicated. If "all is okay," it meant that we landed in the specified quadrangle. The Americans, of course, made a precise determina-

From the author's archives.

This is a rare photograph of an R–7A ICBM on the pad prior to launch at the Scientific-Research and Testing Firing Range No. 5 (Tyuratam) in 1959. Note the unusual nosecone designed for a new series of nuclear warheads. The R–7A was operationally deployed as a Soviet ICBM in September 1960 and carried a single warhead of 3 Mt yield.

tion of the nose cone impact site, but they did not know the estimated splashdown point. Therefore, they could only be guided by the position of the spot relative to our three ships. From the behavior of our ships they had expected the launch on 19 January. The launch was scratched due to fog, but in America they managed to issue a report about a failure that supposedly took place. It turned out like in the film where "for every failure you should be able to fight back"; we launched on the 20th and the communiqué followed. Now people were talking about the success.

The next launch took place on 24 January 1960. This time it really was a failure. The control chamber of strapon Block V exploded, evidently as a result of bad nitrogen purging. A fire started in Block V, the engine "died," and the entire cluster fell apart after 31 seconds. Our ships continued to mess around in the sea, surrounded by American destroyers. Once again a report appeared in the foreign press about the Soviets' latest launch attempt failure. We published no disclaimer, and decided to remain silent until the third and last launch of the program.

The last launch was set for Sunday, 31 January. It was considered so routine that all the senior officers stayed home. A military squad prepared for and executed the launch with minimal participation of industry specialists. Lead designer Kasho reported to us at the OKB from the firing range about the preparation and launch process. Arkadiy Ostashev, Emil Brodskiy, and Yevgeniy Shabarov were on the high-frequency communications line in Korolev's office. In Moscow and at the firing range the cold had settled in—it was –23°C (–9°F) with a breeze.

I was easily persuaded to monitor the last launch into the area of tropical islands in a warm sea without leaving my home. At 8:00 p.m. Arkadiy Ostashev informed me over the phone that, "Everything's okay. Even Mitrofan Ivanovich [Nedelin], who is in his office, has no negative remarks and he's congratulating everyone." An hour and a half later Ostashev telephoned again and asked if I could hear the noise over the telephone line. I confirmed that my ears had picked up a chorus of raucous cheering. Ostashev reported that, "They informed us that everything is a lot better than it has been. We sent a courier to the store and have done everything we were supposed to. We recommend that you also celebrate this occasion without leaving home." I took their wise advice.

The last communiqué in this regard said that the test missions had been accomplished and the region was safe for navigation. The TOGE-4 ships returned to Petropavlovsk-Kamchatskiy. The remaining No. 74 test launches were targeted for Kamchatka.

In September 1960, No. 74 went into service. However, on 7 October 1960, Nedelin wrote to Chairman of the State Committee for Defense Technology Rudnev that according to a TASS report, the U.S. had launched an Atlas intercontinental ballistic missile to a range of 9,000 miles (14,500 kilometers). He requested that they study the possibility of launching an 8K74 missile with a reduced-weight nose cone into the Pacific Ocean to a range of 16,000 to 17,000 kilometers with an azimuth of 45°.

Rudnev readdressed the letter in the form of a directive to Korolev. Bushuyev studied a version of the nose cone containing a warhead that had been designed for the new R-9 missile. It was 1.65 megatons instead of the standard three megatons. The required range was achieved because the warhead had been reduced in weight by 600 kilograms. A nose cone equipped with dye was fabricated, and launches took place during the winter of 1961 to intimidate the Americans. Nedelin's assignment had been completed, but he was no longer able to know about that.[9]

There were just two launch complexes for R-7 rocket and R-7A missile launches at the NIIP-5 firing range in Tyuratam: launch pads at Sites No. 1 and No. 31. Missiles in the Assembly and Testing Buildings were on standby alert; in the event of a stand-to-alert, nonstop work could achieve launch readiness in 12 to 16 hours.

While building the northern firing range in Plesetsk, provisions were made to create four launch complexes for *Semyorkas*. Through all conceivable efforts they reduced the readiness time to 7 or 8 hours. This was acceptable for spacecraft launches, but subsequently it became clear that this amount of time was unsuitable for combat missiles. We understood this as well as the military types did and began the intense development of the new R-9 intercontinental ballistic missile.

During all subsequent launches into the Pacific Ocean, TASS printed official reports warning ships of the danger of being in areas with such-and-such coordinates. These reports served as a signal for the Americans. Their combat and special-purpose ships and their reconnaissance planes appeared in these regions near the time of the scheduled launches. Somehow the Americans had ballpark knowledge of our missile launch schedule. Naval staff representative Oleg Maksimovich Pavlenko recalls that:

"*There was an instance when the American naval ships* Lansing *and* General Arnold *dropped into the nosecone impact area right after a launch. They showed their disregard for the TASS report about this area being closed to navigation. The crews of the American ships were on the open decks and there were signs with Russian text scrawled on them: 'How are things in Moscow?' Admiral Maksyuta communicated with the American ships over the radio and transmitted the following message to the* Lansing*: 'Your presence in this area is dangerous. I request that you leave the area. Please pass on this appeal to the* General Arnold. *Commander.' There was no reply. Immediately after the nosecone splashdown the American ships rushed full steam ahead to the impact point, lowered a launch with scuba divers in protective suits and began to gather everything they possibly could from the surface of the water after the detonation.*

In the early 1970s, three launches were conducted one after the other into the area indicated in the TASS report. The General Arnold *rushed into the impact zone right after the first nosecone splashed down, not imagining that another would follow. The second nosecone fell quite close to its board. The* General Arnold *was lucky. The new nosecones were armed with a warhead that had a 39-kilogram TNT equivalent. Pre-*

---

9. Nedelin died in a massive rocket catastrophe in 1960. See Chapter 32.

vious nosecones were armed with 300-kilogram warheads. Despite the relatively low charge, a 40-meter column of water rose up alongside the ship. The decks filled with curious Americans instantly emptied. The American ships cleared out of the firing area at full steam after sending the Soviet flagship the following message: 'Commander. From now on we will not navigate so dangerously.' After that incident, American naval observers did not get closer than five to six miles away from our ships during launches."

Chapter 25
Lunar Assault

The two years that followed the 1957 satellite successes resembled the war years in terms of pace and intensity. By early 1958, projects were simultaneously under way at OKB-1 in five primary areas:
  • engineering follow-up on the R-7 combat missile to put it into service;
  • updating the R-7 (article 8K71) to achieve a range of 12,000 kilometers (missile R-7A or article 8K74);
  • converting the R-7 from a two stage into a three or even four stage rocket;
  • designing a "heavy satellite" for photoreconnaissance (the future Vostok); and finally
  • projects for the conquest of the Moon, Mars, and Venus.

I have already discussed the first two areas of endeavor. Of the remaining three space projects, the problem of reaching the Moon seemed the most compelling and high-priority.

Each of the possible areas had its own proponents and enthusiasts; there were no opponents. Updating the R-7 missile by augmenting it with successive stages opened such prospects that we wanted to do everything as quickly as possible, to stun the world as often as possible, and to be transfixed with delight at hearing Yuriy Levitan's voice: "Attention! All the radio stations of the Soviet Union are reporting! We are broadcasting a TASS report! Today, pursuant to the space exploration program and preparation for interplanetary flights..."

You can criticize the utopian plans for building communism, the trampling of human rights, and the Communist Party's dictatorship in a totalitarian state all you want. But it is impossible to erase from the history of the Khrushchev era the favorable conditions created for developing cosmonautics and its related sciences. Cosmonautics did not arise simply from militarization, and its aims were more than purely propagandistic. During the first post-*Sputnik* years, the foundations were laid for truly scientific research in space, serving the interests of all humankind. All Soviet people, not just those of us who were directly involved in the missile and space programs, felt proud and were thrilled to be citizens of the country that was blazing the trail for the human race into the cosmos. I am not writing about this out of nostalgia for the "good old days," but because I remember well how people from the most diverse social strata felt about our space successes.

Most historians point to Korolev's genius and capabilities as an organizer to explain the successes of Soviet cosmonautics during that time. There is no doubt that his personality played an enormous role. But the conditions for successful work had been created around Korolev, his inner circle—including the other chief designers and the scientists from academia who had gathered around Keldysh—and the newly spawned missile organizations of Yangel and Chelomey. One would think, why should the Ministry of Defense squander soldiers and officers for a lunar assault? This was clearly detrimental to their primary military missions. Nevertheless, over the entire expanse from Moscow and the sunny Crimea to Kamchatka, at dozens of ground tracking stations and floating tracking stations on the oceans and seas, at Command and Measurement Complex centers, and in all the firing range services, thousands of military service personnel toiled selflessly. Military specialists carried out Korolev's instructions just as fervently as the orders of their commander-in-chief, Chief Marshal of the Artillery Nedelin.

Essentially, our technocratic community was a state within a state, which for the time being did not contradict Communist Party doctrine. High-ranking Party leaders understood that the technocrats needed a certain degree of sovereignty and self-determination. Things were a lot worse for agricultural scientists, biologists, artists, and poets. At that time, despite numerous errors, failures, and severe accidents, the technocrats—nuclear specialists, physicists, and missile specialists—were forgiven everything. Our successes were lauded around the world. Only those directly involved knew about our fiascos and failures.

The history of the conquest of the Moon is an example of this. I was directly involved with all the Moon launches up until 1966. If you were to piece together and describe the entire history of humanity's lunar conquest from our first failures in 1958 until the American manned lunar expeditions, you would get a very informative, fascinating book. It would be full of scientific information, tragic and comic events, and adventures just as riveting as any mystery or science fiction novel.

It bears mentioning that over the 30 years since the six American expeditions to the Moon, various characters have continued their efforts to expose NASA and prove that the presence of the astronauts on the Moon was staged, that is, that it was all Hollywood hocus-pocus. No one doubted our successful moon launches, but fans of big news stories simply knew nothing about our failures. We knew how to hide our failures. During the Cold War, disinformation was fed to the potential enemy as actively as during wartime.

OVER A PERIOD OF JUST ONE YEAR, from 23 September 1958 through 4 October 1959, we undertook seven lunar launches. Of these seven launches, one was partially successful (this was when we announced the creation of the artificial planet *Mechta* [Dream]) and only two fully implemented the tasks assigned them. In the ensuing years, up until 1966, we achieved success in only 1 out of 14 lunar launches. In all, there were 21 lunar launches over a nine-year period. Of these, only three were complete successes! But what hellish, fascinating, risky work it was!

The R-7's two stages were not adequate to reach the Moon with an automatically controlled vehicle loaded with equipment. A third stage, strictly for space, was needed to boost the lunar vehicle to the "second cosmic velocity" of 11.2 kilometers/second.[1] This stage was called the Block Ye.[2] It needed an engine. Mishin, elated with the successes of developing vernier thrusters for the R-7 using the OKB-1 workforce at our factory, persuaded Korolev not to turn to Glushko for help. The department of our chief engine specialist, Melnikov, had pretty good test-stand facilities and sufficient personnel to develop the engine itself, a combustion chamber with a high-altitude nozzle. But we needed a turbopump assembly and we had no experience producing them. We also had no time to learn how. The aviation industry rescued us. I have already mentioned that this industry had gotten on Khrushchev's wrong side. Not only had the factories' production capacity been freed up, but the design bureaus were also looking for interesting work. The very energetic, highly motivated, and talented Semyon Ariyevich Kosberg—chief designer of the Voronezh design bureau for experimental reactive aircraft engines and assemblies—offered his services to Korolev.[3] The son of a blacksmith, Kosberg was short and stout, but very light on his feet. He gesticulated rapidly and animatedly, was always optimistic, and had typically Jewish features. Korolev liked him from the first time they met. Once again I saw Korolev's unique ability to quickly take stock of people, to sense their inner nature from the first encounter.

Collaborative and very productive work began right away. Kosberg set about developing and manufacturing third stage engines running on oxygen/kerosene propellant for the R-7. For the first lunar vehicle, Mishin insisted on dividing the work: we took on the combustion chamber, while Kosberg had the turbopump, gas generator, and fittings.[4] We might have come to an amicable agreement, but one time when the duties and responsibilities were being divided up, Mishin lost his temper and carelessly said to Kosberg: "Why, you obstinate Jew." The latter flew into a rage, shot out of Mishin's office, and flew into Korolev's office across the hall. Kosberg announced to Korolev that he would not work with an anti-Semite. He ran out of the office and commanded his deputy Konopatov, "We're leaving!" Korolev called Mishin into his office. I don't know what transpired between them. But on Korolev's orders they intercepted Kosberg and brought him back. Explanations followed and then peace was restored.

Kosberg's vigorous activity tragically came to an end in 1965. When the *Sovnark-*

---

1. The "second cosmic velocity" is the Russian term for the velocity required to escape Earth orbit.

2. The upper stage was called Block Ye since "Ye" is the sixth letter of the Cyrillic alphabet (A, B, V, G, D, Ye). The first five letters denoted the core and strapon boosters of the R-7.

3. In 1946–66, Kosberg's design bureau was officially known as OKB-154. Today, it is known as the Design Bureau of Chemical Automation (KB Khimavtomatiki).

4. This engine was known as the 8D714 (RO-5).

hozy were dismantled and ministries were restored, they decided to transfer Kosberg's design bureau out of the aviation ministry and over to the Ministry of General Machine Building (MOM). Kosberg flew to Moscow to try to protest, but to no avail. He was extremely distraught by the events in Moscow and flew back to Voronezh. Driving back from the airport over an icy road, Kosberg had an automobile accident and was taken to the hospital with serious injuries. Soon thereafter he was gone. The doctors steadfastly reassured Korolev that Kosberg would pull through, and, when he didn't, Korolev was shaken.[5] After Kosberg's death, Aleksandr Konopatov became KB chief, and the KB was still transferred to the MOM. Today the Voronezh KB is one of the leading design bureaus for the development of oxygen/hydrogen liquid-propellant rocket engines. It developed the engines for the second (hydrogen) stage of the Energiya and third stages of the Proton and Soyuz rockets.

In all fairness, the Soviet scientific community valued the achievements of the school founded by Semyon Ariyevich Kosberg. Aleksandr Dmitriyevich Konopatov, who was in charge of the organization after Kosberg's death, was elected a corresponding member in 1976 and became an academician and active member of the USSR Academy of Sciences in 1991.

When the Block Ye third stage engine with a thrust of almost five metric tons was developed, there was one more difficult task. They needed to determine with complete certainty that ignition and startup would be reliably ensured in space. Up until that time we only knew how to start up engines on the ground with visual control and all kinds of automatic controls, and even then there were misfires. They learned how to start up the first engine of Block Ye reliably on a test rig, but there was no solid assurance that it would start up right away in space.

Pilyugin developed the control system of the third stage using our control surface actuators. The most difficult task was "intercepting" control after separation from the core booster. Large deviations of the gyroscopes could not be allowed. If they were to settle on the limit stops, control would be lost. The new task was to correct the space stage, and then to reliably guide it for almost 6 minutes of acceleration toward the Moon, and shut down precisely when the necessary apparent velocity was achieved. During the acceleration segment, while the control systems of the three stages were operating, one by one, over the course of 725 seconds, we would have to generate the next flight trajectory in order to impact in the center of the Moon's visible disk with a diameter of just 3,476 kilometers.

After the third stage engine shutdown, the flight would be governed only by the laws of celestial mechanics, which in turn, as we used to joke, were governed by our ballistics experts. The ballistics experts, headed by Okhotsimskiy from the Department of Applied Mathematics (OPM) of the Academy of Sciences V. A. Steklov

---

5. Kosberg died on 3 January 1965.

Mathematics Institute, Lavrov from our OKB-1, and Elyasberg from NII-4, performed the calculations on the first computers.[6] One of them was installed at OPM and a second at NII-4 in Bolshevo. The results of their calculations were supposed to be entered into instruments that controlled the flight speed and the moment that the second and third stage engines were shut down.

An error of just one meter per second (i.e., by 0.01% the value of the full velocity) in determining the rocket velocity at engine shutdown would cause a 250-kilometer deviation in the point of contact with the moon. A deviation of the velocity vector from the calculated direction by one angular minute would cause a 200-kilometer shift in the point of contact. A 10-second deviation from the calculated launch time from the Earth would cause a 200-kilometer shift in the point of contact on the Moon's surface. Such strict requirements were new and difficult for us at that time.

When Keldysh presided over meetings, deviation figures, calculations, and selections of orbits, launch dates, and launch times were typically the main subjects of discussions and arguments. He was not a ballistics expert nor a specialist in the field of celestial mechanics, but he quickly grasped the crux of the problem. Keldysh knew how to combine the results of abstract theoretical calculations with common sense and render a verdict for one orbital option or another that no one contested. His authority in this field was indisputable.

Korolev and Keldysh formed a great friendship and mutual understanding that coincided with the era of the first lunar vehicles. Keldysh assumed control of the whole analytical/theoretical portion of the lunar projects. He wanted to land on the Moon, perhaps more than Korolev, especially since research on lunar trajectories was being conducted using the equipment and procedures of Academy scientists. As a result, for the time being, Keldysh wasn't very interested in the human spaceflight projects that Korolev was emphasizing. Unlike Korolev, who was a top-secret figure, Keldysh operated as a partially public figure; he could associate with foreign scientists and travel abroad. Nevertheless, the KGB or the Central Committee forbade Keldysh's name to be linked with space research. His name was also in no way, shape, or form linked with the highly complex mathematical calculations that OPM was performing on the first computers for the nuclear experts.

IT WAS KELDYSH WHO FIRST PROPOSED SEVERAL PROJECTS FOR AUTOMATICALLY CONTROLLED LUNAR VEHICLES. The first, designated Ye-1, made a direct hit on the Moon. The second, Ye-2, flew by the Moon to photograph its invisible far side.[7]

---

6. Academician Dmitriy Yevgenyevich Okhotsimskiy (1921–) was a leading scientist at OPM and one of the most important theoreticians of the Soviet space program. Academy Corresponding Member Svyastoslav Sergeyevich Lavrov (1923–2004) served as head of the ballistics department at OKB-1. Colonel Pavel Yefimovich Elyasberg (1914–88) was a military ballistics expert at NII-4 in 1959–68.

7. *Author's note:* During the design process so many changes were introduced into the Ye-2 automated probe that the vehicle that eventually flew to the Moon was given the designation Ye-2a.

The third mission, Ye-3, was the most exotic; proposed by Academician Zeldovich, its goal was to deliver an atomic bomb to the Moon and detonate it on its surface. The Ye-4 fell through somewhere in our nomenclature. The Ye-5 was a project to take photographs with greater resolution than the Ye-2. Finally, the Ye-6 project, the crown of all our lunar activity, was designed for a soft-landing and transmission to the Earth of a lunar landscape panorama no later than 1964.

The Ye-3 program was concocted exclusively for irrefutable proof of our hitting the Moon. It was assumed that when the atomic bomb struck the Moon, there would be such a flash of light that all observatories capable of observing the Moon at that moment would easily record it. We even fabricated mock-ups of the lunar capsule with a mock-up nuclear warhead. Similar to a naval mine, it was completely covered with detonator pins to guarantee its detonation regardless of the capsule's orientation at the moment of impact. This mission variant was discussed very privately. In one such discussion Keldysh said that he had no desire to alert the world academic community that we were preparing for a nuclear explosion on the Moon. "They won't understand us," he asserted. "However, if we launch a rocket without a preliminary announcement, then there is no guarantee that astronomers will see the flash." In addition, Keldysh asked Korolev not to report this version to Khrushchev until we had discussed everything.

Korolev wavered. I reached an agreement with Pilyugin and Voskresenskiy, and then, on behalf of all the guidance specialists I rather cautiously suggested to him that this variant should be adopted only if there was a guarantee of complete safety in the event of an accident during the powered flight segment after launch. Keldysh added fuel to the fire: "Let the ballistics experts draw all the zones outside our territory in case the stage two or stage three engines don't do the job. Imagine the furor if this thing were to come down on foreign territory, even if it didn't explode."

Soon thereafter, the nuclear specialists themselves abandoned the idea of a nuclear explosion on the Moon. Keldysh paid us a special visit at OKB-1. He was in an excellent mood. As he told us, Zeldovich had rejected his own proposal. After calculating the duration and intensity of the flash in the vacuum of space, he doubted the reliability of photographing it from the Earth. As a result, this project, hazardous both intrinsically and in terms of its political consequences, was laid to rest; the designation Ye-3 was instead assigned to the program following Ye-2 that involved a lunar flyby while performing high-resolution photography.

OF THE 21 R-7 ROCKETS USED ON THE LUNAR PROGRAM FROM 1958 THROUGH 1966, 9 were three stage (known as the 8K72) and 12 were four stage (known as the 8K78) rockets.

The first launches of the lunar version of three stage R-7 rockets (8K72) were conducted on 23 September and 12 October 1958. Both launches ended with identical failures: the cluster broke up during the first stage's final flight segment. This was the first time we had observed this type of failure. The first analysis found no production defects, design errors, or sloppiness on the part of the testers during the rockets' preparation. We suspected that there might be some unknown fundamental flaw in the cluster configuration. The story of the search for the root cause of these failures is very instructive.

The quality of the telemetry recordings was quite adequate for a partial search for signs of failures in the control system or propulsion system assemblies. However, the numerous specialized groups investigating the 23 September crash found no smoking gun. A decision could not be made about the next launch without explaining the cause of the crash and performing some sort of measures. But we had promised Khrushchev a moon shot, so we did not have time for long deliberations and to study telemetry films and recordings without making a decision.

One of those who had lost hope of quickly discovering the secret of the rocket's breakup wistfully remarked that if we wrote it off as sabotage, such as an inconspicuously attached magnetic mine, then no measures other than heightened security would be required and we could continue the launches. In and of itself, the idea of possible sabotage was unacceptable for us, since it involved searching for an enemy among the testers. Over the course of our work, given the burning desire to close our eyes to the true causes, very many crashes could have been attributed to malicious intent. Then the security services would conduct the investigation and the engineers could move on to the next launch with a clear conscience.

Our experience over the first 12 years of work in the rocket field in the postwar era—and to jump ahead, over the ensuing years—showed that if engineers took on the role of private detectives, then we would always achieve success. Not once was a single failure written off as sabotage. Ultimately, even the most baffling incidents were cleared up. But this took time. Our inherent impatience, the pressure from above, and the desire to discover causes using the next launch as a full-scale "reenactment of the crime" were expensive, but on the other hand we were never accused of being idle.

The next astronomical window for hitting the Moon came during the first half of October. If we missed these "lunar" days, then we would miss our opportunity to present a gift in honor of the 41st anniversary of the October Revolution. But it could have been worse.

The biggest headache was the challenge that came from the military. Senior military representative Aleksandr Mrykin declared that, ultimately, the Moon was a matter of prestige, science, and politics. He contended that flight tests on the R-7 ballistic missile would not continue until we had obtained exhaustive explanations as to what caused the missile's breakup and we gave adequate guarantees. "Just imagine an inexplicable breakup of the entire cluster happening after 90 seconds of flight with a missile carrying a real warhead instead of sand!"

But we couldn't imagine such a thing because we had no idea how the automatic controls of the nose cone and the very warhead would behave. In heated debates, one or another person invoked some of the following rationales: "Let's say the missiles are tested in dozens of launches and each one without fail gives us new information that we use to change the designs or structures, ultimately, to increase

441

From the author's archives.

The Ye-1 spacecraft was designed as a simple sphere for direct lunar impact. On the pad, it was installed inside a special shroud built around the new upper stage equipped with a single engine (the RO-5).

reliability. As far as the main problem is concerned—the reliability of the thermonuclear warhead explosion at the target and guaranteeing safety no matter what missile failures occur "en route"—we cannot perform such real tests, much less at full range. Hence, there is a simple conclusion: we must deliver a warhead with an unconditional guarantee that no failure will occur through our fault over the entire route to the target. And if the missile's warhead hits the target, then the nuclear specialists are responsible for everything that happens there. They test our nose cone with the warhead independently, at their firing range. They give the guarantee, and 'may God help them!'"

Apropos of this, Voskresenskiy loved to say that the most reliable guarantee is an insurance policy, but since insurance companies ceased to exist in 1917, the insurance policy should be replaced by an oath signed by all the chiefs. Only Voskresenskiy could take the liberty of saying such things in that highly charged atmosphere. Anybody else risked having Korolev suggest that they catch the first train back to Moscow.

When it already seemed that the best missile detectives had exhausted all their resources to uncover the secret, a cause for first lunar failure in September began to circulate, an explanation that the majority of the chiefs hated. At first, the reasoning seemed purely theoretical, but for the time being it was the only one.

The guidance department of our OKB-1 included a dynamics laboratory. Its engineers analyzed the dynamics of the control processes after each flight, regardless of its results. While analyzing the behavior of the Apparent Velocity Regulation (RKS) system, laboratory chief Georgiy Degtyarenko and Pilyugin's deputy Mikhail Khitrik noticed the strange behavior of the pressure sensors, which acted as feedback devices in this system. These sensors monitored the pressure in the combustion

442

chambers of the strapon boosters. The RKS system's high-resolution sensor showed that the pressure in the chambers pulsed at a frequency from 9 to 13 Hz. This frequency coincided with the missile's normal longitudinal elastic mode frequencies. The amplitude of these vibrations at the moment the recording stopped had reached ±4.5 atmospheres.

If this wasn't stray electrical pickup in the measurement system, then such pressure pulses in the chamber would cause vibrations with a corresponding frequency in the oxygen and kerosene supply system. Indeed, a repeat microanalysis confirmed that the oxidizer pressure at the inlet to the pumps of all the boosters pulsed in the same frequency range. The axial acceleration sensor confirmed the presence of divergent longitudinal acceleration vibrations that had the same frequency as the engines' thrust pulses.

The search was isolated in a loop: from the missile structure, to oxygen pressure pulses at the inlet, to the pumps to engine thrust pulses of the strapon boosters. Vibrations with divergent amplitudes might occur in this closed loop if the natural frequency determined by the missile's structural features coincided with the pressure pulsation frequency in the combustion chamber. Structural deformations and, above all, fuel line deformations at the inlet to the engine pumps would lead to breakdown followed by fire and explosion.

The investigators returned to recordings of these parameters during previous launches and found that, true, there were pulsations of significantly less amplitude on almost all missiles, but no one had attached particular importance to that phenomenon. Usually telemetry system sensors monitored the pressure in the engines' combustion chambers. They were designed for a range from 0 to 50 atmospheres, and therefore the telemetry interpreters did not notice the pulsations on them.

In this case, the logical response would have been to stop flight-tests and switch to a thorough study of the discovered phenomena. But we were like gamblers. The stakes were high, but the payoff was also great—sending an Earth object directly to the Moon. The world's first! No one, certainly not Korolev and Keldysh, wanted to stop for intensive and lengthy investigations and experiments.

After the first reports of the proposed explanation were delivered in private, preventive measures were concocted to avert the cancellation of the next moon shot. Thrust was decreased in the first stage engines beginning at the 85th second, reducing the load on all the structural elements. We suspected that the tank depletion synchronization system might introduce disturbances into the process of feeding oxygen to the pumps. For safety's sake we decided to shut it down during this flight segment and at the same time to shut down the apparent velocity regulation system. We devised and quickly manufactured additional fasteners, hoping to increase the rigidity and thereby increase the natural frequency. We hoped that this modification would keep the pipelines out of a possible resonance zone. These measures were reported at the State Commission, which reluctantly gave the green light for the next launch.

In terms of catastrophic devastation, the second moon shot on 12 October was similar to the preceding one. Analysis of the telemetry recordings showed that the measures were ineffective. Now none of the specialists who had studied the processes giving rise to the destructive vibrations doubted the validity of the initial scenario for the breakup.

At a heated State Commission meeting, Rudnev demanded that Korolev personally head the accident investigation commission and that Keldysh assign scientists to the investigations. The commission took shape as follows: Korolev (chairman), Keldysh, Glushko, Pilyugin, Ishlinskiy, Petrov, Mishin, Akkerman, Narimanov, and Bokov.[8]

In the private conversation that Viktor Kuznetsov and I had with Pilyugin at his cottage at the firing range after all the meetings, Pilyugin grumbled that the guidance specialists had nothing to do with this problem. According to his version, Korolev and his "Mr. Rough" (as Pilyugin defiantly referred to Viktor Gladkiy) weren't looking into the engine's properties, and Glushko couldn't properly explain what he might have going on at the inlet to the oxygen pumps. Kuznetsov, on the other hand, sided with Korolev and Glushko. He didn't think it was right to judge them harshly because they were engineers and not very well versed in theoretical mechanics and oscillatory processes. Instead he asked, "How did Academician Keldysh agree after the first crash to such nonradical measures after having once given the classic explanation for the phenomena of flutter and shimmy in aircraft?" Ishlinskiy, who had just dropped in on us, sided with Keldysh. They shared a "deluxe" room at the new hotel and had the opportunity to discuss the situation in "unofficial" debates. According to *his* version, Keldysh had proposed that Korolev take a break in the launches and conduct serious investigations. But then Korolev and Keldysh would have to report this to Khrushchev and tell him that the next lunar launch attempt would take place at the New Year rather than the anniversary of the October Revolution. Keldysh refused to report to Khrushchev. Then they both decided to take a chance and go to the State Commission with the proposal to launch without dissention. And that's how the launch went off in October.

Now the investigations got under way on a broad scale. Keldysh mobilized the NII-1 theoreticians, Akkerman, Natanzon, and Glikman. They proved analytically that the breakup process was not random, but more likely in keeping with the laws of nature. In their opinion, not only should we increase the structural rigidity, but we should also find ways to preclude the very possibility of oxidizer feed pressure pulsations at the pump inlet. This was precisely the cause of the pressure pulsations in the combustion chamber. The oscillatory process began there and proceeded

---

8. Academician Georgiy Ivanovich Petrov (1912–87) was a prominent scientist at NII-1 before heading the Academy of Sciences Institute of Space Research in 1965–73. Georgiy Stepanovich Narimanov (1922–83) was a deputy director of NII-4 in 1959–65. Vsevolod Andreyevich Bokov (1921–) was chief of the department of analysis at the Tyuratam launch range before becoming a senior official of the Main Directorate of Reactive Armaments (GURVO).

through the entire loop, including the missile structure. You couldn't prevent disturbance processes from occurring solely by increasing rigidity because the pressure pulsation frequency could also rise and then you would have to increase the structural rigidity again.

While NII-1 scientists investigated these processes, the young engineers Degtyarenko, Kopot, and Razygrayev, as yet undistinguished by either awards or academic degrees, were in charge of a parallel investigation at OKB-1 to obtain practical recommendations as to what to do. In our laboratory, one of the first electronic analog simulators was put into operation. Using what were at that time state-of-the-art methods for simulating complex dynamic processes, it was possible to solve high-order differential equations without wasting weeks of work involving numerous analysts using mechanical adding machines.

Degtyarenko received baseline data on structural loads and elastic properties from Gladkiy, a mathematical model of the propulsion system from Natanzon at NII-1, and updates from Glushko's specialists in Khimki. All of this went in to the electronic analog simulator, which made it possible to display the process very graphically on cathode ray tube screens and to record it in the form of oscillograms.

Investigations conducted over many days with no days off and an open-ended workday ended in a proposal to introduce a special hydraulic damper in the oxidizer lines at the inlet to the pumps. Korolev tasked Anatoliy Voltsifer with the design of this damper. Voltsifer was in charge of developing all sorts of engine fittings. The proposed dampers were rather complex and difficult structures that needed to be cut into the oxidizer line. They still had to undergo a cycle of tests on the firing rigs at Glushko's facility and simulate the whole process. Also, the effectiveness of the proposals needed to be verified not only on the simulator, but also on the actual engine.

At the next session of the State Commission, Korolev confirmed the old rule that "no man is prophet in his own country." He thought it more advantageous politically for scientists from the outside—from another very reputable organization—rather than his own subordinates to come up with such a radical idea as a fundamental change in the hydraulic system. Keldysh assigned Natanzon to make a report containing these proposals. All that remained for our comrades to do was to humbly report on the simulation results. Korolev said that the damper design had already been developed, and in any case, it was being manufactured at the factory; plant director Turkov was already organizing round-the-clock work at the factory to produce the dampers.

Subsequently, everything went according to the same optimization plan for new systems that is now classic and universally recognized. Our engineers headed to Khimki along with the dampers. There, firing tests were performed. Using a special device, they set up varying intensities of disturbance at the inlet to the oxidizer line and found that the damper was a splendid shock absorber. Of course, they corrected the damper design and its characteristics several times. But the main thing was achieved. The firing rig tests showed that with the damper, the pressure oscillations in the oxidizer lines at the inlet to the pumps did not cause pressure pulsations in the combustion chambers. Consequently, dampers needed to be immediately installed in all the rockets designated for launches. The danger of the rockets breaking up due to resonance phenomena in the structure-to-engine loop was radically eliminated. This solution was extended to all the missiles developed after the R-7.

I have delved into this story in such detail because it was a consequence of a really fundamental shortcoming in the integration of the missile structure and the engine, which ultimately wasn't recognized until more than a year after flight-development tests began and the announcement to the whole world about the creation of the intercontinental ballistic missile.

At one of the subsequent meetings of the review team, one of the innocents in this story asked why attention hadn't been given to pressure pulsations in the chamber during the many preceding launches. Neither Korolev nor Glushko gave a satisfactory response at that time. Rudnev felt he needed to respond in his own way: "If you add up all the expenditures for each launch, it turns out that we're firing entire cities. Our previous successes have gone to our heads, and we're pressing on for new ones without taking the costs into account. All of us—and I also hold myself responsible—in our race for success have ceased to be vigilant. Indeed, the heroic work that was done in the laboratories, on test rigs, and at the factory after the failures could have been conducted after the first satellite. This is a harsh, but very useful lesson for us all."

Chairman of the State Commission and Chairman of the State Committee on Defense Technology Konstantin Rudnev did not, however, express during this sparsely attended State Commission meeting the idea that he later verbalized to Pilygin, Kuznetsov, and me after the meeting: "With all due respect to the Chief Designer and vice president of the USSR Academy of Sciences, I must admit that the true cause of the failures was discovered by young and as yet quite undistinguished specialists."

"*Today, 14 September, at 12:02:24 a.m. Moscow time, a second Soviet spacecraft reached the surface of the Moon. For the first time in history there has been a space flight from the Earth to another celestial body. In commemoration of this remarkable event, pendants displaying the emblem of the Soviet Union with the inscription 'Union of Soviet Socialist Republics, September 1959,' have been delivered to the surface of the Moon… The Soviet spacecraft's reaching of the Moon is a remarkable success of science and technology. This is the beginning of a new phase in space exploration.*"[9]

THIS IS THE TASS REPORT THAT THE MORNING NEWSPAPERS MANAGED TO PRINT ON 15 SEPTEMBER 1959. At 6:00 a.m. all the radio stations of the Soviet Union

---

9. Many of the original announcements cited by Chertok are collected in A. A. Mikhaylov and V. V. Fedorov, eds., *Stantsii v kosmose: sbornik statey* [*Stations in Space: A Collection of Articles*] (Moscow: AN SSSR, 1960).

broadcast this mind-boggling news throughout the world. There is one inaccuracy in the TASS report cited above, about which Korolev, Keldysh, and the text authors argued bitterly as the text was drafted during the night, the portion with the phrase "A second Soviet spacecraft reached the surface of the Moon..."

Actually only *one* rocket reached the Moon's surface. The preceding lunar probe, which was launched on 2 January 1959, missed the Moon. Its third stage, which carried a lunar capsule housing science equipment and an identical pendant, flew past the Moon and turned into an artificial planet of the Solar System. This *Mechta* (Dream)—it is unclear why it was called "Dream"—was supposed to impact the Moon. In the official history of cosmonautics, 2 January is considered the launch date of *Luna-1* or *Mechta*, the artificial planet, as if that is how it had been conceived. The second Moon shot, *Luna-2*, was officially launched on 12 September.[10]

In reality, the 12 September launch, although the first successful lunar impact, was actually the sixth overall attempt. Despite a year's delay, this event took place just in time for Khrushchev's visit to the United States. On 15 September Nikita Khrushchev departed for the U.S. One simply could not think of a better gift. Coinciding with the meetings of the top-ranking leaders of the U.S. and USSR, this launch could have been the occasion to end the Cold War. Alas, this did not happen. It was not in our power.

American newspapers and radio were abuzz with sensational commentaries.

"President Eisenhower and his chief advisers today were searching for ways to counter-act the new prestige that the Russians' successful Moon shot has created for Premier Nikita Khrushchev for the historic negotiations beginning tomorrow at the White House."

Newspapers around the world justifiably viewed the Moon shot not only from the standpoint of space, but also in social and political terms.

"N. S. Khrushchev arrives in the U.S. bringing the Moon along in his suitcase."

"Unfortunately, it is also true that this successful Moon shot produces complications. A rocket that can hit the Moon proves that other rockets can reach any point on the globe carrying a more deadly cargo and with the same accuracy. A space capsule containing the Soviet pendant is something like a 'flag display' that naval ships used to perform at sea."

Wernher von Braun announced to journalists that Russia had really shot ahead of the U.S. in terms of space projects and that no amount of money could buy the lost time. At a press conference von Braun said, "I am convinced that if Russia were to stop right now, we could catch up in one, two, or three years." After more than 30 years, it is painful and galling to realize that Russia really has stopped. No amount of money can buy lost time—I have to agree with von Braun on that.

Neither von Braun, nor the Americans, nor the Soviet people knew what pains were actually required for this "fantastic achievement," as T. Keith Glennan referred

---

10. The names *Luna-1*, *Luna-2*, and *Luna-3*, were given retroactively after 1963. At the time of the launches, these probes were called *Cosmic Rocket*, *Second Cosmic Rocket*, and *Third Cosmic Rocket*, respectively.

to our triumph.[11] He said, "This is the highest degree of success. No one doubts that the Russians have far surpassed all other peoples in the development of technology for the conquest of space."

On the day of his arrival at the White House, Khrushchev handed President Eisenhower a commemorative gift, a replica of the pendant that our spacecraft delivered to the Moon. This event moved us perhaps as much as the launch of the lunar rocket itself. After all, the pendant was also produced at our OKB-1. It was packed in a wooden case that our best cabinetmakers had toiled over. The case, lined with light blue velvet, contained a gleaming metal ball whose surface was made up of pentagonal cells, each embossed with the emblem of the Soviet Union with the inscription "USSR, September 1959." According to our concept, the pendant's spherical shape symbolized the artificial planet. The pentagonal cells were specially minted from stainless steel. The Mint began producing these historic pentagons back in 1958. The Mint had to produce them again for each new launch date after a failure.

Khrushchev liked this pendant so much that he admired it en route to the U.S. In the airplane Khrushchev took the pendant out of the case to show his American navigator Harold Renegar, who was flying as part of the crew for navigation security in U.S. airspace. "Good thinking!" smirked the navigator. "You launched one of these things at the Moon, and now you're sending another one to us in America."[12]

"The President thoughtfully considered the heavy lunar ball, celebrated in thousands of newspapers, in the palm of his hand. A sunbeam sparkled brightly on its polished facets. The President expressed his deep gratitude to the Soviet government and said that he would hand over the replica of the pendant to the museum in his home town of Abilene, Kansas, so that the people could see it." This is how our correspondents accompanying Khrushchev described this historic event.

Several hours before the solemn ceremony at the White House, a report had come out noting that a Jupiter rocket that was to carry experiments into space had failed to lift off.[13] Two days later an attempt was made to launch a Thor rocket. It also failed.[14] When we learned about these events, we were not gleeful; we knew, and probability theory and prior statistics underscored, that after our triumph we would also have black days.

Khrushchev's talk with American Congressional leaders took place on 16 September. At this meeting, Chairman Richard B. Russell, Jr. of the Senate Armed Services Committee asked Khrushchev the following question: "You have eloquently

---

11. Thomas Keith Glennan (1905–95) served as the first NASA Administrator in 1958–61. Prior to joining NASA, Glennan had been President of Case Institute of Technology in Cleveland, Ohio.

12. M. A. Kharlamov, ed., *Litsom k litsu s Amerikoy [Face to Face with America]* (Moscow: Politizdat, 1959), p. 51.

13. This was possibly the launch of a Jupiter IRBM on 15 September 1959 with biomedical experiments (known as Bioflight 3).

14 .This was Transit 1A launched on 17 September 1959 by a Thor Able II booster.

told us about sending the Soviet rocket to the Moon. We have had failures launching rockets. Have you?"

"Why are you asking me about this?" responded Khrushchev grinning. "You should ask [Vice President Richard M.] Nixon. He already answered this question when he announced that we have allegedly had three failed moon shots. He knows better how things stand with us. Nixon said that he had information from a secret source, but of course, he didn't reveal what that source is. He can't reveal such a secret; after all, it's a fabrication."

"But if you want, I will also respond to this question. Of course, launching rockets into space is not a simple matter. A lot of work goes into this. I'll tell you a secret; our scientists proposed launching a rocket at the moon a week ago. The rocket was prepared and delivered to the launch site, but when they started to check out the equipment, they found that it didn't work quite right. Then, in order to eliminate any possibility of risk, the scientists replaced the rocket with another. This second rocket was the one that we launched. But the first rocket is intact, and if you like, we can launch it, too. That's the situation. I can put my hand on the Bible and swear to this, but let Nixon do the same. (General laughter, applause.)"[15]

After reading this transcript, we noted with satisfaction that Nixon's secret source really was unreliable. In actuality, before 12 September 1959, there had been not three, but *five* moon launch attempts. Only the sixth launch resulted in complete triumph.

I WROTE ABOUT THE FIRST TWO IN DETAIL ABOVE. These were the rockets that broke up in September and October 1958 due to resonance disturbances during the first stage powered flight phase. After installing dampers in the oxygen lines and confirming the effectiveness of these modifications, we succeeded on 4 December 1958 in conducting yet another, third, lunar rocket launch attempt. The failure took place during the second stage flight segment. The accident investigation commission determined with a high degree of reliability that during the 245th second of flight, the reduction/step-up gear driving the hydrogen peroxide pump failed. Subsequently, the precise cause was identified: the breakdown of a gearwheel in the step-up gear due to a lubricant feed failure. Engine thrust dropped fourfold, the control chambers lost effectiveness, the rocket lost stability, and after deviating by more than seven angular degrees, the emergency engine shutdown (AVD) system shut down the engine.[16]

---

15. M. A. Kharlamov, *Zhit v mire i druzhbe!* [*To Live in Peace and Friendship!*] (Moscow: Politizdat, 1959), pp. 95–96.

16. *Author's note:* The measures taken after the crash of the lunar launch in December 1958 proved insufficient. The same defect reoccurred during the launch of standard R-7 missile number IZ-30 on 31 September 1959. On this missile the pump failure occurred 5 seconds later than on the lunar launch. This crash alone compelled the engine specialists to redo the lubrication system and strengthen the step-up gear.

AND SO, it was these three failures (September, October, and December 1958) out of a total of five that U.S. intelligence was able to report to Nixon. Evidently, the U.S. intelligence services were not able to figure out our next two failures. But now we have the opportunity to bring complete clarity into this story.

With the assistance of our powerful propaganda apparatus, we were able to convert the fourth failure on 2 January 1959 into the next brilliant triumph of Soviet science and technology. Among other things, reliably hitting the Moon depended on two factors: the accuracy of the second stage (or core booster) engine shutdown time and the third stage startup time in relation to planned schedules. Possible errors in the automatic system for shutting down engines of the second stage engine—from the longitudinal acceleration integrator—were, however, higher than tolerable. Therefore, to Ryazanskiy's delight, from the very beginning we decided to use a radio-control system to shut down the engine based on velocity and coordinate measurements. But on the January launch the radio command was late! Later, of course, we figured out that the radio-control ground stations (RUPs) were the culprits. The third stage with the lunar capsule containing the pendant missed the Moon by 6,000 kilometers, that is, approximately one and a half times the Moon's diameter. The rocket went into its own independent orbit around the Sun and became a satellite, the world's first artificial planet of the solar system.

Instead of the anticipated debacles—or at least tongue lashings—we were flooded with a deluge of greetings and congratulations. On 5 January, the CPSU Central Committee and USSR Council of Ministers issued a special message, which celebrated: "Glory to the workers of Soviet science and technology who are paving new paths to the discovery of nature and to harness its forces for the good of humanity!"

The January launch was a very good rehearsal and training session for us. For the first time, the third stage operation was completely checked out. It proved to be very beneficial to check out the radio communications system, the reception of telemetry from the capsule, and the processing of the results from the real-time determination of its coordinates and to adjust the interaction between the orbital tracking service's instrumentation complex and the computation centers. All of the onboard equipment worked well, providing the opportunity on 12 January to publish a detailed description of the scientific investigations. The most sensational discovery was the lack of a magnetic field around the Moon. The press gave extensive coverage to the use of an artificial sodium comet formed 113,000 kilometers from the earth to observe the flight of the third stage. The makers of the artificial comet were counting on visual observation by foreign observatories, primarily so that they would acknowledge that the rocket really was flying to the Moon. My departments developed a special timer to ignite this comet. Sixty-two hours after launch "according to the program" the onboard storage batteries, designed for 40 hours of operation, were completely discharged, and "the spacecraft tracking program and program of scientific investigations were completed."

AFTER JANUARY 1959, there was a brief hiatus in the lunar program. The firing range had to return to the R-7 flight-development test program. During this period, nine missiles were launched. Each of them had glitches that needed to be taken into account for the upcoming lunar launches as well.

The fifth attempt to hit the Moon was undertaken during the hot summer of 1959. The launch on 18 June ended in a failure during the operation of the second stage. But we were still plugging away, and the factories continued to produce new Moon rockets.

For our next lunar assault we produced two rockets and two lunar capsules with two "September" pendants. To be on the safe side, we also delivered a third capsule to the engineering facility. This time we decided to err on the side of caution since it was imperative that we hit the Moon. Now Khrushchev wasn't the only one demanding it; our egos had also been wounded. We wouldn't allow ourselves to even think about further failures. At the engineering facility and the launch sites everyone worked with a frantic desire for success. Work went on around the clock. There were relatively few glitches and modifications.

Despite the fact that it was September, the days at the firing range were hot. The nights were warm, still, and clear. The first launch attempt took place on 6 September at 3:49 a.m. in accordance with the flight assignment. It was permissible to err from the launch time by no more than 10 seconds. If the error were greater, the launch would have to be postponed by 24 hours or more, after recalculating the time accordingly. The launch failed on the first attempt. An automatic "circuit reset" occurred. We hunted for the cause for more than 2 hours and found a very stupid operational error when the circuit was assembled at the launch site. As usual, an error analysis revealed a glitch in the electrical circuit. One of the connectors wasn't shown in the electrical diagram, and we failed to connect it during the final assembly of the cables at the launch site. We put the circuit back together, retested it, and made sure that everything was in order, but 24 hours had been lost.

At dawn we reported to the State Commission that it was impossible to reattempt the launch on 7 September. This was because from the very start, we had ordered gyro horizons from Kuznetsov that determined the rocket trajectory inclination angle during the powered flight segment assuming possible launches at 48 hour intervals rather than every 24 hours. For 8 September, the launch time fell at 5:40:40 a.m.

We conducted checkouts all night, continued to fuel the rocket with oxygen, and checked and rechecked the readiness of the ground services. Over the phone I reassured the Command and Measurement Complex colonels who were standing by with their numerous radio specialists nationwide "from Moscow to the very fringes." Everything was going according to plan until it came to the "Drain" command. This command prompts the pressurization of all the tanks with compressed nitrogen. All the tanks were pressurized to the normal pressure, except for the oxidizer tank in the core booster. There was, however, still time in reserve. Upon receipt of the command from the console, pressure was released; the drainage valves opened and we made a second attempt at pressurization, monitoring it using the telemetry system pressure sensor. Golunskiy reported from the first tracking station (IP-1) that according to the visual observation unit the pressure in the tank was 40% of the scale. But what did that actually mean? We needed a precise interpretation, but the contact pressure gauge in the tank wouldn't allow the process to proceed in automatic mode. We missed the launch time again. In the heavy silence that filled the bunker, Voskresenskiy, who never stayed in a bad mood for long in such situations, proposed that we make a third attempt. "Most likely there's an ice plug in the pipe running from the tank to the sensor," he said. "If we knock it out with pressure, the rocket will be ready for launch."

On the third attempt the oxygen tank pressurized, but the launch process had to be halted. We had already missed the launch time. Once again, it turned out that Voskresenskiy was right; his intuition had not deserted him. We needed to decide how to proceed from there. The rocket had already been standing for three days fueled with oxygen. Should we drain the propellant and remove the rocket for drying or make one more attempt?

At that time, after pushing through the crowd of launch team members shrouded in silence, Lavrov reported in his quiet, calm voice that after reviewing the gyroscope programs, they, that is, the ballistics specialists, were clearing the rocket for launch with those very same instruments on 9 September.

"Where were you before?" asked Korolev indignantly, but he did not fly into a rage.

Without any recriminations, we made the only possible decision: leave the oxygen in the tanks for another 24 hours. Meanwhile, we would have to regularly shut down and warm up the control surface actuators and run performance checks on the onboard systems. The electrical firing and fueling departments remained at their work stations. People had already gone two days without sleep. Now they were allowed to sleep right in the bunker, for an hour or two, taking turns. It was decided to blow warm air through all the rocket's instrument compartments and measure the temperature regularly. Under these circumstances the State Commission and chief designers also established a 24-hour duty schedule.

The computing facility, that is, all the [ballistics] theoreticians, received very strict orders to repeatedly recheck everything and give the precise time for the launch on 9 September. The next night at T-minus 4 hours, sleep-deprived and exhausted, everyone once again gathered at the launch site. The launch time was 6:39:50 a.m.

The slanting rays of the sun already illuminated the steppe through large breaks in the light cloud cover. The meteorologists had promised a warm day with no wind. The rocket should finally lift off, and we would catch up on our sleep—if only a bit—before it reached the Moon. At first everything again went according to plan. We had ignition. The roiling flame swirled under all the assemblies as the engines built up to the first intermediate level and ... the "Main" command failed to go through! Because of a problem with the core booster, the circuit was reset, and the fire gradually died out under all the engines. An oppressive silence hung over the bunker. Then the weary voices of Voskresenskiy and Yevgeniy Ostashev issued the prescribed commands for such emergency cases. Fire trucks rolled up to the launch pad. The launch team cautiously inspected the sooty aft compartments. Everyone was tired to the point of indifference. Nevertheless, Korolev ordered the telemetry films to be developed immediately and the findings to be given. Glushko was named chairman of the accident investigation commission. Pilyugin proposed that first they decide what to do next and then investigate. For some reason Korolev suddenly shouted at Pilyugin, "You figure out what your circuitry experts have done!" Voskresenskiy found the cause right away: "Rocket No. 6 is the culprit," he noted. "It already failed at the launch pad once before it was renovated. It shouldn't have been reissued."[17] Everyone was so tired that no one even cracked a smile.

Nevertheless, everyone sighed with relief when they heard the call, "Drain everything right away! Remove the rocket from the launch site! Haul out the next one and prepare for launch on 12 September."

So a new rocket, serial number 43-7b, was delivered to the launch site early in the morning. Khrushchev had spoken of this operation in his reply to Senator Russell (based, of course, on a report that he received from Korolev, Keldysh, or Rudnev) when he said, "In order to eliminate any possibility of risk, the scientists replaced the rocket with another." We learned what Khrushchev had said many days later from the newspapers, and now that we had caught up on our sleep somewhat and relaxed, we vented our spite and had a good laugh. But we could allow ourselves that now "he who laughs last, laughs best."

The rocket launch on 12 September at 9:39:26 a.m. proceeded without a single glitch. The error relative to the calculated launch time was just 1 second. This was the sixth moon shot. I no longer remember who it was (it might have been Colonel Nosov) who announced loudly at the gathering right after the telemetry experts reported that the stage three engines had shut down at precisely the calculated time, "If you don't sleep at all for a week before each launch, then there won't be any failures." Actually, beginning on 6 September, members of the launch team only took cat naps, didn't shave out of superstition, and left the launch site for Site No. 2 only to perform a "hot food input operation." Officers who had served at the front said that even during the war they had more time for sleeping, eating, and shaving. After the launch, almost all the officers headed to Site No. 10 to their families. We gathered in a cramped room at Site No. 2 to receive the latest news via high-frequency communication and then give instructions.

Our first task after launch was to edit the TASS report and transmit it to Moscow. Our second task was to obtain permission to immediately notify Professor Bernard Lovell, director of the British Jodrell Bank Observatory about the impending

---

17. See Chapter 19.

impact.[18] In all of Europe, only this observatory had a large antenna capable of tracking our rocket on its path to the Moon and confirming that we had actually hit it.

Keldysh requested the permission of the State Commission to immediately notify the Briton. Korolev hesitated. What if we miss one more time? No one would believe that we wanted to put one more "artificial planet" into the Solar System. Ultimately, Keldysh prevailed, telephoned the Academy of Sciences, and gave the order to contact Lovell immediately and tell him the projected time of impact with the Moon and the current ephemerides so that he would have time to locate the emitting capsule among all the space noise and crackling.[19] We had some concern that they wouldn't believe our reports; we needed foreign witnesses of the Moon landing in addition to our own. We did not doubt that the Americans were also trying to track our second moon shot, but we had no contacts with American scientists. We figured that they would come up with the idea of turning to Lovell for help on their own. That's just what happened.

NASA's Deputy Administrator Dr. Hugh L. Dryden announced to Soviet correspondents on 14 September that, "We did not have the capability to visually track its lunar impact. But we received *Luna-2* signals on U.S. territory. We maintained constant contact with Professor Lovell from the Manchester observatory, who reported to us about every 'step' of the Soviet lunar rocket. Our scientists calculated the rocket's flight trajectory based on Professor Lovell's data."[20] Thus, NASA confirmed that the Russian lunar spacecraft had indeed hit its target—it had reached the Moon.

The ironies of the Cold War exemplified this episode in the fullest: our scientists did not have the right to communicate directly with American scientists, even for the sake of such a prestigious goal as establishing proof that we had impacted on the Moon.

On the afternoon of 12 September, after receiving preliminary reports that the flight trajectory was very close to the calculated trajectory, Korolev, Keldysh, Rudnev, Glushko, and Ryazanskiy flew out to Moscow. They needed to get to the capital before the lunar impact in order to report to Khrushchev before his departure for the U.S. In addition, Korolev had to personally check the condition of the gift pendant and case.

---

18. Sir Alfred Charles Bernard Lovell (1913–) is a physicist and astronomer who founded and directed (in 1945–81) the world famous Jodrell Bank Observatory near Manchester, England. In the 1950s, the Jodrell Bank complex included the world's largest steerable radio telescope

19. "Ephemerides" (plural of "ephemeris") constitute a set of data providing the locations of planetary bodies at given moments in time.

20. Hugh Latimer Dryden (1898–1965), a prominent aerodynamicist, played an important role in advancing aerospace research in the United States in the postwar era. He served as director of the National Advisory Council on Aeronautics (NACA) in 1947–58 and then deputy administrator of NASA in 1958–65.

Since the evening of 13 September, those of us who had stayed behind at the firing range had occupied the communications room so we wouldn't miss messages about the end of radio communication with the lunar spacecraft. This happened at midnight, and then we no longer felt like sleeping.

The flight of our sixth lunar rocket lasted 38 hours, 21 minutes, and 21 seconds. A flight from the firing range with the traditional stopover in Uralsk took more than 12 hours. For us, 14 September was virtually a holiday. But no one begrudged this, and no one blamed anyone for their boisterous behavior the night before. We received gleeful reports from Moscow that Professor Lovell had tracked the lunar spacecraft and confirmed that they stopped receiving emissions 1 second later than our forecast. After some confusion, we learned that our ballistics experts' forecast did not take the radio wave propagation time into account, thus being off by 1 second. This launch that I have described in such detail was certainly an important event in the history of cosmonautics and international relations.

Chapter 26
Back at RNII

In late 1958, after the first unsuccessful launch attempts for a direct hit on the Moon, S.P. called in Tikhonravov, Bushuyev, and me and announced that Keldysh had invited us to visit Likhobory (i.e., NII-1) and to familiarize ourselves with the proposals for an attitude control system for satellites and lunar vehicles. Tikhonravov said that he had heard about these developments. Boris Viktorovich Rauschenbach was conducting this project at NII-1 and in the opinion of our colleagues Ryazanov and Maksimov, the proposals were very interesting.

I should remind the reader that in 1933 at the initiative of Marshal Tukhachevskiy, the Reactive Scientific-Research Institute (RNII) was established in Likhobory. Korolev and Glushko worked at this institute until their arrests in 1938. Tikhonravov also worked at RNII beginning in 1933. In 1937, RNII was renamed NII-3; later, in 1944, the institute was renamed NII-1 and transferred to the aviation industry. I worked at NII-1 at that time and until my assignment to Germany. After returning from Germany I was transferred from NII-1 to NII-88, that is, from Likhobory to Podlipki. Mishin, Bushuyev, Voskresenskiy, Chizhikov, and several other compatriots from Germany also transferred along with me from Likhobory to Podlipki. Later, in 1948, Isayev's entire team completed this same resettlement routine. Earlier, in 1946, the young Academician Mstislav Keldysh was appointed NII-1 scientific director in place of General Bolkhovitinov.

Korolev recalled that he knew Rauschenbach well from his work at RNII. At the beginning of the war, despite his contributions, Rauschenbach was interned like all Germans.[1] He sat in some concentration camp and by chance stayed alive. After his release he returned to his once familiar institute.

S.P. said that Keldysh was evidently experiencing a "crisis of genre." The intercontinental cruise missile projects that he had overseen were going to be discon-

---

1. During World War II, the Soviet government arrested and then interned many Soviet citizens of German origin (like Rauschenbach) in special labor camps. Most were not released until several years after the end of the war.

From the author's archives.

Boris Rauschenbach (1915–2001) (left) was one of the pioneers of orientation systems for Soviet spacecraft. Rauschenbach, shown here with Chertok, was a man of many interests and wrote a number of books on such diverse topics as painting, philosophy, and history.

tinued.[2] Keldysh was giving more and more attention to our field, that is, ballistic missiles and spaceflight. "During our conversation at Likhobory keep in mind that Keldysh is our ally, not our competitor," said Korolev.

Korolev added that it was time for us to start working seriously on satellite guidance. He had already talked with Pilyugin and Kuznetsov about this. Both were so immersed in projects dealing "purely" with missile systems that they considered involvement with exotic satellites to be a frivolous amusement. Korolev, however, disagreed. He said, "Keldysh has serious proposals, and we shouldn't waste time. And you, Boris, don't be offended. We won't manage all this work with your kids, even with Pilyugin. We need to seek out new cooperation for space."

Tikhonravov supported these thoughts of Korolev's. His designers had already tried to collaborate with the "Pilyuginites" on attitude control systems for satellites, but as yet nothing good had come of it. I must confess that Tikhonravov with his inherent gentleness had already approached me requesting that I go with him to our alma mater NII-1 to see what Rauschenbach was doing. But caught up with

---

2. In the 1950s, Keldysh's NII-1 institute directed work on the Burya and Buran intercontinental cruise missiles through the design bureaus of S. A. Lavochkin and V. M. Myasishchev, respectively.

temporary assignments at the firing range and accident investigation commissions, I simply hadn't gotten around to it.

HERE I FEEL I NEED TO INTERRUPT THE NARRATIVE TO RECALL AND ELABORATE ON THE HISTORY OF THE M. V. KELDYSH RESEARCH CENTER (THE FORMER RNII), which turned 90 years old in 2003. The chronology cited below clarifies my original Russian-language memoirs. The book *M. V. Keldysh Research Center: Seventy Years on the Frontiers of Rocket-Space Technology* aided me in making these elaborations.[3] The editor-in-chief of this scientific work, director of the M. V. Keldysh Research Center A. S. Koroteyev, included the following inscription in it: "To Boris Yevseyevich, with deep respect and gratitude for many years of fruitful collaboration with the Center."

In the early 1930s in Moscow and Leningrad there were two rocket organizations in operation. In Moscow it was the Group for the Study of Reactive Motion (GIRD) and in Leningrad, the Gas Dynamics Laboratory (GDL). The directors of these organizations persistently raised the issue before the Red Army military leadership of merging the two entities. The first military leader to appreciate the prospects of missile armaments was M. N. Tukhachevskiy, deputy commissar for military and naval affairs, chief of armaments of the Workers' and Peasants' Red Army (RKKA), and future (beginning in 1935) marshal of the Soviet Union.[4]

On 21 September 1933, Tukhachevskiy signed "Order No. 0113 of the USSR Revolutionary Military Council" establishing the Reactive-Scientific Research Institute of the RKKA in Moscow. The Moscow Council of People's Deputies confiscated three small buildings from the All-Union Institute of Agricultural Machine Building (VISKhOM) on Likhachevskoye Highway to house the new institute.[5] Another order signed on the same day appointed I. T. Kleymenov (former GDL chief) chief of the new institute and S. P. Korolev (former GIRD chief) as his deputy. At the initiative of Commissar K. Ye. Voroshilov, a Council of Labor and Defense decree dated 31 October 1933 transferred the new institute, dubbed RNII, to the People's Commissariat of Heavy Industry headed by G. V. (Sergo) Ordzhonikidze. During its initial period, four subject areas incorporated into four departments formed the institute's foundation.

In January 1934, Korolev was removed from his post as deputy chief due to production discrepancies and a conflict with Kleymenov. He started working as a senior engineer in Ye. S. Shchetinkov's sector. Subsequently, G. E. Langemak held the position of chief engineer, actually deputy chief for scientific work. Korolev's removal

---

3. A. S. Koroteyev et al., eds., *Issledovatelskiy tsentr imeni M. V. Keldysha: 70 let na peredovykh rubezhakh raketno-kosmicheskoy tekhniki* [*M. V. Keldysh Research Center: 70 Years on the Frontiers of Rocket-Space Technology*] (Moscow: Mashinostroyeniye, 2003).

4. RKKA—*Raboche-krestyanskaya krasnaya armiya* (Workers' and Peasants' Red Army)—was the full and official name of the Red Army in the interwar years.

5. VISKhOM—*Vsyesoyuznyy institut selskokhozyaystvennogo mashinostroyeniya.*

from his high-ranking position was a heavy blow to his morale, but in keeping with the principle that "every cloud has a silver lining," this demotion saved his life since his replacement was later arrested and shot.

During its first years of work, the institute achieved crucial results in the development of rocket-propelled projectiles with solid-propellant engines. In collaboration with NII-6 of the People's Commissariat of Munitions, they developed a new propellant powder compound for rocket-propelled projectiles. This research played an important role in supporting the large-scale production of projectiles on the eve of and during World War II. Troop trials took place 1936–38, and 82- and 132-mm rocket-propelled projectiles (RS-82 and RS-132) were put into service on aircraft. In August 1939, Soviet airmen used RS-82 projectiles for the first time in combat action during the conflict with Japan in the Khalkin-Gol region.

A disgraced Korolev meanwhile was named director of the fifth department and worked simultaneously on several designs:
- the class "212" surface-to-surface cruise missile with a flight range up to 50 kilometers and a payload mass of 150 kilograms, equipped with a liquid-propellant rocket engine with a thrust of 150 kgf;
- a rocket-propelled fighter interceptor with conventional propeller engine system and equipped with a rocket engine enabling it to reliably intercept enemy aircraft;
- a surface-to-air missile for rapid interception of enemy aircraft;
- an air-to-air cruise missile for aerial combat.

In early 1937, the People's Commissariat of Defense Industry split off from the People's Commissariat for Heavy Industry. RNII transferred to the new People's Commissariat and was renamed NII-3. It became a closed organization under a strict regime of secrecy. Unfortunately, during this period internal conflicts fed by a top-down campaign targeting spies and participants in "anti-Soviet Trotskyite subversive organizations" tormented the NII-3 organization. In November 1937, Kleymenov and the institute's chief engineer Langemak were arrested. After a brief inquest, the Military Collegium of the USSR Supreme Court sentenced them and on January 10 and 11, 1938 they were shot dead. Glushko and Korolev were arrested on 23 March and 27 June 1938, respectively. Their fate after arrest has been described in numerous historical works.[6]

On 14 October 1937, B. M. Slonimer was named the new chief of NII-3, and on 15 November 1937, A. G. Kostikov (chief of the liquid-propellant rocket engine development department) began to perform the duties of chief engineer. He was

---

6. For the best Russian-language account, see Yaroslav Golovanov, *Korolev: fakty i mify* [*Korolev: Facts and Myths*] (Moscow: Nauka, 1994), pp. 223–328. See also the two volume work by Korolev's daughter: N. S. Koroleva, *Otets: knigi pervaya i vtoraya* [*Father: Books One and Two*] (Moscow: Nauka, 2001–2002). For an English-language work, see Asif A. Siddiqi, "The Rockets' Red Glare: Technology, Conflict, and Terror in the Soviet Union," *Technology and Culture* 44 no. 3 (2003): 470–501.

confirmed in this post in September 1938. In January 1939, NII-3 was transferred to the just-established People's Commissariat of Ammunition, which had split off from the People's Commissariat of Defense Industry. The Leaders of the new Commissariat considered projects on rocket-propelled projectiles to be NII-3's primary mission. All the projects on liquid-propellant rocket engines and aircraft were actually conducted at the time *only* owing to the enthusiasm of the staff and "available resources." Groups were formed at the institute to develop ground-based launchers and solid-propellant projectiles for them. Engineer I. I. Gvay proposed the first design of the future, the later-famous *Katyusha*. The design of the multiple-launch rocket truck-mounted launcher was updated and modified up until 1940.

In autumn 1939, the new system underwent official firing range tests near Leningrad and was recommended for service. Nevertheless, the system faced a drawn out process to gain the acceptance of the Main Artillery Directorate's leadership. It was difficult to place orders for the series production of 132-mm projectiles at the Vladimir Ilich Factory in Moscow in 1940; the Komintern Factory in Voronezh, meanwhile, produced only two launchers before the war began. The fate of the future *Katyushas* was decided by their splendidly successful firing demonstration at the Sofrino firing range, where the nation's top military brass were attending a review of military technology. On 21 June 1941, just 24 hours before Russia entered World War II, the government decided to put the system into service in the Red Army.[7]

During the first days of the war (from 28 June through 1 July) an order of the People's Commissariat of Defense spurred the formation of the first independent experimental rocket artillery battery under the command of Captain I. A. Flerov. On 2 July, a battery manufactured through the efforts of NII-3 was sent to the Western front, and on 14 July it conducted history's first combat salvo against Nazi troops near the Orsh station. The first combat salvos of the Independent Experimental Battery made a stunning impression not only on the enemy but also on the Western front command. Decisions were made in Moscow to drastically increase the production scales. Berlin, meanwhile, gave orders to uncover the Russian secrets and capture samples of the new weapon. The State Defense Committee (GKO) started keeping a "Special file on reactive technology" that was maintained until the collapse of the Soviet Union!

As far as manufacturing, the Special Design Bureau of the Kompressor Factory in Moscow was entrusted with the leading role in preparing for the series production of the launchers and developing their new modifications. V. P. Barmin was appointed chief designer. That is why later he wound up as chief engineer of the Institute Berlin in 1945 and soon thereafter became chief designer of ground launching equipment for the first domestic R-1 missile and then general designer for many

---

7. See Chertok, *Rockets and People.* Vol. 1, pp. 167–171.

surface-launched missile systems.

Despite the fact that the institute was concentrating its primary efforts on developing a multiple rocket launching system, during the period from January to March 1939 experiments continued with Korolev's 212 cruise missile, which used Glushko's ORM-65 liquid-propellant rocket engine. Having been arrested in the summer of 1938, the authors of these projects knew nothing about the flight-test results.

In addition, on 28 February 1940, at the airfield near the suburban-Moscow Podlipki station where NII-88 (now TsNIIMash) had been located since 1946, and where Korolev would begin working in 1947, the RP-318-1 rocket-glider of his design completed its first flight. The RDA-1-150 liquid-propellant rocket engine designed by Dushkin powered it. This was the first flight in the USSR of a piloted vehicle with a rocket engine. Although the rocket glider took off towed by a Po-2 aircraft, nevertheless, history must not forget the pilot, V. P. Fedorov. He controlled the first piloted flying vehicle, designed by Korolev.[8]

During the first year of the war, the institute's work was completely subjugated to the interests of the front and to the fulfillment of the decree calling for the development of a liquid-propellant rocket engine for the BI fighter-interceptor. It was on this second project that I continued to deal with Dushkin, Shtokolov, and Pallo until the evacuation from Moscow in October 1941.

IN CONNECTION WITH THE ASSIGNMENT TO DEVELOP A LIQUID-PROPELLANT ROCKET ENGINE FOR THE BI FIGHTER-INTERCEPTOR, in 1942, NII-3 began to develop the USSR's first propellant feed systems using turbopump assemblies. Kostikov understood that the capabilities of the rocket-propelled interceptor were determined solely by the sources of thrust, that is, the liquid-propellant rocket engine. Realizing that a basic shortcoming of that aircraft was the very limited flight range, Kostikov decided, independently without Bolkhovitinov, to begin developing an interceptor armed with cannons and rockets. It seemed that this design, under the code number "302," had tremendous advantages over our BI fighter-interceptor.

I used the word "seemed" because in addition to the liquid-propellant rocket engine that enabled takeoff and acceleration when maneuvering, the 302 aircraft had a ramjet engine (PVRD) installed in it for the cruising phase.[9] The primary developers were Kostikov, Tikhonravov, Dushkin, and V. S. Zuyev (the PVRD developer). Kostikov's prestige and that of all of NII-3, achieved thanks to the *Katyusha's* effectiveness, was so great that the government accepted the proposal to develop the new aircraft.

At the same time, the institute's status was changed. NII-3, formerly in the Commissariat of Ammunition, was transformed into the State Institute of Reac-

---

8. *Author's note:* The continuation of this project was the installation of Dushkin's engine not on a glider, but on the BI fighter-interceptor that took off on its own on 15 May 1942. I described this in detail in my first book. See Chertok, *Rockets and People.* Vol. 1, Chapters 12 and 13.

9. PVRD—*Pryamotochnyy vozdushno-reaktivnyy dvigatel.*

tive Technology and subordinated directly to the Council of People's Commissars. In addition to the primary field of endeavor, the development and optimization of rocket-propelled projectiles and multiple rocket launching systems, the institute was tasked with developing reactive aircraft, torpedoes, and various types of reactive engines.

Military Engineer First Class A. G. Kostikov was named the institute's director and chief designer. The State Committee of Defense, on the director's advice, approved the institute's mission and program of operations. Not a single institute or defense technology KB in the Soviet Union had such status. A State Committee of Defense decree dated 26 July 1942 tasked the institute with developing the 302 aircraft in collaboration with the People's Commissariat of Aviation Industry under the supervision of Chief Designer Kostikov and presenting it for flight tests in March 1943. It should be noted that this ambitious timeframe for the development of the aircraft was accepted at the suggestion of Kostikov himself.

The July 1942 decree increased the institute's territory and production area several times. It obtained VISKhOM's main building, which was the institute's "face" for decades to come, and also complexes of facilities later retrofitted with production shops and test rigs. Despite the extreme wartime difficulties, the institute's manpower quota rose significantly. A special decree dated 4 March 1943 classified institute construction among "especially vital construction projects."

Despite all-hands rush jobs, the institute wasn't able to develop the 302 aircraft within the prescribed timeframe. Flight-Research Institute test pilots S. N. Anokhin and M. L. Gallay tested the glider in the air. This was their first contact with piloted rocket-propelled technology, even if it didn't yet have the main element, a rocket engine. As fate would have it, both of these remarkable pilots later participated directly in preparing the first human flights into space. Problems that cropped up during PVRD development prevented its actual production in the foreseeable future; instead, Kostikov made the decision to manufacture the first aircraft with only liquid-propellant rocket engines. This decision stripped the 302 design of whatever advantages it had over the BI. In early 1943, a government commission was sent to the institute headed by Deputy Commissar of Aviation Industry A. S. Yakovlev, also chief designer of the Yak fighter aircraft series. The commission concluded that the institute had not fulfilled the government's assignments. After Yakovlev's report to Stalin on 18 February 1944, Kostikov was removed from the director's post and arrested. The aircraft's development was halted. The State Institute of Reactive Technology under the Council of People's Commissars was renamed NII-1 and transferred to the Commissariat of Aviation Industry.

Over the 10 years since the establishment of the Reactive Scientific-Research Institute, it had been subordinate to five agencies (People's Commissariats for Military and Naval Affairs, of Heavy Industry, of Defense Industry, of Ammunition, and the USSR Council of People's Commissars). Five directors had headed the institute. Of these, the first, Kleymenov, was executed, the next three were just administrators, and the last was Kostikov. After gaining fame for the principal achievement of that period, the development of the *Katyusha* multiple rocket launching system, he was punished relatively mildly for failing to fulfill the clearly unrealistic obligations that he'd taken on.[10]

But what was going on among our allies and enemies in the field of rocket technology during this time? In the U.S., Robert H. Goddard was the only devotee who had really attempted to create a liquid-propellant guided missile. He was the first to use a turbopump assembly to feed propellant into the liquid-propellant rocket engine combustion chamber and the first to use jet vanes and gyroscopes for missile control. However, he conducted all of his work with the help of subsidies from private organizations and foundations. Before 1945, no state organization for developing rocket technology existed in the U.S.[11]

Germany began to set up a scientific and research center for rockets, shifting the energies of its Society for Space Travel *(Verein für Raumschiffahrt)* to the development of real liquid-propellant rockets in the early 1930s. In both Germany and in the USSR, the military took the initiative to bring together spaceflight enthusiasts and to create an experimental base. Almost simultaneously with the establishment of RNII under Tukhachevskiy's patronage in the USSR, an experimental facility was established in Kümmersdorf near Berlin, under the Army's weapons directorate. Walter Dornberger was appointed military director and 20-year-old Wernher von Braun, its technical director. Dornberger and von Braun were permanent leaders of the creation of the world's first long-range ballistic guided missile weapon. From 1936 through 1945 they were in charge of the world's largest scientific-production and rocket testing center in Peenemünde.

As the Soviet Union's fate was being decided in the Battle of Stalingrad in October 1942, in Peenemünde the first successful launch of the A4 (later called the V-2) missile took place, for the first time reaching the altitude where space begins. Beginning in 1943, the development of long-range missile weaponry had top priority in the German defense industry.

The Soviet Union also placed high priority on multiple rocket launching systems. The firepower of the *Katyusha* along with that of classic conventional artillery was decisive in the Battle of Kursk and in subsequent large-scale operations up until the final victory.[12] On the other hand, the massive bombardment of London and other European cities with A4 missiles did not have a substantial effect on the course of the war. However, after Germany's surrender, A4 missile technology served

---

10. Kostikov was arrested in February 1944 and remained in prison until February 1945.

11. There were a number of very important private organizations engaged in extremely innovative work on solid and liquid-propellant rockets during the war, all of whom worked independently of Goddard. These included Reaction Motors, Aerojet, and the Guggenheim Aeronautical Laboratory (GALCIT) at the California Institute of Technology (Caltech). GALCIT would later be reorganized into the famous Jet Propulsion Laboratory (JPL).

12. The Battle of Kursk in July–August 1943 was one of the most important battles on the Eastern Front during World War II. It still holds the record as the largest armored engagement in military history.

as the basis for the creation of new types of missile weaponry in the USSR and in the U.S.

AFTER ITS REORGANIZATION IN FEBRUARY 1944, NII-1 changed its priorities. Projects involving aviation reactive engines moved to the forefront. NII-1 was given a secret name, the Scientific Institute of Reactive Aviation (NIRA).[13] V. I. Polikovskiy was named the first NII-1 director, serving simultaneously as chief of the Central Institute of Aviation Engine Building (TsIAM). Gas dynamics scientist G. N. Abramovich was appointed the deputy director. The institute was energetically staffed with scientific personnel from the aviation industry and related branches of science. During this time N. A. Pilyugin, who would later become an academician, and L. A. Voskresenskiy, who would become Korolev's deputy for missile testing, transferred to the institute from the Central Aero-hydrodynamics Institute (TsAGI).

The GOKO issued a decree in February making Factory No. 293 in Khimki the institute's production and aircraft design facility and renamed it Branch No. 1. There, Isayev, Bushuyev, Mishin, Bereznyak, and I worked under the leadership of Bolkhovitinov. Our "patron" Bolkhovitinov was appointed first deputy director of the institute for scientific-research projects.

Other organizations transferred to NII-1 in addition to Factory No. 293 included M. M. Bondaryuk's ramjet engine design bureau and A. M. Lyulka's gas-turbine engine design bureau (from TsIAM, where he had ended up after returning from Bilimbay). A combined order dated 18 April 1944, issued by two Commissars, L. P. Beriya of Internal Affairs and A. I. Shakhurin of the Aviation Industry, transferred the design bureaus located on the grounds of Factory No. 16 in Kazan to NII-1; the imprisoned Glushko was the chief designer of this KB, while his deputy for testing was his fellow inmate Korolev. Thus, Glushko and Korolev officially returned to their former RNII-NII-3 after a break of six years.

In May 1944, Major General P. I. Fedorov replaced Polikovskiy as chief of NII-1. Until then Fedorov, who held Bolkhovitinov in high esteem, had served as chief of the Air Force NII. With the help of Shakhurin, Fedorov sped up the process to free Glushko, Korolev, and other prisoners of the "special" OKB attached to Factory No. 16 and to have their convictions rescinded. The USSR Supreme Soviet decided in favor of the early "release" of the employees of the "special" rocket engine OKB in Kazan in July 1944.

My sector at the new NII-1 proved to be one that all the others needed. The automatics and control department from the old "Kostikov" institute, as we referred to it, became part of my sector. M. A. Shmulevich directed this department. This very erudite 40-year-old electrical engineer managed to take advantage of favorable

13. NIRA—*Nauchnyy institut reaktivnoy aviatsii.*

conditions during Kostikov's ascent and acquired a lot of extremely scarce electronic measuring instruments. Having become the proprietor of multipurpose research equipment, he rendered invaluable assistance, introducing electronic measurement methods into the experimental and testing operations of the institute's various fields of endeavor.

Shmulevich gave me a lot of help in getting on friendly terms with the old RNII-NII-3 personnel. He had started working there under NII-3's first director, Kleymenov, who was later executed. He knew Langemak, Glushko, and Korolev, although he did not share his memories about his repressed colleagues. The atmosphere of fear that had been established in the NII-3 organization since 1937 had still not lifted. When Shmulevich died after getting hit by a car on the street under strange circumstances, it was a great blow for me.

L. A. Voskresenskiy and A. P. Pleshko, who had recently come on board at the institute, were the chiefs of hydraulic automatic controls laboratories. They had been developing all sorts of electrically controlled valves for engine power systems. Pleshko left the institute after Glushko began staffing his engine design bureau OKB-456 in Khimki after the war. R. I. Popov, A. I. Buzukov, and M. I. Sprinson had also come along with me to the institute from Factory No. 293 in Khimki. I made each of them a department chief. Roman Popov, a brilliant and talented radio engineer worked on a radio guidance system for a jet interceptor aircraft and a radio navigation system to bring it into the landing airfield. He was one of the first to begin developing the idea of radio control for an anti-aircraft guided missile. Popov was younger than I, but we became very good friends. I had faith in his talent, and he believed luck was on his side.

After Shmulevich's mysterious death, an even heavier blow for me and the entire team was Roman Popov's death. On 7 February 1945, he and a group of NII-1 colleagues and institute Director Fedorov were flying to the Western front to retrieve "trophy" materials captured at German firing ranges, including the remains of A4 missiles and ground-based radio equipment. The airplane crashed while landing for a stopover near Kiev, and everyone on the plane was killed. The entire institute was in mourning. Once again the institute was headless.

I appointed V. N. Milshteyn head of the special department for the development of electrical measurement methods and instruments. I had persuaded him to transfer from NII-12, the institute of aircraft equipment. Milshteyn was a specialist not yet well versed in the fine points of rocket technology, but with a brilliant command of the theoretical bases of electrical engineering. Before long, having gained widespread recognition, he published a book on electrical measurement systems design and research methods.[14] Milshteyn found allies among very young devotees of electrical measurements in researching many of the processes that occupied the institute's very thematically motley collective.

---

14. V. N. Milshteyn, *Energetishkiye sootnosheniya v elektroizmeritelnykh priborakh* [*Power Correlation in Electrical Measurement Instruments*] (Moscow, 1960).

Pilyugin was new to everyone in my sector No. 3. Bolkhovitinov had sent him to me; on his own initiative—and with Bolkhovitinov's prompting—he was assigned to supervise the "special automatic controls" group. The project involved developing autopilot systems for reactive aircraft.

THE ESTABLISHMENT OF SCIENTIFIC-RESEARCH DEPARTMENTS AND LABORATORIES took place under extremely complex wartime conditions. Many colleagues faced very difficult living conditions after returning from evacuation to the east. For example, Katya, our five-year-old son, and I were given a nine square meter room in a communal apartment on Novoslobodskaya Street with very unpleasant neighbors. Katya's mother, Kseniya Timofeyevna, came to our rescue in the summer and took Valentin to stay with her in Udelnaya.

While Kostikov was still in charge, despite the war, the institute was granted the opportunity to begin building living quarters for its employees. After Kostikov was taken off the job, the Communist Party leadership at the institute was also revamped. While filling out my Party paperwork for my transfer to the institute from Khimki, I met the new VKP(b) Central Committee Party organizer, Ye. A. Shchennikov. During our conversation, I mentioned my unsettled living situation. In autumn 1944, he invited me for a talk, claiming that he wanted to study the work my sector was conducting. In the course of our conversation he mentioned that the institute was going to add three stories onto an old building on Korolenko Street in Sokolniki. In view of the very difficult housing shortage, there was no hope of getting a separate apartment in this building expansion, but I could expect to get half of a four-room apartment. For me this news was a gift. He gave me a second gift in asking if I would object if his family occupied the other half of the apartment. "Our families are evenly matched," added Yevgeniy Abramovich. "We have three and you have three. Our sons are the same age."

We moved into the new apartment practically simultaneously in February 1945. On 10 February, my family grew from three to four with the birth of our second son, who we named Mikhail. And although the Shchennikov family also grew, from 1945 until 1958, we shared an apartment that didn't have such basic amenities as a bathroom or even a shower. We only had cold water, a wood-burning stove for the first years, a single sink for washing in the kitchen, and a small communal toilet. Nevertheless, we not only maintained civil relations, but our wives and sons became friends and the friendship continued after living together for 13 years, when we moved to separate apartments in different areas of Moscow.

THERE WERE A LOT OF "CRITICAL" PERIODS IN RNII (NII-1)'s HISTORY. One of these was the summer of 1944. In my first book I mentioned Stalin's correspondence with British Prime Minister Winston Churchill regarding studying German rocket technology left behind on the territory of the German firing range in Poland.[15]

---

15. Chertok, *Rockets and People*. Vol. 1, pp. 258–259.

These remains of "trophy" rocket technology enabled us for the first time to assess the scale of operations in Germany.

Our familiarization with German achievements beginning in 1944, and reports about the rocket bombardment of London affected the mindset not only of the institute's specialists, but also that of leaders of the Commissariat of Aviation Industry. On 30 October 1944, an order of the Council of People's Commissars established the Main Directorate of Aviation Reactive Technology, also known as the 18th Main Directorate of the Commissariat of Aviation Industry (NKAP). P. V. Dementyev was named chief of this new main directorate, serving simultaneously as the first deputy commissar.

During the first months after the end of the war, NII-1, the Commissariat of Aviation Industry leadership, and the nation's high-ranking political leaders believed that all projects on reactive technology would be concentrated in the Commissariat of Aviation Industry system. However, aviation industry leaders, having assessed the prospects of jet aircraft, decided that they could not handle two fields of endeavor, jet aircraft and automated long-range missiles.

On July 1945, in a letter to Politburo member G. M. Malenkov, Commissar of Aviation Industry A. I. Shakhurin wrote:

"I am reporting to you the results of an investigation of the German scientific research institute of missile armaments in Peenemünde conducted by NII-1 deputy chief Professor G. N. Abramovich...

From the investigation materials it is evident that production work on V-2 and other types of rocket projectiles is artillery-related. Therefore, it is advisable to assign this work to the People's Commissariat of Ammunition, after handing over to it all the equipment preserved in Peenemünde."[16]

Fortunately for rocket technology, at that time Malenkov did not accept the proposal to transfer us to the ammunition agency.

At the initiative of interdepartmental commission chairman L. M. Gaydukov, on 17 April 1946, Beriya, Malenkov, Bulganin, Vannikov, Ustinov, and Yakovlev signed and sent Stalin a memorandum on the topic of long-range missiles. Less than one month later, on 13 May 1946, the historic decree of the USSR Council of Ministers was composed, presenting a detailed program of the operations and duties of all the branches involved in developing rocket technology. This decree actually determined the birth of the Soviet Union's missile and space industry. (See Chapter 1.)

Surprisingly, the leadership of the aviation industry rejected a leading role in the

---

16. This letter that Chertok cites was first published in V. I. Ivkin, "Raketnoye nasledstvo fashistkoy germanii" ["The Rocket Contribution of Fascist Germany"], *Voyenno-istoricheskiy zhurnal* [*Military-History Journal*], no. 3 (1997): 31–41.

development of rocket-related subject matter.[17] The Ministry of Aviation Industry and its subordinate NII-1 proved to be off the beaten track of rocket technology development. Instead, the Ministry of Armaments took the leading role, even though aviation industry scientists were the best trained for this specific work.

As a result of the events described, Pilyugin, Mishin, Voskresenskiy, and I, as well as Korolev, who was part of NII-1, did not return to our old home, NII-1 (the former RNII), after we returned from Germany. Only Isayev's team remained there for a brief period of time, but it too moved to NII-88 in 1948.

On 2 December 1946, a new NII-1 chief, the ninth since 1933, was appointed: the young (35 years old), recently elected (November 1946) Academician Mstislav Vsevolodovich Keldysh. Keldysh quickly got into the swing of things in the essentially shattered rocket organization. Exhibiting a firm will and formidable capacity for work, he had the knack for swaying even those who had lost perspective. He succeeded in overcoming the somber mood of the scientific employees that had been brought on by numerous reorganizations and in formulating the main principles for the institute's work under the new conditions. He recommended that NII-1 be considered the head institute for liquid-propellant rocket engines and compressorless jet engines.

By 1948, the institute's primary thematic focus included problems of gas dynamics and heat exchange, thermal characteristics of airborne vehicles and fundamental research in the fields of thermodynamics, combustion theory, and the theory of the stability of working processes in engines.

In 1954, the Soviet government issued two historic decrees, one for the development of an intercontinental ballistic missile at NII-88 in Korolev's OKB and the other for intercontinental cruise missiles at the design bureaus of S. A. Lavochkin and V. M Myasishchev. Keldysh was appointed scientific director for the development of the intercontinental cruise missiles Burya (Storm) under Lavochkin and Buran (Snowstorm) under Myasishchev. An astronavigation system was the only thing that could provide flight control and navigation for these missiles. The astronavigation laboratory that I had set up in 1947 at NII-88 was moved to NII-1 to solve these problems. A special branch for the development of cruise missile control systems was also set up at the institute. R. G. Chachikyan, an experienced leader of aviation instrument construction was appointed branch chief. Former NII-88 astronavigation laboratory chief I. M. Lisovich finally received the title and status of chief astronavigation system designer at Chachikyan's design bureau. This design

17. *Author's note:* On 4 April 1946, aviation industry Minister Shakhurin was arrested and sentenced to seven years in prison. M. V. Khrunichev was appointed the new commissar. A Council of People's Commissars decree dated 26 February 1946, entitled "On the Work of the People's Commissariat of Aviation Industry" stated that the NKAP had "permitted a serious lag in the development of new aviation technology, and its leaders A. I. Shakhurin and P. V. Dementyev had showed short-sightedness and narrow-mindedness, failing to use all resources available to the aviation industry to solve the problems of new aviation technology."

bureau produced the first real and flight-tested astronavigation system for the Burya cruise missile.

During the development of the R-7 missile, NII-1 was tasked with ensuring process stability in liquid-oxygen rocket engines, conducting research on gas dynamics, heat exchange, and thermal protection of the nose cones (jointly with the Physics and Technical Institute and NII-88) and developing methods and equipment for measuring pressure pulsations in engines.

I revere Keldysh's memory because I am indebted to him that my proposed idea for using astronavigation to control missiles was not stifled by bureaucrats. Still director at NII-1, Keldysh headed the Department of Applied Mathematics (OPM) at the Steklov Mathematics Institute of the Academy of Sciences. This department was actively involved in the development of problems of missile dynamics and ballistics and research on the theory of flight and the orbital tracking of the first artificial satellites and interplanetary flight programs. Keldysh's extensive network of interests allowed him to lend support to Rauschenbach's idea to develop attitude control systems for spacecraft right at NII-1.

Keldysh's scope of interests was extraordinarily broad. At his initiative, long before the first satellite launch, fundamental research had been conducted on the mechanics of spaceflight and on analyzing and selecting the optimal configurations for staged rockets. These operations helped our designers in the final selection of the cluster configuration for the R-7 rocket. For the first time, NII-1 and OPM jointly studied what was for us the extremely important influence of shifting fluid in missile tanks on the processes of stabilization and control. The work of NII-1 in 1958 to escape the "resonance dead-end" contributed to the subsequent rapprochement of Korolev and Keldysh. By that time Keldysh was respected not only as a scientist, but also as a very capable science organizer who had that practical grasp sometimes lacking in theoreticians who think only in the abstract.

Examining the proposals for new flying vehicles, Keldysh always considered their feasibility. He already had a wealth of experience collaborating with industry and understood very well that any proposal he made regarding the creation of a fundamentally new cruise or ballistic missile required the participation of dozens of scientific-research institutes, design bureaus, and plants and tremendous organizational work. Keldysh viewed Korolev as a man who would deliver him from the most difficult organizational engineering concerns. He considered his own task to be basic research and the organization of scientific teams that would generate ideas. His were top-notch ideas. When a proposal originated in the form of a report or other document with Keldysh's signature on it, it emerged as a result of strict analysis, thorough calculations, and the most nit-picking deliberations in seminars and scientific-technical councils.

In 1954, together with Korolev and Tikhonravov, Keldysh put forth a proposal for the development of an artificial satellite and participated in the preparation of a memorandum to the government on this subject. The next year he was appointed chairman of a USSR Academy of Sciences special commission on artificial satellites. Later, Keldysh became chairman of expert commissions in all the space projects that required highly qualified evaluation.

After the launch of the first artificial satellite, Keldysh became an indispensable participant in the Council of Chiefs, although not all the issues discussed in the council required his involvement. More than once I had the occasion to observe during protracted meetings how Keldysh closed his eyes and withdrew into himself. Everyone figured that Keldysh had dozed off, but few knew his amazing ability in this somnolent state to take the necessary information into his consciousness. To everyone's surprise, he would suddenly toss out a retort or ask a question that "hit the nail on the head." It turned out that Keldysh had caught all the interesting information and his interjection assisted in making the best decision.

Right after the launch of the first artificial satellites, at Keldysh's initiative our work turned to tracking the flights of spacecraft and predicting their orbits. At OPM they established a small but very capable group, which for the first time developed a computerized procedure for determining orbits. Group members included Okhotsimskiy and Eneyev (who would later become Russian Academy of Sciences academicians), Beletskiy, Yegorov, Lidov, and others. The ballistics computing center that was soon created on the basis of this work collaborated closely with the Ministry of Defense NII-4 spaceflight operations facility and with the ballistics experts of our OKB-1 and NII-88. Later this cooperation evolved into a system of Soviet spaceflight operations centers receiving general information from the ground-based Command and Measurement Complex managed by the Ministry of Defense. A coalition of these centers under Keldysh's scientific and procedural leadership participated in all the ballistics design operations and in operations for the ballistics and navigational support of lunar and planetary exploration. Okhotsimskiy at OPM, Elyasberg and Tyulin at the NII-4 computer center, and Lavrov and Appazov at OKB-1 developed methods and programs to determine the optimal launch dates, total control errors, and optimal conditions for correcting flight trajectory via radio transmissions to the spacecraft.

Keldysh's staff bore as much responsibility for the computation results related to orbital correction and prediction of spacecraft trajectories as their colleagues did at NII-4 and OKB-1. In this case, collective responsibility did not lead to irresponsibility. The ballistics experts always covered for each other.

With Keldysh's consent and support, in 1954, future Academician Rauschenbach assembled a small group at NII-1 that began to develop a satellite stabilization and attitude control system. Two of the first to join this team were MVTU graduate Viktor Legostayev and member of the first graduating class of the Moscow Physics and Technical Institute (MFTI) Yevgeniy Tokar. In 1956, Keldysh approved the first basic report by Rauschenbach and Tokar entitled "An Active Stabilization System for an Artificial Satellite of the Earth."[18] In this treatise, the authors conceptualized quite specific equipment, analyzed the difficulties of accomplishing the objective, and presented proposals that would subsequently form the basis for spacecraft control systems design and that still have not lost their relevance to this day.

Our designers Maksimov and Ryazanov, who worked for Tikhonravov, inherited the concepts presented in this report. Tikhonravov reported to Korolev, and they both decided to support this initiative; for the time being, however, they did not involve me, my powerful design team, or the instrument production facility at my disposal. Nor did they involve our colleagues Pilyugin and Kuznetsov, who had at their disposal engineering capabilities for realizing any new ideas in metal and electronics on a wholly different scale than NII-1.

Perhaps they did the right thing. Small independent groups or small laboratories not burdened by ties to the cumbersome structures of production giants and a multitude of day-to-day headaches, not watched over from above with constant control of deadlines, schedules, and all manner of indices of socialist competition, were sometimes capable of creating technical innovations within fantastically short periods of time. In so doing, they were able to implement ideas that would have been rejected in a large firm based on the principle "we can't do this because it can never be done." In the best case they would say: "We can do this, but to do this we need a government decree to build a special building, to obtain the right to increase manpower, to install 30 more telephones with access to the Moscow automatic telephone exchange, to obtain five more service vehicles and a certificate of domicile quota for at least a hundred persons in Moscow and Leningrad."

We called such a list "the standard gentleman's assortment," which in various versions usually accompanied the Central Committee and Council of Ministers draft decrees for the production of new models of military technology as an attachment. Omnipotent clerks in the upper echelons of power thoroughly edited the government's draft decrees. Part of their task was to issue the text of the decree in a form so that all the projects were concisely entered according to deadlines and specific administrators with a minimum number of attachments granting material support, which we called "hay" (*seno-soloma*). When the next decree came out, the administrators were primarily interested in what remained of the "hay." Bitter disappointment set in when they realized that the work had been allotted and assigned, but the "hay" had been thrown out. It was impossible to search out those directly responsible for editing out the "hay" from the text of the decree. The powers-that-be knew how to keep their corporate secrets.

---

18. This report that Chertok cites has been published as M. V. Keldysh, B. V. Rauschenbach, and Ye. N. Tokar, "Ob aktivnoy sistemye stabilizatsii iskusstvennogo sputnika zemli" ["An Active Stabilization System for an Artificial Satellite of the Earth"] in V. S. Avduyevskiy and T. M. Eneyev, eds., *M. V. Keldysh: izbrannyye trudy: raketnaya tekhnika i kosmonavtika* [*M. V. Keldysh: Selected Works: Rocket Technology and Cosmonautics*] (Moscow: Nauka, 1988), pp. 198–234.

RAUSCHENBACH, LEGOSTAYEV, AND TOKAR GRADUALLY INCREASED THE RANKS OF THEIR TASK FORCE, SCREENING PERSONNEL THOROUGHLY. Tokar, who would later become a professor and prominent authority in the field of mechanics and gyroscopic systems theory, acted as personnel officer. He selected staff according to the strict principle that he needed smart, enterprising people, not obedient ones. Thus, Vladimir Branets, Dmitriy Knyazev, Boris Skotnikov, Anatoliy Patsiora, Yevgeniy Bashkin, Igor Shmyglevskiy, Ernest Gaushus, Vadim Nikolayev, Larisa Komarova, Aleksey Yeliseyev, Vladimir Semyachkin, and many others ended up in the task force and later in Rauschenbach's department.

The team that had assembled at NII-1 with Keldysh as its patron, did not know what insurmountable design, production, and organizational difficulties it would need to overcome in order to create a reliable flying vehicle control system; to do this, the team would have to use the academic works of classic automatic control theory and the experience of missile guidance systems that had actually been developed. Simply and unpretentiously, they proposed and developed spacecraft attitude control systems proceeding from the basic laws of mechanics, electrical engineering, and optics. In those days, the developers of control systems loved to boast about the extraordinary complexity of their instruments and the very difficult engineering processes and to show off their wealth of laboratory equipment, never missing the opportunity to reiterate how insufficient it was for new challenges!

At first, what Rauschenbach's task force had proposed required meticulous theoretical study and painstaking calculations. But, for all of this, in the end the proposal looked extraordinarily simple. However, it took Keldysh's initiative and Korolev's will for all of this to be implemented rapidly and at the proper technical level. In the following example I would like to show how remarkably they complemented one another.

In January 1958, Keldysh personally sent Korolev a letter stamped "secret," in which he wrote that the successful launch of two artificial satellites would enable them to move on to solving the problem of sending a rocket to the Moon. This letter proposed just two scenarios:

1. Hitting the Moon's visible surface. When the spacecraft reaches the Moon's surface an explosion takes place that can be observed from Earth. One or more launches can be conducted without an explosion, using telemetry equipment to record the rocket's movement toward the Moon and to confirm that it hit.

2. A lunar fly-by, photographing its dark side and transmitting images to Earth. It is proposed that images be transmitted to the Earth via television when the rocket approaches Earth. Returning observation materials to Earth is a more difficult task. Its solution cannot yet be worked out.

Accomplishing the aforementioned tasks requires overcoming a number of serious technical difficulties.[19]

This passage was followed by a detailed list of tasks that had to be solved to overcome these difficulties. In conclusion Keldysh wrote that, "Working very strenu-

---

19. The original was published as "O zapuske rakety na lunu" ["On the Launch of a Rocket to the Moon"], pp. 241–243.

ously and with constant help on all fronts, the development, design, and construction of a lunar rocket could be completed within the next two to three years."

Corroborated by fundamental theoretical research, Keldysh's intuition abruptly accelerated the practical implementation of new ideas thanks to Korolev's enthusiasm. The timeframe outlined in Keldysh's letter didn't frighten Korolev. The first test launches attempting a direct hit on the Moon's visible surface began in 1958. In September 1959 a direct hit took place, and in October we obtained photographs of the far side of the Moon.

Scrupulous historians can argue who is more prominent in the development of the first lunar programs. To me, such research is purely academic in nature. Beyond Keldysh and Korolev, many dozens more scientists and engineers collaborated closely with one another, ardently deliberated all conceivable options, and exchanged ideas unselfishly without giving a thought to their future fame. Therefore, the prominence of an idea in this case cannot be prescribed to a single individual, not even to the great Korolev or Keldysh.

AND SO—RETURNING TO THE BEGINNING OF THIS CHAPTER—at Keldysh's invitation, Korolev and I drove from Podlipki to Likhobory. As we were driving along in Korolev's ZIM automobile, I found myself musing and reminiscing about working at NII-1. The previous time I was at that institute was to process papers for my transfer to NII-88 more than 10 years before, after returning from Germany. Korolev hadn't even been there since 1938, for over 20 years! What feelings were coming over him now that we were about to enter the building that for him was associated with those years of frustrated hopes and life's tragedies? Usually, in the car, Korolev didn't waste time and when he was traveling with one of his deputies; he discussed current issues or asked him to liven things up with some funny story. This time he sat next to the driver, lost in his own thoughts, without turning around.

At that time there was not yet an overpass over the complex tangle of railroad lines near the Severyanin platform, and we were delayed for a long time at the crossing gate. This wasn't the first time I'd ridden with Korolev, and during long waits at this crossing, he always used to express his indignation in colorful terms when they announced over the railroad loudspeaker, "Train on the belt line." After this message came the next one, "Train bound for Moscow," and then again, "Train on the belt line." It was hard to maintain one's composure and not glance at one's watch. This time Korolev was silent and pretended to snooze.

Only when we approached NII-1 did he snap to and direct our attention to the well-preserved inscription on the façade of the main building, which announced, "All-Union Institute of Agricultural Machine Building." "Look, this masquerade is still going on," he said. "They took this building away from agriculture long ago but left the sign. And now, evidently, they won't allow Keldysh to remove it."

Keldysh met our group very cordially and immediately led us to Rauschenbach's laboratory. Here, laid out on simple tables, were functioning mock-ups of the attitude control system for the automatic unit, which, according to the designers' conception, was supposed to point the cameras and television equipment at the far side of the Moon.

Rauschenbach told us about these principles. Bashkin and Knyazev, two engineers who already had production experience, demonstrated the operation of solar and lunar orientation sensors using simulators. The dramatic actuation of the "whooshing" pneumatic gas-reaction nozzles was supposed to impress the guests. Knyazev and his assistants fidgeted with the high-pressure tanks, opening and closing something. Compressed air was whistling out of a leaky connection somewhere—the inevitable "visit effect" had kicked in. But on the whole the demonstration went well.

Keldysh was very pleased. Korolev said, "The system needs to be refined. I am ready to help with production. But hurry. We need to receive everything and get set up at our facility this year. If you need assistance, Chertok and Bushuyev here are the guys to turn to. If they can't help, call me personally." He didn't praise, but made demands and assigned tasks; such behavior had a mobilizing effect, and people understood that everything was ready and now it was all up to them.

On the trip back Korolev was very animated. "I like these guys. If we help them, they'll do it. We need to pick them up [for our own design bureau]. But, Boris, I can't trust them to you[r management]. You'll probably leak it to your friend Pilyugin, and the two of you will start to prove that nothing will come of these craftsmen. I can't hand them over to Pilyugin either. They'll smother them there or switch to other things. But if we take them on, then we can let them be with Kostya [Bushuyev] to begin with. He doesn't understand instruments and won't bother them. But you, Boris, are going to support them with your KB, electricians, production, and experience. After all, they are still quite green." I was about to protest, but Bushuyev gave me a shove and said, "Sergey Pavlovich, Chertok and I amicably agree. But in order to transfer them, we need to look into how many apartments will be needed in Podlipki. If they don't get living quarters, then eventually they'll run off or they simply won't come here."

In the end, our visit to NII-1 had far-reaching consequences; it influenced Rauschenbach's fate and that of his team. In early 1960, a special government decree transferred Rauschenbach's entire group from NII-1 to OKB-1. Many were provided with living quarters, despite the obvious displeasure of the local union authorities, who had a waiting list of more than 1,000 people in need of housing.

THE OKB-1 ORGANIZATION HAD A WEALTH OF ENGINEERS, among them vibrant personalities. It gave me great pleasure to associate with these people. Working with this group was difficult precisely because they were not docile. They worked furiously, passionately, and selflessly.

In the ensuing years I had a great deal of contact with all of them in complex situations working nonstop on new problems, during days spent investigating serious failures, and during hours of triumph. They not only knew how to work but also how to have fun in skits, to publish hilarious newsletters, and to surround themselves with a feeling of good humor at the right place and the right time. The transfer of Rauschenbach's group was one of the events that to a great extent determined the future success of Soviet cosmonautics. The other was the merger of Vasiliy Grabin's organization with OKB-1.

Chapter 27

The Great Merger

In March 1959, after assembling his closest deputies, Korolev informed us of Ustinov's proposal to annex the neighboring TsNII-58 to OKB-1. Territorially, railroad tracks were all that separated us. Ustinov gave us just three days to mull it over. Ustinov's proposal put an end to all of Korolev's complaints to the government and ministry about the need to significantly strengthen the production base and increase the staffing of engineering and design units at our OKB-1. I recall that at this time Dmitriy Ustinov was not simply a minister, but a Council of Ministers deputy chairman as well as chairman of the Commission on Military-Industrial Issues under the USSR Council of Ministers. S.P. went on to discuss this unexpected and very attractive proposal; but first, having thoroughly prepared himself, he read the memorandum aloud, accompanying its dry text with his own comments.

The Central Artillery Design Bureau (TsAKB) was created in Podlipki in 1942.[1] At the time of its formation, Vasiliy Gavrilovich Grabin was the KB head and its chief designer. In 1945, TsAKB attained scientific-research institute status, after which it was called the Central Scientific-Research Institute for Artillery Armaments (TsNIIAV).[2] After the Ministry of Armaments was transformed into the Ministry of Defense Industry, TsNIIAV was renamed NII-58, and beginning in 1956, it was called Central Scientific-Research Institute-58 (TsNII-58).[3] The personnel of the chief designer's department at Factory No. 92 (or *Novoye sormovo*) in Gorky formed the primary creative nucleus of TsAKB. For a long time the plant director was Amo Sergeyevich Yelyan.

At our meeting, Korolev turned to Turkov: "Roman Anisimovich, you must know Yelyan well.[4] Is he the one who was director of KB-1 by the Sokol metro station?"

---

1. TsAKB—*Tsentralnoye artilleriyskoye konstruktorskoye byuro.*

2. TsNIIAV—*Tsentralniy nauchno-issledovatelskiy institut artilleriyskogo vooruzheniya.*

3. In 1953, the Ministry of Armaments (which oversaw the main rocketry institute, NII-88) was renamed the Ministry of Defense Industry.

4. Roman Anisimovich Turkov was the director of the experimental factory attached to Korolev's OKB-1.

"He's the one," answered Turkov. "During the war he and Grabin started a revolution in artillery production technology in Gorky. Stalin decorated them for good reason. Grabin's famous 76-mm cannons helped defeat the Germans outside Moscow; they were designed on a fast track, preparing for production at the same time. During the war there were more Grabin cannons at the front than any other model, and it was primarily Yelyan's plant that sent them to the front."

When Turkov digressed from the current cares of rocket production and reminisced about the heroic days of the wartime artillery factories, he smiled warmly. He could go on and on about extraordinary episodes during the production of cannons in wartime, underscoring that "Yes, in our time there were people … *Bogatyri*, not like you.[5]"

We had great respect for Turkov. At our plant he enjoyed quite well-deserved prestige among the workers and managers. Only various swindlers, schemers, and slackers disliked him because of his honesty, candor, and integrity. Korolev did not make any decisions concerning the plant without Turkov's approval; each of Korolev's deputies strove to work in close contact with Turkov. From wartime artillery production he brought his experience of working in a single creative surge that went from project conception to design to process development to production to testing.

We all went crawling to Roman Anisimovich when, after discovering a design error, we had to make modifications or even stop manufacturing "articles." In such cases, Turkov, studying the causes and need for the changes in great depth, sought a compromise with the developers and shop managers that would enable the changes to be made with a minimum slippage of the deadlines. The very process of seeking a solution in what would seem a dead-end situation in production gave him pleasure. Once he confessed, "If designers don't suddenly make changes at the very last moment, it means they've overlooked something. This always makes me suspicious."

Korolev continued reading, and we found out that, in addition to awards, Grabin achieved high military ranks for developing artillery weapons systems in the prewar years and during the war. At that time, he was a Colonel-General.[6] Before World War II he was awarded the title Hero of Socialist Labor; through his life, he received the Stalin prize four times and the Order of Lenin four times, as well as many other awards. Grabin's organization was also awarded the Order of Lenin. In late 1948, the total work force at TsNII-58 exceeded 5,000, of which more than 1,500 were engineers. Among the engineers and workers, many were awarded orders and medals, and Grabin's closest associates also received Stalin prizes. "And they received

---

5. A *Bogatyr* is a legendary figure in Russian folklore comparable to fairy-tale knights or the mythical American Paul Bunyan.

6. This rank is roughly equivalent to a U.S. four-star general, that is, above a "Lieutenant-General" and on level with a full "General" in the U.S. Army.

them for good reason," commented Korolev.

Grabin's design bureau designed 13 types of division, tank, and antitank ordnance at Factory No. 92 and then in Podlipki at TsAKB. During the war years, the ZIS-3 division gun was the most famous and largest. The fire power of the legendary T-34 tank came from Grabin's tank guns. While arming our tanks with guns, Grabin also developed antitank guns. Antitank artillery was armed with 57- and 100-mm guns, the latter of which artillerymen called *zveroboy* (hunter). This gun pierced the armor of the heavy German Tiger and Panther tanks and the Ferdinand self-propelled gun.[7] Beyond successful weapons designs, the introduction and use of so-called "rational technology" into mass or "gross" output also contributed to the successes of Grabin's organization.

"I must admit," said Turkov, "that Grabin's revolutionary proposals were sometimes opposed by plant directors and higher-ranking managers."

During the postwar years Grabin worked on automatic antiaircraft guns. In 1953, the 76-mm gun with a rate of fire of 100 rounds per minute was put into service in the air defense forces.

"Imagine," interrupted Korolev, "One hundred rounds a minute, and that caliber to boot! When something went wrong with the production of this gun at the Krasnoyarsk Factory, Stalin ordered the arrest of Marshal Yakovlev and GAU Chief Volkotrubenko. Thank God, they are now free."[8]

Turkov again interrupted and cautiously pointed out that Yakovlev, Volkotrubenko, and a number of other prominent managers were arrested on charges of sabotage, specifically the massive failure of automatic antiaircraft guns of Grabin's design during the Korean War. But Stalin didn't touch Grabin and Ustinov.

According to Turkov, Grabin was unquestionably a very talented designer and at the same time a splendid production engineer. He was also a very commanding, tough, strong-willed leader. His knowledge of production was brilliant. Back before the war, Grabin was the first to propose a fast-track method of design, technological support, and production. He boldly made what were sometimes very risky decisions. Even before the war, Stalin considered Grabin the highest authority on artillery technology. But while Stalin was alive, Grabin showed a blatant disregard for Ustinov. Referring to numerous friends and acquaintances from his days working in artillery production, Turkov confirmed that Ustinov would not forgive such an attitude toward him.

Both Grabin and Ustinov were obliged to Stalin for their high military ranks.

---

7. The Tiger and Panther tanks were two of the best-known and most effective tanks used by the Germans during World War II. The Ferdinand was a huge armored and mobile self-propelled gun that was outwardly similar to the Tiger.

8. Artillery Marshal Nikolay Dmitriyevich Yakovlev (deputy minister of defense) and Colonel-General Ivan Ivanovich Volkotrubenko (chief of the Main Artillery Directorate) were arrested in December 1951 on trumped up charges of obstructing the production of armaments. They were released in April 1953 by Lavrentiy Beriya after Stalin's death.

Stalin first noticed Grabin in 1935 during an inspection of artillery ordnance and since then rendered him effective patronage, was very receptive toward Grabin's new proposals, and appreciated what were for those times revolutionary moves to restructure the ordnance design and production process. During the prewar and war years, Grabin, who held the chief designer post at an artillery plant, was Stalin's "unofficial" consultant. At Stalin's initiative, Grabin was brought in to develop designs for the selection, acceptance, and production startup of field artillery ordnance and tank guns.

As for Ustinov, Stalin first became acquainted with him right before the war, having appointed him to the high post of commissar of armaments at Andrey Zhdanov's recommendation to replace the disgraced Boris Lvovich Vannikov, who was arrested at the behest of the same Zhdanov.[9] Ustinov was asked to set up the mass production of armaments, probably the most difficult task for the first years of the war. The wartime difficulties increased when the Moscow and Leningrad factories were evacuated to the east. Ustinov, who was in charge of all artillery plants, was obliged to carry out Stalin's decisions. There were no discussions; he staked his life on the quantity and dates of delivery to the army of all types of artillery armament.

During the war, factories of the Commissariat of Armaments headed by Ustinov produced 490,000 artillery pieces of all calibers and mortars, of which 188,000 were artillery pieces. Of this number, more than 100,000 were produced in the city of Gorky, where Grabin was the chief designer and Yelyan was the director. Another 30,000 field artillery pieces were manufactured at other factories using chief designer Grabin's designs and technical documentation. Specialists, including Germans, rated the ZIS-3 division gun as an engineering masterpiece.

THE MEMORANDUM THAT KOROLEV READ TO US, and Turkov's reminiscences were new and interesting information for those gathered there in Korolev's office. However, it was not until much later that I learned the details of the stirring story of the postwar career of the very colorful Grabin. Jumping ahead, I will say that I worked over 30 years in that very office where Grabin worked as chief of TsNII-58, and until 2004, my work station was located in the engineering building, which in 1942 was the first home for Grabin's TsAKB.

Three books were published between 2000 and 2002 that recovered the undeservedly forgotten role of Grabin and his organization during the war.[10] I have sup-

---

9. Boris Lvovich Vannikov (1897–1962), who would go on to an illustrious career as one of the senior managers of the Soviet atomic bomb program, was briefly arrested and incarcerated at the beginning of World War II in 1941–42. He had been minister of armaments at the time of his arrest.

10. V. G. Grabin, *Oruzhiye pobedy*, izd. 2-ye, ispr. [*Weapons of Victory*, 2nd ed.] (Moscow: Respublika, 2000); A. P. Khudyakov, *V. Grabin i mastera pushechnogo dela* [*V. Grabin and Masters of Ordnance*] (Moscow: Patriot, 2000); A. B. Shirokorad, *Geniy sovetskoy artilleriy: triumf i tragediya V. Grabina* [*Genius of Soviet Artillery: The Triumph and Tragedy of V. Grabin*] (Moscow: AST, 2002).

plemented the information that Korolev read from his memo with details from these publications and other sources.

In 1954, NII-58 did not receive production assignments worthy of its capabilities from its own ministry. Before the war, Grabin had developed good relations with the commissar of armaments at that time, Boris Lvovich Vannikov. Now in 1954, Vannikov was first deputy minister of medium machine building, in actuality deputy chief of the nuclear empire. Grabin appealed to Vannikov, requesting an engineering assignment worthy of his organization. A government decree tasked NII-58 with designing and manufacturing a fast neutron reactor with liquid-metal coolant and a 5,000-kW output for the Physics and Technical Institute in the city of Obninsk.[11] USSR Academy of Sciences Corresponding Member Aleksandr Ilich Leypunskiy was appointed scientific consultant. All the work for the reactor was completed on time. Moreover, one of Grabin's cannon designers was among those awarded a Lenin prize for the nuclear reactor!

After such successes in the nuclear field, Vannikov proposed that Grabin transfer from the Ministry of the Defense Industry to the Ministry of Medium Machine Building. However, when the government decision on this matter was issued, Council of Ministers Deputy Chairman Vyacheslav Aleksandrovich Malyshev, who was also minister of medium machine building, appointed Academician Anatoliy Petrovich Aleksandrov, who was Kurchatov's deputy at the Institute of Atomic Energy, to be the new NII-58 director. Aleksandrov occupied Grabin's office, while his (Grabin's) deputy Renne made room for the colonel-general. It is now difficult to say who actually initiated such a blow to the great designer's ego. One can only assume that Malyshev did this with Ustinov's approval.

Once again Korolev digressed from his notes and commented:

"All the same, good for Kurchatov. He got such resources from the artillerymen! And after all, they're making those reactors and even sent them to Egypt, Hungary, and who know where else. Look, Kostya," S.P. turned to Bushuyev, "If this incredible proposal goes through, you'll be making spacecraft there instead of fast-neutron reactors and all sorts of ordnance!"

The all-knowing Turkov again supplemented Korolev's words with details that embellished the dry memorandum. After occupying Grabin's office, Aleksandrov started the very rigorous process of restructuring NII-58 from an artillery enterprise to a nuclear technology research and development facility. He hired many new specialists in nuclear physics, measurement technology, and automation. He recruited dozens of graduates from the Moscow Engineering and Physics Institute (MIFI), the

---

11. The first Soviet nuclear power plant to provide energy (mainly electricity) for civilian purposes was opened in the town of Obninsk, about 100 kilometers southwest of Moscow, in June 1954. One of the largest scientific-research networks dedicated to the study of atomic energy is now located at Obninsk.

primary facility for training specialists in the field of nuclear technology.[12] During Aleksandrov's administration, NII-58 was restructured for the series production of nuclear reactors at a rate typical of wartime ordnance production. The old Grabin staff together with the newly arrived young nuclear specialists developed the first monitoring and automatic control system for the new reactors. Integrating nuclear science specialists' experience with artillery technology proved very fruitful.

I heard about the events involving Grabin's struggle to regain his directorship much later. Back in 1959, it was known that the collective appeals of many distinguished artillerymen to the Party Central Committee, and to Khrushchev personally, resulted in a new decree. In early 1956, NII-58 was transferred from the Ministry of Medium Machine Building back to the Ministry of Defense Industry. At the same time, the institute was renamed, not simply as an NII, but Central NII-58 (or TsNII-58). Grabin was appointed director and chief designer of TsNII-58 and Aleksandrov, his deputy. By this time Aleksandrov did not want to be a deputy to the imperious Grabin, and he returned to his old Atomic Energy Institute to work for Kurchatov. TsNII-58 continued to design both ordnance and new nuclear reactors.

Ustinov hadn't forgiven Grabin for snubbing him in days gone by. And despite regaining his post, Grabin didn't change his attitude toward Ustinov either. "I heard," Turkov related, "that when Ustinov paid an unannounced visit to TsNII-58 and went straight to the production site, Grabin didn't meet with him. He stayed in his office despite the fact that the minister of armaments was familiarizing himself with production."

This behavior was in stark contrast to the deferential receptions that we had held for Ustinov right next door on the other side of the railroad tracks at NII-88 since Gonor's time and now under Korolev. Ustinov's relationship with Korolev was also far from smooth. Under Stalin, Korolev never contradicted Ustinov. Now, under Khrushchev, Korolev's prestige had increased immeasurably after the successes in space. Khrushchev also often turned directly to him as Stalin had turned earlier to Grabin.

But Korolev was much more cautious. He always reported all that was necessary to Ustinov and would ask, if only pro forma, for his advice. Ustinov reckoned that if he was going to support anybody, then better that it be Korolev than Chelomey, the new rising star in the missile and space industry. Chelomey enjoyed Khrushchev's support and, like Grabin, also blatantly refused to acknowledge Ustinov's authority.

In 1959, Ustinov was presented with a very convenient opportunity to kill two birds with one stone. He could finally settle the score for all of Grabin's insults, showing him once and for all "who was who," *and* satisfy Korolev's urgent and justified demands for the expansion of his design and production facilities. In other words,

---

12. MIFI—*Moskovskiy inzhenerno-fizicheskiy institut.*

he would terminate all contracts to Grabin's organization and put it at Korolev's disposal. Khrushchev, who was very keen on developing missile weapons at the cost of conventional artillery and aircraft, would certainly support this proposal. He promised to help Korolev and instructed Ustinov to prepare a proposal to this effect. Ustinov did not like to procrastinate. There were also other alternative ideas on the fate of TsNII-58 and of Grabin, so time was of essence. Therefore, he gave Korolev just three days to mull it over.

AND THAT'S HOW THE SITUATION STOOD WHEN KOROLEV WAS READING HIS MEMO TO US AT THE MEETING IN MARCH 1959. "What are we going to do?" asked Korolev. The proposal was not a surprise. There had been talk of merging the enterprises before. With no trouble at all, we would immediately get specialists with ready workstations and workers with machine tools and a large, well-tuned operation with all the auxiliary services.

Grabin's production facilities were equipped with unique, state-of-the-art machine tools. Grabin was also a lot better off than our factory in terms of the scarcest and most sought after professional machine operators. He knew each skilled worker personally. When Grabin visited the main production shops and meetings and spoke with foremen and workers right at their machine tools stations, he was not condescendingly showing his democratic nature, but rather engaging in a custom from the war years that he considered a vital requirement for efficient work. Still young, resourceful, and healthy at the time, he proved that it was possible to develop new artillery system designs in three to four months instead of the usual two to three years.

After a pause, the veteran Turkov took the floor once again. He reiterated that he valued Grabin's wartime contribution very highly. Grabin was a distinguished individual and strong organizer. His team loved and respected him and regarded him as more than just a boss. For the artillerymen he was a real "chief designer." If we took on the role of aggressors who had taken advantage of a situation, that is, Ustinov's settling of old scores with Grabin, it would be dishonest and would generate hostility toward us among his staff.

Korolev understood all this very well himself. Everyone agreed with Turkov and decided that in response to Ustinov, S. P. should announce that he was prepared to comply with the decree, but on the condition that, first, under no circumstances would it contain wording such as "accept the proposal of Chief Designer Korolev" or anything along those lines and, second, Grabin's fate should be decided with consideration of all of his merits.

When the meeting was over, after dismissing everyone, S.P. asked Bushuyev and me to stay. "So here's the deal, my dear boys." This form of address indicated he was in a good mood and was feeling very confident. "I hardly know Grabin at all. I've just met him at municipal conferences a couple of times. I simply feel sorry for him as a person. To lose such a job and such a team after so many years! After all, around here they have a knack for forgetting a person right away and trampling him. I know from my own experience. I don't need to explain this to you. They've probably already told Grabin that Korolev wants to take away everything and bar him from entering the premises. Uncle Mitya [Ustinov] will be seen as innocent while I will be the bad guy who took advantage of Nikita Sergeyevich's good will. I can't meet with Grabin for preliminary explanations. I am entrusting you two to do that. Take your time. Think up some pretext for going to him and talking about possibilities for joint work on spacecraft. Explain that we don't have enough manpower and we are prepared to hand over this project or even the entire spacecraft to him lock, stock, and barrel for development and production. Instead of artillery and nuclear reactors!"

After receiving this assignment from Korolev, Bushuyev and I decided first to conduct deep reconnaissance on the whole situation at TsNII-58, and then ask for a meeting with Grabin. But events interrupted our unhurried preparation for this complicated diplomatic mission. In early May Bushuyev and I received a message through Lelyanov, Korolev's information officer, a former KGB employee, that Grabin had invited us to see him at 11 o'clock the following day. We were told that we wouldn't have to go to the pass office since we'd be on a list for admission.[13]

A contact was already waiting for us in the entryway and immediately led us to a spacious office. Grabin was sitting in his full general's uniform behind a large desk topped with green felt. We introduced ourselves. We were somewhat taken aback that Grabin did not stand up and did not shake hands. True, it was difficult to do that over a broad desk. Nodding toward the heavy, uncomfortable chairs, he gestured for us to take a seat. As we had arranged beforehand, Bushuyev began to speak about new automatically controlled spacecraft designed for a flight to Mars and suggested that Vasiliy Gavrilovich have a look at the project. He asked whether it might make sense to manufacture it here at the pilot plant.

In official portraits, artists had given Grabin a stately bearing. The heavy features of his face expressed pride, haughtiness, and authoritativeness, a true god of war in full regalia. But the face of the man who sat before us was completely different from the portrait on display. He sat in silence and looked first at Bushuyev, then at me, perplexed. Why all this talk? His massive head tended to sink into his shoulders as if retreating from danger. There was an expression of impending doom on his tired face. So many years have passed since then, but even now I recall the mixed feeling of uneasiness and pity that I experienced sitting opposite Grabin.

As Bushuyev spoke, I had time to glance around the spacious office; There was a large conference table, chairs, unpretentious sofas, occasional table by the writing desk, and armchairs devoid of decorative carving, everything was made of light Karelian birch. On the wall above Grabin hung a portrait of Stalin in a gilded frame. When we were preparing for our meeting, someone from Korolev's office—I believe

---

13. In Soviet (and Russian) times, all visitors to industrial enterprises require a pass *(propusk)* to enter the premises; passes were waived for special visitors who would be "on the list" *(po spisku).*

it was that same all-knowing Lelyanov—said to us: "Take note of the furniture in Grabin's office. It was produced in government furniture workshops that were housed in the Butyrskaya prison on Stalin's personal instructions."

The walls of Grabin's office were painted from top to bottom with a profusion of climbing plants, their stems abundant with leaves and huge pale lilac flowers. We scrutinized this mural later. The artist depicted some sort of hybrid of liana, lotus, lilacs, and magnolias. Presumably the artist intended for our host and all who visited his office to feel like they were in a garden. The plaster molding that decorated the ceiling around its entire perimeter and the elegant light fixtures with bronze hanging chandeliers were also unusual. Pilasters with gilded scrolled capitals supported the ceiling.

The architectural and artistic style of the office contrasted with the visage of its owner. He was not the least bit interested in Bushuyev's speech, and for him, our very visit was the result of someone's arm-twisting. Most likely it was a phone call from the Central Committee office. He already knew that there "at the top" Ustinov had arranged everything and it wouldn't be long before a Central Committee and Council of Ministers decree would appear that would be the kiss of death for his career. They would euphemistically propose some post in the Ministry of Defense, in the so-called "heavenly group." Such a group had been instituted for marshals and high-ranking generals who went into retirement due to age or who had fallen out of favor with the Communist Party leadership.[14] Now he would have to say farewell to the team he had gone through the war with, that he had done so much for, say farewell to the design halls with the drafting tables on which drawings of the new assemblies were tacked, to the production bays and their inimitable machine smell, the humming machine tools, the foreman and shop chief rushing to meet him...

Rudnev, minister and chairman of the State Committee of Defense Technology, signed the order for merging TsNII-58 with OKB-1 in June 1959. Grabin convened the managerial staff and leading specialists in the "red hall" and appealed to them with a testimonial speech.

"I believe," he said, "that the right decision has been made. The question about our future fate was posed long ago, and now it has been resolved correctly. Your fate is very important to me. I believe that out of all the possible scenarios in this plan, reunification with our neighbor is the best. Don't ever forget that you are Grabinites. We have traveled a path of glory together, and we can face our nation with a clean conscience. I instruct you to work so that our traditions will never under any circumstances be lost."

This was Grabin's last speech. After giving it, he left the premises never to return. I have reproduced the speech from the words of someone who attended this farewell

---

14. This level of position was called the "Inspectorate" of the Ministry of Defense. Officially it was at a senior level but technically, former high-ranking officials in the Inspectorate had little or no authority.

meeting.

Korolev and Turkov showed the highest degree of scrupulousness in determining the fate of each TsNII-58 employee. Korolev announced that he was prepared to talk personally with each KB and laboratory employee and Turkov would talk with any production worker. Grabin, meanwhile, received an appointment to a consultative group in the Ministry of Defense. This did not keep him very busy. He started to work as a professor in a department at MVTU and to teach a course on artillery ordnance. But he wasn't content doing this, either. Grabin set up a new OKB at MVTU and became its chief designer. He rode a commuter train and municipal public transportation from Podlipki to MVTU and back as long as his health allowed. He brought his work in the development of artillery science commendably to a close by transferring his priceless experience to a new generation.

Many old career artillerymen who didn't wish to change their specialty left TsNII-58 for other defense industry enterprises. But the main TsNII-58 staff and all the young staffers stayed. Together we began the process of reorganizing, expanding our old departments, creating new ones, and selecting managers with the mutual agreement of both sides—"ours," that is, Korolev's staff, and "theirs," that is, Grabin's staff. By mid-1960, the restructuring process was largely completed. Just two weeks after the minister's order, many of the Grabinite specialists had joined in on what was for all of us a new project: the development of solid-propellant rockets.

ACCORDING TO KOROLEV'S CONCEPTION, the TsNII-58 premises were supposed to become an OKB-1 branch. Korolev initially entrusted the duties of deputy chief designer for all space-related projects to Konstantin Bushuyev and moved him into Grabin's office. Bushuyev received not only Grabin's office, but also his Kremlin "hotline" telephone. Consequently, Grabin's name disappeared from the Kremlin automatic telephone system phonebook and Bushuyev's appeared. Bushuyev kept the historic office furniture intact, but he was forced to remove the large portrait of Stalin that hung behind the desk where Grabin formerly presided. At Korolev's insistence some back rooms were remodeled; the personal shower and toilet were remodeled as work spaces for private conferences. The walls of the main office were repainted, covering the renderings of creeping subtropical foliage. TsNII-58 was renamed OKB-1's "second territory" or Branch No. 1. Grabin's personal service vehicle, a ZIS-110, was transferred not to Bushuyev, but to Korolev. Sixty-year-old Colonel General Grabin, having enjoyed 25 years of government-owned cars with personal drivers, had neither his own car nor a driver's license.

For me—Korolev's deputy for guidance systems and, as he sometimes liked to joke—his "rusty electrician," the merger of OKB-1 with TsNII-58 had much greater significance than for Korolev's other deputies. Two events in the years 1959 and 1960, the merger of OKB-1 with TsNII-58 and the transfer of Rauschenbach's team from NII-1 to OKB-1, led to the creation of our country's first, and perhaps the world's first, scientific and technical school for spacecraft guidance systems. The organizational restructuring for my field lasted three years, but I will write about this in greater detail in the next chapter.

Over the course of two years at the new territory, Bushuyev, Tikhonravov, and Tsybin—who moved with him—really organized the space branch of OKB-1. Somewhere in the "upper managerial circles" there was a rumor going around about converting what was now no longer Grabin's, but Bushuyev's space-related territory into an independent "P.O. Box." But one way or another, Korolev decided once and for all to prevent the possible spinoff of the space branch from the main OKB-1 facility into a completely independent organization out of his control.

In late 1962, Korolev carried out another reorganization. He brought back Bushuyev and part of the space designers into his building 65 at the home base; he then sent me to Grabin's former territory with the assignment to convert it mainly into a branch for the development of guidance systems and to set up a worthy instrument production facility. As a result, I settled into Grabin's office as deputy chief of science operations and chief of Branch No. 1 with the added initials "D.T.N.," that is, Doctor of Technical Sciences.[15] One of my first instructions to the local administration was to categorically forbid them to replace the historic furniture in the office, despite their attempts to oblige the new director with something more contemporary. After Bushuyev, I worked in this historic office until 1997, that is, for just under 30 years. I only permitted the furniture to be repaired; there were no replacements.

In 1997, Oleg Igorevich Babkov became the proprietor of the office and in 2002, Vladimir Nikolayevich Branets. They also remained true to the tradition of preserving Grabin's furniture, the very heavy multiple-light chandelier and the wall sconces. But in 2004, the builders presented an ultimatum. The four-story building, which since 1942 had housed the office of Colonel General Grabin, followed by future president of the USSR Academy of Sciences A. P. Aleksandrov, then once again Grabin, next Bushuyev, followed by 30 years of Chertok, then Babkov, and finally Branets, was to undergo a major overhaul. Apparently, it was dangerous to stay any longer in the building where six Heroes of Socialist Labor, more than 20 Lenin and Stalin and then State prize laureates, and hundreds of recipients of many medals had worked! No one knows what will happen now to the office that, as far as I can see, should be converted into a memorial to chief designer Grabin.

As OKB-1 formed its own academic councils, candidates regularly defended their dissertations and diploma projects in Grabin's old office. As chairman of the academic council or chairman of the State examination commission, I congratulated each individual who successfully defended their dissertation or diploma project, conferring on them the appropriate academic degree or title of engineer.

In June 2004, in this same office, as chairman of the State examination commission, I presided over proceedings during which students of the Moscow Physics

---

15. DTN—*Doktor tekhnicheskikh nauk*—is roughly equivalent (and typically higher) than the Western notion of a Ph.D.

and Technical Institute defended projects to obtain bachelor's degrees. Listening to the students, I couldn't help but think that just 40 years ago, less than the lifetime of one scientific generation, the men who worked in this office could not have imagined the problems that these 21-year-old kids were now discussing. They all received grades of "excellent." Congratulating the new degree holders, I said that they could be justifiably proud of the fact that after them no one could boast that they had defended their projects in the office of the legendary creator of the "Victory Weapon."

THE MERGER OF KOROLEV'S OKB-1 AND GRABIN'S TsNII-58, the first "great merger" of the space era, made it possible to expand a common front of projects in the field of space technology. In particular, the union expedited programs to develop spy satellites and the first crewed space ships, the implementation of designs that until then had seemed to be in the distant future. One other outcome of this historic merger was the development of the first RT-2 (8K98) intercontinental solid-propellant missile. After joining forces, the rocket specialists of OKB-1 and the artillerymen of TsNII-58 put into service the intercontinental solid-propellant missile, which stood on alert for 15 years![16] Of all the artillerymen who did not deserve to be debased and insulted, the most distinguished and greatest was Grabin. Korolev was right when he said that it's very easy for us to trample a man for nothing.

The biggest fan of the development of nuclear missiles in the late 1950s and early 1960s was Premier Nikita Sergeyevich Khrushchev. Without plunging into the strategic research of military theoreticians, Khrushchev proceeded from simple considerations such as the country's inability simultaneously to develop fundamentally new branches of the armed forces and substantially modernize classic ones under the conditions of a bitter Cold War. He chose in favor of (ground-based) nuclear-tipped missiles and submarine-launched missiles at the expense of the surface naval fleet, the air force, and conventional artillery.

The second "great merger" took place with a slight time lag, substantially increasing the design and production potential of the Soviet missile and space industry. Aviation industry OKB-52 in suburban Reutov acquired Chief Designer Myasishchev's large-scale OKB-23 as its Branch No. 1 and the aviation Factory No. 23 in Fili, the premises of which had once been Factory No. 22.[17]

As a result, distinguished Chief Designer Myasishchev lost the design team that

---

16. The RT-2 was the first operational Soviet ICBM that used solid propellants. Developed by OKB-1, its initial version was put into service duty in December 1968. Western agencies referred to the missile as SS-13 (U.S. DoD) or Savage (NATO).

17. Myasishchev's OKB-23 was made Branch No. 1 of Chelomey's OKB-52 in October 1960.

had developed what was at that time the best long-range bomber in the world.[18] Myasishchev's design bureau and the Khrunichev Aviation Factory in Fili were handed over to Chelomey.[19] However, aviation industry minister Petr Dementyev did not leave Myasishchev unemployed. He transferred him to the perfectly honorable post of TsAGI chief. As a result of the two "great mergers," in one year's time at the expense of the artillery and aviation industries, the missile and space industry gained more than 3,000 engineers and a state-of-the-art industrial production base numbering more than 15,000 workers and employees.

Grabin and Myasishchev united and trained excellent cadres, who fulfilled the slogan of the first five-year plans that "cadres that have mastered technology solve everything."[20] We should be grateful to Grabin and Myasishchev not only for what each of them developed in their own fields, but also for the enormous contribution that their people made after merging with the organizations of Korolev and Chelomey.

---

18. Myasishchev developed the M-4 (known by NATO as Bison) jet-propelled strategic bomber in the mid-1950s. An improved version, the 3M (Bison-B) with longer range was introduced in the late 1950s, but neither bomber performed very well in service duty due to high costs and middling performance.

19. The factory was originally known as Factory No. 23 but renamed Factory Named After M. V. Khrunichev (*Zavod imeni M. V. Khrunicheva* or ZIM) in 1961 after the death of Mikhail Vasilyevich Khrunichev (1901–61), one of the senior managers of the Soviet aviation industry in the postwar era.

20. During the early "five-year plans" for economic development in the late 1920s and 1930s, the Soviet Party and government promoted the use of modern technology as a solution to many social and economic problems. Such sayings as "technology for the masses" and "cadres that have mastered technology solve everything" were popular at worker meetings and on inspirational posters of the period.

Chapter 28
First School of Control in Space

The generally accepted date for the beginning of the space age is 4 October 1957, the day the world's first artificial satellite was launched. However, if you examine the technical nature of this event, then 4 October and then 3 November, the date of the second satellite launch, were actually proof that the Soviet Union had developed the science and technology of rocketry. *Space* technology is substantially different from rocket technology. In 1957 and up until 1959, space technology did not exist in the Soviet Union or in the U.S.

We had moved ahead of the Americans and taken the leading position in cosmonautics beginning in 1957 because, relying on our rocket technology and with the effective support of the nation's top political leadership and governmental institutions, we had rapidly organized a broad front of operations to develop our own space technology. During those first years there was still no store of knowledge that would enable us to formulate a concise set of requirements as we had produced for rockets, airplanes, and cannons.

The merger of the rocket specialists of OKB-1 and the artillerymen of TsNII-58, and the transfer a year and a half later of Rauschenbach's team from NII-1 to OKB-1 created conditions conducive to developing independent space technology. I dare say that Korolev was perhaps the first to understand that space technology required a new organization. At the time, the science of systemic approaches in its current formalized form had not been developed. The successes achieved by prominent military leaders and government officials were all the more outstanding in that, without textbooks and sometimes despite dogma, they made strategic decisions in the interests of the "big system."

For Korolev, his deputies, and close associates, this gigantic new system came about because of a broad view of space technology, by combining fundamental research, applied science, specific design, production, launches, flight, and flight control, rather than from specific spacecraft. This single-cycle setup began to operate in 1959 and 1960. The mastery of this cycle by hundreds and later by many thousands of scientists and specialists made it possible for humankind to begin the Space Age in the 20th century. The spacecraft themselves were a tool, a means for achieving an end. This restructuring or, more accurately, the creation of a completely new organizational setup began in 1959 after the merger of TsNII-58 with OKB-1.

A unique feature of this period was that no special time was set aside for any organizational restructurings. The goals we had conceived, now supported by government decrees, did not allow us to stop designing, producing, and launching combat missiles and spacecraft. The situation might be compared with the strategy of large military operations. During a victorious military blitz, the troops advance without stopping to destroy the routed enemy units remaining in the rear.

IN 1959, THE COMMUNIST PARTY CENTRAL COMMITTEE AND THE USSR COUNCIL OF MINISTERS ISSUED FIVE DECREES DIRECTLY AFFECTING OUR OPERATIONS. These included decrees:

- on 14 March calling for the creation of and beginning of flight-tests of the R-7A missile;

- on 13 May calling for the development of the new R-9 intercontinental missile;

- on 22 May calling for the development of an orbital spacecraft for reconnaissance and human spaceflight;

- on 20 November calling for the development of the solid-propellant RT-2 missile; and

- on 10 December calling for the further development of outer space research, which was also the first to set the goal of human spaceflight.

Through 1959, we had made significant advances. In February 1959, the first R-11FM naval ballistic missile went into service and in December we began the first launches of the R-7A missile. Also, in January 1959, *Luna-1* was launched, followed by *Luna-2* in September (delivering the pendant to the moon), and *Luna-3* in October, which photographed the far side of the moon. The following year, in 1960, there was no time to take a breather; four more decrees came out that year! The last of them ordered a piloted spaceflight to be carried out in December 1960. On 15 May 1960, the first Vostok orbital spacecraft was launched.

The restructuring of our new organization and the performance of each of these missions took place under the constant monitoring of governmental organizations. The Special Committee of the Council of Ministers that existed before December 1957 and managed the primary fields of new defense technologies was eliminated; instead, its structure was used as a basis to create the Commission on Military-Industrial Issues (VPK) of the Presidium of the USSR Council of Ministers. Dmitriy Fedorovich Ustinov was appointed a deputy chairman of the USSR Council of Ministers and simultaneously VPK chairman. Meanwhile, the ministries of the various branches of the defense industry were reorganized into State Committees. In 1957, all industry was transferred according to territorial status to regional, *oblast-*

level, or *kray*-level Councils of National Economy (*Sovnarkhozy*).[1]

The Ministry of Armaments, renamed the Ministry of Defense Industry in 1953, had direct supervision over the development of missile technology before 1953. D. F. Ustinov headed the ministry until 1957.

Many were opposed to the governmental decision—made at Khrushchev's initiative—to create the *Sovnarkhozy*. In 1965, after Khrushchev's ouster, the industrial branch ministries were reinstated and the State Committees and the *Sovnarkhozy* were eliminated. Although there were many enemies of the *Sovnarkhoz* system, for us the creation of *Sovnarkhozy* was very favorable. I will discuss this below.

With the *Sovnarkhoz* reforms in 1957, the Ministry of Defense Industry was transformed into the State Committee for Defense Technology. All defense industry series-production factories were transferred to *Sovnarkhozy* on the basis of territorial status. The head missile technology developers, head institutes, and pilot production plants remained in the State Committee. Responsibility for developing the appropriate technology, conducting long-range scientific research work, and creating experimental prototypes fell on the State Committee. The *Sovnarkhozy* factories were supposed to carry out series production since the State Committees were relieved of responsibility for the series production plan. As a result of Ustinov's new appointment to the VPK, former NII-88 Director K. N. Rudnev was appointed the new chairman of the State Committee for Defense Technology (GKOT).[2] The main directorate within the GKOT for the development of missile technology was its Seventh Directorate, headed by Lev Arkhipovich Grishin in 1957. Until then he had worked as director of the rocket engine Factory No. 456 in the suburban Moscow area of Khimki. Our reorganization was part of the full-blown process of building a new space industry sector led by Ustinov, Rudnev, and Grishin.

The VPK coordinated projects in the branches of industry, above all by assigning administrators and deadlines, specifying phases for conducting operations, and distributing government funding. In addition to the missile and space fields, the VPK and State Committee for Defense Technology supervised a broad range of armaments for all branches of the armed forces. This was one of the reasons why Korolev's OKB-1 was given a great deal of independence in drawing up the space programs.

The Ministry of Defense, that is, the military and our customer, was also extensively engaged in developing new space technology. It was still responsible for building and operating firing ranges, including future cosmodromes, and developing the network of tracking stations on Soviet territory and the first control centers.

I would like to turn the reader's attention to what I consider one

---

1. The *oblast* and *kray* are geographical administrative units roughly equivalent to a province common to all regions of the former Soviet Union.

2. GKOT—*Goskomitet po oboronnoy tekhniki.*

MORE VITAL CIRCUMSTANCE THAT CONTRIBUTED TO THE SOVIET UNION'S DRAMATIC BREAKTHROUGH INTO SPACE. In the late 1950s and early 1960s, 15 years after the end of the war, the spirit of victory unifying the people got its second wind after society was liberated from the suffocating atmosphere of repressions.[3]

The people who had come together in OKB-1 under the leadership of Chief Designer Korolev either of their own will, the government's will, or that of history were very diverse. They came to OKB-1 from organizations that had their own history and traditions. But the majority of the specialists brought together in OKB-1 had gone through a scientific-technical "school." These were the schools of Bolkhovitinov, Keldysh, Tikhonravov, Grabin, Aleksandrov, Korolev himself, and the first Council of Chief Designers.[4]

In terms of their origins, initial assignments, and makeup, the "schools" were quite varied. But after their merger, success came because they all had one thing in common. The leaders of the schools believed in themselves and in their intuition, and they believed that they were the only ones capable of achieving in the nearest future the goals that they had set. They transferred this belief in their mission to their respective organizations, allowing them to achieve common goals.

Our system of higher education, which maintained a high standard despite very heavy losses due to the war, helped to establish the human intellectual potential in each of the schools. The breadth of our educational system, which had moved away from the narrow pragmatism characteristic of the Western (and particularly American) higher education system) played a decisive role in the formative stage of our cosmonautics.

Traditionally, or as the situation panned out by virtue of the peculiarities of Russian history, the schools that emerged and developed in the Soviet Union were technological rather than purely scientific. Scientific schools existed only in Germany, but the military defeat put an end to them, and after the war they simply never recovered. Soviet scientific schools were a community of people who rallied around a talented leader or organizer who supported the community not for the sake of a pure idea, but for the sake of advancing the idea to the point of *practical* application. In a recent book, Academician Nikita Nikolayevich Moiseyev aptly identified the characteristics of our scientific schools. He noted that, "A community evolves into

---

3. The Khrushchev era, especially from 1956 to 1964, is typically known as a "thaw" period of social and cultural liberalization after the unimaginable repressions during the Stalin times. The first major marker of liberalization was Khrushchev's famous speech at the 20th Communist Party Congress in 1956, where he openly denounced Stalin's many crimes.

4. Representatives of all of these schools converged in OKB-1 over the years. They represented the following leaders and organizations: Bolkhovitinov (from OKB-293), Keldysh (from NII-1), Tikhonravov (from NII-4), Grabin (from TsNII-58), Aleksandrov (from NII-58), and Korolev himself (from OKB-1). Anatoliy Petrovich Aleksandrov (1903–94) was not well-known among the missile and space program community but he was deeply involved in the development of Soviet nuclear weapons. He served as the director of the Institute of Physical Problems and Kurchatov Institute of Atomic Energy. He later served as president of the USSR Academy of Sciences in 1975–86.

a school only when a sense of mutual responsibility emerges within it."[5] This sense of responsibility must be profoundly personal rather than official. After 1945, the nation not only recovered from the ravages of war remarkably swiftly, but it made itself the world's second power in the field of science and technology.

Through the text, I shall attempt to recall the structure of the new missile and space school and mention many of those involved in our missile and space programs who were members of its community. All of them were proud of their work. Each one felt that his country needed him. The fact that government leaders supported an atmosphere of collective creative euphoria was very important. It wasn't just the most outstanding scientists, engineers, workers, and later cosmonauts who received the highest government awards, but entire organizations. For example, NII-88 and then OKB-1 were awarded Orders of Lenin.

I was one of the organizers of the nation's first—and perhaps the world's first—scientific and technical school on spacecraft motion control and a broad range of space electrical engineering, radio electronics, and data transmission projects. For that reason I am devoting the greatest space in my memoirs to the area of work that was closest to me.

Due to its merger with TsNII-58 and its liberal recruitment policy, the ranks of OKB-1 had increased by 5,000, of whom 1,500 were engineers. We received a well-developed tract of land with a large building for the design and laboratory facilities, a closed-cycle pilot production plant, and various auxiliary services. A large orchard, a birch grove, and flower beds adorned the grounds. In the summer you would have thought you were in a park rather than the premises of a weapons production enterprise. All the newly acquired territory was called "the second territory" or "the second production facility." The organizational issues involved in the restructuring in connection with the addition of "the second territory" required constant attention. Korolev and all of his deputies, with the help of the primary managerial staff of TsNII-58, devoted a great deal of attention to the placement of personnel.

After visiting Grabin's office for the first time, Korolev told Bushuyev, the new head of the "second territory" that the flowers on the walls had to go. "And, in general, be less ostentatious. Keep the lounge at the back of the office, but get rid of the general's private bathroom with the shower. You will be using the regular restroom." All this was done. Meanwhile, the vestibule of the central entrance and the broad stairway to the third floor were laid with marble. This refined the modest interior of the engineering building and showed all the workers that the new director was showing the proper attention to even the external appearance of their working environment. The full-tilt construction of new buildings also began.

For three years, Bushuyev was the proprietor of the office that contained Karelian birch furniture. Beginning in May 1963, at Korolev's decision I occupied that

---

5. Nikita Moiseyev, *Put' k ochevidnosti* [*Pathway to Evidence*] (Moscow: Agraf, 1998).

office for more than 30 years. Little had changed there since Grabin's time. An even coat of green paint had replaced the exotic flowers. In place of Stalin's portrait there was now a portrait of Tsiolkovskiy, the work of our [in-house] artist. My job was to organize a cluster of departments that encompassed radio engineering, electrical equipment, spacecraft motion control, dynamics, and rocket guidance.

I ALREADY CONSIDERED MYSELF AN OLD HAND BECAUSE MY TRACK RECORD IN THE ROCKET FIELD COULD BE TRACED BACK TO 1940, when at the suggestion of Bolkhovitinov, Isayev, and Bereznyak, I first began to develop liquid-propellant rocket engine control automatics for the BI airplane. I received the first combat order of the Red Star in 1945 for developing the liquid-propellant rocket engine automatic control system. Everyone who worked in Germany at the Institutes RABE, Nordhausen, and Berlin, felt to some extent like they belonged to a superior "rank," regardless of where they were working now and their departmental affiliations. Rather than encumbering us, this feeling linked and united us and ultimately helped us to solve many problems.

The general volume and complexity of the tasks, which on the whole we had set for ourselves, required an increase in the efficiency of the entire research and development system. As our ranks increased drastically, the problem of achieving the optimal structure to accommodate the influx of new people was vital. Based on my experience, I believed (and this was later confirmed repeatedly) that no structure, no matter how carefully it was thought out, was capable in and of itself of creating and sustaining the creative work of engineers and scientists of various specialties at a high level if there was no friendly contact established between them.

Given the situation at the time, in order for a worthy specialist to be appointed head of a project in line with his capabilities, it was necessary not only to have a request from me, his agreement, and Korolev's approval; appointment to key positions also required the support of the Party Committee and no objections from the personnel and security departments. It is true that during this period background information was not as crucial as it had been during the Stalin years, but individuals who might be suspected of having family or other compromising ties with foreign nationals were not cleared for managerial positions.

The managers who achieved the greatest success were those who learned to understand and appreciate above all the role of people, and then, the role of inanimate technology. For me, the main problem remained uniting the efforts of specialists and managers who differed in terms of character, single-mindedness, culture, experience, and age. The best ways to instill a sense of joining forces and teaching the art of one-on-one contact was participating in hardware preparation at the firing range, conducting launches, and analyzing the results of flight-tests. There was no way that theoretical coordination of specialists in the fields of ballistics, electrical engineering, or control dynamics—as well as designers, production engineers, and many others—could instill a collective approach to work as did joint work at the firing range. Despite the harsh living conditions, the exacting nature of the work, and the endless stressful situations that always accompanied the launch preparation process and flight control, there was an atmosphere at the firing range that inspired each participant and motivated everyone to give their all.

In general, for an ideal merger, as we hoped the union of TsNII-58 with OKB-1 would be, the organization of operations required clear-cut territorial and structural divisions. The first step in this direction was the creation of the OKB-1 space division at the second territory. At Korolev's instruction spacecraft designers headed by Tikhonravov and Tsybin moved over there once again. They reestablished space design departments while specialists transferred from Rauschenbach's department at NII-1 settled in there as well. We also moved the radio engineering and electrical equipment departments from our old campus. The ranks of these departments had increased dramatically, primarily due to the number of specialists recruited by Academician Aleksandrov during his time as NII-58 director.

We had the opportunity to create a unique complex of departments at OKB-1 for the development of spacecraft control systems. Government decrees had already specified head organizations and chief designers for the development and manufacture of missile guidance systems. These were Pilyugin and Ryazanskiy at NII-885; Semikhatov, who split off from them to work on naval missiles in Sverdlovsk; and Konoplev in Kharkov, who split off to work on the missiles developed by Yangel. For some reason, the consensus in the corridors of the authorities was that sooner or later, in their spare time, they would also deal with new spacecraft control systems. This was wishful thinking, merely success-induced giddiness.

Having gathered at "second production facility," the specialists decided that the time had come to seize the initiative to develop fundamentally new systems for spacecraft. Our subsequent successes in space can be explained to a significant extent by the fact that, from the very beginning, the development of space technology was organized in the form of an integrated system process. Research, laboratory development, design work, production of the first experimental flight models, flight-development tests and the incorporation of their results, introduction of modifications during the production process—all of this flowed into a single goal-oriented project that was common for many thousands of participants.

Discussing the immediate plans and prospects with my comrades, we came to the conclusion that the field of spacecraft control systems was a field related to missiles in terms of the technology of instrument production, but a new field in terms of technical principles. None of the venerable chief designers would tackle it in full measure. Our historical mission was to take this entire problem into our own hands at OKB-1. Surprisingly, Korolev expressed no objections or misgivings to my proposals. Ultimately, as the sole chief designer at OKB-1 he took on one more heavy burden of responsibility for the fate of plans for the space program. He very actively supported all my proposals and even went further.

"You aren't going to create anything with all your departments if we don't have our own state-of-the-art instrument production," Korolev decided. "I suggest that you prepare proposals specifying which instrument factories will work on our orders. We're going to begin organizing and building our own instrument factory right away. Your beloved Shtarkov's instrument shop No. 2 is after all just a shop, and we need a production facility that is powerful, very broad-based, and versatile. I have already found someone to head this future plant."

Indeed, soon Isaak Borisovich Khazanov was named chief of instrument production and simultaneously deputy chief engineer of the factory. Before the merger he worked for Grabin as chief of the experimental science division. At first I was surprised because Korolev assigned a nonspecialist to the instrument plant, but he assured me that Khazanov would not fail. Once again Korolev took the opportunity to remind me of my past mistakes in personnel placement, telling me that he was a good judge of character. Korolev had first seen Khazanov in 1959, after the consolidation of Grabin's team with ours. There were two factors that might have come into play for Khazanov's appointment, either Turkov's recommendation or the legendary feats of Khazanov's father, whom Ustinov had thrown into the most cutting-edge weapons production sectors during the war. Or perhaps it was Korolev's innately unique ability to accurately assess people from his first encounter with them. He did not make a mistake in his selection with Khazanov.

Under his supervision, Khazanov brought together random production sections and shops, including shops that manufactured control surface actuators, cables, ground control consoles, and antennas. At the same time we also began new construction. In order to rapidly expand the production areas for instrument production, four three-story buildings were built over a period of several months at the second territory. The construction timeframes were the shortest on record because they used generic designs and standard units designed for school buildings. At that time in Moscow and the surrounding area, school construction was the result of assembly line production. Schools literally sprang up in three or four months. Khazanov also took advantage of this situation with the help of Georgiy Vasilyevich Sovkov, Korolev's enterprising construction assistant. They also began designing a special state-of-the-art, six-story building, an actual instrument factory. The design called for air conditioning, a clean zone for microelectronics manufacturing, and special laboratories for testing instrument reliability under exposure to potential and even seemingly improbable external mechanical, climatic, and space effects.

While construction of our own plant was under way—making use of the proprietary interests of the regional *Sovnarkhozy* and factories, which had gained great independence during the Khrushchev reforms—Khazanov and I tried to place as many orders as possible with instrument-building and radio electronics factories.

The directors of factories subordinate to the *Sovnarkhozy* had acquired the right to accept orders and conclude contracts without waiting for instructions from above. In 1965, however, the *Sovnarkhozy* were eliminated. Once again a centralized command and administrative system prevailed. After that point, maintaining the cooperation that we had organized during the times of the *Sovnarkhozy* was quite a challenge.

Korolev encouraged the expansion of our production base in every way. Here is a typical episode. Once, with Khazanov and myself in tow, Korolev flew out to Kiev, where Ukrainian Central Committee Secretary Petr Shelest received us. Soon, our proposals passed from office to office in the defense department of the Ukrainian Central Committee and the Kiev *Sovnarkhoz*. The eagerness of the factory directors, who showed great interest in our proposals, managed to neutralize the ill will of the higher management of the Ukrainian Party office. The thoughtful directors weren't so much enthralled with the issue of the work load in the days to come, but rather with the prospect of mastering new products and modernizing equipment, building new shops, and obtaining new benefits for their teams under the banner of missiles and space. After roaming the corridors of authority in Kiev for many long hours and attending tedious meetings during which they explained to us that the most important thing for Ukraine at that point in time was the iron and steel industry and not satellites, we departed for Moscow. Still, we had obtained an agreement for the use of two factories, the Kiev Radio Factory (KRZ) and Kievpribor.[6] Both of these factories subsequently had a leading role in the production of complex radio electronic equipment for rocket-space technology.[7]

Korolev wasn't able to accompany us to all the factories that Khazanov and I intended to bring into the orbit of space instrumentation technology. However, he always helped, even without leaving his office. Before setting us loose on our solo expeditions to "colonize" others' factories, Korolev would make arrangements with the Central Committee, *Gosplan,* and the VPK. They immediately issued instructions to Communist Party *oblast* committee secretaries. When we arrived in the appropriate town on our own airplane, we were received as high-ranking guests. Before setting out for the factory, we visited the *oblast* committee defense department. As a rule, representatives of the *oblast* committee and *Sovnarkhoz* accompanied us during all our talks with factory directors, even attending the farewell banquets. Sometimes those situations were funny, bordering on the absurd.

After arriving in Kazan, we found out that the factories there were not suited to our needs, but could be used by our colleague OKB MEI Chief Designer Aleksey Bogomolov for the production of transponders for the orbit radio monitoring (RKO) system.[8] Bogomolov had a small pilot plant right at MEI, but it could not meet our demands for deliveries in terms of quantities and deadlines. Given this situation, where I could, I tried to make arrangements not only for production based on the direct orders from our OKB-1, but also for tasking appropriately specialized factories with production for other chief designers who had been working on our projects. We were able to do this in Kazan, where OKB MEI obtained a good

---

6. KRZ—*Kievskaya radiozavoda.* Kievpribor, founded in 1947, was originally known as Factory No. 7 before becoming Kievpribor in 1956.

7. The Kiev Radio Factory, for example, produced instrumentation for the military Almaz space station in the 1960s and 1970s.

8. RKO—*Radiokontrol orbit.*

production facility for many years.

We were initially unable to offer direct orders for our OKB-1 to anyone in the Tatar *Sovnarkhoz*, so we quickly left to manage matters there. The director of one of the factories located on the bank of the Kama "kidnapped" us and whisked us off to his place. Over the course of two days he organized picnics and fishing on picturesque islands with one goal in mind—to land an order for the production of space instrumentation. He didn't release us until we had reassured him that we would explore this possibility in the next few days. Alas, this was a mass production factory, and there was no way our scientific production could satisfy the appetite of a factory set up to produce batches consisting of many thousands of articles.

Our "raids" on the Rostov and Bashkir *Sovnarkhozy* were considerably more successful. Despite the limited success of our fishing on the Azov Sea and the Belaya River in the Urals, we established lasting friendly contacts with the staff of the Azov Optico-mechanical and Ufa Instrument Building Factories. Soon, the Azov Factory monopolized the production of universal test stations and docking assemblies that we developed.[9] The Ufa Factory, meanwhile, mastered the production of onboard computers and a wide array of switching devices for manned vehicles right up to the Soyuz spacecraft. From the early 1960s until the end of the 20th century, it was one of the primary factories delivering instruments for piloted spacecraft. Additionally, the Sarapul Aviation Parts Factory succeeded in setting up the mass series production of control surface actuators, freeing our pilot plant of this labor-intensive production.

We did not forget Moscow and Leningrad on our trips. During and after the war the Moscow Plastik Factory specialized in manufacturing the most complex fuses for various types of shells and rockets.[10] In late 1959, Plastik's Chief Engineer Boris Zaychenkov showed extraordinary courage in accepting our highly risky proposal for space instrumentation production. According to our targets, by mid-1960, his factory would have to manufacture and refine what was—even by present-day standards—a complex program timing device and a computer for Mars missions. These instruments performed control functions that are now handled by microelectronic digital computers. At that time we did not yet possess this technology and had just barely mastered circuit design with semiconductor triodes, that is, transistors combined with conventional relays, magnetic core matrices, and magnetic amplifiers.

Laboratory chief German Noskin, who worked in Petr Kupriyanchik's department, set about developing these instruments within unthinkably tight deadlines. Among other engineers on his team was Nikolay Rukavishnikov, the future cosmonaut, two-time Hero of the Soviet Union, and president of the Federation of Cosmonautics. One time in the early 1990s, at dinner in our dining hall, Rukavishnikov took me back to those distant days and nights. Rukavishnikov, his boss Noskin, and

---

9. The Azov Optico-mechanical Factory, founded in 1944, was originally known as Factory No. 318.
10. The Plastik Factory, founded in 1932, was originally known as Factory No. 571.

their comrades had been working around the clock for days on end in the shops of the Plastik Factory trying to debug the program timing device by the deadline. Chief Engineer Zaychenkov thought that, even during the war, they had not felt as stressed and slept so little. At one point, he telephoned me at night and said that his foremen were doing everything they could, but my engineers had botched things up completely with the troubleshooting. He asked me to come immediately and to decide on the spot what to do next. I arrived and Zaychenkov and I went to the shop. Once I saw the unshaven faces of the testers, ashen from exhaustion and sleep deprivation, I didn't feel very optimistic. One of them had his face buried in an instrument soldering something; another was clicking toggle switches on the control console; a third was looking for something under the workbench. I took the plunge and asked in a loud, upbeat voice, "How's it going, guys? Tomorrow's the very last day!"

No one raised their head except for the guy who was stooped down under the work bench. He stood up, turned a blank gaze at the big shot who had arrived on the scene and quietly said, "You jerks can go to ..." You can guess the precise destination to which the worn out workhorse would send anyone who interfered with the completion of this crucial work. "OK, fine, we won't bother you," was my simple reply as Zaychenkov and I made our way out.

Two days later, the first electronic instrument, the program timing device for the first automatically controlled interplanetary spacecraft, was delivered. Thirty-two years later cosmonaut Rukavishnikov reminded me of this event with obvious pleasure. At that time the young engineer and his comrades were creators and felt they had complete ownership of their creations. Back then the joy of the creative fire and the sense of doing one's duty may have given the young engineers greater satisfaction than the medals and high ranks conferred in subsequent years.

After that, for many years the Plastik Factory manufactured program timing devices for automatically controlled interplanetary spacecraft, even after this field was handed over to Babakin at the S. A. Lavochkin OKB.[11] Thirty-five years later, despite the hardship of recent years, the Plastik Factory remains a subcontractor in the production of space instrumentation.[12]

In Leningrad, the Instrument Building Factory was loaded with orders to produce semiautomated test hardware. However, when the authority of the *Sovnarkhozy* was phased out, this factory returned to the Ministry of Aviation Industry and our orders were transferred from there.

I have mentioned only a few of the main factories that were supposed to produce the most diverse onboard and ground-based equipment.

---

11. Korolev's OKB-1 handed over the development of all automated lunar and interplanetary to the Lavochkin OKB headed by G. N. Babakin in 1965.

12. In 1977, the Plastik Factory combined with the Delta Scientific-Production Association (NPO Delta), known since 1992 as NPP Delta.

WE NEEDED TO PROVIDE THE NEW PRODUCTION FACILITIES IMMEDIATELY with technical documentation and design "escorting" resources. We also needed to set up for deliveries of systems elements and materials. Every day dozens of questions needed to be answered over the telephone and telegraph, and when complications arose it was necessary to travel in person to resolve the problems on site. We also conducted this work with Khazanov, chief of instrument production at the factory. Three years later Khazanov was appointed chief engineer of our factory. In this role he revealed his brilliant organizational skills to the full extent.

According to the Main Artillery Directorate's traditional rules and laws adopted for instruments installed on combat missiles, the developmental cycle for a complex instrument from design conception to clearance for the first flight took from one to three years. First we developed the idea, performed theoretical calculations, and did laboratory research. Next, we manufactured the laboratory mock-up, and put it through testing, reengineering, and modification. After this, the developer drew up design specifications for the design department, which issued drawings for the fabrication of the prototype. The prototype was manufactured with many deviations from the rigid norms that the military representatives enforced. The drawings needed to be reconciled with the fabricated model as quickly as possible. All the changes needed to be inserted, taking production experience into consideration, and permission needed to be granted to use the new documentation to begin manufacturing the first production units. By this time, in addition to the drawings, full-fledged instructions for the verification and acceptance tests, that is, the testing documentation, would have become available. Issuing them was often more labor-intensive than developing the drawing documentation. I can't remember an instance when the testing documentation drawn up by an instrument developer was suitable for the acceptance and release of instruments without serious corrections being performed "on the fly."

The first instruments that passed verification tests proceeded to design development tests (KDI).[13] The instruments were heated, frozen, shaken on vibration stands, placed in vacuum and humidity chambers, and checked for supply voltage limit tolerances. And inevitably defects surfaced that required revamping, retesting, and the replacement of some parts. When serious defects were found, production came to a halt for a thorough investigation to determine the causes and to coordinate all subsequent measures with the "customer," that is, the military representative.

Finally, when everything had been agreed upon, production would move heaven and earth to meet the delivery deadline for the first instruments cleared for installation on the spacecraft. If an instrument had been delayed, instead of being installed at the assembly shop, it would be installed at the factory control and testing station

---

13. KDI—*Konstruktorsko-dovodochnyye ispytaniya.*

(KIS) where the entire spacecraft would undergo tests.[14] This was the last stage before delivery to the firing range. Here, suddenly they would often find troubles related to the instrument's electromagnetic incompatibility; it was either interfering with adjacent units or vice versa. And sometimes, because of mistakes, the myriad cable connections would actually start smoking! In such situations, developers—of the instrument, onboard circuitry, cable designs—and process engineers would be sent over the edge. The spacecraft could not move on to the next set of tests until the mistake had been found. From the very beginning we managed to train all the developers and testers according to the principles that, first of all, one should find the cause, find a solution to eliminate the defect, perform all the modifications, and retest, and then, once one had determined that the modifications were successful, find the guilty party.

Relationships with the military officers specializing in military acceptance played a vital role in the "development–manufacture–testing–delivery" process. Colonel Pavel Trubachev and his deputy Colonel Pavel Aleksandrov headed our military representation.[15] I knew them very well from our joint work at the Institutes RABE and Nordhausen. We had established good business relations. The acceptance officers (we called them "Trubachevites") could have taken a formal approach and worked "by the book." This would have been most hazardous in our business. In our joint work we managed to avoid this. In 1961, Trubachev was named chief of control issues in the Strategic Rocket Forces system. We always found a common language with Colonel Oleg Zagrevskiy, who replaced him as military representative at OKB-1, and then with Colonel Aleksandr Isaakyan as well.

Conflicts with the military reps that cropped up were resolved in the interests of programs and deadlines. There were usually conflicts between stipulated deadlines and the formal cycle of instrument production described above. In addition to absolute technical competency, all development supervisors from the deputy chief designer down to the engineer developer needed to have a knack for finding compromises. This art is not described in any textbooks, nor is it an engineering discipline taught in institutions of higher learning.

Finding a compromise between the demands of the strict sequence in the instrument production process and deadlines that were totally incompatible with this prolonged optimization cycle was very difficult. Usually we arranged for a parallel cycle, that is, production began long before the optimization of the first laboratory models. This was, however, risky; sometimes we had to throw out a large production stockpile. But on the whole, this method, which later spread to other enterprises,

---

14. KIS—*Kontrolno-ispytatelnaya stantsiya.*

15. During the Soviet era, every design bureau and factory specializing in products for the defense industry (such as OKB-1) was staffed with a few people representing the interests of the customer, that is, the armed forces. These military representatives (or acceptance officers) helped to ensure that military specifications for particular systems were being met.

proved worthwhile.

Today's developers, who can make use of personal computers, simulators, and computer-aided drafting, even for drafting large-scale integrated circuits, have trouble dealing with the whole software optimization cycle. The computerization of control systems has revolutionized the development and hardware fabrication process. During the 1960s we could not imagine that just 20 years later, a mathematician developing software would determine the deadlines for a system's production rather than a designer and the production facility. But we had begun working on this future even back then.

AFTER BRIEF DELIBERATIONS, Korolev agreed to hand over Department No. 27, which Rauschenbach had run after its transfer from NII-1, to my instrument cluster. Bushuyev readily agreed with this. After all these mergers and changes, my deputies responsible for all motion control and "radio electricity" problems in space in the 1960s were Boris Viktorovich Rauschenbach, Viktor Aleksandrovich Kalashnikov, and Igor Yevgenyevich Yurasov. Rauschenbach was in charge of all "theoretical" fields, or what we called "dynamics" in our slang. Kalashnikov was in charge of the design and testing departments and all manner of electromechanics. Yurasov very enthusiastically accepted responsibility for "all electricity" and for conducting flight-tests on the spacecraft control systems. The life of each of these three would make an interesting story in itself. I was as candid as I could be with each of my deputies and felt their support. In short, we were a strong, tight-knit foursome ... for the time being. The end came suddenly.

Soon after Korolev's death, Rauschenbach left OKB-1, and took over a department head post at MFTI.[16] He became engrossed with research in the art of icon painting and cultural history and with restoring German autonomy in the USSR. Rauschenbach was a full member (or Academician) of the USSR Academy of Sciences; with his talent for simplicity in scientific and autobiographical writing, Rauschenbach wrote a riveting account of the main events of his life and work with Korolev.[17] One of Rauschenbach's first published works was a book written in collaboration with Tokar on principles of spacecraft attitude control. Even now, this work can be considered a classic in its field.[18] Rauschenbach's subsequent literary works contain not only new glimpses into the history of art and memoirs, but also interesting worldviews and philosophical opinions developed after he left the team

---

16. By then, OKB-1 had been renamed the Central Design Bureau of Experimental Machine Building (*Tsentralnoye konstruktorskoye byuro eksperimentalnogo mashinostroyeniya* or TsKBEM).

17. B. V. Rauschenbach, *Postskriptum* [*Postscript*] (Moscow: Pashkov dom, 2000).

18. B. V. Rauschenbach and Ye. N. Tokar, *Upravlenie orientatsiei kosmicheskikh apparatov* [*Controlling the Orientation of Space Apparatus*] (Moscow: Nauka, 1974).

that he had founded.[19]

Kalashnikov, despite all my pep talks, left TsKBEM after succumbing to illness, and became a teacher at MVTU. However, his transition to teaching did not save him from cancer, which got him in the end. Yurasov received the title Hero of Socialist Labor at the same time I did in 1961. He was an inveterate smoker. Because of rapidly developing gangrene, both of his legs, one after another, had to be amputated; his condition eventually led to his natural demise.

Saying goodbye to my deputies at their funerals and many other friends and comrades-in-arms in the breakthrough into space who have passed on, I have felt that they took with them a bit of my life. We were all united by the euphoria of the romantic period of magnificent results in cosmonautics' historic infancy.

From the author's archives.

Igor Yurasov was one Chertok's principal deputies in developing control systems for spacecraft.

I cannot take sole credit for creating the first scientific and technological school of space systems control. It's very honorable to show off on the tip of the iceberg during anniversary celebrations, but I relied not only on those first deputies of mine such as Rauschenbach, Kalashnikov, and Yurasov, but also on many department heads and their deputies, each of whom we had carefully selected and whose appointments we defended, first of all before Korolev, and then before the Party Committee and personnel officers.

I would like to briefly describe the main departments of the first school of space control during the 1960s. Viktor Pavlovich Legostayev headed the main theoretical department, that is, all the "dynamics specialists." Now he is a full member of the Russian Academy of Sciences, a member of the International Academy of Astronautics, and first deputy general designer and president of the Energiya Rocket-Space Corporation. Way back in the 1960s, having brought together all the dynamics specialists from OKB-1, NII-1, and TsNII-58, Legostayev created a sort of corporation of space theoreticians who solved many fundamental problems of spacecraft motion

---

19. Rauschenbach died on 27 March 2001. His non-technical works include *Prostranstvennyye postroeniya v zhivopisi: ocherk osnovnykh metodov* [*Spatial Construction in Painting: Notes on the Basic Methods*] (Moscow: Nauka, 1986); *Sistemy perspektivy v izobrazitelnom iskusstve: obshchaya teoriya perspektivy* [*The System of Perspective in the Fine Arts: A General Theory of Perspective*] (Moscow: Nauka, 1986); *Pristrastiye* [*Bias*] (Moscow: Agraf, 1997).

control for our cosmonautics. The then-young engineers Yevgeniy Tokar, Vladimir Branets, Ernest Gaushus, Leonid Alekseyev, Oleg Voropayev, Aleksey Yeliseyev, and Larisa Komarova are now doctors of science and professors at prestigious institutions of higher learning.

This department valued good theory highly and proved that mathematical tools were necessary not only for dissertations, but also for the solution of very practical problems. V. N. Branets and I. P. Shmyglevskiy's monograph *Quaternion Application in Problems of Solid-state Orientation*, published in 1973, is a classic example of the use of a mathematical tool to create actual gimballess inertial navigation systems.[20]

While developing flight control dynamics problems, Aleksey Stanislovich Yeliseyev decided that he personally needed to master the technique of spacecraft control. He went into space three times, twice received the title Hero of the Soviet Union, and until 1986 directed the flight-control service as a deputy general designer at Energiya. Larisa Ivanovna Komarova is the generally recognized authority on the development of spacecraft navigation and motion control systems for the descent phase. To this day she holds the title of professor in my home department of motion control at MFTI. Among my other deputies, Branets was bestowed the honor of being Yuriy Semyonov's deputy general designer and was the last person to occupy the office of the great artilleryman, Vasiliy Grabin.[21]

The practical implementation of the theoretical research of Legostayev's department was entrusted to another of Rauschenbach's NII-1 associates, department chief Yevgeniy Aleksandrovich Bashkin. Bashkin's department developed the actual hardware and specific electrical circuitry of the motion control system. We developed the control instruments ourselves at OKB-1, or we were fully responsible for ordering and keeping track of them at subcontracting organizations.

Earlier I expressed what I thought was the historical significance of the merger of the schools of Korolev, Bolkhovitinov, Grabin, and Keldysh. The personnel of the departments of Legostayev and Bashkin are illustrative in this respect. Legostayev's deputy, Oleg Nikolayevich Voropayev, was one of the first rocket theoreticians and dynamics specialists of Korolev's school, while Bashkin's deputy, Oleg Igorevich Babkov, came to us from Grabin's TsNII-58. When I turned 80, Oleg Babkov assumed leadership over the entire complex of control problems at RKK Energiya. By right of succession, after me he became the proprietor of Grabin's historic office.

Bashkin's department was also assigned the challenge of providing optical instruments for spacecraft orientation and navigation. Stanislav Savchenko, who came

---

20. V. N. Branets and I. P. Shmyglevskiy, *Primeneniye kvaternionov v zadachakh orientatsiy tverdogo tela* [*Quaternion Application in Problems of Solid-State Orientation*] (Moscow: Nauka, 1973).

21. Yuriy Pavlovich Semyonov (1935–) served as general designer and director of the Energiya Rocket-Space Corporation (RKK Energiya) in 1989–2005, only the fourth man to head the organization (after Korolev, Mishin, and Glushko).

from the artillerymen to the space field was, and remains to this day, the chief optical specialist. After working under his leadership, young engineer Viktor Savinykh flew into space three times and, together with Vladimir Dzhanibekov, saved the *Salyut-7* space station.[22]

Everything related to electricity on board a spacecraft was combined into a single onboard complex control system (SUBK). At first this task was assigned to Grabin's chief electrician Boris Pogosyants. However, we soon reorganized, dividing all electricity into "onboard" and "ground" segments. The "heavyweight" among the young engineers, Yuriy Karpov, was put in charge of the power distribution system for all the power-consuming devices on the spacecraft—which was the logical interconnection between all power "consumers" in a single electrical network. He was also responsible for the issuance of commands determining the control logic and flight program and for short-circuit and off-nominal situation protection in the "ground-to-space" system.

Two graduates of the Taganrog Radio Engineering Institute, Yuriy Karpov and Vladimir Shevelev, were the electricians who developed the electrical system of the world's first space laboratory, the third artificial satellite, launched on 15 May 1958. Karpov's deputy was originally Boris Pogosyants, and then "hot war" veteran and participant Isaak Abramovich Sosnovik. All three are gone now. Karpov's doctoral dissertation, which he defended in 1989, was the sum total and synthesis of the systemic method for designing a complicated spacecraft onboard equipment control complex. I believe the development of the *Mir* space station's onboard control complex is the pinnacle of the Karpov collective's creative work.

One of the problems determining the reliability of any spacecraft is the ground-testing hardware, which makes it possible to simulate the work program of the actual onboard equipment in flight and to check the correct operation of all the electrical connections and of each instrument. End-to-end tests, or as we referred to them, "general integration tests," were conducted at the factory control and testing station (KIS) and at the cosmodrome engineering facility. It was necessary to create a single test station instead of the individual "suitcases" that the developer of each system used to connect the bundles of cables to "their" onboard instrument. And that task was anything but easy. There were so many arguments, conflicts, and accommodations.

Petr Nikitovich Kupriyanchik headed the ground-testing equipment department. He had gone through the school of control and measurement, not just Grabin's artillery school, but also Aleksandrov's fast neutron nuclear reactor school. His deputy, radio engineer Anatoliy Aleksandrovich Shustov was supposed to use his radio engineering experience to create a "ground-space-ground" multiplexed communication

---

22. In one of the most dramatic missions of the history of the Soviet space program, in 1985, *Soyuz T-13* cosmonauts Dzhanibekov and Savinykh docked with the "dead" *Salyut-7* space station, and over a period of several months, revived the station to full operation.

and control channel for spacecraft tests at the KIS and cosmodrome. Kupriyanchik proved himself not only as a talented and highly knowledgeable electronics engineer, but also to be extremely skilled at defusing conflicts between developers. Very often, when we needed to find a quick compromise between incompatible proposals or to figure out what caused an off-nominal situation, I put Kupriyanchik in charge of the appropriate commission. He found a way out of dead-end situations surprisingly fast. And it is even more surprising that he is still at it today. During crucial crewed launches in recent years, General Designer Yuriy Semyonov used to demand Kupriyanchik's participation in spacecraft preparation at the cosmodrome. That is what 45 years' experience plus the tradition of a school created during Korolev's time is all about. Kupriyanchik's department, chock full of electronics specialists, was also assigned to develop onboard electronic instruments, including sequencers. One of the electronic control enthusiasts was young engineer Nikolay Rukavishnikov, a graduate of the Moscow Engineering and Physics Institute, a fan of motorcycle racing, and a future cosmonaut.

One of the many new problems for us was radio communications and data transmission at interplanetary distances. We first faced this problem in 1959, when transmitting photographs of the far side of the moon from the *Luna-3* spacecraft. Seeing the photographs, we snidely remarked, "It has been proven that the moon is round." It was very difficult to distinguish other details. The "space-to-ground" radio-link power was determined by a range of factors, including onboard transmitter power, the antenna patterns, and the active area of the onboard and ground antennas.

Increasing the radio-link power by raising the onboard transmitter power required increasing its mass and dimensions and substantially increasing the capacity of the onboard storage batteries, measures which threatened to disrupt the new interplanetary spacecraft projects. Another approach to solving this problem was to construct large antennas on Earth and "good" antennas on board. The first large ground antenna with an acute antenna pattern was erected near Simferopol at NIP-10. It had a diameter of 32 meters, which in those pioneering space years was the very height of ambition. But it would not do the trick if the space-based feeder antenna system converted the greater part of the transmitter's power into heat rather than radio waves carrying relevant data.

In those early years, the chief designer of the radio-links was Mikhail Ryazanskiy, the NII-885 deputy director and scientific chief. He explained to Korolev that his institute would be fully responsible for ground-based antennas; onboard antennas, one other hand, were organically associated with the design of spacecraft, including its attitude control system. After a few disagreements in the Council of Chief Designers, Korolev announced that, yes, OKB-1 would take on the development of that part of the radio-link called the onboard antenna feeder system (AFU) and that I would bear the responsibility for this problem.[23] I did not object because back

---

23. AFU—*Antenno-fidernoye ustroystvo*.

when we were developing antennas for R-7 missile warheads I began to set up an antenna group in the radio department and then a special laboratory. Several years later our antenna laboratory became an independent department.

We were fortunate in terms of talent and enthusiasts. Artillery Captain Mikhail Vasilyevich Krayushkin, who was decorated with combat ribbons and medals when he joined us after demobilization from the army, was a specialist in love with antenna engineering. With four years experience at the front, he was an astonishing combination of Old Russian selfless intellectuality and love for classical music and Maxwell's equations. Krayushkin venerated Academician Pistolkors, the patriarch of the domestic school of antenna science, and attended all of his seminars.[24] He established working contact with the MEI and MFTI antenna engineering departments. In later years his orientation toward young and talented theoreticians led to conflicts with subordinates who had been his pupils but considered themselves veterans in engineering space antenna feeder systems.

Together with Krayushkin, we made sure that a special building for the antenna department was constructed. This building housed what was for those times a unique anechoic hall. The antenna team was one of the first in my control complex to master the methods of electronic mathematic simulation, making it possible to find optimal solutions to ensure the reliability of super long-range space communications, and to create antennas with a large gain factor and minimal reduction of noise level.

However, not one of the most up-to-date theories of electromagnetic processes and conversion of electric power into radio waves helped in the development of a mechanical device for antenna deployment. The synthesis of antenna engineering and electromechanics proved possible because we already had a strong electromechanical design core that had mastered the engineering of control surface actuators. However, the specific character of space electromechanics was new to the already "old" technology of the rocket control surface electric actuator. Back in 1947, the electromechanical duties for missile control systems were clearly divided between Pilyugin and me. Ten years later, we ran into first a trickle and then a tidal wave of problems that were relevant only to spacecraft. Projects on rocket and space electromechanics were split off from the general instrument design department. Lev Borisovich Vilnitskiy, another demobilized army captain, was put in charge of them.

While space antenna engineering became former captain Krayushkin's calling, space electromechanics was the calling of Vilnitskiy, another former captain. Vilnitskiy took up the baton for the development of all types of control surface actuators and zealously made sure that his department held sway over all the new problems of electromechanics on rockets and spacecraft. From the first days of his activity in my

---

24. Academician Aleksandr Aleksandrovich Pistolkors (1896–1996) was one of the most famous Soviet scientists specializing in antenna theory and developed several basic principles in electromagnetics including the so-called "Pistolkors Duality Principle."

branch, Vilnitskiy established close relations with production. Together with Isaak Khazanov, who had been appointed chief of instrument production, he actively engaged in the organization of a special shop for control surface actuators and precise electromechanics. The complex engineering of docking assemblies, electric pumps for thermal control systems, antenna deployment drives, solar arrays, and control of the most powerful liquid-propellant rocket engines in the world got its start in his department.

I also collaborated very closely on theoretical issues with the chief electromechanical scientists in the space program, director of the All-Union Scientific-Research Institute of Electromechanics Andronik Gevondovich Iosifyan and his deputy, future Academician Nikolay Nikolayevich Sheremetyevskiy. The midlevel engineers who worked on actually developing electric motors under Iosifyan and Sheremetyevskiy respected Vilnitskiy very highly. There were tremendous disputes, but they eventually agreed to our requirements for the production of various electric power converters.

In 1983, Lev Vilnitskiy retired. He handed off the actuator design baton to Vadim Vasilyevich Kudryavtsev. At the same time, the volume of electromechanical design projects rapidly increased. We needed to develop complex electromechanical and electrohydraulic simulators to reproduce on Earth the processes taking place in space.

Kudryavtsev's talent and irrepressible enthusiasm contributed to the development of unique simulators. The special *Konus* (Cone) building was constructed for them. Digitally controlled control surface actuators for the Energiya launch vehicles were tested on this unique simulation stand. Kudryavtsev was a pioneer in the digital control of powerful actuator units. Until the last days of his life he heroically struggled with the cancer that had consumed him, trying to prolong his life by devoting himself completely to his work. The valediction that he left this life as a talented engineer and manager at the zenith of his creative powers fully applies in his case.

In the field of electromechanics, problems of docking system dynamics and design took on an international scale. In the early 1970s, the Soviet and U.S. governments began negotiating for a rendezvous in space between the Apollo and Soyuz spacecraft. In the joint project with the Americans we aimed to "not give in" and use our own docking assembly design. Doctor of Technical Sciences, Professor, and member of the International Academy of Astronautics Vladimir Sergeyevich Syromyatnikov supervised this project; his team performed this historic mission honorably. For his work, Syromyatnikov has been well known to specialists in the field of cosmonautics in Russia, the U.S., and Europe.

The field of dynamics and docking hardware design required the organization of an independent department and widespread cooperation with other branches. Automatic docking in space of two multiton masses was possible only with organic interlocking, a joint design project for mechanics and electric-automation engineers. Viktor Kuzmin's special department developed electric automatic controls for docking assemblies. This same department developed instrumentation for the fuel tank depletion monitoring systems and, consequently, also for monitoring the operation of the N-1 lunar rocket engines.

FROM THE FIRST DAYS OF THE SPACE AGE we understood that we needed to devote a lot more attention to a source of electrical power for spacecraft than ones for rockets. It turned out that even a reliable storage battery for a rocket was absolutely unacceptable for a spacecraft in terms of weight and size. A special department of electrical power systems was staffed with "pure" electrical engineers. This department ordered and oversaw the development of storage batteries and solar arrays at subcontractor organizations and developed a monitoring system for buffer battery charge, voltage stabilization, and hazardous discharge protection. The problems of a reliable spacecraft power supply became more complicated as we ventured further into the "forest of space."

Boris Mikhaylovich Penek headed the power engineering department for two decades. A team of very enterprising engineers gathered in the department. Young specialists who had come to TsNII-58 during the mastery of nuclear reactors made up its core. Leonard Petrovich Kozlov, Nikolay Semenovich Nekipelov, and Aleksandr Ivanovich Shuruy monitored the work of the Scientific-Research Institute for Current Sources (whose chief designer was Nikolay Stepanovich Lidorenko), the Leningrad Institute of Current Sources, and other organizations. They developed the electrical circuits of the onboard electric power supply, were active as testers at the factory KIS and at the firing range, and during the flight-control process they kept watch over the expenditure of every ampere-hour. During the early stages of the spacecraft design, Bushuyev and I came to a compromise over conflicts between the designers and "rusty electricians" as to the buffer batteries' weights and dimensions and the solar arrays' area and efficiency.

This same power engineering department inherited a number systems for dealing with off-nominal situations, systems that were not unduly sought after; these included the emergency rocket (or spacecraft) destruction (APR) system, the emergency engine unit shutdown (AVDU) system, the electric automatic controls for the emergency rescue system (SAS), and the electric automatic controls for landing.[25] All the departments that were developing systems needed designers who converted electrical diagrams into instrument drawings suitable for production. From my very first days at NII-88 in 1947, I entrusted the design department to Semyon Gavrilovich Chizhikov. We had started out together equipping the DB-A bomber with instrumentation back in 1935 at Bolkhovitinov's OKB.

Modern-day computer technology enables a designer to draw an instrument

---

25. AVDU—*Avariynoye vyklyucheniye dvigatelnoy ustanovki*; SAS—*Sistema avariynogo spaseniya*.

without touching a pencil or drafting pen, and without using a drafting table or a drafting set. He or she has no need for Whatman paper, tracing paper, or blueprints for the reproduction of drawings. A designer's idea takes shape in a digital file and is emailed to the manufacturing shop or transferred on a disk loaded into a machine tool station, or displayed on an assembler's monitor screen. This high technology did not become commonplace until the very end of the 20th century. For the entire preceding century, a designer issuing a working document had to master the difficult art of technical drawing, know the production process, possess an artist's imagination while complying with hundreds of standards and materials manuals, and keep in mind the actual manufacturing capabilities and deadlines. In this respect, the aviation and artillery systems for issuing drawing documentation were different.

Synthesizing the documentation was also a concern of the design department chief. The volume of design projects grew exponentially, and I had to make the decision to divide them into "onboard" and "ground" projects. The onboard instruments stayed in Chizhikov's department while all of the "ground" instruments, including the ground testing station, known in the history of cosmonautics under code number "11N6110," were transferred to the ground design department. Two "Grabinite" designers, Ivan Ivanovich Zverev and Boris Grigoryevich Pogosyants, headed the "ground" department. Chizhikov was also given a deputy, designer Grigoriy Ivanovich Muravyev of the Grabinite school who rapidly made himself at home in this new sphere. The best measure of the quality of a designer's work were the evaluations he received from the chiefs of the shops where his ideas had materialized rather than those from his immediate supervisors. Chizhikov and Muravyev by rights won authority and respect at the factory not only among the workers but also from Turkov, the very demanding factory director.

The production cycle of any article in our technology ended with rigorous testing. This required a large amount of complex, scarce, and expensive test equipment. But that wasn't all. We called the tests conducted on prototypes "design development tests." Igor Fedorovich Alyshevskiy initally headed the laboratory and soon thereafter ran the special testing department.

BY THE MID-1960S, together with the instrument production that Isaak Borisovich Khazanov managed, we had transformed ourselves into an instrument and electric kingdom, a "kingdom" within Korolev's "empire." I had more than 1,300 engineers and technicians working directly under me. At the production plant, Khazanov had a staff of more than 1,700 workers and process engineers under him. In all there were around 3,000 instrument specialists. If we had been an independent enterprise producing all the same items, we would have needed at least 1,000 more employees for all the services supporting our activity: a commercial unit, procurement, book-keeping, library, guard service, transport, personnel, office of security, internal security, administrative support, technical archives, blueprint copying facility, and so on. In addition, Khazanov and I made broad use of the materials science, instrumentation, and main metallurgist departments, and also of the tooling, casting, and electroplating shops that OKB-1 and the factory shared. Despite our thematic independence, we were organically connected with the entire OKB-1 structure. This was one of the reasons why after Korolev's death I suppressed all sorts of internal and external ideas for creating an independent NII or KB for spacecraft control systems.

Our great advantage in the first years of the space race was the integrated systemic approach. A single organization created the launch vehicle, designed and produced the spacecraft, developed its control system, tested all the components of the large system, and controlled the flight. I would call the regime within the organization a totalitarian democracy.

Chief Designer Korolev was the dictator. Rather than submit to him, we listened to him and we argued with him. But we carried out his decisions and instructions without question—that was taken for granted. During the early years, the Americans expended a lot of energy coordinating specifications between companies. A single project involving hundreds of contracts demanded time that we didn't need to spend.[26]

The hundreds of changes that cropped up during projects—major and minor adjustments—were quickly resolved during daily personal contact between the interested parties who called in all the consultants that were needed. Whereas the Americans wasted weeks coordinating complex issues between companies and drawing up protocols, OKB-1 settled them during hours of productive arguments and business meetings at the workstations. A great deal was resolved at the lowest levels so quickly that managers higher up the line found out about a problem only when they signed the fully thought-out document that dealt with that problem.

One more positive aspect of this integrated method of developing space technology was having powerful production facilities at the chief designer's disposal that were capable not only of manufacturing a rocket or spacecraft, but that, from the very beginning of the design process, enabled the work force of the factory and complex of design departments to manufacture experimental units: mock-ups for developmental testing of the configuration, thermal modes, dynamics, pressure integrity, onboard systems complex, antenna parameters, and so on. These experi-

---

26. For management histories of NASA during the Apollo era, see for example, Arnold S. Levine, *Managing NASA in the Apollo Era* (Washington, DC: NASA-SP-4102, 1982); Stephen B. Johnson, *The Secret of Apollo: Systems Management in American and European Space Programs* (Baltimore, MD: Johns Hopkins University Press, 2002). See also the essays by Robert C. Seamans, James E. Webb, and other Apollo-era NASA managers including Robert R. Gilruth, Wernher von Braun, George M. Low, Rocco A. Petrone, Samuel C. Phillips, and George E. Mueller in Edgar M. Cortright, ed., *Apollo Expeditions to the Moon* (NASA: Washington, DC, 1975).

mental units were specialized mock-up stands. They made it possible to introduce essential changes during the hardware design process.

Even after Korolev's death, special testing facilities created for the entire field at the Scientific-Research Institute of Chemical Machine Building (NIIKhimmash) and at the Central Design Bureau of Machine Building (TsKBM) in Reutov continued to develop and implemented the lessons of his school.[27]

OUR RADIO ELECTRONIC DEPARTMENTS SEETHED WITH PASSIONS THAT WERE MORE THAN PURELY CREATIVE. The desire to touch space "with our own hands" overcame the fear of the unknown. The engineers who had created the rocket and spacecraft systems understood better than anyone the dangers that human spaceflight entailed. And, nevertheless, from the teams in my departments alone, five men flew into space!

I am repeating myself, but I feel compelled to mention them again in alphabetical order. Vladimir Viktorovich Aksenov began working as a designer in Chizhikov's department in 1957. He was promoted through all the levels of engineering work up to laboratory chief. In 1973, he became a member of TsKBEM's cosmonaut corps. He participated in two spaceflights as flight engineer; in 1976 on *Soyuz-22* and in 1980 on *Soyuz T-2.*

After demobilization from the army in 1964, two-time Hero of the Soviet Union Aleksandr Pavlovich Aleksandrov began work as a technician in Bashkin's department while at the same time attending the MVTU night school department at our enterprise. He took a class of my lectures, received a higher education, and was space station shift flight director. He completed two spaceflights as flight engineer, in 1983 on *Soyuz T-9* and the *Salyut-7* space station, and in 1987 on *Soyuz TM-3* and the *Mir* space station. In 2003, I congratulated Aleksandrov on his 60th birthday and recalled that his father Pavel Sergeyevich and mother Valentina Vasilyevna had been employees of the Moscow GIRD under the direction of Fridrikh Tsander and then Sergey Korolev. I met his father, a lieutenant colonel, in Bleicherode, Germany, and later worked with him as the OKB-1 military acceptance representative.

After graduating from MVTU, Aleksey Stanislavovich Yeliseyev began graduate school at MFTI and in 1962, simultaneously began to work as a senior technician in Legostayev's department, where he went on to attain the rank of senior engineer. He completed three spaceflights as flight-engineer, including two flights in 1969, on *Soyuz-5*, executing a transfer in open space to *Soyuz-4*, and a flight on *Soyuz-8*. In 1971 he flew on *Soyuz-10*. He is a two-time Hero of the Soviet Union, was NPO Energiya deputy general designer, and served as president of MVTU from 1986 to

---

27. NIIKhimmash, formerly the NII-229, was the primary rocket-engine testing facility during Soviet times. TsKBM was the name of Vladimir Chelomey's OKB-52 organization from 1966 to 1983.

1991.

Viktor Ivanovich Patsayev was already an engineer when he came to OKB-1. Before transferring to the group of cosmonauts he worked as a designer in Krayushkin's antenna department. He completed a space flight in 1971 as test-engineer on *Soyuz-11* and on the long-duration *Salyut* space station. He perished during the return to Earth and was awarded the Hero of the Soviet Union title posthumously.

Engineer-physicist Nikolay Nikolayevich Rukavishnikov was involved for two years in the development and full-scale testing of automatic control systems and nuclear reactor shielding. In 1960, after the merger of TsNII-58 and OKB-1, he was appointed senior engineer and soon thereafter leader of a group in Kupriyanchik's department. He developed electronic instruments, the precursors of modern-day computers, for the first automatic interplanetary spacecraft and also instruments for the manual control of the L-1 spacecraft for the lunar fly-by.[28]

Rukavishnikov completed his first spaceflight as test-engineer on *Soyuz-10*, along with Shatalov and Yeliseyev. Docking for transfer to the long-duration Salyut space station failed to take place due to equipment malfunction. He completed his second spaceflight in December 1974 as flight-engineer of Soyuz-16 on a 7K-TM spacecraft for the Apollo-Soyuz Experimental Flight (EPAS) in preparation for a rendezvous with the U.S. Apollo spacecraft.[29] His third spaceflight in April 1979 as commander of *Soyuz-33* on a mission to visit the long-duration *Salyut-6* space station almost ended in tragedy. The spacecraft's main braking engine failed, and the crew executed an emergency ballistic landing using a backup engine. In 1981, two-time Hero of the Soviet Union Rukavishnikov was elected president of the USSR Federation of Cosmonautics. His activity in this post often intersected with my work as director of "Korolevian" academic readings.[30]

In 1992, I agreed on behalf of the firm to fly to Tomsk to take part in an anniversary celebration and to strengthen business ties with a Tomsk electrical engineering firm. Nikolay Nikolayevich eagerly accepted my proposal to keep me company. Tomsk was his home town; he was an honorary citizen of this town. The CPSU regional committee, which in such cases had been in charge of receiving honorary guests, no longer existed, but the traditions of Party hospitality had been preserved.

Rukavishnikov and I stayed in a large regional committee hotel on the bank of the Tom, a tributary of the Ob River. As we strolled along the high bank admiring the wide river, Rukavishnikov said, "Just think, I could have been way over there

---

28. The L-1 spacecraft was better known in public as the Zond series of vehicles. These were designed to send a dual-cosmonaut crew on a circumlunar flight in the late 1960s.

29. The 7K-TM was a special variant of the Soyuz spacecraft designed for ASTP.

30. The "Korolev Readings" are annual sessions devoted to papers on the Russian space program. Similar "readings" are also held in honor of many other Soviet scientists and cosmonauts (including Yuriy Gagarin and Konstantin Tsiolkovskiy).

instead of here," and he pointed to some barely visible tall smokestacks and build-ings of another town on the horizon.

"I studied to be a nuclear engineer, and over there beyond the horizon is a Tomsk quite different from the old Siberian Tomsk. It's an ultramodern nuclear Tomsk that produces plutonium or something else for nuclear weapons."

"Do you have regrets, Nikolay Nikolayevich?"

"No, no regrets. Even when the failsafe engine developed by Isayev's design bureau failed [on *Soyuz-33*] and I could have gotten stuck in orbit, I had no regrets and wasn't even afraid."

Ten years later, tired from fatigue, I sat down on a playground bench in an Ostankino courtyard. My memory clearly replayed the tranquil conversation about human destinies that Rukavishnikov and I had had on the bank of the Tom. I plodded out of the space housing complex on Khovanskaya Street after the memo-rial service—a ceremony at the gravesite at the Ostankino cemetery—after hearing reminiscences about cosmonaut Rukavishnikov at his funeral.[31]

As a student of the Moscow Engineering Institute of Geodesy, Aerial Survey-ing and Cartography (MIIGAiK) Viktor Petrovich Savinykh spent his graduate residency in Bashkin's department.[32] After defending his diploma project, he stayed there to work, but then transferred to the newly formed instrument department headed by chief optics specialist Stanislav Andreyevich Savchenko. Savinykh com-pleted his first space flight in March 1981, as flight-engineer of *Soyuz T-4* in a mis-sion to the *Salyut-6* station. He completed his second space flight in 1985, also as a flight-engineer on Soyuz T-13 on a mission to the *Salyut-7* station. Together with Vladimir Dzhanibekov, Savinykh performed heroic work to restore the operating capability of the "dead station." He completed his third flight in 1988 as flight-engi-neer on *Soyuz TM-5* to the *Mir* station. His total time spent in space is more than 252 days! Two-time Hero of the Soviet Union Viktor Savinykh has been president of MIIGAiK since 1989 and president of the Russian Association of Institutions of Higher Learning since 1990.

Space wasn't the only place we sent our control systems specialists. By voluntary-compulsory agreement of the parties involved, our specialists bolstered the ranks of high governmental and Communist Party offices. Radio engineers Aleksey Alek-seyevich Shananin, Aleksandr Ivanovich Tsarev, and Oleg Genrikhovich Ivanovskiy moved from Podlipki to the Kremlin to positions of importance in the Military-Industrial Commission. Radio engineer Viktor Alekseyevich Popov received the office of CPSU Central Committee "instructor" in its Defense Department on Old Square.

In our history, both mergers and divisions have been successes that have contrib-uted to the successes of space technology. In 1950, still in my post as NII-88 deputy

---

31. Rukavishnikov died on 18 October 2002.
32. MIIGAiK—*Moskovskiy institut inzhenerov geodezii, aerofotosyemki i kartografii.*

chief engineer, I set up a sensors and measuring systems laboratory. When OKB-1 was formed, this laboratory moved to Department No. 5, headed first by Yangel and then by me. During the great merger of 1960, laboratory chief Ivan Ivanovich Utkin convinced Korolev that it would be expedient to transfer his laboratories to NII-88 (now TsNIIMash). There he set up a specialized department, on the basis of which the Scientific-Research Institute of Measurement Technology was formed in 1966. The founders and managers of the nation's leading governmental center of measuring instrumentation never failed to remind me that they were of "Korolevian" descent and that in their distant, hazy youth I had been their direct boss. Alas, today not a single one of them is still alive.

For our body of work in the development of control systems in the field of cosmonautics, three individuals—Rauschenbach, Legostayev, and myself—were elected full members, that is, academicians of the Academy of Sciences. Besides the three of us, the Russian Academy of Sciences has only one other scientist with this specialization, Corresponding Member Gennadiy Petrovich Anshakov, organizer of the school of control systems for reconnaissance and remote Earth sensing spacecraft at the Progress State Special Design Bureau (GSKB) in Samara.[33]

---

33. GSKB—*Gosudarstvennoye spetsialnoye konstruktorskoye byuro.*

Chapter 29

Ye-2 Flies to the Moon and We Fly to Koshka

In September 1959, we proved to the whole world that Block Ye, the third stage of the R-7A intercontinental missile, was capable of achieving escape velocity and delivering a payload to the surface of the Moon. The world did not know, though, that of the six three stage missiles with the code number 8K72, the Block Ye had only managed to work twice. In four launches, the mission fell short of Block Ye firing.

However, our enthusiasm, reinforced by the government decree dated 20 March 1958, demanded that we move on to the next phase, a lunar flyby to photograph the far side of the Moon, invisible from the Earth. As was already our custom, no advance publicity about this was allowed.

Compared with a direct Moon shot, the mission to photograph its far side was immeasurably more complex. For the first time in the history of cosmonautics a spacecraft had to be created that was controlled both autonomously and by commands from the ground. A photo-television unit (FTU) was installed on the automatic station (AS), or Ye-2.[1] When the AS reached the lunar region, the orientation system was supposed to turn the station so that the camera lenses were pointed at the far side of the Moon not visible from the Earth. Meanwhile the control system would have to stabilize the AS, switch on the FTU, and shut it down after 40 to 50 minutes.

According to the joint calculations conducted by Okhotsimskiy's mathematical group at OPM, Lavrov's at OKB-1, and Elyasberg's at NII-4, the distance from the station to the lunar surface during the photography process would be around 7,000 kilometers. They selected an extremely elongated elliptical orbit encompassing the Moon and the Earth.

To set up the requisite orbit skirting the Moon's far side, the "celestial mechanics" from OPM proposed using the Moon's gravitational pull. The trajectory of the flyby was calculated so as to obtain the maximum amount of information during the first orbital pass. There was supposed to be enough film on board for a second

---

1. FTU—*Fototelevizionnoye ustroystvo*; AS—*Avtomaticheskoy stantsii.*

orbital pass of the Moon and Earth, but in reality would there be a second orbital pass? There were many disputes about the trajectory selection. The problem was further complicated by the fact that, in order to downlink the photographic results successfully via radio-link during the return to Earth, the AS would have to be over the northern hemisphere since the nation's first and (at the time) only interplanetary communications facility had been built in the Crimea on Mount Koshka in the area of Simeiz.

While discussing the trajectory proposed by the ballistics specialists, we demanded that they solemnly swear that during the station's first orbital pass while coming back toward the Earth it would not graze the atmosphere and burn up. We fiercely debated over the station's possible life expectancy. These arguments affected me directly because, depending on the length of its life cycle and the number of communications sessions, the designers and I would have to address a number of issues: determine the parameters for the power supply system and the program timing devices, arrange with Ryazanskiy and Boguslavskiy about the resources and number of commands in the radio system, and resolve a plethora of other issues that we would be facing for the first time. It was damn interesting to puzzle over and work on all of these cases, which are now textbook classics.

Systems production and testing took place in 1959. I already had a great deal of experience in the developmental testing of control system instruments for combat missiles, and I tried in every way possible to carry it over to the Ye-2 systems. My skepticism toward the issue of reliability was very strong and well founded. Using today's standard of reliability theory, if you were to calculate the probability of photographing the far side of the Moon with equipment produced back then, the chances for success would be no higher than 20% to 30%.

Besides the stabilization and attitude control system developed in Rauschenbach's department at NII-1, the item that caused the most trouble was the *Yenisey* photo-television unit that everyone called the "bath and laundry trust company." NII-380 in Leningrad, later known as the All-Union Scientific-Research Institute of Television, developed this FTU according to our specifications. A team of enthusiasts headed by Director Igor Rosselevich and engineers Pyotr Bratslavets and Igor Valik developed the self-adjusting photo-television unit within a period of time that was quite fantastic even by today's standards. A dual-lens camera took pictures while automatically changing the exposure. The process began only upon receipt of a command indicating precise targeting on the moon. When the photographic session was completed, the film went to the automatic processing unit, where it was developed, fixed, dried, rewound into a special cassette, and prepared for transmission of the images.

I had been an amateur photographer since my childhood. Perhaps this is why I especially sympathized with the team of the photo-television unit specialists who took the brunt of the higher-ups' wrath and the testers' rebukes for the numerous failures and continuous disruptions of the preparation schedule during *Yenisey* testing at the firing range.

Cathode ray tubes and a photomultiplier were used to convert the negative image captured on film into electrical signals. The process involved the electronics of scanning, amplifying, forming a signal, and other processes necessary to feed the data to the radio-link. A new aspect of the process was the broad use of semiconductors—transistors—instead of vacuum tubes. At that time, this was considered exotic and was quite risky.

The spacecraft transmitted images to Earth via a radio-link, which also helped to measure the motion parameters of the spacecraft itself and to transmit telemetry parameters. The same radio-link was used for radio commands for onboard systems' and for receiving response acknowledgements. This was a complex integrated radio system developed at NII-885 under Boguslavskiy's supervision. During work on this system I had many rather friendly debates with him over the selection of the radio transmission principle.

While still in Germany studying German radio control and telemetry expertise, Boguslavskiy had criticized the Germans for using continuous-wave radiation instead of the pulse radiation that was widely used in radar. Developing new systems independently, Boguslavskiy pushed through pulse ideas in every way possible. I supported him in this. I had become used to working with pulse methods back in 1943 while working with Popov on an aircraft positioning system.

Despite his previous preference for pulse methods, Boguslavskiy started to develop an integrated continuous-wave radio-link for the Ye-2. We radio specialists, and there were quite a few of us at OKB-1 at that time, tried to influence Boguslavskiy in reembracing his pulse "world view," but he stood his ground.

S.P. got wind of our differences of opinion. He demanded explanations from Ryazanskiy, who was responsible for the radio system as a whole. The matter was brought up in a private conference. Boguslavskiy forthrightly declared that he was not retreating from his commitment to pulse methods, but that with this deadline it was only possible to develop a reliable system using tested continuous-wave methods. With that we settled our differences in the interests of deadlines and reliability.

As a rule, victors don't have to justify themselves, but the faint and fuzzy image that was first transmitted was due to the radio-link's insufficient power. Many years later, still good friends, Boguslavskiy and I discussed this episode after communications sessions during evening strolls around the grounds of the Simferopol and Yevpatoriya space communications radio centers.

Boguslavskiy was also responsible for the philosophy behind the entire complex of ground radio equipment, command unit, powerful radio transmitters, receiving and recording units, and antenna systems. Military unit 32103 and NII-885 were able to successfully construct and prepare the first space communications station on Mount Koshka in the Crimea for such a crucial job as the far side photography mission because of the smooth working relationship between them.[2] The southern side of the mountain where the station was built faced the sea. There was virtually

---

2. Military Unit 32103 ran the Command and Measurement Complex or the ground communications segment of the Soviet space program. NII-885 was the primary research institute devoted to developing guidance systems for Soviet missiles.

no industrial radio interference. The Crimean climate made it possible to work year-round without a break.

The communications center was part of a large Command and Measurement Complex (KIK) system. At that time the KIK was still under the authority of NII-4 and General Sokolov. The practice we had gained during our 1958 launch failures confirmed that every cloud has its silver lining. When we finally achieved failure-free performance and pulled off a successful Moon shot, the long-range radio communications system had been thoroughly developed and tested.

From the author's archives.

A Ye-2A spacecraft shown on a dolly. Such a vehicle flew by the Moon in October 1959 and captured the first photographs of the farside of the Moon. Later known as Luna-3, the vehicle passed over the southern lunar polar cap at a range of about 7,900 kilometers on 6 October. During its flyby, it took 29 photographs which gave humanity its first view of the farside.

THE ASSEMBLY AND TESTING OF THE AUTOMATIC STATION AT THE FACTORY HADN'T BEEN COMPLETED BY THE REQUIRED DEADLINE. Considering that all of the most qualified testers were at the firing range all the time, with Korolev's agreement, Turkov sent the vehicle to the firing range for final testing and adjustment in August 1959. By that time a system for preparing unfinished articles had already been set up at the engineering facility.

Arkadiy Ostashev and I shared the duties of constantly supervising and monitoring the testing. He generously agreed to be at the Assembly and Testing Building primarily at night, leaving me with the day to work, and also deal with the numerous higher ups who still preferred to catch up on their sleep at night, or to make progress reports to Moscow to quite high-ranking leaders. During this testing, we were simultaneously conducting launch preparations for the Ye-1 Moon shot, the one that carried the historic pendant to the Moon.

From the very start, the first spacecraft tests differed fundamentally from aircraft tests. A test pilot tests an aircraft. The aviation chief designer and his close associates would usually stand on the airfield, fret, and wait for the landing and the pilot's report. In the case of a spacecraft at the firing range, before a launch, the testers and developers tested a spacecraft together. They formed such a close-knit team that you could not always tell who the developer was and who the tester was. Usually a spacecraft arrived at the firing range without having been completely finished and tested at the factory. The systems developers knew about many of their mistakes before testing had begun in the Assembly and Testing Building at the engineering facility, and many more were discovered afterward.

The Ye-2, the first spacecraft equipped with a motion control system and a complex radio system, was a typical example of this process. As usual, testing took place under conditions of constant stress. Time flew before the launch window; it accelerated at an uncanny rate. The closer you got to the deadline, the more snags you found; there were more unforeseen defects, failures, and systems affecting each other for reasons unknown. Sometimes it seemed we were absolutely overwhelmed with obstacles with no end in sight and would have to report that "It is impossible to prepare the vehicle by the deadline. The launch must be canceled!" But this didn't happen. We all believed in success and supported this belief in each other.

During Ye-2 preparation in September and October 1959, the developers of the world's first spacecraft attitude control system, Ye. A. Bashkin, D. A. Knyazev, Ye. N. Tokar, V. P. Legostayev, Yu. V. Sparzhin, V. A. Nikolayev, A. I. Patsiora, M. M. Tyulkin, and A. V. Chukanov, who were all members of the same Rauschenbach team that we first saw at NII-1 with Keldysh, won me over with their engineering fanaticism. They found solutions to what seemed like the most hopeless problems.

After transferring to us at OKB-1 from NII-1, Bashkin soon became one of our leading specialists, the chief of the large spacecraft control systems department. Although he possessed invaluable experience with the space program, in his quest for new areas to apply his talents, he later switched over—to my regret—to work at the television center. Knyazev succeeded in organizing projects in a new field with us at OKB-1, specifically microthruster actuator systems. His tragic death in a plane crash was a terrible blow for all of us.

Bashkin and Knyazev each very clearly explained the problems they found in their operations to Keldysh and Korolev, who were quite anxiously monitoring the testing process on the system, especially since it had been produced by a team of nonprofessionals. A general feeling of optimism seasoned with a good portion of humor were usually enough to reassure them after another restless night.

It was much more difficult to understand problems with the radio equipment. If the radio commands failed, they caused onboard equipment to malfunction. But most often the culprit turned out to be the testing station rather than the system itself. At the very beginning of the space age an American rocket specialist very aptly stated, "If everything goes well during testing, it means that you missed something." Usually that's just what happened.

The *Yenisey* gave us the most trouble during preparations. During integrated real-time tests all the commands were executed, but the film came out first with spots and then overexposed and cloudy. We formed and exchanged all sorts of hypotheses and solutions. Valik and Bratslavets went without sleep for countless nights. One night a phone call from Arkadiy Ostashev woke me up. Almost shouting for joy

Closeups of the Ye–2A (Luna–3) spacecraft shows the "top" and "bottom" of the vehicle. The left image shows the lens of the imaging system (Yenisey) in the center surrounded on the sides by various scientific instruments and sensors mounted on thermal shielding. The right image shows gas jet nozzles and a solar sensor for attitude control.

he reported, "Boris Yeseyevich, the alchemists finally did it. The film is perfect. I request permission to give the order not to change anything and to prepare the last complex by morning."

This was a week before the historic pendant on *Luna-2* impacted on the Moon. After that historic event we flew home for a few days for "a change of underwear" and a breath of air in Moscow and Podlipki. The day after we returned to the firing range I appeared before Korolev for a progress report on Ye-2 preparations and to coordinate the program for the near future. He was very excited by the response to *Luna-2*—the international successes, the nationwide celebration, and the obvious regard of Khrushchev, whose return from America we were expecting on 28 September.

"Well, we won't be meeting with him in Moscow," remarked Korolev with visible regret. "We have to fly out to the firing range and prepare for a launch on October 3rd or 4th. No later! Don't be long. You and Ostashev make sure you leave in a couple of days and, look, we absolutely must not disgrace ourselves now. Lovell and the Americans will be tracking the launch. Keldysh wants the mission objective to be announced immediately after the vehicle goes into orbit. So if we don't see the far side of the Moon, it will be a tremendous disgrace. Report to me as soon as you return to Site No. 2."

On 17 September I returned to Site No. 2 and immersed myself in the continuous, round-the-clock stream of testing concerns. By 25 September we had received relatively solid assurance that, it appeared, all the *bobiki* were gone and we could move on to mate the automatic station with the third stage and then to the assembly and final testing of the entire cluster.

Soon thereafter I had the opportunity to admire the work of Captain Sinekolodetskiy. In soft slippers he deftly moved along the rocket boosters suspended beneath the roof and gave commands using gestures comprehensible only to him and the crane operator. These were signs similar to those used by the deaf, but the heavy-duty overhead traveling crane followed all his commands very precisely. The spectacle of the rocket cluster's nighttime assembly was a real pleasure.

On 28 September at the Sport Palace in Luzhinki, a meeting took place on the occasion of Khrushchev's return from America. Khrushchev was greeted by an auto worker, a collective farm brigade chief, an MVTU coed, and, on behalf of the scientific community, Academician Leonid Sedov. With all due respect and goodwill to Leonid Ivanovich, a great scientist and mechanical engineer of our day, I shared Korolev's hard feelings. To this day, abroad some still call Sedov the "father of the Soviet sputnik." The true creators never got their 15 minutes of fame to boost their spirits.

All those who spoke at the meeting, including Sedov, praised the achievements of "scientists, engineers, and workers who fulfilled humankind's ancient dreams, those who led the way into space and to interplanetary flight." Khrushchev's speech genuinely elated all present at the meeting and the millions in the radio audience. And indeed, he was sincere when he said:

"Our time can and must be a time for the fulfillment of great ideals, a time of peace and progress. The Soviet government recognized this long ago… From this high podium, standing before Muscovites, before all my people, before the government and the Party, I must say that President of the United States of America Dwight Eisenhower displayed statesmanlike wisdom in his assessment of the current international situation. He displayed courage and will…

At the same time, I have gotten the impression that in America there are forces that are not acting in concert with the President. These forces advocate a continuation of the Cold War and the arms race…"[3]

At that time we all had not only underestimated these forces in the U.S., but we had not even thought that such forces also existed among us. They brought the world to the brink of catastrophe just three years later.[4]

AND WHILE OVATIONS WERE THUNDERING ACROSS THE NATION, we were preparing the Ye-2. Preparation at the launch site was proceeding relatively calmly. As I left the launch pad at T-minus 30 minutes to go to the tracking station, in keeping with tradition I told Leonid Voskresenskiy and Yevgeniy Ostashev to "break a leg." Together, they replied, "Go to hell."[5]

The rocket carrying the new lunar spacecraft lifted off just 20 days after the first lunar impact. On 4 October, the second anniversary of the beginning of the space age, Yuriy Levitan announced to the world that a "third cosmic rocket" had been

---

3. *Zhit v mire i druzhbe!* [*To Live in Peace and Friendship!*] (Moscow: Politizdat Publishing House, 1959), pp. 415–16.

4. Here, Chertok is alluding to the Cuban Missile Crisis, which he describes in Volume 3.

5. Russian superstition requires "Break a leg" to be answered with "Go to hell." Not responding is considered bad luck.

successfully launched. Despite promises, the hyper-cautious authors of the TASS report tossed out all references to the flight's primary objective, that of photographing the far side of the moon.

By midday on 4 October, the State Commission was informed that the Control Center on Mount Koshka was monitoring and communicating "using all means." Everything was OK on board and work was continuing according to the program. Early on the morning of 5 October, we left the firing range. The "bath and laundry trust company" team flew to the Crimea and the rest of us flew to Moscow. We celebrated the second anniversary of the first satellite launch on the Il-14 airplane en route to Vnukovo.[6]

After arriving in Moscow, on 6 October I convened a meeting hoping to determine, first of all, the status of operations on future spacecraft scheduled to reach Venus. Celestial mechanics determined the Venus launch dates, and a delay of even one week meant the dates would be postponed for at least a year. In the first half hour of conversation I realized that preparation of the Venus spacecraft was in a catastrophic state. However, my intentions to switch from the Moon to Venus proved to be obviously premature. The telephone rang; an unexpected call from Korolev:

"Boris, get here right away! Don't bring any papers with you. Bear in mind, you won't be going back to your office today."

"Sergey Pavlovich, what about Mars and Venus? The situation is critical!"

"No, did you understand what I said?! You have enough deputies. Get here right away!"

When I arrived, S.P. was on the Kremlin hotline making arrangements with Vladimirskiy, then with Keldysh and Ryazanskiy about when we would take off from Vnukovo.[7] Ostashev, who was summoned right after me, was trying to say something, but S.P. wasn't listening.

"Radio communications with the spacecraft are very poor," he said. "The telemetry is not coming through; radio commands are not getting on board. We are flying to the Crimea and have to be in place before the communications session starting at 4:00 p.m.—that's the time of radio coverage from the Crimea. Two cars are already parked down at the entrance. Figure out who takes which one. Stop by your homes, grab what you need, and drive to Vnukovo. A Tu-104 is waiting for us there—a charter flight. They'll let you go straight to the plane. Departure is at 12:00. We need to arrive there ahead of time to look at the situation and decide what to do."

We both understood that there was no time for inquiries and discussion. On the way to Vnukovo I stopped by my home at 3 Ostankinskaya, and, with a speed now familiar to my wife Katya, I repacked the carry-on bag I had brought with me the

---

6. Vnukovo, Moscow's first international airport, opened in 1941.
7. Sergey Mikhaylovich Vladimirskiy was a deputy chairman of the State Committee for Radio-Electronics (GKRE), the ministry that oversaw the development of the lunar probe's telemetry and communications systems.

day before from the firing range for my new assignment. At the entrance to the airfield the duty attendant simply asked, "Charter flight? Your party has already gone through. Hurry," and he pointed in the direction of the airplane. The Tu-104 was the first jetliner of our civil aviation.[8] It was still a great rarity for domestic flights. It was easy to find this airplane on the airfield.

After climbing on board the aircraft, to my surprise, I saw Keldysh, Vladimirskiy, and Ryazanskiy—all smiling—and S.P.'s disgruntled, worried face. He pounced on me:

"Where's Ostashev? I gave you two cars!"

"But, Sergey Pavlovich, two cars can't make the road any shorter and can't make us go twice as fast," I objected. "Arkadiy will be here any minute now."

In such instances it would have been useless to make excuses or to object. S.P. could not bear to wait idly if he was in a big hurry. He couldn't rail against Keldysh. As I later found out, he had already blown up at Vladimirskiy and Ryazanskiy for "the failed transmission of the radio commands." Now his deputy Chertok was late and Ostashev wasn't there at all! And in a situation like this Keldysh still allows himself to smile! S.P. became more and more incensed, and about 10 minutes after I showed up, he commanded the crew to taxi and take off. S.P.'s agitation had reached the limit. To calm himself he went into the cockpit.

"We can't wait any longer."

They took away the boarding ramp and battened down the doors. The jet engines roared, and the airplane began to taxi to the takeoff strip. Suddenly, a car came careening across all the concrete runways on a course to intercept the taxiing airplane. Out jumped Ostashev, desperately flailing his carry-on bag. The airplane stopped. They quickly let down the onboard ladder and admitted the tardy passenger on board. S.P. came out into the passenger cabin, shook his fist at Ostashev and uttered words whose meaning one could only guess in the roaring crescendo of the engines.

For those times, the Tu-104 was a comfortable, top-of-the-line, high-speed aircraft. Instead of a little over a hundred passengers, there were only six of us. Except for Keldysh, this was the first time any of us had been on board such an aircraft. Smirking good-naturedly, he continued to joke that this flight was Korolev's extravagance. Since we were already here, we should enjoy the "world class" accommodations and service. Having only flown in our own cargo-and-personnel Il-14 or Li-2 service airplanes, we were not accustomed to well-dressed stewardesses. The aircraft and crew had been suddenly taken off of an international flight, and so the cute young ladies had the opportunity to serve a single table and treat us to a delicious dinner.

Soon S.P. was in a good mood. Responding to everybody's praise for the air-

---

8. The Tupolev Tu-104, the first Soviet jet airliner, began regular scheduled civilian flights in September 1956 between Moscow and Irkutsk in the Soviet far east.

plane, the dinner, and stewardesses, he declared, "Ah well, soon we'll be getting these planes too and we'll lure these young ladies away. But remember, we'll only be allowed in such an airplane for good behavior. And if your radio commands fail to go through, Mikhail," he said turning to Ryazanskiy, "you'll be flying on the Li-2 and it will be a long time before you see stewardesses like these again."

"But now, my darling boys," continued Korolev, "keep in mind that we will be landing at a military airfield. A helicopter is waiting to take us to Ay-Petri. Crimean officials will meet us there and take us straight to the control center. For relaxation, if there is any, they've reserved suites for us at the *Nizhnyaya Oreanda* (Lower Oreanda).[9]

Korolev had decided to fly our group to the Crimea just that morning. In a little over an hour he had managed to organize this surprise expedition supported by Aeroflot, the Air Force, the Crimean regional committee of the CPSU, and the USSR Council of Ministers Directorate of Affairs. His brilliant organizational skills came out even in problems that seemingly had nothing to do with systems engineering.

Our flight to the Crimea showed that Korolev knew how to maintain good relations with high officials of the Party and government hierarchy. Korolev's name was no secret to them. They knew perfectly well who had actually delivered lunar pendants to two addresses, and they took into account Khrushchev's regard for Korolev.

At the military airfield we exchanged a warm farewell with the hospitable Tu-104 crew. Air Force commanders greeted us as we came down the boarding ramp, and we squeezed into a helicopter, its rotor already spinning. After crossing over the Crimean Mountains the helicopter flew along the shoreline. There I saw Koktebel and Karadag, the *Zolotyye vorota* (Golden Gates).[10] The last time I was here was with Katya, Isayev, and a team from Bolkhovitinov's OKB the year before the war.[11] Unable to resist, under the racket of the helicopter's engine I recited Pushkin:

> *How beautiful you are, o shores of Tauris*
> *When seen from a ship at sea*
> *As the Morning Star is shining*
> *When you first appeared to me.*[12]

---

9. *Nizhnaya oreanda* was (and still is) a plush resort hotel on the southern coast of Crimea frequented by the noble families during the Tsarist era.

10. Koktebel was (and still is) a beach resort on the eastern end of the Crimean coastline on the Black Sea. It is located near the strikingly beautiful Kara-Dag volcanic mountains, remnants from the Jurassic Era. The area was declared a national reserve in 1979. The *Zolotyye Vorota* (Golden Gates), named after one of Ukraine's oldest monuments created in the 11th century, are beautiful rock formations in the Black Sea.

11. Chertok, *Rockets and People*. Vol. 1, pp. 108, 153–155.

12. This is an excerpt from Aleksandr Sergeyevich Pushkin's *Yevgeniy Onegin* [*Eugene Onegin*]. For a recent translation, see Alexander Pushkin, *Eugene Onegin: A Novel in Verse*, trans. Charles Johnston (London: Penguin, 2003). Pushkin spent some time at Tavrida in Crimea in 1820 which he memorialized in *Eugene Onegin*.

"There goes Boris!" laughed S.P. He was clearly pleased that everything was going precisely according to schedule. He even had time for an unscheduled look at Koktebel, the site of his romantic glider-borne youth.[13] My poetic mood was interrupted by the helicopter's commander. He entered the passenger cabin and, having correctly recognized Korolev as the boss, reported: "Wet snow is falling in the Ay-Petri area. Visibility is virtually zero. Landing is not advisable."

Korolev understood that the decision was up to him.

"We're in a big hurry. Cars are waiting for us at Ay-Petri. Perhaps we can risk it?"

The commander agreed that they could take a chance on landing, but he held his ground.

"It's unwise to drive down from Ay-Petri in weather like this. It's a big risk."

Everyone agreed that there was no sense in us getting into an automobile accident. The commander suggested that we land on a helicopter pad in the mountains near Yalta. Korolev agreed. The commander got on the radio with the Yalta CPSU municipal committee *(gorkom)* and asked them to send cars to pick us up.[14] He wasn't authorized to tell them who we were and why we were landing near Yalta over the radio. According to the regulations of the security services, none of the local authorities were supposed to know about our flight into the Crimea. Nevertheless, when we climbed out of the helicopter and said farewell to our pilots, Yalta Party leaders had already driven out to meet us in ZIM and Pobeda automobiles.

The secretary of the Yalta municipal committee was obviously pleased that we were surprised: "You thought that you were here illegally? The cars couldn't have been sent to Ay-Petri without my involvement. As you can see, news travels fast here. We tracked the helicopter. We are prepared to provide you and your companions everything you'll need to relax after your stressful work. We find this more pleasant than indulging the whims of various high-ranking officials' wives."

Korolev thanked him on behalf of us all and said that he was sorry that we did not have even an hour for recreation and strolling.

"We are very pressed for time. Please take us to the control center in Simeiz."

The Yalta boss was clearly disappointed. He had hoped that he might provide all the pleasures that the best of the resort palaces had to offer these highly classified developers of the secret lunar rockets and at the same time join them for some wining and dining. We squeezed into the ZIM and took off at top speed along the narrow winding Crimean road toward Simeiz. After leaving home at 11:00 a.m., switching from an automobile to a jetliner, then to a helicopter, and back to an automobile, at 2:30 p.m. we were on Mount Koshka towering over Simeiz, a famous resort on the southern shore of Crimea.

The control center was located next to a branch of the Pulkovo Observatory. The

---

13. Korolev visited Koktebel many times in his youth in the 1920s during his days as a glider pilot.

14. *Gorkom—Gorodskoy komitet* (literally, city committee)—was one of the local levels of the Communist Party structure.

main structure was a flat rotary antenna with an area of 120 square meters. Transmitters and receivers were housed in trucks. The control center itself was crammed into a temporary wooden barrack. The photo-recording gear was installed in one of its tiny rooms. The image of the far side of the Moon was supposed to appear on the heat-sensitive paper of these instruments, which required no development process. Simultaneously the image was also recorded on conventional movie film, which required a prolonged chemical treatment process. It was not possible to develop the movie film on site. It was assumed that this would be done in Moscow.

The control center personnel—military and civilian specialists—lived in tents. A conventional wartime field kitchen was smoking nearby. Everything indicated that the military were in charge of all the operations here. They were already doing major construction on new control centers near Simferopol and Yevpatoriya. The center on Mount Koshka was temporary, so everything had a camp-like quality.

At our first operational gathering, Boguslavskiy, who was considered our technical director, reported that the bad antenna patterns of the spacecraft's onboard antennas were evidently the main cause of the unstable communications during the first sessions. What must be must be. You can't correct the patterns. Korolev wanted to speak in person with the operators directly responsible for radio communications. Among all his other rare qualities, as we used to say, S.P. also possessed a seventh sense for detecting "irregularities and sloppiness." Right away he noticed and grasped that there were three people at once in command at the center, his favorite, Boguslavskiy (future doctor of technical sciences, future Hero of Socialist Labor, future Lenin Prize laureate), and Colonels Sytsko and Bugayev (also future laureates and chiefs of the new deep-space communications centers).

During the communications sessions the operators were turning the myriad control and adjustment knobs without a great deal of coordination. Not everyone understood when to execute a command and whose command to execute. They all respected Boguslavskiy, but any operator viewed the colonel standing over him as the more real authority.

"Attention!" commanded Korolev. "During the communications sessions I request that all reports go to Yevgeniy Yakovlevich Boguslavskiy. And I ask that all operators execute only his commands."

A very simple command, it would seem, but a new order was established immediately at the control center. Boguslavskiy felt like he was in charge and took over all responsibility for "communication with the Moon." It turned out that the colonels had quite enough to do. They stopped duplicating Boguslavskiy's actions.

At 4:00 p.m. on 6 October, the telemetry receiving session took place. To everyone's surprise, gradually, as the data was processed, it became clear that everything was working properly on board.

When the general tension had subsided after the session, Boguslavskiy and I stepped out for a smoke. A cold wind was blowing. From the observation deck a wonderful view opened up on the verdant resort coast below. The setting sun illuminated the azure bay. A lone motor boat putted along over the choppy sea.

"Do you see that motor boat?" asked Boguslavskiy. "I requested that. A boat from the Black Sea Fleet is patrolling the bay. It's carrying equipment to hunt for interference sources. In addition, during the communications sessions, as per our notification, the Black Sea Fleet 'quieted down' radio chatter and, when possible, stopped it completely. And down below, the State Traffic Patrol (GAI) isn't allowing vehicles onto the mountain road.[15] Interference has been reduced to a minimum. To tell you the truth, the power of the transmitters is low. But I think that if the "bath and laundry trust company" doesn't let us down, everything will be OK. However, we'll be receiving the picture from a range no greater than 50,000 [kilometers]!"

While telling me all this, he was eagerly smoking a Belomor, having turned down the Kazbek I offered him. When we returned from our smoking break, Korolev had already gotten himself worked up again. He demanded a report of the precise schedule for the next communications sessions and the actions to take in the event of failures.

Orientation on the far side of the Moon and then the activation of the FTU were supposed to begin early on the morning of 7 October. Bratslavets suddenly voiced his apprehension that from prior experience in simulations with the FTU the photography could take more than an hour. It turned out that here at the center the supply of special magnetic tape for the next recording of lunar landscape images from the far side had been used up. Well, that sent S.P. into a rage. I understood him. After all, if they'd let us know we could have grabbed some of that critical tape and brought it with us from Moscow. He let Ryazanskiy, Boguslavskiy, and Bratslavets have it with both barrels.

But you can't fix anything simply by shouting about it. S.P. found satisfaction only in the concrete action resulting from the tongue-lashing. He called Moscow at once, found Minister Konstantin Rudnev, explained the situation, and asked for help. Then he made some more calls to our OKB-1 and explained everything to the staff there. After many conversations with Moscow he calmed down, and turning to Colonel Bugayev, he said, "A Tu-104 is coming into Simferopol on flight such-and-such. The pilot will have a box containing the film. I'm arranging for a helicopter to be at the airport as soon as the plane lands. You must bring that film back here on that helicopter. I'm sorry, but this trouble is the fault of my comrades."

The film incident was soon resolved, and everything worked according to Korolev's schedule.

Already late in the evening, glancing over at Keldysh peacefully snoozing at some console, S.P. gave the last marching orders: "Ostashev will spend the night here, and we'll go check out the *Nizhnyaya oreanda*. Don't expect a peaceful morning. We'll be back early."

From Mount Koshka swept by the cold October winds we drove down in the

---

15. GAI—*Gosudarstvennaya avtomobilnaya inspektsiya.*

State Committee ZIM to the warm resort area and rushed to the *Nizhnyaya oreanda*. Despite the late hour, in the fashionable government health resort, the disciplined staff assigned us each a luxurious suite and announced that dinner was served. At the table spread with delicacies and wines S.P. sternly warned, "No drinking! Tomorrow we're leaving at 6:00 a.m."

We only got 4 hours of sleep that night.

ON 7 OCTOBER AT 6:30 A.M. THE FTU ON BOARD THE LUNAR SPACECRAFT WENT INTO OPERATION. At this point the spacecraft was located on a straight line between the Moon and the Sun. During the communications session on Mount Koshka the team feverishly deciphered the telemetry, which contained glitches. I couldn't restrain myself and said, "It's the Moon that's interfering with data transmission."

We needed to conserve electric power so that the storage batteries wouldn't discharge during FTU operation, so we shut down the telemetry. The photography was completed within the allotted 40 minutes. On the spacecraft, which was now flying toward the Earth, the crucial developing and fixing process began in the "bath and laundry" compartment.

The altitude from which the photography was conducted was extremely interesting for us. The trajectory measurements were processed in tandem at the NII-4 ballistic center and OPM. Now Keldysh was sitting at the telephone, while Korolev was showing his impatience. In his calm voice Keldysh said, "They've recalculated for the third time, but that's just in case. Meanwhile they assure me that we passed no more than 7,000 kilometers above the Moon's surface and so it seems as if everything is going according to schedule. Now we need to make sure that the spacecraft doesn't plow into the [Earth's] atmosphere. The Moon was "disturbed" that we were having a look at its forbidden area, and now the ballistics specialists are finding out how this disturbance will affect the spacecraft's trajectory toward the Earth."

Hours of agonizing waiting passed, during which Ostashev and I incessantly pestered Bratslavets to reassure us with telemetry data that the FTU had functioned without a glitch. Astronomer Andrey Severnyy, director of the Crimean Solar Observatory, arrived on Mount Koshka at Keldysh's invitation. He tried to add panic into the tense atmosphere of anticipation. In his words, there were absolutely no reason to fret over the successful operation of the FTU since, theoretically, we would not be able to receive any images for the simple reason that the film had been exposed to space radiation. It could have been saved only with a lead shield at least five to six centimeters thick.

We'll wait to see!

I joined Boguslavskiy by the device that recorded directly onto electrochemical paper. A report was coming in from the receiving center: "Range—50,000. Signal stable. We have reception!"

They gave the command to reproduce the image. Again it was up to the FTU. On the paper, line by line a gray image was emerging. It was a circle on which you could distinguish details if you had a sufficiently active imagination. Korolev couldn't stand it and burst in on us in the cramped room.

"Well, what have you got there?

"We have determined that the Moon is round," I replied.

Boguslavskiy pulled the paper with the recorded image out of the machine, showed it to Korolev, and calmly tore it up. S.P. didn't even lose his temper.

"Why so soon, Yevgeniy Yakovlevich? After all, this is the first, you see, the first!"

"It's bad. All sorts of junk. We'll clean up the interference and the next frames will come out right."

Gradually, one after the other the frames appeared more and more distinctly. We cheered and congratulated one another. Boguslavskiy assured us that everything would be much better on the film we would develop in Moscow. Already quite late, after parting with our fellow participants in the "campaign" on Mount Koshka, we departed once again for "our health resort." This time Korolev allowed Ostashev to ride with us. I shared a luxurious suite with him. At supper the ban was lifted on the consumption of wines from the government wine cellars.

AT OUR EARLY BREAKFAST KOROLEV PROPOSED THAT WE SEE HOW CONSTRUCTION WAS GOING ON THE NEW DEEP-SPACE COMMUNICATION CENTER NEAR YEVPATORIYA. We set off as a foursome—Korolev, Keldysh, Ryazanskiy, and I—on the drive from Simeiz to Yevpatoriya. After 3 hours on the road through the Crimea we were met by the deputy commander of Military Unit 32103, Colonel Pavel Agadzhanov; military unit 32103 was the organization that was actually in charge of the entire Command and Measurement Complex.

A military work force was building the Yevpatoriya center, which was simply known as NIP-16.[16] Civilian specialists were involved in assembling and debugging the systems equipment, which had been developed at many different organizations, including NII-885, SKB-567, TsNII-173, and MNII-1. The antenna system, which was colossal for those times, was erected in the immediate vicinity of the magnificent Black Sea beaches. There were very few people in that area of the Crimean coast. During the peak resort seasons the sandy beaches that stretched for tens of kilometers seemed deserted.

According to preliminary calculations, for stable communications with spacecraft within the solar system, a dish antenna with a diameter of around 100 meters needed to be built on Earth. Optimists estimated that it would take five to six years to build these unique facilities. But the antenna specialists had less than a year before the first Mars launches! By that time the dish antenna for NIP-10 in Simferopol had already been built. This 32-meter diameter antenna was erected for future lunar programs. It had been hoped that it would begin operating in 1962.

---

16. NIP-16 was one of many Scientific Measurement Stations (*Nauchno-izmeritelnay stantsiya*, NIP) located across the Soviet landmass that made up the ground communications network for the space program, collectively known as the Command and Measurement Complex.

From the author's archives.

Shown here are the technical leaders of the farside photography mission (Luna-3) during a break in their work on 8-9 October 1959 at Simeiz, Mount Koshka in Crimea. From the left are I. I. Pikovskiy, M. V. Keldysh, Kuznetsov, S. P. Korolev, Ye. Ya. Boguslavskiy, and M. S. Ryazanskiy.

SKB-567 Chief Designer Yevgeniy Gubenko accepted engineer Yefrem Korenberg's bold proposal: instead of one large dish, combine eight 12-meter "cups" into a single structure on a common support and rotary mechanism. The production of such medium-sized dish antennas had already been well mastered. Now they needed to learn to synchronize and combine into the necessary phase the kilowatts radiated by each of the eight antennas during transmission. They would have to combine signals of thousandths of a watt reaching the Earth from distances of hundreds of millions of kilometers.

Developing the metal structures of the mechanisms and drives for the support and rotary mechanisms was another problem that could require several years. Not having lost his sense of humor, Agadzhanov explained that Khrushchev's ban on the construction of state-of-the-art heavy ships for the Navy had rendered vital assistance to cosmonautics. Ready-made support and rotary mechanisms for large caliber gun turrets built for a battleship were quickly redirected, shipped to Yevpatoriya, and installed on concrete foundations built for two antenna systems, one for receiving and the other for transmitting.

The defense industry's Gorky Machine Building Factory manufactured the 12-meter dish antennas; the Scientific-Research Institute of Heavy Machine Building (NII Tyazhmash) assembled the metal structure to connect them; TsNII-173 of the State Committee of Defense Technology debugged the drive systems; MNII-1 of the shipbuilding industry developed the electronics for the antenna guidance and control system using their naval experience; the Ministry of Communications provided the communications lines inside NIP-16 and its outlet to the outside world; Krymenergo (the Crimean energy authority) ran power transmission lines; and military builders laid concrete roadways and built office facilities, hotels, and a full-service military post. The scale of the operation was impressive. They were on such a broad front that the deadlines mentioned by Agadzhanov hardly seemed realistic.

While we were talking, Gennadiy Guskov drove up. He was Gubenko's deputy and supervised all the radio engineering work here; when necessary he also got involved in construction problems.

"Both the ADU-1000 units, receiving and transmitting, will be delivered on time! We won't let you down," he cheerfully reported.

"Why 1,000?" asked Keldysh.

"Because the total effective area of the antenna system is 1,000 square meters."

"You don't have to boast," interjected Ryazanskiy. "The total area in your antenna won't be more than 900!"

This was an argument between followers of different ideas, but now a mere hundred square meters didn't matter.

For Agadzhanov and Guskov, work at NIP-16 served as a launching pad for careers that became part of the history of cosmonautics. Agadzhanov directed flights for many years and simultaneously headed a department at MAI. In 1984, by then a professor, he was elected a corresponding member of the USSR Academy of Sciences. At that time he was working on the development of large computer systems to manage the branches of the armed forces.[17]

Guskov also switched from pure radio engineering to its merger with computers. The NII that he organized in Zelenograd on the outskirts of Moscow (later it became NPO Elas) developed onboard computers for the flight control of spy satellites, the *Salyut* and *Mir* orbital stations, space communications systems, and many others. In 1984, he was also elected a corresponding member of the USSR Academy of Sciences.[18]

Back then, in October 1959, tanned by the Crimean sun, Agadzhanov, Guskov, and those surrounding us who were in charge of the construction of various systems, had unwavering faith that everything would be up and running by the scheduled dates. To me it seemed that it would be a miracle if the NIP-16 builders would keep their promises. However, Korolev declared, scowling, "These deadlines are not satisfactory at all. NIP-16 should be in turnkey condition in the first quarter of 1960."

The gathering dispute was interrupted by the announcement that a plane departing for Moscow was waiting for us at the naval airfield in Saki. Colonel Sytsko

---

17. Pavel Artemyevich Agadzhanov (1923–2001) served as deputy chief of the Command and Measurement Complex in 1957–71, during which period he effectively served as the "flight-director" for many important Soviet robotic and human space missions.

18. Gennadiy Yakovlevich Guskov (1918–) served as director and general designer of NPO Elas, the organization that designed many of the onboard digital computers for Soviet and Russian spacecraft.

proposed that we have lunch before we departed; Keldysh agreed. Only when he'd had a good lunch did Korolev finally relax, and, turning to the officers, he said, "Do you have any idea what a tremendous future this center has?"

"Come here in May, Sergey Pavlovich. It will be one of the best resorts in the Crimea!" said one of the officers.

"All you care about is your resorts! This one, of course, isn't bad, but the main thing is that you don't forget the deadlines!"

Upon his return to our OKB in Podlipki, to our surprise S.P. didn't throw himself into daily business, but began inviting astronomers to his office to join him in examining photographs of the far side of the Moon. But more than that, with them, he discussed

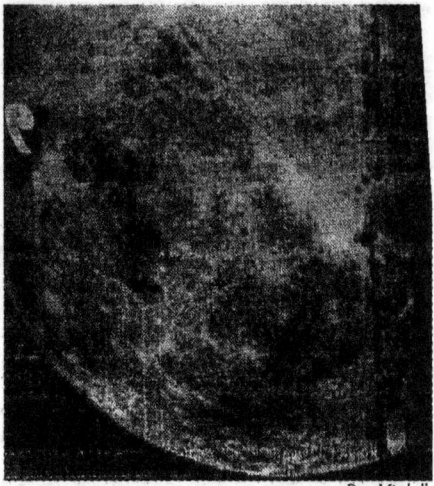

Don Mitchell.

This image from the Luna-3 mission was originally published by the USSR Academy of Sciences as part of a lunar atlas in the mid-1960s. The Luna-3 images showed for the first time that the farside of the Moon lacked the large mare areas present on the side visible to the Earth. The Soviets identified and named a number of features on the farside including the Tsiolkovskiy crater on the lower right which appears as a sea with an island in it.

possible names for the newly discovered formations on the far side. Again and again, when we tried to enter S.P.'s office, his secretary Antonina Alekseyevna warned us, "He asked not to be bothered. Shklovskiy is in there now."

Shklovskiy was already a well-known astronomer at that time.[19] But was it *our* business to think up names for the newly discovered craters on the Moon? Korolev was a strategist. He hurried to take the initiative in his own hands, fearing that those who might get better photos in the future would seize it. You needed to take everything you could from each space success.

On 27 October, newspapers published a photograph of the far side of the Moon. It seemed that the triumph was complete, but there was a misfire with the naming process. The CPSU Central Committee intervened, and this crucial work was

---

19. Iosif Samuilovich Shklovskiy (1916–85), head of the radio-astronomy department at the Shternberg Astronomical Institute, was one of the most prominent Soviet astronomers of the 20th century. His memoirs were published posthumously in English as *Five Billion Vodka Bottles on the Moon: Tales of a Soviet Scientist* (New York: W. W. Norton, 1991).

entrusted to a special commission of the Presidium of the Academy of Sciences. After long arguments, proposals for names were handed over to the Central Committee for approval. They took their time.

Finally, Keldysh's commission got the go-ahead and obtained the Academy Presidium's decision to name the craters and cirques after prominent scientists and cultural figures such as Giordano Bruno, Jules Verne, Heinrich Hertz, Igor Kurchatov, Nikolay Lobachevskiy, James Maxwell, Dmitriy Mendeleyev, Louis Pasteur, Aleksandr Popov, Marie Curie, Tsu Ch'ung Chi, and Thomas Edison.

From reliable sources, we heard that officials debated over Tsu Ch'ung Chi the most.[20] This mathematician who lived in the fifth century was supposedly famous in China, but none of my mathematician friends could explain why he was famous. But we couldn't offend China, a great and friendly nation. A Central Committee directive stipulated that an American and a Chinese should be on the list. Well, they found an easy way out with the American; everyone was happy with Edison. But for a Chinese candidate they had to coordinate with the Chinese embassy. The embassy, in turn, asked Beijing, and that's how Tsu Ch'ung Chi came to be on the list.

T.V. Prygichev.

The Yenisey imaging system used on Luna-3 was a relatively sophisticated instrument that used two lens systems, one a 200mm, f/5.6 wide-angle lens (for widescale images) and a 500mm, f/9.5 lens (for high resolution photos). The system was developed by the Leningrad-based NII-380, later the All-Union Scientific-Research Institute of Television under Igor Rosselevich and Petr Bratslavets.

After all the consultations, the Academy Presidium's decision was finally published on 18 March 1960. In the first draft of the names there was no crater named for Kurchatov. After his death in February, Keldysh and Korolev managed to have him included on the list. Now his name is next to Giordano Bruno's on the lunar map.

IT WOULD SEEM THAT NOW THE TIME HAD COME FOR US TO ENGAGE IN OTHER BURNING ISSUES. Next on the agenda were Venus and Mars. But Keldysh wasn't satisfied with the quality of the lunar photos. He consulted with Boguslavskiy's competitors, who had impressed on him that the images could be substantially

---

20. Tsu Ch'ung Chi (Zhu Chongzhi) was a Chinese mathematician of the fifth century who calculated the value of pi to the seventh decimal place.

improved if the "air-to-ground" radio-link margin were increased. And it wasn't difficult to do. The space communications center on Mount Koshka had done its part, and it was time to move near Simferopol or to Yevpatoriya. Construction of the new large-area, low-noise ground antennas had been completed there and a 10-fold increase in signal power at the ground receivers was possible.

It was difficult to argue against obvious truths corroborated by simple calculation. But no one wanted to repeat all the work that went into photographing the Moon using the same onboard equipment. Not even Korolev. I remember that Bushuyev and I—and even Tikhonravov—persuaded him to work on Keldysh and not force this job on us. Korolev hesitated. Under the astronomers' pressure, Keldysh was uncompromising and managed to have a decree issued, which obligated us to one more launch to obtain high-quality photographs of the far side of the Moon in April 1960.

The schedule in 1960 was already supersaturated with combat and space launches. Preparation for the Vostok launches for human spaceflight, involving automated and dog launches—was under way at full speed. Two Mars spacecraft were also being prepared for the fall, and there was no time for them. And now the Moon was back in our sights again.

"We'd better concentrate our efforts on the soft-landing project. We'll be performing it in two years. This is a lot more glamorous than repeating the photography," I said at various meetings, adding that the onboard radio equipment for the next Moon shot would also not be ready soon. But we couldn't avoid this lunar far side photography mission. As a result, two more hastily assembled automatic stations similar to the Ye-2 were sent to the firing range in early March 1960. Two new three stage 8K72 launch vehicles also arrived there.

Chapter 30

The Beginning of the 1960s

On 31 December 1959, Korolev assembled his inner circle at OKB-1 for the traditional end-of-the-year wrap-up and New Year's celebration.[1] S.P. presented the Ye-2 launch participants copies of the atlas *Pervyye fotografii obratnoy storony Luny* (*First Photographs of the Far Side of the Moon*) that the Academy of Sciences printing office had just issued. My copy was inscribed, "To my dear Boris Yevseyevich Chertok in fond memory of our many years of work together. 31.12.59. S. Korolev." He enclosed a replica of the lunar pendant's ribbon in the atlas.

The detailed description of the automatic station's structure, its flight, and the technology for photographing and transmitting the images of the far side of the moon did not contain a single name of the authors of this project. Only the foreword, signed by Academy of Sciences President A. N. Nesmeyanov, cited the names Galileo and Newton, and the words of N. S. Khrushchev: "How can we not rejoice and be proud of such feats of the Soviet people as the successful launch, in 1959 alone, of three cosmic rockets, which have won the admiration of all of humankind. All the Soviet people celebrate the men and women of science and labor who blazed the trail into space."

Humankind admired us, and the entire Soviet population was proud of us without knowing our names. But we didn't grumble over that. "It wasn't just humankind that appreciated our achievement," said S.P., "but also a wealthy French winemaker. He announced that he would give a thousand bottles of champagne to the ones who reveal the far side of the Moon. He was certain that we wouldn't come up with anything and wasn't afraid of the risk. But once he lost, he kept his word. Of course, there's been a hitch. The vintner asked the embassy in Paris to let him know where to send the champagne. The embassy was at a loss and asked our Ministry of Foreign Affairs. After multilevel coordination, the ministry gave instructions to send the bottles to the Academy of Sciences presidium. So, now, we have the honor of receiving several dozen bottles of champagne from the Academy's stock. You'll

---

1. Under Communism, New Year's Eve took on all the secular attributes of Christmas festivities and was the biggest holiday of the year.

snag a couple of bottles each, and the rest will be dispersed among the Party bigwigs and others who weren't involved." We sniped a lot about that. But still, you have to admit that French champagne received as a lunar congratulatory gift isn't something that everyone gets to bring home to a family New Year's party.

HAVING ENJOYED OURSELVES, we moved on to discuss goals for 1960. We were in a preholiday mood; everyone was in a hurry, even S.P. Nevertheless, discussing the list of future projects took an hour or an hour and a half. I can't quote Korolev because I didn't take verbatim notes, but I will give an account of the gist of his comments and his assessment of the goals for the year ahead.

Our **first** urgent task was to successfully launch 8K74s (or R-7A ICBMs) into the Pacific Ocean. This was not going to please Eisenhower, but might make him more accommodating at the upcoming meeting with Khrushchev. "The meeting will be in May, perhaps in June," said S.P. "I hear that they are rushing to build two cottages on the shore of Lake Baykal, one for Eisenhower and the other for Khrushchev." As far as the cottages are concerned, I can attest that they really did exist. On vacation in 1972, Katya and I had a stroke of luck. While touring around Lake Baykal, the two of us spent an entire week in one of those fashionable cottages.

Khrushchev and Eisenhower, however, never met in those fabulously beautiful sites. Perhaps history would have turned out otherwise if the budding cooperation between the two national leaders had not been destroyed. On 1 May 1960, our S-75 antiaircraft missile system designed by Petr Grushin, and aided by a guidance complex developed by Aleksandr Raspletin, shot down an American U-2 reconnaissance aircraft over the Urals.[2] More than anything else, this spy plane destroyed the hopes for rapprochement between the USSR and the U.S.

The episode with the U-2 aircraft was a striking example of the primacy of military over civilian policy, which soon became an integral trait of U.S. policy during the Cold War years. Those who supported such a dreadful policy for the next 25 years contributed to the stark militarization of public opinion and politics in the U.S., which in turn, fortified the similarly hard-headed positions of individuals in the Soviet Union. I concur completely with the claim from George F. Kennan, the former American ambassador to the Soviet Union, who wrote: "The more America's political leaders were seen in Moscow as committed to an ultimate military rather than political resolution of Soviet-American tensions, the greater was the tendency in Moscow to tighten the controls by both party and police, and the greater the braking effect on all liberalizing tendencies in the regime."[3]

But let's return to our meeting in Korolev's office. The **second** task was to speed up work on the new R-9 ICBM in every way possible. According to Korolev, RVSN Commander-in-Chief Nedelin attached exceptional value to this missile. Glushko was in a very difficult situation; during rig tests on his engines for the R-9, "high

---

2. This is a reference to the shooting down of CIA pilot Francis Gary Powers.
3. Kennan quoted in A. M. Filitov, *Kholodnaya voyna* [*The Cold War*] (Moscow: Nauka, 1991).

frequency" instabilities had occurred and the engines had failed. Glushko was also busy with developmental testing on an engine for Yangel's R-16 missile. Nedelin believed that it might be possible to begin flight-tests on the R-16 this year. Then we would be at a real disadvantage with the R-9. Korolev was quite right; successful testing of the R-16 could be the kiss of death for the R-9, considering the campaign that Yangel was waging, arguing the impracticality of liquid-oxygen propellant missiles for long-term combat duty.

For the **third** task, we needed to once again prepare a couple of rockets and spacecraft to photograph the far side of the Moon. Keldysh had held out for that. With obvious irritation Korolev spoke of his dispute with Keldysh and how he asked him not to insist on re-photographing the far side of the Moon. "But Keldysh believes," he added, "that science will not forgive us if we pass up the opportunity to take better pictures with the Sun illuminating the Moon at an angle, when there will be great contrast between the shadows and light."

Now we were finding possibilities for our *Semyorka* that we had never even thought of during its initial development. By building a third and then even a fourth stage onto the two stage combat cluster, we were making the *Semyorka* into a launch vehicle for spacecraft to do fundamental research on the solar system. "It's difficult to argue with Keldysh," continued Korolev. "He's vice president of the Academy, I'm an academician, and we should enrich science with really fundamental discoveries, especially if they fall right in our laps."

S.P. loved to talk on this subject somewhat tongue-in-cheek. He was trying to show us his supposedly casual attitude toward the Academy scientists. In actual fact—and I saw this on more than one occasion—this was his way of concealing his romantic dreams about really fundamental scientific discoveries from the pragmatists that surrounded him.

It was difficult to prove to marshals, generals, Communist Party leaders, and ministers that for the happiness of the Soviet people it was necessary to spend tens of millions of rubles to explore the Moon, Venus, and Mars. In this regard, cosmonautics was fortunate. It turned out that the main Party leader, Khrushchev, was perhaps a bigger romantic about space exploration than Korolev and Keldysh. Therefore, support from the very top was ensured for the most daring and still half-baked space programs.

And Khrushchev wasn't the only cosmonautics fan. Chief Marshal of the Artillery Nedelin also preferred attention and goodwill toward space projects. Back then no one was thinking about the potential for the military use of planetary exploration programs. Nedelin showed a breadth of thinking that was unusual for his boss Minister of Defense Marshal Malinovskiy, as well as his replacement, Marshal Andrey Grechko.[4]

For the **fourth** task, right after the Moon shot, we were supposed to prepare at

---

4. Marshal Andrey Antonovich Grechko (1903–76), a famous World War II veteran, succeeded Malinovskiy and served as minister of defense in 1967–76.

least two four stage rockets to launch spacecraft to Mars in October 1960. "As far as I know," said Korolev, addressing Turkov and me, "we've never gotten manufacturing and testing of the 1M into gear."[5]

"It's been in gear for a long time now, Sergey Pavlovich, but we haven't had a chance to move on it yet," said Turkov. He didn't think he needed to hold his tongue, and, switching to the offensive, he declared that he was still lacking a lot of drawings for manufacture and, as far as he knew, there was no hope of obtaining the items needed in time for the existing timetable. "The rocket's fourth stage—the Block L with Melnikov's engine—is still just in the preproduction shops," concluded Turkov.[6] In the face of such disrespectful and panicky declarations, S.P. usually traded his amicable, businesslike tone for a furious and accusatory one, but this time he restrained himself. He understood that this fourth task for October was practically unrealistic, but as far as deadlines were concerned, he didn't want to hear any proposals. He replied, "If we aren't ready for the Mars launch in October, we'll have to wait a year for the next launch window! Buck up. Besides, here, my friends, is the most important **fifth** task: we must manufacture, perform developmental testing on the ground, and launch at least four or five habitable spacecraft with recovery of their descent vehicles. It is essential for us and the space photo-reconnaissance experts to optimize the descent phase."

The terms "piloted vehicle" and "spacecraft" were not yet used in 1959. We said simply "object" or "habitable object," meaning that dogs would be flying, or we used the drawing identification numbers "article 1-KP" or "1K." Korolev had already enlisted all of his deputies to develop a crewed spacecraft. But until the first experimental launches began, we didn't have a great deal of faith that this event—the flight of a human being into space—would take place in the next two years. In late 1959, a two-year deadline seemed to verge on the impossible to us. After hearing the goals for 1960, Voskresenskiy ventured to say that, "That works out to at least 10, and, if we have reserves, then 12 launches! That means, Sergey, we'll only be going from the engineering facility to the launch site and back. There won't even be time to look at the films and attend accident investigation commissions."

Korolev wasn't about to get involved in arguments about this on New Year's Eve. He wished everyone good health and told us to enjoy ourselves at our New Year's parties. Despite feelings of doubt, the end-of-the-year meeting described above ended on an optimistic note. Shaking hands with his closest associates in parting, Korolev had something special to say to each of us. Looking into my eyes and smiling like the cat that swallowed the canary, he said, "Don't forget to pass on my New Year's greetings to Katya today!"

Korolev's four deputies—Sergey Okhapkin, Konstantin Bushuyev, Leonid

---

5. The first generation of interplanetary spacecraft designed to fly to Mars were known as the 1M series.

6. Mikhail Vasilyevich Melnikov (1919–96), a deputy chief designer at OKB-1 in 1960–74, headed all rocket engine development work under Korolev. He was the designer of the S1.5400 engine for the Block L of the four-stage version of the R-7.

Voskresenskiy, and I—each having received a personal send-off and a bottle of French champagne, drove away in an excellent mood in Korolev's old ZIM. We drove through 3rd Ostankinskaya Street, which would later become Academician Korolev Street. Korolev himself left in a ZIS-110, the most prestigious automobile of that time, to ring in the New Year at the Kremlin. We all went our separate ways in a good mood; so much interesting work lay ahead of us! Since then gatherings on 31 December just before each New Year's became a tradition with us.

WE SPENT ALL OF JANUARY 1960 DISCUSSING THE FUTURE SPACE PROJECTS. I met often with Mikhail Klavdiyevich Tikhonravov. With his innately subtle and refined sense of humor, he told me how in 1932, when Korolev, Pobedonostsev, and he were working at GIRD in Moscow, the universally respected Fridrikh Tsander would arrive every morning in the basement on Sadovo-Spasskaya Street and, before sitting down at his desk, would exclaim, "Onward to Mars!.." Back then, such exhortations would bring ironic smiles to everyone's faces. "Now a little less than 30 years later, Sergey Pavlovich, who snickered at Tsander's enthusiasm over Mars more than anybody, will soon begin his own briefings with this Tsander-like slogan. I don't think any of us will be smiling ironically," concluded Tikhonravov. He and I had this conversation in late 1959, when our infatuation with Mars really began.

The lunar successes of 1959 made planetologists in academic circles confident of the prospects for exoatmospheric astronomy. We were flooded with proposals to develop spacecraft for the exploration of Mars and Venus, to photograph the Moon again, and to execute a soft lunar landing. This hype was furthered by intra-academic competition between astronomers and geophysicists of various schools and fields. Lunar specialists rejected proposals to send spacecraft to Mars. Proponents of Mars explorations asserted that there was nothing to do on the Moon and the newly discovered capabilities of rocket technology should be used to explore the closest planets. The foreign press also contributed to the hype by reporting that America would not tolerate our supremacy and had already begun work on several designs for automatic interplanetary stations.

Indeed, the U.S. had begun launching the Pioneer series of spacecraft. In 1958 and 1959, these spacecraft used a launch vehicle consisting of a first stage with liquid-propellant rocket engines adopted from the Jupiter combat missile and three upper stages with solid-propellant engines. The first launches were failures, but we knew that the American rocket specialists were right on our heels.[7] The Jupiter rocket had been developed in the U.S. under von Braun's supervision. In this regard, Korolev noted with satisfaction that the Americans still couldn't get along without

---

7. There were six attempted Pioneer launches in 1958–59, none of which achieved their primary objectives, although *Pioneer 4* became the first American spacecraft to reach escape velocity. Only *Pioneer 3* and *Pioneer 4* were launched by the Juno II launch vehicle derived from the Jupiter intermediate range ballistic missile. Of the remainder, three used the Thor Able I and one used an Atlas-Able launch vehicle. For a complete list, see Asif A. Siddiqi, *Deep Space Chronicle: A Chronology of Deep Space and Planetary Probes, 1958-2000* (Washington, DC: NASA SP-2002-4524, 2002).

the Germans and were still quite green.

TIME AND AGAIN KELDYSH AND KOROLEV WERE SUMMONED TO KHRUSHCHEV, who attached exceptional importance to the political side of space successes. In actual fact, Khrushchev not only supported Korolev's and Keldysh's space-related interests, but also demanded that Minister of Defense Rodion Malinovskiy and his deputy Nedelin support Yangel's projects on high-boiling component combat missiles. Our friends from Dnepropetrovsk told us that Brezhnev—a native of Dnepropetrovsk, and now Central Committee Secretary for Defense Industries—had direct instructions to monitor Yangel's OKB and the Dnepropetrovsk missile factory and to assist them.[8] The folks in Dnepropetrovsk boasted that they now had their man in the Central Committee Presidium.[9]

Work on the already flying R-7 and R-7A missiles and on new designs was exceptionally demanding. The military blamed us, and rightfully so, for their insufficient reliability, the long launch preparation cycle, and limited accuracy. We understood these shortcomings all too well. When the rocket was used as a space launch vehicle, a third stage was added to the two primary rocket stages, and in the future, a fourth, needed only for operation in space. Partly because of the use of multiple stages, a spacecraft launch vehicle proved to be more complex and less reliable than a missile delivering a nuclear warhead.

The R-7 rocket was not trusted in its original two stage version to carry the first satellite into orbit until its sixth launch. In its three stage version it was thoroughly tested and had flown numerous times with mock-ups and dogs before it was trusted with the first human being. The four stage version of the launch vehicle, under the code number 8K78, was immediately stacked with the 1M automatic interplanetary station, whose historic mission was to fly past Mars. We had a fervent desire to beat the Americans and be the first in the world to answer the question, "Is there life on Mars?" With the new launch vehicle, we promised to bring just as much glory by revealing the secrets of Venus. What was hidden under Venus' veil of clouds, which was impenetrable for Earth-bound astronomers? We were in a hurry, a desperate hurry.

Before going to Korolev with specific proposals, Mishin, Tikhonravov, Bushuyev, Rauschenbach, and I discussed the possibility of rapidly producing automatic interplanetary stations and an associated fourth stage for the R-7. Tikhonravov and designers Ryazanov and Maksimov studied possible layouts and mass constraints. Rauschenbach, Legostayev, Bashkin, and Knyazev invented—actually invented— attitude control systems to make corrections and to aim cameras at planets and the

---

8. From 1957 on, the person occupying the position of secretary of the Central Committee for defense industries was the de facto governmental head of the Soviet space program. Brezhnev served in this position in 1957–60 and 1964–65.

9. During the Khrushchev era, the Politburo was called the Presidium.

high gain antenna at the Earth.

Tearing myself away from the overwhelming stream of routine matters involving the R-9 rocket, satellites, and multiple Moon shots, I often talked with Ryazanskiy and Boguslavskiy at NII-885 about versions of the radio system for communicating and receiving information from distances of hundreds of millions of kilometers. We had basked in setting the record for long-distance communications range at just over 300,000 kilometers, and now we needed to guarantee 300,000,000 kilometers. Among the electrical engineers there were two enthusiasts, Aleksandr Shuruy and Vitaliy Kalmykov. I tasked them and the conceptual designers with examining the problem of a power supply system for a year-long flight; I also gave them an ultimatum to design a single integrated power network for the entire AMS.[10] I assigned German Noskin and Nikolay Rukavishnikov with devising a PVU (sequencer) that would make it possible to rapidly send various command time sequences on board.[11] Unfortunately, we introduced this instrument only after the sequencer developed by SKB-567 had failed on *Venera-1*.[12]

After the faltering communications during the transmission of photographs of the far side of the Moon on *Luna-3*, Mikhail Krayushkin and his group of antenna fanatics—who believed that the future of radio engineering was in antennas—dreamed of creating the first high gain parabolic dish antenna for outer space.

Mishin and Bushuyev tasked Svyatoslav Lavrov and Refat Appazov to come up with optimal flight plans for interplanetary missions. At OPM, Dmitriy Okhotsimskiy started similar work at the request of Tikhonravov. Very quickly it became apparent that not one of the versions of the three stage R-7 available in the near future was capable of sending a decent-sized payload to Mars or Venus. And even then it was clear to us that it would be necessary to boost at least half a metric ton to escape velocity!

Mishin was the first to hatch the idea of placing one more stage, a fourth, on the three stage *Semyorka*. Thus, we got the idea to use a new oxygen-kerosene engine for this stage. We considered Sergey Okhapkin, another one of Korolev's deputies, to be the most level-headed among us; he was responsible for the work of the design departments, for issuing the main working production documentation, and was directly involved in issues of the rocket's structural integrity. Even he agreed with the idea of a fourth stage without hesitation.

RIGHT AFTER NEW YEAR'S, ON 2 JANUARY, Khrushchev summoned Keldysh, Korolev, Glushko, and Pilyugin. Khrushchev was in a very forceful mood and said that success in space was now just as important to us as the production of combat

---

10. AMS—*Avtomaticheskaya mezhplanetnaya stantsiya* (Automatic Interplanetary Station).

11. PVU—*Programmno-vremennoye ustroystvo*—literally means "programmed-timing device."

12. *Venera(-1)* was launched on 12 February 1961 toward Venus. Communications with the spacecraft failed after last contact at a distance of 1.9 million kilometers from Earth. See Chapter 31.

missiles. He was upset and threatened them saying, "Your work is going rather badly. Soon we will have to punish you for falling behind in space. Work is under way on a broad scale in the U.S. and they could beat us." S.P. quoted Khrushchev from his notes at the meeting on 3 January, to which Keldysh, all the chief designers, and Korolev's deputies were invited. A chaotic discussion began on the program of work on space for that year and the next few years. Keldysh insisted on one more lunar probe, the Ye-2F, which would use more advanced equipment for capturing and transmitting images of the far side of the Moon. I opposed this project in view of the workload for the Mars and Venus program. We had assigned this new program the acronym "MV." Korolev added, "Don't forget that we also have the Vostok." And so, we all left the meeting, having failed to reach an agreement on anything.

On 7 January, Keldysh convened a large interdepartmental council on the Ye-2F and MV. For the Ye-2F, it was agreed that the objectives would be limited strictly to photography. The deadline for approving the mission was extended, but the launch was scheduled for April 1960. As for the MV project, for the first time we began to seriously sort out what was what. Okhotsimskiy, Lavrov, Kryukov, Rauschenbach, Khodarev, Ryazanskiy, and Pilyugin reported each on their own work, and for the time being, only with preliminary considerations. After the meeting, S.P. seated Kryukov and me in his car.[13] In the strongest terms he reprimanded us, saying that we, his deputies, up until now had not sorted out who was responsible for what in the MV program. We were not coordinating the work, and those "idealists in Keldysh's department" want the launch to take place in September of this year.

On January 9, Ustinov held a meeting of the Military-Industrial Commission and presented our report about the status of work on the Vostok and the heavy photo-reconnaissance satellite. The future photo-reconnaissance satellite had already been named Zenit (Zenith).[14] Bushuyev and factory Director Turkov presented reports. The deadlines of the schedule approved by Ustinov would slip by three to four months. Although our subcontractors were largely to blame for the failure to meet the deadlines, OKB-1 bore the brunt of the relentless criticism.

Referring to Zenit, Ustinov noted that, "This is a crucial intelligence tool. There is no mission more vital at this time." Here he clearly castigated Korolev for his interest in the human spaceflight program. Korolev sat glowering in silence. Ustinov had openly attacked Bushuyev, Turkov, and me, but it was understood that the fire was directed at Korolev who could not cope with his deputies himself.

After a break, Ustinov tasked Georgiy Pashkov with preparing a report in a week with proposals for the MV program. At this juncture, military representative Aleksandr Mrykin felt he needed to get involved. In the highly charged atmosphere of

---

13. Segey Sergeyevich Kryukov (1918–) was one of the top-ranking deputies at Korolev's design bureau. He later succeeded Georgiy Babakin as head of the Lavochkin design bureau in 1971.

14. The Vostok piloted spacecraft and the Zenit military photo-reconnaissance satellite shared the same design layout but had entirely different mission goals.

Ustinov's meeting, Mrykin's speech had a sobering ring. "It seems to me that this complex problem cannot be solved by conventional means," he said. "We need to concentrate all of our forces and enlist new cooperation. The VPK must make prompt decisions and not berate the designers from one meeting to another. OKB-1 and its subcontractors need real help and continuous monitoring." Before dismissing everyone, Ustinov warned that very soon Khrushchev would personally review our plans regarding space and wanted to do this right at OKB-1.

S.P. left for several days to think things over and to relax at the government vacation hotel Sosny, having tasked Bushuyev and me with drawing up a draft plan for the MV project and to come see him on 12 January. "But don't put the launch dates beyond September," he instructed.

As usual, the most difficult thing proved to be coordinating the dates with the factory. The deadlines for the development of the drawings and manufacture of the space probes seemed unrealistic to us. But when we arrived at Sosny, S.P. studied our schedules, scowled, and mercilessly set about correcting them, shifting the deadlines "to the left" by two and sometimes by three months. At the same time, he proposed increasing the number of spacecraft being fabricated from two to three. S.P. proposed that we simplify the version headed for Venus, removing any thermal shielding. "We'll fly to Venus, that goddess of love, in the nude," he said. "There isn't time to optimize thermal shielding. If there is a failure in the last stage it'll burn up in the Earth's atmosphere anyway. But we'll be able to prove that we are launching spacecraft, not combat missiles."

On 15 January, after returning from Sosny, S.P. convened an all-hands briefing and announced inconceivable deadlines for the production and launch of three MVs in 1960. Few believed that these deadlines were realistic. S.P. delivered a speech full of threats against those who might be guilty of failing to meet these completely unrealistic deadlines.

WE HAD SOME MAJOR ISSUES ON OUR HANDS. For example, What to do about the control system that must operate continuously for an entire year in space orienting the solar arrays toward the Sun, the parabolic dish antenna toward the Earth, and the entire spacecraft toward Mars or Venus? Each of our main developers took their stand. Having realistically evaluated the situation, Rauschenbach backed out of developing the solar array orientation system and the gyroscopes for the orientation of the entire spacecraft. Clearly he did not want to get involved in projects with risky deadlines. Pilyugin announced that, God willing, if he were really lucky, he could cope with the control of two more upper stages for the R-7.

Ryazanskiy proposed entrusting the entire radio communications problem to SKB-567, where Anatoliy Belousov had been appointed director in place of Yegeniy Gubenko, who had suddenly died, and Khodarev had been appointed chief engineer. Only this young company and also Vladimir Khrustalev—chief designer of optical instruments at the TsKB Geofizika—cheerfully declared, "We'll do it."

Soon thereafter, Andronik Iosifyan invited me to his luxurious mansion by the

Krasnyye vorota (Beautiful Gates).[15] He presented me with his book *The Problem of a Unified Theory of Electromagnetic and Gravitational Inertial Fields*.[16] This work directly contradicted Einstein's theory of relativity. If everything he said was true, then Andronik certainly deserved a Nobel Prize. But the theoretical physicists of our Academy of Sciences did not recognize Iosifyan's scientific treatise. As is commonly known, Einstein spent the last years of his life attempting to develop a unified field theory. To this day, such a unified field theory has not been developed.

I requested that he stoop to the needs of "rusty electricians," setting lofty and pure science aside; Iosifyan assured me of his full support for all our MV projects. He set up a "strike force" headed by Nikolay Sheremetyevskiy. This was the beginning of space-related work for the future academician and director of the All-Union Scientific-Research Institute for Electromechanics (VNIIEM), Nikolay Nikolayevich Sheremetyevskiy.[17] Unfortunately, the team of top-notch electrical engineers that had gathered at NII-627 could not implement a single one of their ideas to meet Korolev's fantastic deadlines, and instead limited themselves to the reliable but routine development of current and voltage converters.

Mrykin's speech at the meeting in Ustinov's office about the "concentration of all forces" had left its mark. On Ustinov's instructions Rudnev assembled Kalmykov, Shokin, and the chiefs of the main directorates—all leaders of the radio electronics industry—in his office. The most erudite among all those present was State Committee on Radio Electronics (GKRE) Chairman Valeriy Kalmykov.[18] When he first heard the mission statement that "Today, in January, we start from scratch, and in September, we launch," he smiled, but he did not take issue with it. He had already been through Beriya's school of deadlines with air defense missiles.[19] In those days, taking issue could lead to arrest. In the best case scenario, you would be taken off the job. He had been in such situations more than once, and like many other ministers, he believed that, as a rule, official wrath came down not on the guilty parties, but the last ones in line. In a multitude of missed deadlines, the important thing was not to be at the very end.

Ustinov informed Korolev that at his request, Khrushchev had personally instructed Kalmykov to help us implement the MV program with the intention of

---

15. This is a reference not to the ruins outside of Kiev, but the area in Moscow where the *Krasnyye vorota* metro station is located.

16. A. Iosifyan, *Voprosy yedinoy teorii elektromagnitnogo i gravitatsionnogo inertsialnogo poley* [*The Problem of a Unified Theory of Electromagnetic and Gravitational Inertial Fields*] (Yerevan, 1959).

17. VNIIEM, formerly known as NII-627, later developed Soviet weather and remote sensing satellites such as Meteor.

18. GKRE—*Gosudarstvennyy komitet po radioelektronike*—was the ministry in charge of developing most of the electronics and guidance systems for the Soviet missile and space industry.

19. In 1951–53, Kalmykov served as chief engineer of the Third Main Directorate (TGU) of the USSR Council of Ministers when he was a senior manager over the development of the first Soviet air defense missile system. Security services supervisor Lavrentiy Beriya oversaw the development of this project.

conducting two launches in September or October of that year. "The whole field of radio electronics is terribly excited," said Korolev, having called me into his office. He instructed me to attend all the assemblies and meetings that Kalmykov and Shokin held and to report to him daily.

After Rudnev's meeting with the GKRE staff and the institute directors, plans were worked out at a feverish pace, assignments were distributed, and questions were asked, for which no one had any answers. Many chiefs called me up directly, trying to understand what they needed to do. When I mentioned the deadlines, rather than argue, they politely said goodbye.

On 22 January, Kalmykov assembled all the potential participants in radio electronic projects in the GKRE conference room. I reported on the MV objectives, the main features of the flight program, orbits, and requirements for the radio communications system. The head of NII-4, General Sokolov, announced the military's proposals for the creation of Crimean and Far Eastern control posts. During the discussion, Kalmykov turned over the meeting to his First Deputy Aleksandr Shokin because he was suddenly summoned regarding a message that an unknown aircraft had violated our air space. One of the meeting participants commented, "That's what we need to be working on instead of this Martian science fiction."

Shokin tried to pin me down, demanding proposals for the distribution of work between the leading organizations for near and far space. I proposed that there be two separate leading organizations. One would be assigned artificial satellite problems and the other, lunar and deep space exploration. In the debate Shokin accused me and OKB-1 as a whole of imposing our will on the different organizations. In his opinion, we were doing this haphazardly, randomly, based on our own sympathies and who our friends were. "We shouldn't have to stand at the beck and call of OKB-1 anymore and wait to see what they will require of us. We should take the initiative ourselves and propose designs, keeping pace with or even moving ahead of OKB-1 requirements," he said. "Wise words," remarked Boguslavskiy, who was seated next to me.

Shokin was keyed up and abruptly cut off television institute (VNII-380) Director Igor Rosselevich and radio-communications institute (NII-695) Director Leonid Gusev, both of whom had spoken in support of my proposals. In this highly charged atmosphere the resilient Aleksey Bogomolov declared that if all the capacity of GKRE wasn't enough, then OKB MEI was ready to take on designing and producing 30- and 64-meter diameter ground-based antennas, and not in the faraway Crimea, but here in the Moscow area, on the Medvezhiye Lakes. Hearing this, everybody laughed and some responded with caustic remarks. The directors of the main radio electronics institutes sensed the boldness of the young OKB MEI organization and clearly felt threatened by its promising proposals.

Sokolov brought everyone back from Martian orbits to Earth. "To build long-range communications tracking stations," he noted, "we'll need to concentrate 10,000 workers in the Crimea alone. And then there's Ussuriysk, from where we must monitor the third stage and, to a certain extent, back up the Crimean tracking stations! At the same time, we still don't have a decree and the construction sites haven't even been finalized. Is it possible in seven months to build these antennas, the likes of which the world has never seen? It appears that, with exceptional effort, everything having to do with the onboard radio complex can be produced. But it's difficult to say how it will be with the ground since we don't have clear-cut specifications from GKRE."

At the end of the meeting Kalmykov reappeared. He informed us that air defense radar stations were tracking an aircraft that had crossed our border from Iran at a very high altitude, but while they were weighing the issue of whether or not to shoot it down with missiles, the plane wisely turned around and left. The meeting was then adjourned with general, vague instructions.

In complex situations dealing with radio-electronics, I preferred to consult with Boguslavskiy. Ever since we'd worked together in Bleicherode I'd had faith in his decency, common sense, and objectivity, regardless of his departmental or company interests. About three years later, in 1963—I no longer remember the circumstances—Korolev said to me, "Of all of your friends and subcontractors in the radio field, the only ones whose objectivity I absolutely trust are Boguslavskiy and Bykov.[20] Even Mikhail (he was referring to Ryazanskiy) can't rise above the interests of his own company."[21] In a man-to-man conversation, Boguslavskiy said, "I don't believe it will be possible to build a reliable multifunctional radio complex for the MV spacecraft in seven months. We'll have to take a completely unjustified risk. Under these conditions, it's impossible to do any serious laboratory study or testing of the components. There is neither the time nor the equipment to run service life or durability tests. I don't want to start a rat race when there's no hope for success, and I will try to dissuade Mikhail. Let Belousov, Khodarev, and Malakhov's organization try to tackle that task. They have a new company. They need to win their 'place in the sun.' If they flub up the job, they'll be forgiven out of consideration for their youth." Despite his stance, Boguslavskiy was at least prepared to persuade Mikhail Ryazanskiy to take on the development of antennas for the Crimean tracking stations; "we shouldn't give up such 'morsels' to Bogomolov," he noted.

And subsequently, the work in these fields was allocated along these lines right into the mid-1960s: SKB-567 produced the multifunctional radio complex and NII-885 took on the job of developing antennas for the deep space tracking station in Crimea. In general, this process also allowed radio-electronics to become an integral part of space technology. As the head OKB dealing with spaceflight, we had a stake in the existence and development of radio electronics systems. In contrast to

---

20. Yevgeniy Yakovlevich Boguslavskiy (1917–69) served as deputy chief designer and then first deputy chief designer at NII-885, the main guidance systems institute, in 1950–69. Yuriy Sergeyevich Bykov (1916–70) served as chief designer of NII-695 (or MNII Radiosvyazi) in 1959–70 during which time he supervised the development of communications systems for several different piloted spacecraft.

21. *Author's note:* I will write about Yuriy Sergeyevich Bykov later.

From the author's archives.

Chief Designer Aleksei Bogomolov (1913–) served as head of OKB of the Moscow Power Institute (OKB–MEI) during a span of nearly thirty years. During this period, he was responsible for a wide range of telemetry, communications, and data recording systems for Soviet missiles and spacecraft.

many other chiefs, at his OKB, Korolev drove home the notion that this was not "support equipment," like automobiles and telephones, but just as organically fused with the overall mission as the engine and the rocket itself!

ON THE MORNING OF 29 JANUARY 1960, Tikhonravov asked me to go with him to see S.P. to work out our general course of action at the next meeting with Keldysh on the lunar program. Recalling Tikhonravov's story about Tsander, I proposed that, "When we walk into S.P.'s office, let's both shout, 'Onward to Mars!'"

Tikhonravov smiled his sweet smile, but declined to indulge in such rowdiness.

Korolev felt wretched. He had just returned the day before from Kuybyshev and had a difficult landing at Vnukovo. There was evening fog, and they didn't want to clear the airplane to land. They were redirected to Leningrad, but Korolev got permission to land through the Air Force command. He noted, "Unfortunately, not everyone at the top understands us. They don't want to try to understand technology at all. They think that's strictly our business. That's why they can't understand our difficulties. And those few that do understand our difficulties don't have the necessary clout. We had a good relationship with Nikita Sergeyevich. But during the last meeting, even he demanded new space triumphs and he laid out our MV mission like this: 'Tell me, is it theoretically possible to do this?' Well, what is one supposed to answer? Of course, it's all theoretically possible. 'Then just don't drag us into the technical details,' said Khrushchev. 'This is your business. Tell me what you need and do it.' That's the whole story. Then they don't give us what we need, but we've still got the assignment that we need to carry out within an insane time frame."

Despite Tikhonravov's insistence, S.P. refused to discuss the lunar program. Instead, he asked which of the designers was doing work on MV. Tikhonravov responded that he had entrusted the project to Gleb Yuryevich Maksimov, but that he was monitoring the work himself and would enlist Ryazanov and other seasoned designers. I liked Gleb Maksimov for his thoughtful, constructively critical attitude toward design work. I supported Tikhonravov. S.P. grumbled that most of the people on Tikhonravov's team had never even seen production and were afraid of factory problems.

Korolev then switched over to me and demanded a report on the latest events in radio-electronics. I began to speak, but he interrupted me, saying "You and Mikhail Klavdiyevich [Tikhonravov] don't know everything. I had a very angry exchange with Kalmykov and Ryazanskiy. I told them that any day now Nikita Sergeyevich [Khrushchev] was going to visit us and we were going to present our proposals. They both promised to give it some more thought, but it's still not clear what they'll come up with."

When Tikhonravov and I left Korolev's office, still without a specific plan of action, I said, "Now, Mikhail Klavdiyevich, when Khrushchev visits us you will get the opportunity to greet him with Tsander's slogan 'Onward to Mars!'"

In anticipation of the "big visit," there was a flurry of activity at OKB-1 and at the factory, which prepared a show of our achievements and future plans. S.P. personally supervised this process. The exhibition was set up in the factory's assembly shop No. 39, the cleanest, brightest, and most spacious shop. The R-7A, also known as the 8K74, was assembled in its complete cluster. Its top-secret specifications were displayed on a placard. Shop No. 39 Chief Vasiliy Mikhaylovich Ivanov confessed that he had not been able to fully rig an authentic missile cluster. The nose cone was made partially of cardboard, the instrument compartment was completely empty, and main Blocks A and B had been temporarily removed from the 8A72.[22] "But, who's going to figure that out?" grinned Ivanov. In addition, full-fledged 8K74 booster components had been laid out at work stations for horizontal tests. A row of warheads was ceremonially displayed, from the now seemingly innocuous R-1, R-2, and R-11 to the formidable intercontinental nuclear missiles. The display placards did not even hint at the warheads' actual TNT equivalent. None of us was trusted to know this. Only the mass was indicated.

The most beautiful and impressive part of the display were the R-11, R-1, R-2, and R-5M missiles standing in order of size, the future R-9 missiles, the global 8K713, the brand-new solid-propellant RT-1 missile, and a mock-up of a "potbellied" micro-missile propelled by high-boiling components, which surprised everybody.[23]

The solid-propellant RT-1 was a three stage missile designed to have a range on the order of 2,500 kilometers. This design was developed under the leadership of Igor Sadovskiy, whom Korolev had appointed as his deputy for solid-propellant missiles in August 1959. This was the first real ballistic missile design in our country using propellant powders manufactured using a new process. This work, which had been actively supported by Korolev for some time, stood as one more testimony to what many viewed as his inscrutable intuition. The "potbellied" liquid-propellant

---

22. The 8A72 was a three stage version of the R-7, designed for launching the early Zenit-2 reconnaissance satellite.

23. The 8K713 was a proposed "global missile" (also known as the GR-1) designed to carry nuclear bombs into Earth orbit and potentially capable of targeting any location on the planet's surface.

missile was displayed at Mishin's insistence as an alternative to the solid-propellant concept, which he did not support.

Space vehicle technology was presented by several displays, including the future Vostok crewed spacecraft with a special winch-driven sliding pilot's seat, a heat shield–coated descent sphere that was prepared for ejection from an aircraft, and a launch vehicle for the future Vostoks along with the third stage (the Block Ye) and its external payload fairing. Interplanetary stations for missions to Mars and Venus had not yet really been designed, but here in the assembly shop it was already possible to touch them; they were shown off in the form of full-scale mock-ups. On the Mars probe, the solar arrays turned smoothly, orienting themselves toward a spotlight. The landing version of the Venus probe was also on display. Of course, we did not forget backup models of the first three satellites and first three lunar probes. We walked through this exhibit with the wonderment of explorers; we had done so much and in just 13 years! Without a doubt, our S.P. had done a brilliant job making everyone move heaven and earth to demonstrate our past, present, and future.

The visit was set for 4 February. Suddenly on 3 February we were informed that Khrushchev would not be there. CPSU Central Committee Secretary Brezhnev would visit us. According to the allocation of duties in the CPSU Central Committee Presidium, Brezhnev was in charge of the entire defense industry and missile technology. Korolev was quite upset that Khrushchev would not be there. Someone warned Sergey Pavlovich that, "Brezhnev is a very shrewd, smart man. Watch what you say." S.P. passed on this warning to the briefers who were supposed to stand by the exhibits.

In the morning the top brass—Ustinov, Serbin, Rudnev, Grishin—and the primary chief designers gathered in the shop. They waited for a long time in the shop chief's office, which had been set up for the meeting. The brass decided to meet Brezhnev at the facility entry gates. When everyone was already weary with anticipation, he appeared accompanied by Ustinov, Serbin, Korolev, and only one bodyguard.

Korolev announced the program for the day, and Brezhnev approved it. The tour of the exhibition began. He walked along, looking and listening attentively, without interrupting and without asking questions. From time to time he raised his extraordinarily bushy eyebrows in surprise. Korolev conducted the narrative very calmly, without losing his train of thought and without repeating himself. One could see that he was in fine form. Only when they reached the RT-1 did Korolev yield the floor to Sadovskiy.

After the tour they went up to the shop chief's office, where tea had been prepared. During tea Korolev said that we would take a break and walk over to the OKB to have a roundtable discussion. Brezhnev perked up and told a relevant joke. "A middle-aged man was being carried around Moscow by his arms. He was holding his feet up in the air, afraid to step on the ground. Passersby were astonished. The people carrying him explained, 'This is our director. They've taken away his private car and he's forgotten how to walk. So we have to carry him to work and back home.'" The joke wasn't new, but everyone laughed. It was a sensitive issue. Khrushchev was trying to reduce the number of service vehicles and transfer them to taxi stations. He had received reports that his decree was being successfully executed. Actually, the transfer of cars to taxi stations was trumped up. On paper the automobiles had been transferred, but in fact in the morning the taxi stations sent the cars back to their old owners, for which they were compensated at the going rates. This suited both sides. After that icebreaker, someone got up the nerve to say that one can laugh, but it's difficult to work without cars. Managers can't drive themselves, and we don't have American-style services yet. The complaints were graciously received.

Next we entered the library building, where posters of future developments were on display. Korolev strolled briefly past the military technology and devoted the majority of the time to space. It was a very convenient opportunity to hint at the unrealistic deadlines for MV, but S.P. did not do that. The graphics on the posters was not the work of professional artists, but of the conceptual designers of Department No. 9, the space department.[24] (When our guest had departed, Grishin rebuked Korolev, saying that any American magazine would have had more colorful pictures).

The general impression was that we still lacked a well-conceived, long-range plan of operations regarding cosmonautics—a *kosmoplan*, as we called it. The more substantive part of the report was when Korolev spoke about the optimization of the *Semyorka* and its conversion into a three stage and then a four stage launch vehicle.

When we had seated ourselves around the large round table, Glushko asked to speak. His speech was in stark contrast with S.P.'s report and was delivered with an aggressive, bristling demeanor. He proposed that we immediately switch to designing and producing a heavy launch vehicle using the RD-111 engine developed for the R-9. "We shouldn't wait for a closed-cycle engine with exhaust gas afterburn in the combustion chamber, as certain incompetent comrades from OKB-1 propose," Glushko said, as always, very convincingly in his soft-spoken voice.

Despite the speech's accusatory tone toward OKB-1, Glushko's face showed no emotion. When he said that certain individuals among those present chided him for being conservative, Mishin snapped and asked, "And who would that be?" Glushko, without missing a beat, responded, "That's your guilty conscience speaking." This brief skirmish was symptomatic of the technical disagreements that had intensified between Glushko and Korolev. As for the relationship between Mishin and Glushko, it deteriorated more and more, beyond recovery. Mishin stopped trying to compromise. On the contrary, he pitted Korolev against his old comrade-in-arms from the very infancy of rocket technology.

---

24. Korolev established his so-called "space department" or Department No. 9 at OKB-1 in 1957 under Mikhail Tikhonravov.

The subsequent presentations of Pilyugin and Ryazanskiy were dull. They spoke in generalities about merging and consolidating institutes and strengthening the production base. Barmin, suddenly and for reasons unknown, supported the "potbellied" nitrogen-oxygen propellant pygmy that Mishin proposed instead of a solid-propellant version. Ending the roundtable session, Korolev made no specific proposals as to organization and future plans, but in guarded terms he rebuffed Glushko for being intolerant of criticism from other specialists.

I have kept my notes of Brezhnev's closing comments. "It is very good that you 'lured' me here," he said. "But, of course, I myself cannot make any decisions. Your proposals need to be discussed in the Central Committee Presidium. You should prepare yourselves, and a bit more thoroughly. In my opinion, the material still needs work. You've got 10 to 15 days to get ready and present a concrete plan. But it would be good if you launched one of these 'bugs' *(zhuchki)* to cause a bit more of a stir."

By mentioning this "bug," Brezhnev immediately dashed the hope for a mutual understanding between the government and our company over deadlines. Not even Ustinov smiled. This attitude toward space technology grated on everyone. On that note, Brezhnev bade us farewell.

When the top brass had left, Grishin the wisecracker turned to us and said, "I've heard that Tikhonravov has a butterfly and bug collection. So tell him to pick out a 'bug' that will cause a bit more of a stir in space."

Mishin couldn't stand it any longer. "He didn't understand anything! These 'bugs' cost us a lot! I can't see any good in this conversation."

"Well, you should still be a bit more careful how you speak!" warned Grishin.

AFTER THE "ROYAL" VISIT, IT WASN'T 10 TO 15 DAYS THAT WENT BY, but almost two months before we had drawn up, coordinated, and sent a draft of the great *kosmoplan* to the GKOT and VPK. S.P. assigned Mishin, Kryukov, and myself to thoroughly edit the section on launch vehicles. We argued a great deal, and it got to the point of shouting. We even reverted to addressing each other formally.[25]

Our most amazing proposal was for a heavy launch vehicle with a 1,600-metric ton launch mass and a nuclear engine in the second stage. In those days the idea of a nuclear rocket engine had only been discussed and there had not yet been any experimental work confirming physicists' optimistic calculations. But for some reason we believed that a nuclear reactor could be put on a rocket. It was a very alluring idea. For two weeks running Korolev spent all his time working on the plan. He got into heated arguments and debates. With Korolev's input, Kryukov and the conceptual designers revamped various layouts for multistage launch vehicles with tandem and parallel staging designs. In response to "Valentin's outburst," S.P. set a

---

25. The Russian language has two forms of the pronoun "you"; friends and family address each other using *ty*, while *vy* is formal and connotes a certain interpersonal distance.

goal of coming up with a three stage launch vehicle that would be capable of insert-ing a 30- to 40-metric ton satellite into Earth orbit by the end of 1961. During the course of the arguments, S.P. realized that this objective wasn't feasible and finally settled for late 1962.

The plan contained a lot of everything: a heavy launch vehicle described in detail, electric rocket engines, automatic and crewed space vehicles, and proposals for their in-orbit assembly and construction. With Korolev's approval, under pressure from Mishin and with objections from Kryukov, the proposals for the new launch vehi-cles stipulated using N. D. Kuznetsov's engines for the first and second stages.[26]

When Glushko paid us a visit to study the plan, he did not sign it, of course, and promptly set off for Dnepropetrovsk to see Yangel to develop counterproposals for a heavy launch vehicle. He offered Yangel high-boiling component engines using the engine that he had already developed for the R-16 missile. By this time, Yangel's OKB had already put into service the R-12 missile with a range up to 2,400 kilome-ters, equipped with a separable nuclear warhead.[27] Its indisputable advantage over our R-5M was its greater range and the fact that its operators didn't have to face the constant hassles of having to replace losses from evaporating liquid oxygen. Yangel had already started a modification of the R-12 for combat duty in a silo version. In this case, the missiles were maintained in launch readiness for long periods of time. At the firing range in Kapustin Yar they were successfully testing the medium-range R-14 missile, now with a range up to 4,500 kilometers. It was also equipped with a nuclear warhead and had a completely automatic control system.[28] Yangel was preparing at a feverish pace to begin flight-tests on his own first two stage intercon-tinental missiles, designated R-16.

Glushko's engines were used on all the missiles. The R-16 missile used hyper-golic propellants (the oxidizer was a mixture of nitrogen oxides and nitric acid, and the fuel was unsymmetrical dimethylhydrazine). The first stage engine generated a thrust of 150 metric tons near the Earth's surface and could lift a 140-metric ton rocket. It was a real competitor to our R-9.

With such work in progress, Yangel could engage in the battle for primacy in the production of the heavy launch vehicle. The high-boiling component engines that Glushko had developed according to his own specific characteristics were infe-rior to similar oxygen-propellant engines that we had anticipated obtaining from Kuznetsov. But Glushko's engines already existed, while Kuznetsov was just on the verge of beginning work in a field that was completely new for him. This was the

---

26. Nikolay Dmitriyevich Kuznetsov (1911–95), a famous designer of aircraft jet engines, served as chief and then general designer of OKB-276 (later NPO Trud) based in Kuybyshev in 1953–94. In the early 1960s, Korolev invited Kuznetsov—who was not a specialist in rocket engines—to participate in several launch vehicle projects including the N-1.

27. The early version of the R-12 (or SS-4 Sandal) was declared operational in March 1959. A silo-capable version came on line in January 1964.

28. Flight-tests of the R-14 (or SS-5 Skean) began in July 1960.

incontestable advantage of Glushko's position. From then on, Korolev's and Mishin's disagreements with Glushko had grave consequences for Soviet cosmonautics.

The great *kosmoplan* got thoroughly bogged down in the offices of the Central Committee and VPK. Korolev often visited the "higher-ups," argued with Ustinov, and was understandably impatient and edgy. Evidently at Ustinov's prompting and with Brezhnev's approval, the VPK staff decided to teach us a lesson for our rebelliousness and "conceit."

We were counting on bonuses and awards for putting the R-7 into service and for the three Moon shot successes. After a lot of official red tape the Council of Ministers issued a decree calling for the payment of so-called governmental graduated bonuses. Basically, these bonuses were designated for the chief designers. The bulk of the creators, despite their extraordinary work, on average could count on a bonus of from 300 to 1,000 rubles. On the other hand, the workers at Dnepropetrovsk Factory No. 586 and Yangel's OKB boasted that their bonuses were twice as big as ours. They were showered with medals, and 23 individuals became Lenin Prize laureates. In our organization, only 15 people were awarded the Lenin Prize for the Moon shots. People grumbled and privately seethed, but they could only vent among themselves.

Almost all my work time during the first months of 1960 was devoted to the Moon and Mars projects. Whereas the day-to-day lunar tasks were primarily organizational and routine—completing units, testing, assembling, and eliminating glitches and defects—unresolved problems continually cropped up on the Mars project. Every day there were new problems.

We managed to use a minimum of already seasoned people for the new Moon shots. For the most part a new contingent was involved with Mars: electronics experts who had transferred from TsNII-58, guidance experts from Rauschenbach's department who had transferred with him from NII-1, and our old cadre of radio specialists.

We had no experience setting up radio communications at distances of millions of kilometers. By the end of the year, we would not be calculating signal power input to receivers using classic formulas, but ensuring the actual transmission of commands on board and receiving information contaminated with interference from the interplanetary station. Designing the antennas, solar arrays, sequencer systems, and an "ideology" for the attitude control computing devices required the continuous interaction of conceptual designers, radio specialists, design engineers, and our subcontractors, who were grappling for the first time with creating a radio-link 150 million kilometers long. I barely managed to scrape some time together to delve into the development of a general concept and configurations for a manned spacecraft. My deputy Yurasov and the young chief of the onboard complex control systems department, Karpov, were in this then-breakthrough field.

In this tower of Babel of systems, instruments, circuits, and cables, Yurasov and Karpov attempted to establish order and a minimum of standardization. "These new 'passengers' are like children," complained Yurasov. "Each one hangs on to his favorite toy and is afraid to let go of it."

I argued the need for an elementary systemic approach until I was blue in the face. But it was too late. Production did not allow us to introduce serious changes. It was very difficult for us to finally drive home the idea that we needed to tightly integrate the onboard systems into a single onboard control complex with interconnected software and hardware. With such an abundance of tasks, bringing order and harmony and reconciling disagreements between dozens of systems developers, conceptual designers, design engineers, subcontractors, and manufacturers faced with impossibly short deadlines required heroic efforts. Many disagreements were resolved more quickly and easily at the firing range during strolls on the concrete runway, in conversations at the hotels, or even at the launch site during the many hours of launch preparation.

On 7 April, I flew out to the firing range with most of the members of the State Commission and technical management to prepare and launch the Ye-2F, which had now been designated Ye-3, a designation previously intended for the lunar craft carrying a nuclear warhead. The Uralsk and Aktyubinsk airfields were waterlogged, and we flew into Tyuratam through Astrakhan. The lower Volga was still hemmed in by the spring flood. From the airplane, the thousands of tributaries of the famous Volga delta looked like a fantastic piece of art. Gradually, dry and barren steppes replaced this abundance of water. Soon the Aral Sea was shimmering in the sunlight, and a half hour later our Il-14 landed in our dear Tyuratam.

At the engineering facility they were busy around the clock preparing the first of the two recently arrived although not fully factory-tested Ye-3 lunar spacecraft. As was the case the year before, the most critical item proved to be the *Yenisey* photo-television unit. The familiar NII-380 engineers Valik and Bratslavets, their faces unshaven and pallid from overwork, but yet optimistic, repeated cycle after cycle of tests, removing one blotchy test film after another. Once again Korolev had to arrange rapid air transport for new developing solution to be carried on a Tu-104 from Leningrad to Moscow and then on an Il-14 to Tyuratam. The fresh photo chemicals were put to the test right away, and the *Yenisey* began to crank out films in excellent condition.

Korolev and Keldysh held a rowdy meeting to demonstrate the use of unsuitable photo chemicals and poor quality photo materials. They decided to schedule the first launch on 15 April and not to let up on the hard work in preparation for the second launch under any circumstances. On the night of 12 April, the first Ye-3 was mated with the cluster and sealed with a fairing, and the entire rocket was assembled and made ready for transport to the launch site.

Meanwhile, Boguslavskiy and I had completely worn ourselves out searching for defects in the radio complex on the second Ye-3. Recalling the shortcomings in the radio-link from our experience working on Mount Koshka, we were trying to achieve the maximum traveling-wave coefficient (KBV), which to a great extent determined the performance factor of the "space-to-ground" radio-link.[29] One of the theoreticians who had flown in with Keldysh had mentioned that the traveling-wave coefficient decreases due to ionization of the area around the antenna.

One night during check-out tests at the MIK, Deputy Ministers Aleksandr Shokin and Lev Grishin came by. I was discussing the situation with Ryazanskiy and Boguslavskiy. Grishin recommended that to get rid of the ionization, the testers should be issued alcohol "to flush out the area."

"In general, my faith in the engineering intuition of designers and testers has been shaken," declared Grishin. Providing an example, he continued, "A main oxygen valve, which passed the inspection sampling tests with flying colors, was routinely dismantled, and it turned out that a part was missing. After that the military rep rejected the tests. If that part had been there the valve might not have passed the tests. They installed the part, repeated the tests, and sure enough, there was a glitch." He provided another example. "And now your people have discovered a 'minus' on the hull; they've found which cable it's in and have decided to throw out the cable and send commands from the ground. What's more, they've discovered a breakage in the temperature sensor. Because there's no time to mess around with it, they decided to chuck it." We could make all kinds of excuses, but wisecracking Grishin hit us where it hurt the most.

On 13 April the State Commission Chairman and Chief Marshal of Artillery Nedelin conducted the first meeting before the launch. Keldysh delivered a general report on the goals of the experiments. Bushuyev, Vernov, and Severnyy presented joint reports. Ryazanskiy, Rosselevich, and I reported on the readiness of the Ye-3 systems, while Colonel Nosov reported on the firing range readiness (I underscore that in 1960 the contemporary term "cosmodrome" was not yet in use) and Colonel Levin reported on the readiness of all the Command and Measurement Complex services. At the launch site all the tests proceeded without incident. Meanwhile, at the MIK work was under way around the clock in preparation for a backup launch.

By morning we had replaced several systems, including the entire radio unit and the failed traveling-wave coefficient sensor that had prompted Grishin to snicker at us. We also repaired the *Yenisey* photo-television unit, which managed after all the tests to receive a "minus on the hull." Soon after, the spacecraft was being mated with the rocket. Sinekolodetskiy's installers were working like artists, balancing on the trusses of the erector and on the rocket boosters, "like circus performers," according to Grishin. At 9 a.m., everyone who had worked at night grabbed some breakfast and headed off to take a nap, in order to be at the launch site at T-minus 4 hours.

The launch took place exactly at the prescribed time, at 18 hours 6 minutes 42

---

29. KBV—*Koeffitsient begushchey volny.*

seconds. I was in a tracking station (IP-1) next to the *Tral* receivers housed in the bodies of vans. It was now commonplace for military operators to sit at the consoles, while our telemetry specialists (Golunskiy, Vorshev, and Semagin) tracked parameters on the monitor screens. OKB MEI engineers Popov and Novikov and their assistants were also standing watch at their stations, ready to replace any system that acted up or come to the aid of the military operators. From a distance of 800 meters, during daylight, the rocket engines' ignition flames can hardly be seen. But then the noiseless lapping flame of preliminary combustion appears, builds to the mounting rumble of main stage combustion, and the rocket is enveloped in flame; the rumbling becomes unbearable, and it smoothly lifts off from the trusses. Now the flame gushes in a clearly defined plume. No matter how many times I have feasted my eyes on a launch, I still can't get used to it. You always fear that something is going to happen and the rocket's soaring flight on its blinding fiery plume will turn into the confused somersaulting of blazing boosters.

The powered flight segment was going right on schedule. We heard reports from the telemetry machines: "Flight normal!"

At 120 seconds the four boosters of the first stage separate in an X-shape. The second stage moves along its trajectory, leaving behind a white vapor trail illuminated by the sun. Now we need to be closer to the telemetry specialists; only they and Bogomolov's *Kama* radar systems could see what was happening with the rocket. Here's the third stage startup report—one can breathe easier now!

And suddenly the news: pressure in the chamber is dropping, the engine has shut down. Well, it should be shut down. Vorshev states that the last stage engine shut down 3 seconds before the designated time. It's over! Our efforts and all the trouble over photo chemicals and eliminating dozens of defects in the Ye-3 had been in vain! "There won't be any movies," said Grishin, who was standing not far away.

The next day, after analyzing the telemetry, the diagnosis was unambiguous and vexing to the point of tears. By all parameters, the flight had proceeded normally. Three seconds before the designated time for engine shutdown, the pressure downstream from the pumps fell by 50%, the pressure in the chamber decreased smoothly, the pressure sensor contact tripped, and the engine shut down. This had caused the spacecraft to be 130 meters per second below final velocity. Where it would end up was not yet clear.

Further investigation showed that there wasn't enough kerosene! The third stage tank had not been completely fueled. I recalled Rudnev's admonition—"We're firing entire cities." Here's one more city gone out the window. This was the sloppiness of the tankers and controllers in Barmin's service! Nedelin, Korolev, and Keldysh sequestered themselves with Barmin, Voskresenskiy, and Nosov for an investigation and report to Khrushchev. Meanwhile, the rest of us, who had no part in this sloppiness, now set our hopes on the second launch—the third for photographing the far side of the Moon.

After three sleepless days, the next rocket and Ye-3 lunar spacecraft were ready for launch on 19 April. This time, taking advantage of the twilight, I decided at T-minus 15 minutes to step out of tracking station IP-1, where many spectators had gathered, into the steppe facing the launch site.

Taking my time, delighting in the fragrance of the steppe, I walked about 300 meters and gazed at the rocket brightly illuminated by flood lights. I could hear the announcement "T-minus 1 minute" over the IP's loudspeakers. Out on the steppe I was engulfed by a feeling of solitude; there was no one nearby, except the image of a beautiful dream embodied in the rocket. I thought that, "If something happens to her now, I and hundreds more of her creators will be powerless to help her." And it happened! I had certainly courted disaster. All the engines of the rocket's main stage produced a deafening roar. Standing 300 meters closer than usual, I felt the difference in the noise level.

But what's going on? I see or surmise that the strapon booster closest to me is not lifting off along with the rest of the cluster and that, belching flame, it collapses. The remaining boosters reluctantly lift off, and when they appear to be directly over my head, they disintegrate. I can't make out what is flying where, but I sense that one of the boosters, its engine roaring, is headed right for me. Run! Just run! To the IP—there are emergency trenches there! Maybe I have time. Back in my *Komsomol* days I was a pretty good 100-meter runner. I once ranked as a champion sprinter at Factory No. 22. Now, on the steppe, brightly illuminated by the plume of a rocket booster flying at me, I probably set a personal record. But the steppe is not a running track. I stumble and fall, banging my knee badly. Behind me an explosion resounds and I feel a blast of hot air. Clods of earth thrown up by the explosion fall around me. Overcoming the pain in my knee, I limp toward the tracking station, away from the enormous bonfire blazing next to the spot where I had been standing. But where are the other boosters!? A blaze shoots up near the MIK. Don't tell me some booster has struck the engineering facility— there are people there!

When I had limped to the trench, suddenly an agitated female scream burst from it, "I said get out!" I recognized the voice of Irina Yablokova, a scientific associate at Lidorenko's institute. We considered her our chief manager of onboard storage batteries. The trench was filled to overflowing with officers of all ranks who had jumped in there. One by one, chuckling with embarrassment and shaking themselves off, they struggled out and ran to their vehicles, searching for the drivers, who had taken cover. Yablokova had a good laugh as she told us how she didn't understand what was going on at first. But suddenly someone shoved her into the trench and then bodies started to pile on from all sides so that it became difficult to breathe. We went up to the *Tral* vans. It turned out that the valiant team of telemetry specialists had jumped out of the vans and taken cover where they could. The accident had caused a great deal of damage, but quite luckily not a single life was lost. The core booster fell and exploded right next to the MIK; the window glass and doors had been blown out, and inside, the plaster was crumbling down. One officer, thrown against the wall by the shock wave, had suffered contusions.

When Voskresenskiy saw how badly I was limping, he couldn't pass up the opportunity to announce that the accident investigation commission protocol would note that, "Among the injured was comrade Chertok, who violated established safety regulations and did not avail himself of the shelter prepared in advance by the firing range authorities."

"Keep in mind," said Voskresenskiy, "Korolev arranged with Nedelin for a special State Commission decree obliging the firing range authorities to evacuate everyone away from the launch site and to herd those remaining at IP-1 into the trenches."

The next morning the State Commission circulated instructions for the immediate restoration of all damaged buildings at the launch site and engineering facility. Keldysh and his entire army of scholars were far more shaken that the others. Despite our resistance, they were the ones who had insisted on these launches. Now there was no hope of repeating a similar experiment in the next few years. Taking stock of our losses from the explosion and fire, we left the firing range for a short while. At the MIK, despite the broken glass, they were already unloading and setting up the next rocket at the work stations.

Chapter 31
"Onward to Mars...and Venus"

The celestial mechanics' calculations confirmed that not every year was suitable for flying to Mars. In 1960, the optimal launch dates fell in late September or the first half of October.

Who could presume to announce to Khrushchev that it was unrealistic to create a rocket-space system for launches to Mars and Venus in the autumn of 1960 and that the plan needed to be postponed for a year until the next launch windows? No one wanted to be the first one to be "thrashed." Now, many years later, I am amazed by the behavior of such sensible high-ranking individuals as Ustinov, Rudnev, and Kalmykov. Unlike Khrushchev, they had studied the technology and understood how unrealistic the missions were. But none of them had the courage to propose realistic dates. It was assumed that such initiative would come from Korolev personally or from the Council of Chiefs. Such initiative could not be viewed as ideological dissent from the Party line. In this case, no one would be threatened with arrest or other repression. But, nevertheless, despite common sense, from ministers to workers, we all devoted all of our efforts to carrying out the latest CPSU Central Committee and government decree.

These decrees usually began with the words, "Adopt the proposal of the USSR Academy of Sciences, Ministry of Defense, State Committee on Defense Technology, State Committee on Radio Electronics ..." and continued with a long list of state committees (after reform in 1965, a list of ministries replaced them), followed by a list of other organizations, then the surnames of the ministers and directors of all the previously listed organizations, and finally, the statement of the mission and the deadlines. The last points of the decree listed the individuals responsible for executing each part of the mission: state committee ministers, organization heads, and the chief designers personally. Thus, from the very beginning, it was established that no one at the top had ordered the flight to the Moon, Venus, or Mars, or to carry out any other space project. The Central Committee and Council of Ministers had simply agreed to the proposals coming from below, and rendered assistance, stipulating not only deadlines, but also making arrangements for financing, bonuses, allocating necessary funds for construction, production facilities in the *Sovnarkhozy*, and so on—everything that the drafters of the decree's text had managed to coordinate with *Gosplan*, *Gossnab*, the Ministry of Finance, and other ministries, that, as they used to say, "didn't give a hoot" about the Moon and Mars.[1]

The events of the first year of the 1960s (now the last century) serve as a very illustrative example of the kinds of decrees issued by the Central Committee and Council of Ministers, which had adopted the proposals generated by the USSR Academy of Sciences, chief designers, and the ministers who patronized them. Here I will cite an example of how the decrees were initiated, prepared, and issued.

On 2 January 1960, Khrushchev summoned Korolev, Keldysh, Glushko, and Pilyugin. Korolev and Pilyugin told us, "Nikita Sergeyevich is very perturbed by the broad scope of space exploration projects under way in the U.S. He believes that now success in space is just as vital as combat missile production." Khrushchev instructed Keldysh and the chief designers to prepare their own proposals for a detailed draft of a Central Committee and USSR Council of Ministers decree. The next day, Korolev and Keldysh held a meeting of the chief designers at OKB-1 summarizing the results of their conversation with Khrushchev the day before. Those attending the meeting decided to prepare, within a week, a detailed long-range research and development program for rocket and space systems. When Brezhnev visited OKB-1 on 4 February, he reiterated Khrushchev's recommendation that we prepare proposals for their serious discussion in the Central Committee Presidium. In private conversations Korolev commented on the month-long delay after Khrushchev's first instruction, saying that Chelomey had gone directly to Khrushchev at the same time and independently of us with long-range proposals. Moreover, Ustinov had also given similar instructions to Yangel.

While an expanded long-range plan covering all conceivable areas of rocket and space technology was being prepared, Council of Ministers Deputy Chairman and VPK Chairman Ustinov supervised all day-to-day operations for all the organizations. Not expecting a general decree, Korolev sent letters to the policy-makers with proposals for speeding up work on the automated lunar stations, including a soft-landing on the Moon, and for work on the 8K78 four stage launch vehicle (which would later be called Molniya [Lightning]).[2]

On 28 February, Korolev approved the schedule of operations for the production and launch of automatic interplanetary stations (AMSs) for Mars missions—the 1M spacecraft. The schedule included deadlines that Bushuyev, Turkov, and I considered impossible to meet, but Korolev made each of us sign this schedule showing

---

1. *Gossnab—Glavnoye upravleniye gosudarstvennogo snabzheniya* (Main Directorate for State Procurement)—was the main governmental agency providing logistical support for the Soviet economy.

2. The Molniya launch vehicle was named after one its primary payloads, the Molniya series of military and civilian communications satellites. During Soviet times, the real designations of boosters (such as 8K78 or 8K72) were kept secret. Instead, the official Soviet media would assign general names to boosters (such as Molniya and Vostok, respectively) that were merely the names of major payloads for those boosters.

our approval.

The schedule stipulated that design documentation and specifications be issued to subcontractor organizations by 15 March, that working drawings be issued in April, that experimental spacecraft and the first 1M spacecraft be manufactured in July, and that the first 1M automatic interplanetary stations be sent to the firing range in mid-August and launched in late September or early October 1960. By contrast, the production cycle for modern (by this I mean late 20th century and early 21 century) spacecraft in the U.S. and Europe for interplanetary flights, including to Mars, is six to eight years! We raised a bit of a ruckus about this in Korolev's office, but then in February 1960, we signed a schedule with a total timeline from conception to launch of eight months!

Not until early April, after deliberating with his main "old" staff of chief designers, did Korolev send to USSR Council of Ministers Deputy Chairman Ustinov, GKOT Chairman Rudnev, Minister of Medium Machine Building Slavskiy, Minister of Defense Malinovskiy, GKRE Chairman Kalmykov, GKAT Chairman Dementyev, GKS Chairman Butoma, and USSR Academy of Sciences President Nesmeyanov a draft of his letter to the Central Committee and a draft of the Central Committee and USSR Council of Ministers decree for the further exploration of space for their approval and signature.[3]

During the period from 1960 through 1962 (three years!), the draft called for the launch of the following:

-Ye-3—a flight around the Moon for higher quality photography of the Moon's far side;

-Vostok—attitude-controlled satellites for photoreconnaissance of the Earth's surface, descent from orbit, launch and landing with a person on board;

-1M and 2M—automatic interplanetary stations for missions to Mars in 1960 and 1962 respectively;

-1V and 2V—automatic interplanetary stations for missions to Venus in 1961 and 1962, respectively;

-Ye-6 and Ye-7—spacecraft for a soft lunar landing and an artificial lunar satellite, 1960–61;

-Elektron—dual satellites for studying the Earth's radiation belts—1960;

-Zond—a spacecraft to study interplanetary space—1961.

Subsequently, over the next three to five years, the draft plan called for the production of the following:

-a new rocket-space system with a launch mass of 1,000 to 2,000 metric tons and a payload mass of 60 to 80 metric tons to be inserted into orbit as an artificial

---

3. GKAT—*Gosudarstvennyy komitet po aviatsionnoy tekhnike* (State Committee for Aviation Technology); GKS—*Gosudarstvennyy komitet po sudostroyeniyu* (State Committee for Ship Building). Like the other state committees (GKOT, GKRE), these bodies were ministry-level institutions supervising various branches of the Soviet defense industry.

satellite;

-powerful, high-performance liquid-propellant rocket engines, including nuclear rocket engines, liquid-hydrogen engines, ion rocket engines, and plasma rocket engines;

-high-precision automatic and radio-based flight control systems and telemetry systems;

-integrated ground systems for the preparation and launch of heavy rockets;

-a system of satellites in geosynchronous orbit to relay radio and television broadcasts and for navigational purposes;

-a system of satellites at various altitudes for the systematic observation of the Earth's surface for meteorological, geophysical, and astronomic purposes;

-a habitable heavy orbital station with a three- to five-man crew and a mass of 25 to 30 metric tons;

-an automatic solar satellite in the closest possible orbit to the Sun;

-a space vehicle with a two- to three-man crew for lunar landing with an approach mass of 10 to 12 metric tons;

-an interplanetary vehicle with a two- to three-man crew with a mission to fly around Mars and Venus with a mass of 10 to 30 metric tons during approach to the planet;

-an interplanetary vehicle with a two- to three-person crew with a mission to disembark on the planets' surface. Plan calls for group flight of three to four vehicles;

-powerful intercontinental rockets having a payload mass of 10 to 40 metric tons and a range of 3,000 to 12,000 kilometers;

-a system for the destruction of satellites, stations, and space rockets from the territory of the USSR;

-photographic and radio-reconnaissance systems aimed at any area of the Earth's territory, military radio communications systems, navigation systems, etc.[4]

When Korolev briefed us, his deputies, about the text of these proposals, we said (Bushuyev quite timidly, and Mishin, Voskresenskiy, and I more emphatically) that it would be better to write up proposals for the next five to seven years, rather than for the next three to five years. Basically, that was the upshot of our internal discord. Korolev did not accept our amendments, arguing that, "If we hold off with a manned Moon landing, the Americans will get ahead of us; if we're another year late with the manned mission to fly around Mars or Venus we'll be forgiven, but if Chelomey or Yangel propose completely crazy deadlines, then these missions will be given to them." And not only did we agree, we believed that this would be achieved—if not

---

4. The April 1960 draft of the *kosmoplan* is reproduced as "Draft of the CPSU Central Committee and USSR Council of Ministers on the Future Mastery of Cosmic Space" in B. V. Rauschenbach and G. S. Vetrov, eds., *S. P. Korolev i ego delo* [*S. P. Korolev and His Affairs*] (Moscow: Nauka, 1998), pp. 289–293. A later version from June 1960 is also reproduced in the same source (pp. 295–301) as "Draft Decree of the CPSU Central Committee and USSR Council of Ministers."

in 5 years, then surely in 10 years. In defense of our optimistic schedule, I should note that in the mid-1960s, even before their Moon landing, American scientists and managers of the aerospace industry were even greater utopians.

But LET'S RETURN TO 1960. Since the time of the first satellite to the present day, any space program begins with a launch vehicle. And that will be the case until humankind comes up with other ways to insert spacecraft into space. During 1959–60 we devoted our primary attention to launch vehicles. For the Mars and Venus missions, the two stage R-7A combat missile was used as the foundation to which we added a third stage, the Block I, and to it a fourth, the Block L.[5] The new four stage rocket was assigned the code number 8K78, and the new interplanetary spacecraft—1M ("the first Mars vehicle"). A lead designer, Vadim Petrov, was appointed for the 1M, and we began to issue schedules. Despite the general hype, there was no documentation for the work of our factories or those of our subcontractors—not in January, not in February, not in March! And the launch was supposed to be in October (the very latest date was 15 October)! A present-day reader, even just slightly versed in technology, would grin and say that only the foolhardy could take on such a mission within such a time frame. But we did not consider ourselves foolhardy. We grumbled that we had very little time, but if we really, really wanted to, we could do it.

And what did we have to do? I will begin with the launch vehicle and the insertion sequence. In early 1960, after two years of studying alternative scenarios for inserting spacecraft into interplanetary trajectories, OPM theoreticians Okhotsimskiy, Eneyev, Yershov, and our ballistics specialists Lavrov, Appazov, and Dashkov agreed on a method for launching spacecraft toward Mars and Venus.

Keldysh devoted a great deal of attention to this problem. At our OKB-1, while keeping a close watch on the theoretical research, Mishin, Okhapkin, and Kryukov made corrections applicable to the specific features of the three stage R-7, which was already flying as the 8K72, later called Vostok. They directly supervised the development of the fourth stage.

Our research showed that the continuous firing regimen using the basic three stages and ending with the insertion of a satellite into an intermediate orbit provided the greatest efficiency in terms of payload mass. Depending on the designated planet and launch date, the fourth stage would be fired at a specific point in the satellite's low intermediate orbit, boosting the interplanetary spacecraft to escape velocity. At the end of the boost segment and engine shutdown, the spacecraft

---

5. In terms of designations, the Soviets typically used "Block" to denote stages of a booster, and "Object" to denote a payload. Of the first 14 letters of the Cyrillic alphabet, Soviet engineers used A, B, V, G, D, Ye, Zh, I, K, L, and M for designating Blocks or Objects, for example, Block A or Object D. The basic R-7A vehicle was the Block A (core) surrounded by four strapon boosters (Blocks B, V, G, and D). The third stage was the Block I while the fourth stages were either the Block L or M. The Block Zh was a paper design never actually built.

would depart on an autonomous voyage into deep space. Its orbit en route to the planet would be monitored from Earth and guided by its own Correction Engine Unit (KDU).[6] The proposed insertion sequence later proved to be all-purpose—it remained in use for all launches to Mars and Venus, for lunar spacecraft soft-landings, and even for the insertion of the Molniya communications satellite. Perhaps that is why all open publications refer to the four stage rocket developed in 1960 as the Molniya, while we simply called it the "seventy-eight"—referring to its design code number 8K78.

Rauschenbach, Yurasov, and myself, as well as all the guidance specialists at OKB-1, were really stirred up by the fever-pitched assignment of projects on the control system for the fourth stage (Block L) and the interplanetary spacecraft. After many arguments, the Council of Chiefs made a decision, which was reinforced by the orders of ministers and state committee chairmen that fourth stage guidance would be considered a continuation of the rocket guidance system, and thus its development would be entrusted to Pilyugin, while OKB-1 would be assigned to develop the guidance systems for the Mars and Venus spacecraft. The decision was an ideological victory for our young guidance team.

The first three stages of the rocket had more or less been tested out, and we were not particularly troubled by their performance. Yet, during every launch, even in hot weather, an uneasy chill would come over me. The fourth stage, however, was unique in that it would require the execution of an in-orbit startup in weightlessness, outside the zone of radio coverage from Soviet territory. A special startup support system (SOZ) was developed for the fourth stage engine.[7] This system contained a solid-propellant engine with a low total thrust.[8] The system imparted the required initial acceleration for the reliable startup of the fourth stage main engine.

Melnikov and his deputies, Raykov and Sokolov, developed the fourth stage's liquid oxygen–kerosene engine under Mishin's strict supervision. They were very proud of the fact that they were producing an engine with a closed-cycle configuration for the first time. After driving the turbine, rather than being ejected into the surrounding space, the generator gas passed into the combustion chamber, where it underwent afterburn, increasing the specific impulse. The primary concerns for the development of the fourth stage fell on Sergey Okhapkin. Sergey Kryukov handled the four stage launch vehicle's general layout and the integration of dozens of design parameters.

Engine production required high standards of metalworking, the mastery of new materials, and very close cooperation with testers and designers. Korolev and Turkov tasked young engineer Vakhtang Vachnadze with introducing the new process to our factory and managing engine production. Once again they made no mistake in

---

6. KDU—*Korrektiruyushchaya dvigatelnaya ustanovka.*
7. SOZ—*Sistema obespecheniya zapuska.*
8. In English, these are typically called ullage motors.

their selection.[9]

Isayev agreed to develop the high-boiling propellant-based KDU for the interplanetary spacecraft, but he asked for assistance from our production facilities. He was not about to mention being loaded down with Makeyev's naval orders.[10] "The journey to Mars is worth the risk," he declared, and threw himself into the general hubbub in creating the AMS.

Gleb Maksimov's group produced the draft plan of the spacecraft itself. Maksimov did not have a great deal of experience in the development of interplanetary spacecraft, but alas, no one had any experience yet. The conceptual designers' imagination needed to be transformed into a specific layout incorporating Isayev's KDU and our own attitude control and stabilization system and control system for all the onboard units. It needed to integrate Lidorenko's solar array, buffer batteries, Belousov's and Khodarev's radio system, a large parabolic antenna, and many more devices of all sorts—each capable, if it failed, of wreaking havoc on the entire undertaking.

OUR DEPARTMENT INHERITED A LOT OF COMPLETELY NEW TASKS. We, the guidance specialists of OKB-1, would have to design from scratch a control system for the world's first spacecraft bound for Mars. The primary task was to develop the logic and hardware for a system that would provide, as instructed from the Earth, virtually any orientation of the automatic Mars station during the operation of the vernier thrusters.

After meetings on Stromynka Street in Moscow at TsKB Geofizika with Vladimir Khrustalev, we arranged to develop sun and star trackers. The newly invented orientation system was multifunctional. Its first task was to provide constant orientation toward the Sun in order to ensure that the solar array was permanently illuminated to the required extent. We developed constant solar orientation (PSO) and approximate solar orientation (GSO) systems.[11] The latter could be used in the event of PSO malfunction to spin the spacecraft about the solar axis. This spinning enabled us to charge the batteries from the solar array. The Sun alone was not enough to correct trajectory. The KDU axis needed to be set in virtually any position in space, depending on the calculations performed on Earth, to generate a correcting pulse. A second optical reference point was needed in addition to the Sun, for which we selected the bright star Canopus; Sirius served as a backup. As for the second task, TsKB Geofizika developed the star tracker with lenses that moved into prescribed angles, depending on the numerical data transmitted from Earth,

---

9. Vakhtang Dmitriyevich Vachnadze (1929–) eventually became director of NPO Energiya, that is, the old Korolev design bureau. He served in this position in 1977–91.

10. Although better known in the West as a designer of rocket engines for spacecraft, Aleksey Isayev's OKB-2 was primarily involved in producing engines for naval, tactical, and air defense missiles.

11. PSO— *Postoyannaya solnechnaya orientatsiya;* GOS—*Grubaya solnechnaya orientatsiya.*

and oriented the KDU axis in space before firing it. OKB-1 and TsKB Geofizika dealt with the development of instruments and a reliable logic for searching for the requisite star. The third task of the orientation system was to aim the narrow beam of the parabolic antenna at the Earth.

It would have been so much easier to solve all of these problems if we could have put a computer on board; but that wasn't developed for another 15 years! In 1960 we had not even dreamed of this. And that is why it was necessary to complicate the radio system apparatus by introducing sequencers into it.

The strategy of flight control, making corrections, and obtaining information was designed to ensure sufficient time to cram in all the necessary operations while the AMS was still in the Yevpatoriya center's coverage zone. In addition to transmitting commands to the craft to control the onboard systems, receive telemetry data, and measure coordinates, the radio system also needed to transmit numerical settings before a correction and perform a back check.

Engineer Vitaliy Kalmykov was expected to develop a unified system for electrical power distribution and the transmission of commands from the radio-link decoders and sequencers. In addition, he needed to create an inhibitor that would permit the vernier thrusters to perform a correction burn only when a star was in the star tracker lens's field of vision.

When designing the onboard automatic controls and general electrical circuitry, one needed to understand the operating logic of each system. Each of the developers created his own "piece" of a complex system. After studying each of these "pieces," the task of an engineer developing the logic and control system for an entire onboard complex was to assemble it all into a unified whole. "Departmentalism" within the cramped quarters of the spacecraft and in the single radio-link might lead to a dangerous situation when a command issued from the Earth would end up at the wrong address, causing an emergency situation on board. The command distribution logic would have to preclude such situations. In 1960, Yuriy Karpov's team was simultaneously developing onboard complex control systems (SUBK) for the first Korabl-Sputniks and AMSs.[12] On the Korabl-Sputnik, each system had "sovereignty," which made it difficult to create a unified power supply system and a common control logic. For the AMS, on the other hand, we had to develop a unified logic and unified centralized electric power supply system. I assigned this task to Yuriy Karpov's newly formed team. The need for systemic integration gradually seeped into the consciousness of each of its engineers. Automatic Mars stations were their first serious test, and it must be said that the general layout that Kalmykov developed passed it.

The creation of an onboard power supply was not a simple task. The core of the system was a system of flat solar arrays that were switched on via the onboard power

---

12. The Korabl-Sputniks were the robotic precursors for both the Vostok crewed spaceship and the Zenit photo-reconnaissance satellites.

source switchboard (BKIP) so as to recharge the buffer batteries.[13] A special ampere-hour meter was installed to protect against overcharging. Working with two subcontracting institutes, the All-Union Scientific-Research Institute of Current Sources (VNIIT) in Moscow and the Scientific-Research Institute of Batteries (NIAI) in Leningrad, Aleksandr Shuruy developed a unified power supply system.[14] Skipping ahead, I will say that this little space power plant did not let us down.

We were at the very beginning of our journey, and we still hadn't tried our hand at systems engineering. One of our mistakes was that we dismissed electromagnetic compatibility problems as inconsequential. Our neglect of these problems soon came back to haunt us.

Outside Yevpatoriya, construction was under way at a furious pace on the Center for Deep Space Communications. Putting this center into operation would mark the real beginning of the Mars program. Agadzhanov, Guskov, and the many creators of the Yevpatoriya center did not let us down. By October 1960, NIP-16 was ready to go to work with a Mars spacecraft. But there was no spacecraft capable of flying to Mars or Venus yet.

OUR FACTORY RECEIVED AN ORDER FOR TWO OF THE FIRST MARS SPACECRAFT code named 1M. Korolev gave Turkov just five months to manufacture them, including testing at the control and test station and shipment to the firing range! Within this same time frame, they also needed to design the fourth stage and conduct ground developmental testing. Our calculations, checked many times, showed that the optimal Mars launch date that year was 26 September. Any delay would force us to reduce the payload mass. We spent an entire year producing the first two "seventy-eights" and the first two Mars spacecraft. By today's standards, that time frame was fantastic. Ignorance truly was bliss.

In my long engineering career, I often had the occasion to deal with situations when a young team would set out any way they could to create a new system within an incredibly tight time frame. You could explain such an approach by a lack of experience, which only comes from many failures. In those days, we were not required to carry out labor-intensive ground developmental testing of individual systems and the entire spacecraft on special mock-ups and rigs. As a result, it was possible to schedule deadlines for the production of a flight-ready flight model while ignoring the protracted cycle of ground developmental testing.

The item that was furthest behind schedule was the radio complex for the 1M. The entire managerial staff of the SKB-567—the developers of the radio complex—comprised former NII-885 employees including Belousov, Khodarev, and the lead developer of the onboard radio unit, Malakhov. Like Belousov's SKB-567, NII-885

---

13. BKIP—*Bortovoy kommutator istochnikov pitaniya.*
14. VNIIT—*Vsesoyuznyy nauchno-issledovatelskiy institut istochnikov toka;* NIAI—*Nauchno-issledovatelskiy akkumulyatornyy institut.*

was subordinate at that time to the State Committee for Radio Electronics. They were responsible for producing the interplanetary radio complex within a fantastically brief period of time.

In August 1960, we returned with Korolev to Podlipki from the firing range after the successful flight of the third Korabl-Sputnik carrying the dogs Belka and Strelka.[15] Despite the hype surrounding the successful landing of Belka and Strelka, I went to the factory to find out the state of affairs with the first Mars spacecraft. The launch was supposed to take place in October—just two months away—and in shop No. 44 the assemblers were fiddling around with a dismantled AMS engineering model. No testing had begun yet; Belousov's radio complex hadn't arrived yet. I burst into Korolev's office. He was shouting loudly over the Kremlin "hotline" that they needed to isolate Belka and Strelka from the company of any other canines. He was suspicions that the medics would pull off some sort of sensational stunt for the sake of glory. Nevertheless, he listened very attentively to what I had to say. Right then and there he called Minister Kalmykov and his deputy Shokin on the Kremlin "hot line." In harsh words he said that Belousov, the new chief designer of the radio complex, was going to completely disrupt the entire schedule; he, Korolev, would have to report to Nikita Sergeyevich [Khrushchev] personally that the Mars launch that had been promised for this year would not take place.

Having finished his loud conversation over the "hotline," S.P. suddenly suggested, "Let's go see Belousov right away. We'll have a look and discuss everything there on site. Tell Bushuyev and Ostashev they should come, too."

At 1 p.m. we were in Belousov's office. Kalmykov and Shokin had arrived there too. Belousov's design bureau and its rather puny pilot-production plant were located next to Moscow's largest new construction project, the Lenin Komsomol automobile factory that was being upgraded and modernized. The managers of this new factory had laid claim to Belousov's work area and had demanded their immediate eviction.

The individual assemblies of the radio complex for the 1M were undergoing debugging and engineering follow-up. They had not yet been checked out together. Integrated testing of the closed communications loop had not even been performed on laboratory mock-ups. The overall picture was depressing. Belousov and his deputies Malakhov and Khodarev did not defend themselves or offer excuses. They had already spent many sleepless nights, but promised to finish everything in just a little while.

After a brief discussion, Korolev suddenly proposed limiting the testing to individual assemblies and sending us the whole complex for installation on board the AMS without performing integrated tests. Such a bold proposal astonished Kalmykov and Shokin. It took away their responsibility for the equipment's reliabil-

---

15. This was a test of the future Vostok crewed spacecraft designed to launch a person into orbit. Chertok includes a full description of the Vostok program in Volume 3.

ity and transferred it to Korolev, who had made this risky decision. I tried to argue, but S.P. gave me such a look that I shut up immediately. "Here's the thing, comrade Belousov, and you all listen. You will perform the integrated testing at *our* facility. Chertok and Ostashev will be in charge. The tested equipment must be sent to the firing range from our shop No. 44 on 28 August." One of the engineers hovering around tugged on my sleeve and whispered, "It'll be at least a week before we debug a single unit. We can't send you semi-finished products right after soldering."

When we took our seats in the spacious ZIS-110 after looking over their tiny factory, Korolev angrily reprimanded me, "Boris, you're incorrigible. You think I don't understand that their operation is an utter failure. But now let them try to tell us that they can't even send us the first unit part by part. I warned Kalmykov a long time ago that he was betting on the wrong horse."

On 30 August, I, the designated chief of engineering operations at the engineering facility, flew to Tyuratam with Arkadiy Ostashev, whom Korolev had appointed as my deputy. Twenty-four hours later an An-12 cargo airplane delivered two semi-assembled Mars spacecraft, 1M No.1 and No. 2. We immediately sent No. 1 for electrical tests and No. 2 to the pressure chamber to check for structural leaks. Pandemonium broke out as we sorted out the equipment that had arrived, the dozens of boxes, cables, and consoles; we tried to identify missing parts, looked for necessary test documentation and even people who had gotten lost somewhere in Moscow and Podlipki. Incoming and outgoing radiograms kept the communications lines busy around the clock. We had a month until the Mars launch.

I must confess that at that time I did not consider the situation hopeless—that was my still-meager space experience showing. In that same year of 1960, we had already conducted successful launches of the Korabl-Sputnik, which had been announced to the whole world. Perhaps we would be lucky here too. In addition, on the threshold of a missed deadline, there would always be one more wicked thought that would come to mind, "I'm not going to be the last in line. The flight doesn't come down to me! After all, the rocket is new!"

Korolev assigned Leonid Voskresenskiy to supervise launch preparation of the four stage 8K78. Voskresenskiy studied the state of affairs with the fourth stage in detail. God had endowed him with the gift of foresight. Having listened to my problems, he advised me, "Forget about that radio unit and all the Mars problems. The first time we won't fly any farther than Siberia!"

We had already become accustomed to working around the clock at the engineering facility. But in terms of sleep deprivation, the number of hourly technical problems, and the deluge of failures, September 1960 set a record. Among all the systems competing for the greatest number of *bobiki*, the record holder was the radio complex.

It started when the radio unit simply proved to be inoperable. At a review debriefing meeting on 9 September, Malakhov, the leading conceptual architect of the onboard radio complex, announced that the situation was far from hopeless and that he needed just a few days for tests—although some of the instruments that

573

had arrived from Moscow were not certifiable and the spares didn't work at all. His announcement provoked an explosion of indignant laughter. I reported the state of affairs by radiogram to Korolev. He responded that he would be flying out in the next few days with Minister Kalmykov who would "dish it out" to this Malakhov and Belousov's entire company.

After Malakhov and Khodarev made transmitters emit and receivers receive commands on the table, I insisted on installing all the units in their proper places in the body of the spacecraft and beginning joint tests with the other systems. We needed to make sure that the commands from the radio unit were not dispersed to false addresses and that the transmitters were capable of emitting the promised wattage through the flight-ready onboard antennas while also consuming no more than the approved number of amperes from the onboard sources.

All hell broke loose! The triodes in the transmitter broke down—we discovered that the wrong triode had been soldered in. The diodes in the transmitter's power converter broke down—it was unclear why. The *Taran* microswitches failed due to their particularly poor quality. The command radio-link electronics burned out due to muddled-up installation. The electronic telemetry switch failed. The transmitter had begun operating, but suddenly it started to smoke! And so on, and so on. The daily list of glitches contained more than 20 items.

Malakhov, appearing after 1 or 2 hours of sleep, crawled halfway into the spacecraft along with his soldering iron. He was the only one who understood or had access to the radio unit. It was difficult to tell whether the smoke was coming from the solder or whether the instruments themselves were smoking.

By 15 September the State Commission headed by Rudnev and Kalmykov had arrived at the firing range. It was their custom to arrive at the MIK at night to make sure that no one was sleeping and that the "soldering" continued. Korolev, Keldysh, and Ishlinskiy were already at the firing range. Meetings on the Korabl-Sputniks took up much of the bigwigs' time. Numerous guests and curious individuals involved in the human spaceflight program started to appear. The Vostok launch was close at hand. The brass kept track of business concerning the Mars launches only at night.

One night Rudnev and Kalmykov arrived at the MIK with Korolev. Rudnev turned to me with a not entirely civil question, "Every night when we come to the MIK I see the same butt sticking out of the spacecraft! Is it going to fly to Mars too?" He said this so loudly that the subject of Rudnev's comment wrenched his other body parts out of the spacecraft, and when he saw the bigwigs, he prepared himself for further censure. However, they weren't in the mood for it now. Malakhov reported that he needed another four hours.

"I'm already used to your needing 28 hours in every day," said Kalmykov. "In a month you've asked me for over a hundred extra hours." If he had elaborated further, we might have ended up in a loud discussion of the real state of things, an outcome undesirable in the presence of State Commission members. The managers soon left us.

Four times we pulled two transmitters out of the spacecraft for "standard repairs";

we pulled out the receivers six times. We modified the logic for issuing commands, resoldered the telemetry circuits countless times, and simply could not make the output of numerical commands match up with the required angular settings for the star and solar trackers. Each new activation simulating one of the onboard operating sessions generated new failures and inexplicable glitches. Each time, the unit had to be opened up and resoldered one more time.

It is unclear when our instrument production facility's two female installers got time to rest. At any time of the day you could see Rimma and Lyuda at the MIK resoldering according to the latest change in an unpredictable instrument's installation chart or producing a new cable. One of the engineers, having received a cable after a resoldering job, tested it with me present to see that it conformed to the diagram drawn in pencil on a scrap of paper. He discovered an error, got mad, and complained, "your installers let me down."

I went up to Rimma to find out what had happened.

"I confess that I made a mistake after soldering for 17 hours without taking a break for dinner and breakfast. We gave up having lunch long ago."

With around-the-clock testing, modifications, resoldering, and rechecking, it wasn't until 27 September that we were finally able to begin full-scale integrated testing. We found such a number of deviations that it became obvious that it would be impossible to launch on the optimal date. Integrated flight-control program compliance tests in communications sessions failed for the most varied reasons. We repeated them until we were slaphappy, striving to get through the simulation of a normal flight at least one time without glitches.

Finally, on 29 September we managed to simulate an image transmission session. To everyone's jubilation, we received some semblance of a test pattern. The photo-television unit was supposed to transmit an image of Mars' surface during a pass from an altitude of around 10,000 kilometers. But, alas, upon repetition we realized that it was unlikely that the FTU would work! Due to a procedural error putting in the preset values, the astro-correction session also went awry. Once again, we tried to repeat and once again something went wrong during another phase.

On 3 October, during a raucous session of the State Commission, so many comments were aimed at Belousov that I began to feel genuinely sorry for him. Ryazanskiy sized up the latest verbal flaying of Belousov, Khodarev, and Makakhov by noting, "It serves them right. It was no use taking on such a project with these deadlines." Agadzhanov, who had made a special trip in from the Crimea, reported that Yevpatoriya was ready for operation, but he requested that the receivers' bandwidth be expanded due to the drifting frequency of Belousov's onboard transmitters. Korolev also made some scathing remarks to Minister Kalmykov, voicing a vote of no confidence toward SKB-567 and Belousov personally. He requested that the SKB be transferred to Ryazanskiy as a branch before continuing work.[16]

---

16. This is actually what happened; SKB-567 was soon absorbed by NII-885.

Meanwhile, the ballistics experts and the conceptual designers calculated the trajectories for each postponed date, reporting that, "We're going to miss the optimal date, so we need to look for weight reserves!"

Without hesitating, the State Commission decreed that the photo-television unit and Professor Lebedinskiy's spectroreflectometer be taken off board.[17] The latter instrument was supposed to determine if there was life on Mars. In order to facilitate making this decision, Korolev proposed that the instrument first be tested on the steppe not far from our site. To everyone's delight, the instrument showed that "there was no life" on Earth in Tyuratam! Lebedinskiy took the State Commission's decision like the death of a close friend. I reassured him, "You were lucky! The chances of making it to Mars are virtually nil. So you'll have time to get your instruments into shape. At the very least, in a year you must prove with your instrument that we do have life here on the steppe."

ON THE EVENING OF 4 OCTOBER, in the cottages, barracks, and hotels we celebrated the anniversary of the first *Sputnik* launch using the gift from the French winemaker. Of the thousand bottles of champagne that he had sent us for photographing the far side of the Moon, a nice round hundred were delivered to us from Moscow for the celebration. Our mood was anything but the best when we marked the anniversary. A year ago we had stunned the world with photography of the far side of the Moon. Last week we were supposed to have launched a spacecraft to Mars to photograph its mysterious canals and transmit the images to Earth. What if some other structures would be discovered there? But a month of round-the-clock work had shown that there wouldn't be any sensation.

By a surprising confluence of random events (or perhaps there is a pattern here), I find myself editing this chapter for republication on 4 October 2004. Forty-four years have passed since humankind's first attempts described in this chapter to discover intelligent life on Mars using automatic spacecraft. Mars proved to be such an intelligent planet that we haven't managed to find even unintelligent, primitive life on its surface in all that time.

The entire month of September 1960, I had worked at the engineering facility with Arkadiy Ostashev on a schedule of 12- to 13-hour shifts. I almost always worked the day shift in order to brief the brass, while Ostashev worked primarily at night. When it became evident that we wouldn't make the optimal date, our morale started to deteriorate—"better a horrible end than endless horror." But promises had been made to Khrushchev about the flight to Mars, and the command "full steam ahead" was still in effect. It really made no sense to postpone the launches until next year. The production of launch vehicles was continuing successfully. Saving

---

17. Andrey Vladimirovich Lebedinskiy (1892–1965) was one of the most prominent aerospace biomedicine specialists in the Soviet Union. In 1954–63, he served as the director of the Institute of Biophysics at the Academy of Medical Sciences.

resources was not a consideration, and extra experience would always be useful.

On 6 October, after three days of uninterrupted testing, modifications, refinements, and authorizations, I reported to Korolev that I was sending 1M No. 1 for assembly and integration with the fourth stage of the launch vehicle and switching all my resources over to the backup—1M No. 2. There was no longer any hope of making a pass close to Mars. The objective remaining was simply to test the fourth stage and perform a test-run of the spacecraft systems operation in a prolonged flight. This, in and of itself, would be a success.

On 10 October, 8K78 No. 1 integrated with spacecraft 1M No. 1 lifted off the launch pad and crashed. Studying the telemetry recordings, we quickly determined the cause. The first two stages operated normally. During the third stage (Block I) segment, the gyro horizon issued a clearly false command approximately 309 seconds into the flight. Evidently, a breakage occurred or a contact was disturbed in the command potentiometer. Because of the false command, the third stage deviated by more than 7°. At the same time, the gyro horizon's terminal contact closed and a command was issued to shut down the engine. The entire Mars-bound stack fell to the Earth and burned up in the atmosphere over eastern Siberia.

The second launch of 8K78 carrying 1M No. 2 took place on 14 October, and it too, failed. This time the engineering defect was in the hydraulic system. A leak in the liquid oxygen line supercooled a kerosene valve that was supposed to open before ignition of the third stage engine while the vehicle was still on the launch pad. As a result, the liquid oxygen–drenched kerosene valve froze up. When the ignition command was issued, the valve failed to open and yet another Mars-bound stack burned up in the atmosphere over Siberia through the fault of the launch vehicle.

Kalmykov had every right to defend himself from the harsh attacks Korolev unleashed on him, yet he did not do so. In both cases OKB-1 was officially the guilty party. Aside from Viktor Kuznetsov, the contractors that we had accused of producing poor-quality equipment and missing deadlines had nothing to do with this failure. We could chalk up the previous crash to Kuznetsov; neither Korolev, nor I, nor my comrades bore any responsibility for the gyro horizon. But the general misery from two crashes in a row after a month and a half of intense, nonstop pressure was so excruciating that no one remembered their former hard feelings. The first launch window for flights to Mars closed. We switched over to Venus.

TWO SPACECRAFT WERE ALSO PREPARED FOR THE FIRST VENUS LAUNCHES UNDER THE CODE NUMBER 1VA. In terms of its control system and onboard equipment makeup, the 1VA spacecraft were very similar to the 1M. The objective of the launches was to gain experience shooting for Venus, to conduct research en route from Earth to Venus and during the approach segment to Earth's mysterious neighbor. It was impossible to design a descent vehicle and equipment to conduct direct research of the planet's atmosphere and surface within this launch window. Although Keldysh tried to bring up the subject, he quickly realized that it was completely unrealistic. A pendant shaped like a small globe with the continents etched on it was placed on the 1VA. Inside this small sphere was a medal depicting the Earth-to-Venus flight path. On the other side of the medal was the emblem of the Soviet Union. The pendant was placed in a spherical capsule with thermal shielding to protect it during entry into Venus' atmosphere at reentry velocity. We weren't very concerned about which of the Venusians would discover this pendant. The important thing was to publicize a description of it and prove that the Soviet Union was the first to touch down on Venus.

The first automatic interplanetary station to be launched toward Venus, the 1VA, was shipped to the firing range from our factory on 1 January 1961. The 8K78 rocket boosters arrived at the firing range four days later. That day Rudnev held a session of the State Commission at the GKOT on the upcoming Venus launches. The session was very preliminary. First they heard Voskresenskiy's report on the failure of the third stage of 8K78 launch vehicle during the two Mars attempts on 10 and 14 October 1960.

I presented the second report. I reported on the objectives and schedule for the AMS launches to Venus. The primary objective was to hit Venus and to test out communications at a range of millions of kilometers and the control system during long-duration flight. The scheduled date of the first launch window was 20 to 23 January; the second—28 to 30 January; and the third—8 to 10 February 1961. Based on the experience of the previous year's Mars launch attempts, after hearing the reports from Voskresenskiy and myself, everyone present was so skeptical that they didn't even ask any questions. Only Rudnev, as State Commission Chairman, asked a question, and it was directed more to himself than to me: "During the Mars launches, we never even determined if the spacecraft itself was reliable. We never got that far. What's the probability that out of three launches we'll send even one to Venus?"

"One out of three will definitely make it to Venus," I cheerfully replied.

On 8 January 1961, I once again flew out to Tyuratam with the main group of engineers and installers. The team was well broken-in after the days and nights of "Martian" preparation. We were already psychologically prepared for the work. The equipment was simpler than that of the Mars shots. Once again the most unreliable element proved to be the radio complex. Debugging it devoured most of the time that remained before the first launch set for 4 February.

During preparation we discovered not only equipment failures, but also obvious mistakes committed during the design process. The setting of the attitude control system sun/star tracker depended on the launch date. The setting was performed at the engineering facility using 4 February as the reference date. After mating the automatic interplanetary station with the unit, the launch vehicle payload was enclosed in the payload fairing. In the event of a launch delay of 24 hours or more, the rocket would have to be removed from the launch site just to reset the sensor because there was no way to access it there.

Apropos this, Voskresenskiy, who supervised operations at the launch site, asked lead designer Vadim Petrov and Gleb Maksimov, "What were you guys thinking when you designed it like this? For this, you designers should have to drop your trousers and get a flogging right here in front of everybody. Then we should either make you modify the sensor or the fairing. Unfortunately, my schedule doesn't include time for a beating demonstration or for modifications. I'm not going to complain to Korolev. But if we don't hit Venus, I'll tell him the reason why."

The endless processes of dismantling and reassembling the payload almost drove us berserk. We would dismantle it, determine the reason for the latest failure, replace the transmitter or find a loss of contact in the feeder cable, assemble it, activate a communications session, and discover a new failure that wasn't there before. On the night of 25 January the fifth such assembly-disassembly cycle took place. This time the high-frequency switch that connected one of the two transmitters with the parabolic antenna failed.

At that time, Korolev had left the firing range for just three days. Now he was flying "home." We had developed the unwritten tradition of driving out to the airfield to meet arriving brass regardless of the current work load. And so, deprived of a good night's sleep, Keldysh, Ishlinskiy, and I drove over to the airfield to meet Korolev. En route Keldysh argued with Ishlinskiy about scientific works submitted as candidates for Lenin Prizes. We warmed up in the car, and as they argued I fell asleep.

It was a sunny day. Korolev, who was the first to deplane and obviously in a good mood, exclaimed, "It's already springtime here! When I left Moscow the temperature was –24°C (–11°F)." I drove back with Korolev and Voskresenskiy. S.P. was not so much interested in Venus as he was in talking about his meeting with Frol Kozlov—the Party's second ranking official after Khrushchev.[18] He complained that our *Devyatka* had been squeezed out in favor of Yangel's *Shestnadtsataya*.[19] He said, "After last October's catastrophe, the bigwigs are sparing no effort to rehabilitate Yangel and his work.[20] Worse than that, Frol told me straight out, 'Yangel gets first priority, Chelomey is the backup, and you're last in line.'"

"I asked Kozlov, 'Does that mean we won't be involved in space?'"

"He replied, 'No, that's not what we're saying; you will absolutely be involved. We are attaching exceptional value to the Venus launch. But don't be in a hurry. We're not rushing you. If necessary, it can wait.'"

The next day and night—with Korolev, Keldysh, and Ishlinskiy attending and with an assemblage of inquisitive onlookers—we once again disassembled the AMS to look for defects in the power supply system automatic controls. We found out

---

18. Frol Romanovich Kozlov (1908–65) served as secretary of the Central Committee for defense industries in 1960–64, that is, the effective Party leader of the Soviet space program.

19. *Devyatka*, literally "niner," and *Shestnadtsataya*, literally "sixteenth," were the nicknames for the R-9 and R-16 missiles, respectively.

20. This is a reference to the terrible disaster during the first attempted launch of the R-16 in October 1960. See Chapter 32.

that a remote control switch was on the blink. While we were at it, we corrected a defect in Konstantin Iosifovich Gringauz' equipment, which was supposed to determine the condition of the interplanetary plasma throughout the journey.

Once again we assembled, tested, and sent the entire AMS to the pressure chamber for a leak check. By morning on 29 January, after the pressure chamber test, I was forced to make the decision to disassemble the AMS instead of handing it over to be integrated with the launch vehicle. We had determined that there was just noise—no valid signals—at the receiver output.

We checked everything in disassembled form. We found the causes. Once again we assembled the vehicle; once again we performed tests on the assembled hardware; and once again there was another leak check in the pressure chamber. In the brief intervals between the incessant testing, the cracking open of the systems, the modifications, and the pressure chamber leak checks, Ostashev and I, taking turns, managed to grab an hour of sleep.

In a state of continuous turmoil, without considering the details of the document, I signed a protocol on equipping the Venus descent vehicle with a pendant of the Soviet Union; running around at the MIK, I asked Korolev to approve it. His attitude toward that document was considerably more serious and he chewed me out. "It was typed sloppily," he said. "Retype a clean copy on good paper. This is a document of national importance. We will sign it together but the State Commission chairman must approve it."

We finally handed over the spacecraft for integration with the launch vehicle. Kirillov designated 7 a.m. on 1 February for the traditional departure from the MIK for the launch site.[21] That night I admired the two launch vehicles. The next 8K78 cluster, the third one now, was lying on the erector. In its nose cone was the elegant, gleaming 1VA with the metallic shine of its thermal insulation foil and blinding white paint of its thermal radiators. Nearby, the final horizontal tests were under way on the fourth 8K78 launch vehicle.

On 31 January, at 5 p.m., a meeting of the Council of Chiefs began on the third floor of the MIK service building. Korolev and Glushko presented proposals for a future heavy launch vehicle. The upshot of their presentations was that the powers that be had issued a directive shifting the project to military objectives. But it was not completely clear which objectives. For the first time, Korolev set the goal of manufacturing the launch vehicle components at the firing range instead of only assembling them. That was the only way to eliminate the problem of transporting the future gigantic rocket from Russia to Kazakhstan. Barmin smirked. Everyone else was silent. The troubles of the next few hours were on everyone's mind. We needed to grab some dinner and meet for a State Commission session at 8 p.m. to make the decision to transport the Venus rocket to the launch site.

---

21. Anatoliy Semenovich Kirillov (1924–87) served as chief of the first directorate at the Tyuratam launch range. In this position, he was responsible for all launch operations at the range.

The Commission convened with the participation of a large number of well-wishers. The proceedings had just begun when an officer ran in and whispered something to Kirillov. He excused himself and dashed for the door, asking permission en route to abscond with me as well. When we ran into the hall out of breath, a smiling Ostashev met us and explained everything. The load-bearing frame holding the 1VA had been brought into the horizontal position and the crane had transferred it to the erector to be mated with the launch vehicle. Everything was going normally, but suddenly all the attitude control valves of the mated spacecraft began to clatter, whistling as they leaked the precious supply of compressed nitrogen. Everyone who had been working on the erector jumped down and dashed for the exit. Memories of the catastrophe at Yangel's adjacent site were still fresh. Everyone knew that the AMS's propulsion system was filled with nitric acid and kerosene. What if an engine started up all of a sudden? Arkadiy Ostashev, who was in the hall, was the first to realize what was happening. He gave the command to quickly de-mate the frame, hook up the ground console, and stop the AMS's premature activity. It turned out that elastic deformation had caused the frame carrying the spacecraft to pull away from the load-bearing ring so much that the limit contacts designed to activate the first near-Earth communications session after separation from the launch vehicle had tripped.

Unable to stand the suspense, the entire State Commission took a break and came down to the hall. I proposed that we cap one of the two limit contacts, put a wider stop on the second one, and put in an electric inhibitor that would be removed once the assembly was vertical at the launch site. The proposals were implemented overnight and rechecked from the ground console.

In the cold morning, as per tradition, everyone drove out to the gates of the MIK for the rocket transfer. The gates swung open and the motor locomotive belching exhaust prepared to push the erector holding the rocket to the launch pad. Suddenly Korolev said to Kirillov:

"Stop the transfer!"

"Why, Sergey Pavlovich?"

"You scheduled this to start at 7 a.m, and it's only 6:50."

Everyone smiled and patiently hopped up and down in the cold for the prescribed 10 minutes. Precisely at 7 o'clock, turning to all those assembled, Rudnev loudly announced, "Sergey Pavlovich has taught us a lesson in precision. I support him and I request that henceforth, none of us do anything ahead of schedule.

These "instructions" had a light-hearted rallying effect.

On the very first day of tests at the launch site, we discovered that the third stage gyro horizon rapidly drifted and moved to a stop position, thus issuing an emergency engine shutdown command. After several tries we still couldn't determine the cause of the defect. Viktor Kuznetsov accepted the blame and proposed replacing the instrument.

At 11 p.m. on 3 February, a 15-minute State Commission meeting was held right in the bunker. On behalf of Chief Designer Pilyugin, Finogeyev reported the readiness of the launch vehicle control system. Ishlinskiy, who had been assigned to determine the possible causes for the gyro horizon's abnormal drift, gave a 3-minute report with his innate professorial flair. The snoozing Keldysh gave a start, and violating the official rules of order, concluded, "Even a person who doesn't know the gyroscope's operating principles can see from your report that it is better to fly without gyroscopes."

Grigoriy Levin reported that all the systems of the Command and Measurement Complex were ready. The ships *Dolinsk* and *Krasnodar* were in the Gulf of Guinea, the *Voroshilov* was standing by near Alexandria, and the *Sibir* and the *Suchan* were in the Pacific Ocean. The weather service reported a temperature of −15° (5° F), with light winds and clear skies. The State Commission members did not want to leave the bunker's warm guest room. "Instead of the Gulf of Guinea, you guys get Tyuratam," remarked State Commission Chairman Rudnev.

Ostashev left and headed to the *Tral* telemetry receiving station at the MIK. I went to IP-1, to the cozy cabin where the *Tral* units that received information from all the stages had been installed. At T-minus 3 minutes I stepped out into the cold darkness. A nighttime launch always made a stronger impression than a launch conducted in daylight. For seconds the steppe, as far as the eye can see, was illuminated by the single flame of five rocket engines. As the thunder receded into the distance, the light gradually faded and once again the steppe became dark, lonely, and bleak. Quickly I returned to the telemetry operators. They didn't detect a single glitch visually. The last messages from Ussuriysk reported that it seemed that the stage four—the Block L—had separated, but things were still not fully clear; they would recheck. Soon everyone was rushing to Site No. 2 for the radiograms. Reports from the ships arrived there via Odessa and Moscow. But no, it wasn't in the cards even on this third launch to test out the fourth stage. But the first three stages had presumably worked normally! We finally went out to the "dotted line," that is, went into orbit as artificial satellites. After that the messages were muddled, but it was already clear that the fourth stage would not depart for Venus at the required time. A commission was immediately created to conduct an in-depth investigation under my chairmanship. I was also tasked with speeding up preparation for the launch of 1VA No. 2.

The process of investigating what caused the failure of the first Venus launch had begun in my commission with a conflict between the "parties under investigation": the launch vehicle control system, which Finogeyev defended; the power supply system, whose reliability Iosifyan vouched for; and our OKB-1 designers, whom Korolev had promised to make "walk to Moscow on the railroad tracks" if they were to blame for the failure of the Block L (fourth stage) to separate from the Block I (third stage).

The telemetry operators rescued everybody. Boris Popov brought graphs constructed per dispatches from the Kamchatka tracking station. The cause was obvious—the PT-200 current converter failed at the end of third stage operation. This converter powered the Block L control system, and this failure completely explained the malfunction.

PT-200 developer Iosifyan asked, "But where did you put my converter?"

After a brief period of confusion it was determined that the PT-200 had been installed on the frame connecting the Block I with the Block L.

"What were you thinking?" the angry Iosifyan blurted out. "This electric device is not designed to work in a vacuum. Evidently the bearings broke down or carbon brushes ground against the commutator all at once the like sandpaper. Most likely, it was a combination of the two. I did not give permission to use this device in a vacuum!" It turned out that the guilty parties were Finogeyev, who had used the PT-200 in his system without coordinating the conditions for its use with the developer, and I, who had supervisory responsibility over the actions of all the "rusty electricians," as Korolev put it.

The cause of the crash was clear, but what could be done in the two days remaining before the last tests on the next launch vehicle? The time had been compressed to the extent that we needed proposals that required very few hours for their implementation. It was impossible to report to Korolev and then to the State Commission unless we had an actual proposal in reserve.

In search for solution, I went to the "working class," to a brigade of our factory workers, to gain an understanding of how much time would be required to manufacture a special pressurized container. On the way I stopped off for a smoke break at the laboratory where our spacecraft control specialists had settled in. There I consulted with my comrades about the problem that had suddenly fallen in my lap. Anatoliy Patsiora, one of the attitude control system developers, pointed to the onboard storage battery that was in the laboratory for some reason.

"Will that work? Dump all the cells out of the pressurized housing and put the PT-200 inside!"

Aleksandr Shuruy was standing nearby. He was an expert on both the design of storage batteries and the PT-200 itself. He immediately checked out the idea on his slide rule; the solution was beyond a doubt. Several hours later the PT-200 was installed inside the pressurized container that had formerly housed the onboard storage battery. The *teploviki*—as we called the thermal control specialists—advised us to wrap the container in vacuum shield insulation and cover it with black and white stripes like a zebra.[22] Iosifyan felt terrible about the failure caused by the PT-200. He approved the improvised packing of the converter into the ready-made storage battery container, but painstakingly checked the thermal control calculations. He didn't stop there. He gathered Korolev, Rudnev, and Kalmykov, and brought them one night to the laboratory where we were finishing up the business of installing the PT-200.

Rudnev and Kalmykov, both chain smokers, gladly sat down for a smoke break.

---

22. *Teploviki* is derived from the Russian word *teplo* meaning "heat."

Cutting me off in the middle of my explanations, Rudnev requested that I say something more entertaining at this time of night. "Tell the esteemed ministers how you and Vasya Kharchev attempted to steal von Braun from the Americans," chimed in Korolev.[23] For a little while my story distracted the crowd gathered in the laboratory—from young engineers to high-ranking industrial leaders—from the everyday routine of our space work. In those days at the firing range a rather democratic style of interaction had developed between those involved in the operations—from young engineer to minister. This was by no means just for show—it was easier to work that way. Rudnev capped off my story saying, "I could listen to you all night, but tomorrow, I mean, today, you have to give a report to the State Commission. Let's go, comrades. We won't distract you any more."

For all the operations, including leak testing, fastening, onboard installation, and the cycle of electrical tests, we spent less than 24 hours. We had not broken the general preparation schedule. It was announced that the next launch would be on 12 February. It turned out that preparing for the next launch was easier than figuring out what went wrong with the previous one.

On the afternoon of 10 February a small group gathered in Korolev's cottage to celebrate Keldysh's 50th birthday. We toasted the birthday boy's health with champagne, while he self-consciously mumbled that the best present for him would be a successful launch to Venus. That same day at 6:00 p.m., birthday boy Keldysh presided over the meeting of the State Commission on behalf of its Chairman Rudnev, who had departed for Moscow. I delivered a brief report on the causes of the previous crash, pointing out that the most probable cause was the failure of the PT-200 DC-to-AC current converter, and announced that the converter for 1VA No. 2 had been installed in a pressurized container. My report was approved. The State Commission believed that the failure of the PT-200 converter was the most probable cause for the failure of the fourth stage engine to start up.

A heavy unguided satellite with a mass of around six tons, not counting the mass of the third stage, ended up in orbit. An argument flared up in the State Commission as to what official communiqué to issue in this regard. Even back then, it wasn't difficult to detect such a satellite in near-Earth space. Korolev argued that we should announce nothing at all. Let the Americans torment themselves trying to figure out the satellite's purpose. Keldysh categorically objected. Glushko proposed a compromise statement: "With the objective of practicing for the launch of a more powerful spacecraft, a satellite was launched. Having transmitted to the Earth all the necessary telemetry data, it fulfilled its mission during its first orbital pass."

To Korolev's displeasure, Glushko's proposal was accepted, and a TASS report was issued noting that, "A Soviet heavy satellite is in orbit. It weighs 6,483 kilo-

---

23. This is reference to Chertok and Kharchev's mini-adventure in trying to "acquire" the services of Wernher von Braun in occupied Germany in 1945. See Chertok, *Rockets and People*. Vol. 1, pp. 294–296, 300–305.

grams... The scientific and technical mission objectives assigned for the satellite launch were accomplished."

According to the forecast, the new heavy satellite that had ended up in near-Earth orbit would quickly plow into the Earth's atmosphere. Given such a low orbit, the ballistics experts could not give a precise answer as to what the impact area might be, but they believed that most probably the heavy satellite would burn up over the ocean after making two to three orbital passes. Keldysh nevertheless was interested in knowing whether there was any information about the orbit of our new *Tyazhelyy sputnik* (Heavy Satellite).[24] Lieutenant Colonel Levin informed him that the telemetry system was ready for its next operation, but only air defense facilities could observe the satellite. However, after receiving the ballistics specialists' forecast, they had not detected anything. "A week has already passed," said Keldysh, "and no one has sent us any protests, so everything must have disappeared in the ocean."

The report by General Nikolay Kamanin, who was attending the State Commission session, cheered everyone up.[25] Air Force Headquarters had communicated to him that after TASS's report about our heavy satellite, Italian and French ham radio operators had supposedly received human calls for help and heard groans over our space frequencies. Based on these reports, some newspapers had surmised that the *Tyazhelyy sputnik* was crewed and that a cosmonaut had died in orbit in horrible suffering. Suddenly, firing range chief Aleksandr Zakharov announced, "All firing range services are ready for operation." The next State Commission meeting was scheduled for 10 p.m. on 11 February.

We forgot about *Tyazhelyy sputnik* for the time being, but it jogged our memories a year and a half later! In the summer of 1963, Korolev asked me to come to his office, having alerted me over the phone. "Don't bring any papers or charts," he added. When I entered the small room of his office, he was smiling like the cat that ate the canary, an indication that he was in a good mood, and he began to unwrap a bundle of crumpled wrapping paper. From a small heap of shapeless pieces of iron he pulled out a slightly deformed, sooty medal and held it out to me, "I received a gift from the Academy of Sciences and decided that by rights it belongs to you," he said.

When I first gazed at the gift, I must have looked pretty silly. This was the pendant from the first 1VA Venus spacecraft. Despite being scuffed up and sooty, you could distinctly make out the inscription:

"*1961* Union of Soviet Socialist Republics *"

At the center of the medal the sun was shining with the orbits of the Earth and

---

24. In all official TASS dispatches from the period, this spacecraft was simply called *Tyazhelyy sputnik,* or Heavy Satellite.

25. Nikolay Petrovich Kamanin (1909–82) at the time (in 1958–62) was the deputy chief of the Air Force General Staff in charge of combat preparations. In this position, he was responsible for coordinating the training of cosmonauts. Although his title changed over the years, he continued to supervise cosmonaut training until 1971.

Venus depicted around it.

Korolev went on to explain that the medal and the remains of the structure in which it was packed were handed over personally to Keldysh from the KGB. The remains of the pendant had fallen into the hands of the KGB, not from space, but from Siberia. While swimming in a river—a tributary of the Biryusa River in eastern Siberia—a local boy hurt his foot on some sort of piece of iron. When he retrieved it from the water, rather than throw it into deeper water, he brought it home and showed it to his father. The boy's father, curious as to what the dented metal sphere contained, opened it up and discovered this medal inside. This took place in a Siberian village, the name of which Korolev had not been told. The boy's father brought his find to the police. The local police delivered the remains of the pendant to the regional department of the KGB, which in turn forwarded it to Moscow. In Moscow the appropriate KGB directorate found no threat to state security in these objects, and after notifying Keldysh as president of the Academy of Sciences, this unique find was delivered to him by courier.

Thus, I was awarded the medal that had been certified for the flight to Venus by the protocol that Korolev and I signed in January 1961. After the launch we were all certain that the *Tyazhelyy sputnik* and the pendant had sunk in the ocean. Now it turned out that it had burned up over Siberia. The pendant had been designed to withstand Venus' atmosphere and therefore it reached the Earth's surface.

According to the ballistics experts' forecasts, the probability of the satellite splashing down in the Pacific Ocean was greater than 90%. The probability of falling on dry land was 10%, of which 3% was the probability for falling on the territory of the USSR. It came down precisely to that 3%. But if, using the theory of random processes, you calculate what the probability would be of finding the pendant on the territory of the USSR, this value would be virtually zero. But it happened! An event occurred, the probability of which was close to zero! Most unfortunately, at that time in the hurly-burly of my daily routine, I didn't bother to find out the names of the boy and his father and the geographical location of the find. Their names deserved to be mentioned in the history of cosmonautics under the heading "Strange but True."

BUT LET'S GET BACK TO 1961. On 11 February at 7 a.m. under clear weather conditions, with an icy Tyuratam breeze, a four stage 8K78 was transported to the launch site for the fourth time. Preparation at the launch site was under way around the clock. Repairing to the warm *bankobus* (our term for the spacious dugout hut 150 meters from the launch pad) to get warm and take a smoke break, the testers knocked on wood and in all seriousness stated, "It's going well; this is the fourth one; it should pan out."

And it did!

On 12 February at 7:04:35 a.m. the fourth 8K78 launch vehicle lifted off. For the first time all four stages worked through their sequences normally. The second 1VA automatic interplanetary station had finally been launched on an interplanetary trajectory. At 9:17 a.m. NIP-16 triumphantly reported from Yevpatoriya that the first long-range communications session was proceeding normally. The second session at 4:23 p.m. confirmed that we really had launched a spacecraft to Venus. After gathering all the data, the ballistics experts from the Moscow ballistics center announced that an orbital correction would be required, and, if it went through, the pendant of the Soviet Union would be on Venus!

Gathering for breakfast at our "deluxe" dining hall after a sleepless night, we all agreed with Voskresenskiy that we had been given the chance to "rob Venus of her virginity." We decided to celebrate such a historic event with a drink. Korolev was in a cheerful mood and announced, "Only the pendant, which has thermal protection, will make it to the surface of Venus. The wrath of Zeus will come down on those who signed the protocol arming the AMS with the pendant. Boris and I signed the document. So let's have another dram so that Zeus will forgive us!"

In all the laughter and joking, everyone gladly joined in the toast. However, Zeus decided to anticipate our trespassing on the honor of the goddess of Love, and not to punish us ex post facto. Our collective jubilation was overshadowed by reports from Yevpatoriya. Telemetry data indicated unstable operation in the continuous solar orientation (PSO) mode, which maintained the required orientation of the solar arrays to charge the storage batteries. The operating logic for the onboard systems was set up so that if a PSO glitch occurred, the equipment would automatically reorient itself toward the Sun and after completing the orientation process, it would spin about its own "solar" axis. In this gyroscopic stabilization mode, rough orientation toward the Sun was maintained. Meanwhile, all systems consuming electric power, except for the thermal control system and sequencer, were shut down. Right then and there we discovered a stupid design flaw. The onboard receivers, which could have received control commands from Earth signaling the beginning of the next session, shut down along with all the other systems. After one "spin" the next communication session was only activated autonomously from the onboard sequencer and not until five days later. We faced five days of complete uncertainty and agonizing waiting.

Nevertheless, unaware of our uncertainties, TASS informed the world about the launch of the interplanetary station *Venera-1*.[26] "The successful launch of a spacecraft to the planet Venus is blazing the first interplanetary trail to the planets of the Solar System." That is how TASS ended its first report about the first attempt to reach Venus.

There was a lot of busy activity during those days; we were planning for an upcoming visit to OKB-1 by Air Force Commander-in-Chief Marshal of Aviation Konstantin Vershinin, and there were unending rush jobs at the firing range to prepare for the launch, scheduled for 10 March, of a Vostok carrying a dummy and

---

26. At the time, TASS simply called the vehicle the "Automatic Interplanetary Station." *Venera-1* was a designation conferred several years later. *Venera* is the Russian word for "Venus."

the next canine crew. Yet, Korolev and Keldysh and all their "Venusians" departed for Yevpatoriya to personally participate in the communication session scheduled for 17 February.

It is difficult to convey the stress that we felt as we waited for 1VA to make radio contact on its own without a prompt from Earth after five days of silence. In the small hall of NIP-16, where field telephones were the primary mode for exchanging information, the triumphant report "We have a signal!" rang out. Everyone broke into applause, but Korolev "stifled" us with one fierce look; silence quickly prevailed. During the session they took the risk of checking the PSO and again came up with a glitch; no other obvious problems were detected on board. Another five days remained before the next session.

On 22 February, the 1VA did not make radio contact. The session on 17 February, from a range of 1.9 million kilometers, was the last one. There was still a glimmer of hope that contact would be restored. After the first communication session, a detailed description of the AMS layout, flight trajectories, and instrumentation and control complex was prepared for publication in the press. An accompanying photograph depicted the same spherical pendant that had been stowed inside the AMS. After heated arguments, on 26 February, *Pravda* published the detailed material about the first flight to Venus without mentioning the radio blackout and without attributing the article to any authors. However, we never restored contact. According to the ballistics experts' calculations, the silent *Venera-1* passed Venus at a distance of approximately 100,000 kilometers in late May 1961.

Once again I was entrusted with a commission comprised of Rauschenbach, Malakhov, Khodarev, Ostashev, Maksimov, and military representatives to investigate the causes of the glitch and the loss of contact after 17 February. We were quickly able to determine the cause for the PSO failure. The optical sensor was not pressurized. Our thermal control specialists had been concerned only about the average temperature of the entire instrument, without performing calculations or conducting experiments to estimate the *local* temperatures of individual elements. The calculations showed that given a permissible average temperature, the maximum temperature reached in a sensitive element might exceed 80°C (176°F). This clearly led to the failure of the PSO system.

After lengthy arguments, we attributed the loss of contact to the failure of the sequencer developed for the radio complex. This had been done to cut down on mass. Korolev upbraided me harshly for giving in to the designers. I vowed that justice would prevail and on all subsequent AMSs we would install reliable event controllers that we had developed and that were manufactured at the Plastik Factory. But the most important action taken as a result of this incident was that henceforth command radio-link receivers would absolutely never be shut down. It was not permissible to save a trickle of energy while risking the loss of an entire spacecraft. We gained the experience of a spacecraft's first interplanetary flight at such a high price.

DURING PREPARATION FOR THESE FIRST TWO VENUS LAUNCHES, almost all of the scientific-technical elite interested in interplanetary flight had gathered at the firing range. Taking advantage of the circumstances, Korolev and Keldysh convened a council during which programs for the future were discussed. Korolev presented his idea for producing a series of standardized automatic spacecraft for interplanetary research, reasoning that series production could cut costs. The council accepted his idea, and straight away Korolev gave the command to start designing a new spacecraft with a maximum degree of structural and onboard systems standardization, taking into consideration the experience gained on the 1M and 1VA. The new spacecraft was assigned the factory index 2MV.

According to the ballistics experts' calculations, the earliest dates for launches of the new spacecraft series were August 1962 for Venus and October 1962 for Mars. The factory received the assignment to start up immediate production of at least six AMSs, three for Venus and three for Mars.

Soon after the decision to develop the 2MV, it became obvious that we first needed to development an analog—a spacecraft model to be used for intensive verification on the ground of flight modes simulating all nominal and potential off-nominal situations. Today, a similar solution is considered a matter of course, and no spacecraft goes into space until all the onboard systems and the complex as a whole have been proven reliable on its analog on the ground. This process increases the total volume of production operations, and in any event, prolongs the production cycle of the first flight model. Such an analog had not yet been provided for the 2MV.

I found myself thoroughly engrossed with the new project only following my return from the firing range after Yuriy Gagarin's flight and the investigation into the R-9 launch crashes.[27] All the teams in Korolev's OKB-1 and all of their subcontractor organizations continued to be in good spirits. No one was losing much sleep over the failures of the interplanetary flights. Gagarin's triumph had overshadowed all other space-related events. Nevertheless, specific schedules had been set up for the 2MV, meetings were being held, and drawings were being issued; we were arguing over each science experiment and presenting reports before ministers and the VPK.

Considering our bitter experience, we insisted on developing a new high-performance radio-link in the SHF range. The onboard equipment of this radio-link operated using a high-gain parabolic dish antenna. In the intervals between the infrequent sessions using this radio-link, it was possible at any time to communicate over the UHF link, which used wide-beam antennas. For communication in an unoriented mode, an "emergency" VHF system was again developed that operated

---

27. Yuriy Gagarin became the first human in space in April 1961 during his Vostok(-1) mission. During this same time period, OKB-1 also conducted a series of launches of the new R-9 ICBM, many of which failed to achieve their goals.

using omnidirectional antennas.

Each of the 2MV spacecraft consisted of two compartments. A standardized orbital compartment contained communications and control equipment that was identical for Mars and Venus. A special compartment housed science equipment determined by the wishes of planetologists. In some cases, instead of a special compartment, equipment intended to land on the planet was installed as part of the descent vehicles, which of course, differed for Venus and Mars. Spacecraft designed for landings were given the code numbers 2MV-1 and 2MV-3, while those designed to conduct research on the planets while flying past them were dubbed 2MV-2 and 2MV-4. The "fly-by" spacecraft had photo-television units installed on them.

To increase reliability and guarantee thermal control, the optical sensors were moved inside the pressurized service compartment from the external vacuum. We took responsibility for the automatic controls for the entire onboard complex away from Malakhov and transferred them as a separate task to the specialists in Karpov's department, where the chief electrician was "our" Kalmykov (we differentiated "our" Vitaliy Kalmykov from Minister Valeriy Dmitriyevich Kalmykov).

We installed sequencers on these new AMSs that we had developed with element-by-element redundancy. The creators of this instrument later took pride in the fact that "chief designers come and go," but their sequencer continued to be used for all subsequent AMS modifications. The unsuccessful experience of transferring the factory assembly and test cycle to the firing range was also taken into consideration. After all, there was more time and they were able to complete the main tests at the factory control and test station.

By the time testing began, an already battle-tested and broken-in team gathered once again at the firing range. It was very important that the people now understood each other much better; personal compatibility contributed to the technical compatibility of the systems. The efforts of two factories—ours in Podlipki and the Progress Factory in Kuybyshev—produced the 8K78 launch vehicles, which were delivered in advance to the firing range, where, rather than "standing in line" to be tested, they lay there already tested.

As per plans, in August 1962, the 2MV launches to Venus began. On 25 August, our fifth four stage 8K78 launch vehicle carrying the automatic interplanetary station 2MV-1 No. 3 with a mass of 1,097 kilograms worked normally through operation of the first three stages. The telemetry operators on the ship in the Gulf of Guinea had learned to quickly diagnose the status of the Block L systems from telemetry. This time we first received the reassuring message that the Block L engine had fired according to the program, but soon there was an alarming message—the engine had operated for just 45 seconds. The Block L failed to stabilize, and the failure was chalked up to the control system.

On 27 August, new State Commission Chairman Leonid Smirnov informed us that the Americans had launched the Mariner-2 spacecraft toward Venus; the list of scientific investigations that Mariner-2 was supposed to conduct was almost identical to ours.[28]

Without waiting for an in-depth investigation into the causes for the failure of the previous launch—we simply physically did not have enough time—we launched our next 2MV, spacecraft No. 4, toward Venus on 1 September. Once again the pendant was not destined to reach Venus' surface. A valve feeding fuel into the combustion chamber of the Block L accelerating engine failed to open.

We launched the last of the three Venus spacecraft, 2MV-2 No.1, on 12 September. The engine of Block L operated for just 0.8 seconds and shut down due to a nonstabilized mode. Once again the blame fell on the control system that Pilyugin had developed. A more in-depth investigation of the last launch showed that upon the issuance of the primary command for the shutdown of the Block I (stage three), there was a violent perturbation and the Block L (stage four) spun sharply. As the fourth stage spun, air bubbles in the tanks moved into the intake ports and the Block L engine failed to start up.

And so, the 1962 season of Venus launches ended in disgrace. All three launches failed through the fault of the fourth stage. We didn't have the opportunity to test out the spacecraft performance during even the first million kilometers of their interplanetary trajectories. How much effort had been invested in the development, manufacture, modification, testing, and retesting of the AMSs—all of it in vain?

But we didn't have the opportunity to grieve for long. The Mars launch windows were approaching. The equipment for the 2MV Mars version was loaded into airplanes and flown, one after the other, to the firing range. Once again the sleepless nights of testing at the MIK at Site No. 2 began. On 15 October 1962 at 11 p.m., I departed Vnukovo airport with the main group of testers for one of our most stressful, interesting, and eventful missions.

After the assault on Venus, all sorts of measures were taken to enhance the reliability of the Block L. However, after thoroughly studying the causes of the failures and the actions taken to correct them, Voskresenskiy told me confidentially, "I recommended to Sergey that we put off work on Mars this year. We're up to our ears in trouble. But he won't listen to me. We didn't conquer the Goddess of Love. I don't think we'll cope with the God of War any better."

"Our mission," I protested, "is to pave the way. Pioneers have not always reached their goals, but those who have come after them have always been grateful."

Smirnov, Keldysh, Ishlinskiy, Ryazanskiy, Kuznetsov, Bogomolov, Rauschenbach, Sheremetyevskiy, Kerimov, and all of our developers, testers, and representatives from subcontractor organizations who had been given a temporary leave of absence "due to family circumstances" flew in to the firing range. Once again for the umpteenth time, despite the string of failures, the now familiar firing range routine

---

28. Mariner-2 was launched on 27 August 1962. The spacecraft accomplished the first successful planetary mission in space history when it passed by Venus on 14 December 1962 at a range of 34,762 kilometers, gathering significant data on the Venusian atmosphere and surface.

was established, involving nothing but continuous work. There were small plea-sures—above all, getting together with friends with whom you had parted company quite recently. There were jokes shared on the job, but most often in the dining hall, or en route to the MIK and to Site No. 10—the town at the range. There had been so much misfortune but none of us felt despondent.

The preliminary schedule called for three launches:
- on 24 October—2MV-4 No. 3 (a flight passing close to Mars);
- on 1 November—2MV-4 No. 4 (a flight passing close to Mars); and
- on 4 November—2MV-3 No. 1 (landing version).

The remedial measures taken on the Block L had required reductions in AMS mass. This was a very painful experience for us, because to a great extent, with these measures we had to reduce the primary objectives of the interplanetary flight.

And so, on 24 October, a Mars launch took place. All the "science" had been removed from the spacecraft, but in exchange, the Block L was equipped with a wealth of monitoring and measurement systems. During the time of radio coverage on the ships located in the south Atlantic, telemetry recorded the normal firing of the Block L engine, but 17 seconds later there was an explosion in the turbopump assembly. That is what the telemetry operators Raykov and Semagin, who were on board the ships, reported. Both groups were experienced enough to be correct in their diagnosis.

There was no connection between the failures of the Block L on the 8K78 and tests of the R-9 (8K75) combat missile. Nevertheless, in line with the axiom "when it rains, it pours," right next to the 8K78 launch pad, at Site No. 51, the turbopump assembly of an R-9 missile exploded on 27 October.

The State Commission held a meeting on 29 October. We listened to a report by the chief engine expert of OKB-1, Mikhail Melnikov, who advanced his explanation of the explosion in the Block L, relying on Raykov's reports and telemetry informa-tion received from the *Dolinsk* and the *Krasnodar* ships. His report was reassuring. He noted that, "In all probability, a foreign particle got into the turbopump assem-bly. The explosion was pure chance. The launches should continue." Oh, these for-eign particles! When necessary, they could be used to explain any crash.

We continued. On 30 October we transported the launch vehicle carrying AMS 2MV-4 No. 4 to the launch pad while in the MIK they were testing the last 2MV-3 No. 1. On the morning of 31 October, I left to attend the State Commission meet-ing. Before this, Vitaliy Kalmykov and his friend Kuyantsev, neither of whom had slept all night, reported that commands did not pass over the VHF (emergency) link to the descent vehicle of the spacecraft. Boguslavskiy remained behind to study the problem with them. During a break in the commission meeting I ran over to the hall and—"hooray!"—they had fixed the defect in the VHF link. The commands were getting through! By lunchtime the tests on the last spacecraft were finished. We sent it to the pressure chamber, and decided to take a 2-hour nap.

November 1 was a clear, cold day. A brisk north wind was blowing. At the launch site, preparations were under way for an evening launch. After lunch I ran over to the cabin, switched on the radio, and made sure it was operating properly over all the bands. At 2:10 p.m. I went outside and waited for the designated time. At 2:15 p.m. under a bright sun a second sun flared up in the northeast. This was a nuclear explosion in the stratosphere—the test of a nuclear weapon code-named K-5. The flash lasted fractions of a second. The nuclear device on the R-12 missile was detonated at an altitude of 60 kilometers to test the capability for terminating all sorts of radio communications. According to the map, the detonation site was 500 kilometers away. After returning quickly to the radio receiver, I realized how effective the nuclear experiment had been. There was complete silence in all ranges. It was a little over an hour before communication was restored.[29]

The Mars launch took place at 7:14 p.m. By that time, the ionosphere had returned to normal after the nuclear explosion. In any event, all stations reported that telemetry monitoring was proceeding without incident. Finally, after all the disasters, the Block L operated according to the program and the AMS departed for Mars. Despite the mishaps with previous publications concerning launches to Venus, on 2 November, *Pravda* and Levitan hurried to report that a rocket bound for Mars had been launched in the Soviet Union.

While 2MV-4 was flying to Mars on a flyby mission, we immediately followed it up on 4 November with the launch of 2MV-3 No. 1 in a Mars-landing version. Alas, evidently, fate or the gods had granted us the preceding launch only for temporary moral support. Soon after launch, we received messages from the Gulf of Guinea that dashed all our hopes. Again there was a breakdown in the propulsion system, and the shutdown command was issued after 33 seconds.

Having successfully left Earth orbit for Mars, the primary objective of 2MV-4, or *Mars-1*, was to photograph the planet in a close flyby. The images were supposed to be transmitted via SHF radio-link using a high-gain parabolic antenna. This process required the reliable operation of the attitude control system. While we were preparing for the next launch, Yevpatoriya, which had begun VHF communications sessions according to the program, was sending optimistic dispatches saying that everything on board was normal, communications were good, but there was one glitch in the attitude control thruster system.

After the mission failure of 4 November, the State Commission agreed that Keldysh would fly to Yevpatoriya to clarify all the circumstances of the ongoing *Mars-1* flight, I would fly with him along with the attitude control and guidance specialists, and Korolev would depart for Moscow. On 5 November, after arriving at NIP-16, we quickly realized that there would be no sensational photographs of Mars. The entire supply of gaseous nitrogen, which was the working medium of the attitude

---

29. Here, Chertok is referring to one of a series of five R-12 missile launches conducted under the codename Operation K designed to test the aftereffects of nuclear explosions (specifically electromagnetic pulse effects) at high altitudes on antiballistic missile system (ABM) radars. Missions K-3, K-4, and K-5 were accomplished in September to October 1962.

From the author's archives.

Shown here is the 2MV-4 type spacecraft, one of which was launched in November 1962 and became the Mars-1 spacecraft. The spacecraft was designed to fly by Mars and take photographs. In the event, the vehicle lost attitude control, and communications were cut off prior to its flyby in June 1963. The two white hemispherical objects with the stripes on each side of the vehicle are thermal control radiators. The contraption at the top is the propulsion system.

control systems, had been lost. How? Telemetry data analysis enabled us to pinpoint that the guilty party was one of the valves in the attitude control system. It had remained open the entire time. Evidently, a large "foreign particle" had gotten under the valve seat and the entire precious gaseous nitrogen supply had whooshed through the open valve.

Right before the November holidays we ruined Korolev's mood and that of all who had flown from the firing range to Moscow with him. Korolev immediately organized a project to analyze the production process of the attitude control system valves, which the aviation industry had manufactured. He even brought in criminal investigators. The cause of the valve failure was determined beyond a doubt. When we soldered the electromagnet winding, we used rosin, crumbs of which might have fallen under the valve seat and prevented the valve from fitting snugly on the seat surface. The resulting gap was quite sufficient for the entire supply of working medium to escape. This event was reproduced at the factory. This incident was discussed in depth in the State Commission and even higher at VPK sessions.

Nevertheless, the AMS flew to Mars, albeit without attitude control, but in every other aspect in very good working order. Communications sessions via the UHF link were conducted regularly, all the "science" that could work en route functioned, and (this was particularly gratifying) all the services of NIP-16, the Center for Deep Space Communications, got a test run and a training exercise. Communication via the UHF radio-link using the semidirectional antenna continued for 140 days. Contact was lost at a range of 106 million kilometers. But back then that was a distance record for space communications. On 15 December, *Pravda* published a description of the spacecraft's trajectory, a photograph of the AMS, and the program of scientific investigations. By that time we already knew that that spacecraft would not reach Mars "alive and kicking."

The *Mars-1* flight gave all of us experience, which increased our optimism. The next phase had begun—the design and manufacture of an improved series of standardized interplanetary spacecraft with the factory index 3MV. The primary action taken to increase the reliability of series 3MV spacecraft was the redundancy in the attitude control system thrusters. We decided to begin series launches of the 3MV automatic interplanetary station by testing the entire complex as an interplanetary probe while taking high-quality photographs of the far side of the Moon en route. The first launch of such a probe was scheduled for November 1963.

Despite the difficulties, misfortunes, and failures, funding for the program to reach Venus and Mars continued. Work on programs for a soft-landing on the moon using the same 8K78 launch vehicle was in progress simultaneously, and there were plans to insert Molniya-1 communications satellites in a highly elliptical orbit. The Mars launches in 1962 overlapped with the Cuban Missile Crisis. This time, Mars, the "God of War," was unable to use missile technology to turn the Cold War into a hot one—World War III. But I'll write of these events in the next book in this series.

Chapter 32
Catastrophes

After two unsuccessful Mars launch attempts in a row, on 18 October 1960, we testers and developers felt gloomy as we left the firing range with Korolev. We had every reason for our somber thoughts. The year had begun with the failure of one of the three 8K74 combat missiles launched to a maximum range over the Pacific Ocean. In April, two launch vehicles carrying Ye-3 lunar spacecraft—designed to take pictures of the far side of the Moon—failed, one after the other. On our very first attempt in 1959, we had gotten photographs of the far side of the Moon. It had caused a stir throughout the world, although the quality of the "far side" images was poor. And it was through the fault of that same R-7A launch vehicle that we just couldn't get the new high-quality photos that Keldysh and the lunar astronomers dreamed of seeing.

It was in the heat of July that the first test descent spacecraft for the future Vostoks was lost with the dogs Chayka and Lisichka on board. And once again the launch vehicle was to blame![1] Now, in October, there had been two more failed launches. These spacecraft hadn't even made it into near-Earth orbit, and we intended to fly to Mars. Over a period of 10 months there had been six failures of a launch vehicle that was officially in service as an ICBM! And in accordance with Korolev's proposal, which all the chiefs and all of his deputies had supported, the Central Committee and the Council of Ministers proposed that we use that very launch vehicle in December 1960, that is, in two months, to launch a man into space![2]

In the airplane on the way back from the firing range, I watched Voskresenskiy and Ostashev busying themselves, pouring the cognac they had on hand in case of a success and spreading out stale sandwiches on newspapers; I tried to say something optimistic: "Well, we had Belka and Strelka! They caused a lot of hoopla in the Hawaiian Islands in July, and our comrades weren't shortchanged with the Lenin

---

1. An attempt to launch a Vostok spacecraft carrying the dogs Chayka and Lisichka ended in disaster on 28 July 1960 when the launch vehicle exploded about 28 seconds into the mission.

2. Chertok will describe the Vostok human space program in Volume 3.

Prizes!"[3]

Voskresenskiy interrupted me and, raising his glass, he proposed a toast, "To the end of failures!"

"Let's drink to that," said Korolev, "But keep in mind that leap year isn't over yet."[4]

And unfortunately, it turned out he was right.

ON THE EVENING OF 24 OCTOBER, Korolev called Ostashev to his office. Shabarov, who had remained at the firing range, had sent a radiogram reporting a serious accident involving Arkadiy's brother, Yevgeniy Ostashev. Korolev recommended that Arkadiy fly out to Tyuratam the next morning. Later, after receiving top-secret information from Moscow sources, Korolev informed only his deputies that a fire and explosion had occurred during preparation of an R-16 missile at Yangel's launch Site No. 41. There had been casualties. How many and who was still unknown. A government commission had already been formed, with Brezhnev himself as chairman.

This was the most horrific disaster in the history of missile and space technology. In the following description of the disaster, I have used the accounts of Yevgeniy Shabarov, who was at the firing range at that time, Arkadiy Ostashev, who arrived the day after the catastrophe, NII-944 Chief Designer Viktor Kuznetsov, OKB MEI Chief Designer Aleksey Bogomolov, and Chief Designer and VNIIEM Director Andronik Iosifyan, all of whom, as luck would have it, survived the accident.

In the second book of the Russian-language version of *Rockets and People,* based on Arkadiy Ostashev's account, I indicated that a total of 126 persons were killed. He had mentioned that number in 1990, citing data obtained at one time in the firing range main office.[5]

According to the official report of Artillery Major General Grigoriy Yerofeyevich Yefimenko who was firing range chief of staff in 1960, 57 servicemen and 17 industrial representatives (or a total of 74 people) died in the explosion and fire right at the launch site.[6] The soldiers and officers who died at the launch site and those who died later in the hospital from wounds, burns, and poisoning were buried in a mass grave in the Baykonur municipal park. Eighty-four soldiers and officers were buried in that grave. The bodies of the industrial representatives were flown back to where they worked. If you accept the official data, there were 17 of them. Thus, the total loss of life was 84 plus 17 or 101 persons.[7]

---

3. Belka and Strelka were the first living beings recovered from orbit after flying for 26 hours in *Korabl-Sputnik-2* in August 1960.

4. Russian superstition holds that leap years are unlucky.

5. B. Ye. Chertok, *Rakety i lyudi: Fili—Podlipki—Tyuratam [Rockets and People: Fili—Podlipki—Tyuratam]* (Moscow: Mashinostroyeniye, 1996), p. 397.

6. This list that Chertok cites was first declassified and published in 1994. See I. D. Sergeyev, ed., *Khronika osnovnykh sobytiy istorii raketnykh voysk strategicheskogo naznacheniya [Chronicle of the Main Events in the History of the Strategic Rocket Forces]* (Moscow: TsPIK, 1994), pp. 248–262.

7. In other words, 57 soldiers died during the accident, 27 soldiers died later of injuries, and 17 civilians (total) died, making a total of 101 fatalities.

Lt.-Col. Yevgeniy Ostashev (1924–60) was one of the victims of the Nedelin Disaster in 1960. His official title was chief of the 1ˢᵗ Directorate at the firing range, i.e., responsible for flight-testing of Korolev's missiles at Tyuratam. Although the disaster was caused by a Yangel missile, Ostashev was at the firing range only to provide moral support to his colleagues. His brother Arkadiy Ostashev was a senior civilian engineer at Korolev's OKB-1.

Despite these figures, one publication cited the total number of deaths as 180! Finally, in 2004, I received from the archives a Xerox copy of the official documents, which for years had been classified "Top Secret." These documents confirmed the data signed by G. Ye. Yefimenko. I did not find any official data about deaths in hospitals, and so I don't feel that I can refer to a total number of deaths. I can only confirm that the first R-16 missile, named "article 8K64," killed, on average, more people without leaving the launch pad than did any 10 V-2 missiles that struck London during World War II.

The R-16 payload container had been filled with inert ballast, that is, it contained no explosives. Nevertheless, right on the launch site the missile killed 74 testers, developers, and the Commander-in-Chief of the Strategic Rocket Forces, Chief Marshal of the Artillery Nedelin.

I WOULD LIKE TO GIVE READERS A LITTLE BACKGROUND TO THE CIRCUMSTANCES SURROUNDING THIS ACCIDENT. OKB-586 Chief Designer Mikhail Yangel was an ardent supporter of missiles using high-boiling components. As far back as his stint as NII-88 director, he had come out against developing intercontinental combat missiles that used liquid oxygen as the oxidizing component. His hard-nosed position severely worsened his relations with Korolev, who had proposed the new, liquid-oxygen R-9 intercontinental missile. As we saw it, the R-9 missile was supposed to enter the strategic armaments arsenal to replace the R-7 and R-7A. After the development of the R-9, *Semyorkas* were supposed to be removed from duty and fully converted to serve cosmonautics.

We had a strong rationale for this line of action. The *Semyorka* launch pads were open on all sides and quite vulnerable to attack. The complexity and duration of their launch preparation, which took at least 7 hours, did not conform to the new doctrines of nuclear missile war. If American nuclear delivery vehicles delivered the first strike, the *Semyorka* launch pads would certainly be destroyed. We would no longer have intercontinental missiles for a retaliatory strike. We needed to develop new intercontinental missiles that would have reliably protected launch pads and would enable us to deliver a retaliatory strike in around 10 minutes. In the 1960s we spoke of 10-minute combat readiness. Today nuclear missile launch readiness is measured in individual seconds.

Which of the intercontinental missiles, Korolev's R-9 or Yangel's R-16, would stand on combat duty in secure silos to protect the country? That is what strained the relations between Korolev and Yangel. I feel compelled to note that Korolev's first deputy, Vasiliy Mishin, was a more vehement defender of liquid-oxygen missiles and greater opponent of the high-boiling component missiles proposed by Yangel than even Korolev himself. There were two reasons for that. First, Korolev began to understand better than Mishin and all of his deputies that for combat purposes the best competitor for both liquid-oxygen and high-boiling missiles would be solid-propellant missiles. Immediately after the merger with TsNII-58, he organized solid-propellant projects.[8] Second, Korolev felt apprehensive, although he did not directly express this to any of his close associates, that in the long term, combining projects involving piloted cosmonautics and intercontinental flights with the development of nuclear missile systems in a single organization would be a very difficult undertaking, even for him.

Nor did Glushko remain on the sidelines. He developed first- stage engines for both the R-9 and R-16 two stage missiles. Over the years of producing R-1, R-2, R-5M, and R-7 missiles, Glushko's design bureau had created an extensive firing test facility and had gained invaluable experience developing liquid-oxygen engines. Despite this, he got into a competition with an obvious trend toward developing high-boiling component engines using nitric acid as oxidizers and unsymmetrical dimethylhydrazine as fuel. Both of these hypergolic components were toxic and explosive. The rank-and-file military testers detested them compared with the "noble" liquid-oxygen, ethyl alcohol, and kerosene engines. However, when it came to maintaining hundreds of missiles in a constant state of launch readiness for months and even years, the high-boiling component missiles had incontrovertible advantages. The intense evaporation of liquid oxygen after a missile was fueled necessitated the constant replenishing of the tanks. As a result of such losses, special storage facilities were designed for the R-9 missile with systems to compensate for evaporation losses. Fueling a missile with oxygen took place right before launch. In principle, high-boiling component missiles could stand by in a fueled state and did not require additional readiness time for the fueling process. This had been proven in the operating experience of the R-12 medium-range missiles that Yangel had developed prior to 1960 using Glushko's engines.

The development of the single stage medium-range R-12 missile had begun under Yangel's leadership in 1951, when he was still Korolev's deputy. The development was transferred to Dnepropetrovsk in 1953. The government decree on the

---

8. The OKB-1 began development of an experimental long-range solid-propellant missile system, the RT-1, in November 1959. The missile first flew in April 1962.

R-12 missile was issued after Yangel had been named chief designer of OKB-586.[9] After flight tests at Kapustin Yar, the R-12 complex was put into service simultaneously with our R-5M and R-7A missiles in March 1959. The missiles went into series production at Factory No. 586 in Dnepropetrovsk, Perm Machine Building Factory No. 172, Orenburg Machine Building Factory No. 47, and Omsk Aviation Factory No. 166. The thermonuclear warhead of the R-12 missile complex had a yield of 2.3 megatons. With a range of 2,100 kilometers, the missile posed a real threat to all NATO countries bordering the Soviet Union. Skipping ahead, I will say that over the years, 2,300 R-12 missiles were produced. The missile was in service for more than 30 years—from March 1959 through June 1989.[10]

Yangel's high-boiling component R-12 missile squeezed out Korolev's liquid-oxygen R-5M missile as a serious contender. However, in order to shore up his success, with the support of the military Yangel proposed one more medium range missile, the R-14. Its development began in 1958, and its flight-tests were conducted in Kapustin Yar in 1960. The R-14 was designed for a maximum range of 4,500 kilometers. Its nuclear warhead also had a yield of 2.3 megatons.

Although Yangel's high-boiling component R-12 and R-14 missiles were capable of destroying all the United States' NATO allies, only Korolev's liquid-oxygen R-7 and R-7A missiles posed a real threat to America itself. However, there were just four very vulnerable launch pads for our *Semyorkas*, two in Tyuratam and two at the new firing range in Plesetsk.[11] A new "massive" intercontinental missile was absolutely necessary, and as soon as possible. That is why Yangel, with the active support of Khrushchev and Nedelin, began to develop the two stage R-16 intercontinental missile. In their drive to develop new missiles, they were motivated by the slogan "our country needs a secure nuclear shield!" And as soon as possible. The reliability of the propulsion systems and control systems proved to be of defining significance in the competition for a new generation of R-9 and R-16 intercontinental missiles. A single chief designer, Valentin Glushko, developed fundamentally different engines for both missiles.[12]

In 1962, the R-12 and R-14 missiles, which had already been put into service, came close to blowing up the world during the Caribbean (or Cuban Missile) Crisis.

---

9. Development of the R-12 was officially approved in August 1955, about 14 months after Yangel's official appointment as chief designer of OKB-586.

10. In the West, the missile was known as the SS-4 (by the U.S. Department of Defense) and Sandal (by NATO).

11. The original missile launch facility, the Scientific-Research and Testing Range No. 53 (NIIP-53), was established in January 1957 near the town of Plesetsk. The range was converted into a testing range and then a future space launch facility in August 1963. Since 1966, more satellites have been launched into orbit from the Plesetsk facility than any other location in the world.

12. Glushko's design bureau produced engines for the first stage of the R-9 (the RD-111) engine and the first and second stages of the R-16 (the RD-218 and the RD-219, respectively). Another organization, the Kosberg design bureau, produced the second stage engine of the R-9 (the RD-0106).

Fortunately for humanity, none of these missiles was launched against the U.S. in 1962. But in and of itself, there were those who cited the journey of the R-12 and R-14 missiles from the USSR to Cuba and back as proof of the superiority of high-boiling component missiles. But this was still in the future, two years after the catastrophe.

When Yangel entered into competition with Korolev for the intercontinental missile, he still only had experience in the experimental development and operation of the intermediate range R-12 missile. These missiles used nontoxic kerosene as fuel. The new toxic unsymmetrical dimethylhydrazine propellant (UDMH) was being used on R-14 missiles for the first time.[13] Viktor Kuznetsov developed a special gyro-stabilized platform for the R-14 missile. Working with Pilyugin, he created an inertial and autonomous guidance system that required no radio correction.

Back then it seemed to us that Glushko did not show the necessary diligence and enthusiasm for the experimental development of engines for our R-9 missile. One of the reasons for this was the "high-frequency" phenomenon that appeared in powerful liquid-oxygen engines when their specific characteristics were augmented. After a series of mysterious breakdowns of liquid-oxygen engines during rig tests, we discovered that high-frequency pressure fluctuations in the chamber preceded the failures. This high frequency resulted in destruction of the combustion chamber or engine nozzle. On engines for the R-9 missile, high frequency proved to be a curse that disrupted their delivery deadlines for the assembly of the first missiles. Neither the theoreticians nor the testers were able to explain why high frequency occurred in liquid-oxygen engines. Skipping way ahead, I'd like to mention that even on the successful *Semyorka*, which has been flying for decades in its modification called the Soyuz launch vehicle, to this day, high frequency will sometimes appear out of nowhere in the core booster.

THE R-16 MISSILE WAS WAY AHEAD OF THE DEADLINE FOR THE BEGINNING OF FLIGHT TESTS. The military builders built Site No. 40 for the R-16 missile on the barren steppe of the Tyuratam firing range. There were two launch sites, the new engineering facility's Assembly and Testing Building, a hotel, and everything else that was needed for flight tests of the new intercontinental missile. In 1960 they built at the same pace as they had on the firing range in 1957, but now with tremendous experience and their own industrial base.

The Strategic Rocket Forces Command and the commander-in-chief himself, Chief Marshal of the Artillery Nedelin supported Yangel with undisguised enthusiasm. The availability of other options made it possible to objectively compare the actual performance specifications of the new generation of intercontinental missiles. The military acceptance staff of the Yuzhmash Factory had a very liberal attitude toward departures from the strict rules of ground experimental development, which

---

13. In Russian, the name of the fuel is abbreviated as NDMG *(Nesimmetrichnyy dimetilgidrozin)* or simply *geptil*.

were tolerated in the interests of saving time.

The first government decree on the development of the R-16 missile (8K64) was issued on 17 December 1956, i.e., before the beginning of flight-tests on our first R-7 intercontinental missile. This decree called for flight-development tests to begin in July 1961. To speed up development, Yangel managed to free his OKB and Factory No. 586 from their naval projects and from the series production of anti-aircraft and air-launched cruise missiles. Despite usual expected delays, the first R-16 missile arrived at the firing range for flight-development tests in September 1960, ten months *ahead* of the deadline set by the government rather than behind schedule.

THE DECISION ABOUT THE CHIEF DESIGNER OF THE MISSILE GUIDANCE SYSTEM WAS UNCONVENTIONAL FOR THOSE TIMES. Yangel counted on traditional cooperation: he recommended that Pilyugin be the chief designer of an autonomous guidance system (with no radio control). However, at the insistence of his then-direct superior Ryazanskiy, who did not want to complicate his relations with Korolev, Pilyugin turned Yangel down. Furthermore, Ryazanskiy believed that an intercontinental range needed a combined guidance system—an autonomous system plus mandatory radio correction to ensure the requisite accuracy. Yangel categorically objected to using radio control. Not without reason, he believed that a combat missile must have a fully autonomous guidance system.

After Pilyugin turned him down, Yangel managed to persuade Viktor Kuznetsov to be the chief designer of an autonomous guidance system with increased accuracy. Though capable of developing and delivering a new gyro-stabilized platform for the inertial system, Kuznetsov did not have the intellectual or the production base to develop the entire ground and onboard electrical complex for the guidance system. Nevertheless, the government decree for the development of the R-16 missile designated V. I. Kuznetsov as chief designer of the guidance system.

The next time we met, I told Viktor Kuznetsov, "In my opinion, you're out of your element here. Not one of your remarkable gyroscope specialists is capable of developing a common electrical circuit for the complex."

To my surprise, he replied, "I have no intention of getting involved with that. Mikhail Kuzmich arranged this with the VPK. OKB-692 is being created in Kharkov. The chief designer of that OKB will be Boris Mikhaylovich Konoplev. And that's where all the onboard and ground electrical circuits will be developed, in Kharkov."

Indeed, by 1957, a high-capacity instrument group had been created in Kharkov to manufacture instrumentation for missiles. One of the leading specialists of NII-885, Abram Markovich Ginzburg, had been appointed chief designer at the Kommunar Factory. He was the same man about whom Serov, Lavrentiy Beria's deputy had said in 1947, "Show me this Ginzburg." At that time Pilyugin had hidden Ginzburg, telling Serov that he was currently at the rig replacing a relay.[14]

---

14. Chertok describes this event in Chapter 2 of this volume.

Ten years later it would have been appropriate to appoint Ginzburg chief designer of the new OKB-692. I have no doubt that he would have done a splendid job of developing a guidance system complex for the new intercontinental missile, provided that Viktor Kuznetsov developed all the gyroscopic command instruments and ensured the required accuracy. Two circumstances, however, interfered with that sensible decision. First, there was item five on Ginzburg's personal history form (ethnicity).[15] Second, Boris Mikhaylovich Konoplev, who had begun to feel cramped at NII-885, came forward out of the blue as an applicant to the post of OKB-692 head and chief designer.

For me, Boris Konoplev was not only a comrade-in-arms from our days preparing for the transpolar flights of 1937, but he was also an authority on radio guidance systems. He came out with a design for a long-range missile radio guidance system back in 1949 for the N-3 theme.[16] We tested the radio system together on the R-2R missile at the Kapustin Yar firing range. His entire engineering career was involved with radio engineering. Even in his personal family life, Konoplev was devoted to radio engineering. Ryazanskiy believed that Konoplev's wife had a better grasp of the theoretical bases of radio engineering than Konoplev himself.

Thus, the R-16 missile had one more fundamental difference from all preceding ones. For the first time since 1946, the Central Committee and government made a decision whereby the missile guidance system was produced without the participation of Ryazanskiy and Pilyugin. Konoplev's talent as an innovator in the field of radio engineering systems was incontestable, although it irritated his radio specialist colleagues. However, I do recall from my encounters with him in 1937 during the preparation of the transpolar flights that Konoplev's working style had a peculiarity that is typical of many talented inventors, but hazardous for a chief designer. He strove to solve a new problem as quickly and with as much originality as possible, without paying a great deal of attention to outside experience. While working with Konoplev on the R-5R, I realized that, first and foremost, he was enthralled with testing the viability of new principles.[17] Konoplev was not interested in who would subsequently conduct all the dirty work of service testing the system and how. Konoplev's obsession with his own new ideas prevented him from objectively embracing much that was already tried and true.

The new electric integrated circuits developed at OKB-692 operated using a different logic from those developed by Pilyugin. Moreover, these circuits required rigorous developmental testing. Pilyugin permitted the delivery of electrical instruments and all cables for Korolev's rockets only after each set had been thoroughly

---

15. Here Chertok is suggesting that Ginzburg's Jewish background was a liability.
16. The N-3 theme, performed in 1948–51, was an exploratory program to research layouts for an intercontinental ballistic missile.
17. The R-5R was an experimental missile launched several times in 1955 to test an experimental radio guidance system for the future R-7 ICBM.

tested, simulating all phases of launch preparation and flight. During rig testing, the behavior of the circuits was also studied during possible fault conditions. Nevertheless, again and again we realized that even on the launch pad, during missile preparation, situations arose that had not been simulated in advance on the test rig, and, therefore, for the next launch we would have to introduce changes, holding up the preparation process. In such situations the schedule was shifted. From the firing range Pilyugin monitored the checkout process in Moscow via radiograms, and he permitted work to continue only after receiving an official radiogram confirming the validity of the decisions made based on the rig test results. Ever since the series of R-7 failures in 1958, this procedure had been introduced for virtually all systems. Korolev demanded that it be meticulously followed and inured the chairmen of the State Commissions to this practice.

Developmental testing of electrical circuits is very tedious and is boring for a creative personality. This was drudgery akin to looking for improperly placed commas and typos in a multivolume work. After a circuit had undergone developmental testing, the final edition of the test instructions was issued. The instructions were supposed to be put together in such a way that during missile preparation, the tester and launch chief would not be intimidated by their ignorance of the subtleties of the circuit's logical connections. Any deviation from the instructions would have to be analyzed and permitted by the system's chief designer after consultation with his specialist, who provided a detailed presentation of all possible consequences of the infraction.

If they had adhered to these rules during the preparation of the guidance system and the R-16 missile itself for the first flight-development tests, it probably would not have been on the launch site in October 1960. Their desire to beat out the R-9 missile that was close on their heels was very great. The first R-16 missile was prepared for launch in October 1960; at that time our R-9 missile was still at the factory in Podlipki waiting for the delivery of an engine from Khimki.

Design errors and production defects caused most of the failures and catastrophic malfunctions of missiles during launch preparation and in flight over the preceding 13 years (beginning in 1947). On the other hand, the majority of failures that are today called "catastrophic" are the result of insufficient knowledge of the operating conditions. As far as the catastrophe—and it truly was a catastrophe, not a failure—that occurred at the firing range on 24 October 1960, one cannot explain it using the terminology or classification system of reliability engineering developed for rocket technology.

Chief Marshal of Artillery Nedelin, the commander-in-chief of the Strategic Rocket Forces, was the chairman of the State Commission on the testing of the R-16 missile. He and Yangel decided to give the nation a gift in honor of the 43rd anniversary of the Great October Socialist Revolution. They would execute the first launch before 7 November! Such was the tradition in our country: to have workers' gifts arrive just in time for holidays celebrating revolutions, significant dates, or the opening of Communist Party sessions. Right off the bat, they were under extreme pressure with tight deadlines preparing to test the new intercontinental missile. The military testers who survived the catastrophe and who had been with us for every sort of all-hands rush job since 1947, said that they had never seen such a violation of testing standards.

The most important cause of the catastrophe was haste, a hurry unjustified by any military or governmental need. In this instance, if the ambition to present a gift in honor of a holiday results in a missile being delivered to the launch pad without undergoing developmental testing on the ground, who is to blame? The first person liable in such a case is the chief designer. But then there is also the military acceptance staff, which knows the weak points at least as well as, and sometimes better than, the chief designer. The regional engineer (the chief military acceptance officer) gave his consent to clear the missile for flight tests. He was the second liable individual. On further investigation, one finds that technically, these first two liable individuals can cite the procurement of a guidance system that had not undergone developmental testing, a system that Boris Konoplev, the system's chief designer and his senior military representative had cleared for flight tests. So now there are at least four who were technically guilty. They would have had every right to say, "We need to do such-and-such—eliminate these particular glitches to obtain the necessary assurance." None of them ventured to do this, even though they were not under the threat of any censure.

Without any special instructions, by early 1960 we in Korolev's team had worked out a style for behaving and for conducting operations at the launch site. No rush jobs were permitted until the missile had been fueled with the rapidly evaporating liquid oxygen. Korolev himself set the example for a cool, unhurried demeanor. Each glitch was analyzed calmly and thoroughly.

Did State Commission Chairman Nedelin know about the breaches in the missile's developmental testing cycle? One can only assume that he had access to the relevant reports. But for each glitch described in these cases a decision followed to "permit" it. The decision was logically valid and backed up by the appropriate authoritative signatures.

Such violations, legally justified by an official clearance for flight tests, typically lead to subsequent violations right on the launch pad. In the case of the R-16, during the prelaunch testing process, glitches occurred one after the other, disrupting the original preparation schedule. In such a situation the primary remedy was to work round-the-clock. The test team did not leave the launch site for 72 hours. I often found myself in such situations while preparing for missile launches, when the launch control team and main staff of testers never had an opportunity for rest. Typically, we faced such situations from the need to launch within a window strictly determined by celestial mechanics.

But in this case, astronomy had nothing to do with it. Not only did State Commission Chairman Nedelin not grant permission to rest, but he appealed for even more self-sacrificing work before the great holiday. Who would dare question the Chief Marshal of Artillery, who for the sake of strengthening the defense capability of the Fatherland was appealing to his men to engage not in battle, but in self-sacrificing work? After all, this wasn't the front. No one was being sent to a certain death. It wasn't like there was some health risk, much less a life-threatening situation.

IN ORDER FOR THE READER TO GRASP THE CIRCUMSTANCES SURROUNDING THE LAUNCH PREPARATION FOR THE FIRST R-16 MISSILE, I have used a literary work by KB Yuzhnoye General Designer Stanislav Konyukhov and Lev Andreyev, *Yangel: Lessons and Legacy.*[18] Of all the literature available on the history of rocket technology, the work by Konyukhov and Andreyev was the first to provide a detailed and credible account of what really happened on 24 October 1960, 41 years after the fact.[19]

The startup of a rocket engine, its operation, and shutdown in flight constitute a complex multiphase process. The hydraulic system feeds the propellant components—oxidizer and fuel from the fuel tanks—according to commands issued by the control system. The chief designer of the

Asif Siddiqi.

In the Soviet military, Marshal Mitrofan Nedelin (1902-60) was the most vociferous advocate of developing and adopting modern strategic missiles. Without his guidance and vision, the Soviets might not have invested enormous resources into developing the early ICBMs such as the R-7 and R-16. For his leading role, Khrushchev appointed him commander-in-chief of the new Strategic Missile Forces in 1959. Some would say his recklessness and hubris was a major contributing factor to the terrible disaster at Tyuratam in late 1960 that killed so many.

---

18. S. Konyukhov and L. Andreyev, *Yangel: uroki i naslediye [Yangel: Lessons and Legacy]* (Dnepropetrovsk: GKB Yuzhnoye, 2002).

19. There are numerous published accounts in Russian of the Nedelin disaster. For books, see M. I. Kuznetskiy and I. V. Strazheva, eds., *Baykonur—chudo XX veka: vospominaniya veteranov Baykonur ob akademike Mikhaile Kuzmichye Yangelye i kosmodrome [Baykonur—Miracle of the 20th Century: Recollections of Baykonur Veterans on Academician Mikhail Kuzmich Yangel and the Cosmodrome]* (Moscow: Sovremennyy pisatel, 1995); M. I. Kuznetskiy, *Baykonur Korolev Yangel* (Voronezh: IPF ·'Voronezh,' 1997). Newspaper and journal accounts include articles in *Ogonek [Spark]* no. 16 (April 15–22, 1989); *Krasnaya zvezda [Red Star]* (October 24, 1990); *Rabochaya tribuna [Working Tribune]* (December 6, 1990); *Krasnaya zvezda* (October 16, 1993); *Istochnik [Source]* no. 1 (1995); *Voyenno-istoricheskiy zhurnal [Military-Historical Journal]* no. 5 (1995); *Nauka i zhizn [Science and Life]* no. 1 (1999); *Novosti kosmonavtiki [News of Cosmonautics]* no. 12 (1999); *Istoricheskiy arkhiv [Historical Archive]* no. 5 (2000). For a published account in English, see Asif A. Siddiqi, "Mourning Star," *Quest* 3 no. 4 (Winter 1994): 38–47.

engine is the developer of the hydraulic system and of the control logic that fires the missile and shuts down the engine. The chief designer of the control system is the developer of the electrical control circuit that starts up and shuts down the missile. But in this case, the automatic controls and all the electric circuits of the control system, including the propulsion systems testing equipment, were developed not by control systems chief designer Kuznetsov, but by OKB-692 Chief Designer Konoplev. The chief designer of each system has his own deputies, who, interacting with one another, are responsible not only for understanding the systems but also for jointly developing a common logic and for reliable performance, including possible off-nominal situations.

The logical interaction of the hydraulic and electrical systems must undergo developmental testing on control system rigs and on firing test rigs during testing of the entire propulsion system complex as per special instructions. Factory tests on each missile check the systems' interaction logic, the functionality of all the elements, and that there are no random production process errors. A mandatory recheck is performed at the firing range at the engineering facility and at the launch site. The missile's chief designer approves all instructions for missile testing and launch preparation.

As a rule, the effector mechanisms in the hydraulic system are valves that have two positions: "open" and "closed." The "open"—"close" process was executed on first generation missiles in two stages: an electrical command was issued to the electro-pneumatic valve, which opened or shut off the flow of high-pressure gas to the hydraulic valve. This system was reusable and reversible: it was possible to open and close the pneumatic and hydraulic valves on an unfueled missile for multiple tests at the factory and at the firing range. In their attempts to streamline and simplify the system, the [R-16] engine specialists and the control specialists did away with the two stage setup, eliminated the electro-pneumatic control valves, and introduced single stage pyrotechnics. An electrical command detonated the explosive cartridges, which were built right into the structure of the hydraulic valves that open or shut off the fuel feed. Pressure from the gases formed during the explosion opened or closed the hydraulic valves. The number of special heavy tanks and pipelines for the high-pressure control gas was reduced. However, the pyrotechnics made the valves nonreusable. After the detonation of the explosive charges the valve had to be replaced; access would have to be provided to it for maintenance and recharging so it could be used again. The integrity of the electrical circuit and of the ignition filaments of the explosive charges were checked by applying "nonfiring" current. Here, there would have to be an absolute guarantee that the nonfiring current was much lower than the current required to ignite the explosive charge...

During all the cycles of factory and firing-range tests on a missile, to avoid setting off nonreusable valves, they were disconnected from the general electrical supply, replaced with an "equivalent" in the form of a signal light indicating that at the proper time the command reached the connector, which would be hooked up to the explosive charge after the tests.

The R-16 first stage sustainer engine constituted three autonomous assemblies, each containing two combustion chambers, connected by a single start system that activated the oxidizer and fuel start tanks and the automatic control assembly.[20] The engine had a ground-level thrust of 226 metric tons and an operating time of 90 seconds. The engine was shut down by control system commands that actuated the pyrotechnic valves, shutting off the flow of oxidizer and fuel into the gas generator of the turbopump assembly. After 90 seconds of flight, the first stage electroautomatic controls turned over the control function to the second stage automatic controls, which started up the second stage engine according to the timeline loaded into the memory of the second stage control system. The second stage sustainer engine consisted of a single assembly with two chambers, oxidizer and fuel start tanks, and their automatic control assemblies. It had a thrust of 90 metric tons.[21] Separate fuel and oxidizer lines ran from the fuel tanks to the engines. The turbopump assemblies, which generated the required pressure to feed the propellant components into each line, ensured the stable operation of the sustainer engines.

Because hypergolic, toxic, and aggressive components were used in R-14 and R-16 missiles, to reliably seal the tanks and pipelines leading into them when the fueled missile was on the launch pad over an extended period of time, and also to prevent the aggressive components from entering into the pump chambers prematurely, special barriers—blowout discs—were installed on the flanges of the fuel and oxidizer pipelines. When the explosive cartridge was triggered, the disc opened up and allowed propellant components to fill the chambers of the turbopump assembly pumps.

The intense evaporation of the oxygen in Korolev's missiles after fueling was a drawback, requiring rapid preparation and launch of the missile after fueling. Similarly, if the discs in Yangel's missiles ruptured, they could not remain on the launch pad for longer than 24 hours. Over a 24-hour period, the aggressive components destroyed the gaskets, leaks developed, and a real fire hazard was created if the nitrogen tetroxide and *geptil* came into contact.

The engine startup process was possible only after the blowout discs were ruptured according to the missile preparation timeline immediately before launch. Before the beginning of flight-development tests, no reliable signaling system had been developed for the startup automatic controls system to confirm that the discs had opened up. This parameter had to be worked into the engine startup automatic control devices, precluding the possibility of subsequent commands passing through until it had been absolutely confirmed that the discs had been blown out. During flight-tests of the R-14 missile, beginning in June 1960 in Kapustin Yar, there were serious glitches with the reliability of the disc opening mechanism. Yangel was briefed on this by his first deputy, Vasiliy Budnik, who supervised the tests. However, they did

---

20. This was the RD-218 engine, made up of three RD-217 modules ("autonomous assemblies").
21. This was the RD-219 engine, which was essentially a high altitude version of the RD-217.

not have time to take any radical measures to increase the reliability of this assembly before the R-16 flight-tests began. Chief Designer Yangel and military acceptance committed a fundamental error by clearing the missile for flight-tests when it was already known that the disc opening mechanism was unreliable.

One more barrier was placed in the path of the components as they left the turbopump assembly—the main control valves, which covered the inlet into the engine combustion chambers. The main control valves automatically opened only when the pressure at their inlets reached a specific value.

The second stage engine startup process after disc blowout proceeded as follows. When the control system issued a command to start up the engine, a special electro-pneumatic valve operated and gas from the onboard high-pressure system was fed into the start tanks containing fuel and oxidizer. The pressure of the gas forced the propellant components into the gas generator, where they combined and ignited. The gas formed during combustion passed into the turbine, on one shaft of which were the oxidizer and fuel line pumps. A powder combustion starter provided the initial "crankup" of the turbopump assembly, and then the turbine gas generator switched to consuming propellant components collected downstream from the fuel and oxidizer pumps. As the turbine spun, the pressure in the chambers downstream from the pumps increased, and when it reached a certain value, the main oxidizer and fuel valves opened. Propellant components rushed into the combustion chambers, combined, ignited, and the engine started.

Remember that **two** separation barriers were provided in the propellant component lines, before the fuel and oxidizer were combined: blowout discs and main valves. The shutdown systems for the first and second stage engines were analogous. A command issued from the control system for both lines—fuel and oxidizer—operated the cutoff pyrovalves. These valves shut off the supply of propellant components, the turbopump assembly shut down, pressure in the system abruptly dropped, and, finally, the cutoff pyrovalve for the line feeding oxidizer to the chambers operated (i.e., closed) in the combustion chamber head.

The process for starting up an R-16 missile does not differ fundamentally from the established startup procedure for other missiles with liquid-oxygen engines. It has been preserved for almost all missiles to this day. After the missile is fueled and made ready for launch, based on all parameters, the missile team leader issues the command from the launch control panel to start up the launch execution program. According to this program, the gyro assemblies of the gyro-stabilized platforms are started up, pyrotechnic devices fire causing the onboard self-activating batteries to fill with electrolyte, the power supply switches from ground sources to onboard buses, and the blowout discs in the oxidizer and fuel lines are detonated.

Having monitored the display lights to see that all preparatory commands were issued, the launch chief presses the "launch" button. The missile is now completely under the control of the launch sequence, that is, the missile is controlled by the series of commands issued by the guidance system in automatic mode to execute the launch without human participation.

After the first stage engine gathers thrust exceeding the weight of the missile, the missile lifts off of the launch pad. As it lifts off, the connectors connecting the onboard electrical network with the ground network pull apart and the liftoff contact closes. From that moment everything that happens on board is under the control of the flight sequence. Signals for the execution of commands according to the flight sequence are strictly timed in relation to the reference point, the moment the liftoff contact tripped. There were no controlling onboard computers at that time, so programmed sequencers (PTRs) generated commands for stage one and two.[22] The timers were camshafts, which were turned by a step motor. The cams closed specific contact groups activating relays in the main distributor of the onboard electrical equipment system. Relays activated by PTR commands or another control system command source actuated corresponding propulsion system controls. A special pulse generator sent pulses to the step motors. These pulses powered the step motor mounted on the gyro-stabilized platform to change the pitch angle depending on the specified flight trajectory.

Before launch, the PTRs of stage one and stage two and the gyro-stabilized platform sequencer are supposed to be set at "zero"—the initial position. Pulses begin to flow to their step motors only after the command arrives from the relay that was actuated at the moment the missile lifted off from the launch pad. However, high-current pulses passing through cables when the explosive cartridges go off or when other onboard power consumers are activated are capable of generating false pulses in common bundles of the cable network; these false pulses can cause the step motors to shift the PTR program and the pitch angle from the initial position. Control system Chief Designer Konoplev and the staff of the missile's chief designer were responsible for checking the control system for this type of parasitic cross-coupling by exposing the onboard network to actual current pulses. They did not conduct such tests on integrated test rigs or on special experimental units.

THE MISSILE WAS TRANSPORTED TO THE LAUNCH SITE ON 21 OCTOBER. During the prelaunch testing process, they detected no substantial glitches preventing them from making a decision to begin fueling. On 23 October, the missile was fueled with the propellant components and compressed gasses and preparation began for its launch, which the State Commission decreed was to take place that very evening.

As soon as the missile appeared at the launch site the missile team worked virtually around the clock. Moscow was putting the pressure on. Khrushchev kept phoning Yangel and Nedelin. State Commission Chairman Nedelin, in order to set an example, was at the launch site almost the entire time overseeing the missile

---

22. PTR—*Programmnyy tokoraspredelitel*—literally represents "programmed current distributor," but they were basically mechanical sequencers which activated systems onboard the rocket in a given and preplanned sequence.

preparation process. A team of military testers who had experience preparing and launching intercontinental *Semyorkas* conducted all the operations, but this was their first time at the new launch site and they had been tasked with rapidly preparing and launching this first new missile without any preliminary training on an engineering model.

I mentioned earlier that we manufactured a special engineering version of the *Semyorka*. The designers needed one of these engineering units for fit checks and interfacing with ground launching and fueling equipment, rather than performing those tasks on a flight unit of the missile. Likewise, the testers and the firing range launching team needed it for training before the first fight unit arrived. In the Council of Chief Designers, Korolev repeatedly harped on the need to fabricate an engineering model. Operations for its manufacture in conjunction with the launch equipment and many days of training for the entire staff were inserted into all the schedules. What prevented Yangel from drawing on this experience? Perhaps it was his confidence that the R-16 launch equipment was much simpler than that of the R-7; the main thing, however, was that there was no time for preliminary training of the launch team. The launch of the first R-16 missile had to be pulled off as the latest gift before the 43rd anniversary of the Great October Revolution.

The majority of the preparatory operations for the first R-16 took a lot more time than had been scheduled. Despite their exhaustion from three days of preparation, the team of testers fueled the missile with all the propellant components by the end of the day on 23 October. This is when the testers encountered their first unforeseen situation. The developers of the missile and propulsion systems were not sure that the propellant lines would maintain their pressure integrity after the blowout discs were detonated. The fact is that back during preparation of the first R-14 flight unit in July 1960, when the blowout disc in the oxidizer line was detonated a leak developed under a flange seal in the oxidizer line. The launch was postponed for this reason, and Budnik, Yangel's first deputy, decided to drain the propellant components and remove the missile from the launch pad.

However, during the draining process, a small fountain of fuel squirted out of the coupling between the drainage hose and the missile drainage line. When it hit the concrete where there were tiny splashes of oxidizer diluted with water, the stream of fuel ignited in the air. The mini-fire was put out with fire extinguishers. After Budnik sent this alarm signal from Kapustin Yar, special attention should have been paid to making sure the fittings containing the blowout discs were intact after the discs were blown out. But they were pressed for time, and this work had not been tested for the R-16. Vasiliy Budnik flew out to the firing range and was present at the R-16 launch site, but did not interfere with the preparation process.

It seemed to me that I had gotten to know Yangel well enough working with him from 1950 to 1953. Why had he not categorically demanded that the blowout discs undergo thorough developmental testing? Everyone was already aware of the precedent set long before the first R-14 missile launch attempt. Our *Semyorka*, already famous by that time, burned up in its powered flight phase due to a fuel line leak during its first launch on 15 May 1957. That time the flight lasted 103 seconds and the missile exploded at high altitude without inflicting any damage except to our morale.

In heated arguments with opponents of the oxygen-kerosene engines, Vasiliy Mishin loved to argue that, "We have no insurance against fires when propellant lines leak near a high heat source, but your missiles will burn and explode for no apparent reason when there is the tiniest hint of a leak of hypergolic components!"

Besides the danger of mechanical seal failure in a blowout disc's fittings themselves, the disc's detonation could adversely affect the electrical system. The explosive charges that blew open the discs contained a metallic filament that acted as an igniter. As voltage was fed to it, the filament became red hot, burned, and ignited a powder charge. It was assumed that a positive determination that the electric circuit powering the explosive charge was broken could prove for the ground monitoring system that the disc had been blown out. In fact, quite often, burned filament remains came into contact with the metallic housing of the explosive charge, thereby indicating a "housing contact," that is, the signal circuit falsely confirmed the integrity of the explosive charge. The OKB-692 electricians and OKB-586 supervisory electricians were unquestionably at fault, since they had developed and approved a system with such a defect. But it wasn't simply a monitoring issue. The contact between the filament remnants and the housing caused the occurrence of high-ampere currents—an immediate short-circuit, which, if it continued for a prolonged period, was capable of damaging the cable network insulation and the wiring in current distribution instruments. They should have called on the experience of NII-885 and OKB-1, which had taken this feature of pyrotechnics into consideration.

Yangel, his two deputies Berlin and Kontsevoy, and Glushko's deputy Firsov made an excessively risky decision: to detonate the blowout discs without sticking to the nominal prelaunch preparation timeline when everyone had already left the launching pad; instead they decided to do this immediately after fueling, when more than 100 people were still at the launch site. And for this, they proposed quite an original method to determine whether the discs had been blown out, not relying on an electrical effect, but on the powerful water-hammer sound and characteristic "gurgling" when the lines fill up with liquid at the moment the discs are blown out. Specialists who had experience testing propulsion systems in Zagorsk were ordered to climb into open hatches and to assess by ear whether the discs had been blown out. And they did in fact crawl in, and even without gas masks, because they would interfere with their hearing! That, in and of itself, was a flagrant violation of safety procedures. If a propellant component leak had occurred, in the best case scenario these "listeners" would have been in danger of severe poisoning and burns.

To begin with, they were supposed to blow out the discs in the first and second stage oxidizer lines, since that component did not pose a fire hazard. They could blow out the discs in the fuel lines only after they had determined by visual inspection that the oxidizer line was leak-tight. The decision to blow out the discs "manually" and to use the subjective method to determine that the disc had been blown out was made and coordinated with State Commission Chairman Nedelin. Based on the "disc" issue alone, the missile should not have been cleared for launch, but the main customer, Nedelin, was in a hurry.

Following instructions, civilian engine specialists from OKB-456 crawled into the first stage aft compartment up to their waists, and when the command was given over the public address system to blow out the discs in the oxidizer line one by one, they started to listen. They were supposed to first give a leak integrity report and then wait for the command to blow out the discs in the fuel line.

Darkness had already fallen. It was pitch black in the aft compartment. The "listeners" attuned themselves to the acoustic signals of the blowout discs in the second stage. Out of the blue, a sound occurred in the first stage aft compartment accompanied by a very strong shock. Several seconds later there was a bright flash in the vicinity of the first stage engine and the aft compartment was filled with the smell of burned powder explosive. Subsequent examination showed that the blowout discs in the *first stage* lines had been detonated instead of those in the second stage. The bright flash was the result of some false command that had detonated the explosive cartridges of the gas generator cutoff valve on the first stage engine. In addition, they found a tiny leak from the fuel line through the turbopump assembly shaft seal.

After analyzing the events, specialists from Yangel's OKB, headed by Deputy Chief Designer V. A. Kontsevoy and lead engineer K. Ye. Khachaturyan and the electrical system developers from Konoplev's OKB came to the conclusion that the main distributor of electrical commands needed to be removed from the missile in order to study the causes for the confusion and transmission of unauthorized commands. And if the operations were to continue, the engine specialists would have to replace the gas generator cutoff valve on the first stage engine.

In addition to these glitches, the control panel operators in the bunker had detected that the PTRs and gyro-stabilized platform pitch angle sensor were not in the "zero" initial position. Analysis of this off-nominal situation showed that it was caused by impulse noise. It was determined beyond a doubt that the system issuing program pulses to the step motors of the pitch command sensors and program timers had insufficient interference protection. It was obvious even to nonspecialists that eliminating such an effect would require special tests for onboard systems cross-coupling. It was absolutely impossible to conduct such operations on the launch pad on a fueled missile. But Konoplev argued it was possible to conduct the first launch of a missile with this fundamental defect, and the State Commission agreed with him.

There was another deviation from the standard missile preparation process: the second stage onboard self-activating batteries became operational prematurely. According to the standard process, the self-activating batteries were supposed to be activated by compressed air pressure after all the checks had been conducted during the launch process when there were no longer any people on the launch pad. In the case of the R-16, compressed air fed into the battery when the command was issued, detonating the pyrotechnic valve, opening up access for the compressed air to enter a rubber bag containing electrolyte. As the bag inflated, a blade cut it open, and the electrolyte flowed into the battery. The second stage ground power supply was disconnected at that time, and, in violation of all the rules, all the electric automatic controls were connected to the onboard power buses. This gross violation of a very logical standard procedure occurred because during the last few hours, controllers discovered that the self-activating battery was at the limit of its capacity given the low temperature that had set in over the past few days. Therefore, the decision was made to keep it in an activated state in a warm place and to put it on board after the announcement of T-minus 30 minutes. Until it was installed on board, the second stage batteries were inside a van, where they were maintained at a temperature of +30°C (86°F) under the immediate control of a representative of the Scientific-Research Institute of Current Sources.

Actually, the second stage onboard battery was installed and hooked up to the onboard network before the official announcement of T-minus 30 minutes. Therefore, the main distributor was replaced and the discs were blown out on the second stage with the onboard battery hooked up.

On the evening of 23 October, the State Commission held a session during which they would have to make a decision covering the whole array of incidents. The meeting participants had to decide what to do with a missile that—after being fueled with hypergolic, toxic, aggressive propellant components—was on the verge of breakdown. During the State Commission meeting, attendees proposed to replace the main distributor, to replace the falsely actuated pyrotechnic cutoff valve, and to continue work. At that fateful meeting, only one State Commission member—firing range department chief Lieutenant Colonel S. D. Titov, argued vehemently against this proposal. After expressing his view that the control system had not undergone sufficient follow-up development, he uncompromisingly proposed to, "Drain the propellant components, neutralize the missile at the firing range, and send it to the factory for modification." Alas! This would have been the only absolutely correct decision.

Lieutenant Colonel Titov understood that his speech contradicted the State Commission chairman's mindset. A lieutenant colonel had dared to disagree with the marshal! Everyone else, on the other hand, including review team members who were more experienced and not subordinate to the marshal or Chief Designer Yangel, decided to continue work at the launch site and to conduct the launch the next day! Closing the State Commission meeting, Nedelin summed up the situation, "We'll modify the missile on the launch pad. The nation is waiting for us!"

I knew Viktor Kuznetsov, Andronik Iosifyan, and Aleksey Bogomolov well and had great respect for them. The people at this meeting were experienced, competent individuals, independent of Nedelin, Yangel, and [State Committee] Deputy Chairman Grishin. More than once I had witnessed them disagreeing with Korolev and taking issue with ministers. Except for Grishin, who was killed, I had the opportunity to ask each of them, "Why did you agree to continue the operation? After all, we all had 13 years of preparation and launch experience. We knew perfectly well what a gamble it was to modify electrical circuits on a fueled missile. What kind of obsession was this?" And no one could give me a clear answer why seemingly sensible, independent, and responsible individuals did not support Titov.

Highly experienced chief designers such as Kuznetsov, Iosifyan, and Bogomolov could well imagine that the decision to replace the main distributor and the pyrotechnic cutoff valves on a fueled missile with its blowout discs detonated *and* the onboard batteries activated was more than a risky undertaking. And, in fact, the vote of either Chief Designer Kuznetsov or Chief Designer Iosifyan at the State Commission meeting could have been decisive!

Modifications were performed all night. The main distributor was removed and opened up. Inspection showed that the insulation on the wires of one of the bundles conducting current to detonate the blowout discs had completely melted and the bare wires were in contact with one another, forming parasitic circuits.

The propulsion system's electrical circuit had been constructed so that the rotary switch on the control panel, from which the detonation commands were issued, needed to be set in the "O-1" position to detonate the blowout discs in the first stage oxidizer line and in the "G-1" position to detonate the discs in the fuel line.[23] Then voltage was fed through the appropriate circuits of the main distributor to the explosive cartridges of the discs. The control panel did not limit the duration of the command and the ensuing short-circuit current.

While the propulsion system engineers were trying to determine by "sound and smell" whether the second stage blowout discs had been actuated, the burned-up explosive cartridges in the first stage shorted the circuits passing through the main distributor with the remains of their own incandescent filaments. The short-circuit current melted the insulation. A false command to actuate the explosive cartridges of the gas generator cutoff valves of one of the first stage propulsion system assemblies actually passed over the bare wires.

As of the morning of 24 October, all the efforts of the civilian specialists and military testers were aimed at eliminating the catastrophic defects. Individuals who already had experience preparing missiles at Kapustin Yar and at Korolev's launch site were, for the first time, confronting a high-ranking military leader's decision to continue with the risky preparation and launch of a missile with a defective control system. But even in the smoking room no one dared to gripe because Marshal Nedelin had taken a seat by the launch pad not more than 20 meters from the missile and was attentively monitoring the actions of the crews on the ground and at all levels of the erector. Industrial representatives joined him on his bench, and he compared notes with them about the progress of the operations.

Several meters from Nedelin stood firing range Chief K. V. Gerchik and his deputy A. I. Nosov, who had been appointed to a new post in Moscow, but had

---

23. The letters O and G stand for *okislitel* (oxidizer) and *goryucheye* (fuel) respectively.

stayed on for the launch. Also present was Chief of the "Korolevian" First Directorate of the firing range Ye. I. Ostashev, who was not supposed to be there at all, but he wanted to give the marshal some documents and also to support his colleague Grigoryants by being there. The latter had been appointed Chief of the "Yangelian" Second Directorate of the firing range.

It was still morning when the NII-4 chief, General A. I. Sokolov approached the marshal and dared to warn him about the danger of being on the launch pad in the immediate vicinity of the fueled missile. The marshal snapped back at him, "If you are a coward, then leave!

The offended Sokolov departed for the airfield and flew to Moscow. By accusing Sokolov of cowardice, Nedelin saved his life. Three months later Sokolov was appointed chairman of the State Commission in place of Nedelin to continue flight-tests on the R-16 missile.

Twilight had descended on the launch pad, and they still needed to perform the operation to blow out the second stage discs. To make sure the operation didn't fail, they decided to perform it manually, disc by disc, rather than using the disc blowout control panel as called for by the instructions.

Engineers K. Ye. Khachaturyan, Ye. A. Yerofeyev, and Senior Lieutenant V. A. Makulenko climbed the ladder of the erector to the upper service platform, opened the hatches in the interstage compartment, disconnected the connector, and fed electric current directly from the technological battery that had been hoisted up there. They determined by the sound that the start tanks had filled. However, a captain who was with them asserted that only one disc had detonated. Later they realized that he was right. It was not clear what was going on with the second stage blowout discs. OKB-456 engine specialists, OKB-586 electricians, and military testers climbed up to the upper service platform to perform the blowout operation. They needed to disconnect and then reconnect the connectors of the second stage main distributor, and, for safety, do the same with the connector to the second stage engine powder combustion starter.

After all the operations had been performed, all the deputy chief designers sent their inspectors, and the launch crew members performed the checks. After T-minus 60 minutes the stairway leading to the upper service platform and the platform itself were swarming with people. Many military and civilian personnel not needed for the preparatory operations had crowded right onto the launch pad near the area where Marshal Nedelin and Deputy Chairman Grishin were sitting.

The postaccident top-secret report (special file) to the Central Committee signed by Brezhnev's commission states that, "At T-minus 60 minutes, besides the 100 individuals needed for the operation, 150 additional individuals were present on the launch pad." Yangel, his deputies Kontsevoy and Berlin, Konoplev, Glushko's deputy Firsov, and Second Directorate Chief Grigoryants were discussing the situation and giving instructions to their specialists right next to the missile.

After completing missile launch preparation, the testers evacuated the area and went to the observation post located approximately one kilometer from the launch site. They were not in a hurry to evacuate. Fate had already divided them all into the living and the dead. Yangel spent almost the entire time on the launch pad. The stressful situation gave him a craving for cigarettes and he gladly accepted Andronik Iosifyan's invitation, literally just a few minutes before the explosion, to go into the smoking room about 150 meters from the launch site. Iosifyan also invited Deputy Chairman Grishin to have a smoke, but for some reason he stayed behind. That cost him his life. The smoking room was also a sort of club, where the members could discuss the situation and temporarily relieve the psychological pressure. Aleksey Bogomolov was already there. He didn't smoke, but he was sensible enough not to be hanging around the launch pad. Viktor Kuznetsov, his deputies Tsetsior and Raykhman, and Yangel's first deputy and formerly also Korolev's deputy Budnik had accepted the invitation of chief designer of ground electrical equipment Goltsman to take cover under the ramp where the diesel generators were located that supplied power to the entire launch pad in the event of a national network power failure.

Second Testing Directorate Chief Grigoryants, the director of operations, announced T-minus 30 minutes at 7:05 p.m. Approximately 1 hour before the accident, OKB-586 lead designer Khachaturyan, who had been busy on the upper service platform of the erector "manually" blowing out the explosive discs, after performing many off-nominal operations, climbed down, reported the situation to Yangel, and was given the following instructions, "There's nothing for you to do here. Go over to the bunker and help Matrenin."

Carrying out Yangel's instructions, Khachaturyan phoned missile crew chief Matrenin in the bunker, informing him about the decision that had been made and telling him to begin setting up the launch circuit. When he left for the bunker, he noted that despite the announcement of T-minus 30 minutes, the ladder on the erector was heavily congested like a main thoroughfare; some were climbing down, others were waiting their turn to climb back up. Yangel's order saved Khachaturyan's life. After the catastrophe he was the first one to discover the connection in the general electrical system that was the direct cause of the command to start up the secondstage engine of the missile standing on the launch pad.

Khachaturyan's recollections cited in the book *Yangel: Lessons and Legacy* also include criticism directed at me. True, using information obtained from Arkadiy Ostashev, and later from the "Kharkovite" Ginzburg, I committed an error in the treatment of the direct circuit error. As the primary immediate supervisory officer over the missile's general electrical system, which was developed by OKB-692, Khachaturyan had the opportunity to give a more credible account in his own memoirs. It is just a shame that the truth was first published 41 years after the catastrophe. Better late than never![24]

---

24. *Author's note:* Teaching a course called "large rocket-space systems" to students at MFTI and MGTU, in the unit on "reliability and safety," I utilize the account of the events of October 1960 as a very instructive example. To one degree or another, the lessons of the past have remained relevant for almost half a century.

Most likely because Khachaturyan went down into the bunker and thus survived, he was the first specialist to discover the reason why the command was transmitted to start up the second stage engine.

Here is his version:

"When I went down into the bunker I found the always calm and collected Matrenin in somewhat of an agitated state, which Aleksandr Sergeyevich explained saying that Grigoryants was putting tremendous pressure on him and always rushing. Continuing our conversation, we stopped by the smoking room and had a cigarette. I started to reassure him, uttering a bunch of platitudes. And suddenly at that moment we heard incomprehensible chaotic, violent noises and explosions. Matrenin and I ran into the control room. Senior Lieutenant V. N. Taran, the preparation and launch control panel operator, and engineers from our design bureau whose responsibilities included monitoring the pre-launch circuit setup, were there at that time.

They looked horrible: ashen and wild-eyed. I dashed to the periscope and saw our missile burning on the launch pad. This hideous conflagration was accompanied by the explosions of the solid-fuel braking engines and the high-pressure tanks."

Forty years later we were able to see on television what happened at the launch site. On that fateful day, operators from the Ministry of Defense film studio had set up equipment to film the launch; everything was ready to roll long before launch time. When the unauthorized startup of the second stage engine occurred, the director of the film crew gave the order to switch on the remote-controlled movie cameras. This enabled many vital moments of the incident to be recorded.

Propellant components splashing out of the tanks soaked the testers standing nearby. Fire instantly devoured them. Poisonous vapors killed them. Of course, the quality of the film frames is not up to today's standards, but when viewed in slow motion you can see how the missile and erector burned and how the frantic people trapped on the service platforms jumped straight into the fire and were instantly consumed. The enormous temperature at a significant distance from the epicenter of the fire burned peoples' clothing, and many of those fleeing who got bogged down in molten asphalt burned up completely. The film chronicle does not show what happened to people who reached a relatively safe area.

Running for their lives, they found themselves in the ditch surrounding the launch site or on sand; instead of throwing off their flaming clothes or falling to the ground to extinguish the flames, like burning torches, they attempted to flee farther from the launch site and got tangled in the barbed wire surrounding it. Rescue workers arriving on the scene attempted to help the people who had run to them. They flung them to the ground and threw sand on them. It was 2 hours before the fire fighters managed to contain the fire and the launch site became accessible to the rescue workers.

According to one account, when Yangel arrived in the smoking room he was surprised that a cigarette lighter could cause a blinding bright flash. That was the moment the fire started. Risking his life, he darted toward the roaring blaze trying to lead frantic people out of the fire. He suffered burns to his hands before he was taken by force to the hospital at Site No. 42. Yangel dispatched a message to the Central Committee, the text of which was published for the first time in the book *Chronicle of the Main Events in the History of the Strategic Rocket Forces* in 1994.[25] The print run was very small, and it was never put on sale.

In the Archives of the President of the Russian Federation (APRF), rocket forces historians sought out Yangel's message, which was transmitted through Nedelin's office over a secure line code-named Purga-3. Here is the text of that message:

MESSAGE

At 6:45 p.m. local time, 30 minutes before the launch of article 8k-64, during the final pre-launch operation a fire broke out causing the destruction of tanks containing propellant components.[26]

The incident resulted in up to 100 or more human casualties, including several dozen fatalities.

Chief Marshal of the Artillery Nedelin was on the launch pad for the tests. They are now searching for him.

Please arrange emergency medical care for those who suffered burns from the fire and nitric acid.

Yangel

Purga-3
Office of Comrade Nedelin[27]

Yangel and Khrushchev spoke to each other. When Yangel reported that they hadn't found Nedelin, and that guidance system Chief Designer Konoplev and his deputy, Glushko's deputy, and two of Yangel's deputies were among the fatalities, Khrushchev asked, "And where was the technical director of testing at this time?"

I have quoted Khrushchev's words from the book by Andreyev and Konyukhov. I think they should be interpreted not as a criticism of Yangel—"And why are you still alive?"—but rather as natural indignation over the fact that the technical director was unable to ensure testing safety.

Khrushchev informed Yangel: "A commission headed by Brezhnev is flying out to you."

I believe that a report sent to Moscow in the line of duty by a KGB representative immediately after the catastrophe prompted the decision to form the State Com-

---

25. Sergeyev, ed., *Khronika osnovnykh sobytiy istorii raketnykh voysk strategicheskogo naznacheniya.* The book was issued by the publishing office of the Russian Strategic Rocket Forces.

26. The code name of the R-16 was the "8K64."

27. All of the documents cited here by Chertok are from the Sergeyev book published by the Strategic Russian Rocket Forces. This document bears the additional text, "Read by the members of the CPSU CC Presidium. V. Malin. 24.10.60. Hold in CC Presidium archives. 20.1.62. V. Malin." The original is in the Archive of the President of the Russian Federation (APRF), *fond* [collection] 3, *opis* [register] 50, *delo* [file] 409, *list* [page] 50. First published in 1994 in Sergeyev, ed., *Khronika osnovnykh sobytiy istorii raketnykh voysk strategicheskogo naznacheniya,* p. 240.

Asif Siddiqi.

Movie cameras recorded the grizzly aftermath of the R-16 explosion at Tyuratam as in desperation men tried to run away from the growing conflagration that melted everything around the rocket.

mission even *before* Yangel's message arrived. The decision calling for the creation of a governmental commission and detailing its makeup was signed on 25 October, after the commission had already arrived at the firing range.

TOP SECRET
(special file)
No. P308/22

To Comrades Brezhnev and Kozlov
Excerpt from Central Committee Presidium session protocol No. 308
dated 25 October 1960
Ministry of Defense issue
Approve the Commission comprising comrades Brezhnev, Grechko, Ustinov, Rudnev, Kalmykov, Serbin, Guskov, Tabakov, and Tyulin to investigate the causes of the catastrophe and take action at [the installation of] military unit 11284.
CENTRAL COMMITTEE SECRETARY[28]

---

28. APRF, f. 3, op. 50, d. 409, l. 49. First published in 1994 in Sergeyev, ed., *Khronika osnovnykh sobytiy istorii raketnykh voysk strategicheskogo naznacheniya*, p. 241

What was the cause of the fire and explosion? Here is the testimony of Khachatu-ryan, who according to Andreyev's and Konyukhov's book, was the first to discover the "direct effector" of the command to start up the second stage engine on the launch pad rather than in flight.

"The morning after the nightmarish evening of 24 October, I was sitting there ana-lyzing the layout of the electrical circuit for the propulsion system control system. I started to feel ill when I saw that when the step motors were reset in the initial state with volt-age present on bus D (the specialists referred to this as having the activated second-stage onboard battery hooked up to the onboard cable network), voltage flowed unimpeded through the working contacts of the timer to the VO-8 electric start tank pressurization valve. Everything was so technically simple in terms of the electrical circuit and so tragi-cally terrible in its consequences! Just then Komissarov entered the room (at that time Boris Alekseyevich Komissarov was the chief of the military delegation at the Yuzhnoye factory and KB; later he became deputy chairman of the VPK). I came to my senses a bit, and showed him on the diagram how the command to start up the second stage engine had been issued yesterday during the control system launch preparation."

Specialists have a right to ask questions: errors in electrical circuits are by no means rare, but why hadn't this specific error been identified during the many cycles of integrated tests at the factory, at the engineering facility, and at the launch site before fueling? The immediate developer of the circuitry at OKB-692 bore the responsibility for the error. The curators of the control system at the KB Yuzh-noye, who approved the circuitry with the error, bore the responsibility for fail-ing to detect it. I would forgive both groups. In a complex circuit it might not be possible to spot errors on paper. But why was the ground developmental testing and multistep testing process set up so that circuitry errors did not come to light before the catastrophic event? The KB Yuzhnoye deputy chief designer, the respec-tive department chief, the testers, and the KB Yuzhnoye military representatives bore the responsibility for this.

A second group of issues: Regardless of who committed the error allowing a false command to be issued to fire the second stage engine on the ground, this command should not have been executed over that circuit because the second stage onboard batteries were supposed to feed power to the automatic controls only *after* launch, and, second, even if this inhibitor had also been removed, that is why blowout discs had been placed in the lines, so that the propellant components would not reach the engine ahead of time.

Both levels of inhibitors had been removed despite elementary safety consider-ations. According to standard procedure, the onboard batteries of the second stage should not have been activated until all checks had been performed, after T-minus 1 minute, when there were no people on the launch pad. And according to standard procedure, the blowout discs were supposed to be detonated immediately before launch.

And, finally, there was the last fateful decision that could only be made by the immediate testing supervisors at the launch site, that is, Chief Designer Konoplev, Yangel's deputies Kontsevoy or Berlin, and Grigoryants and Matrenin, who supervised the actions of the military operators. I have no right to accuse any of them specifically. But based on the logic of the events, one of them gave the operator sitting at the control panels in the bunker the command to set the circuit in the "zero" initial state when there were more than 100 people on the launch pad.

The fact is that after the last cycle of integrated tests was performed, the circuit needed to be set in the initial state before the missile was fueled. The integrated tests at the launch site were performed before fueling. These were the last checks to see that the control system instruments and cable network were in good working order. These tests simulated the whole readiness setup cycle and involved the passage of commands prior to "launch," the actuation of the liftoff contact, the control system's generation of all the commands to the effectors of other systems, all the way down to the passage of the main command to shut down the second stage sustainer and control engines and separation of the nose cone. However, irreversible actuations of effectors, such as the detonation of explosive cartridges, are not typically executed during integrated tests. Therefore, assemblies containing explosive cartridges are disconnected from communications instruments—such as the main current distributors of the control system—and simulators are hooked up instead.

Arkadiy Ostashev, who flew to the firing range because of the death of his brother, Chief of the First Testing Directorate Lieutenant Colonel Yevgeniy Ostashev, undertook his own "private" investigation of all the circumstances surrounding the catastrophe. The version he told me was that Grigoryants, who had been on the launch pad, gave the order to begin preparing the circuit for launch readiness, while Matrenin delivered the specific command to the officers sitting at the control panels in the bunker.

This is at odds with Khachaturyan's recollection; according to him, while the circuit was being reset to the initial state, he and Matrenin stepped out of the control room to have a smoke. The one who executed the fateful operation was an officer operating a control panel, who acted under the supervision of OKB-586 staff members. He was the one who performed the specific act, but there were very many who were actually responsible and culpable for what happened.

Yangel displayed real courage, having declared to the chairman of the State Commission investigating the causes of the catastrophe. "I ask that no one be blamed for what happened. I am culpable for everything as the chief designer who was not able to keep an eye on all the subcontractors."

The chief designer is not capable and not obliged to analyze all the connections in a complex electrical circuit. His deputies, the control system curators, and developers of the missile testing process are supposed to do that at the factory and at the engineering facility. To a great extent Yangel's behavior also determined the behavior of the State Commission that arrived at the firing range on the morning of 25 October.

The following is the Central Committee State Commission report.

TOP SECRET
(special file)

CPSU CC

As instructed by the Central Committee, the commission has conducted an on-site investigation into the circumstances of the catastrophe that occurred on 24 October 1960, at USSR Ministry of Defense NIIP-5 during the testing of an R-16 missile.

With the participation of leading specialists, the following causes of the catastrophe were determined:

As of 26 September, the R-16 missile was located at the firing range in the Assembly and Testing Building. During the missile's technical preparation process, individual defects in control system equipment and cable network were found and corrected by the industrial and military specialists at the firing range.

On 21 October the missile was transported to the launch site and on 23 October the pre-launch tests, which proceeded without incident, were completed. That same day the missile was fueled and its launch preparation began according to the approved procedure.

During the preparation process, when the command was issued to detonate the explosive discs in the second-stage oxidizer lines, the control panel issued a false command, and in fact, the explosive cartridges in the first-stage fuel line were detonated. In addition, the explosive charges of the gas generator cutoff valves in the first assembly of the first-stage sustainer engine detonated spontaneously and the main distributor of the onboard cable network malfunctioned

This circumstance forced the commission to suspend further launch preparation until these defects could be clarified. On the morning of 24 October the missile launch review commission made the decision to continue launch preparation, thereby permitting a departure from the approved procedure.

The violation of the launch preparation procedure consisted in the fact that the step motors of the second-stage control system were reset to their initial position with the engine startup system already fueled and with the onboard power supply activated. As a result of this, the second-stage sustainer engine started up prematurely, its flame burned through the bottom of the first-stage oxidizer tank, and then the second-stage fuel tank disintegrated, which resulted in an intense fire and the complete destruction of the missile on the launch pad. (The technical findings on this matter are attached).

The test directors showed excessive confidence in the operational safety of the article's entire complex. As a result, they made individual decisions hastily without proper analysis of the possible consequences.

During the missile's launch preparation there were serious flaws in the organization of the work and security. At T-minus 60 minutes, besides the 100 individuals needed for the operation, 150 additional individuals were present on the launch pad.

Seventy-four military and civilian workers died in the catastrophe. Among the dead were test commission chairman Chief Marshal of Artillery M. I. Nedelin, guidance system Chief Designer Konoplev, missile Deputy Chief Designers Kontsevoy and Berlin, engine Deputy Chief Designer Firsov, Deputy Chief of the Firing Range Colonel Nosov, and Firing Range Directorate Chiefs Lieutenant Colonels Ostashev and Grigoryants. Fifty-three individuals suffered various degrees of injuries and burns. The injured received immediate medical attention and leading medical specialists were called in to treat them.

Deceased military servicemen were buried in a communal grave on the grounds of the firing range with military honors. Deceased industrial workers were returned to their home towns for burial. Materials concerning assistance and the establishment of pensions for the families of the deceased will be presented at the USSR Council of Ministers.

Numerous conversations with persons directly involved in testing, with eyewitnesses to the catastrophe, and individuals who were injured, attest to the commendable and courageous behavior of those who faced extremely severe conditions. Despite the serious consequences of the event, the firing range personnel and industrial workers are ready and able to correct the flaws that have been revealed and to complete the experimental development of the R-16 missile.

In the interests of recovering from the catastrophe and fulfilling the assignment to produce the R-16 missile, the commission has conducted an investigation involving leading industrial specialists and has held a meeting with the firing range command staff and has outlined the following actions:

- perform additional checks and conduct additional developmental testing on the R-16 missile control system complex;

- review and optimize a procedure for pre-launch preparation and missile launch execution, heighten operational security on the launch pads and intensify safety measures for those involved with testing;

- increase the quality of experimental development and production of assemblies and instruments at KBs, institutes, and factories;

- restore the damaged launch pad within 10-15 days and finish building and rigging a second launch pad with the intention of beginning R-16 missile flight tests in November of this year;

- in view of the death of a number of leading specialists, take action to find qualified staff to strengthen the ranks of the firing range and industry.

Conducting the aforementioned measures will make it possible to fulfill the designated testing program for the R-16 missile.

Attachments:

1. Technical findings – 4 pages.

2. Lists of killed and injured – No. 3386s – 16 pages.

3. Top Secret photographs 5 items/ex. No. 1 from film 680 – 2 items and ex. No. 1 from film 684 – 3 items*

> *L. Brezhnev*
> *A. Grechko*
> *D. Ustinov*

K. Rudnev
V. Kalmykov
I. Serbin
A. Guskov
G. Tabakov
G. Tyulin[29]

The following is the text of the technical findings in its entirety.

<u>TOP SECRET</u>
(special file)

TECHNICAL FINDINGS
*of the commission investigating the causes of the article 8K64 catastrophe
No. LD1-ZT, which occurred during its launch preparation
at military unit 11284 on 24 October 1960*

Article 8K64 No. LD1-ZT was transported to the launch site on 21 October 1960, at 8:00 a.m. Missile launch preparation was conducted without noteworthy incidents until 6:23 p.m. on 23 October, after which operations were halted due to the discovery of the following abnormalities while conducting a routine operation to detonate the blowout discs in the stage two oxidizer lines:

1. Instead of the blowout discs in the stage 2 oxidizer lines, it turned out that the blowout discs in the stage 1 fuel lines were detonated.

2. Several minutes after the aforementioned blowout discs were detonated, the pyro-cartridges of the cutoff valves in the assembly 1 gas generator of the stage 1 sustainer engine detonated spontaneously.

As a result of the subsequent investigation of the causes for the aforementioned abnormalities, on 24 October it was determined that the erroneous execution of a command to detonate the blowout discs and the spontaneous triggering of the gas generator's pyro-cartridges occurred due to design and production defects in the detonation control panel developed by OKB-692 of the State Committee on Radio Electronics (GKRE).

These same circumstances caused the A-120 main distributor to malfunction (the onboard cable network was not damaged).

In accordance with the decision of the testing review team, the gas generator cutoff valves and the A-120 instrument were replaced.

Also, the decision was made to detonate the stage 2 barrier discs using stand-alone

---

29. The document has the following notations: "Reported to members and candidates of the CPSU CC Presidium. V. Malin. Keep in CPSU CC Presidium archives. 20.1.62. V. Malin." APRF, f. 3, op. 50, d. 409, ll. 51-54. First published in 1994 in Sergeyev, ed., *Khronika osnovnykh sobytiy istorii raketnykh voysk strategicheskogo naznacheniya*, pp. 242–244.

circuits from separate power sources rather than from the detonation control panel. After this, the pre-launch missile preparation was continued.

While conducting subsequent missile preparation operations, on 24 October 1960, at 6:45 p.m. local time, a fire broke out on the missile in the area of the stage 2 aft compartment leading to the destruction of the missile and the ground equipment assemblies located at that time on the launch site in the vicinity of the launch pad.

The fire broke out after the announcement of T-minus 60 minutes while the control system step motors were being reset to their initial position. By that time, the barrier discs in the oxidizer and fuel lines of the stage 2 sustainer and control engines had been blown out, a leak check had been performed on the lines, and by order of the review team, the stage 1 and stage 2 self-activating batteries, which had been activated on the ground, were hooked up.

The fire on the missile was caused by the premature actuation of the VO-8 electro-pneumatic start tank pressurization valve, triggered by a command issued by the timer when the control system step motors were reset to the zero (initial) position. The actuation of the VO-8 electro-pneumatic valve, in turn, caused the stage 2 sustainer engine to start up.

One should note that the fire on the missile might not have occurred if, in this instance, the control system step motors had been set in the zero position before the onboard batteries were hooked up, as stipulated in the engineering plan.

The commission identified the actuation of the VO-8 electro-pneumatic valve and startup of the stage 2 sustainer engine by analyzing the technical documentation. The condition of the missile remains unambiguously confirmed this fact (see missile remains inspection protocol).

Additional analysis of the control system integrated circuit showed that the circuit does not preclude the untimely actuation of the VO-8 electro-pneumatic valve during launch preparation operations on the missile, in those cases when the control system might require readjustment after disc detonation and activation of the batteries (for example, when the launch trajectory needs to be changed because of prolonged delays and launch preparation, or when the circuit is dead).

CONCLUSIONS AND PROPOSALS

1. During missile launch preparation a number of incidents occurred indicating that there were abnormalities and defects in the cable network, onboard batteries, blowout disc detonation control panel, and distributor A-120 of the control system.

The testing management team did not attach the proper significance to this, and to eliminate the aforementioned abnormalities and defects, it permitted a number of deviations from the established launch preparation procedure without sufficiently working through and analyzing the consequences.

While conducting the final operations on the fueled missile, a large number of people who were not involved with the execution of any operations were allowed on the launch pad without any justification.

2. The direct cause of the catastrophe was a defect in the integrated circuit of the control system, which allowed the untimely actuation of the VO-8 electro-pneumatic valve

controlling the startup of the stage 2 sustainer engine during launch preparation. This defect was not identified during all the preceding tests.

The fire on article LD1-ZT might not have occurred if the control system step motors had been reset in the zero position before the onboard batteries were hooked up.

3. OKB-692 shall work with NII-944, OKB-586 and VNIIEM to modify the integrated circuit of the control system in order to ensure the complete safety of missile launch preparation and its reliable operation during preparation and launch.

4. OKB-586, NII-944, and OKB-692 shall insert changes into the operational technical documentation based on the results of preparing article No. LD1-ZT at the technical position and launch site, and also based on the results of modifying the integrated circuit.

> *Yangel*
> *Budnik*
> *Tabakov*
> *I. Ivanov*
> *Ishlinskiy*
> *Tretyakov*
> *Kuznetsov*
> *Tyulin*
> *Iosifyan*
> *Medvedev*
> *Tsetsior*
> *Doroshenko*
> *Bokov*
> *Matrenin*
> *Vorobyev*
> *Favorskiy*[30]

These documents, which were declassified in 1994, require further comment. First, not only do the Central Committee commission report and the technical findings *not* mention specific guilty parties by name, but they also do not bring charges against any organization. When he first arrived at the firing range early on the morning of 25 October, addressing the assembled testing participants, Brezhnev said, "Comrades! We do not intend to put anyone on trial; we are going to investigate the causes and take actions to recover from the disaster and continue operations."

The next day, 26 October, at a session of the Central Committee Commission, Brezhnev expressed condolences on behalf of the Central Committee, the government, and Khrushchev himself, on the occasion of the deaths of the firing range testers and industrial specialists and officially announced that all the necessary actions would be taken to render assistance to those who were injured and to the families of

---

30. APRF, f. 3, op. 50, d. 409, ll. 55-58. First published in 1994 in Sergeyev, ed., *Khronika osnovnykh sobytiy istorii raketnykh voysk strategicheskogo naznacheniya*, pp. 245–247.

those who died. He said that because there was no one to hold accountable for the errors and miscalculations that were committed, and since the managers responsible both for the technical side and for operational safety were all killed—except for Yangel and Mrykin—the nation's leadership had decided not to conduct a special investigation, but to let all the participants who survived draw the appropriate conclusions themselves. This was a wise decision.

Second, no one knew why the high commission, without any serious study, had stipulated that flight-tests would continue in November after all defects were eliminated, that is, within one month. One could forgive Brezhnev and Grechko. They were justified in not understanding the amount of work that needed to be done in order to fix the missile. But Ustinov, Rudnev, Kalmykov, Tabakov, and Tyulin—if they didn't understand this, they must have sensed it. However, without hesitation, they put their signatures after Brezhnev's.

Actually, the flight-tests were continued a little more than three months later. And I consider this to be a heroic feat for the organizations of Yangel and Sergeyev (who had been appointed to replace Konoplev), and for all the remaining creators of the R-16.[31]

Third, despite Brezhnev's assurances, one individual was punished. And this one person was a woman. In early November 1960, at the initiative of GKRE Chairman Kalmykov, who was in charge of OKB-692, an extended technical session for R 16 missile control system specialists was held in Kharkov with the participation of control system developers of other missiles. N. A. Pilyugin headed the session, which was attended by VPK Deputy Chairman G. N. Pashkov, GKOT Chairman K. N. Rudnev, Central Committee Defense Industries Department Sector Manager B. A. Stroganov, Chief Designers Kuznetsov, Iosifyan, Lidorenko, and Yangel's first deputy, V. S. Budnik.

According to the recollections of meeting participant K.Ye. Khachaturyan, after calling the meeting to order, the floor was given to I. A. Doroshenko. She reported to the attendees that, in view of the fact that the blowout discs had not undergone thorough developmental testing, a new process operation had been introduced during launch preparation to check their actuation "by sound," and seemed to imply that this had caused the fatalities. At the end of the meeting Kalmykov read an order that dismissed I. A. Doroshenko from her job and barred her from working at defense enterprises in the future.

In fact, Doroshenko was not punished for an error committed in an electrical circuit, but for inappropriate behavior. Instead of being contrite, she attempted to shift the blame for what happened onto the unreliable blowout discs. Due to inexperience or her own pride, she had failed to size up the situation and ended up in the role of the woman who gets thrown overboard during a storm to save the ship.

---

31. Vladimir Grigoryevich Sergeyev (1914–) succeeded Konoplev as chief designer of OKB-692, the organization now known as NPO Khartron.

Actually, this one woman—the developer of the general electrical circuit—was not to blame for what happened.

In and of itself, the situation at the launch site after the missile was fueled was a blatant violation of safety regulations. One could, for the sake of a great goal, compel a dozen testers and electricians to fiddle around with their own connector plugs, test gauges, and portable batteries right on board the missile. But it was the responsibility of the director of testing to clear the launch pad of every single person not involved in that work, regardless of rank or title. First and foremost, the firing range chief was responsible for doing this.[32] But he was subordinate to Nedelin.

Yangel, the chief designer of the missile; Konoplev, the chief designer of the control system; and their deputies for testing should have halted any electrical tests until everyone not needed for the troubleshooting operations had been cleared off of the launch site. They had the right. They did not exercise it.

Sometimes efforts to observe fundamental safety are viewed as cowardice. If a general on the front lines goes from trench to trench under a hail of bullets without cowering, he is commended: "Look what a brave man we have!" Despite the mortal danger, the soldier at his side will not cower either. But in this case the brave men are risking only their own lives.

The testers themselves were so tired that, to a certain degree, they can be posthumously exonerated for their various errors and rash acts. They didn't think things through; didn't grasp the situation; they were rushed. As they say in such situations, "Forgive them, Lord, for they knew not what they did." But it was the duty of the electric circuit developers to know what they were doing. Under conditions when all the electric inhibitors preventing stage two engine startup had been removed, for absolutely unknown reasons, the launch control officer in the bunker decided to conduct the cycle of operations that set the stage two program timer to the initial position. One can only assume that one of Yangel's deputies gave him permission to do so if he requested it over the intercom system. He did not have the right to perform such an operation without authorization, without having coordinated with the test director. The one who gave the OK for that operation must have either forgotten or didn't even know that it needed to be checked against the logic of the circuit—in case anything might happen.

The command to set the PTR into the initial position was the last and fateful error in a long chain of events that set the stage for the biggest catastrophe in the peacetime history of missile technology. While being set in the zero position, the PTR fed power to the stage two engine startup circuit. All the circuit safety inhibitors had been removed beforehand during the process of troubleshooting. And, the engine executed the command.

The technical findings document cited above was signed by 17 people well

---

32. The firing range chief at the time was Konstantin Vasilyevich Gerchik (1918–2001) who served in that capacity in 1958–61.

known to me. They spoke honestly about the "destruction of the missile and the ground equipment assemblies." But for some reason they did not mention that at T-minus 60 minutes, in addition to the 100 persons that were needed to work on the launch pad, as many as 150 other people were also there. And the majority of these 250 individuals were "destroyed" or "damaged" at least as much as the ground equipment.

The firefighting brigades brought in from all the sites and the ambulance crews who rushed to help found a terrible scene. Among those who managed to escape from the missile, some were still alive. They were transported straight to the hospital. The majority of the dead were unrecognizable. Their bodies were laid out in specially designated barracks for identification. Arkadiy Ostashev, who arrived the day after the catastrophe, spent 14 hours in the barracks trying to identify his own brother Yevgeniy. Nedelin was identified by the "Gold Star" medal that had survived. Konoplev's body was identified by size. He had been taller than anyone else at the launch site.

Cigarettes saved the life of Yangel, Iosifyan, and everyone keeping them company in the smoking room, which was located a safe distance from the launch site. Iosifyan had talked Bogomolov, who never smoked, into joining him in the smoking room to discuss the situation. Iosifyan and Bogomolov had experience working at our launch sites. They wanted to persuade Yangel to take the helm, call for a break in missile preparation, let everyone rest, and calmly discuss a further plan of action. Both of them considered the troubleshooting actions of Konoplev and his specialists dangerous. They talked Deputy Chairman Lev Grishin into joining them for a smoke. He promised to catch up with them, but for some reason got delayed. Eleven days later, he died in the hospital in agonizing pain. The firing range chief, General Konstantin Gerchik, managed to move a little farther away from the marshal toward the smoking room. He was taken to the hospital in serious condition; badly burned and poisoned, he survived, spending more than six months in hospitals.

The best burn treatment specialists were summoned from Moscow to attempt to save the survivors. Funeral arrangements needed to be made, the next of kin needed to be notified, and arrangements needed to be made for them to fly to the firing range. Soldiers and officers were buried in a mass grave in the municipal park. It wasn't until three years later that an obelisk bearing the names of those interred there was placed over it. Deceased civilian specialists were placed in zinc coffins and flown back for burial in the towns where they had lived or worked. Konoplev and his staff were buried in Kharkov; Yangel's deputies and staff—in Dnepropetrovsk.

No official reports emerged about the catastrophe at the missile firing range. Relatives, close friends, and all eyewitnesses were urged not to speak of the true scale of the incident. Acquaintances at funerals in other cities were supposed to be told that there had been an accident or an airplane crash.

It was simply impossible to remain quiet about the death of Marshal Nedelin. A brief governmental report was issued about the tragic death of Nedelin in an airplane crash. No mention was made as to the fate of the crew and other passengers.[33] Nedelin's funeral on Red Square took place with the traditional ritual. An urn containing his ashes was placed in the single row in the Kremlin wall columbarium behind Kurchatov's urn. The number one physicist who had supervised the development of the first atomic bomb and the chief military ideologue of the nuclear missile strategy were right next to one another. Lev Arkhipovich Grishin, the most optimistic and quick-witted of all the directors of the missile field at that time, who "died in the line of duty," was buried in Novodevichye Cemetery.[34] A government decree signed by Kosygin called for what were for those times good pensions for the families of the deceased.

Yangel had witnessed the tragedy from beginning to end. His missile destroyed its creators before his eyes. Though his physical injuries were minor, the psychological shock left him unable to work for a month. Iosifyan and Bogomolov, who soon thereafter arrived for joint work with us, spent a long time explaining what a miracle it was that they had come away from there alive and implored Korolev to be prudent. Even without their pleadings, Korolev understood that technical and organizational conclusions needed to be drawn from this tragedy. Although it was the R-16 missile—a competitor of the R-9—that had suffered this catastrophe, we were all subdued by the scale of what had happened. Too many friends, acquaintances, and just plain good people were among the dead. Regardless of who the chief designer had been, the missile was not his, but ours. We were all citizens and patriots of our nation.

New procedures were introduced at Site No. 1. Much more rigorous procedures were put in place for access to what we called "ground zero"—the main concrete pad of the launch site. Special security was established as soon as fueling began. All those involved with preparation wore special color-coded armbands. As launch drew closer, an ever-increasing number of colored armbands were required to leave the launch pad. The last to head for the bunker after the announcement of T-minus 15 minutes were those wearing red armbands.

For the first time, master scheduling of operations at the launch site was initiated. The precise time and place for the performance of each operation were indicated on a master schedule. Once the executive officer of the military unit, the firing range directorate officer supervising him, and the industrial representative had executed their operation, they were supposed to leave the work station until the next call or evacuate to a predetermined area. The total number of operators and inspectors was reduced after many operations were combined and integrated. Each director, crew chief, or testing brigade military and industrial representative was required to thoroughly study his own operations and bore full responsibility for conducting them

---

33. Nedelin's death was reported in the West at the time. See for example, Osgood Caruthers, "Chief of Rockets Killed in Soviet," *New York Times*, October 26, 1960, p. 22.

34. Famous Soviet citizens who had not achieved heroic status were typically buried in the Novodevichye Cemetery instead of the Kremlin Wall.

according to the master schedule and for reporting any discrepancy to the launch director.

No organizational measures could guarantee against possible operational errors and glitches when there were systems failures. Special teams were created to develop "fool-proofing" proposals reducing the number of manual operations as much as possible, increasing the number of automatic functions and introducing an automatic emergency fire-extinguishing system. They did not manage to implement all the sensible solutions right away on R-7 missiles. This would require far-reaching changes. The measures were implemented on a priority basis. Everything that could be devised for safety enhancement was implemented on R-9 missiles, and later—on our new undertaking—the RT-2 solid-fuel intercontinental missile, better known under code number 8K98.

EXACTLY THREE YEARS TO THE DAY AFTER THE CATASTROPHE DESCRIBED ABOVE, on 24 October 1963, at that same firing range in one of the silos of the R-9 missile site, a fire broke out, costing the lives of seven military testers. This time the missile had not been fueled. The testers were conducting routine servicing procedures without having first bothered to check whether the silo had permissible oxygen vapor levels. A second mass grave appeared in the Leninsk municipal park. After that incident, 24 October was considered bad luck at the firing range. Tacitly, it became a day off from work, and military testers even avoided serious domestic chores at home.

Five years later the new firing range chief found this to be harmful superstition and ordered that 24 October be considered a normal work day for all servicemen. On that day, the residents of Leninsk, school children, and relatives who fly in for the occasion manage to scrounge up fresh flowers and place them on the mass graves. The floral selection in Tyuratam is very meager during those cold October days.

HAVING RECOVERED FROM THE SHOCK, Yangel's Dnepropetrovsk team delivered the R-16 missile for flight-development tests three months later.[35] After that, everything took its proper course. Besides the R-9, new competitors emerged for the R-16—missiles developed by Chelomey and then solid-fuel missiles. In his later post as General Secretary and Chairman of the Defense Council, Brezhnev would have to be a peacemaker in the missile "civil war" that had flared up between the schools of the chief designers and the ministers, generals, and Party officials who stood behind each of them. This is a distinct and as-yet little studied field in the history of our missile technology.[36]

---

35. The first successful R-16 lifted off on 2 February 1961.

36. This so-called "little civil war" peaked in 1969 over a decision to select from competing options for a new generation of strategic ICBMs offered by Yangel and Chelomey.

Our missile technology would have shaped up much differently if it had not been for the death of Nedelin. Among the high-ranking military leaders of that time, he was the only marshal and deputy minister of defense who had gained an understanding of our problems. He was a military technocrat, and therefore military and civilian specialists respected him. We really felt Nedelin's loss after Marshals Moskalenko, Biryuzov, and Krylov replaced him one after the other. These men were distinguished military commanders from the World War II era with a great deal of experience managing combined-arms operations. Our Navy and Air Force adhered to the rule that specialists who had served at sea and in the air were appointed commanders-in-chief. After Nedelin's death, when it came to the rocket and space forces, this natural and reasonable order was disturbed.

Having returned to the firing range after the November holidays of 1960, we did not rush from the airfield to our hotel rooms at Site No. 2. Our entire staff visited the fresh mass grave. We placed bouquets of red carnations and roses we had brought with us from Moscow atop the now dried flowers there. Our hats in our hands, we stood in prolonged silence by the wreath-covered mound. At such times, each individual thinks about something quite personal and inevitably something universal. Here lie our comrades-in-arms. They were destroyed by the R-16 missile, the competitor of our R-9. But the R-16 was also our missile. It was created for our nation, to protect us, my family, my Moscow. The road to space was paved with combat missiles.

Little did I imagine back then that 33 years later I would be on a scientific-technical council of the Russian Space Agency defending a project using a Tsiklon rocket to insert six satellites for a communications system developed by us into space. The modern Tsiklon is a launch vehicle developed without Yangel, based on subsequent R-16 modifications.[37] And I certainly could not have imagined that I would be delivering my report in Russia, in Moscow, but not in the Soviet Union.

In closing this chapter I must write of one more catastrophe. The high Party and technical investigative commissions took just two days to determine the causes of the catastrophe of 24 October 1960—the biggest ground disaster in the history of missile technology. A little less than 20 years later, on 18 March 1980, at our northern firing range in Plesetsk a catastrophe occurred during launch preparation of a rocket that had long served as a launch vehicle—the R-7 rocket, produced at the Progress Factory in Kuybyshev (now Samara). Forty-eight people died in that explosion and fire. This time there were no "extras" on the launch pad. All those killed were experienced in missile launch preparation.

Leonid Vasilyevich Smirnov, the Military-Industrial Commission chairman and deputy chairman of the Council of Ministers, was appointed chairman of the State

---

37. Technically, the Tsiklon launch vehicle (and all its various modifications) was based on the R-36 ICBM whose design genealogy can be traced back to the earlier R-16.

Commission investigating the catastrophe. The commission comprised Valentin Petrovich Glushko (at that time General Designer of NPO Energiya), launch complex Chief Designer Vladimir Pavlovich Barmin, TsSKB-Progress General Designer Dmitriy Ilyich Kozlov, Chief of the Main Directorate of Space Assets of the Ministry of Defense Aleksandr Aleksandrovich Maksimov, and Commander of the Strategic Rocket Forces Chief Marshal of Artillery Vladimir Fedorovich Tolubko. Employees of leading firms who might be able to help investigate the causes of the catastrophe were called in to work in the commission.

Vladimir Fedorovich Karashtin, chief of ground systems (our Complex No. 6) from NPO Energiya, took part in the work of one of the groups. When he returned from Plesetsk he was very reluctant to share his impressions. He told me categorically that the military fuel servicers were the culprits. During the fueling process there was an oxygen leak. To stop it they tossed a wet, dirty rag on the spot; soon, *oksilikvid* (a liquid oxygen mixture) explosives formed, resulting in a fire.

I was reminded that when the fourth stage of our *Semyorka* Block L failed, without knowing the cause of the failure and lacking the necessary telemetry information, the engine specialists put forward a version involving *oksilikvidy*. According to the explanations of rocket propellant chemistry specialists, *oksilikvidy* form when liquid oxygen comes into contact with an organic compound, for example, various lubricating compounds. When exposed to heat, shock, or some other influences, supposedly such a component can self-ignite or explode.

In due course, Leonid Voskresenskiy and I conducted experiments to confirm the explosive hazard posed by a mixture of liquid oxygen and used motor oil. The mixture stubbornly refused to self-ignite or explode. As far as the Block L stages were concerned, the *oksilikvidy* theory was soon tossed out because before the next launch the true cause of their failure was always determined. The culprit turned out to be the electrical circuit that supplied power to startup the Block L engine.

"Could it be that this situation also reminded them of *oksilikvidy*, but the culprit is electricity?" I asked.

"No, the electricity hypothesis was rejected right away. There was also a peroxide hypothesis. But there was no proof at all for it."

Without having determined the true cause, after much argument the commission subscribed to the "dirty rag" version as the "most probable." Since all the probable culprits had died, there was also no one to punish. However, time showed that the true culprits were still alive, the "dirty rag" had nothing to do with it, and all those lying in the mass grave under the obelisk had done their duty honorably.

But before I continue, I must explain to the reader why the "peroxide hypothesis" that the commission rejected had come up. The fact was that our distinguished "old lady," the *Semyorka*, still had a birthmark that had been passed on to all Glushko's liquid-oxygen propulsion systems since the German V-2. In addition to the main propellant components—liquid oxygen as the oxidizer and kerosene as the fuel— concentrated hydrogen peroxide was used. It was a sort of a relic from the old days of missile technology. The Germans were the first to use it in the V-2 propulsion system. Under the effect of a catalyst, peroxide violently breaks down into water and oxygen. The gas vapor formed during this process in the closed chamber of a gas generator is directed at the turbine, which then drives the pumps feeding oxygen and kerosene into the engine combustion chambers.

As far back as the tests performed at Kapustin Yar on the German A4 (V-2), everyone involved with that new technology knew that hydrogen peroxide was very temperamental. It could begin to break down when it came into contact with materials that seemed nonhazardous. Therefore, during fueling and when the missile was held on the launch pad, we monitored the temperature in the peroxide tank very closely.

It was Professor Aleksandr Shteynberg, the chief research associate of the Russian Academy of Sciences Semyonov Institute of Chemical Physics, a doctor of physical and mathematical sciences, who provided a conclusive explanation of what happened on 18 March 1980 in Plesetsk. Using the opportunity of my stay in St. Petersburg, he handed me a memo that he had composed, which in my opinion has great value not just for missile technology. Considering its historic value, I am citing it here in a slightly abridged version.

Brief description of the circumstances of the investigation
into the cause of the catastrophe at the Plesetsk cosmodrome on 18.3.80

"On 19 or 20 March 1980, as chief of the Department of Kinetic and Explosive Properties at the State Institute of Applied Chemistry (GIPKh, Leningrad) of the Ministry of Chemical Industry, I was summoned by the institute's first deputy director, Yevgeniy Sivolodskiy, who ordered me to get ready to catch the morning flight to Arkhangelsk with him the following day.[38] *From there we would travel by train to Plesetsk in connection with the disaster that occurred there on the night of 18 March.*

The following morning Sivolodskiy's chauffeur met me at the institute entryway. He said that his boss had gone to the institute polyclinic this morning to get a work release from his doctor. I was ordered to fly alone.

...During launch preparation of the R-7 rocket, an explosion and fire occurred. A large number of cosmodrome employees were killed and the launch pad was completely destroyed. The State Commission is working under the leadership of VPK Chairman and Deputy Premier Leonid Vasilyevich Smirnov. All the leading firms involved with the rocket and with what happened have sent representatives to the commission. General Designers Glushko, Barmin, and Kozlov, Commander-in-Chief of the Rocket Forces Marshal Tolubko, Glavkosmos Chief [sic] Gen. Maksimov, and others are participat-

---

38. GIPKh—*Gosudarstvennyy institut prikladnoy khimii.*

ing in the commission's work.[39] I have been assigned to sub-commission No. 1, which is involved with analyzing the materials and developing scenarios for the cause of the disaster.

Jumping ahead, I will mention that the explosion and fire resulted in the death of 48 individuals. Thus, the Plesetsk disaster is the space industry's second worst after the preceding 'Nedelin' disaster in Tyuratam.

I was taken straight to the main office of the commission where a few people were assembled. All the aforementioned people except for L. V. Smirnov were in the room. Valentin Petrovich Glushko was the first to fill me in on the situation. He knew me personally since I had presented reports on the kinetic and explosive properties of rocket propellants a number of times at his 'Rocket Propellants' Council. Most of these reports had dealt with the relevant properties of HYDROGEN PEROXIDE. Glushko had long wanted to find a way to make peroxide break down at a rate slow enough to enable it to be stored for a fairly long time in ampulized, i.e., pressurized tanks, without danger of the tank exploding. I had several rather heated discussions with Glushko in this regard after presenting reports to his Council. He did, however, greet me quite amiably.

Glushko sized up the situation approximately as follows. 'We have invited you as a chemist, a specialist on the kinetic and explosive properties of propellant components. The cause of the explosion has not yet been ascertained—we're counting on your help. Please pay particular attention to hydrogen peroxide.' At this point in the conversation Academician Barmin interjected: 'But, please, don't concentrate just on peroxide. After all, there's kerosene and oxygen…' Glushko cut him off and in a completely unexpected—and for him almost outrageous—manner, he bellowed, 'Am I going to talk or are you going to prevent me!' As I said, I had observed Glushko many times before and my recollection was that he never even raised his voice—he spoke almost in a whisper. In a word, I was somewhat shocked.

Thus began the work of the commission. There were a lot of iron scraps, but all of them were covered with dirt and soot. The fire lasted a long time; everything got stuck in deep snow. There was no solid data or witness testimony about how the fire started; whether it went from the top down or from the bottom up, etc. Those who were nearby died. Those who escaped—for example, those in the so-called 'octagon' boxes in the lower part of the launch complex— heard the sounds at the beginning of the accident, but they did not see the fire start or how it developed.

I will not go into detail about the hypotheses that were advanced. There were several, but they all had few conclusive facts to back them up. Or there was almost no proof at all—material evidence was missing. As a result, certain investigators began to formulate the so-called upper "oxygen" scenario, relying more and more on the fact that it was a possible cause, if not the real one. They envisioned a model for the formation of so-called

39. Maksimov was, in fact, not chief of *Glavkosmos*, but chief of the Main Directorate of Space Assets (GUKOS). *Glavkosmos*, created in 1985, was the public face of the Soviet space program in the late 1980s.

oksilikvidy—*compounds of liquid oxygen and organic materials—the material of the damp rag that in many cases was wrapped around a joint in the oxygen line when there was a leak. It was a fairly well-known fact that this 'unauthorized' method was used at both cosmodromes.*

...After working extensively on the problem, the Commission issued a statement that the circumstances described above had caused the disaster. It should be stressed that the situation was extremely tense. One could almost physically sense the desire of the majority of the participants not only to find the cause, but also to save the reputation of the organization where this commission member belonged. This accusation did not have the slightest thing to do with the cosmodrome's military staff. It was quite clear that they wanted to know the truth. After all, their safety literally depended on it. I personally worked very closely with them and above all with the chief of the cosmodrome's analysis department, Colonel A. S. Tolstov and the cosmodrome's chief engineer, Colonel L. N. Yashin. A number of others conducted themselves commendably as well. Among them, I should also mention the chairman of sub-commission No. 1, Deputy General Designer A. V. Soldatenkov, and General A. A. Maksimov

But, I repeat, there was little material for the analysis of this [oksilikvidvy-based] *and other hypotheses. There was rather strong pressure from members of the commission who were cosmodrome outsiders. It seemed to me that we needed to look more closely into the peroxide scenario since there was one indisputable fact pointing to its validity. During the so-called equalization operation* [at the time of the accident], *when they monitor the peroxide filling the toroidal tank by checking the level in a glass tube—a vessel connected to the tank—the peroxide level suddenly dropped dramatically. The individual monitoring the process managed to report this over the intercom. The explosion occurred almost immediately thereafter. Nevertheless, this very fact was considered insufficient for the subsequent development of the lower 'peroxide' scenario...*

Sixteen months passed. On 26 July 1981, Sivolodskiy sent his car for me. I was vacationing at a dacha in Sosnovo outside Leningrad. The first deputy director ordered me to fly immediately to Plesetsk, where during launch preparation of a rocket carrying the Bulgaria 1300 *satellite, it seems a similar disaster almost occurred.*[40] *While filling the same R-7 rocket tank with peroxide, the fueling assembly began to heat up. This system comprised a nozzle, similar to one on a fire hose, equipped with a filter—a tightly knit stainless steel mesh. Very fortunately the crew chief (later we learned his name—Konstantin Menyayev) kept his wits about him and gave the command to drain the tank, and as they used to say back then, 'poured the component into the ditch.'*

'I asked to send a radiogram requesting that A. S. Tolstov immediately arrange a special flight to send out fueling accessories from the cosmodromes: rubber hose, nozzle, and filter. I requested that highly-trained specialists who had worked with me be summoned to the institute: L. Ye. Volodina, L. V. Shirokova, S. P. Amelkovich, and S. L.

---

40. *Interkosmos-22* (IK-B-1300), a joint Soviet-Bulgarian satellite was launched into orbit on 7 August 1981 from the Plesetsk Cosmodrome

Dobychin. They all had studied rather extensively both hydrogen peroxide and the appropriate structural materials. These outstanding individuals came right away. Several hours later, A. S. Tolstov and Comrade Istomin (unfortunately I do not remember his first name and patronymic)—an employee of Barmin's company, who was responsible for fueling, arrived with the aforementioned parts. I took all the parts into the laboratory. The colleagues I had summoned and the people who had flown in from the cosmodrome went ...[Sivo]lodskiy remained in his own office to wait.

...The parts had NO traces on them—they were completely clean. I put peroxide in a microburette and began to apply it one drop at a time to the shiny seam of the filter mesh. Three seconds had not passed before a yellow deposit formed on the point of contact with the peroxide. There was smoke and actual loud explosions began accompanied by tongues of yellow flame. The explosions continued as long as even a trace of peroxide remained on the surface of the seam. The spectral analysis that was performed immediately in S.L. Dobychin's laboratory showed that instead of using super pure FOOD-GRADE tin as a soldering material, so-called 'quick solder' was used—radio solder with a 40% lead content. We knew that lead was categorically forbidden in structures in which hydrogen peroxide was used. It is not the metal itself but its oxide that is a very strong catalyst for peroxide breakdown. The latter forms virtually instantaneously when lead comes into contact with such a powerful oxidizer as concentrated hydrogen peroxide. The low durability of lead and its oxide is a property that exacerbates the spontaneous decomposition of peroxide. The effect of the reaction's intense heat release and the powerful gas generation destroys the solid material and disperses the catalyst into the volume of chemically active liquid. If this liquid flows slowly or is allowed to sit even briefly, an explosion is inevitable. (A year before this incident, in August 1980, I managed to determine the cause of an explosion similar in nature in an organization where the catalyst was also a lead oxide—red lead.)

...We wrote a quick procedure for testing ALL filters in the warehouses of Plesetsk and Baykonur. The procedure was put to use immediately and it turned out that both warehouses contained whole batches of such filters waiting to be used. Documentation analyzed by A. S. Tolstov at the Frunze Factory in Sumi where these filters were made showed that in March 1980 the lead filters had already been delivered to Plesetsk. As far as I know, after rejecting all the filters and returning to the use of pure tin solder, there have been no more problems associated with peroxide at the cosmodromes during the past 24 years.

I should add a few words of explanation. Specialists know that the number of materials compatible with hydrogen peroxide is NEGLIGIBLE. Super pure tin is on this short list. The old saying that 'hydrogen peroxide breaks down if you look at it cross-eyed' isn't far from the truth. On the other hand, many know that soldering stainless steel using super pure (food-industry) tin is a pain: the solder doesn't want to flow on steel. It is not surprising that the development engineer in Sumi, who was not well-versed in chemistry, switched to POS solder—a tin and lead alloy.[41] *This solder (radio solder!) flows beautifully on stainless steel. But the fact that such a change in the manufacturing process of peroxide filters was not coordinated with the peroxide chemists/developers (GIPKh, Leningrad), that it was approved at all levels of the MOM, and signed by Academician Barmin is of course a disgrace, a shame, and per se, a crime. Even more shameful is the fact that after the incidents described above in 1981, for a seemingly endless period of time the bureaucracy didn't want to remove the blame from those who had died, particularly Private First Class Yarulla Velikoredchanin, and give the innocent victims their due. In December 1999, a government decree did just that. Dmitriy Viktorovich Ivanov, an officer from Plesetsk who has since passed away, played perhaps the leading role in bringing the affair to this end. He wrote and published a book about the Plesetsk disaster,* Entering the Town of Mirnyy…[42] *And in spring 2000, NTV broadcast the program* Independent investigation *(hosted by Nikolay Nikolayev), which discussed the main circumstances of both the disaster and the investigation on the whole rather truthfully. The program let many of those involved in these events, including the author of this memo, say a few words. NTV sent me a videotape of this broadcast. I was awarded for my work with an Honorary Diploma of the Cosmodrome that said a lot of nice things of which I am proud.*

Chief research associate of the Russian Academy of Sciences N. N. Semyonov Institute of Chemical Physics

Doctor of Physical and Mathematical Science, Professor Aleksandr Shteynberg."

Reflecting on the story of the Plesetsk disaster investigation, I recalled one of my conversations with Vladimir Nikolayevich Pravetskiy. It was during his stint as chief of the Third Main Directorate of the USSR Ministry of Health. He often visited Baykonur while manned flights were being conducted. After one of those sleepless nights I complained to him that a failure had occurred that we hadn't figured out; it had corrected itself, but to be on the safe side we had removed all the instruments and cables related to the failure. During stand-alone tests we found no defects in those instruments and cables.

"In medical practice," said Pravetskiy, "situations occur that we call 'paired incidents.' Sometimes even an experienced physician can't form a diagnosis. If time permits, he waits. Sooner or later a second patient will show up with similar symptoms, but they are so clearly manifested that the diagnosis can be made flawlessly. That's what we call a 'paired incident.'"

I TRIED TO ESTABLISH A CONNECTION BETWEEN THE SCALE OF A ROCKET-SPACE DISASTER AND THE SUBSEQUENT PERIOD OF TIME REQUIRED TO RESTORE THE SYSTEM. Alas! I couldn't find a consistent pattern. After the disaster on 24 October 1960, in which more than 100 people (counting those who died in hospitals) perished,

---

41. POS—*Pripoy olovyanno-svinitsovyy.*

42. D. V. Ivanov, *Pri vyezde v Mirnyy gorodok… [Entering the Town of Mirnyy…]* (Moscow: ZAO izdateskiy dom Gamma, 1997).

R-16 flight tests resumed three months later. After the disaster on 18 March 1980, launches employing the R-7 launch vehicle continued virtually uninterrupted. And after the death of seven astronauts during their return to Earth aboard the Shuttle *Columbia* on 1 February 2003, American Shuttle flights were discontinued for more than two years! Space disasters have occurred for the most diverse reasons and on different scales both in the Soviet Union and in the U.S. In all the cases I'm familiar with, the common factor is that those who were truly to blame for the disasters remained alive and faced no censure.

Index

Ministry of Chemical Industry, 11-15, 636

Ministry of Defense, 9, 30, 177, 200, 209, 314, 323, 403, 416, 425, 429, 436, 471, 486, 563, 619, 624

Ministry of Defense Industries, 258, 275, 381, 481-482, 493

Ministry of Electrical Industry, 11-15, 19

Ministry of Finance, 14, 564

Ministry of Foreign Affairs, 429, 539

Ministry of General Machine Building (MOM), 128, 194, 406, 438, 640

Ministry of Health, 389, 640

Ministry of Higher Education, 15, 20, 332

Ministry of Machine Building and Instrumentation, 11-15

Ministry of Medium Machine Building, 275, 282, 481-482

Ministry of Radio-Electronics Industry, 191

Ministry of Radio Industry, 381

Ministry of Railways, 323

Ministry of Shipbuilding Industry, 5, 11-15, 29, 191, 425-426

Ministry of the Communications Systems Industry (MPSS), 30, 42, 177, 179-180, 534

Mints, Aleksandr L., 212-215, 411-412

Mir publishing house, 39

Mir space station, 507, 514, 535

mischgerät, 37

Mishin, Vasiliy P., 5, 18, 44, 50, 53, 64, 69, 75, 76, 120, 156, 164-165, 166, 168, 226, 232, 247, 259, 261-262, 272, 286, 291, 335, 347, 369, 372, 375, 389, 408, 437, 444, 457, 465, 469, 544-545, 553, 555-557, 566-568, 613; and Hero of Socialist Labor award, 285; and design of R-7 launch pad, 294-295; and development of liquid oxygen systems for R-9 ICBM, 410-411, 600; and election as Corresponding Member of Academy of Sciences, 411-412; and conflict with Glushko, 554-555

MIT (Massachusetts Institute of Technology), 113

Mitenkov, Fedor M., 262

Mitkevich, Olga A., 117, 213

Mnatsakanyan, Armen S., 400

MNII-1 institute, 29-30, 253, 533-534

mobilization economy, 1

Moiseyev, Nikita N., 494-495

Moiseyev, Nikolay D., 62-64, 113

Moisheyev, Igor, 220, 226

Molniya launch vehicle, 420, 564, 568; see also 8K78 launch vehicle

Molniya-1 communications satellite, 595

Molotov, Vyacheslav M., 342

MOM, see Ministry of General Machine Building, 194

Monino, 178

Montania Factory, 43

Moon, see lunar missions

Mordvintsev, Leonid A., 126

Moscow Aviation Institute (MAI), see MAI

Moscow Council of People's Deputies, 459

Moscow Electromechanical Scientific-Research Institute (MNIIEM), see VNIIEM

Moscow Engineering and Physics Institute (MIFI), 481, 508

Moscow Engineering Institute of Geodesy, Aerial Surveying and Cartography (MIIGAiK), 516

Moscow Mashinoapparat Factory, 19

Moscow Physics and Technology Institute (MFTI), 296, 471, 487-488, 504, 506, 509

Moscow Power Engineering Institute, see MEI

Moscow Prozhektor Factory, see Prozhektor Factory

Moscow State University (MGU), 62, 118, 167, 169, 296, 401, 425

Moskalenko, Kirill S., 634

Mossovet (Moscow Municipal Council of People's Deputies), 412

Mount Koshka, 520-522, 526, 529-533, 538

Mozzhorin, Yuriy A., 127-128, 236, 348-349

MPSS, see Ministry of the Communications Systems Industry

Mrykin, Aleksandr G., 128-129, 142, 274, 330, 336, 351-352, 364, 370, 380, 418, 441, 546-548, 629

Mukhanov, Valentin M., 166-167

Mukhina, Vera, 327-328

Müller, Eric, 60

Munich, 394

Munich Technical University, 39

Murayev, Grigoriy I., 127, 512

MV spacecraft (Mars/Venus exploration of 1960), 546-547, 550-551, 554

M. V. Frunze Factory, see Frunze Factory

M. V. Keldysh Research Center, see Keldysh Research Center

M. V. Keldysh Research Center: Seventy Years on the Frontiers of Rocket-Space Technology (book), 459

M. V. Khrunichev Factory, see Khrunichev Factory

www.ingramcontent.com/pod-product-compliance
Lightning Source LLC
Chambersburg PA
CBHW081425170526
45166CB00008B/2107